Intermediate Algebra
with applications

LINDA L. EXLEY
VINCENT K SMITH
DeKalb College

 PRENTICE HALL, Englewood Cliffs, New Jersey 07632

Library of Congress Cataloging-in-Publication Data

Exley, Linda L.
    Intermediate algebra: with applications/Linda L. Exley and Vincent K Smith.
      p.   cm.
    ISBN 0-13-470394-4
    1. Algebra.   I. Smith, Vincent K   II. Title.
QA152.2.E96   1990   89-37979
512.9—dc20   CIP

Editorial/production supervision: Virginia Huebner
Interior design: Anne T. Bonanno
Cover design: Anne T. Bonanno
Cover art: Network Graphics
Manufacturing buyer: Paula Massenaro
Photo editor: Lori Morris-Nantz
Photo research: Rhoda Sidney

Photo credits: 1, Steve Hansen/Stock, Boston • 49, Topham/The Image Works • 95, Photo Source, Ltd. • 119, ProPix/Monkmeyer Press • 153, Bob Daemmrich/The Image Works • 203, NASA • 249, Yann Guichaoua/Photo Researchers • 295, Richard Hutchings/Photo Researchers • 331, Thomas Wanstall/The Image Works • 375, Frank Siteman/Taurus Photos • 439, Robert Goldstein/Photo Researchers • 477, Hugh Rogers/Monkmeyer Press • 525, Jerry Cooke/Photo Researchers

© 1990 by Prentice-Hall, Inc.
A Division of Simon & Schuster
Englewood Cliffs, New Jersey 07632

All rights reserved. No part of this book may be
reproduced, in any form or by any means,
without permission in writing from the publisher.

Printed in the United States of America

10 9 8 7 6 5 4 3 2 1

ISBN 0-13-470394-4

Prentice-Hall International (UK) Limited, *London*
Prentice-Hall of Australia Pty. Limited, *Sydney*
Prentice-Hall Canada Inc., *Toronto*
Prentice-Hall Hispanoamericana, S.A., *Mexico*
Prentice-Hall of India Private Limited, *New Delhi*
Prentice-Hall of Japan, Inc., *Tokyo*
Simon & Schuster Asia Pte. Ltd., *Singapore*
Editora Prentice-Hall do Brasil, Ltda., *Rio de Janeiro*

FORMULAS FOR VOLUME AND SURFACE AREA

Figure		Volume	Surface Area
Rectangular solid	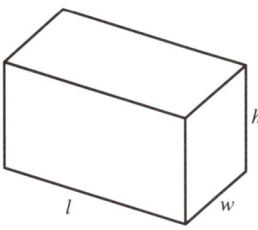	$V = lwh$	$SA = 2wh + 2hl + 2wl$
Cube	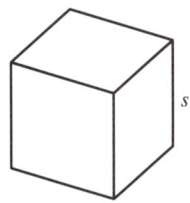	$V = s^3$	$SA = 6s^2$
Right circular cylinder	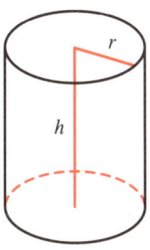	$V = \pi r^2 h$	$SA = 2\pi r^2 + 2\pi r h$
Right circular cone	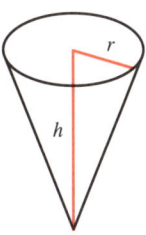	$V = \frac{1}{3}\pi r^2 h$	$SA = \pi r^2 + \pi r \sqrt{r^2 + h^2}$
Sphere	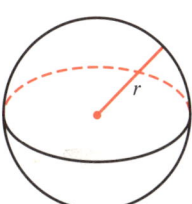	$V = \frac{4}{3}\pi r^3$	$SA = 4\pi r^2$

*Dedicated to Our Spouses
Lois and Charles*

Contents

PREFACE — XI

0. REVIEWING SETS OF NUMBERS AND GEOMETRY — 1

0.1 Sets — 2
0.2 Properties of Real Numbers — 7
0.3 Absolute Value — 16
0.4 Operations with Real Numbers — 21
0.5 Angles, Lines, and Plane Figures — 28
0.6 Perimeter, Surface Area, Area, and Volume — 37

Chapter Summary — *43*
Review Problems — *46*
Chapter Test — *48*

1. POLYNOMIALS — 49

1.1 Exponents — 50
1.2 Definition, Evaluation, and Combining Like Terms — 61
1.3 Addition and Subtraction — 66
1.4 Multiplication — 69
1.5 Pascal's Triangle (Optional) — 79
1.6 Division — 82
1.7 Synthetic Division (Optional) — 88

Chapter Summary — *91*
Review Problems — *93*
Chapter Test — *94*

2. FACTORING — 95

2.1 Greatest Common Factor and Factoring by Grouping — 96
2.2 Factoring Binomials — 104
2.3 Factoring Trinomials — 108
2.4 Summary of Factoring — 113

Chapter Summary — *117*
Review Problems — *117*
Chapter Test — *118*

3. RATIONAL EXPRESSIONS — 119

3.1 Rational Expressions — 120
3.2 Fundamental Principle of Rational Expressions — 123
3.3 Multiplication and Division — 130
3.4 Addition and Subtraction — 136
3.5 Complex Fractions — 143

Chapter Summary — *149*
Review Problems — *150*
Chapter Test — *152*

4. USING LINEAR EQUATIONS IN ONE VARIABLE — 153

- 4.1 Solving Linear Equations in One Variable — 154
- 4.2 Literal Equations and Formulas — 165
- 4.3 Applications — 170
- 4.4 Fractional Equations — 180
- 4.5 Absolute Value Equations — 187
- 4.6 Equations Containing Square Roots — 192

  Chapter Summary — 198
  Review Problems — 200
  Chapter Test — 202

5. EXPONENTS AND RADICALS — 203

- 5.1 Negative Integer Exponents — 204
- 5.2 Roots and Radicals — 217
- 5.3 Operations with Radicals — 223
- 5.4 Rational Exponents — 230
- 5.5 Complex Numbers — 236

  Chapter Summary — 244
  Review Problems — 245
  Chapter Test — 247

6. NONLINEAR EQUATIONS IN ONE VARIABLE — 249

- 6.1 Solutions by Factoring — 250
- 6.2 Applications — 256
- 6.3 Completing the Square — 260
- 6.4 The Quadratic Formula — 267
- 6.5 Equations of Higher Degree — 277
- 6.6 More Fractional Equations — 279
- 6.7 Radical Equations — 287

  Chapter Summary — 290
  Review Problems — 292
  Chapter Test — 294

7. INEQUALITIES IN ONE VARIABLE — 295

- 7.1 Linear Inequalities in One Variable — 296
- 7.2 Compound Inequalities — 305
- 7.3 Boundary Numbers and Absolute Value Inequalities — 310
- 7.4 Higher-Degree and Fractional Inequalities — 319

  Chapter Summary — 326
  Review Problems — 327
  Chapter Test — 329

8. EQUATIONS IN MORE THAN ONE VARIABLE — 331

- 8.1 Cartesian Coordinate System — 332
- 8.2 Graph of $Ax + By = C$ — 340
- 8.3 Slope — 349
- 8.4 Equations of Lines — 357
- 8.5 Variation — 367

  Chapter Summary — 370
  Review Problems — 372
  Chapter Test — 374

9. SYSTEMS OF EQUATIONS — 375

- 9.1 Systems of Equations in Two Variables — 376
- 9.2 Linear Systems in More Than Two Variables — 389
- 9.3 Applications — 397
- 9.4 Linear Inequalities in Two Variables — 405
- 9.5 Systems of Linear Inequalities in Two Variables — 410
- 9.6 Determinants and Cramer's Rule (Optional) — 417
- 9.7 Matrix Methods (Optional) — 425

  Chapter Summary — 433
  Review Problems — 435
  Chapter Test — 438

10. CONIC SECTIONS — 439

- 10.1 The Circle — 440
- 10.2 The Parabola — 447
- 10.3 The Ellipse and the Hyperbola — 458
- 10.4 The General Second-Degree Equation in Two Variables — 470

  Chapter Summary — 472

	Review Problems	473
	Chapter Test	474

11. FUNCTIONS 477

11.1	Definition of Function and Functional Notation	478
11.2	Graphs of Functions	485
11.3	Algebra of Functions	495
11.4	Inverse Functions	499
11.5	Exponential and Logarithmic Functions	504
11.6	Properties of Logarithms	514
	Chapter Summary	519
	Review Problems	520
	Chapter Test	523

12. OTHER TOPICS 525

12.1	Sequences	526
12.2	Summation Notation	532
12.3	Factorials and Binomial Coefficients	535
12.4	The Binomial Theory	539
12.5	Permutations and Combinations	543
	Chapter Summary	550
	Review Problems	551
	Chapter Test	553

ANSWERS TO SELECTED PROBLEMS 555

INDEX 601

OUR COMMITMENT TO EXCELLENCE

Prentice Hall has taken every possible step to ensure the precision and accuracy of this text. Experts reviewed content, checked exercises, and proofread technical material. They assisted the authors and Prentice Hall in producing an outstanding text of the highest quality.

You can help Prentice Hall maintain these high standards by perusing this text and relaying pertinent information found, including errors, to Mathematics Editor, Prentice Hall, Englewood Cliffs, NJ 07632 or to your local Prentice Hall representative. We look forward to serving your text needs now and in the future.
Thank you very much for your support.

Preface

This book is an intermediate algebra textbook with a review of beginning algebra topics. It can be used as a developmental text to prepare students for college algebra, precalculus or decision mathematics. However, it contains sufficient coverage to be used as an introductory college algebra text.

GETTING STARTED

Chapter 0 reviews such fundamental topics as sets, absolute value and the properties of the real number system. The sections in this chapter may be covered in detail, treated as a review, or omitted entirely. If the chapter is omitted, the individual topics can be covered as required with the material of the later chapters. A short review of geometry is included in Chapter 0 and geometric problems are integrated throughout the book.

PEDAGOGY

This book is written in a language that students understand. Wordy statements and mathematical jargon are kept to a minimum while methods and techniques are illustrated with worked examples whenever possible. Although simple language is used and rigor is not stressed, careful mathematics is.

The heart of the book is found in over 6500 carefully selected exercises that develop and illustrate the ideas of algebra in a modern setting. Extensive problem sets occur at the end of each section. Each set starts with a section entitled "Warmups." Warmups are a collection of problems graded in difficulty from easy to medium, which are keyed to worked examples in the text. Warmups are followed by a set of exercises called "Practice Exercises." This longer set of problems contains the drill and practice exercises so necessary for students to develop their manipulative algebra skills. The Practice Exercises are also graded in difficulty but are mixed and not keyed to worked examples. The problem sets also have a few "Challenge Problems" that allow the student to probe into the natural extensions of the ideas presented in the text. The problem sets contain sufficient numbers of problems to provide the instructor considerable flexibility. The pedagogy is never dictated by a limited selection of problems.

The new material and all the teaching points of each section are presented with worked and annotated examples. Problem-solving steps are identified and explained. The students see what is to be done and how to do it. Most examples and all of the teaching points are reinforced with Warmup problems keyed to the examples. The topics are arranged in such a manner as to allow the student to progress through the book developing skills as they are needed. The student is not put into a position of using a concept that has not already been developed.

The topics are developed so that instructors can tailor a course to fit their needs. For example, this book has a separate chapter containing all of the inequalities.

However, if an instructor wishes to teach inequalities with the traditional approach, the individual sections can be taught as disconnected entities.

OTHER FEATURES OF THE TEXT

A Problem Solving Approach: A problem-solving approach is used throughout the book. Worked examples with explanatory text provide the key to an extensive collection of applied problems.

Connections: Each chapter opens with an introductory paragraph putting the upcoming material in context, and showing its connections with other chapters, other courses, and historical development of mathematical ideas.

Key Problems: The Warmups and Practice Exercises contain key problems marked in color. These problems illustrate the necessary teaching points of the section. The instructor can choose to work some or all of them in class as blackboard examples. They can be assigned as hand-in homework to be graded or they can be used for any purpose where the instructor wishes to be sure of a limited but representative cross-section of the material.

Let's Not Forget: The review problems in every chapter, except Chapter 0, end with a novel feature called "Let's Not Forget." This segment contains a few carefully selected problems that repeat earlier themes, particularly certain sticky ideas and problems that historically give students trouble. The Let's Not Forget problems expand with each chapter.

Checkups: Each chapter summary ends with a section called "Checkups." These are worked examples from the text stated as problems. These can be used as a self-test for students to check their understanding of the main points in the chapter. Each problem is keyed to an example number so that students can check their work by referring back to the text.

Be Careful!: Common Student Errors are prominently flagged in the margin with the admonition, "Be Careful!" The adjoining text explains the caution.

In Your Own Words: The problem sets in each section end with a few questions marked, "In Your Own Words." These may be used to encourage good writing in mathematics and to test comprehension of certain ideas.

Calculator Boxes: Calculators are introduced where appropriate and instructional material on the proper use of calculators is included. The calculator boxes are not mere window dressing but provide detailed instructions on meaningful problems for new calculator users. However, the calculator material, including exercises, is segregated and clearly marked for those who wish to omit it. In the calculator boxes and elsewhere in the text, the distinction between exact, approximated and estimated answers is emphasized.

Chapter Summary: Each chapter contains a summary of the material and glossary of new terms.

Chapter Review: Each chapter ends with a set of review problems and a chapter test to reinforce the material.

Inequalities: Traditionally, inequalities are hard to teach, particularly the nonlinear varieties. There are two reasons for this problem. First, in most textbooks, inequalities appear in various places, scattered throughout the books, as the apparently unwanted stepchildren of equations. Linear inequalities are found at the end of the chapter on linear equations, quadratic inequalities at the end of the chapter on quadratic equations and absolute value inequalities stuck in some convenient spot. Secondly, a different technique is taught for each type. In this book all inequalities are gathered together in Chapter 7 for proper emphasis. Next, a single unifying technique is given that solves all of the inequalities of undergraduate mathematics

SUPPLEMENTS

In order to make both the teaching and the learning process easier, the following supplements are available to accompany *Intermediate Algebra with Applications:*

Annotated Instructor's Edition: The Annotated Instructor's Edition is a real time-saving device for the experienced teacher and a near necessity for the new or part-time instructor. It is an exact replica of the student's edition except for certain additions designed to improve the teachability of the text. These additions include five essays that may be of interest to an intermediate algebra teacher, answers next to all the problems, and copious margin notes. The margin notes provide teaching tips and alternate approaches to the material. Historical notes provide the instructor with the background material to liven up the classroom and humanize the material. Common student errors are identified and suggestions given. Places where classroom discussion is appropriate are identified. Points to Emphasize, Discovery Problems, and short True-False Quizzes called "Temptations" are also scattered throughout the margins.

Supplements Guide and Demo Instructor's Disk, (by Joan Dykes)—a unique supplement which facilitates your maximum use of the complete package provided with the book, lists alternative syllabi, possible homework assignments and alternative syllabi, as well as a section-by-section listing of appropriate supplements for use by both the student and the professor. An *"Instructor's Disk"* contains much of this material on disk, so that you can easily edit and adapt it to your own course.

Transparencies and Transparency Masters provide the instructors with an additional teaching tool around which lectures can be built—worked out examples and diagrams.

Instructor's Manual with Tests (by Mary Jean and Shannon Brod) provides 9 tests which were designed specifically for this text and which can be photocopied directly from the manual, thus eliminating typing or cutting and pasting of test questions. 5 tests per chapter are short answer with workspace, 4 are multiple choice tests which have been carefully crafted to include realistic alternate answers.

Computerized Test Generator produces multiple versions of a test with minimum effort. Tests can be randomly generated by the computer to your specifications or you can choose items from a printed *Test Item File,* which lists all the items consecutively. You can add your own test items, as well, by using the editing function.

Instructor's Solutions Manual (by Elizabeth Sirjani) contains fully-worked out solutions to all the even-numbered problems in the book. (The odd-numbered solutions are provided in the Student Solutions Manual.) These can also be photocopied and given to students who need extra help, or made available in the library or math lab; however, they will not be sold to students.

FOR THE STUDENT:

Interactive Algebra Tutor Software provides the student with instant acess to tutorial help, additional practice problems, immediate feedback in the form of diagnostic comments, and sample quizzes. Site License available with qualified adoption.

Algebra Problem Solver Software (by H & N Software) through the use of artificial intelligence, gives students another chance to practice the major topics in algebra,

and provides fully worked-out solutions to problems supplied by either the student or the computer.

Student Solutions Manual (by Virginia Parks) Provides fully worked-out solutions to the odd-numbered problems in the book, and to all Chapter Test questions.

"How to Study Math" (by Helen Burrier) is a unique booklet provided free in quantity to adopting instructors to hand out to each student. Contact your local Prentice Hall rep for more information.

Study Guide keyed directly to the book and software, provides an additional reference for the student who needs more help.

Videotapes keyed to each section of the book, are a reinforcement of the text and lecture, or can be used in a math lab as a lecture substitute. Available with a qualified adoption.

Audiotapes (by Peg Green) Give additional help in the form of worked-out examples, and audiocassettes are available with a qualified adoption.

The supplements package, in combination with an excellent and well-developed text, will provide students and instructors with a state-of-the-art approach to learning Intermediate Algebra and proceed to College Algebra or other courses.

We would like to thank the following reviewers:

Loyde Beam *Garland County Community College*
Dr. B. P. Bockstege *Broward Community College*
Helen Burrier *Kirkwood Community College*
Virginia Carson *DeKalb College*
Margaret Greene *Florida Community College*
Robert L. Hoburg *Western Connecticut State University*
William B. Martin *Pima County Community College*
Phyllis Meckstroth *San Diego State University*
James Peake *Iowa State University*
Lee Price *University of Southwestern Louisiana*
Ronald Rose *American River College*
Herbert Sendek *Community College of Allegheny County*
Edith Silver *Mercer County Community College*
Donald P. Skow *Pan American University*
Ara B. Sullenberger *Tarrant County Junior College*
Bonnie Townsend *DeKalb College*
John Wenger Harold *Washington College*

Preface

xiii

Susan White *DeKalb College*
Jerry Wilkerson *Missouri Western State College*
Brenda Wood *Florida Community College*

Finally, we would like to thank the staff and faculty of DeKalb College for their continued support; Peggy Estes who insisted we do this work; Virginia Huebner who is far too cheerful to be a production editor; designer Anne Bonanno; development editor Christine Peckaitis; and lastly, our senior editor and friend, Priscilla McGeehon.

Linda L. Exley
Vincent K Smith

Advice to the Successful Math Student

Throughout the years, we have seen a parade of algebra students pass through the classroom. Many succeed, but some do not. We would be poor students ourselves if we did not notice the things that make the difference. Three essentials are evident: a will to succeed, homework discipline, and test-taking skills.

Math Success

The first day or two of class, some students freely admit, "I hate math," or "I can't do math," or perhaps, "Math was always my worst subject." We can't make students like math and we realize that many subjects are easier. However, the mental state that such statements indicate leads almost certainly to failure. Regardless of your past experience, you must decide to succeed. You must make up your mind *to do* whatever it takes to succeed. No one can do that for you. You must decide on your own.

Math Homework

Once you have decided to succeed, you must realize that homework is the key to your success. This book has large problem sets to allow plenty of drill problems. Do all of the assigned problems. Don't stop because you already know how to do that kind of problem or because you have figured it out. Working drill problems that you know how to do develops the speed and accuracy you need at test time, and most importantly, gives you the self confidence you need to combat test anxiety.

HOW TO DO HOMEWORK IN MATH

1. Do all homework every day. Plan on 1–3 hours studying for each hour in class.
2. Read the instructions to the problem. Make sure that you understand what it is that you are doing with each problem.
3. Start by working examples in the book. Work them until you can do them without looking at the book.
4. Do homework neatly just as if it were to be turned in. Don't be sloppy and write down things which make no sense. Keep it in a notebook in an organized way.
5. Don't just memorize. Try to understand why you do each step. Don't expect to understand everything the first time you see it. Sometimes you must mimic steps to solve a problem. Understanding may come gradually.
6. Do the problems enough times so that you don't rely on notes, the book, or a friend to give you hints. You must practice until you can work the problems with no help.
7. Develop speed as well as accuracy. This comes with much practice.
8. Don't be afraid of words. Often a word problem is very easy. Follow the procedure outlined in the book for solving word problems and don't just try to figure out the answer.
9. Review old material often.
10. Test yourself by working the keyed problems with no help. Do this after you have done all the problems assigned.
11. Read the material in the book before it is covered in class and again after it is covered.
12. Do the review problems in each chapter.

Math Tests

Test anxiety, indicated by statements like, "I just go blank on tests" or "I understand in class, but can't do the test problems," is usually caused by lack of self confidence. The cure is practice, practice, practice.

HOW TO TAKE A MATH TEST

1. Be rested.
2. Be prepared.
3. Watch the time. Don't rush, but pace yourself so that you can try every problem. If a problem seems hard, skip it. Do problems that you can solve quickly first.
4. Read the instructions for each problem carefully. Make sure that you know *what* you are to do with the problem.

CHAPTER 0

See Problem Set 0.6, Exercise 7.

Reviewing Sets of Numbers and Geometry

- **0.1** Sets
- **0.2** Properties of Real Numbers
- **0.3** Absolute Value
- **0.4** Operations With Real Numbers
- **0.5** Angles, Lines, and Plane Figures
- **0.6** Perimeter, Surface Area, Area, and Volume

CONNECTIONS

In Western Europe for nearly a thousand years after the fall of Rome, it was not fashionable to think. In fact, if caught in the act of any intellectual pursuit, one might become the featured act in the afternoon barbeque! However, the study of mathematics continued in other places, notably along the northern coast of Africa.

The Arabic word *al-jabr,* which means "to transpose," was prominent in the title of a book written by Al-Khowârizmî around A.D. 830. From this word comes our word *algebra.* During this period the Arabs developed most of what we call algebra today.

Algebra is largely arithmetic, with one or more numbers represented by letters. Therein lies its usefulness, and therein lies its difficulty. To be successful with algebra, it is necessary to do arithmetic with letters exactly as if they were numbers.

In this chapter we discuss several topics that should be a review of previous work. It would be wise for the student to master this material before moving on.

0.1 SETS

A set is a collection of objects. The objects are called **elements** or **members.** We often use capital letters to name sets and list the elements in braces. For example, $A = \{a, b, c, d, e\}$, is a set containing the first five letters of the alphabet. $B = \{1, 2, 3, \ldots, 10\}$ is a set containing the counting numbers 1 through 10. (The three dots, called an *ellipsis,* mean "continue on, in the pattern that has been established".) $N = \{1, 2, 3, \ldots\}$ is a set containing all the counting numbers.

If the number of elements in a set is a counting number, the set is called a **finite set.** If there are no elements at all in a set, it is the **empty set.** Otherwise, it is called an **infinite set.** The sets A and B are finite, whereas N is infinite.

The empty set is sometimes called the **null set.** It is written as $\{\ \}$ or $\emptyset$. It would be wrong to write the empty set as $\{\emptyset\}$, as this set would contain one element.

The symbol "$\in$" is used to mean **"is an element of,"** and the symbol $\notin$ is used to mean **"is not an element of."** If $D = \{a, b, c, d, e\}$, then $a \in D$ whereas $2 \notin D$.

Sometimes instead of listing the elements of the set, a notation called **set-builder notation** is used. $\{x \mid x \text{ is a vowel}\}$ describes the "set of all x such that x is a vowel." The bar is read "such that." It is the same set as $\{a, e, i, o, u\}$.

Set A is a **subset** of a set B if every element in A is also an element in B. The symbol "$\subseteq$" means **"is a subset of."** If $A = \{1, 5, 7\}$ and $B = \{1, 3, 5, 7, 9\}$, then $A \subseteq B$.

EXAMPLE 1. List all the subsets of $\{a, b, c\}$.

Solution:

The subsets are: $\{a, b, c\}$ $\{a, b\}$ $\{a\}$ $\emptyset$
                                  $\{a, c\}$ $\{b\}$
                                  $\{b, c\}$ $\{c\}$

Notice that the set is a subset of itself and that the empty set is a subset of every set.

> **Subsets of a Set**
>
> If A is any set, $A \subseteq A$ and $\emptyset \subseteq A$.

Often, sets are combined using the operations of **union** ($\cup$) and **intersection** ($\cap$).

> **Union and Intersection of Sets**
>
> $A \cup B = \{x \mid x \in A \text{ or } x \in B\}$
>
> $A \cap B = \{x \mid x \in A \text{ and } x \in B\}$

[margin note: ∪ = either or both; ∩ = both]

That is, the union of two sets is the set of all elements that belong to *either* of the two sets (or both of them), while the intersection of two sets is the set of all elements that belong to *both* of the two sets.

EXAMPLE 2. If $A = \{1, 2, 7\}$ and $B = \{1, 2, 3, 4, 5\}$, find:

(a) $A \cup B$ (b) $A \cap B$

Solutions:

(a) $A \cup B = \{1, 2, 3, 4, 5, 7\}$
(b) $A \cap B = \{1, 2\}$

Algebra deals with sets of numbers. The symbols used to name the numbers are called **numerals.**

> **Natural Numbers**
>
> $N = \{1, 2, 3, \ldots\}$ *[margin: Counting numbers]*
>
> **Whole Numbers**
>
> $W = \{0, 1, 2, 3, \ldots\}$
>
> **Integers**
>
> $Z = \{\ldots, -3, -2, -1, 0, 1, 2, 3, \ldots\}$
>
> **Rational Numbers**
>
> $Q = \left\{ x \mid x \text{ can be written as } \dfrac{p}{q}, \text{ where } p \text{ and } q \text{ are integers and } q \neq 0 \right\}$ *[margin: $\frac{7}{6}$ or any fraction even $\frac{5}{15}$, can be repeating ∴ the integer is a rational number]*
>
> **Irrational Numbers**
>
> $I = \{x \mid x \text{ is a number but not a rational number}\}$ *[margin: π, √2 non repeating]*
>
> **Real Numbers**
>
> $R = \{x \mid x \text{ is a rational or an irrational number}\}$

The natural numbers are sometimes called the **counting numbers.** Some examples of rational numbers are $\frac{7}{8}$, 5.2, and -3. Rational numbers are often referred to as *fractions*. Some examples of irrational numbers are $\sqrt{2}$, $\sqrt{3}$, and π. As a matter of fact, $\sqrt{x}$, where x is a positive integer but not a perfect square, is an irrational real number.

The following diagram shows how these sets are related to each other.

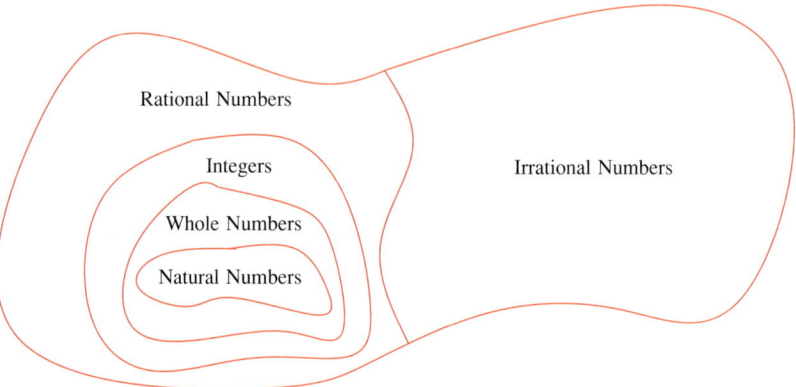

Notice that since $5 = \frac{5}{1}$, the integer 5 is a rational number. Similarly, all integers are rational numbers. As the next example shows, all rational numbers can be expressed as a repeating decimal.

EXAMPLE 3. Express each rational number as a decimal.

(a) $\frac{2}{3}$ (b) $\frac{1}{2}$ (c) $\frac{3}{11}$

Solutions:

(a) $\frac{2}{3} = 0.6666\ldots = 0.\overline{6}$ (b) $\frac{1}{2} = 0.5 = 0.5\overline{0}$

(c) $\frac{3}{11} = 0.272727\ldots = 0.\overline{27}$

Be Careful! Be careful not to confuse exact values with approximations. $\frac{2}{3} = 0.\overline{6}$, which is an exact value. The number 0.67 is an approximation of $\frac{2}{3}$, as is 0.667 and 0.666667. Decimals that do not repeat are irrational numbers.

EXAMPLE 4. Classify each number as rational or irrational.

(a) $\frac{3}{7}$ (b) $0.\overline{78}$ (c) $\sqrt{7}$ (d) 1.7

Solutions:

(a) $\frac{3}{7}$ is rational.

(b) $0.\overline{78}$ is a repeating decimal, thus is rational.

(c) $\sqrt{7}$ is irrational, as 7 is not a perfect square.

(d) 1.7 is rational, as $1.7 = \frac{17}{10}$.

Calculators

In today's world, many calculations with real numbers are done with a calculator. The calculator is not a substitute for mental arithmetic but an aid in carrying out tedious, lengthy computations.

There are two logical systems employed by various calculators. One type uses the common algebraic notation and the other uses reverse Polish notation. Algebraic calculators have a key marked $\boxed{=}$, while reverse Polish calculators have an $\boxed{\text{enter}}$ key. The discussion and examples in this book are written using algebraic notation. Consult an operating manual for instructions and examples for reverse Polish notation. A **scientific calculator** will have the keys necessary for most college courses.

Since a calculator can give only a finite number of decimal places, the user must distinguish between exact values, approximations, and estimates. The exact value of the irrational number $\sqrt{8}$ is a nonrepeating decimal. This value can be approximated with a calculator. Press the keys

$\boxed{8}$ $\boxed{\sqrt{}}$ and read on the display $\boxed{2.828427125}$.

We write $\sqrt{8} \approx 2.828427125$. This answer can be rounded to any desired accuracy. Notice that the symbol $\approx$ means "approximately equal to."

When using a calculator, it is very important to estimate what the answer should be so that we can determine if the answer is reasonable. Since 8 is almost 9, $\sqrt{8}$ is a little less than 3. So the approximation, 2.828427125, is a reasonable answer.

Calculator Exercises

Estimate the decimal value of each real number and approximate each number to the nearest hundredth.

1. (3.12)(4.94)
2. $\frac{3}{7}$
3. $\sqrt{95}$
4. $\frac{11}{12}$
5. $\sqrt{43}$
6. $\frac{1}{9}$

Answers:

1. 15; 15.41
2. 0.5; 0.43
3. 10; 9.75
4. 1; 0.92
5. 6.5; 6.56
6. 0.1; 0.11

Sec. 0.1 Sets

PROBLEM SET 0.1

Warm-ups

In problems 1 through 10, indicate whether each statement is true or false.

1. If $A = \{a, b, c\}$, then $a \in A$.
2. The set of natural numbers is a subset of the set of integers.
3. Every integer is a rational number.
4. $A \subseteq A$, where A is any set.
5. $-\dfrac{1}{2}$ is an integer.
6. The empty set is written as $\{\emptyset\}$.
7. 0 is a natural number.
8. If A is any set, $A \cup \emptyset = A$.
9. If A is any set, then $A \cap \emptyset = A$.
10. The set $\{2, 4, 6, \ldots, 18\}$ is an infinite set.

In problems 11 through 13, $A = \{2, 4, 6, 8\}$ and $B = \{-2, 0, 2\}$. See Examples 1 and 2.

11. List the elements in $A \cup B$.
12. List the elements in $A \cap B$.
13. List all the subsets of B.

In problems 14 through 17, list the elements of A.

14. $A = \{x \mid x \text{ is a natural number less than } 10\}$
15. $A = \{x \mid x \text{ is a color in the U.S. flag}\}$
16. $A = \{x \mid x \text{ is a natural number divisible by } 5\}$
17. $A = \{x \mid x \text{ is one of the first 10 letters in the alphabet}\}$

In problems 18 through 21, write A using set-builder notation.

18. $A = \{w, x, y, z\}$
19. $A = \{2, 4, 6, 8, 10, 12\}$
20. $A = \{\ldots, -5, -3, -1, 1, 3, 5, \ldots\}$
21. $A = \{\text{Monday}, \text{Tuesday}, \ldots, \text{Sunday}\}$

In problems 22 through 26, write each rational number as a repeating decimal. See Example 3.

22. $\dfrac{1}{3}$
23. $\dfrac{3}{4}$
24. $\dfrac{1}{6}$
25. $\dfrac{5}{8}$
26. $\dfrac{2}{7}$

In problems 27 through 32, classify each number as rational or irrational. See Example 4.

27. $\dfrac{2}{3}$
28. 1.25
29. 2π
30. $\sqrt{17}$
31. $0.1\overline{23}$
32. 0.66666

Practice Exercises

In problems 33 through 42, indicate whether each statement is true or false.

33. If $A = \{1, 2, 3\}$, then $3 \notin A$.
34. The set of natural numbers is a subset of the set of whole numbers.
35. Every negative integer is a rational number.
36. $\emptyset \subseteq A$, where A is any set.
37. $\dfrac{1}{2}$ is a real number.
38. The empty set can be written as $\emptyset$.
39. 0 is a rational number.
40. If A is any set, $A \cap A = A$.
41. If A is any set, then $A \cap \emptyset = \emptyset$.
42. The set $\{2, 4, 6, \ldots\}$ is a finite set.

In problems 43 through 45, $A = \left\{\dfrac{1}{2}, 2, \dfrac{9}{4}, 5\right\}$ and $B = \left\{\dfrac{9}{4}, 5\right\}$.

43. List the elements in $A \cup B$.
44. List the elements in $A \cap B$.

45. List all the subsets of B.

In problems 46 through 49, list the elements of A.

46. $A = \{x \mid x \text{ is a natural number less than } 7\}$
47. $A = \{x \mid x \text{ is a day of the week beginning with F}\}$
48. $A = \{x \mid x \text{ is an integer divisible by } 2\}$
49. $A = \{x \mid x \text{ is one of the last 10 letters in the alphabet}\}$

In problems 50 through 53, write A using set-builder notation.

50. $A = \{\text{April, August}\}$
51. $A = \{-2, -1, 0, 1, 2\}$
52. $A = \{\ldots, -6, -4, -2, 0, 2, 4, 6, \ldots\}$
53. $A = \{\text{clubs, diamonds, hearts, spades}\}$

In problems 54 through 59, classify each number as rational or irrational.

54. $\dfrac{2}{5}$ **55.** 1.2 **56.** 4π **57.** $\sqrt{11}$
58. $0.13\overline{7}$ **59.** 0.1111111

In problems 60 through 64, write each rational number as a repeating decimal.

60. $\dfrac{2}{3}$ **61.** $\dfrac{3}{5}$ **62.** $\dfrac{5}{6}$ **63.** $\dfrac{7}{8}$ **64.** $\dfrac{3}{7}$

Challenge Problems

65. Express 0.3 as a quotient of two integers.

66. Express 1.2 as a quotient of two integers.

67. Express 0.655 as a quotient of two integers.

IN YOUR OWN WORDS . . .

68. What is a real number?
69. What is a rational number?

0.2 PROPERTIES OF REAL NUMBERS

It is much easier to visualize sets of numbers by drawing a picture of them. This is done by using a number line. Draw a line and pick a starting point called the **origin** and label it with the number 0. Move to the right of 0 and represent positive integers with a convenient unit of measure. Negative integers are represented to the left of zero. Fractions and irrational numbers are inserted in their place using the same unit of measure.

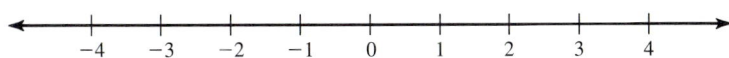

The number associated with a point is called its **coordinate.** Every real number can be located on the number line, and every point on the number line represents a real number.

To **graph** a set of numbers means to locate them on a number line. Looking at the number line, note that if a number, p, lies to the left of another number, q, then p is less than q. We write this using the symbol $<$.

Less Than

$p < q$ means p is less than q. (p is to the *left* of q.)

Similarly, if r lies to the right of s, then r is greater than s. We write this using the symbol $>$.

Greater Than

$r > s$ means r is greater than s. (r is to the *right* of s.)

Note that saying $p < q$ is the same as saying that $q > p$.

Sometimes "equals" is combined with "less than" or "greater than" by using the symbol $\leq$ or $\geq$.

Equality with Less Than or Greater Than

$p \leq q$ means p is less than or equal to q.

$r \geq s$ means r is greater than or equal to s.

The number line helps us to understand some properties of real numbers. If two real numbers, p and q, are graphed, one of three things can occur.

1. p is the same as q.
2. p is to the left of q.
3. p is to the right of q.

This idea is called the Trichotomy Property.

Trichotomy Property

If p and q are real numbers, exactly one of the following statements is true.

$$p = q$$
$$p < q$$
$$p > q$$

If p, q, and r are real numbers with $p < q$ and $q < r$, does p lie to the left or to the right of r? If we look at a number line, we can answer this question. Because we know that $p < q$, p lies to the left of q. Similarly, since $q < r$, we know that q lies to the left of r.

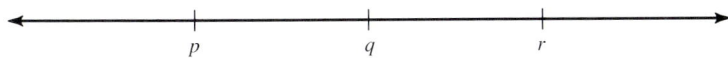

Thus we see that p is to the left of r. That is, $p < r$. This idea is called the Transitive Property.

> **Transitive Property**
>
> Let p, q, and r be real numbers.
>
> If $p < q$ and $q < r$, then $p < r$.

Many properties of the real numbers deal with two numbers being equal, which are called Equality Properties.

> **Properties of Equality**
>
> If p, q, and r are real numbers, then:
>
> 1. $p = p$ — Reflexive
> 2. If $p = q$, then $q = p$. — Symmetric
> 3. If $p = q$ and $q = r$, then $p = r$. — Transitive
> 4. If $p = q$, then we may replace p with q. — Substitution

Arithmetic with real numbers uses the operations of addition, subtraction, multiplication, and division. Symbols are used to indicate which operation to perform.

Operation	*Symbols*	*Name*
Addition	$p + q$	Sum
Subtraction	$p - q$	Difference
Multiplication	pq, $p \cdot q$, $p(q)$	Product
Division	$p \div q$, $\dfrac{p}{q}$, p/q	Quotient

When we express sums, differences, products, and quotients in symbols, we call them **expressions** or **algebraic expressions**. In a sum or difference, the numbers being added or subtracted are called **terms**. In a product, the numbers being multiplied are called **factors**.

In doing arithmetic, we take advantage of properties of numbers without giving them much thought. For example, it is well known that $2 + 3 = 3 + 2$ and that $2 \cdot 3 = 3 \cdot 2$. These ideas are true for all real numbers and illustrate the Commutative Properties of real numbers.

Commutative Properties

If p and q are real numbers, then:

1. $p + q = q + p$ Commutative for addition
2. $pq = qp$ Commutative for multiplication

The next two properties deal with the way in which we group real numbers to add or multiply them. We know that

$$2 + (3 + 4) = (2 + 3) + 4$$

and that

$$2(3 \cdot 4) = (2 \cdot 3)4$$

These ideas are called Associative Properties.

Associative Properties

If p, q, and r are real numbers, then:

1. $p + (q + r) = (p + q) + r$ Associative for addition
2. $p(qr) = (pq)r$ Associative for multiplication

A combination of addition and multiplication such as $2(3 + 4)$ can be done in two ways. In the first way, we add 3 and 4 and then multiply 2 by 7.

$$2(3 + 4) = 2 \cdot 7$$
$$= 14$$

The second way is called the Distributive Property. We multiply each number by 2 and then add.

$$2(3 + 4) = 2 \cdot 3 + 2 \cdot 4$$
$$= 6 + 8$$
$$= 14$$

Thus we see that

$$2(3 + 4) = 2 \cdot 3 + 2 \cdot 4$$

Distributive Property

If p, q, and r are real numbers, then

$$p(q + r) = p \cdot q + p \cdot r$$

There are two real numbers that have special properties. These numbers are 0 and 1. Since the sum of 0 and any real number is the real number itself, we call 0 the identity for addition. Similarly, since 1 times any real number is the real number itself, we call 1 the multiplicative identity.

Identities

If p is a real number, then:

1. $p + 0 = 0 + p = p$ Identity for addition
2. $p \cdot 1 = 1 \cdot p = p$ Identity for multiplication

The **additive identity** is 0 and the **multiplicative identity** is 1.

Looking at a number line, notice that 3 and -3 lie on opposite sides of 0. Also, we note that each number is the same distance from 0. The numbers 3 and -3 have the property that their sum is the additive identity.

$$3 + (-3) = 0$$

The product of 2 and $\dfrac{1}{2}$ is the multiplicative identity. That is,

$$2 \cdot \dfrac{1}{2} = 1$$

Inverses

1. *Additive Inverse (Opposite)*. If p is a real number, there is a real number $-p$ such that

$$p + (-p) = (-p) + p = 0$$

$-p$ is called the **opposite** or **additive inverse** of p. (We read $-p$ as negative p or the opposite of p.)

2. *Multiplicative Inverse (Reciprocal)*. If p is a nonzero real number, there is a real number $\dfrac{1}{p}$ such that

$$p \cdot \dfrac{1}{p} = \dfrac{1}{p} \cdot p = 1$$

$\dfrac{1}{p}$ is called the **multiplicative inverse** or **reciprocal** of p.

Sec. 0.2 Properties of Real Numbers

EXAMPLE 1. Locate each number on a number line.

(a) 5   (b) −5   (c) $-\frac{5}{3}$ (d) $\frac{7}{2}$

(e) p; $p > 0$ (f) $-p$; $p > 0$

(g) q; $q < 0$ (h) $-q$; $q < 0$

Solutions:

(a) through (d) are shown below.

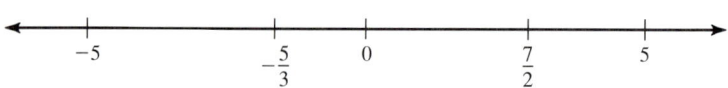

(e) If $p > 0$, then p lies to the right of 0.

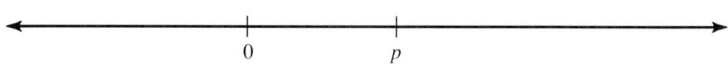

(f) If $p > 0$, then p lies to the right of 0. So $-p$ (the opposite of p) lies to the left of 0.

(g) If q is negative, then q lies to the left of 0.

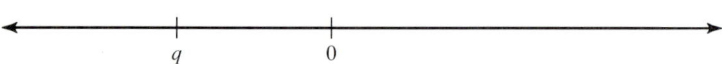

(h) If $q < 0$, q lies to the left of 0. So $-q$ lies to the right of 0.

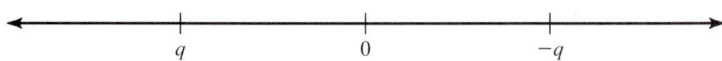

Using the symbol "−" can be very confusing. It has three different meanings. Consider these situations.

$\boxed{1}$ −5 $\boxed{2}$ 4 − 2 $\boxed{3}$ −p

In $\boxed{1}$ it is part of the name of a number, negative five.

In $\boxed{2}$ it indicates subtraction, 4 minus 2 or 4 subtract 2.

In $\boxed{3}$ it indicates the opposite of a number, the opposite of p.

We often interchange the words "minus," "negative," and "opposite" when reading problems that use the "−" symbol.

EXAMPLE 2. Give the reciprocal of each real number.

(a) 2   (b) $\frac{3}{5}$ (c) 8   (d) $\frac{5}{2}$

Solutions:

(a) As $2 \cdot \frac{1}{2} = 1$, the reciprocal of 2 is $\frac{1}{2}$.

(b) As $\frac{3}{5} \cdot \frac{5}{3} = 1$, the reciprocal of $\frac{3}{5}$ is $\frac{5}{3}$.

(c) As $8 \cdot \frac{1}{8} = 1$, the reciprocal of 8 is $\frac{1}{8}$.

(d) As $\frac{5}{2} \cdot \frac{2}{5} = 1$, the reciprocal of $\frac{5}{2}$ is $\frac{2}{5}$.

Example 1 leads us to a discussion of the opposite of the opposite of a number. What is the opposite of the opposite of 3? The opposite of 3 is -3. So the opposite of the opposite of 3 is the opposite of -3, which is 3. Saying this in words is cumbersome; writing it in mathematical symbols makes it easier.

$$-(-3) = 3 \quad \text{(The opposite of negative three is three.)}$$

This can be stated as a property for any real number.

Double Negative Property

If p is a real number, then $-(-p) = p$.

Multiplication by 0

If p is a real number, then $p \cdot 0 = 0 \cdot p = 0$.

EXAMPLE 3. Indicate the property of real numbers that justifies each statement.

(a) $2 + 3 = 3 + 2$
(b) $1(3) = 3$
(c) $-5 + 0 = -5$
(d) $(4 \cdot 3) \cdot 6 = 4 \cdot (3 \cdot 6)$

Solutions:

(a) $2 + 3 = 3 + 2$ Commutative Property for Addition
(b) $1(3) = 3$ Multiplicative Identity
(c) $-5 + 0 = -5$ Identity for Addition
(d) $(4 \cdot 3) \cdot 6 = 4 \cdot (3 \cdot 6)$ Associative for Multiplication

Sec. 0.2   Properties of Real Numbers

It is easy to see that s and $-s$ are opposites because $s + (-s) = 0$. Sometimes it is more difficult to recognize opposites. Consider the following.

$$(p - q) + (q - p) = p + (-q + q) - p \quad \text{Associative Property}$$
$$= p + 0 - p \quad \text{Opposites}$$
$$= p - p \quad \text{Identity (Addition)}$$
$$= 0 \quad \text{Subtraction}$$

Since $(p - q) + (q - p) = 0$, we see that $p - q$ and $q - p$ are opposites. These numbers come up often in algebra.

> **Opposites**
>
> $p - q = -(q - p)$
>
> $p - q$ and $q - p$ are *opposites*.

EXAMPLE 4. Find the opposite of $w - z$.

Solution:

The opposite of $w - z$ is $z - w$.
We say this by writing $-(w - z) = z - w$ or $z - w = -(w - z)$.

■ PROBLEM SET 0.2

Warm-ups

In problems 1 through 7, give the opposite of each number and locate both the number and its opposite on a number line. See Example 1.

1. $\sqrt{3}$
2. -4
3. $-\frac{1}{3}$
4. $\frac{11}{3}$
5. 0
6. $p; \ p > 0$
7. $q; \ q < 0$

In problems 8 through 13, find the reciprocal of each number. See Example 2.

8. 4
9. 7
10. $\frac{5}{3}$
11. $\frac{2}{3}$
12. $\frac{3}{4}$
13. $\frac{8}{5}$

In problems 14 through 28, indicate the property of real numbers that justifies each statement. See Example 3.

14. $2 \cdot 3 = 3 \cdot 2$
15. $4 = 4$
16. If $4 = x$, then $x = 4$.
17. $(3 + 4) + \sqrt{7} = 3 + (4 + \sqrt{7})$
18. $1 + 3 = 3 + 1$
19. $9 + 0 = 9$
20. $7 + (-7) = 0$
21. $1 \cdot (-8) = -8$

22. $4 + (2 + 3) = 4 + (3 + 2)$
23. $2 \cdot \frac{1}{2} = 1$
24. $0 \cdot 6 = 0$
25. $4(3 + 7) = 4 \cdot 3 + 4 \cdot 7$
26. If $a < \sqrt{3}$ and $\sqrt{3} < b$, then $a < b$.
27. If $x + 2 = 5$ and $x = 3$, then $3 + 2 = 5$.
28. $(2 \cdot 9) \cdot 5 = 2 \cdot (9 \cdot 5)$
29. Give a reason for each step. See Example 3.

$$\boxed{1} \quad 3 \cdot x + 7 \cdot x = x \cdot 3 + x \cdot 7$$
$$\boxed{2} \quad = x(3 + 7)$$
$$\boxed{3} \quad = x \cdot 10$$
$$\boxed{4} \quad = 10 \cdot x$$

30. Find the opposite of $x - y$. See Example 4.

Practice Exercises

In problems 31 through 45, indicate the property of real numbers that justifies each statement.

31. $2 + 3 = 3 + 2$
32. $\sqrt{6} = \sqrt{6}$
33. If $y = x$, then $x = y$.
34. $(3 \cdot 4) \cdot 7 = 3 \cdot (4 \cdot 7)$
35. $1 \cdot 3 = 3 \cdot 1$
36. $9 \cdot 0 = 0$
37. $7 + (-7) = (-7) + 7$
38. $1 \cdot (8) = 8$
39. $5 + (-5) = 0$
40. $3 \cdot \frac{1}{3} = 1$
41. $0 \cdot 3 = 0$
42. $4 \cdot (3 + 7) = (3 + 7) \cdot 4$
43. If $x < 1$ and $1 < y$, then $x < y$.
44. If $x + 2 = 4$ and $x = 2$, then $2 + 2 = 4$.
45. $(2 + 9) + 5 = 2 + (9 + 5)$

In problems 46 through 52, give the opposite of each number and locate both the number and its opposite on a number line.

46. $\sqrt{2}$
47. -2
48. $-\frac{2}{3}$
49. $\frac{2}{3}$
50. π
51. $-p; \quad p > 0$
52. $-q; \quad q < 0$

In problems 53 through 58, find the reciprocal of each number.

53. 1
54. 6
55. $\frac{1}{4}$
56. $\frac{2}{5}$
57. $\frac{3}{7}$
58. $\frac{7}{3}$

59. Give a reason for each step.
(1) $3 + (5 + 7y) = (3 + 5) + 7y$
(2) $ = 8 + 7y$

60. Find the opposite of $s - t$.

Challenge Problems

61. Suppose that p and q are positive real numbers with $p > q$. Locate $p - q$ on a number line.

62. Suppose that p and q are positive real numbers with $p < q$. Locate $p - q$ on a number line.

Sec. 0.2 Properties of Real Numbers

IN YOUR OWN WORDS...

63. Explain what opposites are.

64. Explain what the Commutative Properties say.

65. Explain what the Associative Properties say.

66. Explain what the Distributive Property says.

0.3 ABSOLUTE VALUE

The numbers 3 and -3 share a common property. Each number is 3 units from zero on the number line. The measure of this distance is called the **absolute value** of 3 and -3 and is written $|3|$ and $|-3|$. Thus

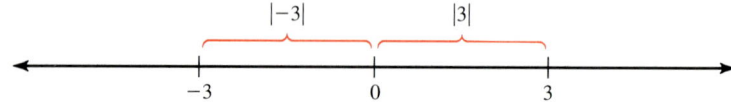

Notice that $|-3| = |3| = 3$.

A similar argument holds for 17 and -17.

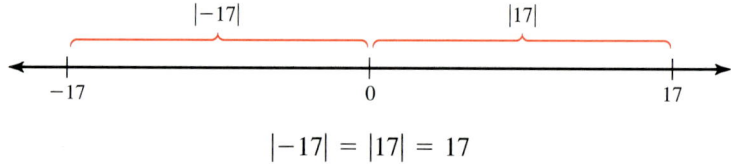

$$|-17| = |17| = 17$$

EXAMPLE 1. Evaluate each absolute value.

(a) $|52|$ (b) $|-7|$ (c) $|\pi|$ (d) $|0|$

Solutions:

(a) The number 52 is 52 units from 0 on the number line, so $|52| = 52$.

(b) Since -7 is 7 units from 0 on the number line, $|-7| = 7$.

(c) π is a positive number (approximately 3.14159), so π is π units from 0 on the number line. Therefore, $|\pi| = \pi$.

(d) As 0 is 0 units from 0 on the number line, $|0| = 0$.   ☐

Suppose that p is a positive number. Then p lies to the right of 0 on the number line.

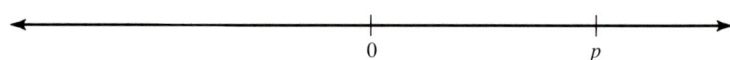

The opposite of p, $-p$, lies the same distance from 0, but to the left of 0.

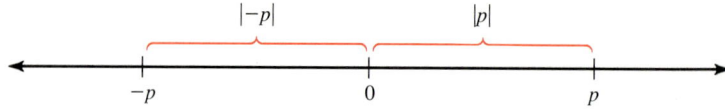

and as before, $|-p| = |p| = p$.

Now suppose that q is a *negative* number. As q is *less* than 0, it lies to the left of 0 on the number line.

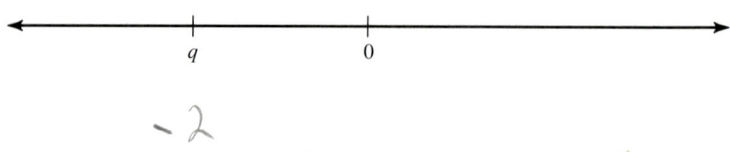

The opposite of q, $-q$, lies the same distance from 0, but to the right of 0.

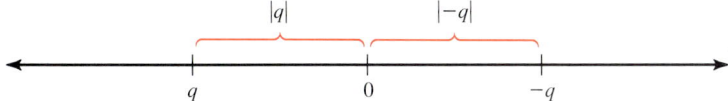

Please note that because q is negative, $-q$ is **positive!** So $-q$ is the measure of the distance and $|q| = |-q| = -q$.

This leads to the definition of absolute value.

Absolute Value

If k is a real number,

$$|k| = \begin{cases} k & \text{if } k \geq 0 \\ -k & \text{if } k < 0 \end{cases}$$

Students starting the study of algebra are often troubled by statements such as "$-q$ is positive" and "$|-q| = -q$." This is understandable, as we have trained ourselves, during years of learning arithmetic, to recognize negative numbers by the minus sign that they wear. Negative numbers in the set

$$\{4, 2, -11, 81, -1, -997\}$$

are easily picked out. Unfortunately, we have taught ourselves a bad theorem!

In algebra, letters are used to stand for numbers, and a letter may be used for a negative number. Suppose that p, q, and r are real numbers. Can you pick out the negative numbers in the set $\{p, q, -r\}$? Try it, and check your choices with the values, 2 for p, -7 for q, and -1 for r.

EXAMPLE 2. Suppose that g, h, j, and k are real numbers with $g > 0$ and $h < 0$. Simplify the following by evaluating the absolute value, *if possible*.

(a) $|g|$ (b) $|h|$ (c) $|j|$ (d) $|-k|$

Solutions:

(a) As $g > 0$, g is positive. So we can write $|g| = g$.
(b) As $h < 0$, h is negative. So $-h$ is *positive*, and we can write $|h| = -h$.
(c) $|j|$ is *either j or $-j$*, depending on whether j is positive or negative. Without that information, $|j|$ cannot be simplified.
(d) $|-k|$ cannot be evaluated for the same reason as in part (c). However, we can write $|-k| = |k|$.

In the discussion above, notice that $|k| = |-k|$ for all values of k. As k and $-k$ are opposites, we see that opposites have the same absolute value. Since $p - q$ and $q - p$ are opposites, we have

$$|p - q| = |q - p|$$

If p and q are any two points on the number line, the distance between them is given by the value of $|p - q|$.

> **The Distance Between Two Numbers on the Number Line**
>
> If p and q are *any* two numbers on the number line, the distance between them is given by
> $$|p - q|$$

Some properties of absolute value are quite useful.

> **Some Properties of Absolute Value**
>
> If p and q are real numbers with $q \neq 0$,
>
> 1. $|p \cdot q| = |p| \cdot |q|$
> 2. $\left|\dfrac{p}{q}\right| = \dfrac{|p|}{|q|}$
> 3. $|p| = |-p|$

EXAMPLE 3. Use the properties of absolute value to simplify each expression.

(a) $|2x|$ (b) $\left|\dfrac{-3}{z}\right|$

Solutions:

(a) $|2x| = |2| \cdot |x|$ Property 1
    $= 2|x|$ 2 is positive

(b) $\left|\dfrac{-3}{z}\right| = \dfrac{|-3|}{|z|}$ Property 2
    $= \dfrac{3}{|z|}$ $|-3| = 3$

EXAMPLE 4. Find the distance on the number line between the points corresponding to each pair of numbers.

(a) 14, 21 (b) −8, 26 (c) 12, −38 (d) −37, −16

Solutions:

In each case, let the first number be p and the second number be q; then calculate $|p - q|$.

(a) $|14 - 21| = |-7| = 7$
    The distance is 7 units.
(b) $|-8 - 26| = |-34| = 34$
    The distance is 34 units.
(c) $|12 - (-38)| = |12 + 38| = |50| = 50$
    The distance is 50 units.
(d) $|-37 - (-16)| = |-37 + 16| = |-21| = 21$
    The distance is 21 units.

EXAMPLE 5. Demonstrate that $|p - q| = |q - p|$ is true with the same pairs of numbers as were used in Example 4.

(a) 14, 21 (b) −8, 26
(c) 12, −38 (d) −37, −16

Solutions:

Again we let the first number be p and the second number be q, but we calculate $|q - p|$.

(a) $|21 - 14| = |7| = 7$, so $|14 - 21| = |21 - 14|$.
(b) $|26 - (-8)| = |26 + 8| = |34| = 34$, and we see that $|-8 - 26| = |26 - (-8)|$.
(c) $|-38 - 12| = |-50| = 50$, and again we see that $|12 - (-38)| = |-38 - 12|$.
(d) $|-16 - (-37)| = |-16 + 37| = |21| = 21$, so $|-37 - (-16)| = |-16 - (-37)|$.

PROBLEM SET 0.3

Warm-ups

In problems 1 through 12, evaluate each expression. See Example 1.

1. $|46|$
2. $|-46|$
3. $|7 - 11|$
4. $|54 - 16|$
5. $|0|$
6. $|7| - |11|$
7. $|54| - |16|$
8. $-|-13|$
9. $-|13|$
10. $|54| - |-16|$
11. $-|32| - |-28|$
12. $-|-17| - |-63|$

In problems 13 through 20, assume that a, b, c, and d are real numbers with $a > 0$ and $b < 0$. Simplify by evaluating the absolute value, if possible. See Example 2.

13. $|a|$
14. $|b|$
15. $|c|$
16. $|d^2|$
17. $|-a|$
18. $|-b|$
19. $|-c|$
20. $|-d|$

In problems 21 through 25, use the properties of absolute value to simplify each expression. See Example 3.

21. $|5y|$
22. $|-14z|$
23. $-|-53k|$
24. $-|2x|$
25. $\left|\dfrac{6}{z}\right|$

In problems 26 through 31, locate each pair of numbers on a number line and find the distance between each pair of numbers. See Example 4.

26. 5 and −3
27. 10 and 2
28. −8 and −6
29. π and $\sqrt{7}$
30. $-\sqrt{5}$ and $-\sqrt{3}$
31. x and y

Practice Exercises

In problems 32 through 43, evaluate each expression.

32. $|16|$
33. $|-16|$
34. $|22 - 9|$

Sec. 0.3 Absolute Value

35. $|19 - 22|$
36. $|31 - 31|$
37. $|22| - |9|$
38. $|19| - |22|$
39. $-|-47|$
40. $-|47|$
41. $|19| - |-22|$
42. $-|19| - |-22|$
43. $-|-27| - |-81|$

In problems 44 through 50, use the properties of absolute value to simplify each expression.

44. $|7z|$
45. $|-21x|$
46. $|8j|$
47. $-|81g|$
48. $-|7xy|$
49. $\left|\dfrac{u}{9}\right|$
50. $\left|\dfrac{9}{u}\right|$

In problems 51 through 62, assume that p, q, r, and s are real numbers with $r > 0$ and $s < 0$. Simplify by evaluating the absolute value, if possible.

51. $|p|$
52. $|q|$
53. $|r|$
54. $|-p|$
55. $|s|$
56. $|-q|$
57. $|-r|$
58. $|-s|$
59. $-|r|$
60. $-|s^2|$
61. $-|-r|$
62. $-|-s|$

In problems 63 through 68, first determine if the statement is true or false, then carefully check the result with the values -3 for p and 7 for q.

63. If $p < 0$, then $|-p| = p$.
64. If $p < 0$, then $|p| = -p$.
65. If $p < 0$, then $|-p| = -p$.
66. $|p - q| = |q - p|$.
67. If $q > 0$, then $|q| = q$.
68. If $q > 0$, then $|-q| = q$.

In problems 69 through 74, locate each pair of numbers on a number line and find the distance between each pair of numbers.

69. 7 and -1
70. 14 and 8
71. -5 and -9
72. π and $\sqrt{10}$
73. $-\sqrt{7}$ and $-\sqrt{5}$
74. $z - w$

75. Find the absolute value of the sum of -8 and -5.
76. Find the absolute value of the product of -2 and 7.
77. Find the quotient when -8 is divided by the absolute value of -4.
78. Find the sum of the absolute value of -7 and the absolute value of -19.
79. Add the absolute value of $-\sqrt{2}$ to the absolute value of π.

Challenge Problems

In problems 80 through 83, rewrite each expression without absolute values.

(HINT: $\pi \approx 3.14$.)

80. $|\pi - 3|$
81. $|3 - \pi|$
82. $|\pi + 3|$
83. $|-\pi - 3|$

■ IN YOUR OWN WORDS . . .

84. What is the absolute value of a number?
85. Explain why we sometimes have $|x| = -x$.
86. Explain why opposites have the same absolute value.

0.4 OPERATIONS WITH REAL NUMBERS

The rules for addition, subtraction, multiplication, and division of real numbers are summarized below.

Addition of Signed Numbers

1. To add two real numbers with **like signs** (both positive or both negative), add their absolute values and keep the common sign.
2. To add two real numbers with **unlike signs** (one positive and one negative), subtract the smaller absolute value from the larger absolute value and keep the sign of the number with the larger absolute value.

Subtraction of Signed Numbers

$$p - q = p + (-q)$$

1. To subtract two real numbers, rewrite the problem as an addition problem.
2. Follow the rules for addition.

Multiplication of Signed Numbers

1. To multiply two real numbers with **like signs** (both positive or both negative), multiply their absolute values. The sign will be positive.
2. To multiply two real numbers with **unlike signs** (one positive and one negative), multiply their absolute values. The sign will be negative.

Division of Signed Numbers

$$p \div q = p \cdot \frac{1}{q}$$

1. To divide two real numbers, rewrite the problem as a multiplication problem.
2. Follow the rules for multiplication.
3. Remember that division by 0 is not allowed.

EXAMPLE 1. Perform the operation indicated.

(a) $5 + (-7)$ (b) $(-6)(-5)$ (c) $-7 - (-3)$ (d) $(-8) \div 4$

Solutions:

(a) $5 + (-7) = -2$
(b) $(-6)(-5) = 30$
(c) $-7 - (-3) = -7 + (+3)$
$= -4$

(d) $(-8) \div 4 = (-8)\left(\dfrac{1}{4}\right)$
$ = -2$

The following products with signed numbers are useful to know.

$$(-p)(-q) = pq$$
$$(-p)q = -pq$$
$$p(-q) = -pq$$
$$-1(p) = -p$$

Fractions are very common in algebra. Operations with fractions are reviewed here. (p, q, r, and s are real numbers.)

Fundamental Property of Fractions

$$\frac{p}{q} = \frac{p \cdot r}{q \cdot r}; \quad q \neq 0 \text{ and } r \neq 0$$

Multiplication of Fractions

$$\frac{p}{q} \cdot \frac{r}{s} = \frac{p \cdot r}{q \cdot s}; \quad q \neq 0 \text{ and } s \neq 0$$

Division of Fractions

$$\frac{p}{q} \div \frac{r}{s} = \frac{p}{q} \cdot \frac{s}{r} = \frac{p \cdot s}{q \cdot r}; \quad q \neq 0; \ r \neq 0; \ s \neq 0$$

Addition of Fractions

$$\frac{p}{q} + \frac{r}{q} = \frac{p + r}{q}; \quad q \neq 0$$

Subtraction of Fractions

$$\frac{p}{q} - \frac{r}{q} = \frac{p - r}{q}; \quad q \neq 0$$

Using zero in a fraction can be troublesome. The following divisions involving 0 are important.

Fractions Containing Zero

$$\frac{0}{p} = 0; \quad p \text{ is any nonzero real number.}$$

$$\frac{p}{0} \text{ is undefined}; \quad p \text{ is any real number.}$$

Signs in fractions must be treated carefully. In working with fractions, a fraction often must be written in another form by manipulating signs. The following is very useful in doing this.

Signs in Fractions
$(q \neq 0)$

$$\frac{-p}{q} = \frac{p}{-q} = -\frac{p}{q}$$

$$\frac{p}{q} = \frac{-p}{-q} = -\frac{-p}{q} = -\frac{p}{-q}$$

Mathematicians have developed a shorthand for writing products with the same number. This shorthand uses exponents. For example, instead of writing $2 \cdot 2 \cdot 2$, we write 2^3 (read "two cubed").

$$2 \cdot 2 \cdot 2 = 2^3$$

The 3 indicates how many times to use 2 as a factor.

Natural Number Exponent

If p is a real number and n is a natural number, then

$$p^n = \underbrace{p \cdot p \cdot p \cdots p}_{n \text{ factors of } p}$$

Thus

$$p^1 = p$$
$$p^2 = p \cdot p \quad (p \text{ squared})$$
$$p^5 = p \cdot p \cdot p \cdot p \cdot p$$

In writing p^n, we call p the **base** and n the **exponent**. The exponent tells how many times to use the base as a factor. We read p^n as p to the nth power.

EXAMPLE 2. Find the value of each expression.

(a) 2^4 (b) $(-3)^3$ (c) $2^2 \cdot 3^3$ (d) $(-2)^2(-5)^2$

Solutions:

(a) $2^4 = 2 \cdot 2 \cdot 2 \cdot 2$
    $= 16$

(b) $(-3)^3 = (-3)(-3)(-3)$
    $= -27$

(c) $2^2 \cdot 3^3 = 2 \cdot 2 \cdot 3 \cdot 3 \cdot 3$
    $= 4 \cdot 27$
    $= 108$

(d) $(-2)^2(-5)^2 = (-2)(-2)(-5)(-5)$
    $= 4 \cdot 25$
    $= 100$

Sec. 0.4 Operations With Real Numbers

Often, more than one operation is involved in simplifying an expression. For example, $2 + 3 \cdot 4$ has multiplication and addition. We must decide which operation to perform first. Grouping symbols such as parentheses ( ), brackets [ ], or braces { } are used to make the order of operations clear.

If we write $(2 + 3) \cdot 4$, the addition is performed first and then the multiplication. So

$$(2 + 3) \cdot 4 = 5 \cdot 4 = 20$$

However, if we write $2 + (3 \cdot 4)$, the multiplication is performed first and then the addition. So

$$2 + (3 \cdot 4) = 2 + 12 = 14$$

If there are no symbols of grouping, we perform multiplications and divisions from left to right before performing additions and subtractions from left to right. So

$$2 + 3 \cdot 4 = 2 + 12 = 14$$

Fraction bars and absolute value bars are also grouping symbols. For example,

$$\frac{2 - 3}{2 + 4} = \frac{-1}{6}$$

This is the same as $(2 - 3) \div (2 + 4)$. The fraction bar can be omitted because of the parentheses.

Absolute value bars are actually grouping symbols that tell where to work first. For example,

$$2|5 - 8| = 2|-3|$$
$$= 2 \cdot 3$$
$$= 6$$

Square root symbols are also grouping symbols. Operations inside the square root symbol must be performed before finding the square root. Consider

$$\sqrt{9 + 16} = \sqrt{25}$$
$$= 5$$

The following rules determine which operation to perform first.

Order of Operations

If grouping symbols are present, perform operations inside them, starting with the innermost symbol, in the following order.

1. Perform any exponentiations.
2. Perform all multiplications and divisions in order from left to right.
3. Perform all additions and subtractions from left to right.

If grouping symbols are not present, perform operations in the order given above.

Chap. 0 Reviewing Sets of Numbers and Geometry

Be careful to note that multiplication does not have to be done before division, nor addition before subtraction. Multiplications and divisions are performed in order as they occur working from left to right. Additions and subtractions are performed in order as they occur working from left to right.

EXAMPLE 3. Simplify each expression.

(a) $-3(5-7)$ (b) $-5(-6)-(-7)$ (c) $(-5)^2 - \dfrac{8}{(-4)}$

(d) $\dfrac{(-10-4)(-2)}{\sqrt{28-12}-11}$ (e) $-\dfrac{2}{3}\left[6(1-5)-|2-8|\right]$

Solutions:

(a) $-3(5-7) = -3(-2)$ Work inside parentheses.
$ = 6$

(b) $-5(-6)-(-7) = 30-(-7)$ Multiply first.
$ = 30+7$
$ = 37$

(c) $(-5)^2 - \dfrac{8}{(-4)} = 25 - (-2)$ Multiply and divide.
$\phantom{(-5)^2 - \dfrac{8}{(-4)}} = 25 + 2$
$\phantom{(-5)^2 - \dfrac{8}{(-4)}} = 27$

(d) $\dfrac{(-10-4)(-2)}{\sqrt{28-12}-11} = \dfrac{(-14)(-2)}{\sqrt{16}-11}$

$\phantom{\dfrac{(-10-4)(-2)}{\sqrt{28-12}-11}} = \dfrac{28}{4-11}$

$\phantom{\dfrac{(-10-4)(-2)}{\sqrt{28-12}-11}} = \dfrac{28}{-7}$

$\phantom{\dfrac{(-10-4)(-2)}{\sqrt{28-12}-11}} = -4$

(e) $-\dfrac{2}{3}\left[6(1-5)-|2-8|\right] = -\dfrac{2}{3}\left[6(-4)-|-6|\right]$

$\phantom{-\dfrac{2}{3}\left[6(1-5)-|2-8|\right]} = -\dfrac{2}{3}[-24-6]$

$\phantom{-\dfrac{2}{3}\left[6(1-5)-|2-8|\right]} = -\dfrac{2}{3}[-30]$

$\phantom{-\dfrac{2}{3}\left[6(1-5)-|2-8|\right]} = 20$

EXAMPLE 4. Subtract -8 from the sum of -5 and 2.

Solution:

We must add -5 and 2 and then subtract -8.

$(-5+2)-(-8) = -3-(-8)$ Work inside parentheses.
$ = -3+8$ Subtract.
$ = 5$

Sec. 0.4 Operations With Real Numbers

CALCULATOR BOX

Operations with Real Numbers on a Calculator

$\boxed{+/-}$, $\boxed{(}$, $\boxed{)}$

The calculator keys will appear in boxes, $\boxed{}$. We will use a single box for numbers instead of boxing each digit of the number. Notice the $\boxed{+/-}$ key. (On some calculators, this key is marked $\boxed{\text{CHS}}$.) This key is used to enter negative numbers. Pressing it changes the sign on the number in the display. Enter a number and try it.

Most modern calculators have the proper order of operations built in. That is, they will multiply and divide before they add and subtract. In the computation

$$-4 + 8 - 3 \cdot 18 \div 9$$

multiplication and division must be done before addition and subtraction. Thus we should get -2 from this computation. Try the following keystrokes:

$\boxed{4}$ $\boxed{+/-}$ $\boxed{+}$ $\boxed{8}$ $\boxed{-}$ $\boxed{3}$ $\boxed{\times}$ $\boxed{18}$ $\boxed{\div}$ $\boxed{9}$ $\boxed{=}$

If the calculator follows the order of operations, the display should read $\boxed{-2}$.

Often we need to override the natural order of operations with grouping symbols. Let's calculate $\dfrac{5 + 17}{3 \cdot 4 - 1}$. If we enter $5 + 17 \div 3 \cdot 4 - 1$, the only division will be 17 by 3! The numerator and the denominator must be grouped together, so we must enter $(5 + 17) \div (3 \cdot 4 - 1)$. Try

$\boxed{(}$ $\boxed{5}$ $\boxed{+}$ $\boxed{17}$ $\boxed{)}$ $\boxed{\div}$ $\boxed{(}$ $\boxed{3}$ $\boxed{\times}$ $\boxed{4}$ $\boxed{-}$ $\boxed{1}$ $\boxed{)}$ $\boxed{=}$

The display should read $\boxed{2}$.

Calculator Exercises

Use a calculator to perform the operation indicated.

1. $-14 + 71 - 11 + 316$
2. $-177 - (311 - 622) + 543$
3. $333 - \dfrac{759}{15} + 17.5$
4. $\dfrac{83 - 38}{9}$
5. $\dfrac{-117.18}{5 - 23}$
6. $\dfrac{(155 - 23)}{[2(8 + 3)]}$

Answers:

1. 362   2. 677   3. 299.9   4. 5   5. 6.51   6. 6

■ **PROBLEM SET 0.4**

Warm-ups

In problems 1 through 24, perform the operation indicated. For problems 1 through 6, see Example 1.

1. $6 - (-3)$
2. $(-7)(-7)$
3. $\dfrac{-16}{4}$

Chap. 0 Reviewing Sets of Numbers and Geometry

4. $-9 + (-9)$

5. $\dfrac{16}{0}$

6. $\dfrac{0}{-5}$

For problems 7 through 18, see Examples 2 and 3.

7. $|-12 + 4| \div 4$

8. $-\dfrac{2}{3}\left(-\dfrac{1}{2} + 3\right)$

9. $\dfrac{1}{6} \div \left(-\dfrac{1}{3}\right)^2$

10. $-2 + 3 \cdot 5$

11. $(2 - 3)^2 \cdot 5$

12. $-(5 - \sqrt{18 - 2})$

13. $15 \div (-3) \cdot \left(\dfrac{-1}{5}\right)$

14. $3|4 - 7| + 2^2(-3)$

15. $\dfrac{2 - 7 + 4}{3(2) - (-1)}$

16. $\dfrac{(-6 - 3)}{(4 - 7)}$

17. $5[6 + 2(-5 + 1)]$

18. $3\left(\dfrac{2}{3} - 2\right) \div (-2)$

For problems 19 through 24, see Example 4.

19. Subtract -7 from the product of -6 squared and $-\dfrac{2}{3}$.

20. Subtract 11 from the sum of -13 and -4.

21. Square the sum of -13 and 4.

22. Find the absolute value of the sum of 6 and -15.

23. Subtract -2 cubed from -2 squared.

24. Find one-half of the sum of $\dfrac{2}{5}$ and $-\dfrac{1}{3}$.

Practice Exercises

In problems 25 through 55, perform the operation indicated.

25. $7^2 - 8^2$

26. $0 - 3^3$

27. $-7 + (-7)^2$

28. $(2)(-7) - (-3)(-6)$

29. $|-15 - (-3)| \div (-3)$

30. $-\dfrac{3}{8} - \left(\dfrac{-1}{4}\right)^2$

31. $\dfrac{\left(\dfrac{1}{8}\right)}{\left(\dfrac{-1}{4}\right)^3}$

32. $\dfrac{0}{0}$

33. $2^2 - 3 \cdot 5$

34. $2 - \dfrac{2|-1 - 9|}{(-2)(-3) - (-2)(7)}$

35. $\dfrac{3^2 - 4^2}{(-2)(-1) - 1}$

36. $\dfrac{0}{-1}$

37. $-2[6 - 3^2(\sqrt{5} + 4)]$

38. $-[\sqrt{50 - 1} - (3 - 7)]$

39. $-12 \div (-3)^3 \cdot \dfrac{-1}{3}$

40. $\dfrac{-3 - 5 - (-1)^2}{2[1 - 3(2 - 5) - 1]}$

41. $(2 - 7)^2 - 7^2$

42. $\dfrac{2 - 3(4 - 7)}{3}$

43. $\dfrac{3}{2} - \left(-\dfrac{1}{4} \div 4\right)$

44. $[6 - (-5) - 11] \div |-3 - 3| + 2$

45. $6\left(\dfrac{-2}{3} - 1\right) \div 3$

46. $5[-6 - 2(5 - 14)]$

47. $\dfrac{2\left[\dfrac{-1 - (-7)}{4 \cdot 2 - 10}\right] - 8 \cdot 3}{-(-5) - 7}$

48. $\dfrac{-3[2(3 - 7) - (4 - 9)]}{3^3 - 2 \cdot 7}$

49. $\dfrac{-4|-9 - 2| + 8 \cdot 5}{2\left(\dfrac{-5 - 9}{5 + 2}\right) - \dfrac{6 - 2 \cdot 3}{0 - 2}}$

50. $\dfrac{-\dfrac{1}{5}[25 - (5 - 15)]}{-7 - 7}$

51. Multiply $\dfrac{3}{4}$ by the sum of -24 and 16.

52. Find the reciprocal of the product of $\dfrac{5}{14}$ and $\dfrac{2}{15}$.

53. Divide the square root of the sum of -12 and 16 by -2.

54. Subtract the product of $\dfrac{1}{3}$ and -63 from the product of $-\dfrac{3}{7}$ and -21.

55. Subtract the sum of 5 and -7 from the absolute value of -11.

Sec. 0.4 Operations With Real Numbers

Challenge Problems

In problems 56 through 58, perform the operation indicated.

56. $\dfrac{\frac{1}{2} - \frac{1}{3}}{\frac{1}{5}}$
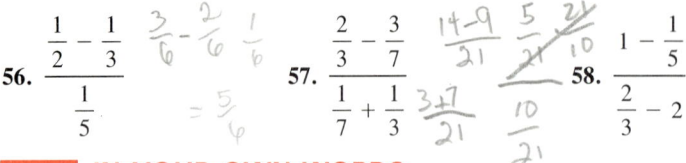
57. $\dfrac{\frac{2}{3} - \frac{3}{7}}{\frac{1}{7} + \frac{1}{3}}$
58. $\dfrac{1 - \frac{1}{5}}{\frac{2}{3} - 2}$

IN YOUR OWN WORDS . . .

59. State the order of operations.

0.5 ANGLES, LINES, AND PLANE FIGURES

This section is a review of some basic ideas and terminology from geometry. Such fundamental ideas as point, line, angle, and length are left undefined. No attempt is made to use precise notation. Angles with the same measure are said to be *equal*. The symbol "∠" indicates the measure of an angle. Lowercase letters are used to designate the length of line segments.

Angles

An **acute angle** is an angle of *less than* 90°.

A **right angle** is an angle of *exactly* 90°. The symbol "⌐" is used to indicate a right angle.

An **obtuse angle** is an angle of *greater than* 90° but *less than* 180°.

A **straight angle** is an angle of *exactly* 180°. Its sides form a line.

Complementary angles are *two* angles whose sum is 90°. Each is called the **complement** of the other.

Supplementary angles are *two* angles whose sum is 180°. Each is called the **supplement** of the other.

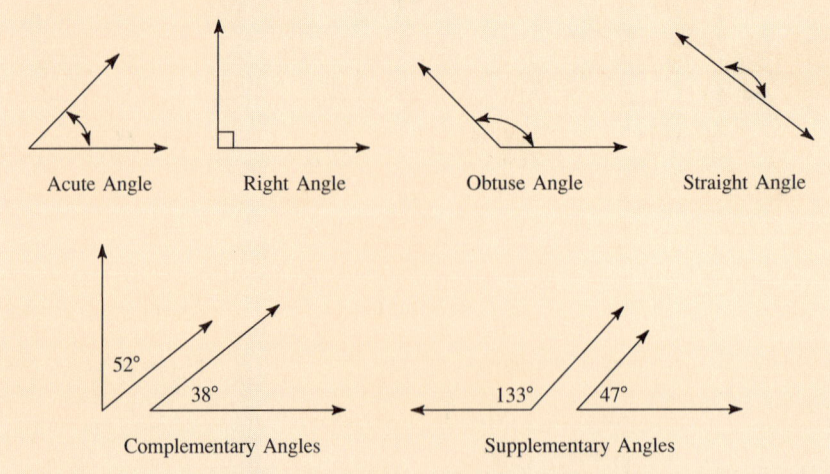

Two intersecting lines form the angles numbered 1, 2, 3, and 4.

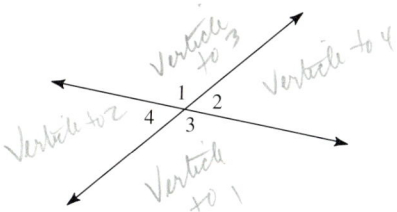

Angles 1 and 3 are **vertical angles,** as are angles 2 and 4. *Vertical angles are equal.* Angles that have a common vertex and a common side are called **adjacent angles.** Angles 1 and 2 are adjacent angles, as are angles 2 and 3, angles 3 and 4, and angles 1 and 4.

EXAMPLE 1. Are the following angles complementary or supplementary?

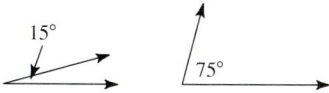

Solution:

The sum of the measures of the two angles is 90°. The angles are complementary.

Lines

Two lines that lie in the same plane and do not intersect are called **parallel lines.**

Two lines that intersect at right angles are called **perpendicular.**

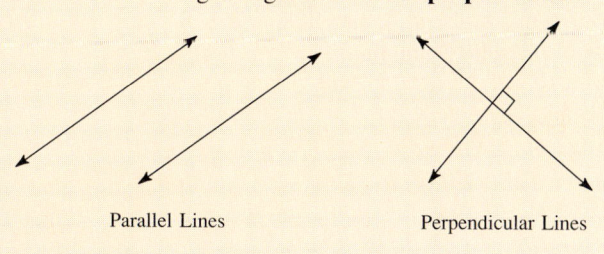

Parallel Lines   Perpendicular Lines

A **transversal** is a line that intersects two or more lines at different points. A transversal of two lines forms some angles that are given special names.

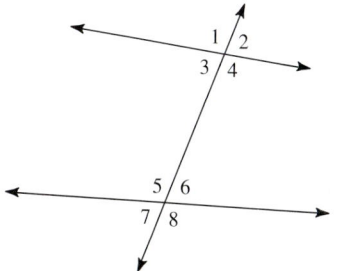

Interior angles: 3, 4, 5, and 6

Exterior angles: 1, 2, 7, and 8

Corresponding angles: 1 and 5, 2 and 6, 3 and 7, 4 and 8

Alternate interior angles: 3 and 6, 4 and 5

Alternate exterior angles: 1 and 8, 2 and 7

Sec. 0.5 Angles, Lines, and Plane Figures

> **Parallel Lines Cut by a Transversal**
>
> If two lines are cut by a transversal, the lines are parallel if any of the following statements are true.
>
> 1. *Corresponding angles* are equal.
> 2. *Alternate interior angles* are equal.
> 3. *Alternate exterior angles* are equal.
>
> If two lines are parallel, all of the statements are true.

EXAMPLE 2. Determine the measure of each angle in the figure. Lines l_1 and l_2 are parallel. $\angle 1 = 120°$.

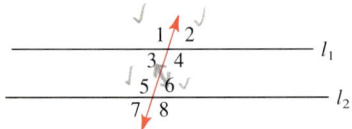

Solution:

Angles 1 and 2 are supplementary angles. So $\angle 2$ is 60°.
Angles 1, 4, 5, and 8 are equal. So each is 120°.
Angles 2, 3, 6, and 7 are equal. Thus each is 60°.

A **polygon** is a closed plane figure bounded by line segments. Triangles, quadrilaterals, pentagons, hexagons, and octagons are polygons with 3, 4, 5, 6, and 8 sides, respectively. If all the sides are of equal length, a polygon is called **equilateral.** If all the angles are equal, it is said to be **equiangular.** A **regular** polygon is one that is both equilateral and equiangular. Several polygons are shown below. The dashed lines are called **diagonals.**

Equilateral   Equiangular   Regular

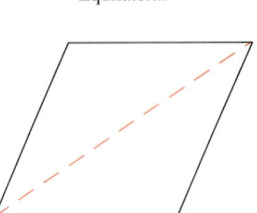 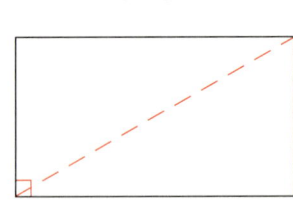 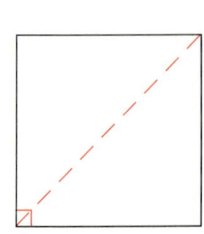

A **triangle** is a polygon with three sides.

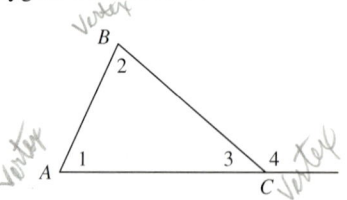

The three points *A*, *B*, and *C* are the **vertices** (singular, *vertex*) of the triangle. The three angles 1, 2, and 3 are the **interior angles** of the triangle, or simply the *angles*

of the triangle. The three sides of a triangle and the three angles are sometimes called the **parts** of a triangle. Angle 4 is called an **exterior angle** of the triangle.

Two important relationships exist for these angles. The sum of the measures of the interior angles is 180°. The measure of an exterior angle is equal to the sum of the measures of the opposite interior angles.

$$\angle 1 + \angle 2 + \angle 3 = 180°$$

$$\angle 4 = \angle 1 + \angle 2$$

EXAMPLE 3. Determine $\angle A$ in each figure.

(a)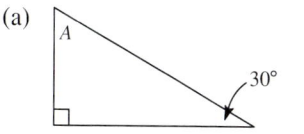

(b)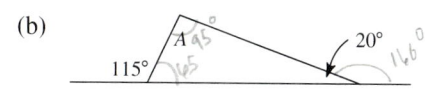

Solutions:

(a) Since this is a right triangle, $\angle A$ must be 60°.
(b) $\angle A + 20° = 115°$
    $\angle A = 95°$

Triangles

An **isosceles triangle** has two sides of equal length. The angles opposite the equal sides are equal.

An **equilateral triangle** has all three sides equal. It is also equiangular.

An **acute triangle** has three acute angles.

An **obtuse triangle** has an obtuse angle.

A **right triangle** contains a right angle. The longest side of a right triangle (the side opposite the right angle) is called the **hypotenuse.** The other two sides are called the **legs** of a right triangle.

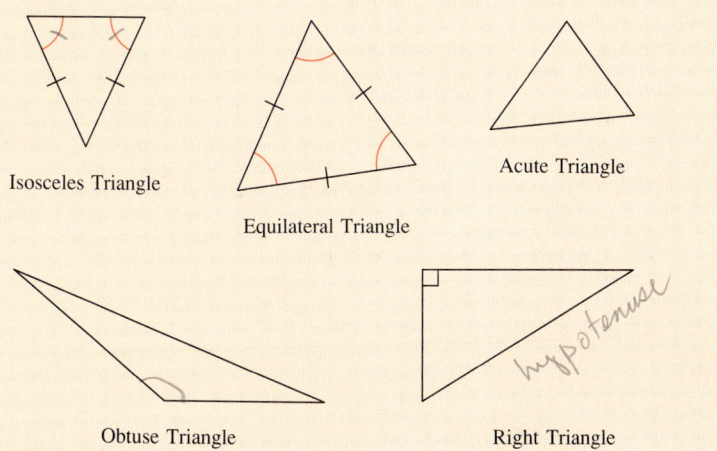

Isosceles Triangle   Equilateral Triangle   Acute Triangle

Obtuse Triangle   Right Triangle

The following result is one of the most useful relationships in all mathematics.

Pythagorean Theorem

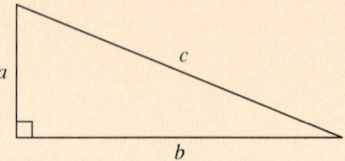

In the right triangle above, the side of length c units is the *hypotenuse* and the sides of lengths a and b units are the *legs*. The square of the length of the hypotenuse equals the sum of the squares of the lengths of the legs. That is,

$$c^2 = a^2 + b^2$$

EXAMPLE 4. Verify the Pythagorean Theorem for a right triangle with sides of 5, 12, and 13 inches.

Solution:

The hypotenuse is the longest side. So it must be 13.

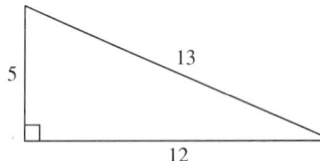

$$13^2 = 169$$
$$5^2 + 12^2 = 25 + 144$$
$$= 169$$

Thus the square of the length of the hypotenuse equals the sum of the squares of the lengths of the two legs. □

A geometric property that is of great practical use involves triangles that are the same shape.

Similar Triangles

Two triangles are similar if any one of the following statements is true.

1. Two angles of one triangle equal two angles of the other.

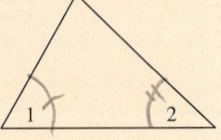 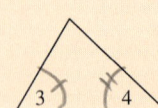

$$\angle 1 = \angle 3$$
$$\angle 2 = \angle 4$$

2. Corresponding sides are proportional.

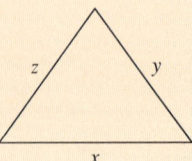 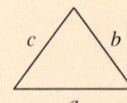

$$\frac{x}{a} = \frac{y}{b} = \frac{z}{c}$$

3. Two corresponding pairs of sides are proportional and the angle between them is equal.

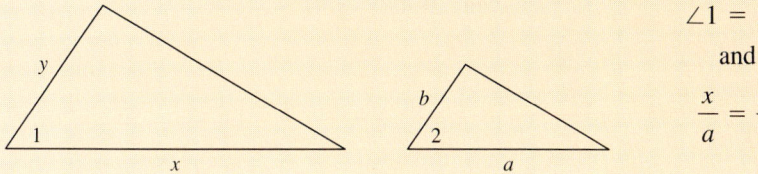

If two triangles are similar, all of these statements are true.

EXAMPLE 5. Are the triangles in the figure similar?

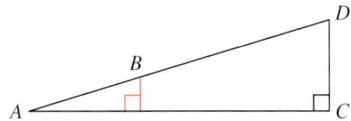

Solution:

As we examine the figure we notice a large triangle and a small triangle. Both are right triangles and they share the angle at the vertex A. This makes the angles at B and D equal. We see that they have two equal angles. Thus, by **1.** (page 32), they are similar triangles. ◻

EXAMPLE 6. The triangles are similar. Set up proportions for corresponding sides.

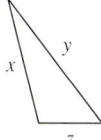 

Solution:

$$\frac{x}{4} = \frac{y}{5} = \frac{z}{2}$$

◻

A **quadrilateral** is a four-sided plane figure.

Quadrilaterals

A **trapezoid** is a quadrilateral with exactly two sides parallel. The parallel sides are called **bases** and the other sides are called **legs**. A trapezoid with equal legs is called an **isosceles trapezoid**.

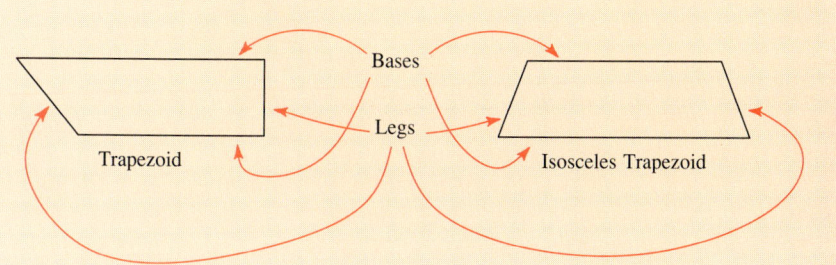

Sec. 0.5 Angles, Lines, and Plane Figures

A **parallelogram** is a quadrilateral in which *both* pairs of opposite sides are parallel. The diagonals of a parallelogram bisect each other, and any two consecutive angles are supplements of each other.

A **rhombus** is a parallelogram with equal sides.

A **rectangle** is a parallelogram whose angles are right angles.

A **square** is a rectangle with equal sides.

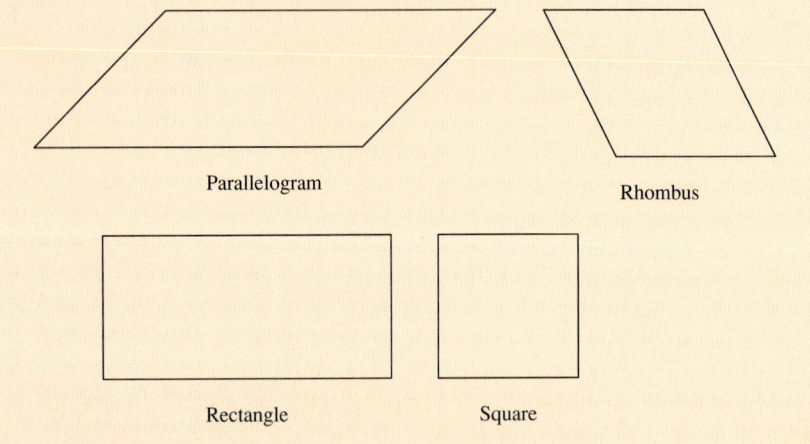

Circles

A **circle** is the set of all points in a plane that are the same distance from a given point called the **center.**

A **radius** of a circle is a line segment connecting the center with a point on the circle.

A **chord** is a line segment connecting two points on the circle.

A **diameter** is a chord that contains the center.

A **tangent** to a circle is a line in the plane of the circle that intersects the circle in exactly one point.

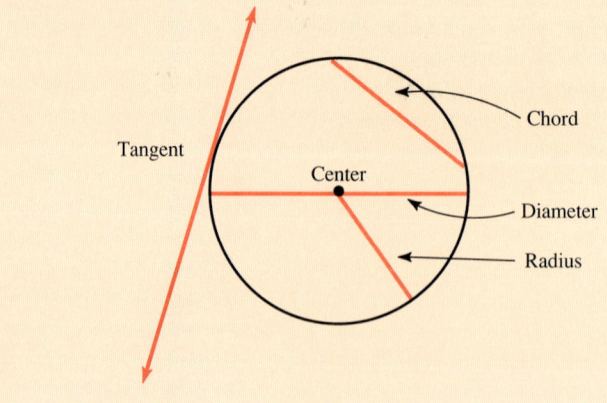

PROBLEM SET 0.5

Warm-ups

1. Find the measure of each angle in the figure. Lines l_1 and l_2 are parallel. Lines l_3 and l_4 are perpendicular. $\angle d = 45°$. See Examples 1 and 2.

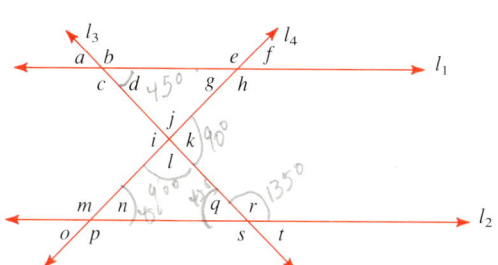

a = 45°   e. 135°   i. 90°   m. 135°   q. 45°
b = 135° f. 45°   j. 90°   n. 45°   r. 135°
c = 135° g. 45°   k. 90°   o. 45°   s. 135°
d = 45°   h. 135°   l. 90°   p. 135°   t. 45°

In problems 2 through 5, find $\angle A$. See Example 3.

2.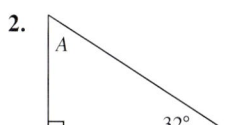

3. 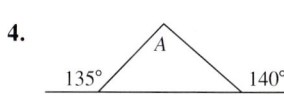 120°

4.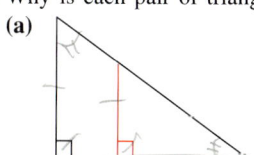

5. A 120° 30°

6. Verify the Pythagorean Theorem for a right triangle with sides of lengths 3, 4, and 5 units. See Example 4.

7. Why is each pair of triangles similar? See Example 5.
   (a)
   (b) $l_1 \parallel l_2$

Practice Exercises

In problems 8 through 19, determine if each statement is true or false. Use the figure that follows. Lines l_1 and l_2 are parallel. $\angle c = 120°$ and $\angle h = 105°$.

8. $\angle i = 75°$
9. $\angle j + \angle n = 180°$ false 75°
10. $\angle g + \angle j = 105°$
11. $\angle g = 45°$ false 75°
12. $\angle m + \angle n = 135°$
13. Angles h and k are supplementary. T
14. $\angle l = 45°$
15. Angle b is an acute angle. no
16. Angle f is obtuse.
17. Angles k and n are complementary. no
18. $\angle b = 105°$
19. $\angle m + \angle n = 105°$ yes

In problems 20 through 25, find $\angle A$.

20. 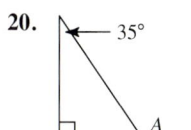 35°

21. 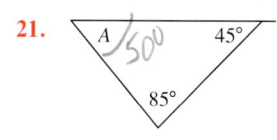 50° 45° 85°
    45
    85
    130
    = 50°

22. 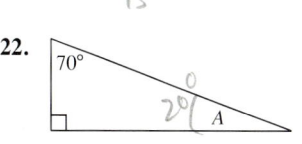 70° 20° A
    90
    70
    160
    = 20°

Sec. 0.5 Angles, Lines, and Plane Figures

35

23. 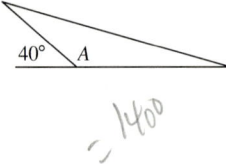   24. 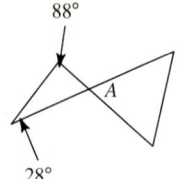   25.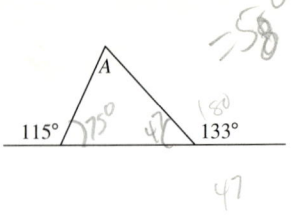

26. Find the measure of each angle in the figure. Lines l_1 and l_2 are parallel. $\angle d = 35°$. $\angle k = 130°$.

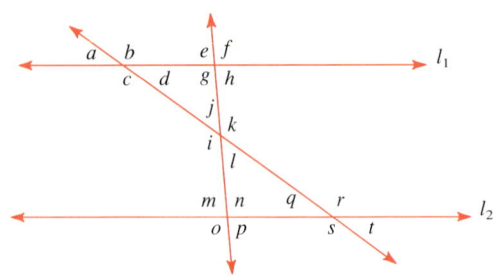

27. Verify the Pythagorean Theorem for a right triangle with sides of lengths 6, 8, and 10 units. $10^2 = 100$; $6^2 + 8^2 = 100$
28. Why is the following pair of triangles similar?   29. Why is the following pair of triangles similar?

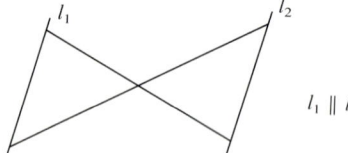   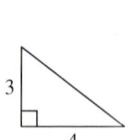   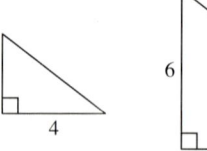   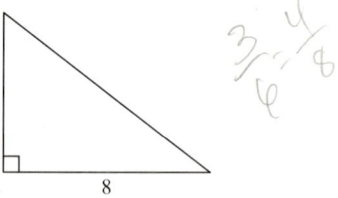

In problems 30 through 40, determine whether each statement is true or false.

30. A rectangle is equiangular.
31. An equilateral triangle is equiangular.
32. An isosceles triangle is an acute triangle.
33. A parallelogram is a quadrilateral.
34. A square is a rhombus.
35. A pentagon is a polygon with 7 sides.
36. A square is a regular polygon.
37. A triangle is a polygon.
38. A right triangle is an acute triangle.
39. The acute angles in a right triangle are complementary.
40. A rhombus is an equilateral polygon.

Challenge Problems

41. The sum of the measures of the interior angles of a triangle is 180°. Figure out a formula for the sum of the measures of the interior angles of a polygon with *n* sides.

(HINT: Divide a polygon into triangles.)

Quadrilateral (4 sides)   Pentagon (5 sides)

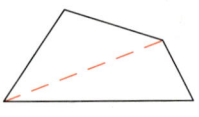   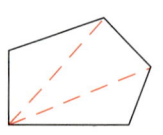

2 triangles   3 triangles

■ IN YOUR OWN WORDS . . .

42. What are similar triangles?
43. What does the Pythagorean Theorem say?

0.6 PERIMETER, SURFACE AREA, AREA, AND VOLUME

This section contains formulas for perimeter, area, surface area, and volume of geometric figures. The **perimeter** of a polygon is the sum of the lengths of its sides. The perimeter of a circle is called the **circumference** of the circle.

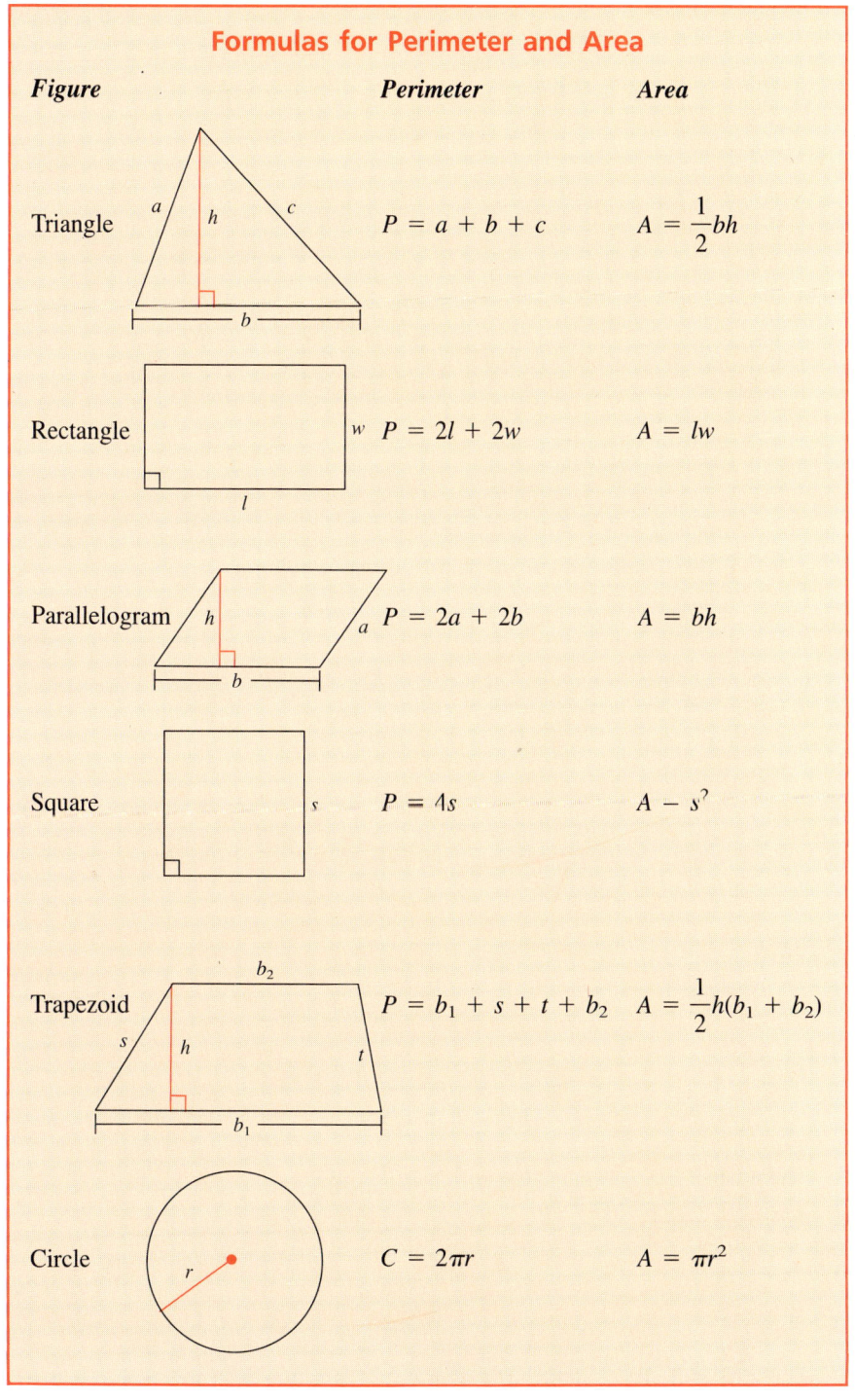

Formulas for Perimeter and Area

Figure	Perimeter	Area
Triangle	$P = a + b + c$	$A = \dfrac{1}{2}bh$
Rectangle	$P = 2l + 2w$	$A = lw$
Parallelogram	$P = 2a + 2b$	$A = bh$
Square	$P = 4s$	$A = s^2$
Trapezoid	$P = b_1 + s + t + b_2$	$A = \dfrac{1}{2}h(b_1 + b_2)$
Circle	$C = 2\pi r$	$A = \pi r^2$

Sec. 0.6 Perimeter, Surface Area, Area, and Volume

EXAMPLE 1. Find the perimeter of each figure.

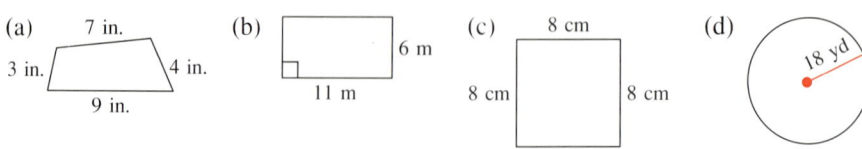

Solutions:

(a) A trapezoid is a quadrilateral, so its perimeter is given by

$$3 + 7 + 4 + 9 = 23$$

The perimeter is 23 inches (in.).

(b) The perimeter of a rectangle is twice the width plus twice the length.

$$P = 2 \cdot 6 + 2 \cdot 11$$
$$= 34$$

The perimeter is 34 meters (m).

(c) The perimeter of a square is four times the length of a side.

$$P = 4 \cdot 8$$
$$= 32$$

The perimeter is 32 centimeters (cm).

(d) The circumference of a circle is 2π times the radius.

$$C = 2 \cdot \pi \cdot 18$$
$$= 36\pi$$

The circumference is 36π yards (yd).

EXAMPLE 2. Find the area of the following figures.

(a) A parallelogram with base 7 in. and height 21 in.
(b) A 3-foot (ft) by 8-ft rectangle
(c) A right triangle with hypotenuse 5 cm and legs of 3 cm and 4 cm
(d) A circle of diameter 8 ft 4 in.

Solutions:

(a) The formula for the area of a parallelogram is

$$A = bh$$
$$= 7 \cdot 21$$
$$= 147$$

The area is 147 square inches (in.^{2}).

(b) The area of a rectangle is

$$A = lw$$
$$= 3 \cdot 8$$
$$= 24$$

The area is 24 ft^2.

(c) In a right triangle, if the base is one leg, the height is the other leg. Therefore,

$$A = \frac{1}{2}bh$$
$$= \frac{1}{2} \cdot 3 \cdot 4$$
$$= 6$$

The area is 6 cm^2.

(d) To find the area of a circle of diameter 8 ft 4 in., we first must write the diameter in one unit of measure. Since 8 ft is $8 \cdot 12 = 96$ in., the diameter of the circle is $96 + 4 = 100$ in. Thus the radius is 50 in. The area is given by

$$A = \pi r^2$$
$$= \pi(50)^2$$
$$= 2500\pi$$

The area is 2500π in.^{2}

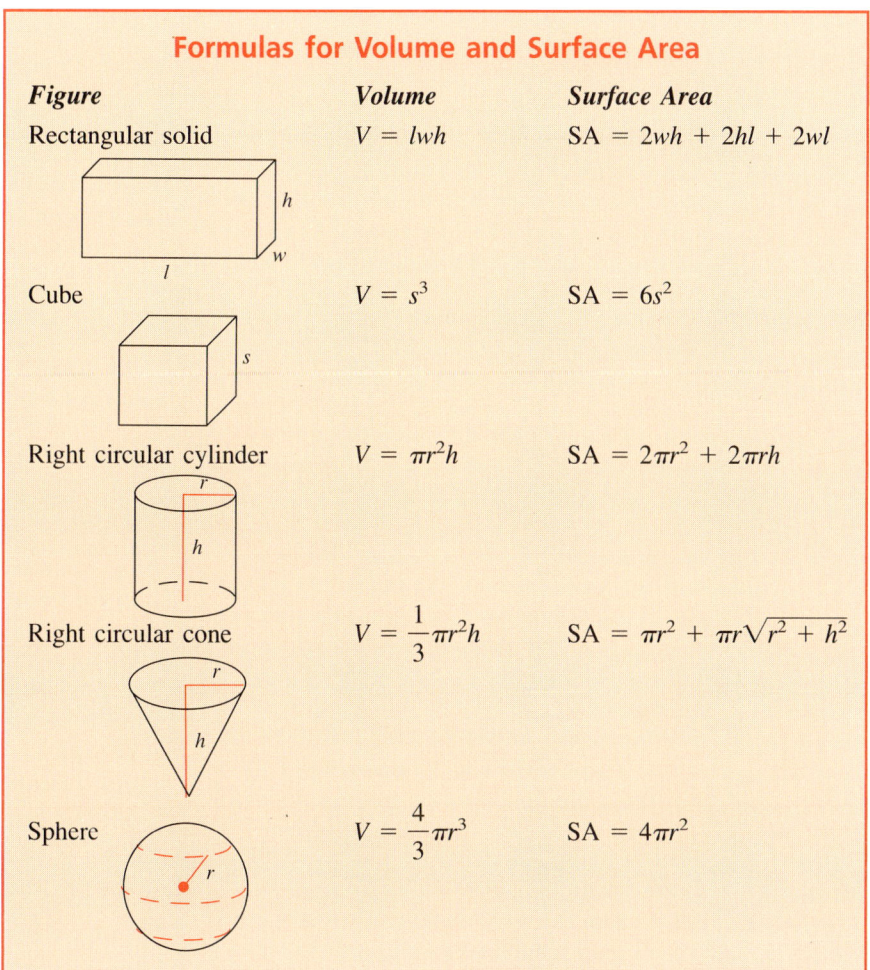

Formulas for Volume and Surface Area

Figure	Volume	Surface Area
Rectangular solid	$V = lwh$	$SA = 2wh + 2hl + 2wl$
Cube	$V = s^3$	$SA = 6s^2$
Right circular cylinder	$V = \pi r^2 h$	$SA = 2\pi r^2 + 2\pi rh$
Right circular cone	$V = \frac{1}{3}\pi r^2 h$	$SA = \pi r^2 + \pi r\sqrt{r^2 + h^2}$
Sphere	$V = \frac{4}{3}\pi r^3$	$SA = 4\pi r^2$

Sec. 0.6 Perimeter, Surface Area, Area, and Volume

EXAMPLE 3. How many cubic meters of water will it take to fill a swimming pool 50 m long, 20 m wide, and 2 m deep?

Solution:

Think of the pool as a rectangular solid.

$$V = lwh$$
$$= 50 \cdot 20 \cdot 2$$
$$= 2000$$

It will take 2000 cubic meters (m³) of water to fill the pool. ▫

EXAMPLE 4. How many cubic feet of silage can be stored in a silo 80 ft high with radius 18 ft?

Solution:

If we assume that the silo is a right circular cylinder, we have

$$V = \pi r^2 h$$
$$= \pi(18)^2(80)$$

The silo can store $25{,}920\pi$ cubic feet of silage. ▫

EXAMPLE 5. Find the surface area of a can of cola that is 5 in. high with a radius of $1\frac{1}{8}$ in.

Solution:

We assume that the can is a right circular cylinder.

$$SA = 2\pi r^2 + 2\pi rh$$
$$= 2\pi\left(\frac{9}{8}\right)^2 + 2\pi\left(\frac{9}{8}\right)(5)$$
$$= 2\pi \cdot \frac{81}{64} + 2\pi \cdot \frac{45}{8}$$
$$= \frac{81\pi}{32} + \frac{45\pi}{4}$$
$$= \frac{81\pi}{32} + \frac{360\pi}{32}$$
$$= \frac{441\pi}{32}$$
$$= 13\frac{25}{32}\pi$$

The surface area is $13\frac{25}{32}\pi$ square inches. ▫

PROBLEM SET 0.6

Warm-ups

In problems 1 through 6, find the perimeter of each plane figure. See Example 1.

1. A triangle with sides of 3, 4, and 5 in.
2. A trapezoid with sides of 17, 6, 19, and 5 m
3. A parallelogram with one side 13 cm and one side 4 cm
4. A rhombus with one side of 7 ft
5. A rectangle with length of 20 yd and width of 17 yd
6. A circle of radius 14 mm
7. Jim Rice hits a home run. How many feet must he run? (The distance from home plate to first base is 90 ft. Assume that his path around the bases forms a square.)
8. How much fencing must Tom Brown buy to enclose his backyard, pictured below?

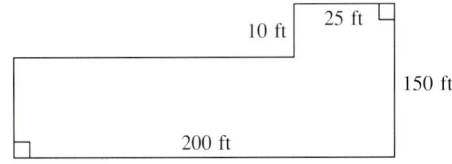

For problems 9 through 12, see Example 2.
In problems 9 through 11, find the area of the shaded part of each figure.

9.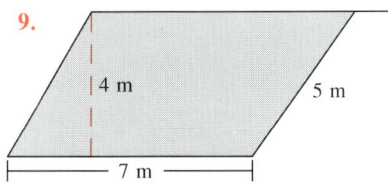

   Opposite sides parallel.

10.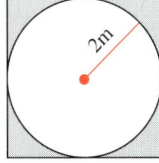

    Quadrilateral is a square.

11.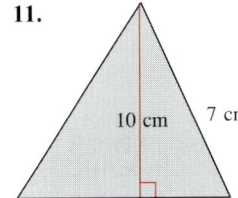

12. Find the area of a circular rug with a diameter of 8 ft 4 in.

For problems 13 through 18, see Examples 3 and 4.

13. Find the volume of a cube 5 units on a side.
14. Find the volume of a basketball whose radius is $5\frac{1}{2}$ in.

    Approximate this to the nearest hundredth.
15. How much water does it take to fill a rectangular swimming pool that measures 35 ft long, 25 ft wide, and 6 ft deep?
16. How much cork is there in a bottle cork that measures 20 mm in diameter and 45 mm in length? (Assume that the bottle cork is a right circular cylinder.)
17. Find the volume of an ice cream cone of height 3.6 in. and a base radius of 1.1 in.
18. How much Coca-Cola is there in a can 5 in. high with a radius of $1\frac{1}{8}$ in.

For problems 19 through 22, see Example 5.

19. Find the surface area of the earth. Use 4000 miles (mi) as the radius. Use $\pi \approx 3.14$ and approximate the answer to the nearest mile.
20. An oil tank is a right circular cylinder of radius 40 ft and height of 100 ft. Find its surface area.
21. A box has dimensions of 14 in. by $6\frac{1}{2}$ in. by $3\frac{1}{2}$ in. Find its surface area.
22. Find the number of square inches of material needed to make 5000 Dole pineapple juice cans if each can is 5 in. tall and 3 in. in diameter.

Practice Exercises

In problems 23 through 28, find the perimeter of each plane figure.

23. An equilateral triangle with sides of 2.5 in.
24. A trapezoid with sides of $17\frac{1}{2}$ m, $6\frac{1}{3}$ m, $19\frac{1}{6}$ m, and 5 m
25. A parallelogram with one side of 23 cm and one side of
26. A rhombus with one side $6\frac{2}{3}$ ft long

27. A rectangle with length of 20.3 yd and width of 17.23 yd
28. A square with one side $5\frac{1}{4}$ ft long
29. John wants to walk around the equator. How many miles would he walk? (The diameter of the earth is 8000 miles.)

In problems 30 through 32, find the area of each figure.

30.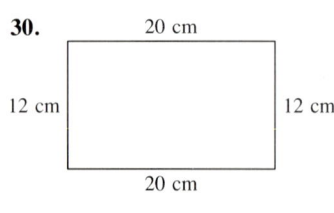
Rectangle

31.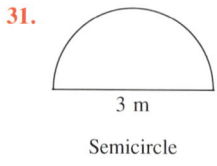
Semicircle

32.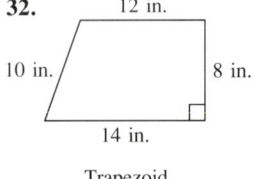
Trapezoid

33. Find the volume of a cube $2\frac{1}{3}$ units on a side.
34. How much air is in a spherical balloon of radius 18 cm?
35. How much Clorox 2 does it take to fill a rectangular box that measures $5\frac{1}{2}$ in. tall, $3\frac{1}{4}$ in. wide, and $1\frac{1}{2}$ in. deep?
36. How much detergent is there in a cylindrical bucket that measures 10 in. in diameter and is 1 ft tall?
37. Find the volume of a conical sand pile of height 5 yd and a base radius of 12 yd.
38. How much trash can be put in a cylindrical garbage can that is 1 yd high and has a radius of 2 ft?
39. Find the volume of the earth. Use 4000 miles as the radius.
40. A water tank is a right circular cylinder of radius 25 ft and height of 30 ft. Find its surface area.
41. Suzanne is making a box from a rectangular piece of cardboard that is 3 ft by $2\frac{1}{2}$ ft by cutting out a 6-in. square from each corner. How many cubic feet of crystals will the box hold?
42. A tool box has dimensions of 3 ft by 18 in. by 16 in. Find its surface area.
43. Find the number of square inches of material needed to make 10,000 Campbell soup cans if each can is 4 in. tall and $2\frac{1}{2}$ in. in diameter.
44. A tennis ball has a radius of $1\frac{1}{4}$ in. If there are three tennis balls in a can $8\frac{1}{2}$ in. tall with a radius of $1\frac{1}{2}$ in., how much space is not used by the tennis balls?
45. JoAnn is making a box from a piece of cardboard that is a 3-ft square by cutting out a 6-in. square from each corner. How many cubic feet of glitter will the box hold?
46. A marble of radius $\frac{1}{2}$ in. is placed inside a cube with an edge of length 1 in. Find the volume of that part of the cube which is outside the marble.

Challenge Problems

47. Find the surface area of the trough shown at the right.

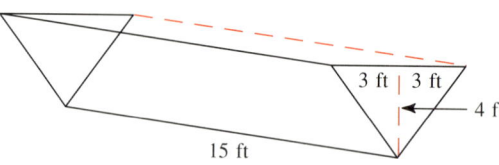

▬ IN YOUR OWN WORDS . . .

48. Make up a problem to find the volume of a sphere.
49. Make up a problem to find the surface area of a cube.

CHAPTER SUMMARY

GLOSSARY

Set: A collection of objects

Element or **member** of a set: The objects in the set. The symbol $\in$ means "is an element of."

Empty set or **null set:** A set with no elements, written as $\{\ \}$ or $\emptyset$.

Subset of a set A: A set with the property that each of its elements is in set A. A is a subset of B is written as $A \subseteq B$.

The **union** of sets A and B is the set of all members that belong to either A or B and is written $A \cup B$.

The **intersection** of sets A and B is the set of all elements that belong to both A and B and is written $A \cap B$.

Natural numbers or **counting numbers:** $\{1, 2, 3, \ldots\}$

Whole numbers: $\{0, 1, 2, 3, \ldots\}$

Integers: $\{\ldots, -3, -2, -1, 0, 1, 2, 3, \ldots\}$

Rational numbers: $\left\{\dfrac{p}{q},\text{ where } p \text{ and } q \text{ are integers and } q \neq 0\right\}$

Irrational numbers: {Numbers on a number line that are not rational}

Real numbers: {Rational and irrational numbers}

Additive identity: The number 0 is the additive identity.

Multiplicative identity: The number 1 is the multiplicative identity.

p is **less than** to q: The number p lies to the left of the number q on a number line. Write this as $p < q$.

p is **greater than** to q: The number p lies to the right of the number q on a number line. Write this as $p > q$.

$p \leq q$: The number p is less than or equal to the number q.

$p \geq q$: The number p is greater than or equal to the number q.

Reciprocal or **multiplicative inverse** of p: The number that when multiplied by p gives a product of 1.

Opposite or **additive inverse** of p: The number that when added to p gives a sum of 0.

Absolute value of a real number is its distance from zero on the number line. We define the absolute value of a number as

$$|b| = \begin{cases} b & \text{if } b \geq 0 \\ -b & \text{if } b < 0 \end{cases}$$

Base of x^n: The base is x.

Exponent of x^n: The exponent is n. The exponent tells how many times to use the base as a factor.

Acute angle: An angle with measure less than 90°.

Right angle: An angle with measure of 90°.

Obtuse angle: An angle with measure more than 90°.

Straight angle: An angle with measure of 180°.

Complementary angles: Two angles whose measures total 90°.

Supplementary angles: Two angles whose measures total 180°.

Parallel lines: Two lines in the same plane that do not intersect.

Perpendicular lines: Two lines that intersect at right angles.

Polygon: A closed figure whose sides are line segments.

Similar triangles: Triangles that are the same shape.

Isosceles triangle: A triangle with two sides of equal length.

Equilateral triangle: A triangle with three sides of equal length.

Acute triangle: A triangle with three acute angles.

Obtuse triangle: A triangle with one obtuse angle.

Quadrilateral: A polygon with four sides.

Trapezoid: A quadrilateral with two sides parallel.

Isosceles trapezoid: A trapezoid with equal legs.

Parallelogram: A quadrilateral with both pairs of sides parallel.

Rhombus: A parallelogram with equal sides.

Rectangle: A parallelogram with right angles.

Square: A rectangle with equal sides.

Sec. 0.6   Perimeter, Surface Area, Area, and Volume

PROPERTIES OF REAL NUMBERS

Properties of Addition

1. $p + q = q + p$ Commutative
2. $p + (q + r) = (p + q) + r$ Associative
3. $p + 0 = 0 + p = p$ Identity
4. $p + (-p) = 0$ Inverse

Properties of Multiplication

1. $pq = qp$ Commutative
2. $p(qr) = (pq)r$ Associative
3. $p \cdot 1 = 1 \cdot p = p$ Identity
4. $p \cdot \dfrac{1}{p} = 1$ Inverse

Distributive Property

$p(q + r) = pq + pr$

OPERATIONS WITH REAL NUMBERS

Addition

1. If the two numbers have the same sign, add the absolute values of the numbers and keep the common sign.
2. If the two numbers have different signs, subtract the absolute value of the smaller number from the absolute value of the larger number and keep the sign of the number with the larger absolute value.

Subtraction

1. Change the subtraction to addition.
2. Follow the rules for addition.

Multiplication and Division

1. If the signs of the two numbers are alike, multiply or divide the absolute values of the numbers. The answer will be positive.
2. If the signs of the two numbers are not alike, multiply or divide the absolute value of the numbers. The answer will be negative.

ORDER OF OPERATIONS

If grouping symbols are present, perform operations inside them, starting with the innermost symbol, in the following order.

1. Perform any exponentiations.
2. Perform all multiplications and divisions in order from left to right.
3. Perform all additions and subtractions from left to right.

If grouping symbols are not present, perform operations in the order given above.

PROPERTIES OF ABSOLUTE VALUE

If p and q are real numbers, with $q \neq 0$:

1. $|p \cdot q| = |p| \, |q|$
2. $\left| \dfrac{p}{q} \right| = \dfrac{|p|}{|q|}$
3. $|p| = |-p|$

$|p - q| = |q - p|$

DISTANCE BETWEEN TWO REAL NUMBERS, p AND q

If two lines are cut by a transversal, the lines are parallel if any of the following statements are true.

1. *Corresponding angles* are equal.
2. *Alternate interior angles* are equal.
3. *Alternate exterior angles* are equal.

If two lines are parallel, all of the statements are true.

PARALLEL LINES CUT BY A TRANSVERSAL

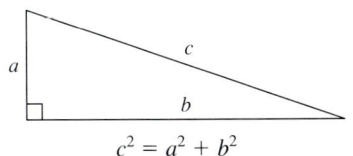

$c^2 = a^2 + b^2$

PYTHAGOREAN THEOREM

Figure	Perimeter	Area
Triangle	$P = a + b + c$	$A = \frac{1}{2}bh$
Rectangle	$P = 2l + 2w$	$A = lw$
Parallelogram	$P = 2a + 2b$	$A = bh$
Square	$P = 4s$	$A = s^2$
Trapezoid	$P = b_1 + s + t + b_2$	$A = \frac{1}{2}h(b_1 + b_2)$
Circle	$C = 2\pi r$	$A = \pi r^2$

PERIMETER AND AREA FORMULAS

Figure	Volume	Surface Area
Rectangular solid	$V = lwh$	$SA = 2wh + 2hl + 2wl$
Cube	$V = s^3$	$SA = 6s^2$
Right circular cylinder	$V = \pi r^2 h$	$SA = 2\pi r^2 + 2\pi rh$
Right circular cone	$V = \frac{1}{3}\pi r^2 h$	$SA = \pi r^2 + \pi r\sqrt{r^2 + h^2}$
Sphere	$V = \frac{4}{3}\pi r^3$	$SA = 4\pi r^2$

VOLUME AND SURFACE AREA FORMULAS

CHECKUPS

1. Classify each number as rational or irrational.
    (a) $\frac{3}{7}$ (b) $0.\overline{78}$ (c) $\sqrt{7}$ (d) 1.7   Section 0.1; Example 4

2. Indicate the property of real numbers that justifies each statement.
    (a) $2 + 3 = 3 + 2$ (b) $1(3) = 3$
    (c) $-5 + 0 = -5$ (d) $(4 \cdot 3) \cdot 6 = 4(3 \cdot 6)$ Section 0.2; Example 3

Sec. 0.6 Perimeter, Surface Area, Area, and Volume

3. Use the properties of absolute value to simplify.

   (a) $|2x|$ (b) $\left|\dfrac{-3}{x}\right|$ Section 0.3; Example 3

4. Simplify $\dfrac{(-10-4)(-2)}{\sqrt{28-12}-11}$.   Section 0.4; Example 3d

5. Find $\angle A$ in the figure.   Section 0.5; Example 3b

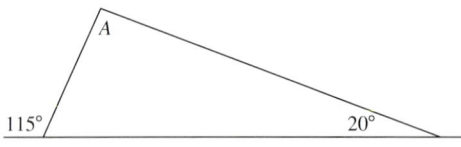

6. Find the perimeter of the figure.   Section 0.6; Example 1d

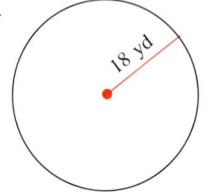

7. Find the area of a right triangle with hypotenuse 5 cm and legs of 3 cm and 4 cm.   Section 0.6; Example 2c

8. How many cubic feet of silage can be stored in a silo 80 ft high with a radius of 18 ft?   Section 0.6; Example 4

9. Find the surface area of a can of cola that is 5 in. high with radius of $1\dfrac{1}{8}$ in.   Section 0.6; Example 5

REVIEW PROBLEMS

In problems 1 through 14, perform the operation indicated.

1. $-(-7)[2-(5-8)]$

2. $\dfrac{3}{7} \div \left(\dfrac{-9}{14}+\dfrac{1}{7}\right)$

3. $\dfrac{-6(-7)}{(-2-19)-\sqrt{30-5}}$

4. $(-2-8)\cdot(-3)^2$

5. $\dfrac{(-3)(-4)-(-4)(-5)}{-5-2+(-1)}$

6. $\dfrac{5}{0}$

7. $-|6+(-11)|-6^2$

8. $\dfrac{-2}{9}\left[-\dfrac{1}{2}\div\dfrac{1}{4}-2(5-15)\right]$

9. $|(-4)(5)|-|6-19|$

10. $-7+(-6)-(-4)$

11. $(6-8)^2+\dfrac{9}{3}$

12. $3\cdot\dfrac{3-8}{2^2-3^2}$

13. $(-9)^2-[17-(4-12)]$

14. $\dfrac{\dfrac{2-3(5-9)}{(-6)}}{3-(-6)}$

In problems 15 through 20, indicate the property that justifies each statement.

15. $2+7=7+2$

16. $(3\cdot 4)\cdot 4 = 3\cdot(4\cdot 4)$

17. $6(7+5)=6\cdot 7+6\cdot 5$

18. $5+(-5)=0$

19. $9\cdot 1=9$

20. $4\cdot\dfrac{1}{4}=1$

In problems 21 through 25, classify each number as rational or irrational.

21. $\sqrt{11}$

22. $\dfrac{3}{8}$

23. 7.68

24. $\sqrt{14}$

25. $3.3\overline{7}$

In problems 26 through 28, A = {a, b, c} and B = {c, d, e, f}.

26. List the elements of $A \cup B$.
27. List the elements of $A \cap B$.
28. List the subsets of A.
29. If $C = \{x \mid x$ is an odd natural number less than 7$\}$, list the elements of C.
30. If $D = \{3, 6, 9, 12, 15\}$, write D using set-builder notation.

In problems 31 through 35, evaluate each expression.

31. $|-5| - |25|$ **32.** $|3x^2y|$ **33.** $-|-25|$
34. $-|6p^2|$ **35.** $-|-4| - |-21|$

In problems 36 through 40, assume that a, b, and c are real numbers with $a > 0$ and $b < 0$.
Simplify, if possible.

36. $|a|$ **37.** $|b|$ **38.** $|c|$ **39.** $|-a|$ **40.** $|-b|$

In problems 41 through 44, find the perimeter and area of each figure.

41.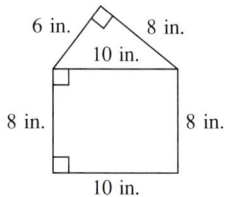

42.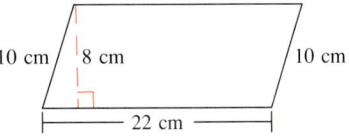

Opposite sides parallel.

43.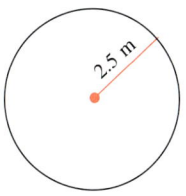

44.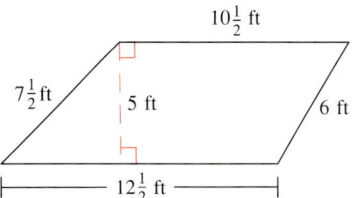

In problems 45 through 47, find the volume and surface area of each figure.

45.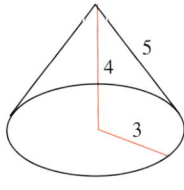

Cone

46.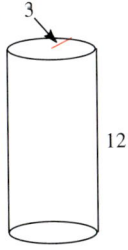

Cylinder

47.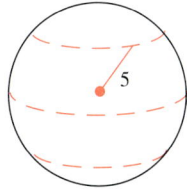

Sphere

48. Find the measure of each angle. l_1 and l_2 are parallel. $\angle b = 65°$.

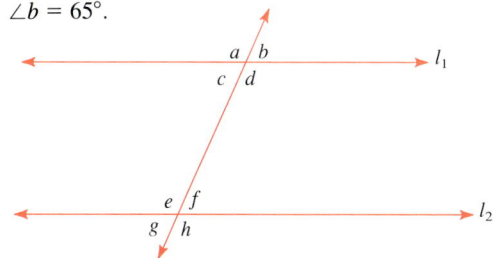

Sec. 0.6 Perimeter, Surface Area, Area, and Volume

In problems 49 through 50, indicate why the pair of triangles are similar.

49. A pair of equilateral triangles

50. A pair of right triangles with an acute angle of 33°

CHAPTER TEST

In problems 1 through 9, choose the correct answer.

1. If A is a real number and $A < 0$, which of the following is true?
   A. $|A| = A$
   B. $|A| = -A$
   C. $|A - B| = |A + B|$
   D. All are true

2. The set of irrational numbers is a subset of (?)
   A. the set of rational numbers
   B. the set of integers
   C. the set of natural numbers
   D. the set of real numbers

3. The rational number $\frac{1}{6}$ expressed as a decimal is (?)
   A. .16   B. .$\overline{16}$ C. .1$\overline{6}$ D. .2

4. If $A = \{1, 2, 5\}$, which statement is true?
   A. $1 \notin A$ B. $1 \subseteq A$ C. $\emptyset \subseteq A$ D. $a \in A$

5. The statement $5 + 4 = 4 + 5$ illustrates which property?
   A. Distributive   C. Commutative
   B. Associative   D. Closure

6. The distance between -5 and p on the number line is (?)
   A. $p - 5$ B. $p + 5$ C. $|p - 5|$ D. $|p + 5|$

7. The opposite of $2 - x$ is (?)
   A. $2 + x$
   B. $-2 - x$
   C. $x - 2$
   D. None of the above

8. If p is a negative real number, which statement is true?
   A. p lies to the right of 0 on a number line.
   B. $-p$ lies to the left of 0 on a number line.
   C. $-p$ lies to the right of 0 on a number line.
   D. None of the above

9. The reciprocal of $\frac{3}{8}$ is (?)
   A. $\frac{3}{8}$ B. $\frac{8}{3}$ C. 1   D. Does not exist

In problems 10 through 16, perform the operations indicated.

10. $(-7)^2 - (-13)$

11. $\dfrac{-5}{9} \div (-15)$

12. $-5^2 + 4 \cdot 6$

13. $3\{2 - 3[5 + 2(8 - 9)]\}$

14. $\dfrac{16 - 4^2}{8}$

15. $-\dfrac{2}{3}\left(3 - \dfrac{9}{2}\right)$

16. $3|11 - 17|$

In problems 17 through 20, rewrite the expression using the property listed.

17. $6 + (-7)$ Commutative Property of Addition

18. $x(6 + y)$ Distributive Property

19. $-5 + 5$ Additive Inverse

20. $2(x + t)$ Commutative Property of Multiplication

21. Find the measures of each angle. $\angle a = 130°$ and $\angle d = 30°$.

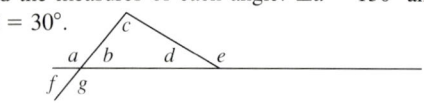

22. Find the volume of a cylindrical tank 12 ft high that has a radius of 4 ft.

23. Find the amount of fencing needed to enclose a rectangular garden with dimensions 28 m by 22 m.

24. Find the surface area of a circular ball of radius $3\frac{1}{2}$ cm.

25. Find the area of the parallelogram at the right.

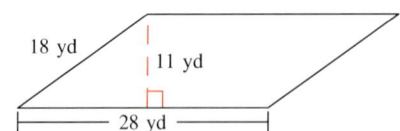

CHAPTER 1

See Section 3, Exercise 53.

Polynomials

1.1 Exponents
1.2 Definition, Evaluation, and Combining Like Terms
1.3 Addition and Subtraction
1.4 Multiplication
1.5 Pascal's Triangle (Optional)
1.6 Division
1.7 Synthetic Division (Optional)

CONNECTIONS

For centuries mathematicians have sought to approximate the value of π. Near the end of the seventeenth century, James Gregory used the expression

$$x - \frac{1}{3}x^3 + \frac{1}{5}x^5 - \frac{1}{7}x^7$$

to approximate π to 100 decimal places. He showed that $\frac{\pi}{4}$ is approximately the value of this expression when x is 1.

$$\frac{\pi}{4} \approx 1 - \frac{1}{3} + \frac{1}{5} - \frac{1}{7}$$

Expressions such as $x - \frac{1}{3}x^3 + \frac{1}{5}x^5 - \frac{1}{7}x^7$ are called polynomials.

Today polynomials are among the most useful expressions in mathematics. For example, they may be used in a business to find the cost of manufacturing calculators, by a biologist to find the volume of a spherical cell; or, by a physicist to find the distance fallen by a free-falling body.

This chapter deals with polynomials. We learn how to add, subtract, multiply, and divide polynomials. In Chapter 2, we will learn how to factor polynomials. In Chapter 3, we will see a connection between algebraic fractions and polynomials, and in Chapter 6 we will see a connection between equations and polynomials.

1.1 EXPONENTS

Exponents were developed as a notation for writing products with the same number. For example, instead of writing $2 \cdot 2 \cdot 2 \cdot 2 \cdot 2$, we write 2^5 (read "two to the fifth power").

Natural Number Exponents

If x is a real number and n is a natural number, then

$$x^n = \underbrace{x \cdot x \cdot x \cdots x}_{n \text{ factors of } x}$$

So

$$x^1 = x$$
$$x^2 = x \cdot x$$
$$x^{10} = x \cdot x \cdot x \cdot x \cdot x \cdot x \cdot x \cdot x \cdot x \cdot x$$

In writing x^n, we call x the **base** and n the **exponent.** The exponent gives the number of times the base is used as a factor. Sometimes the exponent is called the **power.** Read x^n as "x to the nth power."

NOTE: x^n does **not** mean n times x.

EXAMPLE 1. Identify the base and simplify each expression.

(a) 3^2 (b) $(-3)^2$ (c) -3^2 (d) $4\left[\dfrac{1}{3}\right]^2$

(e) $-2x^3$ (f) $(-2x)^3$ (g) $(-2)^3$ (h) -2^3

Solutions:

(a) 3^2. This is 3 squared.
The base is 3.
$$3^2 = 3 \cdot 3 = 9$$

(b) $(-3)^2$. This is -3 squared.
The base is -3.
$$(-3)^2 = (-3)(-3) = 9$$

(c) -3^2. This is the opposite of 3 squared.
The base is 3.
$$-3^2 = -3 \cdot 3 = -9$$

(d) $4\left[\dfrac{1}{3}\right]^2$. This is four times $\dfrac{1}{3}$ squared.
The base is $\dfrac{1}{3}$.
$$4\left[\dfrac{1}{3}\right]^2 = 4\left[\dfrac{1}{3} \cdot \dfrac{1}{3}\right]$$
$$= \dfrac{4}{9}$$

(e) $-2x^3$. This is -2 times the cube of x.
The base is x. $-2x^3$ is simplified.

(f) $(-2x)^3$. This is the cube of $-2x$.
The base is $-2x$.
$$(-2x)^3 = (-2x)(-2x)(-2x)$$
$$= (-2)(-2)(-2)x \cdot x \cdot x$$
$$= -8x^3$$

(g) $(-2)^3$. This is the cube of -2.
The base is -2.
$$(-2)^3 = (-2)(-2)(-2) = -8$$

(h) -2^3. This is the opposite of 2 cubed.
The base is 2.
$$-2^3 = -2 \cdot 2 \cdot 2 = -8$$

As Example 1 indicates, negative signs can be troublesome. Notice that $-p^n$ and $(-p)^n$ are not the same. Compare (b) and (c), (e) and (f), and (g) and (h).

Identifying the Base

$-p^n$	$(-p)^n$
The base is p.	The base is $-p$.
$-p^n = -p \cdot p \cdot p \cdots p$	$(-p)^n = (-p)(-p)\cdots(-p)$
$-4^2 = -4 \cdot 4 = -16$	$(-4)^2 = (-4)(-4) = 16$

Looking closer at $(-p)^n$, we see that if n is odd,

$$(-p)^n = (-p)(-p)\cdots(-p) \quad (n \text{ factors of } -p, \text{ and } n \text{ is odd})$$
$$= -p^n$$

However, if n is even,

$$(-p)^n = (-p)(-p)\cdots(-p) \quad (n \text{ factors of } -p, \text{ and } n \text{ is even})$$
$$= p^n$$

Compare (b) with (g) in Example 1.

This discussion leads to the following result.

Odd and Even Powers

If p is a real number and n is a natural number,

$$(-p)^n = \begin{cases} p^n & \text{if } n \text{ is even} \\ -p^n & \text{if } n \text{ is odd} \end{cases}$$

EXAMPLE 2. Simplify each expression.

(a) $(-x)^5$ (b) $(-y)^6$ (c) $(-4)^3$ (d) $(-3)^4$

Solutions:

(a) This is of the form $(-p)^n$ with n odd.

$$(-x)^5 = -x^5$$

(b) This is of the form $(-p)^n$ with n even.

$$(-y)^6 = y^6$$

(c) $(-4)^3 = -4^3$ 3 is *odd*
$ = -4 \cdot 4 \cdot 4$
$ = -64$

(d) $(-3)^4 = 3^4$ 4 is *even*
$ = 3 \cdot 3 \cdot 3 \cdot 3$
$ = 81$

Not only is the exponent notation convenient to use, but it has some very nice properties. Suppose that we want to multiply $x^3 \cdot x^2$. Use the definition of exponent and write

$$x^3 \cdot x^2 = (x \cdot x \cdot x)(x \cdot x) = x \cdot x \cdot x \cdot x \cdot x = x^5$$

This leads to the first property of exponents:

Product with Same Base

$$x^m \cdot x^n = x^{m+n}$$

To multiply two expressions with the **same** base, keep the base and **add** the exponents.

Look at $(x^2)^3$. Use the definition of exponents and write

$$(x^2)^3 = x^2 \cdot x^2 \cdot x^2$$

Now use the first property and write

$$= x^{2+2+2}$$
$$= x^6$$

This gives the second property of exponents:

Power of a Power

$$(x^m)^n = x^{mn}$$

To raise a power to a power, keep the base and **multiply** the exponents.

Now consider a product raised to a power, such as $(xy^3)^2$. Again we use the definition of exponent and write

$$(xy^3)^2 = (xy^3)(xy^3)$$

Then use the commutative and associative laws and write

$$= x \cdot x \cdot y^3 \cdot y^3$$

Applying the *product with same base* rule gives

$$= x^{1+1}y^{3+3}$$
$$= x^2y^6$$

This leads to the third property of exponents:

Product to a Power

$$(xy)^n = x^n y^n$$

To raise a product to a power, raise each factor in the product to the power.

Sec. 1.1 Exponents

The next property is just like the preceding one except that we use a quotient instead of a product. Consider $\left[\dfrac{x^2}{y}\right]^2$. We apply the definition of exponent and write

$$\left[\dfrac{x^2}{y}\right]^2 = \dfrac{x^2}{y} \cdot \dfrac{x^2}{y}$$

$$= \dfrac{x^2 \cdot x^2}{y \cdot y}$$

$$= \dfrac{x^4}{y^2}$$

This leads to the fourth property of exponents:

Quotient to a Power

$$\left[\dfrac{x}{y}\right]^n = \dfrac{x^n}{y^n}; \quad y \neq 0$$

To raise a quotient to a power, raise both the numerator and the denominator to the power.

For the final property of exponents, let's look at some examples of quotients with the same base. Consider $\dfrac{x^5}{x^3}$, $\dfrac{x^3}{x^5}$, and $\dfrac{x^4}{x^4}$. Again using the definition of exponent, we write

$$\dfrac{x^5}{x^3} = \dfrac{x \cdot x \cdot x \cdot x \cdot x}{x \cdot x \cdot x} \qquad \dfrac{x^3}{x^5} = \dfrac{x \cdot x \cdot x}{x \cdot x \cdot x \cdot x \cdot x} \qquad \dfrac{x^4}{x^4} = \dfrac{x \cdot x \cdot x \cdot x}{x \cdot x \cdot x \cdot x}$$

$$= x^2 \qquad\qquad\qquad = \dfrac{1}{x^2} \qquad\qquad\qquad = 1$$

So we have the fifth property of exponents:

Quotient with the Same Base

$$\dfrac{x^m}{x^n} = \begin{cases} x^{m-n} & \text{if } m > n \\ 1 & \text{if } m = n \\ \dfrac{1}{x^{n-m}} & \text{if } n > m \end{cases} \quad x \neq 0$$

Properties of Exponents

If m and n are natural numbers and x and y are real numbers,

1. $x^m x^n = x^{m+n}$ Product with same base

2. $(x^m)^n = x^{mn}$ Power of a power

3. $(xy)^n = x^n y^n$ Product to a power

4. $\left[\dfrac{x}{y}\right]^n = \dfrac{x^n}{y^n};\ y \neq 0$ Quotient to a power

5. $\dfrac{x^m}{x^n} = \begin{cases} x^{m-n} & \text{if } m > n \\ 1 & \text{if } m = n \\ \dfrac{1}{x^{n-m}} & \text{if } n > m \end{cases}\ x \neq 0$ Quotient with same base

EXAMPLE 3. Use the properties of exponents to simplify each expression.

(a) $-3^2 \cdot 3^3$ (b) $(-x^2)^3 \cdot (-x)^4$ (c) $(-2x^2y^3)^3$

(d) $\left[-\dfrac{x^2}{y}\right]^4$ (e) $\dfrac{-5^5}{5^2}$ (f) $\dfrac{(x+2)^2}{(x+2)^8}$

Solutions:

(a) $-3^2 \cdot 3^3 = -3^{2+3} = -3^5$ Product with same base

(b) $(-x^2)^3 \cdot (-x)^4 = -(x^2)^3 \cdot (x)^4$ Odd and even powers
$ = -x^{2 \cdot 3} \cdot x^4$ Power of a power
$ = -x^{6+4}$ Product with same base
$ = -x^{10}$

(c) $(-2x^2y^3)^3 = (-2)^3(x^2)^3(y^3)^3$ Product to a power
$ = -8x^{2 \cdot 3}y^{3 \cdot 3}$ Power of a power
$ = -8x^6 y^9$

(d) $\left[-\dfrac{x^2}{y}\right]^4 = \left[\dfrac{x^2}{y}\right]^4$ Even power
$\phantom{\left[-\dfrac{x^2}{y}\right]^4} = \dfrac{(x^2)^4}{y^4}$ Quotient to a power
$\phantom{\left[-\dfrac{x^2}{y}\right]^4} = \dfrac{x^{2 \cdot 4}}{y^4}$ Power of a power
$\phantom{\left[-\dfrac{x^2}{y}\right]^4} = \dfrac{x^8}{y^4}$

(e) $\dfrac{-5^5}{5^2} = -\dfrac{5^5}{5^2}$
$\phantom{\dfrac{-5^5}{5^2}} = -5^{5-2}$ Quotient with same base
$\phantom{\dfrac{-5^5}{5^2}} = -5^3$
$\phantom{\dfrac{-5^5}{5^2}} = -125$

(f) $\dfrac{(x+2)^2}{(x+2)^8} = \dfrac{1}{(x+2)^{8-2}}$ Quotient with same base
$\phantom{\dfrac{(x+2)^2}{(x+2)^8}} = \dfrac{1}{(x+2)^6}$

Properties 2, 3, and 4 tell us when to *multiply* exponents. They are sometimes called power rules and are used in problems in the form

$$(\text{expression})^{\text{power}}$$

We must be very careful in simplifying problems of this type. The following two examples illustrate an *important* difference in simplifying an expression raised to a power.

EXAMPLE 4. Simplify each expression.

(a) $(x^4y^3)^2$ (b) $(x^4 + y^3)^2$

Solutions:

(a) $(x^4y^3)^2$

This is an expression raised to a power. Notice that the expression is a **product**. It **can** be simplified by multiplying exponents.

$$(x^4y^3)^2 = (x^4)^2(y^3)^2 \quad \text{Product to a power}$$
$$= x^8y^6 \quad \text{Power of a power}$$

(b) $(x^4 + y^3)^2$

Be Careful!

This also is an expression raised to a power. Notice that the expression is a **sum**. It **cannot** be simplified by multiplying exponents.

$$(x^4 + y^3)^2 = (x^4 + y^3)(x^4 + y^3) \quad not \quad x^8 + y^6$$

EXAMPLE 5. Simplify each expression.

(a) $\left[\dfrac{t^2}{z}\right]^3$ (b) $(t^2 - z)^3$

Solutions:

(a) $\left[\dfrac{t^2}{z}\right]^3$

This is an expression raised to a power. Notice that the expression is a **quotient**. It **can** be simplified by multiplying exponents.

$$\left[\dfrac{t^2}{z}\right]^3 = \dfrac{(t^2)^3}{z^3} \quad \text{Quotient to a power}$$

$$= \dfrac{t^6}{z^3} \quad \text{Power of a power}$$

(b) $(t^2 - z)^3$

Be Careful!

Again, we have an expression raised to a power. Notice that this expression is a **difference**. It **cannot** be simplified by multiplying exponents.

$$(t^2 - z)^3 = (t^2 - z)(t^2 - z)(t^2 - z) \quad not \quad t^6 - z^3$$

The point in Examples 4 and 5 is that an expression raised to a power can be simplified by multiplying exponents *if* the expression is made up of *products and* quotients. It *cannot* be simplified by multiplying exponents if it contains *sums or differences*.

In Section 1.4 we will see how to multiply expressions such as $(x^4 + y^3)^2$ and $(t^2 - z)^3$.

EXAMPLE 6. Simplify each expression.

(a) $\left[\dfrac{-2r^3s}{t^2}\right]^3$ (b) $\dfrac{(-xy^3)^2}{(-x^3y)^3}$

Solutions:

(a) Notice that this is a quotient to a power. We use properties 4, 3, and 2.

$$\left[\dfrac{-2r^3s}{t^2}\right]^3 = \dfrac{(-2r^3s)^3}{(t^2)^3} \quad \text{Quotient to a power}$$

$$= \dfrac{(-2)^3(r^3)^3 s^3}{(t^2)^3} \quad \text{Product to a power}$$

$$= \dfrac{-8r^9 s^3}{t^6} \quad \text{Power of a power}$$

(b) The numerator and the denominator each have a product to a power which can be simplified using properties 2, 3, and 5.

$$\dfrac{(-xy^3)^2}{(-x^3y)^3} = \dfrac{(-x)^2(y^3)^2}{(-x^3)^3 y^3} \quad \text{Product to a power}$$

$$= \dfrac{x^2 y^6}{-x^9 y^3} \quad \text{Power of a power}$$

$$= -\dfrac{y^3}{x^7} \quad \text{Quotient with same base}$$

Consider the quotient $\dfrac{2^5}{2^5}$. This is a number divided by itself and has the value 1. However, if we subtract exponents, we get

$$\dfrac{2^5}{2^5} = 2^{5-5} = 2^0$$

For exponent property 5 to make sense in both situations,

$$2^0 = 1$$

So we define an exponent of 0 as follows:

Zero Exponent

If x is any nonzero real number, then

$$x^0 = 1$$

Sec. 1.1 Exponents

EXAMPLE 7. Simplify each expression.

(a) $3x^0$ (b) $(23x^3y^5)^0$ (c) $(-5)^0$

(d) -5^0 (e) $\left[\dfrac{s}{t}\right]^0$ (f) $(2a + 3b)^0$

Solutions:

(a) $3x^0 = 3 \cdot 1 = 3$ (The base is x.)
(b) $(23x^3y^5)^0 = 1$ (The base is $23x^3y^5$.)
(c) $(-5)^0 = 1$ (The base is -5.)
(d) $-5^0 = -1$ (The base is 5.)
(e) $\left[\dfrac{s}{t}\right]^0 = 1$ $\left[\text{The base is } \dfrac{s}{t}.\right]$
(f) $(2a + 3b)^0 = 1$ (The base is $2a + 3b$.)

CALCULATOR BOX

Exponents with a Calculator

To raise a number (the base) to a power (the exponent), use the $\boxed{y^x}$ key and follow these steps:

1. Enter the base (y).
2. Press $\boxed{y^x}$. (On some calculators this key is marked $\boxed{x^y}$.)
3. Enter the exponent (x).
4. Press $\boxed{=}$.

EXAMPLE Evaluate 6^7.

Solution:

Press the keys,

$\boxed{6}$ $\boxed{y^x}$ $\boxed{7}$

and read $\boxed{279936}$ on the display.

$$6^7 = 279{,}936$$

The exponent, 2, is so common that scientific calculators have a special key, $\boxed{x^2}$, for squaring a number. To find 87^2 by using the $\boxed{x^2}$ key, press the keys

$\boxed{87}$ $\boxed{x^2}$

and read $\boxed{7569}$ on the display.

When using a calculator, it is **important** to decide if an answer is reasonable. To do this, we must learn to **estimate** the result of a calculator computation.

EXAMPLE Estimate the value of each expression, then use a calculator to approximate the value of each to the nearest thousandth.

(a) $(-2.13)^5$ (b) $\frac{4}{3}\pi(1.972)^3$

Solutions:

(a) Since -2.13 is close to -2, the answer should be about $(-2)^5$ or -32. Press the keys

$\boxed{2.13}$ $\boxed{+/-}$ $\boxed{y^x}$ $\boxed{5}$ $\boxed{=}$

and read -43.84277 on the display. This is reasonably close to the estimate, -32.

$$(-2.13)^5 \approx -43.843$$

(Notice the $\approx$ sign. This is an *approximation*.)

(b) Since 1.972 is close to 2 and π is close to 3, we see that an estimate of $\frac{4}{3}\pi(1.972)^3$ is $\frac{4}{3}3(2)^3$ or 32. Press the keys

$\boxed{4}$ $\boxed{\div}$ $\boxed{3}$ $\boxed{\times}$ $\boxed{\pi}$ $\boxed{\times}$ $\boxed{1.972}$ $\boxed{y^x}$ $\boxed{3}$ $\boxed{=}$

and read 32.1225 on the display. Therefore,

$$\frac{4}{3}\pi(1.972)^3 \approx 32.123$$

Calculator Exercises

In problems 1 through 4, find the value of each expression.

1. 7^8 **2.** $(-13)^5$ **3.** 679^2 **4.** $(-55.7)^2$

In problems 5 through 8, estimate the value of each expression, then use a calculator to approximate the value to five decimal places.

5. $(2.14)^3$ **6.** $(-9.1234)^2$ **7.** $(1.01)^5$ **8.** $\frac{\pi^2}{3}$

Answers

1. 5,764,801 **2.** $-371,293$ **3.** 461,041 **4.** 3102.49

5. 8; 9.80034 **6.** 81; 83.23643 **7.** 1; 1.05101 **8.** 3; 3.28987

PROBLEM SET 1.1

Warm-ups

In problems 1 through 44, simplify each expression.
For problems 1 through 8, see Example 1.

1. 2^4 **2.** $(-4)^2$ **3.** -4^2 **4.** $(-4)^3$

5. -4^3 **6.** $3\left[-\frac{1}{5}\right]^2$ **7.** $-3(-2)^2$ **8.** $-(-2)^3$

Sec. 1.1 Exponents

For Problems 9 through 12, see Example 2.

9. $(-x)^8$ **10.** $(-y)^9$ **11.** $(-s)^{11}$ **12.** $(-7)^{10}$

For problems 13 through 24, see Example 3.

13. $2^2 \cdot 2^3$ **14.** $-2^2 \cdot 2^3$ **15.** $-x^2 \cdot x^3$ **16.** $(-x)^2(-x)^3$

17. $(2^3)^2$ **18.** $(3x^2)^3$ **19.** $(-x^2)^5$ **20.** $\left[\dfrac{2}{z^4}\right]^3$

21. $\left[-\dfrac{3}{2}\right]^2$ **22.** $\left[-\dfrac{3}{2}\right]^3$ **23.** $\dfrac{x^{12}}{x^4}$ **24.** $\dfrac{2^2}{2^5}$

For problems 25 through 30, see Example 4.

25. $[(-x)^2]^3$ **26.** $(x^2y)^3$ **27.** $(2xy^2)^3$
28. $(3x^2y^3z^5)^4$ **29.** $[(x+y)^4]^5$ **30.** $[-x^2y(-z)^3]^3$

For problems 31 through 35, see Example 5.

31. $\left[\dfrac{2x}{y^2}\right]^3$ **32.** $\left[\dfrac{-2x}{y^3}\right]^3$ **33.** $\left[\dfrac{x^2}{-y^4}\right]^2$

34. $\left[\dfrac{x^5}{y^3}\right]^4$ **35.** $\left[\dfrac{(x-1)^3}{(x-1)^2}\right]^5$

For problems 36 through 41, see Example 6.

36. $\left[\dfrac{2x^3y}{-z^2}\right]^3$ **37.** $\left[\dfrac{-x^4yz^3}{2s^2}\right]^5$ **38.** $-\left[\dfrac{5xy^2z}{-15x^2yz^3}\right]^4$

39. $\dfrac{(25x^5z)^2}{(5x^3z^3)^3}$ **40.** $\dfrac{(2xy^4z^3)^3}{(-6x^3yz^4)^2}$ **41.** $\dfrac{(-xyz)^3}{(-x^2yz^4)^2}$

For problems 42 through 44, see Example 7.

42. $\left[\dfrac{-2xy^2}{z^{11}}\right]^0$ **43.** $-2^0\left[-\dfrac{1}{2}\right]^4$ **44.** $\dfrac{(-5x^0)^2}{(-5x^2)^0}$

Practice Exercises

In problems 45 through 114, simplify each expression.

45. 2^3 **46.** $(-3)^3$ **47.** -3^3 **48.** $(-3)^2$

49. -3^2 **50.** 2^0 **51.** $3^0\left[\dfrac{1}{2}\right]^2$ **52.** $2\left[-\dfrac{1}{3}\right]^3$

53. $-2(-3)^3$ **54.** $-(-3)^2$ **55.** $(3^2)^2 \cdot (3^2)^3$ **56.** $2^2(3)^2$

57. $-3^0\left[-\dfrac{1}{2}\right]^2$ **58.** $-3^3 \cdot (3^2)^2$ **59.** $(-2)^3 \cdot 3^2$ **60.** $2^2(-3)^2$

61. $2^2(-3^2)$ **62.** $-2^2(-3)^3$ **63.** $(-2)^2(-3^3)$ **64.** $3^2 \cdot 3^3$
65. $-2^3 \cdot (2^3)^2$ **66.** $(-2)^3 \cdot 2^3$ **67.** $2^2(-2)^2$ **68.** $2^2(-2^2)$

69. $(2 \cdot 3)^3$ **70.** $(-2 \cdot 3)^3$ **71.** $(-2 \cdot 3)^2$ **72.** $\left[\dfrac{3}{2}\right]^2$

73. $\left[-\dfrac{3}{2}\right]^3$ **74.** $\left[-\dfrac{3}{2}\right]^2$ **75.** $\left[\dfrac{2 \cdot 3}{5}\right]^2$ **76.** $\dfrac{2^3}{2^6}$

77. $\dfrac{2^3}{2^2}$ **78.** $\dfrac{-2^3}{2^2}$ **79.** $\dfrac{-2}{(-2)^9}$ **80.** $\dfrac{(-2)^7}{(-2)^3}$

81. $\dfrac{-3^4}{3^7}$ **82.** $\dfrac{(-5)^4}{5^5}$ **83.** $\dfrac{(-7)^9}{-7^8}$ **84.** $\dfrac{5^2}{11^0}$

85. -8^0 **86.** $\left[\dfrac{4^9}{13^{15}}\right]^0$ **87.** $x^3 \cdot x^4$ **88.** $(x-2)^3(x-2)^5$

89. $(-x^3)(x^2)$ **90.** $(z+1)^7(z+1)$ **91.** $(-x)^3 \cdot x^2$ **92.** $(-x^5)(-x^4)$

93. $x \cdot (-x)^2$
94. $x \cdot (-x)^4$
95. $(-x^2)^3$
96. $[(-x)^4]^5$
97. $(x^5y)^7$
98. $(3xy^3)^5$
99. $(-4x)^3$
100. $(-2x)^4$
101. $[(x+7)^4]^4$
102. $\left[\dfrac{3x}{y^3}\right]^2$
103. $\left[\dfrac{-2x}{y^2}\right]^5$
104. $\left[\dfrac{x^3}{-y^3}\right]^3$
105. $\left[\dfrac{(x+4)^4}{(x+4)^3}\right]^5$
106. $\dfrac{(5xy^4)^3}{(2x^3)^3}$
107. $\dfrac{(-3x^3y)^4}{(2^3x^3y^4)^3}$
108. $\dfrac{(16xy^0z)^3}{(8x^2yz)^2}$
109. $\dfrac{(14xz)^2}{(7x^2z)^3}$
110. $\dfrac{(4xy^2z^7)^4}{(-8x^2yz^4)^3}$
111. $\dfrac{(-xyz)^2}{(-x^2yz^3)^3}$
112. $\left[\dfrac{6xy^4z}{-18x^5yz^8}\right]^3$
113. $\left[\dfrac{-30x^5yz^8}{45x^5yz^3}\right]^2$
114. $\left[\dfrac{-x^4y^7z^8}{(xy^6z^3)^2}\right]^3$

115. Find the square of $3x^2y^3$.
116. Find the cube of $-5x^2z^4$.
117. Square the quotient when $11x^5z$ is divided by $7y^3$.
118. Cube the quotient when $12x^2y$ is divided by $8uv^3$.
119. What is the area of a square whose side has length $3x$ feet?
120. What is the area of a square whose side has length $3x^3$ meters?
121. What is the volume of a cube whose side has length $4t$ inches?
122. What is the volume of a cube whose side has length $2t^2$ centimeters?

Challenge Problems

In problems 123 through 130, simplify each expression. Assume that n is a natural number.

123. $6^n \cdot 6^2$
124. $\dfrac{3^{3n}}{3^n}$
125. $(8^n)^2$
126. $(5^2)^n$
127. $2 \cdot 2^n$
128. $\dfrac{9^{n+3}}{9^{n+1}}$
129. $x^n \cdot x^{2n}$
130. $x^2 \cdot x^n$

■ IN YOUR OWN WORDS . . .

131. Explain what is squared in each expression.
    (a) $x^2 + y^2$ (b) $(x+y)^2$ (c) $(xy)^2$
132. Explain when the situation (expression)^{power} tells us to multiply exponents.
133. Explain what is meant by -7^2 and by $(-7)^2$.

1.2 DEFINITION, EVALUATION, AND COMBINING LIKE TERMS

The expression bx^n, where b is a nonzero real number and n is a nonnegative integer, is called a **monomial** in one variable. We call b the **coefficient**, x the **variable**, and n the **degree**. The following are monomials in one variable.

① $2x^2$	② $\dfrac{1}{2}y^3$	③ $\sqrt{3}x^5$	④ $-5z$	⑤ -7

	Coefficient	Variable	Degree
①	2	x	2
②	$\dfrac{1}{2}$	y	3
③	$\sqrt{3}$	x	5
④	-5	z	1
⑤	-7	any letter	0

(Note in ⑤ that $-7 = -7x^0$.) Monomials such as -7 are called **constants**.

The expression $bx^m y^n$, where b is a nonzero real number and m and n are nonnegative integers, is called a **monomial in two variables.** Its degree is $m + n$. For example, $-12x^2 y^4$ is a monomial in the two variables x and y. Its degree is 6 since $2 + 4 = 6$.

A **polynomial** is a finite sum of monomials. Its degree is the highest degree of the monomials that make it up. Consider $4x^5 + 3x^3 + (-2x^2) + (-6)$. This is a polynomial in one variable since it is the sum of four monomials in the same variable. However, we would write it in the form

$$4x^5 + 3x^3 - 2x^2 - 6$$

Its degree is 5. This polynomial is written so that the monomial of highest degree comes first and the degrees of the other monomials decrease in order. This is called **standard form.** The coefficient of the first term is called the **leading coefficient.** The monomials are most often called **terms.** The last term, -6, is called the **constant term.**

Consider the polynomial in two variables

$$4x^5 - 6x^4 y + 7x^2 y^3 - 3x + y^4$$

Its degree is 5. Notice that the powers of x are in decreasing order. This is considered standard form with respect to x for a polynomial in two variables.

EXAMPLE 1. Write each polynomial in standard form with respect to x and give the degree of each.

(a) $5 - 3x^2 + 2x$ (b) $2x + 3$
(c) $x^5 y - x^6 + 2y^6$ (d) $5 - 2x^2 y^3 + 3xy + 3y^5$

Solutions:

(a) Standard form is $-3x^2 + 2x + 5$.
   The degree is 2.
(b) Standard form is $2x + 3$.
   The degree is 1.
(c) Standard form is $-x^6 + x^5 y + 2y^6$.
   The degree is 6.
(d) Standard form is $-2x^2 y^3 + 3xy + 3y^5 + 5$.
   The degree is 5.

Polynomials that have one term are called **monomials.** Those that have two terms are called **binomials,** and those that have three terms are called **trinomials.**
Listed below are examples of each.

Monomials	*Binomials*	*Trinomials*
x	$x + y$	$x + y + z$
$5y^4$	$z^3 - 7$	$w^2 - 8w + 6$

Polynomials have many different values, depending on the value of each variable. To **evaluate** a polynomial means to find its value. Use the order of operations discussed in Section 0.4.

EXAMPLE 2. Evaluate each polynomial for the values given for each variable.

(a) $3x^2 - x + 4$ if x is -2 (b) $3x^2 y^4 - xy + y^5$ if x is 2 and y is -1

Solutions:

(a) $3x^2 - x + 4 = 3(-2)^2 - (-2) + 4$
$= 3(4) + 2 + 4$
$= 12 + 2 + 4$
$= 18$

(b) $3x^2y^4 - xy + y^5 = 3(2)^2(-1)^4 - (2)(-1) + (-1)^5$
$= 3(4)(1) - (-2) + (-1)$
$= 12 + 2 - 1$
$= 13$ □

The Distributive, Associative, and Commutative Properties apply to polynomials just as to real numbers.

Properties of Polynomials

If A, B, and C are polynomials, then

1. $AB = BA$ — Commutative for Multiplication
2. $A + B = B + A$ — Commutative for Addition
3. $A(BC) = (AB)C$ — Associative for Multiplication
4. $A + (B + C) + (A + B) + C$ — Associative for Addition
5. $A(B + C) = AB + AC$ — Distributive

These properties enable us to simplify many polynomials. For example, the binomial $7x + 3x$ can be simplified.

$7x + 3x = x(7 + 3)$ Distributive Property
$= x(10)$ Commutative Property
$= 10x$

This would not work with $7x + 3x^2$. We say that $7x$ and $3x$ are **like terms**. $5x^3$ and $-6x^3$ are like terms, while $2x$ and $5x^2$ are **unlike terms**. **Like terms** are "alike" everywhere except the coefficient.

EXAMPLE 3. Combine like terms in each expression.

(a) $3a - 7b + 12a + 4b$
(b) $5x^2 - 5y^2 + 3x^2 - 4y^2$
(c) $3x + 7y$
(d) $c^3d + 6c^2 - 5c^3d - 4c^2 + 6c$

Solutions:

(a) $3a - 7b + 12a + 4b = 15a - 3b$
(b) $5x^2 - 5y^2 + 3x^2 - 4y^2 = 8x^2 - 9y^2$
(c) $3x + 7y$ There are no like terms.
(d) $c^3d + 6c^2 - 5c^3d - 4c^2 + 6c = -4c^3d + 2c^2 + 6c$ □

Sec. 1.2   Definition, Evaluation, and Combining Like Terms

CALCULATOR BOX

Evaluating Polynomials with a Calculator

A polynomial can be evaluated, or at least approximated, with a calculator.

EXAMPLE Evaluate the polynomial $2x^3 + 3x^2 - 7x + 5$ when x is 9.

Solution:

We replace x with 9 in the polynomial

$$2(9)^3 + 3(9)^2 - 7(9) + 5$$

press the keys

| 2 | × | 9 | y^x | 3 | + | 3 | × | 9 | x^2 | − | 7 | × | 9 | + | 5 | = |

and read **1643** on the display. Therefore, the value of the polynomial is 1643 when x is 9.

Notice that we did not write down intermediate results. The calculator keeps the running total as each term is evaluated.

EXAMPLE Estimate the value of the polynomial $5t^7 - t^3 + 6$ when t has the value 1.028, then approximate the value to three decimal places with a calculator.

Solution:

Note that 1.028 is very close to 1, so we estimate the value of the polynomial using 1 for x.

$$5(1)^7 - (1)^3 + 6 = 5 \cdot 1 - 1 + 6$$
$$= 5 - 1 + 6$$
$$= 10$$

To find the approximate value, press the keys

| 5 | × | 1.028 | y^x | 7 | − | 1.028 | y^x | 3 | + | 6 | = |

and read **10.979897** on the display. The polynomial has the *approximate* value of 10.980 when t has the value 1.028. As this is very close to the estimate, it is a reasonable answer.

Calculator Exercises

Approximate the value of each polynomial when x is 2.171 and y is 0.964. Express answers to the nearest thousandth. Make an estimate to see if the answer is reasonable.

1. $8x^5 - 11x^4 + 2x + 3$
2. $13y^{11} + 25y^6 - 9y + 1$
3. $2x^2 + 3xy + y^2 - 11$
4. $12x^5 - 10x^2y + 4y^3 - 2$

Answers

1. 148.805   2. 21.073   3. 5.634   4. 534.884

PROBLEM SET 1.2

Warm-ups

In problems 1 through 6, write each polynomial in standard form with respect to x, and give its degree. See Example 1.

1. $x + 6$
2. $-3x^2y^2 + 4xy^3 - 5 + 2x^3y$
3. $4 - 3x + x^2$
4. $9 - x^3$
5. $\frac{1}{2}x - 3x^3 + 4x^2 - 7$
6. $4x^3 + \frac{2}{3}x - 17x^7 + 89$

In problems 7 through 10, evaluate each polynomial when x is −2. See Example 2.

7. $3x + 7$
8. $4x^2 - 2x + 1$
9. $2x^3 - x - 2$
10. $x^2 - x$

In problems 11 through 14, evaluate each polynomial when x is 1 and y is −1. See Example 2.

11. $xy + 6$
12. $x^2y^2 - xy + 1$
13. $2x^3y - 5x^2y^2 + xy + y^2$
14. $x^3 - y^3$

In problems 15 through 22, combine like terms, if possible. See Example 3.

15. $5x^3 - 7x^2 + 3x^3 - x^2$
16. $6xy + 5x^2y - 2x^2y - 8xy$
17. $3z^2 - 4z - z^2 + 6z$
18. $6z - 3z^2$
19. $17xyz - 11xy - 13z$
20. $7t^3 - 5t^2 + 6t - 5t^3 - t^2$
21. $w^5 - w + 6 - w^5 + w$
22. $6x^2 + x - 3 + 5x^2 + 3x + 1$

Practice Exercises

In problems 23 through 32, write each polynomial in standard form with respect to x, give its degree, and identify those which are monomials, binomials, or trinomials.

23. $x - 4 \quad x - 4;$
24. $-5x^2y^2 + 7x^4y^3 - 8 + 3x^3y$
25. $1 - x + x^2 \quad x^2 - x + 1;$
26. x
27. $\frac{2}{5}x - 2x^5 + 7x^4 - 9$
28. $4x^5 + \frac{1}{2}x^2 - 11x^8 + 19$
29. $x^5 + y^5$
30. $6xy + 2x^2 + 3y^2$
31. $-x^4y^3 + x^7y^2 + y^7$
32. $4 - 2x^6$

In problems 33 through 38, evaluate each polynomial when x is −3.

33. $3x + 7$
34. $4x^2 - 2x + 1$
35. $8 \quad x$
36. $x^2 - x$
37. $1 - x^2$
38. $2x^3 - x - 2$

In problems 39 through 44, evaluate each polynomial when x is −2 and y is 3.

39. $xy + 6$
40. $x^2y^2 - xy + 1$
41. $x^2 + 2xy + y^2$
42. $x^3 - y^3$
43. $x^2 - y^2$
44. $2x^3y - 5x^2y^2 + xy + y^2$

In problems 45 through 52, simplify each polynomial, if possible.

45. $6x - 7 - 5x + 3$
46. $5x^2y^3 - 7xy^2 + 3x^2y^3 + 2xy^2$
47. $7y - 2z - 3y - 2z$
48. $12abc + 4ab + c$
49. $4x^3 - x^2 + 2 - x^3 + 2x^2$
50. $7t - t + 7t^2 +$
51. $-x^3 - w^3 + x^3 + w^3$
52. $-2xy - 2xy$

53. A stone is thrown into the air in a manner such that its height in feet after t seconds is given by the polynomial $64t - 16t^2$. How high is the stone after 3 seconds? After 4 seconds?

54. A ball is thrown down from the top of the Mile High skyscraper. Its height from the ground in feet after t seconds is given by the polynomial $5280 - 74t - 16t^2$. How high is the ball after 5 seconds? After 10 seconds? After 15 seconds and 16 seconds?

55. The sum of the squares of the first n natural numbers is given by the expression $\frac{1}{3}n^3 + \frac{1}{2}n^2 + \frac{1}{6}n$. What is the sum of the squares of the first five natural numbers? The first six?

Challenge Problems

In problems 56 through 59, evaluate each polynomial when x is -1. Assume that n is a natural number.

56. $x^n - 1$
57. $x^{2n} + 1$
58. $x^{2n+1} + 1$
59. $x^{2n-1} - 1$

■ IN YOUR OWN WORDS . . .

60. What is a polynomial?
61. How do we evaluate a polynomial?
62. Explain how to combine like terms.
63. Explain the meaning of an exact answer, an approximation, and an estimate.

1.3 ADDITION AND SUBTRACTION

Addition of polynomials is just a matter of adding up like terms. For example, to add the polynomials $5x^2 - 7x + 6$ and $4x^2 + 4x - 5$, we use the Associative and Commutative Properties to rearrange the terms

$$(5x^2 - 7x + 6) + (4x^2 + 4x - 5) = (5x^2 + 4x^2) + (-7x + 4x) + (6 - 5)$$

and then we add the like terms,

$$= 9x^2 - 3x + 1$$

EXAMPLE 1. Perform the addition in each.

(a) $(4x^3 + 4x^2 - x + 2) + (-7x^2 + x - 5)$

(b) $(3xy^2 + 4x^3y) + (xy^2 - 2x^3y)$

Solutions:

(a) $(4x^3 + 4x^2 - x + 2) + (-7x^2 + x - 5)$
$$= 4x^3 + (4x^2 - 7x^2) + (-x + x) + (2 - 5)$$
$$= 4x^3 - 3x^2 - 3$$

(b) $(3xy^2 + 4x^3y) + (xy^2 - 2x^3y) = (3xy^2 + xy^2) + (4x^3y - 2x^3y)$
$$= 4xy^2 + 2x^3y$$

The Opposite of a Polynomial

If A is a polynomial, $-A$ is called the **opposite** of A and

$$A + (-A) = 0$$

EXAMPLE 2. Find the opposite of each polynomial.

(a) $3x^2$ (b) $x^2 - x - 2$

Solutions:

(a) $-(3x^2) = -3x^2$, since $3x^2 + (-3x^2) = 0$

(b) $-(x^2 - x - 2) = -x^2 + x + 2$, since
$(x^2 - x - 2) + (-x^2 + x + 2) = 0$

Example 2 suggests that to find the opposite of a polynomial, we change the sign of each term in the polynomial.

We define subtraction as we did for real numbers.

Subtraction of Polynomials

If A and B are polynomials, then

$$A - B = A + (-B)$$

To subtract B from A, we add the opposite of B to A.

EXAMPLE 3. Perform the subtraction indicated.

(a) $(3x^3 - 7x + 2) - (x^3 + x - 8)$
(b) $(5x^4y^2 + 6xy^3 - y^4) - (-4x^4y^2 + 2xy^3 - 3y^4)$

Solutions:

(a) $(3x^3 - 7x + 2) - (x^3 + x - 8) = (3x^3 - 7x + 2) + (-x^3 - x + 8)$
$= 2x^3 - 8x + 10$

(b) $(5x^4y^2 + 6xy^3 - y^4) - (-4x^4y^2 + 2xy^3 - 3y^4)$
$= (5x^4y^2 + 6xy^3 - y^4) + (4x^4y^2 - 2xy^3 + 3y^4)$
$= 9x^4y^2 + 4xy^3 + 2y^4$ ☐

We can combine the operations of addition and subtraction together by using grouping symbols.

EXAMPLE 4. Perform the operations indicated in the expression

$$[(x^2 - x) - (2x^2 + 3x + 2)] + (6x^2 - x + 1)$$

Solution:

Start with the inside parentheses.

$[(x^2 - x) - (2x^2 + 3x + 2)] + (6x^2 - x + 1)$
$= [(x^2 - x) + (-2x^2 - 3x - 2)] + (6x^2 - x + 1)$
$= [-x^2 - 4x - 2] + (6x^2 - x + 1)$
$= 5x^2 - 5x - 1$ ☐

When doing additions and subtractions of polynomials, think of them in terms of parentheses preceded by a positive or a negative sign.

Look at an addition.

$(5x^2 + x - 8) + (3x^2 + 4x + 3) = 5x^2 + x - 8 + 3x^2 + 4x + 3$
$= 8x^2 + 5x - 5$

Parentheses preceded by a positive sign can be removed without changing the problem.

Sec. 1.3 Addition and Subtraction

Look at a subtraction.

$$(4x^2 - 3x - 7) - (x^2 + 5x - 2) = (4x^2 - 3x - 7) + (-x^2 - 5x + 2)$$
$$= 4x^2 - 3x - 7 - x^2 - 5x + 2$$
$$= 3x^2 - 8x - 5$$

Thus parentheses preceded by a negative sign can be removed if the sign of each term inside the parentheses is changed.

PROBLEM SET 1.3

Warm-ups

In problems 1 through 4, perform the operations indicated. See Example 1.

1. $(4x - 5) + (3x - 9)$
2. $(y^2 + 6y - 5) + (2y^2 - 7)$
3. $(x^2 + 4xy + y^2) + (4x^2 - 7xy - 4y^2)$
4. $(5x^3 - 6x^2 + 5x - 7) + (10x^3 + 7x^2 - x - 1)$

In problems 5 through 8, find the opposite of each polynomial. See Example 2.

5. $5x^6 - 3x^4 + 3$
6. $-4x^5 + 7x^4 - x^3 - 4$
7. $15x^4y - 12x^3y^2 + 7x^2y^2 - 12$
8. $3x^3y^2 - 5xy^2 + 6$

In problems 9 through 16, perform the operations indicated. For problems 9 through 12, see Example 3.

9. $(5x - 8) - (7x + 4)$
10. $(z^2 - 6z) - (2z^2 - z + 8)$
11. $(2x^2 - 5xy + 3y^2) - (6x^2 + 3xy - 8y^2)$
12. $(6x^4 + 5x^3 - x^2 + x - 4) - (7x^4 - 4x^3 - 4x^2 + 7x - 9)$

For problems 13 through 16, see Example 4.

13. $[(x^2 + 2x + 5) + (4x^2 - 8)] - (5x^2 - x)$
14. $(v^4 + 4v^2 - 8) - [(2v^4 + v - 7) + (3v^3 - 5v^2 + 6v + 5)]$
15. $[(5x^3y - x^2y^2 + 7xy + 5) - (3x^3y + 2x^2y^2 - 4xy)] + (2x^3y + 7)$
16. $(t^3 - 27) - [(t^3 + 27) - (27 - t^3)]$

Practice Exercises

In problems 17 through 48, perform the indicated operations in each expression.

17. $(x - 5) + (23x + 9)$
18. $(7x + 6) - (x - 4)$
19. $(3y^2 + 4y - 8) + (3y^2 - 9)$
20. $(z^2 - 3z) - (4z^2 + z - 5)$
21. $(x^2 - 4xy - y^2) - (3x^2 + 7xy - 5y^2)$
22. $(5x^2 - 2xy + 4y^2) - (7x^2 - 5xy - 11y^2)$
23. $(3x^3 - 4x^2 + 3x - 9) + (12x^3 + 6x^2 + x + 1)$
24. $(2x^4 + 5x^3 + x^2 - x + 4) - (3x^4 - 8x^3 + 5x^2 - 2x - 1)$
25. $(8x^5 - 4x^3 + 7x) - (3x^4 - 17x^3 + x^2 - 12)$
26. $(4x^6 - 2x^4 + 13) - (9x^5 + 4x^4 - x^3 + 3)$
27. $(4x^3y - 2x^2y^2 + 8xy - 12) + (2x^3y - 7x^2y^2 + 7xy - 1)$
28. $(-3x^4y^2 + 6x^3y + x^2 - 2) - (x^4y^2 - 3x^3y + 2x^2 - 9)$
29. $(13x^5 - 4x^3 - 7x) + (11x^5 + 14x^4 - 3x)$

30. $(17x^4y - 18x^3y^2 - 3x^2y^2 + 19) - (2x^3y^2 + 2xy^2 - 13)$
31. $(x^2 - 7x - 5) - (x^2 + 6x - 4) + (x^2 - 2x + 11)$
32. $(9y + 8) + (10y + 6) - (y - 4)$
33. $(z^2 - 3) - (z^2 + 3) + (z^2 - 2z - 1)$
34. $(w^3 - 1) - (w^3 - 2w^2 + 2w - 2) - (w^2 + 4w - 4)$
35. $[(3x^2 - 4x + 7) + (2x^2 + 10)] - (7x^2 - x)$
36. $[(2x^3y - x^2y^2 - 3xy - 7) - (5x^3y + 3x^2y^2 - 6xy)] + (4x^3y - 1)$
37. $(4v^4 - v^2 + 4) - [(2v^4 - v + 8) + (2v^3 + 3v^2 - 5v + 8)]$
38. $(t^3 + 27) - [(t^3 - 27) - (27 - t^3)]$
39. $[(z^2 - 2z) - (2z^2 + 8)] - (6z^2 - 2z - 7)$
40. Find the sum of $6x^2 - 5x + 1$ and $x + 7$.
41. Subtract $3s + 2$ from $s^2 + 7s - 6$.
42. Find the sum of $3x^2 - 7x + 5$ and $3x + 4$.
43. Subtract $x^2 + x + 1$ from the sum of $x^2 + 1$ and $x^2 + 2x + 1$.
44. Subtract $x^2 - x - 1$ from the sum of $x^2 - 1$ and $x^2 - 2x - 1$.
45. Subtract the sum of $3x + y$ and $4x - 8y$ from the sum of $6x - 5y$ and $3x - 13y$.
46. Subtract $x - y$ from $y - x$.
47. Find the sum of $2x^2 + 2x + 2$, $-x^2 - 4x - 4$, and $6x^2 + 7x - 3$.
48. Subtract $-5t^3 - t^2 + t + 3$ from $t^3 + 7t^2 - t - 5$.

49. Find a polynomial that gives the perimeter of a rectangle if the width is given by W units and the length by $2W - 3$ units. What is the perimeter of such a rectangle with a width of 7 ft?
50. Find a polynomial that gives the perimeter of a triangle if the lengths of the sides are $s + 1$ units, $s^2 - 2s + 2$ units, and $s^3 + 2s^2 + s - 9$ units.
51. Two Amtrak passenger trains leave Philadelphia, one traveling north and the other south. If the northbound train travels $65x$ miles while the southbound train travels $54x + 27$ miles, how far apart are the trains?
52. A barbed-wire fence is cut into three sections of lengths $3t$ yards, $2t - 1$ yards, and $t^2 + 4t - 5$ yards. Write a polynomial that gives the original length of the fence.
53. Alex has a collection of 42 Wilson and Penn tennis balls. If x of them are Wilson, how many are Penn?
54. Two angles of a triangle have measure $x + 30$ degrees and $3x - 80$ degrees, respectively. What is the measure of the third angle?
55. Charles and Kitty have saved $10x + 4$ dollars for their summer trip. If Charles saved $6x - 22$ dollars, how much has Kitty saved?
56. A section of cast-iron pipe $4s$ feet long is cut into three pieces. If one piece is 16 ft long and another is $2s - 7$ feet long, what is the length of the third piece?
57. An angle has measure $x^2 - 2x + 30$ degrees. Find the measure of its supplement.
58. One acute angle of a right triangle has measure $x - 30$ degrees. What is the measure of the other acute angle?

Challenge Problems

In problems 59 through 62, perform the operations indicated. Assume that m and n are natural numbers.

59. $(2x^{2n} - x^n + 1) + (5x^{2n} + 2x^n + 3)$
60. $(5x^{2n} - 3x^n - 2) - (3x^{2n} - x^n - 4)$
61. $(x^m - 5x^my^n + y^n) + (2x^m + x^my^n - 5y^n)$
62. $(x^{mn} - y^{mn}) - (2x^{mn} - 2y^{mn})$

■ IN YOUR OWN WORDS . . .

63. How do we subtract polynomials?

■ 1.4 MULTIPLICATION

We multiply polynomials by using the Distributive, Commutative, and Associative Properties for polynomials given in Section 1.2 and the laws of exponents from Section 1.1. To multiply monomials, multiply the coefficients and use the laws of

exponents. For example,
$$(5x^2)(-3x^3)(4x) = (5)(-3)(4)x^2 \cdot x^3 \cdot x$$
$$= -60x^6$$

Use the Distributive Property to multiply a monomial and a binomial.
$$a(b + c) = ab + ac$$

The product of two binomials is one of the most important products that we learn. Let's look at $(a + b)(c + d)$. Using the Distributive and Commutative Properties, write
$$(a + b)(c + d) = (a + b)c + (a + b)d$$
$$= ac + bc + ad + bd$$

Note that there are four terms in the product found by multiplying each term in the first binomial by each term in the second binomial.

Product of Two Binomials
$$(a + b)(c + d) = ac + ad + bc + bd$$

A common way to remember this is:

ac represents the product of the "**f**irst" terms,
ad represents the product of the "**o**utside" terms,
bc represents the product of the "**i**nside" terms, and
bd represents the product of the "**l**ast" terms.

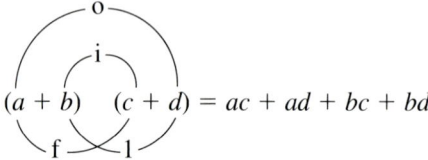

This is often called the "**foil**" method.

EXAMPLE 1. Find the product of $(2x + 3)$ and $(x - 4)$.

Solution:

The product of the **f**irst is	$(2x)(x)$	or	$2x^2$.
The product of the **o**utside is	$(2x)(-4)$	or	$-8x$.
The product of the **i**nside is	$(3)(x)$	or	$3x$.
The product of the **l**ast is	$(3)(-4)$	or	-12.

Thus
$$(2x + 3)(x - 4) = 2x^2 + (-8x) + 3x + (-12)$$
$$= 2x^2 - 5x - 12$$

The distributive property allows other multiplications, such as

$$(a + b)(c + d + e) = (a + b)c + (a + b)d + (a + b)e$$
$$= ac + bc + ad + bd + ae + be$$

Again, each term in the first polynomial is multiplied by each term in the second polynomial. It is important to note that the foil method works only for the product of two binomials.

EXAMPLE 2. Find each product.

(a) $xyz^2(xy^2z)$ (b) $x^3(x - 2y)$
(c) $(x + 2)(x + 1)$ (d) $(x-y)(x^2 - xy + 1)$

Solutions:

(a) $xyz^2(xy^2z) = x^2y^3z^3$

(b) $x^3(x - 2y) = x^3 \cdot x - x^3 \cdot 2y$
$\qquad\qquad\quad = x^4 - 2x^3y$

(c) $(x + 2)(x + 1) = x \cdot x + 2x + 1 \cdot x + 2 \cdot 1$
$\qquad\qquad\qquad = x^2 + 3x + 2$

(d) $(x - y)(x^2 - xy + 1) = x \cdot x^2 - x \cdot xy + x \cdot 1 - yx^2 + y \cdot xy - 1 \cdot y$
$\qquad\qquad\qquad\qquad\quad = x^3 - x^2y + x - x^2y + xy^2 - y$
$\qquad\qquad\qquad\qquad\quad = x^3 - 2x^2y + xy^2 + x - y$ □

The next several examples develop some multiplications that occur very often in algebra. They are **special products** and should be memorized. The first product is called the **square of a binomial.**

EXAMPLE 3. Find each product.

(a) $(p + q)^2$ (b) $(p - q)^2$

Solutions:

(a) $(p + q)^2 = (p + q)(p + q)$
$\qquad\qquad = p^2 + pq + pq + q^2$
$\qquad\qquad = p^2 + 2pq + q^2$

(b) $(p - q)^2 = (p - q)(p - q)$
$\qquad\qquad = p^2 - pq - pq + q^2$
$\qquad\qquad = p^2 - 2pq + q^2$ □

Square of a Binomial
$$(p + q)^2 = p^2 + 2pq + q^2$$
$$(p - q)^2 = p^2 - 2pq + q^2$$

Sec. 1.4 Multiplication

These should be memorized as a pattern and remembered in words rather than symbols. The square of a binomial is the first term squared plus (or minus) twice the product of the two terms plus the last term squared. Notice that these patterns are alike everywhere except for the sign in the middle term.

EXAMPLE 4. Find each product.

(a) $(x + 4)^2$

(b) $(2x - 3y)^2$

Solutions:

(a) $(x + 4)^2$

$$(p + q)^2 = p^2 + 2pq + q^2 \quad \text{Square of binomial}$$

We replace p with x and q with 4.

$$(x + 4)^2 = x^2 + 2x \cdot 4 + 4^2$$

(the first term squared plus twice the product of the two terms plus the last term squared)

$$= x^2 + 8x + 16$$

(b) $(2x - 3y)^2$

$$(p - q)^2 = p^2 - 2pq + q^2 \quad \text{Square of binomial}$$

We replace p with $2x$ and q with $3y$.

$$(2x - 3y)^2 = (2x)^2 - 2(2x)(3y) + (3y)^2$$
$$= 4x^2 - 12xy + 9y^2 \qquad \square$$

Another special product comes from multiplying the sum of two numbers by the difference of the same two numbers. It is called the **difference of two squares.**

EXAMPLE 5. Find the product of $(p - q)$ and $(p + q)$.

Solution:

$$(p - q)(p + q) = p^2 + pq - pq - q^2$$
$$= p^2 - q^2$$

Note that $(p - q)(p + q) = (p + q)(p - q)$. $\qquad \square$

Difference of Two Squares

$$(p - q)(p + q) = p^2 - q^2$$

The product of the sum of two numbers and the difference of the same two numbers is the first number squared minus the second number squared.

EXAMPLE 6. Find each product.

(a) $(x + 2)(x - 2)$

(b) $(4r + 7)(4r - 7)$

Solutions:

(a) $(x + 2)(x - 2)$

$$(p + q)(p - q) = p^2 - q^2 \quad \text{Difference of two squares}$$

We replace p with x and q with 2.

$$(x + 2)(x - 2) = x^2 - 2^2$$
$$= x^2 - 4$$

(the first number squared minus the second number squared)

(b) $(4r + 7)(4r - 7) = (4r)^2 - 7^2$
$$= 16r^2 - 49 \qquad \square$$

The next special product is called the **sum (or difference) of two cubes.**

EXAMPLE 7. Find each product.

(a) $(p - q)(p^2 + pq + q^2)$

(b) $(p + q)(p^2 - pq + q^2)$

Solutions:

(a) $(p - q)(p^2 + pq + q^2)$
$$= p^3 + p^2q + pq^2 - p^2q - pq^2 - p^3$$
$$= p^3 - q^3$$

(b) $(p + q)(p^2 - pq + q^2)$
$$= p^3 - p^2q + pq^2 + p^2q - pq^2 + q^3$$
$$= p^3 + q^3 \qquad \square$$

Sum of Two Cubes

$$(p + q)(p^2 - pq + q^2) = p^3 + q^3$$

Difference of Two Cubes

$$(p - q)(p^2 + pq + q^2) = p^3 - q^3$$

Care must be taken in using these rules. Notice that the first factor is a binomial and the second factor is a trinomial. The middle sign in the trinomial is the opposite of the sign in the binomial. Two examples illustrate the pattern.

Sec. 1.4 Multiplication

EXAMPLE 8. Find each product.

(a) $(x - 3)(x^2 + 3x + 9)$ (b) $(x + 4)(x^2 - 4x + 16)$

Solutions:

(a) $(x - 3)(x^2 + 3x + 9)$

Notice that the first factor is a binomial and the second factor is formed by squaring the first term, the middle term is the product of the two terms, and the last term is the last term squared.

$$(p - q)(p^2 + pq + q^2) = p^3 - q^3 \quad \text{Difference of two cubes}$$

We replace p with x and q with 3.

$$(x - 3)(x^2 + 3x + 9) = x^3 - 3^3$$
$$= x^3 - 27$$

(the first number cubed minus the second number cubed)

(b) $(x + 4)(x^2 - 4x + 16)$

$$(p + q)(p^2 - pq + q^2) = p^3 + q^3 \quad \text{Sum of two cubes}$$

We replace p with x and q with 4.

$$(x + 4)(x^2 - 4x + 16) = x^3 + 4^3$$
$$= x^3 + 64$$

The next special product is called the **cube of a binomial**.

EXAMPLE 9. Find each product.

(a) $(p + q)^3$ (b) $(p - q)^3$

Solutions:

(a) $(p + q)^3 = (p + q)(p + q)(p + q)$
$\qquad\quad = (p + q)(p^2 + 2pq + q^2)$
$\qquad\quad = p^3 + 2p^2q + pq^2 + p^2q + 2pq^2 + q^3$
$\qquad\quad = p^3 + 3p^2q + 3pq^2 + q^3$

(b) $(p - q)^3 = (p - q)(p - q)(p - q)$
$\qquad\quad = (p - q)(p^2 - 2pq + q^2)$
$\qquad\quad = p^3 - 2p^2q + pq^2 - p^2q + 2pq^2 - q^3$
$\qquad\quad = p^3 - 3p^2q + 3pq^2 - q^3$

Cube of a Binomial

$$(p + q)^3 = p^3 + 3p^2q + 3pq^2 + q^3$$
$$(p - q)^3 = p^3 - 3p^2q + 3pq^2 - q^3$$

Notice that the powers of p decrease in each term while the powers of q increase in each term. The coefficients of the two middle terms are both 3.

EXAMPLE 10. Find each product.

(a) $(x + 2y)^3$ (b) $(2x - 3y)^3$

Solutions:

(a) Since $x + 2y$ is a binomial, $(x + 2y)^3$ is the cube of a binomial. Replace p with x and q with $2y$ in the formula

$$(p + q)^3 = p^3 + 3p^2q + 3pq^2 + q^3$$

and get

$$(x + 2y)^3 = x^3 + 3x^2(2y) + 3x(2y)^2 + (2y)^3$$
$$= x^3 + 6x^2y + 12xy^2 + 8y^3$$

(b) Replace p with $2x$ and q with $3y$ in the formula

$$(p - q)^3 = p^3 - 3p^2q + 3pq^2 - q^3$$

and get

$$(2x - 3y)^3 = (2x)^3 - 3(2x)^2(3y) + 3(2x)(3y)^2 - (3y)^3$$
$$= 8x^3 - 36x^2y + 54xy^2 - 27y^3$$

Summary of Special Products

Square of a Binomial

$$(p + q)^2 = p^2 + 2pq + q^2$$
$$(p - q)^2 = p^2 - 2pq + q^2$$

Difference of Two Squares

$$(p + q)(p - q) = p^2 - q^2$$

Sum or Difference of Two Cubes

$$(p + q)(p^2 - pq + q^2) = p^3 + q^3$$
$$(p - q)(p^2 + pq + q^2) = p^3 - q^3$$

Cube of a Binomial

$$(p + q)^3 = p^3 + 3p^2q + 3pq^2 + q^3$$
$$(p - q)^3 = p^3 - 3p^2q + 3pq^2 - q^3$$

We can combine the operations of multiplication, addition, and subtraction along with symbols of grouping. The order of operations from Section 0.4 also still apply.

> **Order of Operations**
>
> If grouping symbols are present, perform operations inside them, starting with the innermost symbol, in the following order.
>
> 1. Perform any exponentiations.
> 2. Perform all multiplications or divisions in order from left to right.
> 3. Perform all additions or subtractions from left to right.
>
> If grouping symbols are not present, perform operations in the order given above.

EXAMPLE 11. Simplify each expression.

(a) $6 - 2[(x + 2)(x - 3) - x(x + 1)]$ (b) $(3x - 2)^2 - (x - 7)(2x + 3)$

Solutions:

(a) Begin on the inside of the grouping symbols.

$$6 - 2[(x + 2)(x - 3) - x(x + 1)] = 6 - 2[x^2 - x - 6 - x^2 - x]$$
$$= 6 - 2(-2x - 6) \quad \text{Combine like terms.}$$
$$= 6 + 4x + 12 \quad \text{Distributive Property}$$
$$= 4x + 18$$

(b) Begin by performing the indicated multiplications.

$$(3x - 2)^2 - (x - 7)(2x + 3) = 9x^2 - 12x + 4 - (x^2 - 11x - 21)$$

Notice that when we multiplied $(x - 7)$ and $(2x + 3)$, we enclosed the product in parentheses. Care must be taken with the minus sign in front of this term. The problem indicates subtraction of the entire product. Each sign in the parentheses must be changed.

$$9x^2 - 12x + 4 - x^2 + 11x + 21 = 8x^2 - x + 25 \quad \square$$

■ PROBLEM SET 1.4

Warm-ups

In problems 1 through 34, find each product.
For problems 1 through 8, see Example 1.

1. $(x + 2)(x + 1)$
2. $(x + 4)(x + 5)$
3. $\left[\frac{1}{2}x + 2\right]\left[\frac{1}{2}x + 1\right]$
4. $(x + 7)(x + 1)$
5. $(x - 6)(x - 7)$
6. $(x - 5)(x - 8)$
7. $(x + 7)(x - 5)$
8. $(x - 9)(x + 7)$

For problems 9 through 18, see Example 2.

9. $(4ab)(-3a^2b)$
10. $(2x^3)(x^2y^3)$
11. $(rs)s^3$
12. $3x(x + 4)$
13. $-3x^2(1 - x)$
14. $x(2x + 5)$
15. $(5x - 7)(3x - 8)$
16. $(3x - 1)(4x + 5)$
17. $(x - y)(x^2 - 2xy^2 + y^3)$
18. $(x + 2y)(x^2 + 3xy + y^2)$

For problems 19 through 22, see Examples 3 and 4.

19. $(x + 5)^2$
20. $(5x + 12)^2$
21. $(3x - 4y)^2$
22. $(6x + 7)^2$

For problems 23 through 26, see Examples 5 and 6.

23. $(x - 2)(x + 2)$
24. $(1 + 4y)(1 - 4y)$
25. $(5x + 7)(5x - 7)$
26. $(8 - 9z)(8 + 9z)$

For problems 27 through 30, see Examples 7 and 8.

27. $(x - 2)(x^2 + 2x + 4)$
28. $(t + 3r)(t^2 - 3rt + 9r^2)$
29. $(x + 5)(x^2 - 5x + 25)$
30. $(2u - v)(4u^2 + 2uv + v^2)$

For problems 31 through 34, see Examples 9 and 10.

31. $(x + 2)^3$
32. $(x - 4)^3$
33. $(3a - b)^3$
34. $(3w + 4x)^3$

In problems 35 through 38, simplify each expression. See Example 11.

35. $(x + 3)(x - 1) + 2x(x - 7)$
36. $(x + 3)^2 - (x - 8)(2x + 5)$
37. $5 - 2[3(x^2 - x) + (x + 1)(x + 3)]$
38. $12 - [5 - (x + 4)^2 + (2x - 1)(3x + 2)]$

Practice Exercises

In problems 39 through 87, find each product.

39. $(2ab)(-4a^3b)$
40. $(4x^5)(x^3y^6)$
41. $(2rs)s^4$
42. $x^3(-7x^4y)$
43. $-2^3xy^3(x^4y)$
44. $2a^4b^5(-3^2ab)$
45. $3b^5(-3^2abc)^3$
46. $(-ab)^2$
47. $(-r^4t)^5$
48. $(-3cd^6)^3$
49. $(rst^5)^3(-3r^7)^2$
50. $(-u^3)(-u)^3(-u)^4$
51. $(abc^5)(a^3bc)^7(-ab^2c)^4$
52. $3x(2x - 5)$
53. $a^3bc(3a + 5ab^3c^6 - 4b)$
54. $x(2x - 1)$
55. $-\dfrac{3}{5}x^5yz(5xz - 3y^5z^7 - 1)$
56. $3a^4(a^5 + a^3 - a)$
57. $\dfrac{1}{3}x^4(3x^7 - 9x^6 + x^3 - 3)$
58. $(x + 3)(x + 5)$
59. $(x + 1)(x + 9)$
60. $(x - 5)(x + 3)$
61. $(x - 3)(x + 7)$
62. $\left[\dfrac{1}{2}x + 3\right]\left[\dfrac{1}{2}x + 4\right]$
63. $(5 + x)(7 + x)$
64. $(x - 5)(x - 7)$
65. $(x - 4)(x - 6)$
66. $(x^2 + 2)(x^2 - 5)$
67. $\left[x - \dfrac{1}{3}\right]\left[x + \dfrac{2}{3}\right]$
68. $(x - 4)(x - 9)$
69. $(3 - x)(2 - x)$
70. $\left[x - \dfrac{2}{5}\right]\left[x + \dfrac{3}{5}\right]$
71. $(7 - x^2)(4 + x^2)$
72. $(2x + 7)(3x + 8)$
73. $(4x + 5)(2x + 5)$

Sec. 1.4   Multiplication

74. $(5x^2 + 7)(4x^2 + 9)$
75. $(11x + 5)(3x + 1)$
76. $(2x - 3)(x - 5)$
77. $(4 - x)(5 - 4x)$
78. $(x - 6)(2x - 7)$
79. $(5x^2 - 6)(x^2 - 7)$
80. $(8x - 5)(3x - 8)$
81. $(10x - 7)(3x - 5)$
82. $[(m + n) + 4][(m + n) + 3]$
83. $[(u - z) - 5][(u - z) + 3]$
84. $(c - d)(c^2 - 6cd + d^3)$
85. $(2x - 3y)(x^2 - 2xy + 3y^2)$
86. $(x + 3)(x - 5)(x + 7)$
87. $(2x + 7)(x - 7)(3x + 1)$

In problems 88 through 115, use special products to find each product.

88. $(x - 3)(x + 3)$
89. $(x - 4)^2$
90. $(x + 6)^2$
91. $(11x + 12)^2$
92. $(9x - 8)^2$
93. $(5x - 4)^2$
94. $(7x + 5)(7x - 5)$
95. $(x - 5y)(x + 5y)$
96. $(5x - 3y)^2$
97. $(7x + 6)^2$
98. $(4x + 7y)^2$
99. $(ax - 2by)^2$
100. $(2x + 5)^2$
101. $(7a + 3b)(49a^2 - 21ab + 9b^2)$
102. $(2s - 5)(4s^2 + 10s + 25)$
103. $(t + 4r)(t^2 - 4rt + 16r^2)$
104. $(x - 3)(x^2 + 3x + 9)$
105. $(x - 6)(x^2 + 6x + 36)$
106. $(2u - 3v)(4u^2 + 6uv + 9v^2)$
107. $(x - 5)^3$
108. $(x + 3)^3$
109. $(x + 2y)^3$
110. $(a - 2b)^3$
111. $(4s - 3t)^3$
112. $(3w + 5x)^3$
113. $[(x - y) + 7][(x - y) - 7]$
114. $[6 + (r + s)][6 - (r + s)]$
115. $[(s + t) + 3]^2$

In problems 116 through 119, simplify each expression.

116. $(x + 4)^2 - (x - 7)(2x - 3)$
117. $(x + 6)(x - 3) - 3x(x - 5)$
118. $22 - [7 - (x + 3)^2 - (2x - 5)(3x - 2)]$
119. $7 - 3[3(x^2 + x) - (x - 1)(x + 3)]$

120. If N is an *even* integer, find the product of N and the next even integer.
121. If N is an *odd* integer, find the product of the next two odd integers.
122. Find a polynomial that gives the area of a circle whose radius is $2x - 5$ feet.
123. Suppose N is an integer. Find the product of the next two consecutive integers.
124. Cube the integer just before the integer N.
125. Find a polynomial that gives the volume of a sphere whose radius is $x - 5$ meters.
126. Barbara Phipps keeps her favorite mare on a pasture in the shape of a triangle whose base is $2W$ yards and whose height is $3W^2$ yards. How many square yards are there in the pasture?
127. Donald Crankshaw planes a block of wood that is W inches wide and $3W^2$ inches long until it is uniformly $\frac{1}{2}$ in. thick. How many cubic inches are there in the block of wood?

Challenge Problems

In problems 128 through 133, find each product. Assume that m and n are natural numbers.

128. $x^n(x^2 - x + 1)$
129. $2x^2(x^{2n} + 3x^n + 1)$
130. $(x^n + 1)^2$
131. $(x^n + 2)(x^n - 2)$
132. $(x^n + y^n)^2$
133. $(x^n + 1)^3$

■ IN YOUR OWN WORDS . . .

134. Explain the special products.
135. Explain how to multiply a binomial by a trinomial.
136. Perform the multiplication indicated.
    (a) $(p - q)^2$ (b) $(q - p)^2$
    Why are these products the same?

1.5 PASCAL'S TRIANGLE (OPTIONAL)

Consider $(p + q)^n$, where n is a whole number.

$$(p + q)^0 = 1$$
$$(p + q)^1 = p + q$$
$$(p + q)^2 = p^2 + 2pq + q^2$$
$$(p + q)^3 = p^3 + 3p^2q + 3pq^2 + q^3$$
$$(p + q)^4 = ?$$

What will be the result of raising $p + q$ to the fourth power? There will be five terms. The first will be p^4, and the last will be q^4.

$$p^4 + _____ + _____ + _____ + q^4$$

Note that in $(p + q)^3$ the powers of p decrease while the powers of q increase, and the exponents total three in each term. So if we follow this pattern, we will have everything but the coefficients.

$$p^4 + \_\_\_p^3q + \_\_\_p^2q^2 + \_\_\_pq^3 + q^4$$

However, we can figure out the coefficients by a particular pattern known as **Pascal's Triangle.** It is named for the famous French mathematician Blaise Pascal.

Removing everything but the coefficients in the pattern,

$$1 \qquad n = 0$$
$$p + q \qquad n = 1$$
$$p^2 + 2pq + q^2 \qquad n = 2$$
$$p^3 + 3p^2q + 3pq^2 + q^3 \qquad n = 3$$

yields the following pattern.

$$1 \qquad n = 0$$
$$1 \quad 1 \qquad n = 1$$
$$1 \quad 2 \quad 1 \qquad n = 2$$
$$1 \quad 3 \quad 3 \quad 1 \qquad n = 3$$

If we could find the next row in the triangle, we would have the coefficients for the expansion of $(p + q)^4$. An interesting and uncomplicated pattern is present. The triangle is made by placing 1's on the outsides. The numbers inside are found by adding the two numbers directly above.

$$1 \qquad n = 0$$
$$1 \quad 1 \qquad n = 1$$
$$1 \quad 2 \quad 1 \qquad n = 2$$
$$1 \quad 3 \quad 3 \quad 1 \qquad n = 3$$
$$1 \quad 4 \quad 6 \quad 4 \quad 1 \qquad n = 4$$

Thus the coefficients we need are 1, 4, 6, 4, and 1.

$$(p + q)^4 = p^4 + 4p^3q + 6p^2q^2 + 4pq^3 + q^4$$

Pascal's Triangle can be used to find the coefficients in expanding $(p + q)^n$ when n is a nonnegative integer. In Chapter 12 we look at a statement of the general result of expanding a binomial, the Binomial Theorem. It is clear that using the triangle would be difficult if the power were very large.

We can expand $(p - q)^n$ in the same manner, but the signs will alternate from positive to negative.

EXAMPLE 1. Expand each binomial.

(a) $(x + 2)^4$ (b) $(2x + 3)^5$

(c) $(x - 3)^6$ (d) $(2x - 3)^4$

Solutions:

(a) There will be five terms. We make the powers of x decrease and the powers of 2 increase.

$$(x + 2)^4 = x^4 + \_\_\_x^3(2)^1 + \_\_\_x^2(2)^2 + \_\_\_x(2)^3 + 2^4$$

Now we use the triangle to find the coefficients.

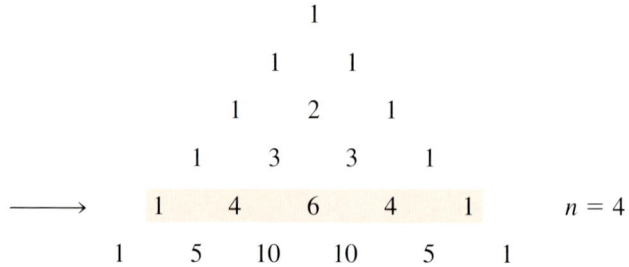

Notice that we can identify the row we need in the triangle by looking at the second number in the row.

The coefficients we need are 1, 4, 6, 4, 1.

$$(x + 2)^4 = x^4 + 4x^3(2) + 6x^2(2)^2 + 4x(2)^3 + 2^4$$
$$= x^4 + 8x^3 + 24x^2 + 32x + 16$$

(b) There will be six terms. We make the powers of $2x$ decrease and the powers of 3 increase. The exponents will total 5 in each term.

$$(2x + 3)^5 = (2x)^5 + \_\_\_(2x)^4(3)^1 + \_\_\_(2x)^3(3)^2$$
$$+ \_\_\_(2x)^2(3)^3 + \_\_\_(2x)^1(3)^4 + (3)^5$$

We use the triangle to find the coefficients.

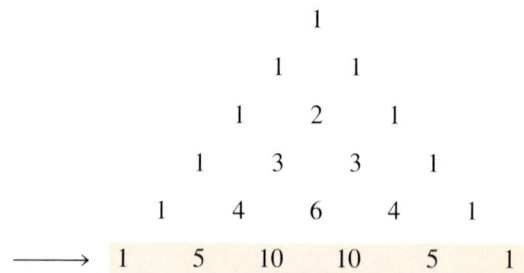

Chap. 1 Polynomials

The coefficients are 1, 5, 10, 10, 5, 1.

$$(2x + 3)^5 = (2x)^5 + 5(2x)^4(3)^1 + 10(2x)^3(3)^2$$
$$+ 10(2x)^2(3)^3 + 5(2x)^1(3)^4 + (3)^5$$
$$= 32x^5 + 240x^4 + 720x^3 + 1080x^2 + 810x + 243$$

(c) There will be seven terms. The powers of x will decrease and the powers of 3 will increase. The exponents will total 6 in each term. The signs will alternate beginning with positive.

$$(x - 3)^6 = x^6 - \_\_\_x^5(3)^1 + \_\_\_x^4(3)^2 - \_\_\_x^3(3)^3$$
$$+ \_\_\_x^2(3)^4 - \_\_\_x(3)^5 + \_\_\_(3)^6$$

We look for the coefficients in the triangle.

```
                    1
                  1   1
                1   2   1
              1   3   3   1
            1   4   6   4   1
          1   5  10  10   5   1
⟶   1   6  15  20  15   6   1
```

The coefficients are 1, 6, 15, 20, 15, 6, and 1.

$$(x - 3)^6 = x^6 - 6x^5(3)^1 + 15x^4(3)^2 - 20x^3(3)^3$$
$$+ 15x^2(3)^4 - 6x(3)^5 + (3)^6$$
$$= x^6 - 18x^5 + 135x^4 - 540x^3 + 1215x^2 - 1458x + 729$$

(d) There will be five terms. The powers of $2x$ will decrease and the powers of 3 will increase. The exponents will total 4 in each term. The signs will alternate beginning with positive.

$$(2x - 3)^4 = (2x)^4 - \_\_\_(2x)^3(3)^1 + \_\_\_(2x)^2(3)^2 - \_\_\_(2x)^1(3)^3 + 3^4$$

We find the coefficients.

```
                  1
                1   1
              1   2   1
            1   3   3   1
⟶       1   4   6   4   1
        1   5  10  10   5   1
```

They are 1, 4, 6, 4, and 1.

$$(2x - 3)^4 = (2x)^4 - 4(2x)^3(3) + 6(2x)^2(3)^2 - 4(2x)^1(3)^3 + 3^4$$
$$= 16x^4 - 96x^3 + 216x^2 - 216x + 81 \qquad \square$$

Sec. 1.5 Pascal's Triangle (Optional)

PROBLEM SET 1.5

Warm-ups

In problems 1 through 9, expand each binomial. See Example 1.

1. $(x + 1)^3$
2. $(x - 2)^3$
3. $(x + 2)^4$
4. $(3x - 2)^5$
5. $(2x + 3y)^6$
6. $(x + y)^6$
7. $(u - 2w)^6$
8. $(x + 2)^8$
9. $(x - 2)^{10}$

Practice Exercises

In problems 10 through 30, expand each binomial.

10. $(x + 2)^3$
11. $(x - 1)^3$
12. $(x + 3)^4$
13. $(y - 2)^4$
14. $(2x + 1)^4$
15. $(b - 3)^5$
16. $(y - 1)^4$
17. $(2x - 1)^4$
18. $(b - 2)^5$
19. $(a + 1)^5$
20. $(3x - 1)^3$
21. $(2x + 3)^4$
22. $(a + 2)^5$
23. $(3x - 2)^3$
24. $(2x + 5)^4$
25. $(3x + 2)^5$
26. $(2x - 3y)^6$
27. $(x - y)^6$
28. $(u - 3w)^6$
29. $(x - 2)^8$
30. $(x + 2)^{10}$
31. Find the fourth term in the expansion of $(3x - y)^{11}$.
32. Find the third term in the expansion of $(3x + 7)^8$.
33. Find the middle term in the expansion of $(5z - 2)^8$.

IN YOUR OWN WORDS . . .

34. Explain how to raise a binomial to the nth power if n is a whole number.
35. How is Pascal's Triangle formed?
36. Explain why $(x + y)^n \neq x^n + y^n$.

1.6 DIVISION

Division of polynomials can be done like division of real numbers. Recall that there are several ways of writing a division problem. If P and D are polynomials, the following statements have the same meaning.

1. P divided by D, $P \div D$, $\dfrac{P}{D}$, P/D, $D\overline{)P}$

We call P the **dividend** and D the **divisor.**

When we divide real numbers, the divisor must be nonzero. For polynomials, the divisor must not have a value of zero.

We divided monomials by monomials in Section 1.1 using the laws of exponents. For example,

$$\frac{4x^5}{2x^3} = 2x^2 \quad \text{and} \quad \frac{3a^2}{6a} = \frac{a}{2}$$

The laws of exponents do not tell how to do other division, such as $\dfrac{6x^3 + 3x^2}{2x}$.

The next result follows from the *properties of real numbers,* found in Section 0.2.

> **Division by a Monomial**
>
> If A, B, and C are monomials, then
> $$\frac{A+B}{C} = \frac{A}{C} + \frac{B}{C}$$
>
> (C must not have a value of 0.)

EXAMPLE 1. Perform the division $\dfrac{6x^3 + 3x^2}{2x}$.

Solution:
$$\frac{6x^3 + 3x^2}{2x} = \frac{6x^3}{2x} + \frac{3x^2}{2x}$$
$$= 3x^2 + \frac{3x}{2}$$
$$= 3x^2 + \frac{3}{2}x$$

If the divisor is not a monomial, we use long division. It is much like long division of numbers. The steps are illustrated in this example.

EXAMPLE 2. Consider $(2x^3 - x^2 + 5x + 4) \div (2x + 1)$.

Solution:

Step 1 Put each polynomial in standard form and write
$$2x + 1 \,\overline{\smash{\big)}\, 2x^3 - x^2 + 5x + 4}$$

Step 2 What is $\dfrac{2x^3}{2x}$? It is x^2. So write
$$\begin{array}{r} x^2 \\ 2x + 1 \,\overline{\smash{\big)}\, 2x^3 - x^2 + 5x + 3} \end{array}$$

Step 3 Multiply x^2 by $2x + 1$ and write
$$\begin{array}{r} x^2 \\ 2x + 1 \,\overline{\smash{\big)}\, 2x^3 - x^2 + 5x + 4} \\ 2x^3 + x^2 \end{array}$$

Step 4 Subtract $2x^3 + x^2$ from $2x^3 - x^2$.
$$\begin{array}{r} x^2 \\ 2x + 1 \,\overline{\smash{\big)}\, 2x^3 - x^2 + 5x + 4} \\ 2x^3 + x^2 \\ \hline -2x^2 \end{array}$$

Be Careful!

Step 5 Bring down the next term.
$$\begin{array}{r} x^2 \\ 2x + 1 \,\overline{\smash{\big)}\, 2x^3 - x^2 + 5x + 4} \\ 2x^3 + x^2 \\ \hline -2x^2 + 5x \end{array}$$

(continued)

Now we start the process over. What is $\dfrac{-2x^2}{2x}$? It is $-x$. So write

$$\begin{array}{r} x^2 - x \\ 2x + 1 \overline{\smash{)}\, 2x^3 - x^2 + 5x + 4} \\ \underline{2x^3 + x^2 } \\ -2x^2 + 5x \end{array}$$

Multiply $-x$ by $2x + 1$.

$$\begin{array}{r} x^2 - x \\ 2x + 1 \overline{\smash{)}\, 2x^3 - x^2 + 5x + 4} \\ \underline{2x^3 + x^2 } \\ -2x^2 + 5x \\ -2x^2 - x \end{array}$$

Be Careful! **Subtract** $-2x^2 - x$ from $-2x^2 + 5x$.

$$\begin{array}{r} x^2 - x \\ 2x + 1 \overline{\smash{)}\, 2x^3 - x^2 + 5x + 4} \\ \underline{2x^3 + x^2 } \\ -2x^2 + 5x \\ \underline{-2x^2 - x } \\ 6x \end{array}$$

Now we begin again. Bring down the 4. Since $\dfrac{6x}{2x} = 3$, write

$$\begin{array}{r} x^2 - x + 3 \\ 2x + 1 \overline{\smash{)}\, 2x^3 - x^2 + 5x + 4} \\ \underline{2x^3 + x^2 } \\ -2x^2 + 5x \\ \underline{-2x^2 - x } \\ 6x + 4 \end{array}$$

Multiply 3 by $2x + 1$.

$$\begin{array}{r} x^2 - x + 3 \\ 2x + 1 \overline{\smash{)}\, 2x^3 - x^2 + 5x + 4} \\ \underline{2x^3 + x^2 } \\ -2x^2 + 5x \\ \underline{-2x^2 - x } \\ 6x + 4 \\ 6x + 3 \end{array}$$

Be Careful! **Subtract** $6x + 3$ from $6x + 4$ and write

$$\begin{array}{r} x^2 - x + 3 \quad \longleftarrow \text{Quotient} \\ \text{Divisor} \longrightarrow 2x + 1 \overline{\smash{)}\, 2x^3 - x^2 + 5x + 4} \quad \longleftarrow \text{Dividend} \\ \underline{2x^3 + x^2 } \\ -2x^2 + 5x \\ \underline{-2x^2 - x } \\ 6x + 4 \\ \underline{6x + 3} \\ 1 \quad \longleftarrow \text{Remainder} \quad \square \end{array}$$

The degree of 1 is less than the degree of $2x + 1$, so the procedure stops. Thus,

$$(2x^3 - x^2 + 5x + 4) \div (2x + 1) = x^2 - x + 3 + \frac{1}{2x + 1}$$

Be careful when carrying out the subtraction step. Most of the errors made in long division result from not subtracting correctly.

The polynomial $2x + 1$ is the **divisor.**

The polynomial $2x^3 - x^2 + 5x + 4$ is the **dividend.**

The polynomial $x^2 - x + 3$ is the **quotient.**

The polynomial 1 is the **remainder.**

To check the division, we multiply the divisor by the quotient and add the remainder. This should give the dividend.

$$\text{Divisor} \cdot \text{Quotient} + \text{Remainder} = \text{Dividend}$$

$$(2x + 1)(x^2 - x + 3) + 1 = (2x^3 - x^2 + 5x + 3) + 1$$
$$= 2x^3 - x^2 + 5x + 4$$

EXAMPLE 3. Perform the division $\dfrac{x^2 + x - 6}{x + 3}$.

Solution:

Both polynomials are in standard form.

$$x + 3 \overline{\smash{\big)}\, x^2 + x - 6}$$

$$\begin{array}{r} x \phantom{{}+3\,)\,x^2+x-6} \\ x + 3 \overline{\smash{\big)}\, x^2 + x - 6} \\ \underline{x^2 + 3x} \\ -2x \end{array}$$
 $\longleftarrow x^2 \div x = x$
 $\longleftarrow x(x + 3)$
 $\longleftarrow x^2 + x - (x^2 + 3x)$

$$\begin{array}{r} x - 2 \\ x + 3 \overline{\smash{\big)}\, x^2 + x - 6} \\ \underline{x^2 + 3x} \\ -2x - 6 \\ \underline{-2x - 6} \\ 0 \end{array}$$
 $\longleftarrow -2x \div x = -2$

$\longleftarrow$ Bring down -6
$\longleftarrow -2(x + 3)$

$$\frac{x^2 + x - 6}{x + 3} = x - 2$$

To check the answer:

$$(x - 2)(x + 3) = x^2 + x - 6$$

EXAMPLE 4. Divide $2 - 2x + x^3$ by $x + 1$.

Solution:

Write each polynomial in standard form first. Notice that there is no x^2 term in the dividend. Indicate this by writing $0x^2$ in its proper place.

$$\begin{array}{r}
x^2 - x - 1 \\
x+1 \overline{\smash{)}\, x^3 + 0x^2 - 2x + 2}\\
\underline{x^3 + x^2 }\\
-x^2 - 2x \\
\underline{-x^2 - x }\\
-x + 2\\
\underline{-x - 1}\\
3
\end{array}$$

$(2 - 2x + x^3) \div (x + 1) = x^2 - x - 1 + \dfrac{3}{x+1}$ ∎

EXAMPLE 5. Divide $(2x^3 - 3x^2y + 2xy^2 + y^3)$ by $(2x + y)$.

Solution:

$$\begin{array}{r}
x^2 - 2xy + 2y^2 \\
2x+y \overline{\smash{)}\, 2x^3 - 3x^2y + 2xy^2 + y^3}\\
\underline{2x^3 + x^2y }\\
-4x^2y + 2xy^2 \\
\underline{-4x^2y - 2xy^2 }\\
4xy^2 + y^3\\
\underline{4xy^2 + 2y^3}\\
-y^3
\end{array}$$

$(2x^3 - 3x^2y + 2xy^2 + y^3) \div (2x + y)$

$= x^2 - 2xy + 2y^2 + \dfrac{-y^3}{2x + y}$ ∎

EXAMPLE 6. Perform the division $\dfrac{x^3 - y^3}{x - y}$.

Solution:

$$\begin{array}{r}
x^2 + xy + y^2 \\
x-y \overline{\smash{)}\, x^3 + 0x^2y + 0xy^2 - y^3}\\
\underline{x^3 - x^2y }\\
x^2y \\
\underline{x^2y - xy^2 }\\
xy^2 - y^3\\
\underline{xy^2 - y^3}\\
\end{array}$$

$\dfrac{x^3 - y^3}{x - y} = x^2 + xy + y^2$ ∎

Chap. 1 Polynomials

PROBLEM SET 1.6

Warm-ups

In problems 1 through 17, perform the division indicated.
For problems 1 through 6, see Example 1.

1. $\dfrac{a^4 b^2}{b}$

2. $\dfrac{xy^3}{x^3 y}$

3. $\dfrac{(x-y)^4 z^2}{(x-y)^2 z}$

4. $\dfrac{4a + 2b}{2b}$

5. $\dfrac{6x^2 + 12y}{6x}$

6. $\dfrac{x^2 - x}{x}$

For problems 7 through 13, see Examples 2 and 3.

7. $\dfrac{z^2 - 4z - 12}{z + 2}$

8. $\dfrac{x^2 - x - 2}{x + 1}$

9. $\dfrac{2x^2 - 5x - 3}{2x + 1}$

10. $\dfrac{x^2 + 3x + 4}{x + 2}$

11. $\dfrac{x^2 + 2x - 4}{x + 3}$

12. $(4x^4 - 3x^2 + 1 + 2x) \div (x + 2x^2 - 2)$

13. $\dfrac{x^5 - 3x^4 + x^3 - 2x^2 + 5x + 4}{x^2 - 2x + 3}$

For problems 14 through 17, see Examples 4 and 5.

14. $\dfrac{x^2 - 2xy + y^2}{x - 3y}$

15. $\dfrac{x^3 + x^2 y - xy^2 + 2y^3}{x + 2y}$

16. $\dfrac{x^3 + y^3}{x + y}$

17. $(x^4 + x^2 - x^3 + 1)/(x^2 - 1)$

Practice Exercises

In problems 18 through 47, perform the division indicated.

18. $\dfrac{-16a^2 c}{8ac}$

19. $\dfrac{27ab}{18b^2}$

20. $\dfrac{(a+b)^3 t}{(a+b)^2 t^3}$

21. $\dfrac{5x^2 + 10x - 15}{10y}$

22. $\dfrac{a^2 b - ac}{ab}$

23. $\dfrac{x^2 - xy + y^2}{xy}$

24. $\dfrac{x^2 - 4x + 4}{x - 2}$

25. $\dfrac{6x^2 + 7x - 3}{2x + 3}$

26. $\dfrac{x^3 + 3x + 3x^2 + 1}{x + 1}$

27. $\dfrac{y^2 + 2y - 3}{y - 1}$

28. $\dfrac{6x^2 - x - 2}{3x + 1}$

29. $(2x^3 - x^2 - 3x + 3) \div (2x + 1)$

30. $\dfrac{2x^3 - x^2 - 5x + 2}{2x - 3}$

31. $\dfrac{x^2 + 6x^3 + x - 5}{3x + 2}$

32. $\dfrac{x^3 - 6x^2 + 12x - 8}{x - 2}$

33. $\dfrac{2x^2 - 5x - 3}{x - 3}$

34. $\dfrac{3x + x^2 - 5}{4 + x}$

35. $(6x^2 + 5xy + y^2)/(2x + y)$

36. $\dfrac{12x^2 + 17x + 12}{4x + 3}$

37. $\dfrac{(3x^4 - 2x^3 - 2x + x^2 + 4)}{(x + 3x^2 - 1)}$

38. $\dfrac{6x^3 + 4x^2 + x - 2}{3x^2 - x - 2}$

39. $\dfrac{21x^3 + 5x^2 + 3x + 8}{7x + 4}$

40. $(2x^4 - x^3 + 5x + 3x^2 + 7) \div (x^2 + 4 + 2x)$

41. $\dfrac{x^4 - x + 2}{3 + x}$

42. $\dfrac{x^4 - 2x^2 + 1}{x + 1}$

43. $\dfrac{x^3 + x^6 - 3}{x^3 - 1}$

44. $\dfrac{(x^4 - y^4)}{(x - y)}$

45. $\dfrac{4x^5 - 6x^3 + 2x^2 + 1}{2x^2 + 1}$

46. $\dfrac{x^5 + x^4 - 4x + 1}{x^2 - 2}$

47. Divide $x^3 + x^2 - 10x + 8$ by $x - 2$.

48. Find the remainder when $4x^3 - 12x^2 - x + 7$ is divided by $2x - 1$.

49. Find the remainder when $6t^3 - 13t^2 + 21t + 17$ is divided by $3t - 2$.

Challenge Problems

In problems 50 through 52, perform the division indicated.

50. $\dfrac{x^2 - 2ax + 2a^2}{x - a}$

51. $\dfrac{x^{2n} - 2x^n + 4}{x^n + 1}$; n is a natural number.

52. $\dfrac{x^n - 1}{x - 1}$; n is a natural number.

53. Find the value of k so that if $x^3 + 2x^2 + x + k$ is divided by $x + 3$, the remainder will be zero.

■ IN YOUR OWN WORDS . . .

54. Explain the procedure for long division.

55. When do we use long division?

1.7 SYNTHETIC DIVISION (OPTIONAL)

The long-division procedure that we learned in Section 1.6 can be shortened if the divisor is of the form $x - r$, where r is a constant. This process is called **synthetic division.**

Let's divide $x^2 + 3x - 5$ by $x - 1$ using long division.

$$\begin{array}{r} x + 4 \\ x - 1 \overline{\smash{\big)}\, x^2 + 3x - 5} \\ \underline{x^2 - x} \\ 4x - 5 \\ \underline{4x - 4} \\ -1 \end{array}$$

Removing all the variables but leaving their coefficients, we have the display

$$\begin{array}{r} 1 \quad\; 4 \\ 1 \;\; -1 \overline{\smash{\big)}\, 1 \quad 3 \; -5} \\ \underline{1 \; -1} \\ 4 \; -5 \\ \underline{4 \; -4} \\ -1 \end{array}$$

Many of the numbers in this display are repetitions of the numbers above them. Removing them and condensing further, we obtain

$$\begin{array}{r} 1 \quad\; 4 \\ 1 \;\; -1 \overline{\smash{\big)}\, 1 \quad 3 \; -5} \\ \underline{-1 \; -4} \\ 4 \; -1 \end{array}$$

If the first number in the dividend is copied below the line, the quotient appears there and we can eliminate the top line.

$$\begin{array}{r|rrr} 1 & -1 & 1 & 3 & -5 \\ & & & -1 & -4 \\ \hline & & 1 & 4 & -1 \end{array}$$

Finally, if we eliminate the leading 1 in the divisor and change the sign of the remaining number and all the signs in the second row (so that we can add instead of subtracting), we get the following display.

$$\begin{array}{r|rrr} 1 & 1 & 3 & -5 \\ & & 1 & 4 \\ \hline & 1 & 4 & -1 \end{array}$$

All the information in the original long-division problem is contained in this condensed form. We only need to know how to make it and read the result.

Suppose that we wish to divide $3x^2 - 2x + 10$ by $x - 2$. **Remember,** we use synthetic division **only** when we are dividing by a binomial of the form $x - r$. Here r is 2. We make the first line of the display by writing r and the coefficients of the dividend.

Be Careful!

$$\underline{2}|\ 3\quad -2\quad 10$$

Next, we draw a line and bring down the 3.

$$\begin{array}{r|rrr} 2 & 3 & -2 & 10 \\ \hline & 3 & & \end{array}$$

Next we multiply 2 by 3 and write

$$\begin{array}{r|rrr} 2 & 3 & -2 & 10 \\ & & 6 & \\ \hline & 3 & & \end{array}$$

Now we *add* -2 and 6.

$$\begin{array}{r|rrr} 2 & 3 & -2 & 10 \\ & & 6 & \\ \hline & 3 & 4 & \end{array}$$

Multiply 2 by 4 and write

$$\begin{array}{r|rrr} 2 & 3 & -2 & 10 \\ & & 6 & 8 \\ \hline & 3 & 4 & \end{array}$$

Add 10 and 8.

$$\begin{array}{r|rrr} 2 & 3 & -2 & 10 \\ & & 6 & 8 \\ \hline & 3 & 4 & 18 \end{array}$$

→ Remainder
→ Constant
→ Coefficient of x

Sec. 1.7 Synthetic Division (Optional)

The last line contains the coefficients of the quotient and the remainder. Therefore, $3x^2 - 2x + 10$ divided by $x - 2$ is $3x + 4$ with a remainder of 18.

Synthetic division can be used only when the divisor is of the form $x - r$, that is, a variable to the first power minus a constant. The dividend **must** be in standard form.

EXAMPLE 1. Use synthetic division to divide
$$4x^4 - 6x^3 - 3x^2 - 4x + 5 \quad \text{by} \quad x - 2$$

Solution:

We note that both polynomials are in standard form.

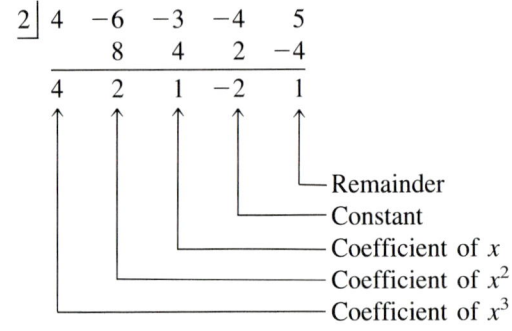

$$\frac{4x^4 - 6x^3 - 3x^2 - 4x + 5}{x - 2} = 4x^3 + 2x^2 + x - 2 + \frac{1}{x - 2} \qquad \square$$

EXAMPLE 2. Use synthetic division to perform the division
$$(x^3 - 8x - 1)/(x + 3)$$

Solution:

We notice that the x^2 term is missing in the dividend. This means that the coefficient of x^2 is 0. Also, the divisor must be written in the form $x - r$, so $x + 3 = x - (-3)$.

$$\begin{array}{r|rrrr} -3 & 1 & 0 & -8 & -1 \\ & & -3 & 9 & -3 \\ \hline & 1 & -3 & 1 & -4 \end{array}$$

$$(x^3 - 8x - 1)/(x + 3) = x^2 - 3x + 1 + \frac{-4}{x + 3} \qquad \square$$

EXAMPLE 3. Find the remainder when $x^3 - x^2 - 5x + 6$ is divided by $x - 1$.

Solution:

The easiest way to find the remainder in such a situation is by synthetic division.

$$\begin{array}{r|rrrr} 1 & 1 & -1 & -5 & 6 \\ & & 1 & 0 & -5 \\ \hline & 1 & 0 & -5 & 1 \end{array}$$

The remainder is 1. $\qquad \square$

PROBLEM SET 1.7

Warm-ups

In problems 1 through 4, use synthetic division to perform the division indicated. See Examples 1 and 2.

1. $\dfrac{2x^2 - 5x - 3}{x - 3}$

2. $\dfrac{y^2 + 2y - 3}{y - 1}$

3. $\dfrac{x^3 + 3x^2 + 3x + 1}{x + 1}$

4. $\dfrac{x^3 + 8}{x + 2} \; x^2$

In problems 5 and 6, use synthetic division to find the remainder when the given polynomial is divided by $x - 4$. See Example 3.

5. $x^3 - 64$

6. $x^3 - 2x^2 - 9x + 3$

Practice Exercises

In problems 7 through 22, use synthetic division to perform the division indicated.

7. $\dfrac{x^4 - 2x^2 + 1}{x - 1}$

8. $\dfrac{3x + x^2 - 5}{4 + x}$

9. $\dfrac{7x^2 + 4x^3 + x + 5}{x + 2}$

10. $\dfrac{3x^3 - 12x^2 - 14x - 6}{x - 5}$

11. $(2x^2 - 9x + 9) \div (x - 3)$

12. $\dfrac{2x + 1 + 3x^3 + 4x^2}{x + 1}$

13. $\dfrac{x^4 - 2x^2 - 5x + 3}{x - 2}$

14. $\dfrac{(x^4 + x^2 + 1)}{(x + 1)}$

15. $(2x^3 - x^2 - 3x - 4)/(x - 2)$

16. $(z^5 - 1) \div (z - 1)$

17. $\dfrac{(y^4 + 1)}{(y + 1)}$

18. $(x^4 - 16) \div (x - 2)$

19. $(x^4 + x^2 - 2) \div (x - 1)$

20. $\dfrac{x^5 - 1}{x - 1}$

21. $\dfrac{x^4 + x^3 + x^2 + x + 1}{x - 1}$

22. $\dfrac{x^4 - x^3 + x^2 - x + 1}{x - 1}$

In problems 23 and 24, use synthetic division to find the remainder when the polynomial is divided by $x - 2$.

23. $2x^4 - 4x^3 + x^2 - x - 2$

24. $x^3 + 3x^2 - 3x + 4$

IN YOUR OWN WORDS . . .

25. Explain how to divide $x^2 - x - 6$ by $x - 3$ using synthetic division.

CHAPTER SUMMARY

GLOSSARY

Base in the expression x^n: The base is x.

Exponent in x^n: The exponent is n. If n is a natural number, it tells how many times to use the base as a factor.

Monomial: An expression of the form bx^n, where n is a whole number and b is a real number.

Coefficient in the expression bx^n: The coefficient is b.

Polynomial: A sum of monomials.

A polynomial written in **standard form:** The monomial of highest degree is written first and other monomials are written in decreasing order of degree.

Binomial: A polynomial with two terms.

Trinomial: A polynomial with three terms.

To evaluate a polynomial: To find the value of a polynomial given the value of its variable(s).

DEFINITIONS
1. If n is a natural number, then $x^n = x \cdot x \cdot x \cdots x$. ($n$ x's)
2. $x^0 = 1$; $x \neq 0$

PROPERTIES OF EXPONENTS
1. $x^m \cdot x^n = x^{m+n}$ — Product with same base
2. $(x^m)^n = x^{mn}$ — Power of a power
3. $(xy)^n = x^n y^n$ — Product to a power
4. $\left(\dfrac{x}{y}\right)^n = \dfrac{x^n}{y^n}$; $y \neq 0$ — Quotient to a power
5. $\dfrac{x^m}{x^n} = \begin{cases} x^{m-n} & \text{if } m > n \\ 1 & \text{if } m = n, \ x \neq 0 \\ \dfrac{1}{x^{n-m}} & \text{if } n > m \end{cases}$ — Quotient with same base

SPECIAL PRODUCTS

Square of a binomial
1. $(p + q)^2 = p^2 + 2pq + q^2$
2. $(p - q)^2 = p^2 - 2pq + q^2$

Difference of two squares
3. $(p - q)(p + q) = p^2 - q^2$

Sum of two cubes
4. $(p + q)(p^2 - pq + q^2) = p^3 + q^3$
5. $(p - q)(p^2 + pq + q^2) = p^3 - q^3$

Cube of a binomial
6. $(p + q)^3 = p^3 + 3p^2q + 3pq^2 + q^3$
7. $(p - q)^3 = p^3 - 3p^2q + 3pq^2 - q^3$

DIVISION OF POLYNOMIALS To divide polynomials, use long division if the divisor is not a monomial. If the divisor is a monomial, divide each term in the dividend by the divisor.

■ CHECKUPS

1. Simplify $\left(\dfrac{-2r^3 s}{t^2}\right)^3$. Section 1.1; Example 6a

2. Evaluate the following polynomial when x is 2 and y is -1.

$$3x^2y^4 - xy + y^5$$

Section 1.2; Example 2b

Perform the operation indicated.

3. $(3x^3 - 7x + 2) - (x^3 + x - 8)$ Section 1.3; Example 3a
4. $(x - y)(x^2 - xy + 1)$ Section 1.4; Example 2d
5. $(2x^3 - 3x^2y + 2xy^2 + y^3) \div (2x + y)$ Section 1.6; Example 5

OPTIONAL SECTIONS

6. Use Pascal's Triangle to expand

$$(x - 3)^6$$ Section 1.5; Example 1c

7. Use synthetic division to perform the division

$$(x^3 - 8x - 1)/(x + 3)$$ Section 1.7; Example 2

REVIEW PROBLEMS

In problems 1 and 2, evaluate each polynomial when x is 1 and y is −3.

1. $x^3 - xy^2 + y^3$
2. $x^2 - y^2$

In problems 3 through 17, use the properties of exponents to simplify each expression.

3. $(-5)^2$
4. -5^2
5. $(-2)^2(-3)^3$
6. $\dfrac{(x - t)^4}{(x - t)^2}$
7. $(xy^3)^4$
8. $(-xy)^3$
9. $(a + b)^4(a + b)$
10. $\left(\dfrac{x^5}{y^7z^8}\right)^5$
11. $\left(\dfrac{-2x}{y}\right)^2$
12. $\left(\dfrac{xz}{-y^2}\right)^3$
13. $\dfrac{-10^0}{10}$
14. $(-2x^2z^3)^3$
15. $\dfrac{(-6)^4}{(-6)^5}$
16. $\left(\dfrac{2x^5}{x^4z^7}\right)^0$
17. $\dfrac{-5^3}{-2^4}$

In problems 18 through 49, perform the operations indicated.

18. $(x^2 - 3x + 8) + (3x^2 - x - 6)$
19. $abc(a^2b^3c - ac^4)$
20. $\dfrac{7t^2 - 7t}{7t}$
21. $(7p^3 - 4p^2 - 7) - (p^3 + 6p - 5)$
22. $(2x - 3)(x + 4)$
23. $\dfrac{(xy^3z^2)}{(x^2yz)}$
24. $(3r - 7) + (r - 1)$
25. $(r^3s^2t)(r^3st^4)$
26. $\dfrac{x^2 + 2x - 3}{x + 3}$
27. $(x^4 - x^3y + 3x^2y^2 - 5xy + y^5) - (2x^4 + 2x^3y - x^2y^2 + 7xy)$
28. $(2x - 3)^2$
29. $\dfrac{x^3 + 5x^2 - 10x + 20}{x + 7}$
30. $(x + 2)(x^2 - 2x + 4)$
31. $(t - 5)^3$
32. $(8p^5 - 6p^4 + 3p^2 - 3) + (6p^4 - 5p^2 + 6p - 2)$
33. $(7r - 2)(2r - 1)$
34. $(x + 3)(x^2 - 2xy + y^2)$
35. $(c + 3d)(c - 3d)$
36. $\dfrac{18r^2t^2 - 12t + 14rt}{6rt}$
37. $(x - y) - (x + y)$
38. $(s + 3)^3$
39. $xy^2(xy - 1)$
40. $(2x - y)(x^2 + 2xy + y^2)$
41. $(8s^3 + 27) \div (2s + 3)$
42. $2x + x(x - 2)$
43. $(x + 3)^2 - (x - 2)^2$
44. $(3x - 7)(5x + 3)$

45. $\dfrac{x^4 + 2x^3 - x^2 + 6}{x^2 - x + 2}$

46. $3a^2b(3ab^3)$

47. $r(r - t) + (2r - t)(2r + t)$

48. $(x^4 - x^3 - x^2 - x + 1) \div (x + 1)$

49. $6 - [x(x - 1) - 2(x + 3)]$

50. If a stone is thrown into the air in such a manner that its height in feet after t seconds is given by the polynomial $5 + 100t - 16t^2$, how high is the stone after 3 seconds? After 6 seconds?

51. Bernard's porch has a triangular roof whose base is twice its height. What is the area of Bernard's porch roof if its height is x feet?

... LET'S NOT FORGET ...

How many terms are in each polynomial?

52. $y^2 - 49$

53. $y^2 + 6y - 7$

54. $y^2 + 14y + 49$

Find each product.

55. $(2x)^2$

56. $2x(x + 2)$

57. $(x - 2)(x + 2)$

58. $(x - 2)^2$

59. $(x + 1)^3$

Simplify each expression.

60. -3^2

61. $(-3)^3$

62. $(-3)^2$

63. $(-b)^2$

64. $(-b)^3$

CHAPTER TEST

In problems 1 through 5, choose the correct answer.

1. $2x^3[(3x)^2 - x^5] = (?)$
   A. $18x^5 - 2x^8$ C. $6x^5 - 2x^8$
   B. $72x^5 - 8x^8$ D. $19x^5$

2. The value of the polynomial $3x^3 - 2x^2y^2$ when x is -1 and y is 3 is (?)
   A. 15   C. 21
   B. -15 D. -21

3. $\left(\dfrac{-2a^5b}{b^4}\right)^2 = (?)$
   A. $\dfrac{4a^{10}}{b^6}$ C. $\dfrac{4a^7}{b}$
   B. $\dfrac{-2a^{10}}{b^6}$ D. $\dfrac{-4a^7}{b^8}$

4. $\dfrac{(-3)^4}{(-3)^6} = (?)$
   A. $\dfrac{1}{6}$ C. $-\dfrac{1}{6}$
   B. $\dfrac{1}{9}$ D. $-\dfrac{1}{9}$

5. $-3^2ab^4(2ab)(-a^4) = (?)$
   A. $-18a^6b^5$ C. $12a^4b^4$
   B. $18a^6b^5$ D. $-12a^6b^5$

In problems 6 through 15, perform the operations indicated.

6. $\dfrac{6x^2 - 4x^3}{x^2}$

7. $(2x - 3)(5x + 7)$

8. $(5x^3y^2 - 4x^2y + 3xy) + (-3x^3y^2 + x^2y - 7xy)$

9. $(4x - 7)^2$

10. $(2x + 3)(4x^2 - 6x + 9)$

11. $(x^4 - 2x^3 + 3x - 2) + (x^2 - 1) - (2x^4 - x^3 + 2x^2 + 3x)$

12. $(x + 3)^3$

13. $(6x^3 - 2x^2 + x - 2) \div (x + 1)$

14. $(5x^3 - x^2 + 4) - (x^3 - 2x - 7)$

15. $7 - [2(x + 4) - (x + 2)(x - 2)]$

CHAPTER 2

See Section 2, Exercise 78.

Factoring

- **2.1** Greatest Common Factor and Factoring by Grouping
- **2.2** Factoring Binomials
- **2.3** Factoring Trinomials
- **2.4** Summary of Factoring

CONNECTIONS

Just as we can multipy 2 and 3 to get 6, we can *unmultiply* 6 to obtain 2 and 3. We call such unmultiplication, *factoring*. Factoring can be harder than multiplication. The product of 3 and 6 is always 18, and is only 18. However, 3 and 6 are not the only factors of 18. Nine and 2 are factors of 18, as are 18 and 1. In fact, the three numbers, 2, 3, and 3 are probably the most useful factors of 18.

The inherent difficulty in factoring large numbers is the basis for a revolutionary new breakthrough in secret codes. Essentially, a message is encoded using a huge number that is the product of two very large prime integers, (perhaps several hundred digits long!) The message is easily decoded (with a computer) if you know the two prime numbers, but practically impossible if you do not. This is one of the latest and most exciting uses of number theory.

Algebra is full of uses of factoring. The Fundamental Principle of Fractions allows us to reduce a fraction by dividing out *factors* common to both the numerator and the denominator. Finding the least common denominator for an addition problem involves using the *factored* form in the denominators.

In this chapter we learn how to factor polynomials. Factored polynomials are among the most useful expressions found in mathematics. The multiplication of polynomials, learned in Chapter 1, will help us to factor. Factoring will be necessary to work with rational expressions in Chapter 3 and to solve equations in Chapter 6. Factoring is among the most useful skills you will learn in algebra.

2.1 GREATEST COMMON FACTOR AND FACTORING BY GROUPING

In algebra it is often necessary to write a number as a product of numbers. This process is called **factoring.** Factoring often deals with natural numbers. A natural number is **prime** if the only way it can be written as a product of natural numbers is as the number itself times 1. For example, 5 is prime because $5 = 5 \cdot 1$. This is the only way to write 5 as a product of natural numbers. Similarly, 2, 3, 7, and 11 are prime. The number 1 is not considered to be a prime number. The smallest prime number is 2.

There are usually several ways to factor a natural number that is not prime. For example,

$$12 = 2^2 \cdot 3 \qquad 12 = 2 \cdot 6 \qquad 12 = 3 \cdot 4 \qquad 12 = 12 \cdot 1$$

Frequently, the most useful of these is $2^2 \cdot 3$ because it is a product of prime numbers. There is only one way to factor a natural number using prime factors. When a number is written as a product of prime numbers, it is **factored completely** or it is written in its **prime factorization.** The numbers in the product are called **factors.**

In factoring large numbers, it is helpful to know when a number is divisible by a prime.

> **Some Tests for Divisibility**
>
> 1. A natural number is divisible by 2 if it is even, that is, if its last digit is either 0, 2, 4, 6, or 8.
> 2. A natural number is divisible by 3 if the sum of the digits of the number is divisible by 3.
> 3. A natural number is divisible by 5 if its last digit is 0 or 5.

To factor a number into prime factors, start with the smallest prime that will divide into the number and continue until all prime factors are found.

EXAMPLE 1. Factor each number into prime factors.

(a) 45
(b) 108
(c) 392

Solutions:

(a) Notice that 45 is not even. So it is not divisible by 2. However, it is divisible by 3. So

$$45 = 3 \cdot 15$$
$$= 3 \cdot 3 \cdot 5$$
$$= 3^2 \cdot 5$$

(b) Since 108 is even, it is divisible by 2.

$$108 = 2 \cdot 54$$

Now 54 is also divisible by 2. So

$$= 2 \cdot 2 \cdot 27$$
$$= 2 \cdot 2 \cdot 3 \cdot 3 \cdot 3$$
$$= 2^2 \cdot 3^3$$

(c) Since 392 is even, divide by 2.

$$392 = 2 \cdot 196$$
$$= 2 \cdot 2 \cdot 98$$
$$= 2 \cdot 2 \cdot 2 \cdot 49$$
$$= 2^3 \cdot 7^2$$

To find the largest factor that two numbers have in common, consider the numbers 15 and 25. The only factor of both is 5. The largest common factor is 5. If the numbers are too large to do this by inspection, the prime factorization of the numbers will help us. Such a factor is called the **greatest common factor.**

Sec. 2.1 Greatest Common Factor and Factoring By Grouping

> **To Find the Greatest Common Factor (GCF)**
> 1. Write the prime factorization of each number.
> 2. List the prime factors that are common to all the numbers.
> 3. The GCF is the product of the factors in step 2 with each factor raised to the smallest power that occurs in any number.

EXAMPLE 2. Find the GCF for each pair of numbers.

(a) 24 and 80
(b) 72 and 378

Solutions:

(a) $24 = 2 \cdot 12$ $80 = 2 \cdot 40$
   $ = 2 \cdot 2 \cdot 6$ $ = 2 \cdot 2 \cdot 20$
   $ = 2 \cdot 2 \cdot 2 \cdot 3$ $ = 2 \cdot 2 \cdot 2 \cdot 10$
   $ = 2^3 \cdot 3$ $ = 2 \cdot 2 \cdot 2 \cdot 2 \cdot 5$
                      $ = 2^4 \cdot 5$

The different prime factors are 2, 3, and 5. The only factor common to both numbers is 2. Since 2 occurs three times in 24 and four times in 80, the greatest common factor is 2^3.

The GCF is 2^3 or 8.

(b) $72 = 2 \cdot 36$ $378 = 2 \cdot 189$
   $ = 2 \cdot 2 \cdot 18$ $ = 2 \cdot 3 \cdot 63$
   $ = 2 \cdot 2 \cdot 2 \cdot 9$ $ = 2 \cdot 3 \cdot 3 \cdot 21$
   $ = 2 \cdot 2 \cdot 2 \cdot 3 \cdot 3$ $ = 2 \cdot 3 \cdot 3 \cdot 3 \cdot 7$
   $ = 2^3 \cdot 3^2$ $ = 2 \cdot 3^3 \cdot 7$

The common prime factors are 2 and 3. Since 2 occurs three times in 72 and once in 378, the GCF will contain 2. Since 3 occurs two times in 72 and three times in 378, the GCF will contain 3^2.

The GCF is $2 \cdot 3^2$ or 18.

EXAMPLE 3. Find the GCF of $108x^3y^6$ and $24x^4y^2$.

Solution:

$$108x^3y^6 = 2^2 \cdot 3^3 x^3 y^6$$
$$24x^4y^2 = 2^3 \cdot 3 x^4 y^2$$

The common prime factors are 2, 3, x, and y. The GCF is $2^2 \cdot 3x^3y^2$ or $12x^3y^2$.

Let's consider writing a polynomial as a product of polynomials. The process of writing a polynomial as a product of polynomials is called **factoring** a polynomial. The polynomials that we factor in this chapter will use only integers as coefficients.

The next type of factoring that we will study is called **factoring out common factors.** It is done by using the Distributive Property to factor out the greatest common factor of the terms in the polynomial. When we multiplied polynomials in Section 1.4, we used the Distributive Property and wrote

$$a(b + c) = ab + ac$$

To factor, turn this around and write

$$ab + ac = a(b + c)$$

Notice the difference in $ab + ac$ and $a(b + c)$. $ab + ac$ is a sum of two terms and *is not factored*, whereas $a(b + c)$ is a product and is *factored*.

We say that we have "taken out" a common factor. If no more factoring can be done, the polynomial is **factored completely.** If no factoring can be done, the polynomial is **prime.** Monomials are considered to be prime polynomials.

To factor out common factors, we must identify the *factors* in each *term*. In the polynomial

$$3x^2 + 5x$$

there are two terms. In the first term, 3, x, and x^2 are factors. In the second term, 5 and x are factors. So x is a common factor. Thus

$$3x^2 + 5x = x(3x + 5)$$

Notice that $3x^2 + 5x$ is a *sum* while $x(3x + 5)$ is a *product*.

Let's consider $x(a + y) - 2b(a + y)$. If we think of it as two terms, the first term is $x(a + y)$ and the second term is $2b(a + y)$. Naming the factors of each term, we have

x and $(a + y)$ are factors of $x(a + y)$.

2, b, and $(a + y)$ are factors of $2b(a + y)$.

Thus $(a + y)$ is the only common factor. So we factor

$$x(a + y) - 2b(a + y) = (a + y)(x - 2b)$$

Notice again that $x(a + y) - 2b(a + y)$ is a *difference*, whereas $(a + y)(x - 2b)$ is a *product*.

Consider the polynomial

$$4(x + y)^3 + 2(x + y)^2 + 6(x + y)$$

If we think of it as a trinomial, then $2(x + y)$ is a factor of each term. So $2(x + y)$ is a common factor and

$$4(x + y)^3 + 2(x + y)^2 + 6(x + y) = 2(x + y)[2(x + y)^2 + (x + y) + 3]$$

We must distinguish between polynomials that are factored and those that are not factored. A polynomial is *factored* if it is written as a *product*.

Sec. 2.1 Greatest Common Factor and Factoring By Grouping

EXAMPLE 4. Factor each polynomial completely.

(a) $ax^2 + ax^3$ (b) $54a^2bc^3 - 72ab^2c^2 + 144abc^2$

Solutions:

(a) $ax^2 + ax^3 = ax^2(1 + x)$ Common factor ax^2 Check by multiplying.

(b) $54a^2bc^3 - 72ab^2c^2 + 144abc^2 = 2 \cdot 3^3 a^2bc^3 - 3^2 \cdot 2^3 ab^2c^2 + 2^4 \cdot 3^2 abc^2$
$$= 2 \cdot 3^2 abc^2(3ac - 2^2 b + 2^3)$$
$$= 18abc^2(3ac - 4b + 8) \qquad \square$$

EXAMPLE 5. Factor $-15x^2y + 5xy^2$ completely.

Solution:
$$-15x^2y + 5xy^2 = -5xy(3x - y)$$

We could have factored out 5 instead of -5.
$$-15x^2y + 5xy^2 = 5xy(-3x + y)$$

Sometimes it is convenient to factor out a negative coefficient. $\square$

EXAMPLE 6. Factor each polynomial completely.

(a) $x(a + b) + y(a + b)$ (b) $3a(m - n)^2 + 2a(m - n)^3$

Solutions:

(a) $x(a + b) + y(a + b) = (a + b)(x + y)$
   Think of this as two terms with a common factor of $(a + b)$.

(b) $3a(m - n)^2 + 2a(m - n)^3 = a(m - n)^2[3 + 2(m - n)]$
$$= a(m - n)^2(3 + 2m - 2n)$$
$$= a(m - n)^2(2m - 2n + 3) \qquad \square$$

The next example contains opposites. Remember that
$$p - q = -(q - p)$$

To handle the opposites, we must look at how to multiply three numbers. Multiplying $(3)(y)(-z)$ gives $-3yz$. The negative sign is usually written in front of the product. Similarly, $3y[-(a - b)]$ is the product of three numbers, 3, y, and $-(a - b)$. So $3y[-(a - b)] = -3y(a - b)$.

EXAMPLE 7. Factor $2x(a - b) + 3y(b - a)$.

Solution:

We will replace $b - a$ with $-(a - b)$.
$$2x(a - b) + 3y(b - a) = 2x(a - b) + 3y[-(a - b)]$$

$$= 2x(a - b) - 3y(a - b)$$

Now a common factor of $(a - b)$ appears.
$$= (a - b)(2x - 3y) \qquad \square$$

Be Careful!

Compare the original problem, $2x(a - b) + 3y(b - a)$, with the step marked with $\boxed{*}$, $2x(a - b) - 3y(a - b)$. This maneuver occurs over and over again in working with polynomials. After doing several problems such as Example 5, it is easy to skip some steps and go directly from the original problem to the $\boxed{*}$ step.

Factoring out the common factors should be done before any other type of factoring is done. Check for common factors *first*.

After a little rearranging, we can factor

$$ax + bx + ay + by$$

Notice that there is no common factor. However, grouping the terms will allow us to factor.

$$\begin{aligned} ax + bx + ay + by &= (ax + bx) + (ay + by) \\ &= x(a + b) + y(a + b) \\ &= (a + b)(x + y) \end{aligned}$$

This type factoring is called **grouping**. It should be used when factoring a polynomial with four or more terms.

EXAMPLE 8. Factor each polynomial completely.

(a) $ax - by + ay - bx$
(b) $rm - rn - sm + sn$
(c) $rx - r + x - 1$

Solutions:

(a) $ax - by + ay - bx = (ax + ay) - (by + bx)$
   Notice the negative in front of the parentheses.

$$\begin{aligned} &= a(x + y) - b(x + y) \\ &= (x + y)(a - b) \end{aligned}$$

Often, there is more than one way to group.

$$\begin{aligned} ax - by + ay - bx &= (ax - bx) + (ay - by) \\ &= x(a - b) + y(a - b) \\ &= (a - b)(x + y) \end{aligned}$$

It does not matter which we use.

(b) $\begin{aligned} rm - rn - sm + sn &= (rm - rn) - (sm - sn) \\ &= r(m - n) - s(m - n) \\ &= (m - n)(r - s) \end{aligned}$

(c) $\begin{aligned} rx - r + x - 1 &= (rx - r) + (x - 1) \\ &= r(x - 1) + (x - 1) \\ &= (x - 1)(r + 1) \end{aligned}$

The 1 in $(r + 1)$ is necessary. Why?

[The coefficient of $(x - 1)$ is 1.]

Grouping is often used to factor completely polynomials with four or more terms. Be careful to look for a common factor before grouping.

Sec. 2.1 Greatest Common Factor and Factoring By Grouping

EXAMPLE 9. Factor each polynomial completely.

(a) $a^2x + ax + abx + bx$ (b) $y^7 - y^6z + y^5 - y^4z$

Solutions:

(a) There is a common factor of x. Factor this out first.

$$a^2x + ax + abx + bx = x(a^2 + a + ab + b)$$

Now use grouping.

$$= x[(a^2 + a) + (ab + b)]$$
$$= x[a(a + 1) + b(a + 1)]$$
$$= x(a + 1)(a + b)$$

(b) $y^7 - y^6z + y^5 - y^4z = y^4(y^3 - y^2z + y - z)$
$$= y^4[(y^3 - y^2z) + (y - z)]$$
$$= y^4[y^2(y - z) + (y - z)]$$
$$= y^4(y - z)(y^2 + 1)$$

In using grouping to factor, we get expressions such as

$$m(x + y) - n(x + y)$$

It is very important to recognize that this is *not* factored. *It is a difference, not a product.* However, $(x + y)(m - n)$ is factored because it is a *product.*

Factoring should always be checked by multiplying the factors to see if we get the original polynomial.

PROBLEM SET 2.1

Warm-ups

In problems 1 through 10, factor each number into prime factors. See Example 1.

1. 64
2. 56
3. 45
4. 196
5. 162
6. 1000
7. 243
8. 900
9. 216
10. 180

In problems 11 through 14, find the GCF. See Example 2.

11. 24; 48
12. 125; 75
13. 36; 54
14. 162; 72

In problems 15 through 21, find the GCF. See Example 3.

15. x^3y^2; x^4y^6
16. a^4b^2; ab^3
17. s^5t^5; st^2
18. $30x^3y^2$; $21x^2y^4$
19. $96x^2y$; $80x$
20. $100x^2yz^4$; $40y^3z^4$
21. $192x^2y^3$; $48x^2y^4$; $72x^2y^3$

In problems 22 through 44, factor out all common factors in each polynomial. For problems 22 through 30, see Example 4.

22. $x^2y - xy^2$
23. $48a^2b^3c + 64a^3b^2c$
24. $21a^2b - 14ab^2 + 7ab$
25. $20p^2q^7 - 28p^5q^5 + 36p^8q^3$
26. $15 - 5a$
27. $15t^3 -$
28. $108x^2y^3z^4 - 96x^2y^4z^2 - 84x^6y^6z^6$
29. $120m^2n^3 - 72m^4n^4 + 48m^3n^5$
30. $162p^8 + 189p^7 - 135p^9$

For problems 31 through 33, see Example 5.

31. $-6x^2 + 3x$ **32.** $-18x^5 - 27x^4$ **33.** $-6uv + 3v$

For problems 34 through 40, see Example 6.

34. $5(a + 2) + b(a + 2)$
35. $m(p - q) + n(p - q)$
36. $(x + 2)(x + 3) - x(x + 3)$
37. $(x + y)^2 + 3(x + y)$
38. $a(z + 2)^3 - b(z + 2)^2$
39. $5(1 - r)^4 + 3(1 - r)^3 - (1 - r)^2$
40. $x(5 + 2t)^2 - (2t + 5)$

For problems 41 through 44, see Example 7.

41. $r(s - t) - u(t - s)$
42. $2x(1 - x) - 3y(x - 1)$
43. $w(3u - v)^3 - 3u(3u - v)^2 + v(v - 3u)$
44. $16(y - x)^3 + 24(y - x)^2 - 32(x - y)$

In problems 45 through 54, factor each polynomial by grouping. See Example 8.

45. $7x + 7y + ax + ay$
46. $xy + x + by + b$
47. $a^2 - ab + 2a - 2b$
48. $6r - 6t - sr + st$
49. $a^2 + ab + a + b$
50. $y^2 - xy - y + x$
51. $5r + s^2 - 5s - rs$
52. $5a + 10 + ab + 2b$
53. $uv + wv - u - w$
54. $z^3 - z^2 + z - 1$

In problems 55 through 60, factor each polynomial completely. See Example 9.

55. $a^3 - a^2c + a^2b - abc$
56. $x^6 + x^5y + x^4 + x^3y$
57. $xz^2 - z^2 + xz - z$
58. $30st^2 + 60t^2 - 24st - 48t$
59. $15abx - 15aby - 25acx + 25acy$
60. $y^3 - xy^2 - y^2 + xy$

Practice Exercises

In problems 61 through 87, factor out all common factors in each polynomial.

61. $42a^4b^2c + 54a^2b^5c$
62. $-8x^3 + 4x$
63. $24r^4s^4 - 48r^3s^3$
64. $12s^6 - 18s^3$
65. $27a^4b - 18ab^5 + 9ab$
66. $25p^5q^9 - 35p^3q^7 + 45p^6q^2$
67. $-16x^3 - 24x^6$
68. $18 - 6a$
69. $12t^4 - 12t^3$
70. $-9uv - 3v$
71. $a^3x - 2a^3y + a^3w$
72. $72x^3y^2z^5 - 108x^4y^3z^3 - 36x^5y^5z^5$
73. $36m^3n^3 - 48m^5n^3 + 60m^4n^6$
74. $66p^5 + 132p^7 - 55p^9$
75. $c(d - 2) + b(d - 2)$
76. $4(r - t) - n(r - t)$
77. $a(s - t) + u(t - s)$
78. $a^4(x - b) + a^3(x - b)$
79. $(x + y)^3 - 5(x + y)^2$
80. $a(z + 2)^3 - b(z + 2)$
81. $3(1 - r)^3 - 3(1 - r)^4 - 4(1 - r)^2$
82. $z(5 + 2x)^2 - (2x + 5)$
83. $2t(a - x) - t(x - a)$
84. $z(4u - 3v)^3 - 3u(4u - 3v)^2 + v(3v - 4u)$
85. $c^2(5 + b)^4 - d^2(5 + b)^5 - 2(b + 5)^2$
86. $-x(3 - 2x)^3 - 2x^2(3 - 2x)^2 - (3 - 2x)^2$
87. $15(a - b)^4 + 45(a - b)^2 - 35(b - a)$

Sec. 2.1 Greatest Common Factor and Factoring By Grouping

In problems 88 through 99, factor each polynomial by grouping.

88. $4x + 4y + zx + zy$
89. $xy + 6y + x + 6$
90. $z^2 + 2z - yz - 2y$
91. $ar - br - as + bs$
92. $t + b + at + ab$
93. $a + b^2 - b - ab$
94. $3x + y^2 - 3y - xy$
95. $4s + 12 - 3b - bs$
96. $2zu + 2zv^2 - 3wu - 3wv^2$
97. $t^5 + t^4 + s^4t + s^4$
98. $am + an - m - n$
99. $q^4 - q^2t + q^2 - t$

In problems 100 through 105, factor each polynomial completely.

100. $b^4z - b^4 - b^2 + b^2z$
101. $ax^2 + b^2x^2 - ax - b^2x$
102. $uz - vz + uz^2 - vz^2$
103. $16c^2r + 32c^2s - 8cr - 16cs$
104. $36a + 12ax - 18a^2 - 6a^2x$
105. $w^3 - w^2t + w^3t - w^4$

Challenge Problems

In problems 106 through 109, factor each polynomial completely. Assume that m and n are natural numbers.

106. $x^{n+2} + x^2$
107. $x^{n+2} - x^{n+3}$
108. $3x^{4n} - 2x^{3n} + x^{2n}$
109. $x^{2mn} + x^{4mn}$

IN YOUR OWN WORDS . . .

110. Explain how to factor by grouping.

111. Make up a polynomial that can be factored by grouping.

2.2 FACTORING BINOMIALS

Special Products in Section 1.4 lead to three types of factoring.

Factoring Rules

1. $p^2 - q^2 = (p + q)(p - q)$ Difference of squares
2. $p^3 - q^3 = (p - q)(p^2 + pq + q^2)$ Difference of cubes
3. $p^3 + q^3 = (p + q)(p^2 - pq + q^2)$ Sum of cubes

The rules above tell how to factor the difference of squares and the difference of cubes as well as the sum of cubes. However, $p^2 + q^2$, the sum of squares, will not factor using real coefficients.

Sum of Squares

$p^2 + q^2$ will not factor using real coefficients.

Let's look at rule 1. To factor the difference of squares, the squares must be recognized. Consider

$$16x^2 - 25$$

If written as
$$(4x)^2 - 5^2$$
factoring rule 1 can be used to identify $4x$ as p and 5 as q. So
$$16x^2 - 25 = (4x)^2 - 5^2$$
$$= (4x + 5)(4x - 5)$$

The factored expression can also be written as $(4x - 5)(4x + 5)$. Why? (The commutative property of multiplication.)

EXAMPLE 1. Factor $4x^2 - 9y^2$ completely.

Solution:

This is the difference of squares.
$$4x^2 - 9y^2 = (2x)^2 - (3y)^2$$
$$= (2x - 3y)(2x + 3y)$$ □

Let's look at rules 2 and 3. They are almost alike. The first factor in each can be obtained by looking at the cubes. The second factor is obtained from the first. The first and last terms in the second factor are the first and last terms of the first factor squared. Notice that the middle term is the product of p and q and that its sign is *opposite* to the sign in the first factor.

Consider $x^3 - 125$. If we look at the cubes
$$x^3 - 125 = x^3 - 5^3$$
we can identify x as p and 5 as q in rule 2.
$$p^3 - q^3 = (p - q)(p^2 + pq + q^2) \quad \text{Rule 2}$$
$$x^3 - 5^3 = (x - 5)(x^2 + 5x + 25)$$

EXAMPLE 2. Factor each expression completely.

(a) $x^3 - 8$
(b) $8x^3 + 27$

Solutions:

(a) This is the difference of cubes. Think of $x^3 - 8$ as $x^3 - 2^3$. Apply rule 2.
$$x^3 - 8 = x^3 - 2^3$$
$$= (x - 2)(x^2 + 2x + 4)$$

(b) This is the sum of cubes. Use rule 3.
$$8x^3 + 27 = (2x)^3 + 3^3$$
$$= (2x + 3)[(2x)^2 - (2x)(3) + 3^2]$$
$$= (2x + 3)(4x^2 - 6x + 9)$$ □

Sec. 2.2 Factoring Binomials

EXAMPLE 3. Factor each expression completely.

(a) $y^2 - (m - n)^2$ (b) $(a + b)^3 - 8y^3$

Solutions:

Be Careful!

(a) Treat $(m - n)$ as one term to use rule 1 for the difference of two squares.
$$y^2 - (m - n)^2 = [y - (m - n)][y + (m - n)]$$
Carefully remove the parentheses.
$$= (y - m + n)(y + m - n)$$

(b) Treat $(a + b)$ as one term and use the difference of two cubes.
$$(a + b)^3 - 8y^3$$
$$= [(a + b) - 2y][(a + b)^2 + (a + b)(2y) + (2y)^2]$$
$$= (a + b - 2y)[(a + b)^2 + 2y(a + b) + 4y^2]$$

Remember always to look for common factors. This should be done before applying rules for difference of squares or sum or difference of two cubes.

EXAMPLE 4. Factor each completely.

(a) $x^4 - 16$ (b) $8x^3 - 64$

Solutions:

(a) This is the difference of squares. Use rule 1. Think of $x^4 - 16$ as $(x^2)^2 - 4^2$.
$$x^4 - 16 = (x^2)^2 - 4^2$$
$$= (x^2 + 4)(x^2 - 4)$$

Be Careful!

Since $x^2 - 4$ will factor, continue factoring.
$$= (x^2 + 4)(x + 2)(x - 2)$$

(b) This one is tricky. It is the difference of cubes, but unless we look for a common factor, we may not factor it completely.
$$8x^3 - 64 = 8(x^3 - 8)$$
$$= 8(x^3 - 2^3)$$
$$= 8(x - 2)(x^2 + 2x + 4)$$

■ PROBLEM SET 2.2

Warm-ups

In problems 1 through 36, factor each polynomial completely, if possible.
For problems 1 through 6, see Example 1.

1. $x^2 - 9$
2. $x^2 - 16$
3. $x^4 - 25$
4. $36 - x^6$
5. $9x^8 - 4y^4$
6. $16x^{10} - 81y^{10}$

For problems 7 through 18, see Example 2.

7. $x^3 + 1$
8. $8 + x^3$
9. $64 - x^3$
10. $x^3 - 125$
11. $1 - 8x^3$
12. $27x^3 + 125$
13. $r^3 - s^3$
14. $t^3 + 8s^3$
15. $x^6 + y^3$
16. $a^9 - b^3$
17. $27a^3 - 8b^3$
18. $64x^6 + 125y^3$

For problems 19 through 27, see Example 3.

19. $(x + y)^2 - 4$
20. $(a + b)^3 - 8$
21. $81 - (r + t)^2$
22. $(x + z)^3 + y^3$
23. $64 - (a - b)^3$
24. $169 - (x + y)^4$
25. $(a + b)^2 - (x + y)^2$
26. $(u + v)^3 - w^6$
27. $(s + t)^3 - (u + v)^3$

For problems 28 through 36, see Example 4.

28. $4x^8 - 64$
29. $a^9 - b^9$
30. $1 - x^8$
31. $25(x + 1) - r^2(x + 1)$
32. $3t^3 + 24s^3$
33. $5(x + y)^2 - 20$
34. $54 - 2(s + t)^3$
35. $4x^2 - 64$
36. $x^3(a + b) + y^3(a + b)$

Practice Exercises

In problems 37 through 74, factor each polynomial completely, if possible.

37. $x^2 - 81$
38. $x^2 - 49$
39. $x^4 - 36$
40. $16 - x^6$
41. $4x^6 - 25y^4$
42. $9x^{12} - 64y^{10}$
43. $x^3 + 27$
44. $64 + x^3$
45. $27 - x^3$
46. $x^3 - 216$
47. $1 - 27x^3$
48. $8x^3 + 27$
49. $r^3 - 8s^3$
50. $t^3 + 64s^3$
51. $x^6 + 8y^3$
52. $8a^9 - 27b^3$
53. $125a^3 - 64b^3$
54. $x^6 + 343y^3$
55. $(x - y)^2 - 16$
56. $(a - b)^3 - 1$
57. $8 - (r - t)^3$
58. $(x + 2z)^3 + 27y^3$
59. $125 - (a - b)^3$
60. $144 - (x + y)^6$
61. $(a - b)^2 - (x - y)^2$
62. $(u + v)^3 + w^9$
63. $8(a + b) + y^3(a + b)$
64. $8x^2 - 128$
65. $3x^4 - 1875$
66. $a^9 + b^9$
67. $1 - x^9$
68. $36(r - 3) - 4r^2(r - 3)$
69. $5t^3 - 625s^3$
70. $8(x + y)^2 - 8$
71. $16 + 2(s - t)^3$
72. $(s - t)^3 + (u - v)^3$
73. $64 + s^2$
74. $a^2 + b^2$
75. Find a value of k so that $z + 4$ will be a factor of $z^2 - k$.
76. Find a value of k so that $3y + 1$ is a factor of $ky^3 + 1$.

Sec. 2.2   Factoring Binomials

In problems 77 and 78, express the volume of the shaded figure as a polynomial in completely factored form.

77.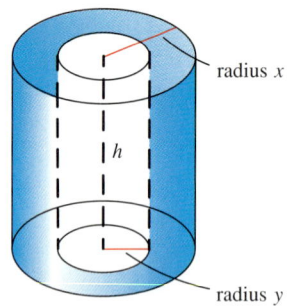

78.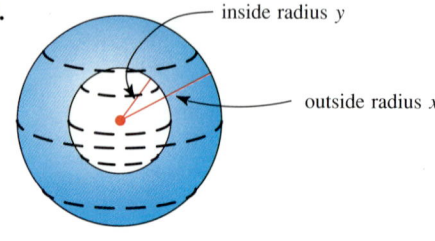

In problems 79 and 80, express the area of the shaded region as a polynomial in completely factored form.

79.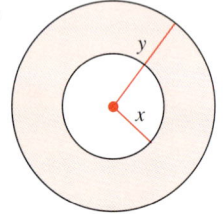

80.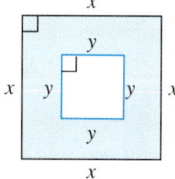

Challenge Problems

In problems 81 through 84, factor each polynomial completely. Assume that p and q are natural numbers.

81. $x^{2p} - 1$
82. $x^{3q} - 1$
83. $x^{3p} + 1$
84. $x^p y^3 - x^p$

In problems 85 and 86, factor each as the difference of two squares and then factor each as the difference in two cubes.

85. $x^6 - 1$
86. $x^6 - 64$

■ IN YOUR OWN WORDS . . .

87. Explain how to factor the sum or difference in two cubes.

■ 2.3 FACTORING TRINOMIALS

The product of two binomials sometimes gives a trinomial. To factor a trinomial, we look for two binomials. The process is trial and error for the most part. That is why multiplication of binomials is such an important skill.

There are some guidelines that we can use to help cut down on the trials and errors. Generally, it is easier to factor a trinomial if it is in standard form.

Suppose that the trinomial has the form

$$x^2 + 6x + 5$$

The factors must multiply to make a first term of x^2, a middle term of $+6x$, and a final term of $+5$. The only way that this can happen is for the signs in the factors to look as follows:

$$(\_\_ + \_\_)(\_\_ + \_\_)$$

The signs in both factors are both positive.

Now suppose that the trinomial has the form

$$6x^2 - 17x + 5$$

The factors must multiply to give a first term of $6x^2$, a middle term of $-17x$, and a final term of $+5$. This forces the signs in the factors to look as follows:

$$(\_\_ - \_\_)(\_\_ - \_\_)$$

The signs are both negative.

What happens if the trinomial has the form

$$x^2 + 7x - 8 \quad \text{or} \quad 2x^2 - x - 3?$$

A little trial and error shows that the signs in the factors must look as follows:

$$(\_\_ + \_\_)(\_\_ - \_\_) \quad \text{or} \quad (\_\_ - \_\_)(\_\_ + \_\_)$$

The signs are different.

If a trinomial is in standard form with a positive leading coefficient, the following guidelines are useful.

Guidelines for Factoring Trinomials

1. If the last term is positive, the factors will have the form

$$(\_\_ + \_\_)(\_\_ + \_\_) \quad \text{or} \quad (\_\_ - \_\_)(\_\_ - \_\_)$$

The $+$ or $-$ sign is determined by the coefficient of the middle term.

2. If the last term is negative, the factors will have the form

$$(\_\_ + \_\_)(\_\_ - \_\_) \quad \text{or} \quad (\_\_ - \_\_)(\_\_ + \_\_)$$

Practice makes factoring trinomials easier. The more practice, the less trial and error. Always check an answer by multiplying the factors.

EXAMPLE 1. Factor each trinomial completely.

(a) $x^2 + 3x + 2$ (b) $-6x + x^2 + 5$

Solutions:

(a) $x^2 + 3x + 2$. The first guideline shows what the factors must look like.

$$(x + \_\_)(x + \_\_)$$

The missing numbers have a product of 2 and a sum of 3. We need 2 and 1.

$$x^2 + 3x + 2 = (x + 2)(x + 1)$$

Check by multiplying.

(b) $-6x + x^2 + 5$. Write the trinomial in standard form first, $x^2 - 6x + 5$. Since the last term is positive and the middle term is negative, the first guideline gives factors of the following type:

$$(x - \_\_)(x - \_\_)$$

The missing numbers have a product of 5 and a sum of 6. We need 5 and 1.

$$x^2 - 6x + 5 = (x - 5)(x - 1)$$

Check by multiplying.

Sec. 2.3 Factoring Trinomials

EXAMPLE 2. Factor each trinomial completely.

(a) $x^2 + 3x - 18$ (b) $x^2 - 3x - 18$

Solutions:

(a) $x^2 + 3x - 18$. The second guideline gives factors in the form

$$(x + \underline{})(x - \underline{})$$

The missing terms have a product of 18. They could be

$$\begin{array}{ll} 6 \text{ and } 3 & \text{or} \\ 9 \text{ and } 2 & \text{or} \\ 18 \text{ and } 1 & \end{array}$$

Since the middle term is $+3x$, the only possibility is 6 and 3. The 6 must go in the factor with the "+" and the 3 in the factor with the "−". So

$$x^2 + 3x - 18 = (x + 6)(x - 3)$$

Check by multiplying.

(b) $x^2 - 3x - 18$. By the second guideline the factors must look as follows:

$$(x + \underline{})(x - \underline{})$$

Note that this trinomial is the same as the one in part (a) except that the sign of the middle term is negative. Just swap the 6 and the 3 so that the middle term will be $-3x$.

$$x^2 - 3x - 18 = (x + 3)(x - 6)$$

Check by multiplying.   ◻

EXAMPLE 3. Factor each trinomial completely.

(a) $6x^2 - 13x + 6$ (b) $3 + 5x - 2x^2$

Solutions:

(a) $6x^2 - 13x + 6$. Notice that

$$6x^2 = (2x)(3x) \quad \text{and} \quad 6x^2 = (6x)(x)$$

So the factors must look as follows:

$$(2x - \underline{})(3x - \underline{}) \quad \text{or} \quad (6x - \underline{})(x - \underline{})$$

We can write 18 as $6 \cdot 3$ or $9 \cdot 2$ or $18 \cdot 1$. There are several possible combinations to try. Some possible factors are:

|1| $(6x - 2)(x - 3)$ |2| $(6x - 3)(x - 2)$
|3| $(2x - 3)(3x - 2)$ |4| $(2x - 2)(3x - 3)$
|5| $(6x - 6)(x - 1)$ |6| $(6x - 1)(x - 6)$
|7| $(2x - 6)(x - 1)$ |8| $(2x - 1)(3x - 6)$

Chap. 2 Factoring

After multiplying each of these, note that ③ is the correct answer. Notice that ① could not be the correct answer because $(6x - 2)$ has a factor of 2, that is, $(6x - 2) = 2(3x - 1)$. This would mean that there was a common factor of 2 in $6x^2 - 13x + 6$. The same idea applies to ②, ④, ⑤, ⑦, and ⑧ as well. This idea can help cut down on trial and error. So

$$6x^2 - 13x + 6 = (2x - 3)(3x - 2)$$

(b) $3 + 5x - 2x^2$. This one is tricky! Do not try to rearrange the terms. It is easier to factor as it is, because the coefficient of x^2 is negative.

$$3 + 5x - 2x^2 = (3 - x)(1 + 2x)$$

Check by multiplying.

Before factoring a trinomial, look for common factors.

EXAMPLE 4. Factor each trinomial completely.

(a) $t^4 - 2t^3 + t^2$ (b) $6x^2 - 6xy - 36y^2$

Solutions:

(a) $t^4 - 2t^3 + t^2 = t^2(t^2 - 2t + 1)$ Common factor
$= t^2(t - 1)(t - 1)$
$= t^2(t - 1)^2$

Be Careful!

(b) $6x^2 - 6xy - 36y^2 = 6(x^2 - xy - 6y^2)$ Common factor
$= 6(x - 3y)(x + 2y)$

EXAMPLE 5. Factor each trinomial completely.

(a) $x^4 + 3x^2 + 2$ (b) $x^6 - 2x^3 - 8$

Solutions:

(a) $x^4 + 3x^2 + 2 = (x^2 + 2)(x^2 + 1)$
(b) $x^6 - 3x^3 - 4 = (x^3 - 4)(x^3 + 1)$

The second factor, $x^3 + 1$, will factor further.

$= (x^3 - 4)(x + 1)(x^2 - x + 1)$

Sometimes the square of a binomial special product can speed up factoring a trinomial. It gives us these factoring rules. If a trinomial can be factored by one of these rules, the trinomial is called a **perfect trinomial square.**

Factoring Perfect Trinomial Squares

$$p^2 + 2pq + q^2 = (p + q)^2$$
$$p^2 - 2pq + q^2 = (p - q)^2$$

Sec. 2.3 Factoring Trinomials

EXAMPLE 6. Determine if each trinomial is a perfect trinomial square.
(a) $x^2 - 6x + 9$ (b) $x^2 + 10x + 25$ (c) $x^2 - 12x - 36$
(d) $x^2 + 8x + 15$ (e) $x^2 - x + 4$

Solutions:

(a) $x^2 - 6x + 9 = x^2 - 2 \cdot x \cdot 3 + 3^2$. So the first and last terms are squares and the middle term is twice the product of x and 3.
$x^2 - 6x + 9 = (x - 3)^2$, so it is a perfect trinomial square.

(b) $x^2 + 10x + 25 = x^2 + 2 \cdot x \cdot 5 + 5^2$. This is also a perfect trinomial square, as
$$x^2 + 10x + 25 = (x + 5)^2$$

(c) The last term must be positive. So
$x^2 - 12x - 36$ is *not* a perfect trinomial square.

(d) The last term is not a square. So
$x^2 + 8x + 15$ is *not* a perfect trinomial square.

(e) $x^2 - x + 4 \neq (x - 2)^2$

The first and last terms are squares, but the middle term is not twice the product of the x and 2.
$x^2 - x + 4$ is *not* a perfect trinomial square.

PROBLEM SET 2.3

Warm-ups

In problems 1 through 38, factor each trinomial completely, if possible.
For problems 1 through 4, see Example 1.

1. $x^2 + 3x + 2$
2. $s^2 + 7s + 6$
3. $x^2 - 14x + 49$
4. $24 - 10x + x^2$

For problems 5 through 12, see Example 2.

5. $r^2 - 2rs - 15s^2$
6. $z^2 + z - 12$
7. $x^2 - x + 2$
8. $u^2 - 5u - 24$
9. $z^2 - 3z - 10$
10. $x^2 + xy - 20y^2$
11. $y^2 - 7y - 60$
12. $t^2 - 10t - 24$

For problems 13 through 22, see Example 3.

13. $2z^2 + 3z + 1$
14. $6x^2 + 11x + 4$
15. $20y^2 + 27yz + 9z^2$
16. $4x^2 + 20x + 25$
17. $8x^2 - 22x + 15$
18. $3x^2 - 26x + 16$
19. $2 - 13x + 21x^2$
20. $18y^2 - 45y + 28$
21. $12z^2 + 4z - 1$
22. $25z^2 - 30z + 9$

For problems 23 through 28, see Example 4.

23. $4a^2 - 4a - 8$
24. $5s^2 + 15s + 10$
25. $-6b^2 + 6bc + 12c^2$
26. $6x^2 + 2xy - 4y^2$
27. $z^3 + 2z^2 + z$
28. $6y^3 + 15y^2 - 9y$

For problems 29 through 38, see Example 5.

29. $x^4 + 6x^2 + 5$
30. $x^4 - 3x^2 - 18$
31. $x^6 + 5x^3 - 14$
32. $3 + 8x^2 + 4x^4$
33. $6y^4 - y^2 - 2$
34. $8z^4 - 2z^2 - 15$
35. $x^4 - 2x^2 + 1$
36. $x^6 - 2x^3 + 1$
37. $x^4 + x^2 - 2$
38. $x^4 - 5x^2 + 4$

Practice Exercises

In problems 39 through 88, factor each trinomial completely, if possible.

39. $x^2 + 9x + 8$
40. $s^2 + 6s + 8$
41. $x^2 + 4x + 4$
42. $y^2 + 9y + 14$
43. $x^2 - 6x + 8$
44. $t^2 - 10t + 21$
45. $x^2 - 2x - 1$
46. $24 - 11x + x^2$
47. $r^2 - 5rs - 14s^2$
48. $z^2 + z - 6$
49. $x^2 - x - 12$
50. $u^2 - 3u - 28$
51. $z^2 - 3z - 54$
52. $x^2 + 4xy - 21y^2$
53. $y^2 - y - 56$
54. $t^2 - 2t - 48$
55. $2z^2 + 5z + 3$
56. $3x^2 + 23x + 14$
57. $6y^2 + 25yz + 4z^2$
58. $4x^2 + 16x + 15$
59. $6x^2 - 17x + 12$
60. $15x^2 - 28x + 5$
61. $35 - 31x + 6x^2$
62. $25y^2 - 10y + 1$
63. $3z^2 - z - 10$
64. $3z^2 + z - 44$
65. $3t^2 - 2t - 1$
66. $24u^2 - 7u - 6$
67. $14s^2 - 11s - 15$
68. $30x^2 + 17x - 21$
69. $28r^2 + r - 15$
70. $6z^2 + 5z - 25$
71. $5a^2 - 5a - 30$
72. $7s^2 + 42s + 63$
73. $-6b^2 - 3bc + 3c^2$
74. $2x^2 + 6xy + 4y^2$
75. $z^4 - 2z^3 - 8z^2$
76. $6y^3 + 6y^2 - 12y$
77. $x^4 - 7x^3 - 60x^2$
78. $x^4 - 3x^3 - 10x^2$
79. $x^4 + x^3 - 12x^2$
80. $x^6 + 6x^3 + 9$
81. $x^6 - 8x^3 + 15$
82. $1 + 3x^2 + 2x^4$
83. $4y^6 - 4y^3 - 8$
84. $z^4 - 3z^2 - 4$
85. $x^6 + 7x^3 - 8$
86. $x^4 - 10x^2 + 9$
87. $x^4 - 2x^2 - 8$
88. $x^6 + 9x^3 + 8$

89. Find all values of k so that $t^2 + kt - 7$ will factor.
90. Find all values of k so that $s - 7$ is a factor of $s^2 - s + k$.
91. Find all values of k so that $x^2 + kx - 3$ will factor.
92. Find all values of k so that $y + 4$ is a factor of $y^2 + ky - 8$.

Challenge Problems

In problems 93 through 96, factor each trinomial completely. Assume that n is a natural number.

93. $x^{2n} + 2x^n - 8$
94. $x^{2n} + 3x^n - 10$
95. $x^{2n} + 2x^n y^m + y^{2m}$
96. $x^{2n} + x^n y^m - 6y^{2m}$

■ IN YOUR OWN WORDS . . .

97. Explain the guidelines used to determine the signs in the factors of a trinomial.

■ 2.4 SUMMARY OF FACTORING

Now that we have looked at several types of factoring, here is a procedure to summarize how to factor a polynomial completely.

Procedure to Factor a Polynomial Completely

1. Factor out the greatest common factor (GCF) if there is one. This should be done *first*.
2. Count the number of terms.
3. If the polynomial is a binomial, check for difference of squares or cubes, or sum of cubes. $P^2 + Q^2$ **does not factor** *with real coefficients*.
4. If the polynomial is a trinomial, look for two binomial factors.
5. If the polynomial has four or more terms, try grouping.
6. Make sure that each factor is prime.
7. Check to see if the product of the factors is the original polynomial.

EXAMPLE 1. Factor each polynomial completely.

(a) $50 - 2x^2$ (b) $t^9 - 1$ (c) $5(a-b)^2 - 5x^2$

Solutions:

(a) There is a common factor.
$$50 - 2x^2 = 2(25 - x^2)$$
There are two terms in $25 - x^2$. This is the difference in squares.
$$50 - 2x^2 = 2(25 - x^2)$$
$$= 2(5 - x)(5 + x)$$
Check by multiplying.

(b) There is no common factor. There are two terms that are the difference in cubes.
$$t^9 - 1 = (t^3)^3 - 1^3$$
$$= (t^3 - 1)(t^6 + t^3 + 1)$$
$$= (t - 1)(t^2 + t + 1)(t^6 + t^3 + 1)$$
Check by multiplying.

(c) $5(a-b)^2 - 5x^2 = 5[(a-b)^2 - x^2]$
Treat $(a-b)$ as one term and the polynomial becomes the difference of two squares.
$$= 5[(a-b) - x][(a-b) + x]$$
$$= 5(a - b - x)(a - b + x)$$ ☐

EXAMPLE 2. Factor each polynomial completely.

(a) $x^4 + x^3 - 2x^2$ (b) $2x^4 - x^2 - 1$ (c) $6r^2t - 2rt^2 + 3rst - st^2$

Solutions:

(a) There is a common factor.
$$x^4 + x^3 - 2x^2 = x^2(x^2 + x - 2)$$
$$= x^2(x + 2)(x - 1)$$
Check by multiplying.

(b) There is no common factor. There are three terms.
$$2x^4 - x^2 - 1 = (2x^2 + 1)(x^2 - 1)$$
This is not factored completely, so continue.
$$= (2x^2 + 1)(x - 1)(x + 1)$$
Check by multiplying.

(c) There is a common factor.
$$6r^2t - 2rt^2 + 3rst - st^2 = t(6r^2 - 2rt + 3rs - st)$$
Use grouping on the polynomial in the parentheses.
$$t[(6r^2 - 2rt) + (3rs - st)] = t[2r(3r - t) + s(3r - t)]$$
$$= t[(3r - t)(2r + s)]$$
$$= t(3r - t)(2r + s)$$ ☐

Grouping can lead to the difference of squares.

EXAMPLE 3. Factor $x^2 - 2xy + y^2 - 4$ completely.

Solution:

The polynomial will not factor by grouping two terms together. However, if grouped as

$$(x^2 - 2xy + y^2) - 4$$

the polynomial will factor.

$$x^2 - 2xy + y^2 - 4 = (x^2 - 2xy + y^2) - 4$$
$$= (x - y)^2 - 4$$

This becomes the difference of two squares.

$$= [(x - y) - 2][(x - y) + 2]$$
$$= (x - y - 2)(x - y + 2)$$

Often, it is necessary to determine if a polynomial is factored. Which of the following polynomials is factored?

$$a^3 - b^3 \quad (a - b)^3$$

The polynomial $a^3 - b^3$ is not factored, whereas the polynomial $(a - b)^3$ is factored. To factor a polynomial means to write it as a *product* of polynomials. Thus $a^3 - b^3$ is a difference, *not* a product, but $(a - b)^3$ *is* a product, as

Be Careful!

$$(a - b)^3 = (a - b)(a - b)(a - b)$$

Factored Form	Not Factored
$(a - b)^3$	$a^3 - 3a^2b + 3ab^2 - b^3$
$(a + b)^3$	$a^3 + 3a^2b + 3ab^2 + b^3$
$(a - b)(a^2 + ab + b^2)$	$a^3 - b^3$
$(a + b)(a^2 - ab + b^2)$	$a^3 + b^3$

■ PROBLEM SET 2.4

Warm-ups

In problems 1 through 38, determine whether or not each polynomial is factored. If not, factor it completely, if possible. For problems 1 through 19, see Example 1.

1. $16x^2 - 16$
2. $(x + 5)^2$
3. $x^2 - x$
4. $w^3 + 125x^3$
5. $x(x^2y - 1)$
6. $49 - 4(r + s)^2$
7. $225a^5b^2c - 90a^2b^2c$
8. $x^2 + 4$
9. $4x^2 - 1$
10. $x^9 - y^6$
11. $16 - 25x^2$
12. $x^2(a + b)^2 - x^2$
13. $(z + t)^3 + 1$
14. $z^2(1 - y) - 4(y - 1)$
15. $8w^3 - 27$
16. $y(z - 4) + (4 - z)$
17. $x^{10} - 16$
18. $ab(y - x) + c(x - y) \quad (y - x)(ab - c)$
19. $a^2b + c$

Sec. 2.4 Summary of Factoring

In problems 20 through 36, see Example 2.

20. $x^2 + x + 1$ **21.** $2ab + 2ax - by - xy$ **22.** $16x^2 + 8x + 1$
23. $16 + 15y^2 - y^4$ **24.** $(3x - 5)^3$ **25.** $2x^2 - 5x - 7$
26. $18a^3 + 36b^3c^3 + 27b^2c^2$ **27.** $x^6 + x^3 - 2$
28. $x^2 - 10x + 25$ **29.** $x^6 + x^5 + x^4 + x^3$ **30.** $a^2z + a^2x - z - x$
31. $x^2(x + y)$ **32.** $y^2 - yz - z + y$ **33.** $9r^2 - 23r - 8$
34. $21x^2 + 24x - 36$ **35.** $2x^2 + x + 1$ **36.** $3x^2 - 30x + 27$

For problems 37 and 38, see Example 3.

37. $x^2 + 6x + 9 - z^2$ **38.** $4 - a^2 - 2ab - b^2$

Practice Exercises

In problems 39 through 76, determine whether or not each polynomial is factored. If not, factor it completely, if possible.

39. $x^2 + 9x + 14$ **40.** $9 - 16x^8$
41. $135r^3t^5 + 225r^2t^4$ **42.** $x^4 - 8x^2 + 16$
43. $1 - 27a^3b^3$ **44.** $x^4 - 64$
45. $x^2 - y^2 + x + y$ **46.** $a^3b(x + 1) + ab^2(1 + x)$
47. $s^3t(2st + u)$ **48.** $x^2 + x$
49. $32 - 2x^2$ **50.** $r^3t(s - 1) + rt(1 - s)$
51. $r^3(x - t) + (t - x)$ **52.** $6s^2 + st - 15t^2$
53. $(z + 4)^2 - x^2y^2$ **54.** $27r^3 + 64$
55. $a^2b + ab^2 + a + b$ **56.** $x(r + s) - y(s + r)$ **57.** $x^4 + 3x^2 + 2$
58. $x^8 - 16$ **59.** $8x^9 + y^3$ **60.** $a(x - y) - b(y - x)$
61. $x^2 + 12x + 36$ **62.** $x^2 - 6x + 5$ **63.** $3x^2 + x + 1$
64. $x^4 - x^3 + x^2 - x$ **65.** $(t + 8)^3 + z^3$ **66.** $2 - 2x^2$
67. $x^4 + x^2 - 20$ **68.** $x^{10} + x^5 - 6$ **69.** $81 - 18x + x^2$
70. $z(a + b)$ **71.** $a^9 + 8$ **72.** $(a + b)^2$
73. $a^7b^5c^4 - a^6b^6c^6 + a^5b^6c^3$ **74.** $(a + 2)^2 - 4b^2$
75. $16 - x^2 - 2xy - y^2$ **76.** $x^2 - y^2 + 2y - 1$

Challenge Problems

In problems 77 through 80, factor each polynomial completely. Assume that n is a natural number.

77. $x^{4n} - 2x^{2n} + 1$ **78.** $x^{5n} + 8x^{2n}$
79. $x^{4n} - x^{2n}$ **80.** $x^{6n} - 8$

81. We can factor certain trinomials by writing the polynomial as the difference of two squares. Consider $x^4 + x^2 + 1$. This will not factor as it is. If the middle term were $2x^2$, it would factor. Let's make the middle term $2x^2$ by adding x^2 and subtracting x^2.

$$x^4 + x^2 + 1 = (x^4 + x^2 + 1) + (x^2 - x^2)$$
$$= (x^4 + 2x^2 + 1) - x^2$$
$$= (x^2 + 1)^2 - x^2$$
$$= [(x^2 + 1) - x][(x^2 + 1) + x]$$
$$= (x^2 - x + 1)(x^2 + x + 1)$$

Factor each polynomial. **(a)** $x^4 + 3x^2 + 4$ **(b)** $x^4 + 2x^2 + 9$ **(c)** $x^4 + 7x^2 + 16$ **(d)** $x^4 - 8x^2 + 4$ **(e)** $x^4 - 11x^2 + 25$

▬ IN YOUR OWN WORDS . . .

82. Explain the steps to follow in factoring a polynomial.
83. What do we mean when we say that a polynomial is factored?
84. What do we mean when we say that a polynomial is factored completely?

CHAPTER SUMMARY

GLOSSARY

To **factor** a polynomial: To write the polynomial as a product of polynomials.

To **factor** a polynomial **completely**: To write the polynomial as a product of prime polynomials.

Greatest common factor of a set of numbers: The largest factor that is a factor of each number in the set.

PROCEDURE TO FACTOR A POLYNOMIAL COMPLETELY

1. Factor out the GCF if there is one. Do this *first*.
2. Count the number of terms.
3. If two terms, try difference of squares, difference of cubes, or sum of cubes.
   NOTE: $P^2 + Q^2$ will not factor with real coefficients.
4. If three terms, look for two binomial factors.
5. If four or more terms, try grouping.
6. Make sure that each factor is prime.
7. Check to see if the product of the factors is the original polynomial.

FACTORING RULES

1. $p^2 - q^2 = (p - q)(p + q)$ — Difference of two squares
2. $p^3 - q^3 = (p - q)(p^2 + pq + q^2)$ — Difference of two cubes
3. $p^3 + q^3 = (p + q)(p^2 - pq + q^2)$ — Sum of two cubes
4. $p^2 + 2pq + q^2 = (p + q)^2$ — Perfect trinomial square
5. $p^2 - 2pq + q^2 = (p - q)^2$ — Perfect trinomial square

CHECKUPS

1. Factor 392 into prime factors. — Section 2.1; Example 1c
2. Find the GCF of $108x^3y^6$ and $24x^4y^2$. — Section 2.1; Example 3
3. Factor $-15x^2y + 5xy^2$ completely. — Section 2.1; Example 5
4. Factor $rm - rn - sm + sn$ completely. — Section 2.1; Example 8b
5. Factor $4x^2 - 9y^6$ completely. — Section 2.2; Example 1
6. Factor $8x^3 + 27$ completely. — Section 2.2; Example 2b
7. Factor $6x^2 - 13x + 6$ completely. — Section 2.3; Example 3a
8. Factor $t^4 - 2t^3 + t^2$ completely. — Section 2.3; Example 4a
9. Factor $5(a - b)^2 - 5x^2$ completely. — Section 2.4; Example 1c

REVIEW PROBLEMS

In problems 1 through 30, factor each polynomial completely, if possible.

1. $2x^2 - 162$
2. $x^2 - 7x - 18$
3. $at - 4z^2 - 2az + 2zt$
4. $64y^3 + 125$
5. $(x + 4z)^2 - 4$
6. $2x^2 + x + 1$

7. $12x^2 - 38x - 14$
8. $2x^2 + x - 1$
9. $12a^2b^2c - 96a^2b^3c + 54abc$
10. $4z^2(s - t) - 2z(t - s)$
11. $x^4 - 15x^2 - 16$
12. $8 + (r + t)^3$
13. $x^6 + 28x^3 + 27$
14. $144 - (x + y)^2$
15. $2a^2b + 4a^2y - 3abz - 6ayz$
16. $30(u - v)^2 + 42(v - u)$
17. $2ax + 4x + 2a + 4$
18. $x^2 + 4$
19. $x^2 + 18x + 81$
20. $3y^2 - 11y + 8$
21. $w^2 - x^2 - 4x - 4$
22. $a^6 - 125$
23. $x^8 - 81$
24. $15 - x - 2x^2$
25. $3x^2 + 4x + 5$
26. $16x^2 + 8x + 1$
27. $27a^3 - 1$
28. $27a^3 - 3a$
29. $x^2 + 4x + 4 - z^2$
30. $x^2 + z^2 - 2xz - y^2$

... LET'S NOT FORGET ...

Identify the expressions that are in factored form. Factor those that are not factored.

31. $(2x + 5)^3$
32. $8x^3 + 125$
33. $(s + t)^3 - a(s + t)^2$
34. $5x(x^2 + y^2)$
35. $12ab^2$
36. $a(1 - b) + (b - 1)$

How many terms are in each expression? Which expressions have $y + 7$ as a factor?

37. $y^2 - 49$ Two terms;
38. $(y + 7)^4$ One term;
39. $y^2 + 6y - 7$ Three terms;
40. $y^2 + 14y + 49$ Three terms;
41. $4(y + 7) - (7 + y)$ Two terms;

Find each product.

42. $(-3x)^3$
43. $(3x - 2)^2$
44. $(x - 3)^3$

Simplify each expression.

45. -5^2
46. $(-5)^3$
47. $(-5)^2$
48. $(-x)^2$
49. $(-x)^3$

CHAPTER TEST

In problems 1 through 6, choose the correct answer.

1. The GCF of $48x^2y$ and $80xy$ is (?)
    A. 16
    B. $16xy$
    C. $8xy$
    D. xy
2. One of the factors of $x^3 - 64$ is (?)
    A. $x^2 - 8x + 16$
    B. $x^2 + 8x + 16$
    C. $x^2 + 4x + 16$
    D. $x^2 - 4x + 16$
3. One of the factors of $x(a - 1) + y(1 - a)$ is (?)
    A. $x + y$
    B. $x - y$
    C. x
    D. $x(a - 1)$
4. $x^2 - 12x + 36 = $ (?)
    A. $(x - 9)(x - 4)$
    B. $(x - 6)(x + 6)$
    C. $(x + 6)^2$
    D. $(x - 6)^2$
5. $x^4 - 16 = $ (?)
    A. $(x^2 + 4)^2$
    B. $(x^2 - 8)(x^2 + 2)$
    C. $(x - 2)^2(x + 2)^2$
    D. $(x^2 + 4)(x - 2)(x + 2)$
6. Which polynomial is factored?
    A. $x^2 + 2x$
    B. $a(r - s) + b(r - s)$
    C. $x^3 - 8$
    D. $(x + 2)^3$

In problems 7 through 15, factor each polynomial completely.

7. $8a^3 - 27$
8. $4x^2 + 18x - 10$
9. $ax - bx + 2a - 2b$
10. $x(p - q) - y(q - p)$
11. $a^4 - 2a^2 - 8$
12. $x^2 + 14x + 49$
13. $(x + y)^2 - 9$
14. $4y^2 - 25$
15. $27a^2b^3 - 36a^3b^2 + 45a^4b^4$

CHAPTER 3

See Problem Set 3.1, Exercise 51.

Rational Expressions

- **3.1** Rational Expressions
- **3.2** Fundamental Principle of Rational Expressions
- **3.3** Multiplication and Division
- **3.4** Addition and Subtraction
- **3.5** Complex Fractions

CONNECTIONS

To find the capacitance of a spherical capacitor, an electrical engineer might use the formula, $C = \dfrac{4kR_1R_2}{R_2 - R_1}$. Expressions such as the right side of this formula are common in science and engineering. They are called rational expressions. Rational expressions are simply polynomials transformed into fractions.

Just as fractions form an important part of arithmetic, rational expressions play an important role in algebra. Rational expressions, like the fractions of arithmetic, indicate a division operation. We will learn how to reduce rational expressions, in much the same way as we simplify common fractions.

In this chapter we define rational expressions, then we learn how to simplify, add, subtract, multiply and divide them. Complex fractions (fractions containing fractions) are introduced and we learn how to simplify them. The factoring skills we acquired in Chapter 2 are a necessary prerequisite to this study, and are essential to our understanding of it. Then, we will use the skills developed in this chapter to solve fractional equations in Chapters 4 and 6, and throughout algebra, as well.

3.1 RATIONAL EXPRESSIONS

In Section 0.1 a rational number was defined as the quotient of two integers, a and b, written as $\dfrac{a}{b}$ with $b \neq 0$, and in Section 1.6 this idea was extended to the division of polynomials. Next we examine quotients of polynomials in more detail.

A **rational expression** is the quotient of two polynomials. Rational expressions are often called *algebraic fractions* or simply *fractions*. We use the terms *numerator* and *denominator* just as we would with fractions in arithmetic. The following are examples of rational expressions.

$$\boxed{1}\ \dfrac{3}{x} \qquad \boxed{2}\ \dfrac{5x^2 + 2x - 1}{x^3 + 3x^2 - x + 1} \qquad \boxed{3}\ \dfrac{x}{(x+1)(x-2)}$$

Each of these is a quotient of two polynomials and each has many different values, depending on the value of x.

$\boxed{1}$ has the value $\dfrac{3}{2}$ when x is 2 and $\dfrac{3}{7}$ when x is 7.

$\boxed{2}$ has the value of -1 when x is 0 and $\dfrac{23}{19}$ when x is 2.

$\boxed{3}$ has the value $-\dfrac{3}{10}$ when x is -3.

Let's consider the rational expression $\dfrac{1}{x-2}$. If x is 4, its value is $\dfrac{1}{2}$. However, if x is 2, the expression takes the form $\dfrac{1}{0}$. This is not a real number. We say that the rational expression $\dfrac{1}{x-2}$ is undefined when x is 2. A rational expression is **undefined** when its denominator has the value 0, regardless of the value of its numerator.

EXAMPLE 1. Determine the values of x for which each rational expression is undefined.

(a) $\dfrac{x}{x+3}$ (b) $\dfrac{x+2}{(x+1)(x-2)}$ (c) $\dfrac{5}{x^2+1}$

Solutions:

(a) If x is -3, the rational expression $\dfrac{x}{x+3}$ becomes $\dfrac{-3}{0}$. So $\dfrac{x}{x+3}$ is undefined when x is -3.

(b) If x is -1, the rational expression $\dfrac{x+2}{(x+1)(x-2)}$ becomes $\dfrac{1}{0}$, and if x is 2, it becomes $\dfrac{4}{0}$. So $\dfrac{x+2}{(x+1)(x-2)}$ is undefined when x is -1 or 2.

(c) The denominator, x^2+1, is never 0 for any value of x. (Try some numbers!) The rational expression $\dfrac{5}{x^2+1}$ is defined for all real numbers. □

From the rules of signed number arithmetic we know that if p and q are real numbers with $q \neq 0$, then

$$\dfrac{-p}{q} = \dfrac{p}{-q} = -\dfrac{p}{q}$$

This idea is also true for rational expressions and is very important.

Sign Property of Rational Expressions

If A and B are polynomials and $\dfrac{A}{B}$ is defined, then

$$\dfrac{-A}{B} = \dfrac{A}{-B} = -\dfrac{A}{B}$$

EXAMPLE 2. Use the Sign Property of Rational Expressions to write each rational expression two different ways.

(a) $\dfrac{-5}{x}$ (b) $-\dfrac{x}{x^2+3}$ (c) $\dfrac{-x+2}{x^2}$

Solutions:

(a) $\dfrac{-5}{x} = \dfrac{5}{-x} = -\dfrac{5}{x}$

(continued)

Sec. 3.1 Rational Expressions

(b) $-\dfrac{x}{x^2+3} = \dfrac{-x}{x^2+3} = \dfrac{x}{-(x^2+3)}$

The parentheses in the denominator are necessary, as $-(x^2+3)$ is not the same as $-x^2+3$.

(c) Be very careful with this one! The leading negative in $-x+2$ **does not** indicate the opposite of $(x+2)$. However, as $-x+2 = -(x-2)$, we can write

$$\dfrac{-x+2}{x^2} = \dfrac{-(x-2)}{x^2}$$

so

$$\dfrac{-x+2}{x^2} = \dfrac{-(x-2)}{x^2} = \dfrac{x-2}{-x^2} = -\dfrac{x-2}{x^2}$$

PROBLEM SET 3.1

Warm-ups

In problems 1 through 16, determine the values of x, if any, for which each rational expression is undefined. See Example 1.

1. $\dfrac{5}{x-3}$
2. $\dfrac{1}{x+1}$
3. $\dfrac{-6}{x-5}$
4. $\dfrac{3}{x}$
5. $\dfrac{x}{6}$
6. $\dfrac{4}{x^2}$
7. $\dfrac{7}{x^2+1}$
8. $\dfrac{x}{x-9}$
9. $\dfrac{x-5}{x-2}$
10. $\dfrac{x+3}{x+8}$
11. $\dfrac{x}{(x+3)(x-1)}$
12. $\dfrac{x^2+2}{(x-6)(x-4)}$
13. $\dfrac{5}{x(x-2)}$
14. $\dfrac{x-7}{x^2-x-6}$
15. $\dfrac{x^2-2x-3}{x^2-4}$
16. $\dfrac{x}{x^2+6x+9}$

In problems 17 through 25, use the Sign Property of Rational Expressions to write each rational expression in two different ways. See Example 2.

17. $\dfrac{-3}{x}$
18. $\dfrac{x^2}{-8}$
19. $-\dfrac{7}{x}$
20. $\dfrac{-(x-1)}{x}$
21. $\dfrac{2x}{-(x+5)}$
22. $-\dfrac{x}{x+3}$
23. $\dfrac{x-1}{-x^2}$
24. $\dfrac{-x-6}{x+3}$
25. $-\dfrac{x-7}{x+4}$

Practice Exercises

In problems 26 through 41, determine the values of x, if any, for which each rational expression is undefined.

26. $\dfrac{3}{x-1}$
27. $\dfrac{5}{x+6}$
28. $\dfrac{-3}{x-2}$
29. $\dfrac{5}{x}$
30. $\dfrac{x}{7}$
31. $\dfrac{6}{x^2}$
32. $\dfrac{1}{x^2+4}$
33. $\dfrac{x}{x-7}$
34. $\dfrac{x-3}{x-4}$
35. $\dfrac{x+5}{x+9}$
36. $\dfrac{x}{(x-3)(x+1)}$
37. $\dfrac{x^2+3}{(x-5)(x-2)}$
38. $\dfrac{5}{x(x-1)}$
39. $\dfrac{x-4}{x^2-x-2}$
40. $\dfrac{x^2-2x-8}{x^2-1}$
41. $\dfrac{x}{x^2-6x+9}$

In problems 42 through 50, use the Sign Property of Rational Expressions to write each rational expression in two different ways.

42. $\dfrac{-6}{x}$

43. $\dfrac{x^2}{-2}$

44. $-\dfrac{1}{x}$

45. $\dfrac{-(x+1)}{x}$

46. $\dfrac{3x}{-(x-5)}$

47. $-\dfrac{x}{x-3}$

48. $\dfrac{x+3}{-x^2}$

49. $\dfrac{-x-5}{x+1}$

50. $-\dfrac{x-4}{x+3}$

51. A Canadian Pacific passenger train leaves Vancouver for Banff and travels $160x$ miles in $x + 12$ hours. Write a rational expression that gives the average rate of this train for that part of the trip.

    (HINT: Rate is distance divided by time.)

52. The acceleration of a certain rocket is given by the relationship, twice the distance divided by the elapsed time squared. Find a rational expression that gives the acceleration of this rocket if it travels $14t^3 - 2t + 77$ meters in $t + 5$ seconds.

53. Sharon runs N times around a 2-mile track at an average rate of $8 - N$ miles per hour. Write a rational expression that gives the time it takes Sharon to run these laps.

54. Power is the time rate of doing work. It is given by the relationship
$$\text{Power} = \dfrac{\text{Work}}{\text{Time}}$$
and work is calculated as force times distance. Write a rational expression that gives the power required for a force of measure $x + 5$ to move through a distance of $2x - 3$ units in $x^2 + 1$ hours.

Challenge Problems

In problems 55 through 57, determine the values of x, if any, for which each rational expression is undefined.

55. $\dfrac{x+1}{x^2 + 4x + 3}$

56. $\dfrac{x-2}{x^2 - 4x + 4}$

57. $\dfrac{x+2}{x^4 - 16}$

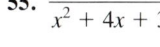 IN YOUR OWN WORDS . . .

58. Explain what we mean when we say that a rational expression is undefined.

3.2 FUNDAMENTAL PRINCIPLE OF RATIONAL EXPRESSIONS

There are many ways of writing the same fraction. For example, $\dfrac{1}{2}$ can be written as $\dfrac{3}{6}, \dfrac{4}{8},$ or $\dfrac{6}{12}$. Which form we use depends on what we are doing with the fraction. The same idea applies to rational expressions, as stated in the following principle.

> **Fundamental Principle of Rational Expressions**
>
> If A, B, and C are polynomials, then
> $$\dfrac{AC}{BC} = \dfrac{A}{B}$$
> where each rational expression is defined.

This principle states that we can divide the numerator and denominator of a rational expression by the same nonzero *factor*. When we write a rational expression so that the numerator and denominator have no common factor, we say that the rational expression is in **lowest terms** or that we have **reduced** to lowest terms.

Consider the rational expression, $\dfrac{2(x+1)}{a(x+1)}$. By the Fundamental Principle of Rational Expressions,

$$\frac{2(x+1)}{a(x+1)} = \frac{2}{a}$$

Notice that $x + 1$ is a factor in both the numerator and the denominator. By dividing both numerator and denominator by the only common factor, we have written $\dfrac{2(x+1)}{a(x+1)}$ in lowest terms.

The Fundamental Principle of Rational Expressions says that each rational expression must be defined. This means that when we divide by $x + 1$, we must assume that $x + 1 \neq 0$ or $x \neq -1$. So it would be proper to write

$$\frac{2(x+1)}{a(x+1)} = \frac{2}{a}; \qquad x \neq -1$$

However, we usually do not write the $x \neq -1$. Instead, we understand that when we divide by a common factor, its value must not be zero.

Be Careful! When using the Fundamental Principle of Rational Expressions to reduce, **factoring** is the key. We divide by common *factors*. So make sure that both the numerator and the denominator are **factored** before reducing.

To Reduce a Rational Expression to Lowest Terms

1. Factor numerator and denominator.
2. Apply the Fundamental Principle of Rational Expressions.

EXAMPLE 1. Reduce each rational expression to lowest terms.

(a) $\dfrac{12x^2}{-18x^3}$ (b) $\dfrac{5x + 5y}{x^2 - y^2}$ (c) $\dfrac{x + 3}{x^2 + 6x + 9}$ (d) $\dfrac{x^2 + xy - xz - yz}{ax - az + bx - bz}$

Solutions:

(a) First factor the numerator and the denominator.

$$\frac{12x^2}{-18x^3} = \frac{(6)(2)x^2}{-(6)(3)x^2 x}$$

Then divide by any common factors. Both 6 and x^2 are factors in both the numerator and denominator. So we divide both the numerator and the denominator by $6x^2$.

$$\frac{12x^2}{-18x^3} = \frac{(6)(2)x^2}{-(6)(3)x^2 x} = \frac{2}{-3x}$$

This answer could be written in various forms:

$$\frac{2}{-3x} = \frac{-2}{3x} = -\frac{2}{3x}$$

(b) $\dfrac{5x + 5y}{x^2 - y^2} = \dfrac{5(x + y)}{(x + y)(x - y)}$ Factor.

$= \dfrac{5}{x - y}$ Reduce.

(c) $\dfrac{x + 3}{x^2 + 6x + 9} = \dfrac{x + 3}{(x + 3)^2}$ Factor.

$= \dfrac{1}{x + 3}$ Reduce.

(d) $\dfrac{x^2 + xy - xz - yz}{ax - az + bx - bz} = \dfrac{x(x + y) - z(x + y)}{a(x - z) + b(x - z)}$

Be Careful!

Although we have factored some *terms* in order to factor by grouping, at this point neither the numerator nor the denominator is factored. So we **cannot** divide out any common factors here. Continuing to factor by grouping gives us

$= \dfrac{(x + y)(x - z)}{(x - z)(a + b)} = \dfrac{x + y}{a + b}$ ☐

EXAMPLE 2. Reduce each rational expression to lowest terms.

(a) $\dfrac{a - b}{b - a}$ (b) $\dfrac{x^2 - y^2}{y^3 - x^3}$

Solutions:

(a) The numerator and the denominator are opposites. Remember that $a - b = -(b - a)$. So we write

$$\frac{a - b}{b - a} = \frac{-(b - a)}{b - a}$$

Now $b - a$ is a factor in both the numerator and the denominator. So

$$\frac{a - b}{b - a} = \frac{-(b - a)}{b - a} = -\frac{b - a}{b - a} = -1$$

(b) $\dfrac{x^2 - y^2}{y^3 - x^3} = \dfrac{(x - y)(x + y)}{(y - x)(y^2 + yx + x^2)}$ Factor.

We replace $x - y$ with $-(y - x)$ in the numerator.

$$= \frac{-(y - x)(x + y)}{(y - x)(y^2 + yx + x^2)}$$

Now $(y - x)$ is a factor of both numerator and denominator.

$$= \frac{-(x + y)}{x^2 + xy + y^2}$$

Notice in the last step we wrote the denominator with the x-terms first. ☐

Sec. 3.2 Fundamental Principle of Rational Expressions

Be Careful! The most common mistake made in reducing fractions is to divide by a *term* and not a *factor*. Make a habit of checking carefully to see that anything being divided out of a fraction is a **factor**.

There are two ways to look at the Fundamental Principle of Rational Expressions. If we write $\frac{AC}{BC} = \frac{A}{B}$, we are reducing. If we write it the other way, $\frac{A}{B} = \frac{AC}{BC}$, we are multiplying the numerator and denominator by C. Later we will need to do this when adding and subtracting rational expressions. This is sometimes called *building up a denominator*.

Let's write the rational expression $\frac{10m}{7n^2}$ with a denominator of $21n^3$. First notice that $21n^3 = 7n^2(3n)$. So use the Fundamental Principle of Rational Expressions and multiply both numerator and denominator by $3n$.

$$\frac{10m}{7n^2} = \frac{10m(3n)}{7n^2(3n)}$$

$$= \frac{30mn}{21n^3}$$

EXAMPLE 3. Write each rational expression with the denominator indicated.

(a) $\dfrac{-xy}{a}$; a^3 (b) $\dfrac{x+2}{x-1}$; $x^2 - 1$

(c) $\dfrac{x}{x+1}$; $x^2 + 2x + 1$ (d) $\dfrac{x-2}{x-3}$; $x^2 - x - 6$

Solutions:

(a) First note that $a^3 = a(a^2)$.

$$\frac{-xy}{a} = \frac{-xy(a^2)}{a(a^2)} = \frac{-a^2xy}{a^3}$$

(b) Notice that $x^2 - 1 = (x - 1)(x + 1)$. So we multiply both numerator and denominator by $x + 1$.

$$\frac{x+2}{x-1} = \frac{(x+2)(x+1)}{(x-1)(x+1)}$$

$$= \frac{x^2 + 3x + 2}{x^2 - 1}$$

(c) As $x^2 + 2x + 1 = (x + 1)^2$, multiply by $x + 1$.

$$\frac{x}{x+1} = \frac{x(x+1)}{(x+1)(x+1)}$$

$$= \frac{x^2 + x}{x^2 + 2x + 1}$$

(d) Notice that $x^2 - x - 6 = (x - 3)(x + 2)$. So we multiply numerator and denominator by $x + 2$.

$$\frac{x-2}{x-3} = \frac{(x-2)(x+2)}{(x-3)(x+2)}$$

$$= \frac{x^2 - 4}{x^2 - x - 6}$$ ☐

EXAMPLE 4. Write each rational expression with the denominator indicated.

(a) $\dfrac{5}{7-x}$; $x - 7$ (b) $\dfrac{3}{2-x}$; $x^2 - 4$

Solutions:

(a) Since $7 - x$ and $x - 7$ are opposites, we replace $7 - x$ with $-(x - 7)$.

$$\frac{5}{7-x} = \frac{5}{-(x-7)}$$

$$= \frac{-5}{x-7}$$

(b) As $x^2 - 4 = (x - 2)(x + 2)$, and $2 - x$ and $x - 2$ are opposites, replace $2 - x$ with $-(x - 2)$ and multiply the numerator and denominator by $x + 2$.

$$\frac{3}{2-x} = \frac{3}{-(x-2)}$$

$$= \frac{-3}{x-2}$$

$$= \frac{-3(x+2)}{(x-2)(x+2)}$$

$$= \frac{-3x - 6}{x^2 - 4}$$ ☐

■ PROBLEM SET 3.2

Warm-ups

In Problems 1 through 28, reduce each rational expression to lowest terms. For problems 1 through 22, see Example 1.

1. $\dfrac{x^7}{x^2}$

2. $\dfrac{y^2}{y^3}$

3. $\dfrac{28t^2}{21t}$

4. $\dfrac{-2^3 xy}{2^2 x^2 y^2}$

5. $\dfrac{(-3)^2 a^2}{3a}$

6. $\dfrac{2^2 3^3 x^3 y}{2^3 3^2 xy^2}$

7. $\dfrac{5x^2(x+y)^2}{10x(x+y)^3}$

8. $\dfrac{-n^2(m-n)}{n^3(m-n)^4}$

9. $\dfrac{(x+y)^5}{(x+y)^2}$

Sec. 3.2 Fundamental Principle of Rational Expressions

10. $\dfrac{(x+1)(3+x)}{(x-1)(x+3)^2}$
11. $\dfrac{12x^2 - 12x}{18x^2}$
12. $\dfrac{x+1}{x^2+4x+3}$

13. $\dfrac{x^2-16}{x-4}$
14. $\dfrac{x+1}{x^3+1}$
15. $\dfrac{x^2-3x-10}{x^2-2x-15}$

16. $\dfrac{x^2+2x-8}{x^2-4x+4}$
17. $\dfrac{s^2-1}{1+s^3}$
18. $\dfrac{8x^3-8}{2x^2-6x+4}$

19. $\dfrac{6x^2+x-2}{3x^2-4x-4}$
20. $\dfrac{a^3+2a^2-15a}{a^4+8a^3+15a^2}$

21. $\dfrac{ac+ad+bc+bd}{ac-ad+bc-bd}$
22. $\dfrac{x^3+xz^2+x^2z+z^3}{x^3+xz^2-x^2z-z^3}$

For problems 23 through 28, see Example 2.

23. $\dfrac{s-1}{1-s}$
24. $\dfrac{(x-2)(x+3)}{(2-x)(1+x)}$
25. $\dfrac{(2x-1)^2}{(1-2x)(x+2)}$

26. $\dfrac{5x^2-10x}{10-5x}$
27. $\dfrac{x^2-3x+2}{3-2x-x^2}$
28. $\dfrac{m^3-n^3}{n^2-m^2}$

In problems 29 through 40, write each rational expression with the denominator indicated.
For problems 29 through 36, see Example 3.

29. $\dfrac{5}{x};\quad 10x$
30. $\dfrac{p}{7};\quad 21(p+1)$
31. $\dfrac{r^2}{6};\quad 6m+6n$

32. $\dfrac{3}{x^2};\quad x^3$
33. $\dfrac{3}{a};\quad 4a$
34. $\dfrac{15b}{2c};\quad 10bc^2$

35. $\dfrac{-3x}{x+1};\quad x^2-1$
36. $\dfrac{a}{a-2};\quad a^2-a-2$

For problems 37 through 40, see Example 4.

37. $\dfrac{x+1}{x-1};\quad x-x^2$
38. $\dfrac{2z}{3-z};\quad z^2-6z+9$

39. $\dfrac{x}{3-x};\quad x^2-9$
40. $\dfrac{3z}{y-2z};\quad 4z^2-y^2$

Practice Exercises

In problems 41 through 96, reduce each rational expression to lowest terms.

41. $\dfrac{x^8}{x^3}$
42. $\dfrac{y^3}{y^7}$
43. $\dfrac{48t^3}{16t}$

44. $\dfrac{-15x^3y}{5xy^2}$
45. $\dfrac{-2x^2y}{-8xy}$
46. $\dfrac{18x^3yz^2}{-27xy^3z^3}$

47. $\dfrac{-3^3xy}{3^2x^3y^3}$
48. $\dfrac{(-5)^2a^2}{5a}$
49. $\dfrac{7^2 5^3 x^2 y}{7^3 5^2 xy^3}$

50. $\dfrac{5x^2(x-y)^2}{15x(x-y)^3}$
51. $\dfrac{-n^5(n-m)}{n^4(n-m)^4}$
52. $\dfrac{(x+y)^2}{(x+y)^7}$

53. $\dfrac{(x+2)(5+x)}{(x-1)(x+5)^2}$
54. $\dfrac{t-1}{1-t}$
55. $\dfrac{(a-3)(a+3)}{(3-a)(1+a)}$

56. $\dfrac{(3x-1)^2}{(1-3x)(x+1)}$
57. $\dfrac{4a^2-8a}{8-4a}$
58. $\dfrac{18x^3-12x}{27x^2}$

59. $\dfrac{5x}{25x - 15x^2}$

60. $\dfrac{xy^3}{xy + y}$

61. $\dfrac{3y + 12y^4}{15y^3 - 25y}$

62. $\dfrac{6x}{24x - 12x^2}$

63. $\dfrac{ab^2}{ab + b}$

64. $\dfrac{5y + 15y^3}{10y^2 - 25y}$

65. $\dfrac{3y^4 + 15y^2}{3y^3 + 9y^5}$

66. $\dfrac{4rs - 8s}{6 - 3r}$

67. $\dfrac{4x^3 - 24x}{x^3 - 6x}$

68. $\dfrac{x + 3}{x^2 + 5x + 6}$

69. $\dfrac{x^2 - 25}{x + 5}$

70. $\dfrac{x - 1}{x^3 - 1}$

71. $\dfrac{x^2 + 8x + 16}{x + 4}$

72. $\dfrac{b^2 - 1}{b^2 - b - 2}$

73. $\dfrac{p^3 - 8}{p^2 - 5p + 6}$

74. $\dfrac{x^2 - 8x + 16}{x - 4}$

75. $\dfrac{b^2 - 1}{b^2 + b - 2}$

76. $\dfrac{p^3 + 27}{p^2 + 5p + 6}$

77. $\dfrac{x^2 - x - 12}{x^2 - 3x - 4}$

78. $\dfrac{x^2 + x - 2}{x^2 + 4x + 4}$

79. $\dfrac{x^2 + 3x - 10}{8 - 2x - x^2}$

80. $\dfrac{s^2 - 1}{1 - s^3}$

81. $\dfrac{3x^4 - 3x}{6x^3 - 6x}$

82. $\dfrac{6x^2 + x - 2}{3x^2 - x - 2}$

83. $\dfrac{x^4 - x^2 - 2}{x^4 + x^2 - 6}$

84. $\dfrac{x^4 - 16}{x^2 - 4}$

85. $\dfrac{9 - 4x^2}{2x^2 - x - 3}$

86. $\dfrac{x^4 - x^2 - 6}{x^4 - 3x^2 - 10}$

87. $\dfrac{x^4 - 25}{x^2 - 5}$

88. $\dfrac{25 - 4x^2}{2x^2 + x - 15}$

89. $\dfrac{m^2 - n^2}{n^3 - m^3}$

90. $\dfrac{a^4 + 2a^3 - 8a^2}{a^5 + 8a^4 + 16a^3}$

91. $\dfrac{am + 3m - 2a - 6}{cm - 2c + 6m - 12}$

92. $\dfrac{a^3 + ab^2 + a^2c + b^2c}{a^3 + ab^2 - a^2c - b^2c}$

93. $\dfrac{ps - pt - qs + qt}{pt + qt - ps - qs}$

94. $\dfrac{2x^2 + 6ax + 4a^2}{x^2 + bx + ax + ab}$

95. $\dfrac{x^3 - xy + x^2y - y^2}{x^3 - x^2y - xy + y^2}$

96. $\dfrac{y^2 + yc + yd + dc}{3y^2 - 3yc + 3yd - 3dc}$

In problems 97 through 124, write each rational expression with the denominator indicated.

97. $\dfrac{5}{x}$; $12x^2$

98. $\dfrac{p}{14}$; $28(p + 1)$

99. $\dfrac{r^2}{3}$; $6x + 6y$

100. $\dfrac{3}{x}$; x^3

101. $\dfrac{3}{a}$; $4a^2$

102. $\dfrac{15b}{2c^2}$; $10bc^2$

103. $\dfrac{-10a}{3xy}$; $30x^2y^3$

104. $\dfrac{p}{p - 5}$; $2p - 10$

105. $\dfrac{a^2}{a + b}$; $4a + 4b$

106. $\dfrac{-10a}{6x^2y}$; $30x^2y^3$

107. $\dfrac{p}{p - 4}$; $2p - 8$

108. $\dfrac{a^2}{a + b}$; $5a + 5b$

109. $\dfrac{-3x}{x - 1}$; $x^2 - 1$

110. $\dfrac{a}{a + 2}$; $a^2 + a - 2$

111. $\dfrac{2z}{z + 3}$; $z^2 + 6z + 9$

112. $\dfrac{x - 1}{x}$; $x^2 + 2x$

113. $\dfrac{x - 2}{x}$; $x^2 - 3x$

114. $\dfrac{5}{m - n}$; $n - m$

115. $\dfrac{x}{2 - x}$; $x^2 - 4$

116. $\dfrac{3z}{y + 2z}$; $4z^2 - y^2$

Sec. 3.2 Fundamental Principle of Rational Expressions

117. $\dfrac{2x}{x+y}$; $x^3 + y^3$

118. $\dfrac{x-1}{x^2 + 2x + 4}$; $x^3 - 8$

119. $\dfrac{2x}{x-y}$; $x^3 - y^3$

120. $\dfrac{x-1}{x^2 - 2x + 4}$; $x^3 + 8$

121. $\dfrac{w^2 + 3w + 9}{w + 3}$; $w^2 - 9$

122. $x + 2$; 7

123. x^3; $2z$

124. $x - 1$; $x + 1$

Challenge Problems

In problems 125 and 126, reduce each rational expression to lowest terms. (n is a natural number.)

125. $\dfrac{(x+2)^{n+1}}{(x+2)^n}$

126. $\dfrac{x^{2n+1} y^{2n-1}}{x^{2n+3} y^{2n}}$

■ IN YOUR OWN WORDS . . .

127. Explain what the Fundamental Principle of Rational Expressions allows us to do with a fraction.

128. The rational expression $\dfrac{x^2 - 4}{x^2 + x - 6}$ can be reduced to $\dfrac{x+2}{x+3}$. Is this reduction true for any value of x? Explain. Why can't we divide the x^2 terms out of the original expression and reduce it to $\dfrac{-4}{x-6}$?

■ 3.3 MULTIPLICATION AND DIVISION

The operations of multiplication and division are performed with rational expressions just as they are with rational numbers.

Multiplication of Rational Expressions

If A, B, C, and D are polynomials,

$$\frac{A}{B} \cdot \frac{C}{D} = \frac{AC}{BD}$$

where each rational expression is defined.

To multiply rational expressions, multiply the numerators and multiply the denominators. This gives a single rational expression that should be reduced, if possible.

Procedure to Multiply Rational Expressions

1. Factor numerators and denominators.
2. Multiply numerators and multiply denominators.
3. Reduce to lowest terms.

EXAMPLE 1. Perform the operation indicated.

(a) $\dfrac{abc^3}{d^4} \cdot \dfrac{d^2}{a^2b^4c}$

(b) $(x+1)^2 \cdot \dfrac{x}{x+1}$

(c) $\dfrac{a+2}{a+1} \cdot \dfrac{a^2-1}{a^2-4}$

(d) $\dfrac{t^2-t-2}{2t-6} \cdot \dfrac{t-3}{t^3-8}$

Solutions:

(a) As the fraction is already factored, we multiply numerators and multiply denominators.

$$\dfrac{abc^3}{d^4} \cdot \dfrac{d^2}{a^2b^4c} = \dfrac{(abc^3)d^2}{d^4(a^2b^4c)} \qquad \text{Multiply.}$$

$$= \dfrac{abc^3 d^2}{a^2 b^4 c d^4}$$

$$= \dfrac{c^2}{ab^3 d^2} \qquad \text{Reduce.}$$

(b) Again we begin by multiplying numerators and multiplying denominators.

$$(x+1)^2 \cdot \dfrac{x}{x+1} = \dfrac{(x+1)^2 \cdot x}{x+1} \qquad \text{Multiply.}$$

We do not want to multiply further in the numerator because our next step is to reduce if possible. Look for common factors to reduce.

$$= x(x+1) \qquad \text{Reduce.}$$

(c) First we factor in each fraction.

$$\dfrac{a+2}{a+1} \cdot \dfrac{a^2-1}{a^2-4} = \dfrac{a+2}{a+1} \cdot \dfrac{(a+1)(a-1)}{(a+2)(a-2)} \qquad \text{Factor.}$$

Next, we multiply numerators and denominators.

$$= \dfrac{(a+2)(a+1)(a-1)}{(a+1)(a+2)(a-2)} \qquad \text{Multiply.}$$

$$= \dfrac{a-1}{a-2} \qquad \text{Reduce.}$$

(d) As multiplication is just a copying step, we can factor at the same time.

$$\dfrac{t^2-t-2}{2t-6} \cdot \dfrac{t-3}{t^3-8} = \dfrac{(t-2)(t+1)(t-3)}{2(t-3)(t-2)(t^2+2t+4)}$$

$$= \dfrac{t+1}{2(t^2+2t+4)} \qquad \text{Reduce.}$$

It is not necessary to multiply further in the denominator. However, it would not be wrong to do so.

Sec. 3.3 Multiplication and Division

EXAMPLE 2. Perform the operation indicated.

$$\frac{m^2 - n^2}{m^5} \cdot \frac{m^2}{n - m}$$

Solution:

$$\frac{m^2 - n^2}{m^5} \cdot \frac{m^2}{n - m} = \frac{(m - n)(m + n)m^2}{m^5(n - m)}$$

Notice that $m - n$ and $n - m$ are opposites. We replace $m - n$ with $-(n - m)$ and then reduce.

$$= \frac{-(n - m)(m + n)m^2}{m^5(n - m)}$$

$$= \frac{-(m + n)}{m^3}$$

$$= -\frac{m + n}{m^3}$$

Next, notice that division is just a form of multiplication.

Division of Rational Expressions

If A, B, C, D are polynomials,

$$\frac{A}{B} \div \frac{C}{D} = \frac{A}{B} \cdot \frac{D}{C}$$

$$= \frac{AD}{BC}$$

where each rational expression is defined.

To divide two polynomials, multiply by the reciprocal of the divisor.

Procedure to Divide Two Rational Expressions

1. Change the division to multiplication by multiplying by the reciprocal of the divisor.
2. Factor numerators and denominators.
3. Multiply numerators and multiply denominators.
4. Reduce to lowest terms.

EXAMPLE 3. Perform the operation indicated.

(a) $\dfrac{6a^3b}{5c^4} \div \dfrac{2ab^2}{c^2}$ (b) $\dfrac{x^2 + 3x + 2}{x^4} \div \dfrac{x^2 + 2x + 1}{x^2}$

Solutions:

(a) We change the division to multiplication.

$$\dfrac{6a^3b}{5c^4} \div \dfrac{2ab^2}{c^2} = \dfrac{6a^3b}{5c^4} \cdot \dfrac{c^2}{2ab^2} \qquad \text{Change to multiplication.}$$

$$= \dfrac{(6a^3b)(c^2)}{(5c^4)(2ab^2)} \qquad \text{Multiply.}$$

$$= \dfrac{3a^2}{5bc^2} \qquad \text{Reduce.}$$

(b) $\dfrac{x^2 + 3x + 2}{x^4} \div \dfrac{x^2 + 2x + 1}{x^2} = \dfrac{x^2 + 3x + 2}{x^4} \cdot \dfrac{x^2}{x^2 + 2x + 1}$

$$= \dfrac{(x+2)(x+1)x^2}{x^4(x+1)^2}$$

$$= \dfrac{x+2}{x^2(x+1)} \qquad \square$$

EXAMPLE 4. Perform the operation indicated.

$$\dfrac{x^3 + 1}{3 - x} \div \dfrac{x + 1}{x^2 - 9}$$

Solution:

We convert the division to multiplication, multiply, and factor.

$$\dfrac{x^3 + 1}{3 - x} \div \dfrac{2(x + 1)}{x^2 - 9} = \dfrac{x^3 + 1}{3 - x} \cdot \dfrac{x^2 - 9}{2(x + 1)}$$

$$= \dfrac{(x + 1)(x^2 - x + 1)(x - 3)(x + 3)}{2(3 - x)(x + 1)}$$

The factors of the numerator include $(x - 3)$, and the factors of the denominator include $(3 - x)$. They are opposites. We replace $(3 - x)$ with $-(x - 3)$.

$$= \dfrac{(x + 1)(x^2 - x + 1)(x - 3)(x + 3)}{-2(x - 3)(x + 1)}$$

$$= \dfrac{(x^2 - x + 1)(x + 3)}{-2}$$

$$= -\dfrac{(x^2 - x + 1)(x + 3)}{2} \qquad \square$$

The key to multiplying or dividing rational expressions is factoring. Make sure that the numerator and denominator are **factored** before reducing.

Be Careful!

Sec. 3.3   Multiplication and Division

PROBLEM SET 3.3

Warm-ups

In problems 1 through 38, perform the operation indicated. Write answers in reduced form.
For problems 1 through 16, see Example 1.

1. $\dfrac{xy^2}{z} \cdot \dfrac{z^2}{x^4 y}$

2. $\dfrac{-8ab^4}{c^4} \cdot \dfrac{c^5}{4a}$

3. $\dfrac{-2^3 m}{n} \cdot \dfrac{n}{(-2)^2 m^3}$

4. $\dfrac{a^4 (bc)^2}{d} \cdot \dfrac{(-cd)^2}{a^2}$

5. $\dfrac{z}{x^2 y} \cdot \left(\dfrac{-xy}{z}\right)^3$

6. $\dfrac{360 s^2}{(-t)^2} \cdot \dfrac{st^2}{270 s}$

7. $\dfrac{17a + 17b}{3} \cdot \dfrac{-39}{51a + 51b}$

8. $\dfrac{p^2 q - p^2 q^2}{p + p^2} \cdot \dfrac{pq + q}{p - pq}$

9. $\dfrac{-4}{a^4 - 5a^2 + 6} \cdot (a^4 - 4)$

10. $\dfrac{s - 2t}{s^2} \cdot \dfrac{s^2 + s^6}{s^3 - 8t^3}$

11. $\dfrac{2x^2 + 2x - 12}{x^2 - x - 12} \cdot \dfrac{x^2 - 3x - 4}{4x^2 - 4x - 8}$

12. $\dfrac{m^2 - n^2}{m^3 + n^3} \cdot \dfrac{m^2 - mn + n^2}{m - n}$

13. $\dfrac{6a^2 - 6b^2}{a^2 + 2ab + b^2} \cdot \dfrac{a^2 - ab - 2b^2}{3a^2 + 3ab - 6b^2}$

14. $\dfrac{ca - da - bc + db}{ca + da + bc + db} \cdot \dfrac{ca + da - bc - db}{ca - da + 2bc - 2bd}$

15. $\dfrac{6x^2 + 7x - 3}{10x^2 - x - 2} \cdot \dfrac{5x^2 - 3x - 2}{4x^2 + 4x - 3} \cdot \dfrac{6x^2 - 5x + 1}{3x^2 - 4x + 1}$

16. $\dfrac{a^2 + 2ab + b^2}{2a^2 - 3ab + b^2} \cdot \dfrac{a^2 - b^2}{a^3 + 3a^2 b + 3ab^2 + b^3} \cdot \dfrac{6a^2 - ab - b^2}{3a^2 - 5ab - 2b^2}$

For problems 17 through 19, see Example 2.

17. $\dfrac{s - t}{t^2} \cdot \dfrac{t}{t - s}$

18. $(x^2 - x) \cdot \dfrac{y^2}{x^2 - x^3}$

19. $\dfrac{16 - x^2}{x^2 + 3x + 2} \cdot \dfrac{x^2 + x - 2}{x^2 - 5x + 4}$

For problems 20 through 34, see Example 3.

20. $\dfrac{abc^2}{xyz^2} \div \dfrac{ab^2 c}{xy^4}$

21. $\dfrac{(2x)^2 y}{15} \div \dfrac{(2x)^3}{75}$

22. $\dfrac{-2^2 x}{ab} \div \dfrac{2^3 x^4 y}{a^2}$

23. $\dfrac{(-2a)^3}{x^5} \div \dfrac{(-2a)^2}{x^2}$

24. $\dfrac{(-ab)^3}{r^2 s} \div \dfrac{a^2 b}{(rs^2)^2}$

25. $\dfrac{(-3)^2 r}{(2st)^2} \div \dfrac{3r^2}{8t}$

26. $\dfrac{(x + a)^2}{4a^2} \div (x + a)$

27. $\dfrac{50 - 2x^2}{x^2 + 2x} \div \dfrac{4x + 20}{4 - x^2}$

28. $\dfrac{x^2 + 3x - 4}{x^2 + 4x + 4} \div \dfrac{x^2 + 2x - 3}{x^2 + 3x + 2}$

29. $\dfrac{2x^2 + 5x - 3}{2x^2 + 5x + 3} \div \dfrac{2x^3 - x^2}{2x^2 + x - 3}$

30. $\dfrac{2x^2 + 3x - 2}{x^2 - 2x - 3} \div \dfrac{x^2 - 3x - 10}{x^2 - 7x + 12}$

31. $\dfrac{2x^2 - 8x - 42}{x^2 + 2x + 1} \div \dfrac{4x^2 + 8x - 12}{x + x^2}$

32. $\dfrac{b^3 + 8c^3}{x^4 - 1} \div \dfrac{ab + 2ac}{x^2 + 1}$

33. $\dfrac{p^2 - q^2}{r^2 - 2rs + s^2} \div \dfrac{pr + ps + rq + qs}{pr - ps - rq + qs}$

34. $\dfrac{ax - ay - bx + by}{cx + cy + dx + dy} \div \dfrac{ax + ay - bx - by}{cx - cy - dx + dy}$

For problems 35 through 38, see Example 4.

35. $(x - y) \div \dfrac{y - x}{y}$

36. $\dfrac{s^2 - r^2}{r^2 + 2rs + s^2} \div \dfrac{r^3 - s^3}{r + s}$

37. $\dfrac{x^3 + 3x^2 + 4x + 12}{x^2 + x - 2} \div \dfrac{x^4 - 16}{1 - x^2}$

38. $\dfrac{2x^2 + x - 10}{3x^2 + 7x + 4} \cdot \dfrac{3x^2 + x - 4}{6x^2 + 13x - 5} \div \dfrac{x^2 + 5x - 6}{3x^2 + 2x - 1}$

Practice Exercises

In problems 39 through 96, perform the operation indicated. Write answers in reduced form.

39. $\dfrac{xy^3}{z} \cdot \dfrac{z^4}{x^2y}$

40. $\dfrac{-6ab^3}{c^5} \cdot \dfrac{c^7}{3a}$

41. $\dfrac{-3^3 m}{n} \cdot \dfrac{n}{(-3)^2 m^4}$

42. $\dfrac{a^5(bc)^3}{d} \cdot (-cd)^2$

43. $\dfrac{z}{x^4 y} \cdot \left(\dfrac{-xy}{z^2}\right)^3$

44. $\dfrac{120s^3}{(-t)^2} \cdot \dfrac{st^3}{320s}$

45. $\dfrac{(ab^4)^3}{c(dx)^2} \cdot \dfrac{(c^3 d)^2}{(-a)^3 b^3}$

46. $\dfrac{(2x)^3}{5y} \cdot \dfrac{15y^4}{4x}$

47. $\dfrac{(abc)^3}{xy} \cdot \dfrac{(xy)^2}{(abc)^2}$

48. $\dfrac{x-1}{x} \cdot \dfrac{x^2}{x-1}$

49. $\dfrac{a-b}{t^3} \cdot \dfrac{t}{b-a}$

50. $\dfrac{13a+13b}{5} \cdot \dfrac{-55}{39a+39b}$

51. $\dfrac{pq+pq^2}{p-p^2} \cdot \dfrac{q-qp}{p^2q^2+p^2q}$

52. $\dfrac{a^4+a^2}{b^8} \cdot \dfrac{b^2}{a^3+a}$

53. $\dfrac{r^4 s^5 - r^3 s^3}{t^2 u^2 + t^4 u^4} \cdot \dfrac{t^3 u^2}{r^2 s^5}$

54. $\dfrac{(z+2)^2}{z^2-1} \cdot \dfrac{z-1}{z+2}$

55. $\dfrac{s+t}{s-t} \cdot \dfrac{t^2-s^2}{t+s}$

56. $\dfrac{c^2-16}{2c-8} \cdot \dfrac{-8}{c^2+3c-4}$

57. $(x^3 - 125) \cdot \dfrac{x^3}{5-x}$

58. $\dfrac{2x^2+6x+4}{x^2-4x+3} \cdot \dfrac{x^2-x-6}{4x^2-4x-8}$

59. $\dfrac{r^3+s^3}{r^2-s^2} \cdot \dfrac{r+s}{r^2-rs+s^2}$

60. $\dfrac{16-x^2}{x^2+7x+10} \cdot \dfrac{x^2+3x-10}{x^2-6x+8}$

61. $\dfrac{3m^2-3n^2}{m^2+2mn+n^2} \cdot \dfrac{m^2+3mn+2n^2}{6m^2-6mn-12n^2}$

62. $\dfrac{ax-az+bx-bz}{ax-2bx+az-2bz} \cdot \dfrac{ax+az+bx+bz}{ax-az-bx+bz}$

63. $\dfrac{as+st+a+t}{ax^2-bx^2+2a-2b} \cdot \dfrac{bx^2-ax^2+2b-2a}{as-2a+st-2t}$

64. $\dfrac{x^2-9}{2sx+2s-tx-t} \cdot \dfrac{2sx-tx+6s-3t}{sx+tx-3s-3t}$

65. $\dfrac{abc^3}{xyz^4} \div \dfrac{ab^2 c}{xy^3}$

66. $\dfrac{(2x)^3 y}{25} \div \dfrac{(2x)^2}{85}$

67. $\dfrac{m^5(n^3 p)^2}{rst^3} \div \dfrac{(mnp)^3}{r^3 s^2 t^2}$

68. $\dfrac{-16a^3}{(bc)^5} \div \dfrac{24a}{b}$

69. $\dfrac{-5^2 x}{ab} \div \dfrac{5^3 x^2 y}{a^3}$

70. $\dfrac{(-2d)^3}{x^4} \div \dfrac{(-2d)^2}{x^3}$

71. $\dfrac{(-ab)^3}{r^3 s} \div \dfrac{a^5 b}{(rs^3)^4}$

72. $\dfrac{(-4)^2 r}{(2st)^3} \div \dfrac{4r^2}{4t}$

73. $\dfrac{p}{p-q} \div \dfrac{q}{q-p}$

74. $\dfrac{(x-r)^3}{4a^3} \div (x-r)$

75. $\dfrac{ax-bx}{ay+by} \div \dfrac{a^2-ab}{2a+2b}$

76. $\dfrac{5u^3+10u^2}{2u-2} \div \dfrac{15u^2+30u}{au-a}$

77. $\dfrac{p^2+pt}{p-p^2} \div \dfrac{pq+qt}{pq-q}$

78. $\dfrac{a^3 b+2a^2 b^2}{c^2 b+cb} \div \dfrac{a^2+2ab}{c^2 d+c^3}$

79. $\dfrac{16x^2-1}{x^2+8x+16} \div \dfrac{64x^3-1}{x+4}$

80. $\dfrac{8-2a^2}{a^2+3a} \div \dfrac{5a+10}{9-a^2}$

81. $\dfrac{x^2-4x-12}{x^2-4x-5} \div \dfrac{x^2-3x-18}{x^2-7x+10}$

82. $\dfrac{2x^2+7x+3}{3x^2-x-10} \div \dfrac{2x^2-x-1}{x^2-4}$

83. $\dfrac{x^2-3x-10}{2x^2+5x+3} \div \dfrac{5+4x-x^2}{2x^2+7x+6}$

84. $\dfrac{8t^3-s^3}{32t^2-2} \div \dfrac{4t^2+2st+s^2}{16t^2-4t}$

85. $\dfrac{x^2 + 10x + 21}{x^2 - 4x + 3} \div \dfrac{x^3 + 7x^2}{x^2 - 2x + 1}$

86. $\dfrac{2x^2 + 16x - 18}{x^3 + 3x^2} \div \dfrac{x^2 + 7x - 18}{x^2 - 4x - 21}$

87. $\dfrac{b^3 - 2b^2 + 3b - 6}{b^2 - b - 6} \div \dfrac{b^4 - 9}{b^2 - 9}$

88. $\dfrac{p^4 - q^4}{p^2 - q^2} \div \dfrac{p^2 + q^2}{p^2 + 2pq + q^2}$

89. $\dfrac{ac + bc + 2ad + 2bd}{a^2 - d^2} \div \dfrac{ac - bc + 2ad - 2bd}{a^2 + 2ab - ad - 2bd}$

90. $\dfrac{rt - ru + st - su}{2rt + ru + 4st + 2su} \div \dfrac{rt - ru - st + su}{r^2 + 3rs + 2s^2}$

91. $\dfrac{2x^2 + 5x - 3}{3x^2 + 2x - 5} \cdot \dfrac{3x^2 - x - 10}{x^2 + 2x - 3} \cdot \dfrac{x^2 - 2x + 1}{2x^2 + x - 1}$

92. $\dfrac{a^2 + ab - 2b^2}{2a^2 + 5ab - 3b^2} \cdot \dfrac{2a^2 + ab - b^2}{2a^2 + ab - 3b^2} \cdot \dfrac{2a^2 + 9ab + 9b^2}{a^2 + 3ab + 2b^2}$

93. $\dfrac{p^2 + pt - 2t^2}{2p^2 - 5pt - 3t^2} \cdot \dfrac{p^2 - 2pt - 3t^2}{p^2 - 3pt + 2t^2} \cdot \dfrac{4p^2 + 4pt + t^2}{2p^2 + 3pt - 2t^2}$

94. $\dfrac{x^2 - y^2}{2x^2 + 3xy - 2y^2} \cdot \dfrac{x^3 + 2x^2y}{x^2 + 2xy + y^2} \div \dfrac{x^2 + xy - 2y^2}{x^3 + x^2y + x + y}$

95. $\dfrac{ac + bc - ad - bd}{2ac + 4ad + bc + 2bd} \cdot \dfrac{c^4 + 2c^3d}{ac - bc - ad + bd} \div \dfrac{ac + ad + bc + bd}{2a^2 - ab - b^2}$

96. $\dfrac{x^2 + x - 6}{2x^2 - x - 1} \cdot \dfrac{x^2 + 2x + 4}{2x^2 + 9x + 9} \div \dfrac{x^3 - 8}{2x^2 + 3x + 1}$

Challenge Problems

In problems 97 and 98, perform the operations indicated. Write answers in lowest terms. Assume that n, s, and p are natural numbers.

97. $\dfrac{x^{2n} - x^n - 6}{x^{2p} + 2x^p - 3} \cdot \dfrac{x^{2p} - 1}{x^{2n} + 4x^n + 4}$

98. $\dfrac{x^{3s} - 8}{x^{2s} + 2x^s - 3} \div \dfrac{x^{2s} - 4}{x^{2s} + 3x^s}$

■ IN YOUR OWN WORDS . . .

99. Explain how to multiply two rational expressions.

100. Explain how to divide two rational expressions.

■ 3.4 ADDITION AND SUBTRACTION

Addition and subtraction of rational expressions are performed using the same procedures that we use for rational numbers.

> **Addition and Subtraction of Rational Expressions**
>
> If A, B, C are polynomials and if each rational expression is defined,
>
> $$\dfrac{A}{C} + \dfrac{B}{C} = \dfrac{A + B}{C} \quad \text{and} \quad \dfrac{A}{C} - \dfrac{B}{C} = \dfrac{A - B}{C}$$

To add or subtract rational expressions we first write each rational expression with the same denominator and then write the sum or difference of the numerators divided by the denominator. This gives us a single rational expression that we reduce to lowest terms.

EXAMPLE 1. Perform the operation indicated.

(a) $\dfrac{x+y}{xy} + \dfrac{1}{xy}$ (b) $\dfrac{a}{a+b} - \dfrac{a-b}{a+b}$

Solutions:

(a) $\dfrac{x+y}{xy} + \dfrac{1}{xy} = \dfrac{x+y+1}{xy}$

Note the parentheses in the next example. Be careful to subtract the entire numerator of the second fraction.

(b) $\dfrac{a}{a+b} - \dfrac{a-b}{a+b} = \dfrac{a-(a-b)}{a+b}$

$= \dfrac{a-a+b}{a+b}$

$= \dfrac{b}{a+b}$

EXAMPLE 2. Perform the operation indicated.

$$\dfrac{p}{p-q} + \dfrac{q}{q-p}$$

Solution:

Notice that the denominators are opposites. By substituting $-(p-q)$ for $q-p$, we make the denominators the same.

$$\dfrac{p}{p-q} + \dfrac{q}{q-p} = \dfrac{p}{p-q} + \dfrac{q}{-(p-q)}$$

$$= \dfrac{p}{p-q} - \dfrac{q}{p-q}$$

$$= \dfrac{p-q}{p-q}$$

$$= 1$$

Often the denominators are not the same. We must figure out an appropriate denominator and write each rational expression with this denominator. It must be a polynomial that is divisible by each denominator. We call this denominator the **least common denominator (LCD)**.

Procedure to Find the Least Common Denominator (LCD)

1. Factor each denominator completely, using exponents.
2. List all different prime factors from all denominators.
3. Write the LCD. The LCD is the product of the factors in step 2 each raised to the highest power of that factor in any single denominator.

Sec. 3.4 Addition and Subtraction

EXAMPLE 3. Find the LCD for each pair of denominators.

(a) $x + y$ and $x - y$
(b) $x^2 + 2x - 15$ and $x + 5$
(c) $x^2 + 6x + 9$ and $x^2 + 3x$
(d) x^7y^3 and xy^4

Solutions:

Step 1	(a) $x + y$ and $x - y$.	
	$x + y$ and $x - y$ are prime.	Factor.
Step 2	The different prime factors are $x + y$ and $x - y$.	List factors.
	Each factor occurs one time in each denominator.	
Step 3	The LCD is $(x + y)(x - y)$.	Write LCD.

(b) $x^2 + 2x - 15$ and $x + 5$.

$$x^2 + 2x - 15 = (x + 5)(x - 3) \qquad \text{Factor.}$$

Prime factors are $x + 5$ and $x - 3$.   List factors.

$x + 5$ occurs one time in $(x + 5)(x - 3)$ and one time in $x + 5$.

$x - 3$ occurs one time in $(x + 5)(x - 3)$ and no times in $x + 5$.

So we must use each factor one time in the LCD.

The LCD is $(x + 5)(x - 3)$.   Write LCD.

(c) $x^2 + 6x + 9$ and $x^2 + 3x$.

$$x^2 + 6x + 9 = (x + 3)^2 \text{ and } x^2 + 3x = x(x + 3) \qquad \text{Factor.}$$

The different prime factors are $x + 3$ and x.

$x + 3$ occurs two times in $(x + 3)^2$ and one time in $x(x + 3)$.

x occurs no times in $(x + 3)^2$ and one time in $x(x + 3)$.

We must use $x + 3$ two times and x one time.

The LCD is $x(x + 3)^2$.   Write LCD.

(d) x^7y^3 and xy^4.

Both are factored.

The different prime factors are x and y.

x occurs seven times in x^7y^3 and one time in xy^4.

y occurs three times in x^7y^3 and four times in xy^4.

The LCD is x^7y^4.

Procedure to Add or Subtract Rational Expressions

1. Find the least common denominator (LCD).
2. Rewrite each rational expression with the LCD as its denominator.
3. Write the sum or difference of the numerators divided by the LCD.
4. Reduce to lowest terms.

EXAMPLE 4. Perform the operations indicated and reduce answers to lowest terms.

(a) $\dfrac{5}{a^2b} - \dfrac{3}{ab^4}$ (b) $\dfrac{8}{m+n} - \dfrac{1}{m-n}$ (c) $\dfrac{-28}{x^2 - 2x - 3} + \dfrac{7}{x-3}$

(d) $\dfrac{t}{t-2} - 1$ (e) $\dfrac{1}{x^2 + 4x + 4} + \dfrac{2}{x^2 + 5x + 6}$

Solutions:

(a) The LCD is a^2b^4.

$$\frac{5}{a^2b} - \frac{3}{ab^4} = \frac{5(b^3)}{a^2b(b^3)} - \frac{3(a)}{ab^4(a)}$$

$$= \frac{5b^3}{a^2b^4} - \frac{3a}{a^2b^4}$$

$$= \frac{5b^3 - 3a}{a^2b^4}$$

(b) The LCD is $(m+n)(m-n)$.

$$\frac{8}{m+n} - \frac{1}{m-n} = \frac{8(m-n)}{(m+n)(m-n)} - \frac{m+n}{(m-n)(m+n)}$$

$$= \frac{8(m-n) - (m+n)}{(m+n)(m-n)}$$

Notice that neither the numerator nor the denominator is factored at this point. So **do not** try to reduce! *Be Careful!*

$$= \frac{8m - 8n - m - n}{(m+n)(m-n)}$$

$$= \frac{7m - 9n}{(m+n)(m-n)}$$

We could multiply in the denominator, but it is not necessary.

(c) $\dfrac{-28}{x^2 - 2x - 3} + \dfrac{7}{x-3} = \dfrac{-28}{(x-3)(x+1)} + \dfrac{7}{x-3}$ Factor.

The LCD is $(x-3)(x+1)$.

$$\frac{-28}{(x-3)(x+1)} + \frac{7(x+1)}{(x-3)(x+1)} = \frac{-28 + 7(x+1)}{(x-3)(x+1)}$$

$$= \frac{-28 + 7x + 7}{(x-3)(x+1)}$$

$$= \frac{7x - 21}{(x-3)(x+1)}$$

$$= \frac{7(x-3)}{(x-3)(x+1)}$$

$$= \frac{7}{x+1}$$ *(continued)*

Sec. 3.4 Addition and Subtraction

(d) $\dfrac{t}{t-2} - 1 = \dfrac{t}{t-2} - \dfrac{t-2}{t-2}$

When subtracting, *use parentheses*.

$= \dfrac{t - (t-2)}{t-2}$

Be Careful!

Be careful when removing parentheses!

$= \dfrac{t - t + 2}{t-2}$

$= \dfrac{2}{t-2}$

(e) $\dfrac{1}{x^2 + 4x + 4} + \dfrac{2}{x^2 + 5x + 6} = \dfrac{1}{(x+2)^2} + \dfrac{2}{(x+2)(x+3)}$ Factor.

The LCD is $(x+2)^2(x+3)$.

$\dfrac{x+3}{(x+2)^2(x+3)} + \dfrac{2(x+2)}{(x+2)^2(x+3)} = \dfrac{x + 3 + 2(x+2)}{(x+2)^2(x+3)}$

$= \dfrac{x + 3 + 2x + 4}{(x+2)^2(x+3)}$

$= \dfrac{3x + 7}{(x+2)^2(x+3)}$ ☐

EXAMPLE 5. Perform the operation indicated.

$$\dfrac{3}{x^2 - 4} + \dfrac{1}{2 - x}$$

Solution:

$\dfrac{3}{x^2 - 4} + \dfrac{1}{2-x} = \dfrac{3}{(x-2)(x+2)} + \dfrac{1}{2-x}$

We replace $2 - x$ with $-(x - 2)$, which will make two of our factors the same.

$= \dfrac{3}{(x-2)(x+2)} + \dfrac{1}{-(x-2)}$

$= \dfrac{3}{(x-2)(x+2)} + \dfrac{-1}{x-2}$

Now the LCD is $(x-2)(x+2)$.

$= \dfrac{3}{(x-2)(x+2)} + \dfrac{-(x+2)}{(x-2)(x+2)}$

$= \dfrac{3 - (x+2)}{(x-2)(x+2)}$

$= \dfrac{1 - x}{(x-2)(x+2)}$ ☐

PROBLEM SET 3.4

Warm-ups

In problems 1 through 33, perform the operation indicated. Write answers in lowest terms.
For problems 1 through 8, see Example 1.

1. $\dfrac{7}{x^2} - \dfrac{x+2}{x^2}$

2. $\dfrac{2a}{t} - \dfrac{a+b}{t}$

3. $\dfrac{x+y}{xy} + \dfrac{x-y}{xy}$

4. $\dfrac{x-y}{x+y} + \dfrac{y-x}{y+x}$

5. $\dfrac{x}{x+y} + \dfrac{x+y}{x+y}$

6. $\dfrac{2}{r-2} - \dfrac{1}{r-2}$

7. $\dfrac{3}{m-2} - \dfrac{2-m}{m-2}$

8. $\dfrac{2}{p+1} - \dfrac{5-p}{p+1}$

For problems 9 and 10, see Example 2.

9. $\dfrac{3y}{x-5} - \dfrac{2y}{5-x}$

10. $\dfrac{x}{z-t} + \dfrac{3}{t-z}$

For problems 11 through 29, see Example 4.

11. $\dfrac{p}{pq^2} + \dfrac{q}{p^2q}$

12. $\dfrac{1}{x^2y^3} - \dfrac{3}{x^4y}$

13. $\dfrac{1}{r+2} + \dfrac{2}{r-3}$

14. $\dfrac{a}{a+2} - \dfrac{1}{a-3}$

15. $\dfrac{2}{5a+10} + \dfrac{7}{3a+6}$

16. $\dfrac{-3}{x^3-x^2} + \dfrac{2}{x^2-x}$

17. $\dfrac{x}{x^2-4x+4} + \dfrac{2}{x^2-4}$

18. $\dfrac{w}{w^2+w-12} - \dfrac{w}{w^2-2w-3}$

19. $\dfrac{x+5}{x^2-2x-15} - \dfrac{x}{x^2-6x+5}$

20. $\dfrac{s}{s^2-5s-24} - \dfrac{s-1}{s^2-10s+16}$

21. $\dfrac{3y+6}{8-y^3} + \dfrac{4}{4-y^2}$

22. $\dfrac{2y}{y^2-9y+14} - \dfrac{y-1}{y^2-8y+7}$

23. $x + \dfrac{2}{x}$

24. $\dfrac{u^2}{u-1} - u$

25. $a + \dfrac{3}{a+b}$

26. $\dfrac{t}{t-1} - 1$

27. $\dfrac{1}{a} - \dfrac{1}{a+1} + \dfrac{1}{a^2+a}$

28. $\dfrac{3}{z+1} - \dfrac{1}{z-1} + \dfrac{1}{z+2}$

29. $\dfrac{p}{p+2} + \dfrac{p+1}{p+3} + \dfrac{2}{p^2+5p+6}$

For problems 30 through 33, see Example 5.

30. $\dfrac{10}{4-2a} - \dfrac{12}{3a-6}$

31. $\dfrac{3}{p^2-p} + \dfrac{7}{1-p}$

32. $\dfrac{x}{x^2-6x+8} - \dfrac{2}{2+x-x^2}$

33. $\dfrac{2}{x-2} + \dfrac{3}{x+2} - \dfrac{5}{4-x^2}$

Practice Exercises

In problems 34 through 82, perform the operations indicated. Write answers in lowest terms.

34. $\dfrac{8}{y^3} - \dfrac{x+2}{y^3}$

35. $\dfrac{x+2}{x+1} - \dfrac{x+4}{x+1}$

36. $\dfrac{a}{b^3c} + \dfrac{b}{bc^2}$

37. $\dfrac{2}{r^2t} - \dfrac{1}{rt^4}$

38. $\dfrac{2}{x^2y} - \dfrac{4}{xyz} + \dfrac{1}{x}$

39. $\dfrac{1}{x} + \dfrac{1}{y} - \dfrac{1}{z}$

Sec. 3.4 Addition and Subtraction

40. $\dfrac{2}{y-1} + \dfrac{1}{y}$

41. $\dfrac{r}{r+2} + \dfrac{1}{2} - \dfrac{3}{2r+4}$

42. $\dfrac{u}{u-v} + \dfrac{2v}{u+v}$

43. $\dfrac{3}{s+1} - \dfrac{1}{s-1}$

44. $\dfrac{5}{a+3} - \dfrac{3}{a+1}$

45. $\dfrac{5}{p+5} - \dfrac{7}{p-7}$

46. $\dfrac{q}{q-4} + \dfrac{q}{q+6}$

47. $\dfrac{v}{v-3} - \dfrac{1}{v+7}$

48. $\dfrac{a}{x-1} + \dfrac{4a}{1-x}$

49. $\dfrac{4}{x-y} - \dfrac{2}{y-x}$

50. $\dfrac{2}{3r+6} + \dfrac{5}{4r+8}$

51. $\dfrac{x}{x^2-2x} + \dfrac{4}{2x^2-x^3}$

52. $\dfrac{3}{5m-5n} + \dfrac{4}{2n-2m}$

53. $\dfrac{3r+1}{1-r^2} + \dfrac{2}{r-1}$

54. $\dfrac{2t}{t-1} - \dfrac{t^2-4t+3}{t^2-2t+1}$

55. $\dfrac{a}{b+a} - \dfrac{2b^2}{a^2-b^2}$

56. $\dfrac{d}{d+3} + \dfrac{5d+9}{d^2+4d+3}$

57. $\dfrac{2x}{x-5} - \dfrac{14x+20}{x^2-x-20}$

58. $\dfrac{x^2-x}{x^2-9} + \dfrac{1}{3-x}$

59. $\dfrac{y}{y-5} + \dfrac{y^2}{5+4y-y^2}$

60. $\dfrac{1}{x^2-7x+12} + \dfrac{x}{x^2-12x+32}$

61. $\dfrac{2t}{t^2-2t-3} - \dfrac{t}{t^2-8t+15}$

62. $\dfrac{z+2}{2z^2-21z+10} + \dfrac{1}{2z^2-7z+3}$

63. $\dfrac{3b-23}{b^2+8b-9} + \dfrac{35}{b^2+11b+18}$

64. $\dfrac{-3y+9}{y^3+27} + \dfrac{4}{y^2-9}$

65. $\dfrac{14x+63}{x^2+9x+14} + \dfrac{8x}{7-6x-x^2}$

66. $\dfrac{3z+17}{2z^2+z-3} + \dfrac{3z+7}{2z^2+7z+6}$

67. $\dfrac{3t+19}{t^2+t-2} + \dfrac{3t-46}{t^2+4t-5}$

68. $\dfrac{y}{y^2-10y+25} - \dfrac{y+1}{y^2-25}$

69. $\dfrac{5}{p+1} + p$

70. $\dfrac{a}{a+b} - 2$

71. $r + \dfrac{rt}{r+t}$

72. $\dfrac{-2w}{w^2-1} - 1$

73. $\dfrac{2}{x} - \dfrac{3}{x+2} + \dfrac{4}{x^2+2x}$

74. $\dfrac{3}{a+b} + \dfrac{4}{a-b} + \dfrac{6a}{b^2-a^2}$

75. $\dfrac{1}{x+4} - \dfrac{1}{x+3} - \dfrac{1}{x+2}$

76. $\dfrac{2}{y-1} - \dfrac{1}{y+1} - \dfrac{1}{y-2}$

77. $\dfrac{4x}{x^2+2x-3} + \dfrac{1}{1-x} - \dfrac{2}{x+3}$

78. $\dfrac{qt}{q^2-qt} + \dfrac{1}{q-t} + 1$

79. $\dfrac{n^3}{m^3+m^2n} + \dfrac{n}{m} - 1$

80. $\dfrac{7}{x^2+5x+6} - \dfrac{49}{x^2-x-12} + \dfrac{x}{x^2-2x-8}$

81. $\dfrac{3(r+1)}{r^2-9} - \dfrac{40}{r^2-4r-21} + \dfrac{16}{r^2-10r+21}$

82. $\dfrac{15u}{125-u^3} + \dfrac{6}{25+5u+u^2} + \dfrac{1}{u-5}$

83. If f is the focal length of a lens, then

$$\dfrac{1}{f} = \dfrac{1}{\text{object distance}} + \dfrac{1}{\text{image distance}}$$

Find a rational expression that gives $\dfrac{1}{f}$ for a lens if the object distance is x centimeters and the image distance is $x + 5$ cm.

84. Find a rational expression that gives $\dfrac{1}{f}$ for a lens if the object distance is $x^2 + 6x + 9$ meters and the image distance is $x^2 - 9$ meters. (See problem 83.)

85. When lights, toasters, and other appliances are plugged into wall outlets at home, they are connected in *parallel*. If R_1 and R_2 are two resistances connected in parallel, the total resistance, R_t, can be found from the relationship

$$\frac{1}{R_t} = \frac{1}{R_1} + \frac{1}{R_2}$$

If a coffee maker with a resistance of $x + 1$ ohms and a toaster with a resistance of $2x - 3$ ohms are plugged into a kitchen circuit, find a rational expression for $\frac{1}{R_t}$.

86. If a lamp with a resistance of k ohms and a steam iron with a resistance of $k^3 - k^2$ ohms are plugged into the same household circuit, find a rational expression for $\frac{1}{R_t}$. (See problem 85.)

Challenge Problems

In problems 87 through 89, perform the operations indicated. Write answers in lowest terms. Consider whether or not to work inside the parentheses first.

87. $(x + 2)\left(\dfrac{3}{x + 2} + \dfrac{1}{x - 1}\right)$

88. $\left(\dfrac{1}{x} + \dfrac{1}{y}\right) \div \left(\dfrac{1}{x} - \dfrac{1}{y}\right)$

89. $\left(\dfrac{3}{x - 3} - \dfrac{x}{x + 3}\right)\left(\dfrac{3}{x + 3} + \dfrac{x}{x - 3}\right)$

■ IN YOUR OWN WORDS . . .

90. Explain how to find a least common denominator.

■ 3.5 COMPLEX FRACTIONS

A fraction that contains a fraction is called a **complex fraction.** There are two methods that are used to simplify complex fractions. The first is to think of the fraction as a division problem, and the second is to use the Fundamental Principle of Rational Expressions.

EXAMPLE 1. Simplify $\dfrac{\frac{1}{x}}{\frac{2}{x^2}}$.

Solution:

We will work this by the two methods.

METHOD 1—Division method:

This complex fraction is $\dfrac{1}{x}$ divided by $\dfrac{2}{x^2}$.

$$\frac{\frac{1}{x}}{\frac{2}{x^2}} = \frac{1}{x} \div \frac{2}{x^2}$$

$$= \frac{1}{x} \cdot \frac{x^2}{2}$$

$$= \frac{x^2}{2x}$$

$$= \frac{x}{2}$$

(continued)

METHOD 2—Multiplication method:

The LCD for x and x^2 is x^2. We use the Fundamental Principle of Rational Expressions and multiply the numerator and denominator by x^2.

$$\frac{\frac{1}{x}}{\frac{2}{x^2}} = \frac{\frac{1}{x} \cdot x^2}{\frac{2}{x^2} \cdot x^2}$$

$$= \frac{\frac{x^2}{x}}{\frac{2x^2}{x^2}}$$

$$= \frac{x}{2}$$

Procedures to Simplify Complex Fractions

Division Method:

1. Write the fraction as a division problem.
2. Perform the division.
3. Reduce to lowest terms.

Multiplication Method:

1. Find the LCD of all denominators.
2. Multiply numerator and denominator by the LCD.
3. Reduce to lowest terms.

EXAMPLE 2. Simplify $\dfrac{\dfrac{a}{x+1} + 2}{\dfrac{b}{x+1} - 1}$.

Solution:

Again we show both methods.

DIVISION METHOD

Before writing the fraction as a division problem, we make a single fraction in the numerator and the denominator. The numerator becomes

$$\frac{a}{x+1} + 2 = \frac{a}{x+1} + \frac{2(x+1)}{x+1}$$

$$= \frac{a + 2x + 2}{x+1}$$

The denominator becomes

$$\frac{b}{x+1} - 1 = \frac{b}{x+1} - \frac{x+1}{x+1}$$

$$= \frac{b-(x+1)}{x+1}$$

$$= \frac{b-x-1}{x+1}$$

Now we begin the simplification of the complex fraction.

$$\frac{\dfrac{a}{x+1} + 2}{\dfrac{b}{x+1} - 1} = \frac{\dfrac{a+2x+2}{x+1}}{\dfrac{b-x-1}{x+1}}$$

Notice that there is now a single fraction (or one term) in the numerator and in the denominator. Now we write the division problem.

$$= \frac{a+2x+2}{x+1} \div \frac{b-x-1}{x+1}$$

$$= \frac{a+2x+2}{x+1} \cdot \frac{x+1}{b-x-1}$$

$$= \frac{(a+2x+2)(x+1)}{(x+1)(b-x-1)}$$

$$= \frac{a+2x+2}{b-x-1}$$

MULTIPLICATION METHOD

The LCD is $x + 1$. So we multiply numerator and denominator by $x + 1$.

$$\frac{\left(\dfrac{a}{x+1} + 2\right)(x+1)}{\left(\dfrac{b}{x+1} - 1\right)(x+1)} =$$

(Notice the distributive property.)

$$\frac{\dfrac{a}{x+1}(x+1) + 2(x+1)}{\dfrac{b}{x+1}(x+1) - 1(x+1)} = \frac{a+2x+2}{b-x-1} \qquad □$$

A complex fraction may be simplified by either method. However, as Example 2 indicates, one method might be better than the other. In general, it is usually better to use the division method when the complex fraction contains a single term in its numerator and one term in its denominator. If the numerator *or* denominator of a complex fraction contains more than one term, it is usually better to use the multiplication method.

Sec. 3.5 Complex Fractions

EXAMPLE 3. Simplify $\dfrac{\dfrac{p}{p-2}+p}{\dfrac{p}{2-p}-p}$.

Solution:

We choose the multiplication method, as there are two terms in the numerator and also in the denominator. Notice that $p - 2$ and $2 - p$ are opposites.

$$\dfrac{\dfrac{p}{p-2}+p}{\dfrac{p}{2-p}-p} = \dfrac{\dfrac{p}{p-2}+p}{\dfrac{p}{-(p-2)}-p}$$

$$= \dfrac{\dfrac{p}{p-2}+p}{\dfrac{-p}{p-2y}-p}$$

(The LCD is $p - 2$.)

$$= \dfrac{\left(\dfrac{p}{p-2}+p\right)(p-2)}{\left(\dfrac{-p}{p-2}-p\right)(p-2)}$$

$$= \dfrac{\dfrac{p}{p-2}(p-2)+p(p-2)}{\dfrac{-p}{p-2}(p-2)-p(p-2)}$$

$$= \dfrac{p+p^2-2p}{-p-p^2+2p}$$

$$= \dfrac{p^2-p}{p-p^2}$$

$$= \dfrac{p(p-1)}{p(1-p)}$$

$$= \dfrac{p-1}{1-p}$$

$$= \dfrac{p-1}{-(p-1)}$$

$$= -1 \qquad \square$$

PROBLEM SET 3.5

Warm-ups

In problems 1 through 21, simplify each complex fraction. Write answers in lowest terms. For problems 1 through 9, see Example 1.

1. $\dfrac{\dfrac{2}{a}}{\dfrac{4}{a^2}}$

2. $\dfrac{\dfrac{xyz^2}{t^4}}{\dfrac{x^3y}{t^2}}$

3. $\dfrac{\dfrac{-3r^3}{s^4}}{\dfrac{18r^4}{s^6}}$

4. $\dfrac{\dfrac{32a^3b}{m^2n}}{\dfrac{-48ab^2}{mn^3}}$

5. $\dfrac{\dfrac{xy}{x+1}}{\dfrac{x}{x+1}}$

6. $\dfrac{\dfrac{4s^2}{p-2}}{\dfrac{12s}{p-2}}$

7. $\dfrac{\dfrac{4}{s-1}}{\dfrac{8}{1-s}}$

8. $\dfrac{\dfrac{z-2}{24}}{\dfrac{z-2}{36}}$

9. $\dfrac{\dfrac{-12}{5r+5s}}{\dfrac{18}{r+s}}$

For problems 10 through 21, see Examples 2 and 3.

10. $\dfrac{\dfrac{1}{y}-1}{\dfrac{1}{y}+1}$

11. $\dfrac{1+\dfrac{1}{r+1}}{1-\dfrac{1}{r+1}}$

12. $\dfrac{\dfrac{1}{a+2}-1}{\dfrac{1}{a+2}+1}$

13. $\dfrac{\dfrac{1}{m}+\dfrac{1}{n}}{\dfrac{1}{m}-\dfrac{1}{n}}$

14. $\dfrac{\dfrac{1}{a}-\dfrac{1}{b}}{\dfrac{a^2-b^2}{ab}}$

15. $\dfrac{\dfrac{2t^2-st-s^2}{st}}{\dfrac{2}{s}+\dfrac{1}{t}}$

16. $\dfrac{\dfrac{1}{b-1}+2}{3-\dfrac{1}{1-b}}$

17. $\dfrac{\dfrac{2}{m+n}+\dfrac{1}{m-n}}{\dfrac{1}{m+n}-\dfrac{2}{m-n}}$

18. $\dfrac{\dfrac{2x-1}{x^2-x}}{\dfrac{2x}{x-1}+\dfrac{1}{x}}$

19. $\dfrac{\dfrac{w+18}{4-w^2}+\dfrac{w}{w+2}}{\dfrac{2}{w+2}-\dfrac{1}{w-2}}$

20. $\dfrac{\dfrac{m+3}{m-3}+\dfrac{m-3}{m+3}}{\dfrac{m+3}{m-3}-\dfrac{m-3}{m+3}}$

21. $\dfrac{\dfrac{a-b}{a+b}-\dfrac{a+b}{a-b}}{\dfrac{a-b}{a+b}+\dfrac{a+b}{a-b}}$

Practice Exercises

In problems 22 through 58, simplify each complex fraction. Write answers in lowest terms.

22. $\dfrac{\dfrac{3}{a}}{\dfrac{9}{a^3}}$

23. $\dfrac{\dfrac{xyz^3}{t^5}}{\dfrac{x^2y}{t^7}}$

24. $\dfrac{\dfrac{-5r^4}{s^3}}{\dfrac{15r^7}{s^8}}$

25. $\dfrac{\dfrac{64a^4b}{m^3n}}{\dfrac{-48ab^5}{mn^2}}$

26. $\dfrac{\dfrac{xy}{x-2}}{\dfrac{ax}{x-2}}$

27. $\dfrac{\dfrac{7s^3}{p-3}}{\dfrac{28s}{p-3}}$

28. $\dfrac{\dfrac{z+2}{64}}{\dfrac{z+2}{48}}$

29. $\dfrac{\dfrac{15}{t-1}}{\dfrac{35}{1-t}}$

Sec. 3.5 Complex Fractions

30. $\dfrac{\dfrac{-14}{3r+3s}}{\dfrac{21}{r+s}}$
31. $\dfrac{\dfrac{x}{bc+bd}}{\dfrac{4x}{ab}}$
32. $\dfrac{\dfrac{m-1}{m^2-1}}{\dfrac{3}{m+1}}$
33. $\dfrac{\dfrac{16}{u-v}}{\dfrac{16}{u+v}}$

34. $\dfrac{\dfrac{x}{abc+abd}}{\dfrac{2x}{ab}}$
35. $\dfrac{\dfrac{m+1}{m^2-1}}{\dfrac{3}{m-1}}$
36. $\dfrac{\dfrac{6}{u-v}}{\dfrac{2}{u+v}}$
37. $\dfrac{\dfrac{x}{x^2-x-6}}{\dfrac{x}{x-3}}$

38. $\dfrac{\dfrac{24}{x-5}}{\dfrac{128}{125-x^3}}$
39. $\dfrac{\dfrac{3}{y}-2}{\dfrac{1}{y}+4}$
40. $\dfrac{1-\dfrac{1}{r-1}}{1+\dfrac{1}{r-1}}$

41. $\dfrac{\dfrac{1}{a-2}+1}{\dfrac{1}{a-2}-1}$
42. $\dfrac{\dfrac{1}{m}-\dfrac{1}{n}}{\dfrac{1}{m}+\dfrac{1}{n}}$
43. $\dfrac{\dfrac{2}{a}+\dfrac{2}{b}}{\dfrac{a^3+b^3}{ab}}$

44. $\dfrac{\dfrac{3t^2+2st-s^2}{st}}{\dfrac{3}{s}-\dfrac{1}{t}}$
45. $\dfrac{\dfrac{1}{2-b}+1}{1-\dfrac{1}{b-2}}$
46. $\dfrac{\dfrac{2}{m-n}-\dfrac{3}{m+n}}{\dfrac{1}{m+n}+\dfrac{2}{m-n}}$

47. $\dfrac{\dfrac{t+5}{t^2-16}}{1+\dfrac{1}{t+4}}$
48. $\dfrac{\dfrac{2}{x-2}+\dfrac{1}{x+4}}{\dfrac{x+2}{x^2+2x-8}}$
49. $\dfrac{\dfrac{x}{x^2-x-12}}{\dfrac{x}{x-4}}$

50. $\dfrac{\dfrac{11}{3-x}}{\dfrac{33}{x^3-27}}$
51. $\dfrac{\dfrac{2x+1}{x^2+x}}{\dfrac{2x}{x+1}-\dfrac{1}{x}}$
52. $\dfrac{\dfrac{w}{w-3}-\dfrac{108}{w^2-9}}{\dfrac{2}{w+3}-\dfrac{1}{w-3}}$

53. $\dfrac{\dfrac{m+3}{m-3}-\dfrac{m+3}{m-3}}{\dfrac{m+3}{m-3}+\dfrac{m+3}{m-3}}$
54. $\dfrac{\dfrac{a+b}{a-b}+\dfrac{a+b}{a-2b}}{\dfrac{a-b}{a-2b}-\dfrac{a+b}{a-b}}$
55. $\dfrac{1+\dfrac{1}{x}-\dfrac{1}{x+1}}{\dfrac{x^2+1}{x+1}-\dfrac{1}{x}}$

56. $\dfrac{\dfrac{3}{w-2}+\dfrac{1}{w}+2}{\dfrac{w}{w-2}+\dfrac{1}{w}}$
57. $\dfrac{\dfrac{3}{x}-\dfrac{2}{y}-\dfrac{4}{z}}{\dfrac{1}{x}-\dfrac{1}{y}-\dfrac{1}{z}}$
58. $\dfrac{1-x+\dfrac{12}{x+3}}{1+x-\dfrac{8}{x+3}}$

Challenge Problems

In problems 59 through 61, simplify each complex fraction.

59. $1+\dfrac{1}{1+\dfrac{1}{1+1}}$

60. $x+\dfrac{x}{x+\dfrac{x}{x+x}}$

61. $\dfrac{1+\dfrac{1}{1-\dfrac{1}{x}}}{1-\dfrac{1}{1+\dfrac{1}{x}}}$

■ IN YOUR OWN WORDS . . .

62. Explain the two methods that we use to simplify a complex fraction. When do we use each method?

CHAPTER SUMMARY

GLOSSARY

Rational expression: The quotient of two polynomials. A rational expression is **undefined** when its denominator has a value of zero.

To **reduce** a rational expression to lowest terms: To write the expression so that there are no common factors in the numerator and the denominator.

$$\frac{A}{B} \cdot \frac{C}{D} = \frac{AC}{BD}$$

MULTIPLICATION OF RATIONAL EXPRESSIONS

$$\frac{A}{B} \div \frac{C}{D} = \frac{A}{B} \cdot \frac{D}{C} = \frac{AD}{BC}$$

DIVISION OF RATIONAL EXPRESSIONS

$$\frac{A}{B} + \frac{C}{B} = \frac{A+C}{B} \qquad \frac{A}{B} - \frac{C}{B} = \frac{A-C}{B}$$

ADDITION AND SUBTRACTION OF RATIONAL EXPRESSIONS

1. Factor each denominator completely, using exponents.
2. List all different prime factors from all denominators.
3. Write the LCD. The LCD is the product of the factors in step 2 each raised to the largest power of that factor in any single denominator.

PROCEDURE TO FIND THE LEAST COMMON DENOMINATOR

There are two techniques to use in simplifying a complex fraction.

1. Treat the complex fraction as a division problem.
2. Multiply numerator and denominator by the same nonzero factor.

COMPLEX FRACTIONS

■ CHECKUPS

1. Determine the values of x for which $\dfrac{x+2}{(x+1)(x-2)}$ is undefined.   Section 3.1; Example 1b

2. Reduce $\dfrac{5x+5y}{x^2-y^2}$ to lowest terms.   Section 3.2; Example 1b

3. Write the rational expression $\dfrac{x-2}{x-3}$ with a denominator of $x^2 - x - 6$.   Section 3.2; Example 3d

4. Perform the operation indicated.

$$\frac{t^2 - t - 2}{2t - 6} \cdot \frac{t - 3}{t^3 - 8}$$

Section 3.3; Example 1d

Sec. 3.5 Complex Fractions

5. Perform the operation indicated.

$$\frac{x^2 + 3x + 2}{x^4} \div \frac{x^2 + 2x + 1}{x^2}$$

Section 3.3; Example 3b

6. Perform the operation indicated.

$$\frac{3}{x^2 - 4} + \frac{1}{2 - x}$$

Section 3.4; Example 5

7. Simplify $\dfrac{\dfrac{a}{x+1} + 2}{\dfrac{b}{x+1} - 1}$.

Section 3.5; Example 2

REVIEW PROBLEMS

In problems 1 through 5, determine the values of x, if any, for which each rational expression is undefined.

1. $\dfrac{14}{x}$
2. $\dfrac{x}{x - 8}$
3. $\dfrac{x^2}{67}$
4. $\dfrac{x + 4}{(x + 2)(x - 8)}$
5. $\dfrac{x - 1}{x^2 + 4x + 3}$

In problems 6 through 10, express each rational expression with the denominator indicated.

6. $\dfrac{3}{x^4 y^2}$; $x^8 y^3$
7. $\dfrac{4}{3p}$; $6p^2(p - 2)$
8. $\dfrac{x}{x - 1}$; $x^2 - 7x + 6$
9. $\dfrac{x + 3}{x - 4}$; $x^2 - 8x + 16$
10. $\dfrac{4}{x + 2}$; $x^2 + 2x$

In problems 11 through 15, reduce each rational expression to lowest terms.

11. $\dfrac{m^5 n}{m^7 n^3}$
12. $\dfrac{4 - x}{x^2 - 16}$
13. $\dfrac{x^2 - 4x - 5}{x^3 + 1}$
14. $\dfrac{r^2 + 6r + 9}{r^2 - 9}$
15. $\dfrac{s^3 t - st}{s^2 + s}$

In problems 16 through 35, perform the operations indicated. Write answers in lowest terms.

16. $\dfrac{3}{x^2 y} - \dfrac{x + y}{x}$
17. $\dfrac{12 - 3x^2}{2x^2 + x - 15} \cdot \dfrac{2x^2 - 3x - 5}{x^2 - 4x + 4}$
18. $\dfrac{4}{t - 7} + \dfrac{3}{t + 7}$
19. $\dfrac{a^2 - 3a + 9}{a^3 - ab^2} \div \dfrac{a^3 + 27}{a^4 + 2a^3 b + a^2 b^2}$
20. $\dfrac{2x - 2}{x^2 - 2x - 8} - \dfrac{1}{4 - x}$
21. $\dfrac{r^2 s t^3}{u^4 v} \cdot \dfrac{-t^5}{u} \cdot \dfrac{u^2 v^3}{rst}$
22. $\dfrac{2}{x + 1} + \dfrac{3}{x^2 + 2x + 1}$
23. $\dfrac{u^2 - uv + 2uw - 2vw}{u^3 + 8w^3} \div \dfrac{u^2 + 4uw + 4w^2}{u^2 - 2uw + 4w^2}$
24. $\dfrac{4}{x + 3} - \dfrac{x + 1}{x + 7}$
25. $\dfrac{1 - 2y + y^2}{6y^2 - y - 1} \cdot \dfrac{y^2 + 7y + 12}{3y^2 + y - 4} \cdot \dfrac{6y^2 + 5y - 4}{y^2 + 2y - 3}$
26. $\dfrac{15x}{x^2 + x - 6} - \dfrac{11x - 3}{x^2 + 2x - 3}$
27. $\dfrac{12x^2 - 5x - 2}{x^2 + 2xy + y^2} \div \dfrac{12x^2 + x - 6}{x + y}$

28. $\dfrac{1}{xy} + \dfrac{3}{y} - \dfrac{4}{xyz}$

29. $\dfrac{12t^3 - 27t}{t^8 - 16} \cdot \dfrac{t^4 + 2t^2}{2t^2 - 3t - 9}$

30. $\dfrac{25x}{x^2 - 3x - 4} - \dfrac{21x - 64}{12 - 7x + x^2}$

31. $\dfrac{96t^7}{375s^2} \div \dfrac{-72t^4}{125s^7}$

32. $\dfrac{1}{x+2} - \dfrac{3}{x-1} + \dfrac{5}{x-3}$

33. $\dfrac{a^3 + a^2b + ab^2 + b^3}{32a^7b^3} \cdot \dfrac{-768a^2b}{a^4 - b^4}$

34. $\dfrac{7}{11 - 2x} + \dfrac{3}{2x - 11}$

35. $\dfrac{m^2 - 16n^2}{6m^4 + 3m^2} \div \dfrac{64n^3 - m^3}{12m^2}$

In problems 36 through 45, simplify each complex fraction.

36. $\dfrac{\dfrac{2x^3z^2}{y}}{\dfrac{-48x}{y^6}}$

37. $\dfrac{\dfrac{x^2y}{x+1}}{\dfrac{x}{x+1}}$

38. $\dfrac{\dfrac{5x}{p-1}}{\dfrac{30x}{p-1}}$

39. $\dfrac{\dfrac{1}{x} - \dfrac{2}{xy}}{\dfrac{2}{x} + \dfrac{1}{xy}}$

40. $\dfrac{\dfrac{2}{a+b} - \dfrac{3}{a-b}}{\dfrac{3}{a-b} - \dfrac{2}{a+b}}$

41. $\dfrac{\dfrac{3t^2 + 5t}{t^2 - 25}}{\dfrac{2}{t-5} + \dfrac{1}{t+5}}$

42. $\dfrac{\dfrac{t^2 - 4z^2}{zt}}{\dfrac{1}{z} - \dfrac{2}{t}}$

43. $\dfrac{\dfrac{1}{x-5} - \dfrac{1}{x+3}}{\dfrac{16x^2 + 16}{x^2 - 2x - 15}}$

44. $\dfrac{\dfrac{1}{x+2} + x + 2}{\dfrac{1}{x+2} - x - 2}$

45. $\dfrac{\dfrac{1}{s-5} + \dfrac{s+5}{s^2 + 5s + 25}}{\dfrac{2s + 5}{s^3 - 125}}$

. . . LET'S NOT FORGET . . .

Identify the expressions that are in factored form. Factor those that are not factored.

46. $(3x - 1)^3$
47. $27x^3 - 1$
48. $x(a - b) + y(a - b)$
49. $3s^2t(st + 1)$
50. $3x^4y$

How many terms are in each expression? Which expressions have $x - 1$ as a factor?

51. $x^3 + 1$
52. $(x - 1)^3$
53. $x^2 - x - 2$
54. $x^2 - 2x + 1$
55. $3(x - 1) - (x - 1)^2$
56. $a(x - 1) + b(1 - x)$

Find each product.

57. $(x + 2y)^2$
58. $(c - 2)^3$

Simplify each expression.

59. -2^2
60. $(-2)^2$
61. $(-2)^3$
62. $(-a)^4$
63. $(-a)^5$

Reduce, if possible.

64. $\dfrac{3(p + q) - 2(p - q)}{(p + q)(p - q)}$

65. $\dfrac{a(y + z) - b(y + z)}{(y + z)(y - z)}$

The next two problems can be worked by using a least common denominator. Follow the directions in each and notice how the LCD is used.

66. Perform the operation indicated: $\dfrac{1}{x - 2} + \dfrac{4}{x + 2}$.

67. Simplify $\dfrac{\dfrac{4}{x - 2} + 1}{\dfrac{2x}{x + 2} - 1}$.

CHAPTER 3 TEST

In problems 1 through 5, choose the correct answer.

1. The rational expression $\dfrac{x-2}{x(x+3)}$ is undefined if x has the value(s) of (?)
   A. 2
   B. -2
   C. 0 and -3
   D. 3

2. If the rational expression $\dfrac{3x}{x-7}$ is written with a denominator of $x^2 - 14x + 49$, it becomes (?)
   A. $\dfrac{3x^2 - 7}{x^2 - 14x + 49}$
   B. $\dfrac{3x^2 - 21x}{x^2 - 14x + 49}$
   C. $\dfrac{3x}{x^2 - 14x + 49}$
   D. $\dfrac{3x^2 + 21x}{x^2 - 14x + 49}$

3. When reduced to lowest terms, the rational expression $\dfrac{x^3 - 8y^3}{x^2 - 4y^2}$ becomes (?)
   A. $\dfrac{x^2 + 2xy + 4y^2}{x + 2y}$
   B. $\dfrac{(x - 2y)^2}{x + 2y}$
   C. $x - 2y$
   D. $x + 2y$

4. $\dfrac{x}{x+4} - \dfrac{x-3}{x-2} = (?)$
   A. $-\dfrac{x + 12}{(x+4)(x-2)}$
   B. $\{4\}$
   C. $\dfrac{-3(x-4)}{(x+4)(x-2)}$
   D. $-3(x+4)$

5. When simplified, the complex fraction $\dfrac{\dfrac{3}{x} + \dfrac{1}{y}}{\dfrac{9y}{x} - \dfrac{x}{y}}$ becomes (?)
   A. $\dfrac{4}{9y - x}$
   B. $\dfrac{1}{3y - x}$
   C. $\dfrac{1}{3y + x}$
   D. $\dfrac{3x + y}{9y - x}$

In problems 6 through 11, perform the operation indicated. Write answers in lowest terms.

6. $\dfrac{a}{ab - b^2} + \dfrac{b}{a^2 - ab}$

7. $\dfrac{(a+2)^2}{a^2 + 6a + 8} \cdot \dfrac{a^2 - 2a - 8}{(a+2)^3}$

8. $\dfrac{x^3 + 3x^2}{x^3 - 6x^2 + 9x} \div \dfrac{x^2 + 2x - 3}{x^2 - 9}$

9. $\dfrac{y+1}{y-2} - \dfrac{y^2 + 5y + 1}{y^2 + y - 6}$

10. $\dfrac{6x^2 + 13x + 6}{9x^2 - 4} \cdot \dfrac{9x^2 - 12x + 4}{2x^2 + x - 3}$

11. $\dfrac{ar + at + 2br + 2bt}{a^2 - ab + b^2} \div \dfrac{r^2 + rt}{a^3 + b^3}$

12. Simplify the complex fraction.
$$\dfrac{\dfrac{2t}{s+t} + 1}{\dfrac{4t}{s+t} - 1}$$

13. Reduce the rational expression to lowest terms.
$$\dfrac{x^2 - 4x + 4}{16x^2 - 4x^4}$$

14. Write the rational expression $\dfrac{w}{5}$ with a denominator of $15w^3$.

15. Determine the values of x, if any, for which each rational expression is undefined.
    (a) $\dfrac{5}{x^2 - x - 2}$
    (b) $\dfrac{x-1}{8}$

CHAPTER 4

See Problem Set 4.3, Exercise 40.

Using Linear Equations in One Variable

- **4.1** Solving Linear Equations in One Variable
- **4.2** Literal Equations and Formulas
- **4.3** Applications
- **4.4** Fractional Equations
- **4.5** Absolute Value Equations
- **4.6** Equations Containing Square Roots

CONNECTIONS

During the Renaissance, the Hindu–Arabic notation for numbers became popular throughout Western Europe. This compact number system was soon followed by the development of the common symbols we use today for addition, multiplication, and equality. These symbols became the key that unlocked algebra.

Using Hindu–Arabic numerals and the new symbols simplified complicated arithmetical questions to stating the equality of two expressions. We call such statements "equations." Equations are one of the central ideas of algebra.

A large class of equations of the form $Ax + B = 0$ are called linear equations. In this chapter we study linear equations and their applications and continue with the examination of other equations that lead to linear equations.

4.1 SOLVING LINEAR EQUATIONS IN ONE VARIABLE

An **equation** is a statement that two numbers are equal. This statement may or may not be true. For example:

$$5 = 5$$
$$17 = 31$$

are two such statements. The first is true and the second false.

Equations are interesting when they contain letters, such as x, which stand for numbers whose value is unknown. For example, the equation

$$x + 4 = 11$$

is true if x is 7 and false otherwise. We call x a **variable**. Any number that replaces the variable and makes the equation true is called a **solution** or **root**. The set of all solutions is called the **solution set**. To **solve** an equation means to find the solution set.

Two equations that have the same solution set are called **equivalent**. The three equations

$$2(x - 3) = x - 12$$
$$2x - 6 = x - 12$$
$$x = -6$$

are equivalent, as each has $\{-6\}$ as its solution set.

Equations such as these which can be written in the form

$$Ax + B = 0$$

are called **linear equations** in one variable.

The equations

$$x^2 + 3x - 7 = 0$$

$$\sqrt{x} + x = 13$$

$$\frac{1}{x} - \frac{2}{x} = 1$$

are examples of *nonlinear* equations.

Here are two tools used in solving equations.

Two Tools Used to Solve Equations

Addition Property of Equality
If the same number is added to (or subtracted from) both sides of an equation, the resulting equation is equivalent to the first equation.

Multiplication Property of Equality
If both sides of an equation are multiplied (or divided) by the same nonzero number, the resulting equation is equivalent to the first equation.

The equations

$$x - 21 = -4 \quad \text{and}$$

$$x = 17$$

are equivalent, as the second was obtained by adding 21 to both sides of the first. Also, the equations

$$3x = -24 \quad \text{and}$$

$$x = -8$$

are equivalent, as the second was obtained by dividing both sides of the first by 3 $\left(\text{or multiplying by } \dfrac{1}{3}\right)$. In both of these cases, the solution set in the second equation is easier to see than in the first. These are simple illustrations of the general techniques to be developed to find the solution sets of linear equations in one variable.

The general approach used to solve linear equations is to apply our two tools, forming equivalent equations until an equation of the form

$$x = \text{a constant}$$

is reached. The solution set consists of the set containing the constant. Examples 1 and 2 that follow illustrate the technique.

Checking Equations

Linear equations can be checked for arithmetic errors. When checking a possible solution, **be sure to use a correct procedure.**

Sec. 4.1 Solving Linear Equations in One Variable

Be careful not to just put the number being checked back into the equation to see if it "works out." This involves a subtle, but *important*, error in logic. A checking procedure that evaluates both the left and right sides of the original equation separately is recommended. If the value obtained from the left side is the same as the value obtained from the right side, the number being checked is in the solution set. Otherwise, the number is not in the solution set.

EXAMPLE 1. Solve $3x + 2 = 10$.

Solution:

Subtract 2 from each side to obtain

$$3x = 8$$

Then divide both sides by 3 to get

$$x = \frac{8}{3}$$

To check $\frac{8}{3}$ in $3x + 2 = 10$, replace x with $\frac{8}{3}$ in the left side.

Left side (LS): $3 \cdot \frac{8}{3} + 2 = 8 + 2 = \boxed{10}$

Right side (RS): $\boxed{10}$

The value of each side is 10, so $\frac{8}{3}$ is the solution. Write the solution set.

$$\left\{ \frac{8}{3} \right\}$$

EXAMPLE 2. Solve $2x - 5 = 5x - 2$.

Solution:

Subtract $2x$ from each side to get all the terms containing x on the same side of the equation.

$$2x - 5 - 2x = 5x - 2 - 2x$$
$$-5 = 3x - 2$$

Next we add 2 to each side to obtain

$$-5 + 2 = 3x - 2 + 2$$
$$-3 = 3x$$

Divide both sides by 3 to get
$$-1 = x$$
Check -1 in $2x - 5 = 5x - 2$.

LS: $2(-1) - 5 = -2 - 5 = \boxed{-7}$

RS: $5(-1) - 2 = -5 - 2 = \boxed{-7}$

Write the solution set.
$$\{-1\} \qquad \square$$

In the first step of Example 2 all the terms containing x are collected on one side of the equation by subtracting $2x$ from both sides. Notice that the same thing could have been accomplished by subtracting $5x$.

$$2x - 5 = 5x - 2$$
$$2x - 5 - 5x = 5x - 2 - 5x$$
$$-5 - 3x = -2$$
$$-3x = 3$$
$$x = -1$$

The solution set is the same. Either way is correct, but it is common to gather terms containing the variable so that the resulting coefficient is positive.

A Procedure for Solving Linear Equations in One Variable

1. Remove all grouping symbols and combine like terms, where possible, on each side of the equation.
2. Clear the equation of fractions, if necessary, by multiplying both sides by the least common denominator (LCD) of all the fractions in the equation.
3. Use the Addition Property to collect all the terms containing the variable on one side of the equation and all constant terms on the other side.
4. Combine like terms.
5. Use the Multiplication Property to divide each side by the coefficient of the variable.
6. Check the solution in the original equation, if required.
7. Write the solution set.

In some problems it is necessary to repeat step 3 and sometimes it is convenient to interchange steps 1 and 2.

Sec. 4.1 Solving Linear Equations in One Variable

EXAMPLE 3. Solve $5(2x + 3) = 3 - 2(3x - 5)$.

Solution:

$$5(2x + 3) = 3 - 2(3x - 5)$$

Step 1 Remove all grouping symbols.

$$10x + 15 = 3 - 6x + 10 \qquad \text{Distributive Property}$$

Step 3 Collect all the terms containing x on one side and constant terms on the other side.

$$10x + 15 + 6x = 13 \qquad \text{Collect } x \text{ terms on left.}$$
$$10x + 6x = 13 - 15 \qquad \text{Collect constant terms.}$$

Step 4 Combine like terms.

$$16x = -2$$

Step 5 Divide both sides by the coefficient of the variable.

$$x = -\frac{2}{16} \qquad \text{Divide both sides by 16.}$$

$$x = -\frac{1}{8}$$

Step 6 Check $-\frac{1}{8}$ in the original equation.

Step 7 Write the solution set.

$$\left\{ -\frac{1}{8} \right\}$$

Often, the solution set for linear equations with one variable contains one element. Note that some equations have all real numbers as solutions, and some have empty solution sets.

EXAMPLE 4. Solve $x + 7 = x + 2$.

Solution:

$$x + 7 = x + 2$$
$$x - x = 2 - 7$$
$$0 = -5$$

Be Careful! As $0 = -5$ is *never* true, the equation $x + 7 = x + 2$ will not be true for any value of x. Thus the solution set is the empty set, which is written as { } or ∅ (*never* as {∅}).

EXAMPLE 5. Solve $2x + 2 = 2(x + 1)$.

Solution:

$$2x + 2 = 2(x + 1)$$
$$2x + 2 = 2x + 2$$
$$2x - 2x = 2 - 2$$
$$0 = 0$$

As $0 = 0$ is *always* true, the equation $2x + 2 = 2(x + 1)$ will be true for all values of x. Thus the solution set is all real numbers.

$$\{x \mid x \text{ is a real number}\}$$

This set is often written as $\mathbb{R}$. □

Equations that are never true, such as in Example 4, are sometimes called **contradictions,** and equations that are true for all real numbers, such as in Example 5, are sometimes called **identities.**

If some of the terms in an equation are fractions or have fractional coefficients, it is usually best to remove the fractions first. This can be done by multiplying both sides by the LCD of all the fractions involved.

EXAMPLE 6. Solve the following equation:

$$\frac{1}{3}x - 4 = \frac{1}{2}x + 1$$

Solution:

Multiply both sides by 6 to clear fractions.

$6\left(\frac{1}{3}x - 4\right) = 6\left(\frac{1}{2}x + 1\right)$ Clear fractions.

$2x - 24 = 3x + 6$ Distributive Property

$-24 - 6 = 3x - 2x$ Collect x terms and constant terms.

$-30 = x$ Combine like terms.

Check -30 in original equation.

Replace x with -30 in the *left* side.

LS: $\frac{1}{3}(-30) - 4 = -10 - 4 = \boxed{-14}$

Replace x with -30 in the *right* side.

RS: $\frac{1}{2}(-30) + 1 = -15 + 1 = \boxed{-14}$

The left and right sides have the same value. So -30 is the solution.

$$\{-30\}$$ □

Sec. 4.1   Solving Linear Equations in One Variable

EXAMPLE 7. Solve $\dfrac{2x + 1}{2} = \dfrac{1}{3} + \dfrac{x - 3}{4}$.

Solution:

Multiply both sides by 12, the LCD of the three fractions.

$12\left(\dfrac{2x + 1}{2}\right) = 12\left(\dfrac{1}{3} + \dfrac{x - 3}{4}\right)$ Clear fractions.

$6(2x + 1) = 4 + 3(x - 3)$ Distributive Property

$12x + 6 = 4 + 3x - 9$ Remove parentheses.

$12x + 6 = 3x - 5$ Combine like terms.

$12x - 3x = -5 - 6$ Collect x terms and constant terms.

$9x = -11$ Combine like terms.

$x = -\dfrac{11}{9}$ Divide.

Check $-\dfrac{11}{9}$ in the original equation.

$\left\{-\dfrac{11}{9}\right\}$ Write solution set.

EXAMPLE 8. Solve $0.8x + 1.7 = 0.2x + 10.0$.

Solution:

Remember that decimals are just fractions, and we can write the equation as

$$\dfrac{8}{10}x + \dfrac{17}{10} = \dfrac{2}{10}x + 10$$

Multiply both sides by 10.

$10\left(\dfrac{8}{10}x + \dfrac{17}{10}\right) = 10\left(\dfrac{2}{10}x + 10\right)$ Clear fractions.

$8x + 17 = 2x + 100$ Distributive Property

$8x - 2x = 100 - 17$ Collect terms.

$6x = 83$ Combine like terms.

$x = 13.8\overline{3}$ Divide.

$\{13.8\overline{3}\}$ Write solution set.

The solution set is written as a decimal rather than a fraction since the original equation used decimals. Be careful in writing $\dfrac{83}{6}$ as a decimal. $\dfrac{83}{6} = 13.8\overline{3}$. This is

Chap. 4   Using Linear Equations in One Variable

an *exact* answer and should not be confused with *approximations* of $\frac{83}{6}$, such as 13.8 or 13.83.

Equations are used in the real world to solve problems as diverse as finding interest and designing rocket boosters. Equations are the heart of engineering and the sciences. In an algebra course we use word problems to imitate such real-world applications.

We solve a word problem by carefully translating the situation described into the language of algebra. Often this leads to a linear equation. This equation is called a *mathematical model* of the problem. We then solve the equation and answer the question. It is important to make sure that we answer the question asked and that the answer we give makes sense.

EXAMPLE 9. Mary Adair Pittman bought a Persian rug on her trip to China. She has forgotten its dimensions. She remembers that the length is 4 ft more than its width and that its perimeter is 40 ft. Find the dimensions of the rug.

Solution:

Choose a variable to represent the width of the rug.

Let w be the width of the rug in feet. Since the length is 4 ft more than the width, the length is $(w + 4)$ feet.

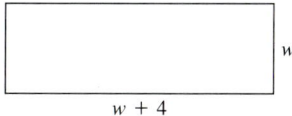

Form an equation.

Remember that the perimeter of a rectangle is

$$P = 2L + 2W$$
$$40 = 2(w + 4) + 2w$$

Solve the equation.

$$40 = 2w + 8 + 2w$$
$$40 = 4w + 8$$
$$32 = 4w$$
$$8 = w$$

Thus $w + 4 = 12$.

Check these values in the original word problem.

Using 8 and 12 gives a perimeter of $2 \cdot 8 + 2 \cdot 12$, which is 40. So a width of 8 ft and length of 12 ft makes sense and answers the question.

Answer the question.

The dimensions are 8 ft by 12 ft.

Sec. 4.1 Solving Linear Equations in One Variable

Consecutive integers is a commonly used phrase in word problems. The numbers 5, 6, 7 are consecutive integers, as are $-10, -9, -8, -7$. The numbers 11, 13, 15 are consecutive *odd* integers, and 6, 8, 10, 12 are consecutive *even* integers.

Consecutive integers: $\quad n, n + 1, n + 2$
Consecutive odd integers: $\quad n, n + 2, n + 4$
Consecutive even integers: $\quad n, n + 2, n + 4$

EXAMPLE 10. Find three consecutive odd integers whose sum is 45.

Solution:

Let x be the smallest of the three integers, then $x + 2$ is the next odd integer, and $x + 4$ is the largest of the three.

As the sum of the three is 45, we write

$$x + (x + 2) + (x + 4) = 45$$

We solve this equation.

$$3x + 6 = 45$$
$$3x = 39$$
$$x = 13$$

Since x is the *smallest* of the three, the other two must be 15 and 17.

Since 13, 15, and 17 are three consecutive odd integers and $13 + 15 + 17 = 45$, they answer the question and make sense.

The three integers are 13, 15, and 17.

CALCULATOR BOX

Solving Equations with a Calculator

To use a calculator to solve an equation, the idea is to write the equation in the form $x =$ a constant, without doing any of the arithmetic.

EXAMPLE Solve $23.56 - 0.87x = 11.02x - 19.04$.

Solution:

Collect the x-terms on the left side and the constants on the right side.

$$-0.87x - 11.02x = -19.04 - 23.56$$

Instead of carrying out intermediate steps of combining like terms, factor an x from the terms on the left side.

$$(-0.87 - 11.02)x = -19.04 - 23.56$$

Now divide both sides by the coefficient of x.

$$x = \frac{-19.04 - 23.56}{-0.87 - 11.02}$$

Enter these keystrokes and let the calculator work!

(19.04 +/− − 23.56) ÷ (0.87 +/− − 11.02) =

The display should read 3.582842725 .
The solution is approximately 3.582842725. Rounding to the nearest hundredth gives us

$$x \approx 3.58$$

Calculator Exercises

Use a calculator to approximate the solution of the equations below correct to two decimal places. Be sure to indicate that answers are approximations.

1. $66.75 - 23.47x = 2.22 + 1.13x$
2. $23.56 - 0.87x = 11.02x - 19.04$
3. $0.987x - 9.87 = 18.95 + 1.22x$
4. $2.37(1.23t - 3.09) = 2.01t - 18.0$

Answers:

1. 2.62
2. 3.58
3. −123.69
4. −11.80

PROBLEM SET 4.1

Warm-ups

In problems 1 through 40, find the solution set.
For problems 1 through 6, see Example 1.

1. $x - 4 = 2$
2. $x + 3 = 5$
3. $2x = 10$
4. $6x = -18$
5. $3x - 5 = 4$
6. $4x - 10 = 2$

For problems 7 through 12, see Examples 2, 4, and 5.

7. $5 + 3x = 5x - 1$
8. $x + 5 = x - 3$
9. $5x - 3 = 3x + 7$
10. $6x + 5 = 6 + 4x$
11. $6 + 3x = 3x + 6$
12. $2 - 3x = 2x - 8$

For problems 13 through 22, see Examples 3, 4, and 5.

13. $2(x + 1) = 3x - 4$
14. $4(x - 6) = 5x + 2$
15. $2(x - 6) - 3(2x - 2) = 0$
16. $7(2t + 3) = -(4 + t)$
17. $-(1 - 2x) = 3 + 5x$
18. $4(3 - x) = 3(4 - x) - x$
19. $1 + 3x = 3(2 + x)$
20. $2(x - 1) + 2 = 3(1 + x) - (x + 3)$
21. $2(x + 3) - 3[2(x - 3) + 4] = 0$
22. $3[6 - (2x + 5) + (4 - x)] = -3(3x - 1)$

For problems 23 through 28, see Example 6.

23. $\frac{1}{2}x - 1 = \frac{1}{3}x$
24. $\frac{1}{4}s + 1 = \frac{1}{3}s$
25. $\frac{8}{9}x - \frac{1}{3} = \frac{2}{3}x + 2$
26. $\frac{3x}{5} + \frac{5x}{2} = \frac{31x}{10}$
27. $\frac{x}{6} = \frac{5}{12}(x + 2)$
28. $\frac{2}{3}(x - 4) = \frac{1}{2}x$

Sec. 4.1 Solving Linear Equations in One Variable

For problems 29 through 34, see Example 7.

29. $\dfrac{x+1}{2} = 2$

30. $\dfrac{7x+1}{10} + 17 = x$

31. $\dfrac{1-3x}{2} - 3x = 5$

32. $\dfrac{x+5}{2} = \dfrac{x+3}{3}$

33. $\dfrac{7x-3}{7} - \dfrac{5x+7}{5} = 1$

34. $\dfrac{y-3}{3} + \dfrac{2y-1}{2} = \dfrac{y-19}{4}$

For problems 35 through 40, see Example 8.

35. $x + 0.4 = 1.5$

36. $2x - 5.66 = 2.02$

37. $1.3x - 6.1 = 2x + 5.8$

38. $0.1y + 1.9 = 0.8 - 0.2y$

39. $1.1x = 2.3(x + 2)$

40. $0.25t + 0.85 = 1.35(1 - t)$

For problems 41 and 42, see Example 9.

41. Ed Wiley is fencing a rectangular strawberry patch in his yard. He wants to use all 78 m of wire that he has. The length of the patch is to be 5 m more than the width. Find the dimensions of the strawberry patch.

42. The width of a rectangle is 5 m less than its length. If half the perimeter of the rectangle is 17 m, what is the length?

For problems 43 and 44, see Example 10.

43. The sum of three natural numbers is 24. The second number is five times the first number and the third number is 10 more than the first. Find the three numbers.

44. The sum of two consecutive odd integers is 32. Find the integers.

Practice Exercises

In problems 45 through 95, find the solution set.

45. $x + 5 = 9$

46. $x - 4 = 0$

47. $2x - 12 = 7$

48. $3x + 14 = 0$

49. $5 - 3x = 2$

50. $4x - 8 = 3x$

51. $5x + 8 = 11x$

52. $3x - 2 = 2x + 13$

53. $x + 7 = x + 3$

54. $11 - 7x = x + 5$

55. $2y + 3 = -y - 6$

56. $-7y - 1 = 3y - 2$

57. $5x - 22 = 17x - 16$

58. $6 + x = x + 6$

59. $7x + 15 = 5x + 31$

60. $3(x - 9) = 2(x + 1)$

61. $2(3 - x) = 5(x + 4)$

62. $7(t + 11) = 3(t - 1)$

63. $(1 + s)3 = 7(-1 - s)$

64. $2(x - 5) - (3x + 1) = 0$

65. $4(2x + 1) - (4 - x) = 18$

66. $9(3s - 1) - 2(3 - 4s) = 10$

67. $3(x + 1) - 3 = 2(x - 1) + (x + 2)$

68. $6 + 3[2(x - 7) + x] = 2(x - 4)$

69. $2[1 - 3(x + 2)] = 0$

70. $2(1 - x) + 3x = 4(3 - x)$

71. $-6\{x + 3[x + 3(2x - 5)]\} = 5 - 3(44x + 8)$

72. $\dfrac{1}{9}x + \dfrac{1}{3} = \dfrac{1}{3}x$

73. $\dfrac{1}{2}x - \dfrac{1}{4} = \dfrac{1}{3}x$

74. $\dfrac{1}{2}(3x - 2) = \dfrac{1}{5}(12 - x)$

75. $\dfrac{1}{6} - \dfrac{1}{3}x = \dfrac{1}{9}(2x + 1)$

76. $\dfrac{2}{3}x - \dfrac{1}{5} = \dfrac{2}{5}x + 1$

77. $\dfrac{2x}{7} + \dfrac{x}{2} = 11$

78. $\dfrac{x}{3} = \dfrac{3}{5}(x - 2)$

79. $\dfrac{3}{4}(x - 3) = \dfrac{5}{8}x$

80. $\dfrac{2x + 2}{2} = 3$

81. $\dfrac{x - 2}{3} = 4$

82. $1 - \dfrac{2x - 1}{11} = 0$

83. $\dfrac{x + 3}{5} - \dfrac{2 - x}{10} = 7$

84. $\dfrac{2y - 3}{3} = \dfrac{y + 4}{2}$

85. $\dfrac{3t-7}{7} = \dfrac{t+1}{2}$

86. $\dfrac{5-8x}{6} - \dfrac{2x+3}{4} = 1$

87. $\dfrac{5x-10}{3} + \dfrac{x-8}{2} = -3$

88. $\dfrac{x+1}{3} - \dfrac{2x-1}{4} = \dfrac{x+2}{5}$

89. $\dfrac{2x-3}{6} = \dfrac{3x}{2} + \dfrac{6-x}{3}$

90. $x - 1.4 = 6.8$

91. $3x + 6.03 = 3.42$

92. $1.4x - 1.4 = 2.0x + 2.2$

93. $1.4x + 0.5 = 1.1x + 1.5$

94. $1.25x - 0.44 = 1.24(x-1)$

95. $1.5(x+2) = 2.8 + x$

96. What are the dimensions of the largest picture frame that can be constructed of 60 in. of molding if the width is to be two-thirds of the length?

97. The length of a rectangle is 8 ft more than the width. If the perimeter is 104 ft, find the dimensions of the rectangle.

98. The sum of two consecutive even integers is 50. Find the integers.

99. The sum of three natural numbers is 29. The second number is 5 more than the first number and the third number is twice the first. Find the three numbers.

100. Find the measure of each angle in a triangle if the angles are equal.

101. Find the measure of each acute angle in a right triangle if the measure of one angle is twice the other.

102. Find the measure of two supplementary angles if the measure of one of the angles is four times the other.

Challenge Problems

In problems 103 through 105, find the solution set; a and b are constants.

103. $x - a = b$

104. $2x - a = 4b$

105. $\dfrac{x-a}{2} = \dfrac{4-x}{3}$

IN YOUR OWN WORDS . . .

106. How do we solve a linear equation in one variable?

107. Describe a valid procedure for checking an equation.

4.2 LITERAL EQUATIONS AND FORMULAS

Equations frequently contain more than one variable. Such equations are often called **literal equations.** In this section we solve some of them.

Notice the similarity in the steps below. Equation  contains only one letter, the variable x, whereas equation ② contains several letters, the variable x, and constants b, c, and m.

$$\boxed{1}\ \ 2x + 1 = 7 \qquad \boxed{2}\ \ mx + b = c$$
$$\phantom{\boxed{1}\ \ }2x = 6 \qquad \phantom{\boxed{2}\ \ }mx = c - b$$
$$\phantom{\boxed{1}\ \ }x = 3 \qquad \phantom{\boxed{2}\ \ }x = \dfrac{c-b}{m};\ \ m \neq 0$$

If we were solving equation ①, we would continue and write the solution set, {3}. Set notation is seldom used in literal equations. We simply write the equation in an equivalent form.

It is *important* to note that literal equations are solved in the same way that other equations are solved.

Formulas provide useful examples of literal equations. Consider the formula for the volume of a box:

$$V = lwh$$

where V is the volume, l is the length, w is the width, and h is the height. This form says that it is solved for V. Often, the formula will be solved for another variable, such as h.

$$V = lwh$$

$$\frac{V}{lw} = h \quad \text{Divide by } lw \text{ (the coefficient of } h\text{)}.$$

This can also be written $h = \dfrac{V}{lw}$.

EXAMPLE 1. Solve $A = 2\pi rh + 2\pi r^2$ for h.

Solution:

To isolate the term containing h, *subtract $2\pi r^2$ from both sides.*

$$A - 2\pi r^2 = 2\pi rh$$

$$\frac{A - 2\pi r^2}{2\pi r} = h \quad \text{Divide by } 2\pi r \ (r \neq 0).$$

EXAMPLE 2. Solve $A = \dfrac{1}{2}h(b_1 + b_2)$ for h.

Solution:

$$A = \frac{1}{2}h(b_1 + b_2)$$

$$2A = h(b_1 + b_2) \quad \text{Clear fractions.}$$

$$\frac{2A}{b_1 + b_2} = h \quad \begin{array}{l}\text{Divide by the coefficient of } h. \\ (b_1 + b_2 \neq 0)\end{array}$$

EXAMPLE 3. Solve $T = 2ab + 2bc + 2ac$ for c.

Solution:

$$T = 2ab + 2bc + 2ac$$

$$T - 2ab = 2bc + 2ac \quad \begin{array}{l}\text{Collect the } c \text{ terms on} \\ \text{right and others on left.}\end{array}$$

Be Careful!

We are solving for c, so we *factor it out* of the terms containing c.

$$T - 2ab = (2b + 2a)c \quad \text{Distributive Property}$$

$$\frac{T - 2ab}{2b + 2a} = c \quad \begin{array}{l}\text{Divide by coefficient} \\ \text{of } c. \ (2b + 2a \neq 0)\end{array}$$

Chap. 4 Using Linear Equations in One Variable

> **A Procedure for Solving Literal Equations**
>
> 1. Remove all symbols of grouping and combine like terms where possible.
> 2. Clear the equation of fractions, if necessary.
> 3. Use the Addition Property of equality to collect all terms containing the variable being solved for on one side of the equation and all other terms on the other side of the equation.
> 4. Combine like terms where possible.
> 5. If the variable being solved for appears in more than one term, use the Distributive Property to factor it out.
> 6. Use the Multiplication Property of equality to divide both sides of the equation by the coefficient of the variable being solved for.

Sometimes it is convenient to interchange steps 1 and 2.

EXAMPLE 4. Solve $cx - 3 = b$ for x.

Solution:

$$cx - 3 = b$$
$$cx = b + 3 \qquad \text{Collect } x \text{ term on left.}$$
$$x = \frac{b + 3}{c} \qquad \text{Divide by } c. \ (c \neq 0)$$

EXAMPLE 5. Solve $bx - 5 = 3bx - 7$ for x.

Solution:

$$bx - 5 = 3bx - 7$$
$$bx - 3bx = -7 + 5 \qquad \text{Collect } x \text{ terms on left and other terms on right.}$$
$$-2bx = -2 \qquad \text{Combine like terms.}$$
$$x = \frac{1}{b} \qquad \text{Divide by } -2b. \ (b \neq 0)$$

EXAMPLE 6. Solve $ax - b(x + a) = 6 + 2bx$ for x.

Solution:

$$ax - b(x + a) = 6 + 2bx$$
$$ax - bx - ab = 6 + 2bx \qquad \text{Distributive Property}$$
$$ax - bx - 2bx = 6 + ab \qquad \text{Collect } x \text{ terms on left and others on right.}$$
$$ax - 3bx = 6 + ab \qquad \text{Combine like terms.} \qquad \textit{(continued)}$$

Sec. 4.2 Literal Equations and Formulas

The next step is common in solving literal equations. *Two* terms contain x, the variable being solved for, and just one is needed. Factor out x.

$$(a - 3b)x = 6 + ab \qquad \text{Factor.}$$

$$x = \frac{6 + ab}{a - 3b} \qquad \text{Divide by } a - 3b.$$
$$(a - 3b \neq 0)$$

EXAMPLE 7. Solve $\dfrac{x - a}{5} = x(a + b)$ for x.

Solution:

$$\frac{x - a}{5} = x(a + b)$$

$$\frac{x - a}{5} = ax + bx \qquad \text{Distributive Property}$$

$$x - a = 5ax + 5bx \qquad \text{Clear fractions.}$$

$$x - 5ax - 5bx = a \qquad \text{Collect } x \text{ terms on left and others on right.}$$

$$(1 - 5a - 5b)x = a \qquad \text{Factor.}$$

$$x = \frac{a}{1 - 5a - 5b} \qquad \text{Divide. } (1 - 5a - 5b \neq 0)$$

PROBLEM SET 4.2

Warm-ups

In problems 1 through 7, solve for the variable indicated.
For problems 1 and 2, see Example 1.

1. $C = 2\pi r$, for r

2. $V = \pi r^2 h$, for h

For problems 3 through 5, see Example 2.

3. $F = \dfrac{9}{5}C + 32$, for C

4. $\dfrac{m_1}{m_2} = \dfrac{d}{D}$, for D

5. $A = P\left(1 + \dfrac{i}{m}\right)$, for i

For problems 6 and 7, see Example 3.

6. $A = P + Prt$, for P

7. $S = 2(lh + hw + lw)$, for w

In problems 8 through 30, solve each for x.
For problems 8 through 17, see Example 4.

8. $x + a = 5$

9. $2x - 7 = a$

10. $2ax + b = 3$

11. $\dfrac{2}{3}x - a = \dfrac{1}{3}$

12. $ax - ab = a + b$

13. $b - ax = 4b$

14. $\dfrac{x-a}{5} = b$

15. $\dfrac{x-b}{3} = \dfrac{2}{5}$

16. $a(6-x) = b$

17. $0.1x + b = 1.0$

For problems 18 through 20, see Example 5.

18. $5ax - 3 = 2ax$

19. $ax + b = bx$

20. $ax + 3 - a^2 = bx$

For problems 21 through 30, see Examples 6 and 7.

21. $a(x-b) = b(x+b)$

22. $3a - (a-x) = 2b(x+3)$

23. $\dfrac{3bx}{4} - a - b = \dfrac{x}{3}$

24. $(x-a)(x+b) = x^2 + 2ab$

25. $\dfrac{7}{x} - \dfrac{a}{3} = \dfrac{b}{2}$

26. $b(x-b+1) = x+b$

27. $(x-b)^2 - (x+b)^2 = 2$

28. $\dfrac{1}{3}(ax-b) = bx + b^2$

29. $\dfrac{3}{2}(bx+a) = \dfrac{3}{2}ax + a^2$

30. $\dfrac{5}{x} - \dfrac{b}{7} = \dfrac{1}{7}$

Practice Exercises

In problems 31 through 48, solve for the variable indicated.

31. $A = lw$, for l

32. $A = \dfrac{1}{2}bh$, for b

33. $V = lwh$, for w

34. $d = rt$, for r

35. $I = Prt$, for t

36. $V = \dfrac{1}{3}Bh$, for B

37. $\dfrac{1}{R} = \dfrac{1}{R_1} + \dfrac{1}{R_2}$, for R

38. $C = \pi d$, for d

39. $A = P + Prt$, for t

40. $C = \dfrac{5}{9}(F - 32)$, for F

41. $P = 2l + 2w$, for l

42. $V = \dfrac{1}{3}\pi r^2 h$, for h

43. $S^2 = 1 - \dfrac{a}{r}$, for r

44. $S = 2\pi rh$, for h

45. $E = IR$, for R

46. $S = \dfrac{n}{2}[2a + (n-1)d]$, for a

47. $S = \dfrac{a}{1-r}$, for a

48. $S = 2(lh + hw + lw)$, for h

In problems 49 through 54, use the formula in the problem number indicated to find each of the following.

49. The area of a rectangle is 46.7 ft² and its width is 6.6 ft. Find its length. (Use problem 31.)
50. The area of a triangle is 6 yd² and its height is 4 yd. Find the length of its base. (Use problem 32.)
51. A jogger runs 10 km in 55 minutes. What is his rate? (Use problem 34.)
52. The circumference of a circle is 21.2 m. Find the length of the diameter of the circle correct to two decimal places. Use $\pi \approx 3.14$. (Use problem 38.)
53. Change 100°C to degrees Fahrenheit. (Use problem 40.)
54. The perimeter of a rectangle is $110\dfrac{1}{4}$ ft and its width is $27\dfrac{1}{2}$ ft. Find its length. (Use problem 41.)

Sec. 4.2 Literal Equations and Formulas

In problems 55 through 80, solve each equation for x.

55. $x - b = 6$

56. $3x + 5 = a$

57. $4bx + a = 7$

58. $\dfrac{3}{4x} + b = \dfrac{1}{4}$

59. $ax + ab = b$

60. $a - bx = 6a$

61. $\dfrac{x + a}{3} = 1$

62. $\dfrac{x - a}{4} = \dfrac{3}{2}$

63. $b(2 - x) = a$

64. $0.1x - 1.0a = 2.0a$

65. $3(5 + ax) = 1$

66. $2ax + 3 = 4ax$

67. $ax - 2 = bx$

68. $a(x + b) = b(x - a)$

69. $2b - (x - a) = a(x - 2)$

70. $(x + a)^2 - (x - a)^2 = 1$

71. $bx + 4 - a^2 = ax$

72. $a(x + a - 1) = x - a$

73. $(x - b)(x + a) = x^2 + ab$

74. $\dfrac{1}{3}bx - a = ax + b$

75. $\dfrac{2}{3}bx + a + b = \dfrac{2}{5}x$

76. $\dfrac{4}{7}(ax - b) = 2 \quad x =$

77. $\dfrac{5}{6}(bx + 2a) = \dfrac{1}{6}ax + b^2$

78. $\dfrac{7}{x} + \dfrac{a}{3} = \dfrac{1}{3}$

79. $a(x - 2) = a + x$

80. $\dfrac{1}{x} + \dfrac{1}{2x} = \dfrac{b}{2}$

Challenge Problems

81. Sometimes we must be careful to note restrictions when reducing. Consider

$$x = \dfrac{(b - 2)(b + 2)}{b - 2}$$

Notice that if we reduce by dividing out the common factor $(b - 2)$, we will have *divided by zero if b* has the value 2. In cases like this we alert the formula user with the restriction $b \neq 2$.

$$x = b + 2; \quad b \neq 2$$

Solve each equation for *x*, and note any restrictions.
(a) $a(a - x) = x + 1$
(b) $2(x + 2) = b(b + x)$
(c) $(b + 2)(x - b + 3) = 0$
(d) $a(x - a) = x - 1$

82. Solve $\dfrac{x + 2}{x - a} = \dfrac{x + a}{x - 1}$ for *x*. (Be careful to note where denominators are zero.)

IN YOUR OWN WORDS . . .

83. Describe how to solve the equation $Ax + Bx = C$ for *x*.

84. What does solving a formula for a variable mean?

4.3 APPLICATIONS

In Section 4.1 we saw that equations could be used to solve real-world problems. Following certain steps will help organize our efforts. In this section we form a procedure that is very effective for solving applications.

> **A Procedure to Solve Word Problems**
>
> 1. Read the problem to determine what must be found.
> 2. Assign a variable, such as x, to represent one of the quantities to be found. (This is the "let" statement.)
> 3. Express all other quantities to be found in terms of x (or the variable chosen).
> 4. Draw a figure or picture, if possible. Label it.
> 5. Reread the problem and form an equation.
> 6. Solve the equation and find the values of all the quantities to be found.
> 7. Check the values in the original word problem. They should answer the question and make sense.
> 8. Write an answer to the original question.

Mixture

These problems contain percents. Remember that percents represent hundredths. To do arithmetic with them, we replace them with their value. That is, we would replace 7% with 0.07 or $\frac{7}{100}$.

EXAMPLE 1. Jack Glover wishes to add enough 50% antifreeze solution to 16 gallons of a 5% antifreeze solution to obtain a 20% antifreeze solution. How much of the 50% solution should Jack add?

Solution:

Step 1 Read the problem and determine what is to be found.

The problem asks for the number of gallons of 50% antifreeze solution to be added.

Step 2 Assign a variable to represent the quantity to be found.

Let x be the number of gallons of 50% solution added.

Step 4 Draw a figure.

Mixture problems become clearer with a picture.

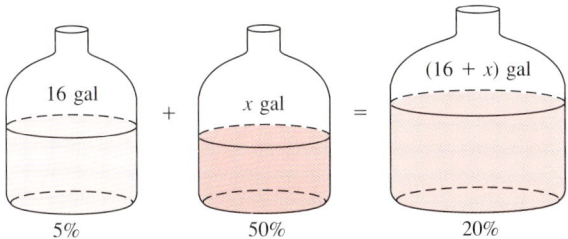

The amount of antifreeze in the first solution plus the amount of antifreeze in the solution added must equal the amount of antifreeze in the final solution. $\Big($The amount of antifreeze in 16 gallons of a 5% solution is $16 \cdot \frac{5}{100}$ gallons.$\Big)$

(continued)

Sec. 4.3 Applications

Step 5 Read the problem again and form an equation.

$$16 \cdot \frac{5}{100} + x \cdot \frac{50}{100} = (16 + x) \cdot \frac{20}{100}$$

Antifreeze in + Antifreeze in = Antifreeze in
5% solution   50% solution   20% solution

Step 6 Solve the equation.

Multiply both sides by 100 to clear fractions.

$$16 \cdot 5 + x \cdot 50 = (16 + x) \cdot 20$$

$$80 + 50x = 320 + 20x$$

$$30x = 240$$

$$x = 8$$

Step 7 Check this value in the original word problem to see if it makes sense and answers the question.

Adding 8 gallons of antifreeze answers the question and makes sense.

Step 8 Write an answer to the question.

Jack should add 8 gallons of the 50% solution.

EXAMPLE 2. A chemist has 10 gallons of a 3% alcohol solution. How many gallons of pure alcohol should she add to the 10 gallons to obtain a new solution that is 4% alcohol?

Solution:

Determine what is to be found and assign a variable to represent this quantity. Let x be the number of gallons of pure alcohol added. Draw a figure.

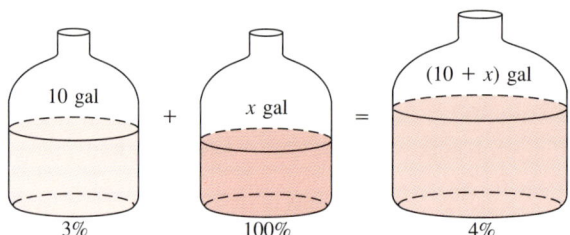

Read the problem again and form an equation.

The amount of alcohol in the original mixture plus the amount of alcohol added must equal the amount of alcohol in the final solution.

$$10 \cdot \frac{3}{100} + x \cdot \frac{100}{100} = (10 + x) \cdot \frac{4}{100}$$

Solve the equation.

We multiply both sides by 100 to clear fractions.

$$10 \cdot 3 + x \cdot 100 = (10 + x) \cdot 4$$

$$30 + 100x = 40 + 4x$$

$$96x = 10$$

$$x = \frac{10}{96} = \frac{5}{48}$$

Check this value in the original word problem to see if it makes sense and answers the question.

Adding $\frac{5}{48}$ of a gallon makes sense and answers the question.

Write an answer to the question.

The chemist should add $\frac{5}{48}$ gallon of pure alcohol. ☐

Work

EXAMPLE 3. Pipe A alone can fill a tank in 2 hours, and pipe B alone can fill the same tank in 3 hours. If both pipes run at the same time, how long will it take to fill the tank?

Solution:

Let x be the number of hours it takes both pipes, working together, to fill the tank. As pipe A alone fills the tank in 2 hours, it must fill $\frac{1}{2}$ of a tank in 1 hour. So in x hours it must fill $x \cdot \frac{1}{2}$ of a tank. Similarly, pipe B must fill $\frac{1}{3}$ of a tank in 1 hour or $x \cdot \frac{1}{3}$ of a tank in x hours. Both pipes working together for x hours fill one tank, so we have the equation

$$\frac{x}{2} + \frac{x}{3} = 1$$

We multiply both sides by 6 to clear fractions.

$$3x + 2x = 6$$

$$5x = 6$$

$$x = \frac{6}{5}$$

As $\frac{6}{5}$ of an hour answers the question and makes sense, we write the solution:

It takes $\frac{6}{5}$ of an hour for both pipes to fill the tank. ☐

Sec. 4.3 Applications

Distance

The next example is a distance problem. We must remember the formula

$$\text{Distance} = \text{Rate} \times \text{Time}$$

which we usually write $D = RT$. Often it is useful to solve this equation for rate or time, giving us the three forms of the distance formula,

$$D = RT \qquad R = \frac{D}{T} \qquad T = \frac{D}{R}$$

EXAMPLE 4. Two trains, 780 miles apart, are heading toward each other at rates of 85 mph and 110 mph. When will they crash?

Solution:

Let x be the number of hours until they crash.

A picture or diagram is very helpful in solving this type of problem. Usually, there are two pieces in a distance problem. In this problem there are the two trains.

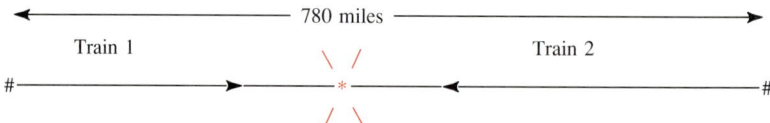

Next we find D, R, and T for each train. One of these should be found from the "let statement" and one should be given in the problem statement. In this problem the let statement gives the time for each train, as they each travel x hours before they crash. The problem statement gives both rates.

	DISTANCE	RATE	TIME
Train 1		85	x
Train 2		110	x

We find the third member of the DRT triple using the appropriate form of the distance formula. In this case, $D = RT$.

	DISTANCE	RATE	TIME
Train 1	$85x$	85	x
Train 2	$110x$	110	x

With D, R, and T for each train, we must find an equation. In this problem the trains started out 780 miles apart and one went $85x$ miles and the other $110x$ miles. Thus

$$85x + 110x = 780$$
$$195x = 780$$
$$x = 4$$

The trains will crash in 4 hours, answers the question and makes sense.   □

Chap. 4 Using Linear Equations in One Variable

EXAMPLE 5. Frank McComb rides his bike from home to the stadium at a constant rate of 6 mph. He returns along the same route at a rate of 10 mph. If the round trip took $1\frac{3}{5}$ hours, how far is the stadium from his house?

Solution:

Let x be the distance from Frank's house to the stadium (in miles).
There are two pieces in this problem, going and returning.

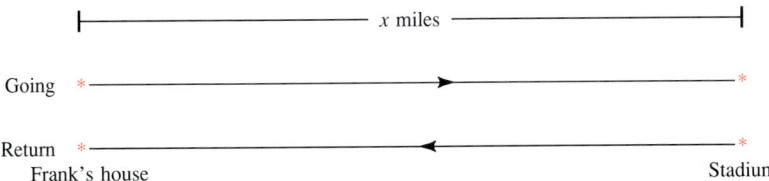

Find D, R, and T for each piece. The distance comes from the let statement and each rate is given in the problem itself.

	DISTANCE	RATE	TIME
Going	x	6	
Returning	x	10	

The missing part of the chart can be found using the formula $T = \dfrac{D}{R}$.

	DISTANCE	RATE	TIME
Going	x	6	$\dfrac{x}{6}$
Returning	x	10	$\dfrac{x}{10}$

To find an equation, we note that the round trip took $1\frac{3}{5}$ hours. So the time going added to the time returning is $1\frac{3}{5}$ hours or $\frac{8}{5}$ hours.

$$\frac{x}{6} + \frac{x}{10} = \frac{8}{5}$$

$$30\left(\frac{x}{6} + \frac{x}{10}\right) = 30 \cdot \frac{8}{5} \qquad \text{Clear fractions.}$$

$$5x + 3x = 48$$

$$8x = 48$$

$$x = 6$$

The stadium is 6 miles from Frank's house. □

Sec. 4.3 Applications

Interest

Investment or interest problems usually use the simple interest formula

$$I = PRT$$

where I is the amount of interest, P is the principal, R is the interest rate, and T is the time. It is *important* that the units of time be the same in both R and T.

EXAMPLE 6. Charles invested part of his savings at 7% interest and the remainder at 8%. In one year he receives $1415 in interest. If his total savings were $19,000, how much money did he have invested at each rate?

Solution:

Let x be the amount he has invested at 7%. Then he has $19{,}000 - x$ invested at 8%. The total interest on the $19,000 is the sum of the interest on each of the two parts, so using the formula $I = PRT$, we form the equation

$$1415 = x \cdot \frac{7}{100} \cdot 1 + (19{,}000 - x) \cdot \frac{8}{100} \cdot 1$$

We clear fractions and solve.

$$141{,}500 = 7x + (19{,}000 - x)8$$
$$141{,}500 = 7x + 152{,}000 - 8x$$
$$-10{,}500 = -x$$
$$10{,}500 = x$$

Thus $19{,}000 - x$ is $19{,}000 - 10{,}500$ or 8500.

Charles had $10,500 invested at 7% and $8500 at 8%.

Geometry

EXAMPLE 7. In a right triangle one leg is 3 cm longer than the other leg. If the shorter leg is increased by 6 cm, the area is increased by 24 cm². Find the length of each leg in the original triangle.

Solution:

We are to find the lengths of the two legs of a given right triangle.

Let x be the length of the shorter leg. Then $x + 3$ is the length of the other leg. Reread the problem and see that it involves the area of the original triangle and the area of a modified triangle. The area of a right triangle is one-half the product of the lengths of the two legs.

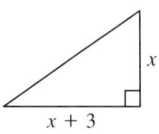

Area $= \dfrac{1}{2}x(x + 3)$

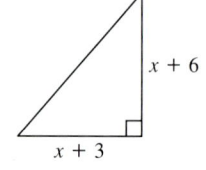

Area $= \dfrac{1}{2}(x + 6)(x + 3)$

Chap. 4 Using Linear Equations in One Variable

The problem also states that the area of the *new* triangle is 24 cm² greater than the area of the *original* triangle. So if we add 24 to the area of the original triangle, it will equal the area of the new triangle.

$$\frac{1}{2}x(x+3) + 24 = \frac{1}{2}(x+6)(x+3)$$

This is the equation we were seeking.

Multiply by 2 to clear fractions, then solve.

$$2\left[\frac{1}{2}x(x+3) + 24\right] = 2\left[\frac{1}{2}(x+6)(x+3)\right]$$

$$x(x+3) + 48 = (x+6)(x+3)$$

$$x^2 + 3x + 48 = x^2 + 9x + 18$$

$$3x + 48 = 9x + 18$$

$$30 = 6x$$

$$5 = x$$

The shorter leg must be 5 cm and the longer leg 5 + 3 or 8 cm.

The area of the original triangle is $\frac{1}{2} \cdot 5 \cdot 8$ or 20 cm², while the area of the *new* triangle is $\frac{1}{2} \cdot 11 \cdot 8$ or 44 cm². As 44 is 24 greater than 20, these dimensions make sense and answer the question. The lengths of the legs of the original triangle are 5 cm and 8 cm.

PROBLEM SET 4.3

Warm-ups

For problems 1 through 4, see Examples 1 and 2.

1. How many gallons of a 60% solution should Christine add to 30 gallons of a 10% solution to produce a 40% solution?

2. How many gallons of 50% lemon juice should Orval add to 20 gallons of a 5% lemonade punch to produce a 10% lemonade punch?

3. A chemist has 10 gallons of a 30% alcohol solution. How many gallons of pure alcohol should he add to obtain a mixture of 40% alcohol?

4. Linda has 10 gallons of a 5% iodine solution but wishes to lower the concentration to 2.5%. How many gallons of water should she add?

For problems 5 through 7, see Example 3.

5. Pipe A alone can fill the "Little Squirt" pool in 2 hours, and pipe B alone can fill the pool in 4 hours. If both pipes run at the same time, how long will it take to fill the pool?

6. Kimberly leaves the hot and cold water faucets on her kitchen sink turned on full blast. The hot water faucet alone can fill the sink in 2 minutes and the cold faucet alone can fill it in 3 minutes. The drain is open and it can drain the sink in 4 minutes. How long will it take for the sink to fill up?

7. Rowland can mow his yard in 21 minutes. It takes Don 28 minutes to mow the same lawn. How long should it take both of them, working together, to mow the yard?

Sec. 4.3 Applications

For problems 8 through 11, see Examples 4 and 5.

8. A TWA 747 flying 720 km/hr travels 161 km farther than a Sabina 737 flying 650 km/hr. If departure and arrival times were the same for both planes, how far did each plane fly?

9. Frank, who lives in College Park, starts walking toward Chuck's house in Stone Mountain, at a rate of 2 mph. Chuck starts at the same time toward Frank's house at a rate of 3 mph. If their houses are 20 miles apart, in how many hours will they meet? How far has each walked?

10. Martin sets off for Tampa, a distance of 244 miles, in his Trihull Special at a constant rate. After 4 hours he has to reduce his speed by 12 mph due to a malfunctioning spider valve. If he finishes the trip in 3 hours, what was his original speed?

11. Lois and Joan walk in opposite directions, Lois walking at the rate of $6\frac{2}{3}$ mph and Joan walking 6 mph. In how many hours will they be 38 miles apart?

For problems 12 through 15, see Example 6.

12. Peggy's Card Shop invests $25,000 for 1 year. Part of it is to be invested at 5.5% and the rest at 6.5%. If a total of $1470 in interest is earned, how much money is invested at each rate?

13. Jane Batten wishes to invest $10,000 for 1 year. Part of it is to be invested at 5% and the rest at 6%. If $560 in interest is earned, how much money is invested at each rate?

14. Marion has $6500 invested at 9%. How much should she invest at 10.5% to yield 10% on her investments?

15. Alice needs a return of 12% on her investments. She has $14,000 in bonds that yield 8% and $16,000 to invest. What yield should she get on the $16,000?

For problems 16 through 19, see Example 7.

16. If the sides of a square are increased by 4 cm, the area is increased by 40 cm^2. Find the length of a side of the original square.

17. In a right triangle the length of one leg is twice the length of the other leg. If each of the legs is increased by 2 m, the resulting triangle has an area of 17 m^2 more than the area of the original triangle. Find the length of each leg in the original triangle.

18. The length of a rectangle is twice its width. If its length and width are each decreased by 2 cm, its area is decreased by 38 cm^2. Find the original dimensions of the rectangle.

19. The length of a rectangle is 5 ft. If the length and width are each increased by 2 ft, the area is increased by 24 ft^2. Was the original rectangle a square?

Practice Exercises

20. How many liters of a 65% solution should be added to 35 liters of a 12% solution to produce a 15% solution?

21. Virginia Carson has 30 ounces of 15% silver alloy that she wishes to upgrade to 20% silver. How many ounces of 50% silver alloy should she add?

22. How many gallons of pure alcohol should be added to 20 gal of an 8% alcohol solution to produce a 10% solution?

23. Virginia Parks has 10 milliliters (ml) of a 6% iodine solution, but wishes to lower the concentration to 3%. How many milliliters of water should she add?

24. How many grams (g) of an amalgam containing 20% silver should be added to 100 g of an amalgam containing 32% silver to obtain an amalgam containing 28% silver?

25. How much pure lead should a glass manufacturer add to 532 kg of a molten mixture of 20% lead crystal to make a batch of 24% lead crystal?

26. A vat contains 20 liters of a 40% detergent solution. How much should be drained and replaced by pure detergent to obtain 20 liters of a 64% solution?

27. Pipe A alone can fill a tank in 5 hours and pipe B alone can fill the tank in 6 hours. If both pipes run at the same time, how long will it take to fill the tank?

28. Pipes A and B fill a tank. Pipe A alone can fill it in 2 hours and pipe B alone can fill it in 5 hours. The drain can empty it in 4 hours. If both pipes and the drain are open, how long will it take to fill the tank?

29. Jose can fill a sound tire with air in 3 minutes. An irate motorist brings Jose a tire that will go flat in 5 minutes if filled with air. How many minutes will it take Jose to fill the faulty tire with air so that he can locate the leak?

30. J.R. can dig a certain length of ditch in 2 hours working alone, while John takes 3 hours to dig the same length of ditch if he is working alone. How long does it take John and J.R. to dig the length of ditch working together?

31. The cold water tap can fill a sink to the overflow level in 3 minutes. The hot water tap can fill the sink to the same level in 4 minutes. How long will it take both taps, running together, to fill a half-full sink to the overflow level?

32. Pipe A alone can fill a tank in 2 hours, pipe B alone can fill the same tank in 3 hours, and pipe C alone can fill it in 6 hours. How long will it take all three pipes working together to fill the tank?

33. The left valve can fill a tank in 5 hours. The right valve can fill the same tank in 10 hours. The drain can empty a full tank in $2\frac{1}{2}$ hours. If the tank is full and both valves are opened and the drain is opened, how long will it take to empty the tank?

34. If Barney's boat were full of water, it would take the bilge pump 2 hours to pump it dry. Barney could bail the water out in 3 hours if the pump were not working. A leak develops that takes 12 hours to fill the boat. How long does it take Barney and the bilge pump working together to empty the leaking boat so that Barney can repair it?

35. Jo can chop a batch of slaw in 30 minutes working alone, while JoAnne can chop the same batch in 42 minutes if she works alone. If they work together, they each work half as fast, due to the intensity of their conversation. How long does it take Jo and JoAnne, working together, to chop a batch of slaw?

36. Two planes leave Daytona at the same time traveling in opposite directions. If they are 2480 km apart after 2 hours and one plane travels 40 km/hr faster than the other plane, what are their average rates?

37. If Henry drives his old truck from home to Clarks Hill at 50 mph and returns immediately at 60 mph, his total elapsed driving time is 5.5 hours. How far is it from Henry's home to Clarks Hill?

38. Charlotte sets off on her bicycle at a rate of 15 mph to the local K-mart, a distance of 5 miles. After traveling part way, a low tire forces her to reduce her speed. She pedals at a rate of 10 mph for the rest of the trip. If the total trip takes 24 minutes ($\frac{2}{5}$ of an hour), how far did Charlotte pedal at 15 mph?

39. The Midnight Express leaves the station at midnight traveling east. At 2 A.M. a slow freight train leaves the same station traveling west. At 4 A.M. the distance traveled by the Midnight Express is 210 miles more than the distance traveled by the slow freight. If the average speed of the Midnight Express is twice the average speed of the slow freight, how far has the slow freight traveled at 4 A.M.?

40. Carolyn and Irma jog the same track. Carolyn jogs at a rate of 0.15 mile/min and Irma at 0.10 mile/min. If Carolyn starts 1.5 minutes after Irma starts, how long will it take Carolyn to overtake Irma?

41. John and Jim run the same track. John runs at the rate of $\frac{1}{6}$ mile/min and Jim at $\frac{1}{8}$ mile/min. Jim starts first and in 3 minutes John overtakes Jim. How long after Jim started did John start?

42. Mary and Susan jog together. Mary jogs at the rate of $\frac{1}{5}$ mile/min and Susan at the rate of $\frac{1}{6}$ mile/min. If Mary starts 1 minute after Susan starts, how long will it take Mary to overtake Susan?

43. Leonard White wishes to invest $12,000 for 1 year, part of it at 5.5% and the rest at 6.5%. If a total of $700 in interest is earned, how much money has Leonard invested at each rate?

44. A total of $33,500 is to be invested, some of it at 6.5% and the remainder at 7.3%. If a total interest of $2273.50 is earned in 1 year, how much money is invested at each rate?

45. Ken and Margaret won a lottery prize of $100,000, after taxes. They invest part of it in a second mortgage paying 18% and the rest in utility stock paying 12%. If their total yearly income from these investments is $16,200, how much was invested at each rate?

46. Beverly wants to get a return of 12% on her total investments. If she has $15,000 invested at 10%, how much must she invest at 15% to realize her goal?

47. Jack wishes to divide $100,000 between AAAA bonds that yield 8.25% and AA bonds that yield 9.25%. How should he divide his investment to obtain an annual return of $8500?

48. Peggy has $12,000 invested at 8%. How much of this investment should she convert to another which yields 8.5% to increase her yearly return by $10?

49. Gerald invested $240,000 in two real estate deals last year. He made a profit of 20% on one deal, but lost 5% on the second one. If Gerald's net profit was 13.75%, how much did he invest in each deal?

50. Barbara has invested $25,000 in two accounts. One is a money market account paying 9.5% annual interest and the other is a CD that pays 12% yearly. If Barbara receives $2700 from these two investments, how much has she invested in each?

51. If the sides of a square are increased by 2 in., the area is increased by 48 in.^{2}. Find the length of a side of the square.

52. One leg in a right triangle is twice the other leg. If each leg is decreased by 4 ft, the area of the resulting triangle is 28 ft^2 less than the area of the original triangle. Find the length of each leg in the original triangle.

53. Glenn gave away tickets to the Redskins game. Bonnie took half of them and Jackie took one-fifth of them. If Bonnie has three more tickets than Jackie, how many tickets did Glenn give away?

54. Find three consecutive even integers such that one-third of the first is 2 more than the second subtracted from the third.

55. The sum of one-half a number and two-thirds of the number is 3 more than the number. Find the number.

Sec. 4.3 Applications

56. Mom left a plate of brownies on the kitchen table. Amanda ate one-third of them and Chuck ate one-half of them. If Chuck ate one more than Amanda, how many brownies were on the table?

57. When one-fifth of a number is subtracted from one-third of the same number, the result is 1 more than one-tenth of the number. Find the number.

58. A 1-acre field is in the shape of a right triangle. If the shortest side measures 110 yd, how many yards of fencing are needed to fence the two legs? (Hint: 1 acre = 4840 yd^2.)

Challenge Problems

59. A boat travels 20 km upstream from the shack to Long Bridge and then returns to the shack. The boat can travel at a rate of 4 km/hr in still water. It takes the boat twice as long to make the trip upstream as downstream. What is the speed of the current?

60. A model airplane flies 5 miles from the treehouse to the barn against the wind and back to the tree house with the wind. The wind is blowing at the rate of 1 mile/hr and it takes twice as long to fly against the wind as it does to fly with the wind. What is the airplane's speed in still air?

▮▮▮ IN YOUR OWN WORDS . . .

61. Why do word problems seem difficult?

62. What should we try to determine when we first read a word problem?

▮ 4.4 FRACTIONAL EQUATIONS

In Section 4.1 linear equations in one variable were solved. Some of the coefficients were fractions, but none of the equations contained variables in denominators. An equation that contains variables in a denominator is called a *fractional equation*. In this section we learn how to solve some fractional equations.

When studying rational expressions in Section 3.1, we noted that rational expressions are undefined when a denominator has a value of zero. The same situation exists with fractional equations. For example, in the equation

$$\frac{1}{x-3} = 3$$

the fraction $\frac{1}{x-3}$ is not defined when x is 3. Thus 3 *cannot* be in the solution set of the equation, because it cannot make the equation true. Remember this as we solve fractional equations.

To solve a fractional equation, use the two tools from Section 4.1. The general strategy is to clear the equation of fractions, solve the resulting equation, then check the proposed solution(s) to see if they make any denominator have a value of zero.

To clear the equation above of fractions, multiply both sides by the LCD, $(x - 3)$.

$$\frac{1}{x-3} = 3$$

$$(x-3)\frac{1}{x-3} = 3(x-3)$$

$$1 = 3x - 9$$

$$10 = 3x$$

$$\frac{10}{3} = x$$

Since $\dfrac{10}{3}$ does not make a denominator in the original equation have the value zero, the solution set is

$$\left\{\dfrac{10}{3}\right\}$$

EXAMPLE 1. Solve $\dfrac{1}{x} + \dfrac{1}{2x} = \dfrac{3}{2}$.

Solution:

To clear the fractions, multiply both sides of the equation by $2x$, the LCD of all the denominators in the equation.

$$2x\left(\dfrac{1}{x} + \dfrac{1}{2x}\right) = 2x\left(\dfrac{3}{2}\right)$$

$$2x \cdot \dfrac{1}{x} + 2x \cdot \dfrac{1}{2x} = 2x \cdot \dfrac{3}{2} \qquad \text{Distributive Property}$$

$$2 + 1 = 3x$$

$$3 = 3x$$

$$1 = x$$

Since 1 does not make any denominator in the original equation have the value zero, write the solution set,

$$\{1\}$$

A Procedure for Solving Fractional Equations

1. Multiply both sides by the least common denominator of all denominators in the equation.
2. Solve the resulting equation for possible solutions.
3. Check to see if any of the possible solutions make a denominator have the value zero. If so, **do not** include them in the solution set.
4. Write the solution set.

EXAMPLE 2. Solve $\dfrac{7}{x-5} + 2 = \dfrac{x+3}{x-5}$.

Solution:

We multiply both sides by the LCD, which is $(x - 5)$.

$$(x-5)\left(\dfrac{7}{x-5} + 2\right) = (x-5) \cdot \dfrac{x+3}{x-5} \qquad \text{(continued)}$$

Sec. 4.4 Fractional Equations

Now apply the distributive property.

$$(x-5)\frac{7}{x-5} + (x-5)2 = (x-5)\frac{x+3}{x-5}$$

$$7 + 2(x-5) = x+3$$

$$7 + 2x - 10 = x+3$$

$$2x - 3 = x + 3$$

$$x = 6$$

Since 6 does not make any denominator have the value 0, we write the solution set,

$$\{6\}$$

When multiplying both sides of an equation by a number, the distributive property allows us to multiply *each* term in the equation by the number. This is a convenient shortcut when clearing an equation of fractions.

EXAMPLE 3. Solve $\dfrac{2}{x-2} - \dfrac{1}{x+1} = \dfrac{5}{x^2 - x - 2}$.

Solution:

We must factor so that we can find the LCD.

$$\frac{2}{x-2} - \frac{1}{x+1} = \frac{5}{(x-2)(x+1)} \qquad \text{Factor.}$$

Multiply each term by the LCD, $(x-2)(x+1)$.

$$(x-2)(x+1) \cdot \frac{2}{x-2} - (x-2)(x+1) \cdot \frac{1}{x+1}$$

$$= (x-2)(x+1) \cdot \frac{5}{(x-2)(x+1)}$$

$$2(x+1) - (x-2) = 5$$

Carefully remove the parentheses!

$$2x + 2 - x + 2 = 5 \qquad \text{Distributive Property}$$

$$x + 4 = 5$$

$$x = 1$$

Since 1 does not make a denominator in the original equation have a value of 0, 1 is the solution.

$$\{1\}$$

Be Careful!

EXAMPLE 4. Solve $\dfrac{2}{x+2} - \dfrac{1}{x} = \dfrac{-4}{x(x+2)}$.

Solution:

The LCD is $x(x+2)$. Multiply each term by $x(x+2)$.

$$x(x+2) \cdot \dfrac{2}{x+2} - x(x+2) \cdot \dfrac{1}{x} = x(x+2) \cdot \dfrac{-4}{x(x+2)}$$

$$2x - (x+2) = -4 \qquad \text{Multiplication}$$
$$x - 2 = -4$$
$$x = -2$$

Since -2 makes a denominator in the original equation have the value zero, it cannot be in the solution set. Thus the solution set is

$$\emptyset$$

EXAMPLE 5. Solve $\dfrac{3}{x-2} - \dfrac{2}{4-x^2} = \dfrac{1}{x+2}$.

Solution:

To compute the LCD we need to factor.

$$\dfrac{3}{x-2} - \dfrac{2}{(2-x)(2+x)} = \dfrac{1}{x+2}$$

We replace $(2-x)$ by $-(x-2)$.

$$\dfrac{3}{x-2} - \dfrac{2}{-(x-2)(2+x)} = \dfrac{1}{x+2}$$

$$\dfrac{3}{x-2} + \dfrac{2}{(x-2)(2+x)} = \dfrac{1}{x+2}$$

Now we see that the LCD is $(x-2)(x+2)$. We multiply each term by the LCD.

$$3(x+2) + 2 = 1(x-2)$$
$$3x + 6 + 2 = x - 2 \qquad \text{Distributive Property}$$
$$2x = -10$$
$$x = -5$$

Since -5 does not make a denominator in the original equation have the value zero, we write the solution set,

$$\{-5\}$$

Word problems can have fractional equations as models.

Sec. 4.4   Fractional Equations

EXAMPLE 6. Mary Kay drives 1 mile in the same amount of time that Lettie drives 1.5 miles. If Lettie's rate is 10 mph more than Mary Kay's rate, how fast does Mary Kay drive?

Solution:

Let r be Mary Kay's rate in mph. Then $r + 10$ is Lettie's rate in mph.

We make a chart. Using the distance–rate–time formula, $t = \dfrac{d}{r}$. We complete the chart.

	D	R	T
Mary Kay	1	r	
Lettie	1.5	$r + 10$	

	D	R	T
Mary Kay	1	r	$1/r$
Lettie	1.5	$r + 10$	$1.5/(r + 10)$

Mary Kay's time is the same as Lettie's time. We form an equation and solve.

$$\frac{1}{r} = \frac{1.5}{r + 10}$$

Multiply by $r(r + 10)$.

$$r(r + 10) \cdot \frac{1}{r} = r(r + 10) \cdot \frac{1.5}{r + 10}$$

$$r + 10 = 1.5r$$

$$10 = 0.5r$$

$$20 = r \qquad \text{Divide by 0.5.}$$

20 does not make a denominator have a value of 0 and 20 makes sense and will answer the question.

Mary Kay drives at a rate of 20 mph. □

EXAMPLE 7. The reciprocal of twice a number is equal to the reciprocal of one more than the number. Find the number.

Solution:

Let x be the number.

Form an equation and solve.

$$\frac{1}{2x} = \frac{1}{x + 1}$$

$$2x(x + 1) \cdot \frac{1}{2x} = 2x(x + 1) \cdot \frac{1}{x + 1} \qquad \text{Multiply by LCD.}$$

$$x + 1 = 2x$$

$$1 = x$$

1 does not make a denominator in the equation have value 0. 1 is the solution. The number is 1. □

EXAMPLE 8. The ratio of wins to losses for the Texas Rangers in one season was 7 to 4. How many games did they lose if they played 143 games that season?

Solution:

Let x be the number of games lost. Since a total of 143 games were played, the number of games won is $143 - x$.

We represent a ratio of 7 to 4 as $\dfrac{7}{4}$.

$$\frac{7}{4} = \frac{\text{wins}}{\text{losses}}$$

$$\frac{7}{4} = \frac{143 - x}{x}$$

$$4x \cdot \frac{7}{4} = 4x \cdot \frac{143 - x}{x} \quad \text{Multiply by } 4x.$$

$$7x = 4(143 - x)$$

$$7x = 572 - 4x$$

$$11x = 572$$

$$x = 52$$

The Rangers lost 52 games.

PROBLEM SET 4.4

Warm-ups

In problems 1 through 27, find the solution set. For problems 1 through 8, see Example 1.

1. $\dfrac{1}{x} + 2 = 3$
2. $\dfrac{2}{x} + 3 = 5$
3. $\dfrac{3}{x} - 1 = \dfrac{1}{2}$
4. $\dfrac{7}{s} - 2 = \dfrac{1}{3}$
5. $\dfrac{1}{2x} + \dfrac{1}{x} = \dfrac{1}{2}$
6. $\dfrac{1}{3x} - \dfrac{1}{x} = \dfrac{1}{2}$
7. $\dfrac{2}{3t} + \dfrac{1}{4} = \dfrac{3}{4t}$
8. $\dfrac{1}{5} + \dfrac{7}{10x} = \dfrac{3}{5x}$

For problems 9 through 14, see Example 2.

9. $\dfrac{3}{x-1} + 2 = \dfrac{5}{x-1}$
10. $\dfrac{4}{x+1} - 3 = \dfrac{-2}{x+1}$
11. $\dfrac{5}{2x+3} = \dfrac{1}{2x+3} + 1$
12. $\dfrac{4}{3x-2} = \dfrac{-4}{3x-2} - 1$
13. $\dfrac{1.4}{x} + \dfrac{3.2}{2x} = -1.2$
14. $\dfrac{0.3}{x} + 1.2 = \dfrac{1}{0.2x}$

For problems 15 through 21, see Example 3.

15. $\dfrac{1}{x+2} + \dfrac{1}{x} = \dfrac{12}{x^2+2x}$
16. $\dfrac{2}{x} - \dfrac{1}{x+1} = \dfrac{3}{x^2+x}$
17. $\dfrac{2}{x-3} = \dfrac{4}{x+4}$
18. $\dfrac{3}{x-1} - \dfrac{1}{x+3} = \dfrac{8}{x^2+2x-3}$
19. $\dfrac{6}{x-3} - \dfrac{3}{x+2} = \dfrac{12}{x^2-x-6}$
20. $\dfrac{2}{x-5} + \dfrac{1}{x-1} = \dfrac{2}{x^2-6x+5}$
21. $\dfrac{x-3}{x+4} = \dfrac{x-2}{x+3}$

For problems 22 through 24, see Example 4.

22. $\dfrac{1}{z+1} + \dfrac{1}{z-1} = \dfrac{2}{z^2-1}$

23. $\dfrac{-4}{5(x+2)} = \dfrac{3}{x+2}$

24. $\dfrac{3}{x-2} + \dfrac{1}{x-5} = \dfrac{-9}{x^2-7x+10}$

For problems 25 through 27, see Example 5.

25. $\dfrac{4}{1-x} + \dfrac{1}{x} = \dfrac{5}{x^2-x}$

26. $\dfrac{1}{x} + \dfrac{2}{1-x} + \dfrac{1}{x+1} = 0$

27. $\dfrac{3}{2+x} + \dfrac{2}{2-x} = \dfrac{2x}{x^2-4}$

For problems 28 through 30, see Example 6.

28. Jeri swims 10 miles in the same amount of time as Douglas swims 12 miles. If Douglas's rate is 5 mph faster than Jeri's rate, find Jeri's rate.

29. John Phillip averages 2 mph faster than Kenneth when riding a bicycle. John Phillip gives Kenneth a 7-mile headstart, then catches him after going 56 miles. What rate does John Phillip average on his bike?

30. A United jet flies 400 miles in the same amount of time that a Delta jet flies 320 miles. If the rate of the Delta plane is 50 mph slower than the United plane, find the speed of the United jet.

For problems 31 and 32, see Example 7.

31. One number is twice another number. If the sum of their reciprocals is $\dfrac{1}{2}$, find the two numbers.

32. The reciprocal of 1 more than a number is twice the reciprocal of the number. Find the number.

For problems 33 through 36, see Example 8.

33. The ratio of boys to girls in the algebra class is 4 to 5. If there are 32 boys in the class, how many girls are in the class?

34. The ratio of bluebirds to woodpeckers in a certain area is 3 to 7. If 600 of both kinds of birds have been counted, how many of them are bluebirds?

35. If 1 inch is 2.54 centimeters, how many centimeters are in 1 foot?

36. If 1 kilogram is 2.2 pounds, how many pounds are in 5 kilograms?

Practice Exercises

In problems 37 through 66, find the folution set.

37. $\dfrac{1}{x} + \dfrac{1}{3} = \dfrac{1}{4}$

38. $\dfrac{1}{x} - \dfrac{2}{3} = \dfrac{1}{2}$

39. $\dfrac{2}{y} - \dfrac{1}{5} = -\dfrac{1}{4}$

40. $\dfrac{1}{x} + \dfrac{1}{3} = -\dfrac{1}{6}$

41. $\dfrac{2}{x} - \dfrac{3}{4} = \dfrac{1}{x}$

42. $\dfrac{5}{x} - \dfrac{2}{7} = \dfrac{1}{x}$

43. $\dfrac{1}{x+1} - \dfrac{2}{x} = 0$

44. $\dfrac{2}{x-1} + \dfrac{1}{x} = 0$

45. $\dfrac{4}{2-x} - 1 = \dfrac{3}{2-x}$

46. $\dfrac{7}{2x-1} - 1 = \dfrac{9}{2x-1}$

47. $\dfrac{3}{x} = \dfrac{5}{x+4}$

48. $\dfrac{7}{x} = \dfrac{8}{2x-6}$

49. $\dfrac{14.3}{w} - \dfrac{3.1}{2w} = 1.1$

50. $\dfrac{0.5}{x} + 1.2 = \dfrac{1}{1.2x}$

51. $\dfrac{16.4}{x} + 12.2 = 1.4$

52. $\dfrac{2}{z-2} + \dfrac{3}{z} = \dfrac{1}{2-z}$

53. $\dfrac{4}{x} + \dfrac{3}{5-x} = \dfrac{-20}{x^2-5x}$

54. $\dfrac{1}{x+3} = \dfrac{x}{x+1} - 1$

55. $\dfrac{3}{2(x+2)} + \dfrac{3}{(x+2)^2} = 0$

56. $\dfrac{w}{w+2} = 1 - \dfrac{1}{w-3}$

57. $\dfrac{2}{x+1} + \dfrac{3}{x+2} = \dfrac{-3}{x^2+3x+2}$

58. $\dfrac{x}{x-3} = \dfrac{x+1}{x-1}$

59. $\dfrac{u-2}{u+3} = \dfrac{u-1}{u+2}$

60. $\dfrac{x}{x+4} - \dfrac{2}{x-3} = 1$

61. $\dfrac{3}{x-1} + x = \dfrac{x^2}{x-1}$

62. $\dfrac{3}{x-1} - 1 + \dfrac{x^2-5}{x^2+2x-3} = 0$

63. $\dfrac{1}{x-5} + \dfrac{x}{25-x^2} = 0$

186   Chap. 4   Using Linear Equations in One Variable

64. $\dfrac{x-2}{x+1} = \dfrac{x+2}{x-1}$

65. $\dfrac{3}{x-1} - \dfrac{1}{x+2} = \dfrac{9}{x^2+x-2}$

66. $\dfrac{2}{x-2} + \dfrac{1}{x-5} = \dfrac{-6}{x^2-7x+10}$

67. Irene jogs 6 miles in the same amount of time that Joyce jogs 8 miles. If Irene's rate is 2 mph slower than Joyce's rate, find Joyce's rate.

68. Manuel lives 10 miles from Riverfront Stadium and Maria lives 16 miles away from the stadium. They plan to meet at the stadium before the game. If each travels at the same rate, Manuel will arrive $\frac{1}{2}$ hour earlier than Maria. How long will it take Maria to get to the stadium?

69. Bill Anderson drives from home to work on Memorial Drive every morning. This is a distance of 15 miles. He uses I-285 to come home. Although this is 10 miles farther, he can average 20 mph faster. If he drives to work in the same amount of time that he drives home from work, find his speed going to work.

70. One number is three times another number. If the sum of their reciprocals is $\dfrac{1}{4}$, find the two numbers.

71. The reciprocal of one less than a number is twice the reciprocal of the number. Find the number.

72. The reciprocal of three times a number is equal to the reciprocal of 2 more than the number. Find the number.

73. The ratio of Democrats to Republicans in the state legislature is 3 to 8. If 45 of them are Democrats, how many Republicans are there?

74. The ratio of pine trees to oak trees in the park is 5 to 9. If 42,000 of both kinds of trees have been counted, how many of them are pine trees?

75. If 1 gallon is 3.8 liters, how many liters are in 5 gallons?

76. If 1 foot is 30.5 centimeters, how many centimeters are in 10 yards?

Challenge Problems

In problems 77 through 80, find the solution set; a, b and c are constants.

77. $\dfrac{x}{x+4} - 1 = \dfrac{x}{x+4}$

78. $\dfrac{1}{x} + \dfrac{1}{x+2} = \dfrac{2}{x+1}$

79. $\dfrac{a}{x} + \dfrac{b}{2x} = c$

80. $\dfrac{a}{x-1} - \dfrac{2a}{x+1} = 0$

IN YOUR OWN WORDS . . .

81. What must be checked when solving fractional equations, and why?

4.5 ABSOLUTE VALUE EQUATIONS

In this section equations containing absolute values will be examined. The absolute value of a number is its distance from zero on the number line. The absolute value of a number k was defined as

$$|k| = \begin{cases} k & \text{if } k \geq 0 \\ -k & \text{if } k < 0 \end{cases}$$

Let's see how we might solve the equation $|x| = 3$. Its solution set would be the set of all numbers whose distance from zero is 3 units.

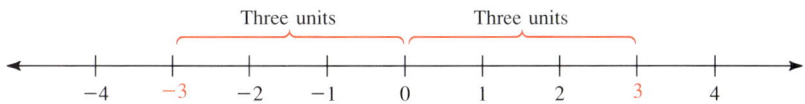

There are two such numbers, -3 and 3, so the solution set is the set $\{-3, 3\}$.

The following property of absolute value equations states this in a useful manner.

> **Absolute Value Equations: Property 1**
>
> If $q \geq 0$, then the equation $|X| = q$ is equivalent to the pair of equations
>
> $$X = q \quad \text{or} \quad X = -q$$

IMPORTANT! This means that if q is not negative, *both* q and $-q$ are solutions to the equation $|X| = q$.

EXAMPLE 1. Solve each equation.

(a) $|t| = 4$
(b) $|x + 9| = 4$

Solutions:

(a) $|t| = 4$

By property 1, this equation is equivalent to the pair of equations

$$t = 4 \quad \text{or} \quad t = -4$$

so the solution set is

$$\{-4, 4\}$$

(b) $|x + 9| = 4$.

Let the X of property 1 be $x + 9$.

$$x + 9 = 4 \quad \text{or} \quad x + 9 = -4$$
$$x = -5 \quad \text{or} \quad x = -13$$
$$\{-13, -5\}$$

Now let's solve the equation $|x| = -2$. Are there any numbers whose *distance* from zero is -2? Of course not, so the solution set is the empty set.

> **Absolute Value Equations: Property 2**
>
> If $q < 0$, the solution set for the equation $|X| = q$ is the empty set.

In other words, if q is negative, the solution set for $|X| = q$ is the empty set, since $|X|$ is never negative.

EXAMPLE 2. Solve $|x| = -3$.

Solution:

As $-3 < 0$, the solution set is the empty set by property 2.

$$\emptyset$$

EXAMPLE 3. Solve $|3x + 7| = 0$.

Solution:
$$|3x + 7| = 0$$

Since 0 is *not* negative, use property 1. However, -0 and 0 represent the same number, so we just get one equation.

$$3x + 7 = 0$$
$$3x = -7$$
$$x = -\frac{7}{3}$$
$$\left\{-\frac{7}{3}\right\}$$
□

EXAMPLE 4. Solve $\left|\dfrac{3 - 2x}{5}\right| - 7 = 0$.

Solution:
$$\left|\frac{3 - 2x}{5}\right| - 7 = 0$$

Before using any of the properties, we *must* isolate an absolute value. Add 7 to both sides.

$$\left|\frac{3 - 2x}{5}\right| = 7$$

We let the X of property 1 be $\dfrac{3 - 2x}{5}$ and we have

$$\frac{3 - 2x}{5} = 7 \quad \text{or} \quad \frac{3 - 2x}{5} = -7$$
$$3 - 2x = 35 \quad \text{or} \quad 3 - 2x = -35$$
$$-2x = 32 \quad \text{or} \quad -2x = -38$$
$$x = -16 \quad \text{or} \quad x = 19$$
$$\{-16, 19\}$$
□

Property 3 provides a method to solve an equation where an absolute value equals an absolute value.

Absolute Value Equations: Property 3

The equation $|U| = |V|$, where U and V are expressions, is equivalent to the pair of equations

$$U = V \quad \text{or} \quad U = -V$$

Sec. 4.5 Absolute Value Equations

EXAMPLE 5. Solve each equation.

(a) $|2x| = |x + 1|$ (b) $|x - 1| = |x + 5|$

Solutions:

(a) $|2x| = |x + 1|$

By property 3, this equation is equivalent to the pair of equations

$$2x = x + 1 \quad \text{or} \quad 2x = -(x + 1)$$
$$x = 1 \quad \text{or} \quad 3x = -1$$
$$x = -\frac{1}{3}$$

so the solution set is

$$\left\{1, -\frac{1}{3}\right\}$$

(b) $|x - 1| = |x + 5|$

This is equivalent to the pair of equations

$$x - 1 = x + 5 \quad \text{or} \quad x - 1 = -(x + 5)$$
$$-1 = 5 \quad \quad x - 1 = -x - 5$$
$$\quad \quad 2x = -4$$
$$\quad \quad x = -2$$

There are no solutions from the first equation. So the solution set is

$$\{-2\}$$

A Procedure for Solving Absolute Value Equations

1. Isolate an absolute value.
2. Apply property 1, 2, or 3.
3. Solve the resulting equations.
4. Check, if required.
5. Write the solution set.

PROBLEM SET 4.5

Warm-ups

In problems 1 through 30, find the solution set. For problems 1 through 10, see Example 1.

1. $|x| = 3$
2. $|x + 5| = 5$
3. $|2x| = 5$
4. $\left|\frac{1}{2}x\right| = \frac{1}{4}$
5. $|1 - 2x| = 4$
6. $|x - 2| = 4$
7. $\left|\frac{1}{4}x\right| = 1$
8. $|1.5x - 5| = 10$

9. $\left|\dfrac{3x}{2} - \dfrac{1}{2}\right| = \dfrac{1}{4}$

10. $\left|\dfrac{7y - 1}{3}\right| = 3$

For problems 11 and 12, see Example 2.

11. $|x| = -2$

12. $|x + 2| = -39$

For problems 13 through 16, see Example 3.

13. $|x| = 0$

14. $|x - 7| = 0$

15. $|2x - 5| = 0$

16. $\left|\dfrac{1}{3} - \dfrac{2}{3}x\right| = 0$

For problems 17 through 22, see Example 4.

17. $|2x - 3| - 5 = 0$

18. $|4z + 1| + 6 = 0$

19. $|5 - 2x| + 2 = 5$

20. $3 - |x + 4| = 3$

21. $7 + |2x + 3| = 5$

22. $8 - |5x - 2| = 0$

For problems 23 through 30, see Example 5.

23. $|x + 2| = |x|$

24. $|2 - x| = |x - 2|$

25. $|x + 1| - |x - 3| = 0$

26. $|x + 9| = |x + 10|$

27. $|2x - 1| = |x + 3|$

28. $|4x + 3| = |x + 7|$

29. $\left|\dfrac{x + 1}{3}\right| - |x| = 0$

30. $\left|\dfrac{2x}{5}\right| = |4x|$

Practice Exercises

In problems 31 through 66, find the solution set.

31. $|x| = 6$

32. $|x| = -1$

33. $|2x| = 0$

34. $|y + 3| = 1$

35. $|x - 4| = 2$

36. $|x + 11| = 0$

37. $|x - 6| = -7$

38. $|x - 7| = 7$

39. $|x + 13| = 7$

40. $|2x + 7| = 3$

41. $|4x + 5| - 4 = 4$

42. $|5x + 1| = 9$

43. $|10r - 3| = 0$

44. $|12t + 15| + 5 = 1$

45. $|5x + 11| = 9$

46. $\left|\dfrac{1}{4}x + 3\right| - 2 = 0$

47. $6 - |2.5x - 4| = 0$

48. $\left|\dfrac{2}{7}v - \dfrac{1}{3}\right| = \dfrac{1}{7}$

49. $|2.3 - x| = 4.5$

50. $\left|\dfrac{3}{4} - \dfrac{1}{2}x\right| = \dfrac{1}{8}$

51. $|2s| = |s + 3|$

52. $|x| = |x - 4|$

53. $|x + 7| = |x - 4|$

54. $|x - 8| = |x - 1|$

55. $|3x + 4| = |x - 2|$

56. $|x + 6| - |2x - 3| = 0$

57. $|x + 1| = |2x + 1|$

58. $|x - 9| = |x + 9|$

59. $|2x - 5| = |4x + 3|$

60. $|x - 3| - |3 - x| = 0$

61. $|2x + 9| = |7 - x|$

62. $|6 + x| = |6 - x|$

63. $|9x - 5| = |x + 3|$

64. $|2x + 5| = |7 - 2x|$

65. $|x + 1| = \left|\dfrac{1}{2}x\right|$

66. $\left|\dfrac{1}{3}x\right| = |x + 2|$

67. Find all numbers such that the absolute value of the number is 3.

68. Find all numbers such that the absolute value of 1 less than the number is $\dfrac{1}{2}$.

Sec. 4.5 Absolute Value Equations

69. Find all numbers such that the absolute value of 3 less than four times the number is 1.

70. Find all numbers such that the absolute value of 3 minus $\frac{1}{5}$ of the number is 2.

Challenge Problems

In problems 71 through 75, find the solution set.

71. $|x - 5| = a$, $a > 0$

72. $|x - a| = 1$

73. $|x - a| = b$, $b > 0$

74. $|x + a| = |x|$

75. $|2x - a| = |x|$

In problems 76 through 79, use properties 1 and 2 to solve each equation. Be sure to check all solutions. Why must these equations be checked?

76. $|2x + 1| = x$

77. $|2x - 3| = x + 1$

78. $|x - 2| = x + 3$

79. $|x + 5| = x + 2$

IN YOUR OWN WORDS . . .

80. What do we mean by the absolute value of a number?

81. Describe how to solve equations of the form |variable| = a constant.

4.6 EQUATIONS CONTAINING SQUARE ROOTS

The equation $\sqrt{x + 1} = 2$ cannot be solved by adding a number to both sides or by multiplying both sides by the same nonzero number. Somehow we must undo the square root.

Remember that the square root of 4 is 2 and that the square root of 9 is 3. That is,

$\sqrt{4} = 2$ because $2^2 = 4$, and 2 is positive

$\sqrt{9} = 3$ because $3^2 = 9$, and 3 is positive

In general, for any positive number q, $\sqrt{q}$ is that positive number that when squared is q.

Square Roots

If q is not negative,

$$(\sqrt{q})^2 = q$$

How do the solutions of the equation $P = Q$ compare to the solutions of the equation $P^2 = Q^2$? What happens if both sides of an equation are squared?

Squaring Property

If the equation $P = Q$ has solutions, these solutions are found in the solutions of the equation $P^2 = Q^2$.

Consider the equation

$$x = 1$$

Its solution set is

$$\{1\}$$

Now if both sides of the equation are squared, it becomes

$$x^2 = 1$$

and we see by inspection that the solution set of the "squared" equation is

$$\{-1, 1\}$$

Note that 1 is a solution to $x = 1$ but -1 is not.

Notice that the solutions of the "squared" equation, $P^2 = Q^2$, are *possible* solutions of the equation $P = Q$, but some solutions of the "squared" equation *may not* be solutions of the original equation. Each solution of the "squared" equation *must* be checked in the original equation.

The general strategy to solve an equation with variables under a square root is to solve the "squared" equation to find all possible solutions and then *check each* possible solution in the original equation.

In Section 4.1 we discussed checking possible solutions. Be sure to use a correct procedure for checking. Be careful to evaluate the left side and the right side separately.

EXAMPLE 1. Solve $\sqrt{z + 1} = 2$.

Solution:

[1] $$\sqrt{z + 1} = 2$$

Square both sides.

[2] $$(\sqrt{z + 1})^2 = 2^2$$
$$z + 1 = 4$$
$$z = 3$$

3 is a solution to equation [2]. It may or may not be a solution to equation [1]. We **must** check to see if it is.

Check 3 in equation [1]. Evaluate each side when z is 3.

LS: $\sqrt{z + 1} = \sqrt{3 + 1} = \sqrt{4} = $ [2]

RS: [2]

Since the values of the LS and the RS are the same, 3 is a solution to equation [1] and the solution set is

$$\{3\}$$

□

Sec. 4.6   Equations Containing Square Roots

EXAMPLE 2. Solve $\sqrt{x+1} + 2 = 0$.

Solution:

The first step is to isolate the radical. Subtract 2 from both sides.

$$\sqrt{x+1} = -2$$

Now square both sides and solve.
$$(\sqrt{x+1})^2 = (-2)^2$$
$$x + 1 = 4$$
$$x = 3$$

We check 3 in the original equation.

LS: $\sqrt{x+1} + 2 = \sqrt{3+1} + 2 = 2 + 2 = \boxed{4}$
RS: $\boxed{0}$

Since we get different values in the left side and the right side, 3 is not in the solution set of $\sqrt{x+1} + 2 = 0$. Thus the solution set is the empty set,

$$\emptyset$$

Notice that it could be seen that the solution set would be empty at the step marked $\boxed{*}$, as a square root cannot be negative. ▭

Possible solutions that do not check are often called **extraneous** solutions.

A Procedure for Solving Equations Containing Square Roots

1. Isolate one radical.
2. Square both sides.
3. Solve the resulting equation for all possible solutions.
4. Check all possible solutions in the original equation.
5. Write the solution set.

EXAMPLE 3. Solve each equation.

(a) $\sqrt{3x+1} - \sqrt{2x+6} = 0$ (b) $\sqrt{x+1} = \sqrt{x+4}$

Solutions:

(a) $\sqrt{3x+1} - \sqrt{2x+6} = 0$

The first step (isolate one radical) is very important. Notice what happens if both sides of the original equation are squared. The left side becomes

$$(3x+1) - 2\sqrt{3x+1} \cdot \sqrt{2x+6} + (2x+6)$$

Chap. 4 Using Linear Equations in One Variable

[Remember, $(p - q)^2 = p^2 - 2pq + q^2$.] This is certainly no improvement over the original! So the first thing we do is isolate one radical.

$$\sqrt{3x + 1} = \sqrt{2x + 6}$$

Square both sides and solve.

$$(\sqrt{3x + 1})^2 = (\sqrt{2x + 6})^2$$
$$3x + 1 = 2x + 6$$
$$x = 5$$

Check 5 in the original equation.

LS: $\sqrt{3x + 1} - \sqrt{2x + 6} = \sqrt{15 + 1} - \sqrt{10 + 6} = 4 - 4 = \boxed{0}$

RS: $\boxed{0}$

Since we get the same value in the LS and the RS, the solution set is

$$\{5\}$$

(b) $\sqrt{x + 1} = \sqrt{x + 4}$

$\qquad (\sqrt{x + 1})^2 = (\sqrt{x + 4})^2 \qquad$ Square both sides

$\qquad x + 1 = x + 4 \qquad$ Solve.

$\qquad 1 = 4$

Since this statement is false no matter what the value of x, the solution set is the empty set,

$$\emptyset$$ ▢

EXAMPLE 4. Solve $5\sqrt{x - 3} = 3\sqrt{2x + 1}$.

Solution:

$\qquad 5\sqrt{x - 3} = 3\sqrt{2x + 1}$

$\qquad (5\sqrt{x - 3})^2 = (3\sqrt{2x + 1})^2 \qquad$ Square both sides.

$\qquad 25(x - 3) = 9(2x + 1)$

$\qquad 25x - 75 = 18x + 9 \qquad$ Solve.

$\qquad 7x = 84$

$\qquad x = 12$

Check 12 in the original equation.

LS: $5\sqrt{x - 3} = 5\sqrt{12 - 3} = 5\sqrt{9} = 5 \cdot 3 = \boxed{15}$

RS: $3\sqrt{2x + 1} = 3\sqrt{24 + 1} = 3\sqrt{25} = 3 \cdot 5 = \boxed{15}$

$\qquad\qquad\qquad \{12\} \qquad$ Write solution set. ▢

Sec. 4.6 Equations Containing Square Roots

EXAMPLE 5. Solve $\sqrt{x^2 + 8} - 4 = x$.

Solution:

$$\sqrt{x^2 + 8} - 4 = x$$
$$\sqrt{x^2 + 8} = x + 4 \qquad \text{Isolate radical.}$$
$$(\sqrt{x^2 + 8})^2 = (x + 4)^2 \qquad \text{Square both sides.}$$
$$x^2 + 8 = x^2 + 8x + 16$$

Don't forget the middle term!

$$8 = 8x + 16 \qquad \text{Solve.}$$
$$-8 = 8x$$
$$-1 = x$$

Check -1 in the original equation.

LS: $\sqrt{x^2 + 8} - 4 = \sqrt{(-1)^2 + 8} - 4 = \sqrt{9} - 4 = 3 - 4 = \boxed{-1}$

RS: $x = -1 = \boxed{-1}$

$\qquad\qquad\qquad\qquad\qquad \{-1\} \qquad$ Write solution set. ☐

CALCULATOR BOX

Square Root Equations with a Calculator

[MR]   [M IN]

Equations involving square roots are common in many engineering applications where approximate solutions are acceptable.

The period of a compound pendulum is given by the formula $T = 2\pi\sqrt{\dfrac{I}{32Md}}$, where T is the period in seconds, I is the moment of inertia, M is the mass, and d is the distance from the center of mass to the axis of rotation in feet.

Find the distance d to the nearest hundredth of a foot, for a compound pendulum of mass 11.42 and moment of inertia 339, if the period is 3.71 seconds.

$$3.71 = 2\pi\sqrt{\dfrac{339}{32(11.42)d}} \qquad \text{Substitution}$$

$$(3.71)^2 = \left[2\pi\sqrt{\dfrac{339}{32(11.42)d}}\right]^2 \qquad \text{Square both sides.}$$

$$(3.71)^2 = 4\pi^2\dfrac{339}{32(11.42)d}$$

$$(3.71)^2(32)(11.42)d = 4\pi^2(339) \qquad \text{Multiply by } 32(11.42)d.$$

$$d = \dfrac{4\pi^2(339)}{(3.72)^2(32)(11.42)}$$

Chap. 4 Using Linear Equations in One Variable

Press the keys

$4 \times \pi \; x^2 \times 339 \div (\; 3.72 \; x^2 \times 32 \times 11.42 \;) =$

and read $\boxed{2.6464121}$ on the display. To check this result, first we save the answer in memory. This key may be $\boxed{\text{STO}}$ or $\boxed{\text{M IN}}$. Press

$339 \div (\; 32 \times 11.42 \times \text{MR} \;) = \sqrt{} \times 2 \times \pi =$

where $\boxed{\text{MR}}$ is the memory recall key, and read $\boxed{3.72}$ on the display. Although this is not exactly equal to the left side of the original equation, it is close enough to confirm the result. Remember, the answer is an approximation.

$$d \approx 2.65 \text{ ft}$$

Calculator Exercises

Find x to the nearest thousandth in each equation.

1. $55.321 = \sqrt{14.007x}$
2. $22.822 = \sqrt{x - 61.190}$
3. $0.00456 = 3\sqrt{\dfrac{x}{123456}}$
4. $8.631 = 15.889\sqrt{6.331 - x}$
5. $\sqrt{\dfrac{x+1}{x-1}} = \pi$
6. $\sqrt{2x + 1.133} - \sqrt{x + 4} = 0$

Answers:

1. 218.492
2. 582.034
3. 0.285
4. 6.036
5. 1.225
6. 2.867

■ PROBLEM SET 4.6

Warm-ups

In problems 1 through 31, find the solution set.
For problems 1 through 10, see Examples 1 and 2.

1. $\sqrt{x + 1} = 2$
2. $\sqrt{x - 3} = -4$
3. $\sqrt{2t + 3} = 5$
4. $\sqrt{3x - 1} = 5$
5. $\sqrt{2 - x} = 3$
6. $\sqrt{4 - x} = 2$
7. $\sqrt{x} = 0$
8. $\sqrt{x + 1} = 0$
9. $\sqrt{x - 3} = -8$
10. $\sqrt{2x + 1} = 4$

For problems 11 through 17, see Example 3.

11. $\sqrt{w + 1} = \sqrt{3w - 1}$
12. $\sqrt{2x - 3} = \sqrt{6 - x}$
13. $\sqrt{3x + 2} + \sqrt{2x + 7} = 0$
14. $\sqrt{3x + 3} - \sqrt{4x + 2} = 0$
15. $\sqrt{x + 1} = \sqrt{2x - 2}$
16. $\sqrt{2x - 1} - \sqrt{x} = 0$
17. $\sqrt{x - 1} = \sqrt{x + 1}$

For problems 18 through 21, see Example 4.

18. $2\sqrt{x - 3} = \sqrt{x}$
19. $12 = 3\sqrt{x - 1}$
20. $2\sqrt{2x + 5} = 5\sqrt{x - 6}$
21. $\sqrt{2x + 3} - 3\sqrt{x - 2} = 0$

Sec. 4.6 Equations Containing Square Roots

For problems 22 through 31, see Example 5.

22. $\sqrt{x^2 + 2} = x$
23. $\sqrt{x^2 + 3} = x + 1$
24. $\sqrt{x^2 - 2} = x + 2$
25. $\sqrt{y^2 - 2y} = -y$
26. $\sqrt{x^2 + 24} - x = -4$
27. $\sqrt{x^2 + x} = x$
28. $\sqrt{x^2 - 9} = 3 - x$
29. $\sqrt{x^2 - 4} = x + 2$
30. $\sqrt{3 + x^2} + x = 3$
31. $\sqrt{4x^2 + 9} + 1 = 2x$

Practice Exercises

In problems 32 through 61, find the solution set.

32. $\sqrt{x + 3} = 4$
33. $\sqrt{z - 5} = -7$
34. $\sqrt{2x - 5} = 15$
35. $\sqrt{3x + 4} = 4$
36. $\sqrt{3 - x} = 3$
37. $\sqrt{1 - x} = 5$
38. $\sqrt{2 - 3x} = 3$
39. $\sqrt{3 - 5x} = -2$
40. $\sqrt{x - 1} = 0$
41. $\sqrt{2 - t} = 0$
42. $\sqrt{x + 4} = -5$
43. $\sqrt{2x - 1} = 3$
44. $\sqrt{x - 3} = \sqrt{2x - 7}$
45. $\sqrt{2 - x} = \sqrt{x + 3}$
46. $\sqrt{x - 7} - \sqrt{2x - 18} = 0$
47. $\sqrt{1 - 3s} + \sqrt{2s + 11} = 0$
48. $\sqrt{x + 3} = \sqrt{2x - 1}$
49. $\sqrt{1 + 3x} = \sqrt{3x}$
50. $2\sqrt{1 - 3x} - \sqrt{x + 17} = 0$
51. $3\sqrt{6 - x} = 2\sqrt{x + 46}$
52. $\sqrt{x^2 + 2x} = x$
53. $2\sqrt{x} = \sqrt{3x + 1}$
54. $2\sqrt{2x + 1} - \sqrt{7x + 8} = 0$
55. $\sqrt{x^2 - 1} = 1 - x$
56. $\sqrt{x^2 + 3x} = x + 3$
57. $\sqrt{x^2 - 3x} = x - 3$
58. $\sqrt{x^2 + 4} = x - 2$
59. $\sqrt{x^2 + 11} = x + 1$
60. $\sqrt{x^2 + 12} - x = 6$
61. $\sqrt{x^2 - 1} = x + 1$

62. The period of a simple pendulum is given by $T = 2\pi\sqrt{\dfrac{L}{32}}$, where T is the period in seconds and L is the length of the pendulum in feet. Find the length of a pendulum whose period is 2 seconds.

63. The velocity of a wave in shallow water is given by the formula $v = \sqrt{32h}$, where v is the velocity of the wave, h is the wave height. Find the wave height in feet if the velocity is 4 ft/sec.

■ IN YOUR OWN WORDS . . .

64. Why do we isolate a radical before we square both sides of an equation containing square roots?

65. Why is checking part of the procedure for solving an equation containing square roots?

CHAPTER SUMMARY

GLOSSARY

Equation: A statement that two numbers are equal.

Variable: A letter representing a number.

Solution or **root**: A number that makes an equation a true statement when it replaces the variable.

Solution set: The set of all solutions of an equation.

Solve an equation: To find the solution set.

Equivalent equations: Equations that have the same solution set.

Linear equation: An equation that can be written in the form

$$Ax + B = 0$$

Contradiction: An equation that is always false.

Identity: An equation that is always true.

Fractional equation: An equation that contains a fraction with a variable in its denominator.

Literal equation: An equation that contains more than one letter.

PROPERTIES OF EQUALITY
1. Add or subtract the same number to both sides.
2. Multiply or divide both sides by the same nonzero number.

FRACTIONAL EQUATIONS
1. Clear the equation of fractions (multiply both sides by the least common denominator).
2. Solve the resulting equation for possible solutions.
3. Possible solutions that make any denominator have a value of zero *cannot* be included in the solution set.

ABSOLUTE VALUE EQUATIONS

Equations with absolute values can be solved by using these properties:

1. If $q \geq 0$, then $|X| = q$ is equivalent to the pair of equations $X = q$ or $X = -q$.
2. If $q < 0$, there are no solutions of the equation $|X| = q$.
3. $|U| = |V|$ is equivalent to the pair of equations $U = V$ or $U = -V$.

EQUATIONS CONTAINING SQUARE ROOTS
1. Isolate one radical.
2. Square both sides.
3. Solve the resulting equation.
4. *Check* all possible solutions in the original equation.
5. Write the solution set.

APPLICATIONS

Word problems can be solved by following these steps:

1. Determine what is to be found.
2. Assign a variable, such as x, to represent one of the quantities to be found.
3. Express all other quantities to be found in terms of x (or the variable chosen).
4. Draw a figure or picture, if possible. Label it.
5. Reread the problem and form an equation.
6. Solve the equation and find the values of all the quantities to be found.
7. Check the values in the original word problem. They should answer the question and make sense.
8. Write an answer to the original question.

■ CHECKUPS

In problems 1 through 5, solve each equation.

1. $5(2x + 3) = 3 - 2(3x - 5)$ Section 4.1; Example 3
2. $\dfrac{2}{x+2} - \dfrac{1}{x} = \dfrac{-4}{x(x+2)}$ Section 4.4; Example 4
3. $\left|\dfrac{3-2x}{5}\right| - 7 = 0$ Section 4.5; Example 4
4. $\sqrt{x^2 + 8} - 4 = x$ Section 4.6; Example 5
5. $ax - b(x + a) = 6 + 2bx$ for x.   Section 4.2; Example 6

Sec. 4.6 Equations Containing Square Roots

6. Jack Glover wishes to add enough 50% antifreeze solution to 16 gallons of a 5% antifreeze solution to obtain a 20% antifreeze solution. How much of the 50% solution should Jack add?

Section 4.3; Example 1

7. Two trains, 780 miles apart, are heading toward each other at rates of 85 mph and 110 mph. When will they crash?

Section 4.3; Example 4

REVIEW PROBLEMS

In problems 1 through 80, find the solution set.

1. $6(x - 5) - (2 - x) = 2x + 1$
2. $2[3 - 2(x - 2)] = 6(x - 4)$
3. $|2z - 3| = 4$
4. $\sqrt{x - 2} = 5$
5. $\dfrac{1}{x - 3} + \dfrac{1}{x + 3} = \dfrac{4}{x^2 - 9}$
6. $|1 - x| = 0$
7. $\sqrt{x} = -3$
8. $\dfrac{1}{2}x - \dfrac{1}{3} = \dfrac{1}{4}$
9. $\dfrac{1}{x + 1} + 1 = \dfrac{x - 2}{x}$
10. $\sqrt{2x - 1} = \sqrt{x + 3}$
11. $\left|\dfrac{3 + y}{2}\right| = 5$
12. $\dfrac{5}{x - 2} = \dfrac{2}{x + 2}$
13. $\dfrac{3}{2}(x + 2) = \dfrac{1}{2}(x - 2)$
14. $|3x - 1| = |2x + 1|$
15. $\sqrt{1 + x^2} = x + 1$
16. $|3x + 4| = \dfrac{1}{4}$
17. $\dfrac{1}{x} - \dfrac{1}{3x} = \dfrac{1}{2}$
18. $6x + 3 = 2(3x + 1)$
19. $\sqrt{3s + 2} = 2$
20. $|6 - x| = 4$
21. $\dfrac{5}{2}(1 - x) = \dfrac{1}{3}(2x + 1)$
22. $\dfrac{3}{x} - 1 = \dfrac{1}{2}$
23. $\sqrt{3x + 4} = \sqrt{x + 2}$
24. $3(2x - 1) = 2(3x + 1) - 5$
25. $\dfrac{3}{x + 1} - 2 = \dfrac{2x}{1 - x}$
26. $\dfrac{1}{4x} + \dfrac{1}{2} = \dfrac{1}{2x}$
27. $\sqrt{1 - x} = \sqrt{2 - x}$
28. $\left|\dfrac{1}{2}x + 3\right| = \dfrac{1}{4}$
29. $6x + 5 = 2x - 3$
30. $\sqrt{x} = 2$
31. $\dfrac{1}{4 - x} + \dfrac{8}{16 - x^2} = \dfrac{1}{4 + x}$
32. $|x| = |2x + 1|$
33. $\sqrt{x^2 + 3} = x + 3$
34. $|3x + 5| = 6$
35. $3(4 - (2x + 7)) = 6(x + 4)$
36. $|1.5x| = 4$
37. $\dfrac{2}{3(w + 3)} - \dfrac{1}{3} = \dfrac{2}{w + 3}$
38. $1.2x + 3.4 = 0.2(x - 1)$
39. $\sqrt{1 - x} = 4$
40. $3x - (1 - x) = 0$
41. $\dfrac{3x}{x + 2} = \dfrac{3}{2}$
42. $\left|\dfrac{1 - x}{4}\right| = \dfrac{1}{2}$
43. $\sqrt{x + 4} = \sqrt{2x}$
44. $\dfrac{1}{2}x + \dfrac{1}{3}(x + 1) = \dfrac{1}{6}$
45. $\dfrac{2}{1 - x} = \dfrac{2x}{1 - x^2}$
46. $|3x - 5| = 0$
47. $\sqrt{x^2 - 15} = x + 5$
48. $3 - (2 + x) = 6x - 7$
49. $|2x - 7| = -3$
50. $\sqrt{6x - 2} = \sqrt{3x + 1}$
51. $\dfrac{1}{x + 2} - 1 = \dfrac{x^2}{4 - x^2}$
52. $\dfrac{2}{x} - \dfrac{1}{3} = \dfrac{4}{3x}$
53. $3[2 - (6 - x) + 3(x + 2)] = 3(x - 2)$

54. $7(2x - 3) - 15 = 6(2 + x)$
55. $|t - 17| = 14$
56. $\sqrt{x + 4} = 8$
57. $\dfrac{1}{x} - \dfrac{2}{3} = \dfrac{1}{2}$
58. $|1 - 2x| = 5$
59. $\sqrt{1 + x} = 3$
60. $\dfrac{1}{2}x - \dfrac{3}{7} = \dfrac{1}{7}x$
61. $\left|\dfrac{1 - x}{2}\right| = 2$
62. $\sqrt{0.5x} = 5$
63. $\dfrac{1}{x - 2} = \dfrac{2}{x - 2}$
64. $\sqrt{x + 2} = \sqrt{2x - 5}$
65. $0.2s - 0.9 = 0.3$
66. $|2x + 5| = |x + 1|$
67. $\sqrt{x + 2} = \sqrt{x - 5}$
68. $\dfrac{2x}{x + 1} - 2 = \dfrac{1}{x}$
69. $\dfrac{3}{4}(x - 2) + (x - 1) = 1$
70. $\sqrt{3x - 1} = \sqrt{5x - 3}$
71. $\dfrac{1}{2x - 1} + \dfrac{1}{2x + 1} = \dfrac{2}{1 - 4x^2}$
72. $|1 - 4x| = 0$
73. $6x - (1 - x) = 6$
74. $\sqrt{x + 2} = -7$
75. $\left|\dfrac{1}{2}x + 1\right| = 4$
76. $4.4x + 2(x + 1.2) = 0$
77. $\dfrac{5}{x + 2} = \dfrac{6}{2x + 3}$
78. $\dfrac{5x}{2x - 3} = \dfrac{3}{2}$
79. $|6x - 7| = 0$
80. $\sqrt{x^2 + 10} = x + 5$

81. Find three consecutive odd integers such that the first plus twice the second plus three times the third is 94.

82. Martin can mow the lawn in 2 hours and Tom can mow the lawn in 6 hours. If they both mow at the same time, how long will it take to mow the lawn?

83. Alice has 15 gallons of 8% Clorox solution, but she wishes to lower the concentration to 5%. How many gallons of water should she add?

84. Irene has $14,000 invested at 9.5% and 11%. If her yearly interest is $1405, how much does she have at each rate?

85. The perimeter of a rectangle is 86 m. If the length is two less than twice the width, find the dimensions of the rectangle.

86. Two cars are 760 km apart. They are approaching each other at speeds of 110 km/hr and 80 km/hr. In how many hours will they meet?

▬ . . LET'S NOT FORGET . . .

Identify the expressions that are in factored form. Factor those that are not factored.

87. $(x - 3)^3$
88. $x^3 - 27$
89. $(x + y) - 4(x + y)^2$
90. $2xy(x^2 + y^2)$
91. x^3yz

How many terms are in each expression? Which expressions have $x + 2$ as a factor?

92. $x^2 - 4$
93. $(x + 2)^2$
94. $x^3 + 8$
95. $xy(x + 2)$
96. $(x + 2)^2 - y(x + 2)$
97. $x^2 + 4x + 4$

Simplify each expression.

98. -8^2
99. $(-8)^2$

Find each product.

100. $(x - 2y)^2$
101. $(a + 2b)^3$
102. $(x^2y)^2$
103. $x^3(x^2y - xy)$

Reduce, if possible.

104. $\dfrac{2(x - 3) - 3 + x}{x^2 - 9}$
105. $\dfrac{2(x - 3) - (3 + x)}{x^2 - 9}$

The following problems can be worked by using a least common denominator. Follow the directions in each and notice how the LCD is used.

106. Solve $\dfrac{1}{2x} - \dfrac{1}{3} = \dfrac{1}{x}$.
107. Perform the operation indicated: $\dfrac{1}{2x} - \dfrac{1}{3}$.
108. Simplify $\dfrac{\dfrac{15t}{2x}}{\dfrac{t^2}{3}}$.

Sec. 4.6 Equations Containing Square Roots

Label each as an equation or an expression. Solve the equations and perform the operations indicated on the expressions.

109. $\dfrac{1}{x+1} - \dfrac{x}{x+2}$

110. $\dfrac{3}{7}(x-1) = \dfrac{1}{3}x$

111. $\dfrac{1}{x+3} - \dfrac{2}{x-2} = \dfrac{3}{x^2+x-6}$

112. $3(x-4) - (x-1) = 6$

113. $|x+3| = 6$

114. $\sqrt{x+2} = 4$

115. $\dfrac{4}{x-y} - \dfrac{8}{y-x}$

CHAPTER 4 TEST

In problems 1 through 5, choose the correct answer.

1. The solution set for the equation $3(2x-1) - 4x = 3(5-x)$ is (?)
   A. $\left\{\dfrac{14}{5}\right\}$ B. $\{-12\}$ C. $\left\{\dfrac{18}{5}\right\}$ D. $\{6\}$

2. The solution set for the equation $\dfrac{2x}{x-1} - \dfrac{1}{x+2} = 2$ is (?)
   A. $\{-5\}$ B. $\{-3\}$ C. $\left\{-\dfrac{3}{2}\right\}$ D. $\emptyset$

3. When solved for x, the literal equation $b(1-6x) = 2ax$ becomes (?)
   A. $x = \dfrac{b(1-6x)}{2a}$ B. $x = \dfrac{b}{2a+6b}$
   C. $x = \dfrac{b}{2a-3b}$ D. $x = \dfrac{1}{2a+6}$

4. The solution set for the equation $\sqrt{x+1} = \sqrt{2x-2}$ is (?)
   A. $\emptyset$ B. $\{-3\}$ C. $\{1\}$ D. $\{3\}$

5. If x is the smallest of three consecutive odd integers, which of the following is the sum of the first plus twice the second plus four times the third?
   A. $x + 2x + 2 + 4x + 4$ B. $x + 2x + 4x$
   C. $x + 2(x+2) + 4(x+4)$ D. $x + 2(2x) + 4(4x)$

In problems 6 through 10, find the solution set.

6. $|2x+3| = 7$

7. $\sqrt{x+3} + \sqrt{2x-3} = 0$

8. $\dfrac{1}{3x} - \dfrac{3}{2x} = \dfrac{1}{3}$

9. $\dfrac{2}{3}(x-1) = \dfrac{1}{6} + x$

10. $|1 - 2x| = -3$

11. Solve for x.
    $$\dfrac{a-x}{2} + y = ax$$

12. Solve for C.
    $$F^2 = \dfrac{1}{LC} - \dfrac{R}{4L^2}$$

13. Little Red Riding Hood is going from her house to her grandmother's house, which is a distance of 600 miles. She rides her Honda for 5 hours and then rides 9 hours on a bus the rest of the way. If the bus went 20 mph faster than the Honda, how fast did the Honda go?

14. Sarah has 134 in. of picture-frame material. What are the dimensions of the largest frame that she can make if the length is to be 5 in. more than the width?

15. Shirley Ann Exum has 7 quarts of a 5% vinegar solution. How many quarts of a 20% vinegar solution should she add to obtain a 13% vinegar solution?

CHAPTER 5

See Problem Set 5.1, Exercise 139.

Exponents and Radicals

- **5.1** Negative Integer Exponents
- **5.2** Roots and Radicals
- **5.3** Operations With Radicals
- **5.4** Rational Exponents
- **5.5** Complex Numbers

CONNECTIONS

The way we write algebra today is mainly due to the French mathematician René Descartes (1596–1650). Descartes popularized the use of superscripts, which we call exponents, to indicate a number multiplied by itself. For example, we write x^3 to mean $x \cdot x \cdot x$. Soon the properties of exponents were discovered and this concise notation was expanded to include zero, negative integers, and rational number exponents.

An important consideration in the design of an auditorium is the time difference in sound traveling to a seat along different paths. If the time difference is too great, poor acoustics will result.

For this study, the formula $t = \dfrac{d}{\sqrt{grT}}$ gives the time for sound to travel a distance, d. Formulas that contain square roots, such as the one above, are common in all engineering fields.

In this chapter we will see that exponents are related to square roots and other radicals. We will investigate this relationship. This study and the complex numbers also developed in this chapter, are necessary to solve quadratic equations in Chapter 6.

5.1 NEGATIVE INTEGER EXPONENTS

In Chapter 1 we learned that exponents were developed as shorthand for multiplications with the same number, such as $x^3 = x \cdot x \cdot x$ or $a^6 = a \cdot a \cdot a \cdot a \cdot a \cdot a$. From this simple idea the five properties that exponents share were discovered.

Properties of Exponents

If m and n natural numbers and x and y real numbers,

1. $x^m \cdot x^n = x^{m+n}$ — Product with same base
2. $(x^m)^n = x^{m \cdot n}$ — Power of a power
3. $(x \cdot y)^n = x^n y^n$ — Product to a power
4. $\left(\dfrac{x}{y}\right)^n = \dfrac{x^n}{y^n}; \quad y \neq 0$ — Quotient to a power
5. $\dfrac{x^m}{x^n} = \begin{cases} x^{m-n} & \text{if } m > n \\ 1 & \text{if } m = n \\ \dfrac{1}{x^{n-m}} & \text{if } n > m \end{cases} \quad x \neq 0$ — Quotient with same base

---

The following definition is consistent with the properties listed above and allows us to use zero as an exponent.

$$x^0 = 1 \quad \text{for all values of } x \text{ (except } x = 0)$$

If we have a definition for 3^{-4} such that the five properties of exponents are valid, then the following must be true.

$$3^{-4} \cdot 3^4 = 3^{-4+4} \quad \text{Product with same base}$$
$$= 3^0$$
$$= 1 \quad \text{Definition of zero exponent}$$

So we see that

$$3^{-4} \cdot 3^4 = 1$$

But if we divide both sides by 3^4,

$$\frac{3^{-4} \cdot 3^4}{3^4} = \frac{1}{3^4}$$

$$3^{-4} = \frac{1}{3^4}$$

Since this is true in general, we make the following definition.

Definition of Negative Exponent

For x any nonzero real number and n a natural number,

$$x^{-n} = \frac{1}{x^n}$$

This definition is consistent with all five of the exponent properties.

EXAMPLE 1. Rewrite each expression without negative exponents, and simplify.

(a) 5^{-3} (b) $2^4 x^{-5}$

(c) $-3^{-2} z^4$ (d) $4^{-1} x^{-11}$

(e) $(-5)^{-2} xy^{-1}$ (f) $7x^{-3}$

Solutions:

(a) $5^{-3} = \dfrac{1}{5^3}$

$$= \frac{1}{5 \cdot 5 \cdot 5}$$

$$= \frac{1}{125}$$

Notice that the negative exponent did not make the answer negative.

(continued)

Sec. 5.1 Negative Integer Exponents

(b) $2^4 x^{-5} = 2^4 \cdot \dfrac{1}{x^5}$

$= \dfrac{2^4}{x^5} = \dfrac{16}{x^5}$

(c) $-3^{-2} z^4 = -\dfrac{1}{3^2} \cdot z^4$

Be Careful! Notice that the base for -2 is 3, *not* -3.

$= -\dfrac{z^4}{3^2} = -\dfrac{z^4}{9}$

(d) $4^{-1} x^{-11} = \dfrac{1}{4^1} \cdot \dfrac{1}{x^{11}}$

$= \dfrac{1}{4} \cdot \dfrac{1}{x^{11}} = \dfrac{1}{4x^{11}}$

In the next example be very careful with the factor x.

(e) $(-5)^{-2} xy^{-1} = \dfrac{1}{(-5)^2} \cdot x \cdot \dfrac{1}{y^1}$

$= \dfrac{1}{5^2} \cdot \dfrac{x}{y^1}$

$= \dfrac{x}{5^2 y^1} = \dfrac{x}{25y}$

The next example is similar but is even more treacherous, as there is a tendency to let the 7 stick to the variable, particularly when working in a hurry.

Be Careful! (f) $7x^{-3} = 7 \cdot \dfrac{1}{x^3}$

$= \dfrac{7}{x^3}$

Notice that the base for the exponent, -3, is x, not $7x$. ▫

Including this last definition, the properties of exponents can now be written.

Properties of Exponents

For m and n any *integers* and x and y real numbers,

1. $x^m \cdot x^n = x^{m+n}$ Product with same base
2. $(x^m)^n = x^{m \cdot n}$ Power of a power
3. $(x \cdot y)^n = x^n y^n$ Product to a power
4. $\left(\dfrac{x}{y}\right)^n = \dfrac{x^n}{y^n}; \quad y \neq 0$ Quotient to a power
5. $\dfrac{x^m}{x^n} = x^{m-n}; \quad x \neq 0$ Quotient with same base

Compare this form of property 5 with that of Section 1.1 and notice the simplification that results when negative exponents are allowed.

EXAMPLE 2. Use the properties of exponents to simplify each expression. The answers should not contain negative exponents.

(a) $5^5 \cdot 5^{-3}$ (b) $(a^{-2})^{-3}$

(c) $\left(\dfrac{c^{-2}}{d}\right)^{-2}$ (d) $(3x^{-4})^{-2}$

Solutions:

(a) $5^5 \cdot 5^{-3} = 5^{5+(-3)}$ Product with same base

$\quad\quad\quad\quad = 5^2$

$\quad\quad\quad\quad = 25$

(b) $(a^{-2})^{-3} = a^{(-2)(-3)}$ Power of a power

$\quad\quad\quad\quad = a^6$

(c) $\left(\dfrac{c^{-2}}{d}\right)^{-2} = \dfrac{(c^{-2})^{-2}}{d^{-2}}$ Quotient to a power

$\quad\quad\quad\quad = \dfrac{c^{(-2)(-2)}}{d^{-2}}$ Power of a power

$\quad\quad\quad\quad = \dfrac{c^4}{d^{-2}}$

$\quad\quad\quad\quad = \dfrac{c^4}{\dfrac{1}{d^2}}$ Definition of negative exponent

$\quad\quad\quad\quad = c^4 \cdot \dfrac{d^2}{1} = c^4 d^2$

(d) $(3x^{-4})^{-2} = 3^{-2}(x^{-4})^{-2}$ Product to a power

$\quad\quad\quad\quad = 3^{-2} x^{(-4)(-2)}$ Power of a power

$\quad\quad\quad\quad = 3^{-2} x^8$

$\quad\quad\quad\quad = \dfrac{1}{3^2} \cdot x^8$ Definition of negative exponent

$\quad\quad\quad\quad = \dfrac{x^8}{9}$ ☐

In Chapter 1 we saw that if n is a natural number,

$$(-x)^n = \begin{cases} x^n & \text{if } n \text{ is even} \\ -x^n & \text{if } n \text{ is odd} \end{cases}$$

The same idea is also true for negative exponents.

$$(-x)^{-n} = \begin{cases} x^{-n} & \text{if } n \text{ is even} \\ -x^{-n} & \text{if } n \text{ is odd} \end{cases}$$

Therefore,

> ### Odd and Even Powers
> If t is an *integer*,
> $$(-x)^t = \begin{cases} x^t & \text{if } t \text{ is even} \\ -x^t & \text{if } t \text{ is odd} \end{cases}$$

Notice how this property is used in the next example.

EXAMPLE 3. Write each expression without negative exponents.

(a) $\left(-\dfrac{2}{5}\right)^{-2}$ (b) $(-4^{-1})^{-3}$

Solutions:

(a) $\left(-\dfrac{2}{5}\right)^{-2} = \left(\dfrac{2}{5}\right)^{-2}$ Even power

$\qquad\qquad = \dfrac{1}{\left(\dfrac{2}{5}\right)^2}$ Definition of negative exponent

$\qquad\qquad = \dfrac{1}{\dfrac{4}{25}}$

$\qquad\qquad = \dfrac{25}{4}$

(b) $(-4^{-1})^{-3} = -(4^{-1})^{-3}$ Odd power

$\qquad\qquad = -4^3$ Power of a power

$\qquad\qquad = -64$

Many problems with negative exponents can be simplified by applying some consequences of the definition. First, we note the following:

$$\dfrac{1}{x^{-n}} = \dfrac{1}{\dfrac{1}{x^n}} = 1 \cdot \dfrac{x^n}{1} = x^n$$

Combining this result with the definition of negative exponents, we get the following result.

> ### Simplifying Negative Exponents
> For x any nonzero real number and n any natural number,
> $$x^{-n} = \dfrac{1}{x^n} \quad \text{and} \quad \dfrac{1}{x^{-n}} = x^n$$

Let's simplify $\dfrac{a^{-2}b}{c^{-3}}$ using this result.

$$\dfrac{a^{-2}b}{c^{-3}} = a^{-2}b \cdot \dfrac{1}{c^{-3}}$$

$$= \dfrac{1}{a^2} \cdot b \cdot c^3$$

$$= \dfrac{bc^3}{a^2}$$

Now, if all the middle steps are left out, we have that

$$\dfrac{a^{-2}b}{c^{-3}} = \dfrac{bc^3}{a^2}$$

Notice that a^{-2} was *a factor* of the numerator and that a^2 ends up *as a factor* of the denominator. Also, c^{-3} was *a factor* of the denominator and c^3 ends up *as a factor* of the numerator. This leads directly to the following results.

1. If p^{-n} is a **factor** of the numerator of a fraction, the fraction may be rewritten with p^n as a factor of the denominator.

2. If q^{-m} is a **factor** of the denominator of a fraction, the fraction may be rewritten with q^m as a factor of the numerator.

In other words, a **factor** of the numerator (denominator) of a fraction may be moved to the denominator (numerator) of the fraction by changing the sign of its exponent.

It is **very** important to remember that these rules apply **only to factors!**

Be Careful!

EXAMPLE 4. Simplify the following expressions. The answers should not contain negative exponents.

(a) $\dfrac{x^{-2}y}{z^{-3}}$ (b) $\left(-\dfrac{p}{q}\right)^{-3}$ (c) $\dfrac{2x^{-1}}{y}$

(d) $\left(-\dfrac{xy^{-1}z^2}{3k^3}\right)^{-2}$ (e) $\dfrac{3^{-1} + 3^{-2}}{1 - 3^{-2}}$

Solutions:

(a) $\dfrac{x^{-2}y}{z^{-3}}$

As x^{-2} is a *factor* of the numerator we can apply the negative exponent rules and write,

$$\dfrac{x^{-2}y}{z^{-3}} = \dfrac{y}{x^2 z^{-3}}$$

Also, z^{-3} is a *factor* of the denominator.

$$= \dfrac{yz^3}{x^2}$$

(continued)

Sec. 5.1 Negative Integer Exponents

(b) $\left(-\dfrac{p}{q}\right)^{-3} = -\left(\dfrac{p}{q}\right)^{-3}$ Odd power

$= -\dfrac{p^{-3}}{q^{-3}}$ Quotient to a power

$= -\dfrac{q^3}{p^3}$

(c) This is a simple example of negative exponents but watch out for a sticky -2!

$$\dfrac{-2x^{-1}}{y} = \dfrac{-2}{x^1 y}$$

Be Careful!

The base for the exponent, -1, is x, not $-2x$! Don't let the -2 *stick* to the x!

$$= \dfrac{-2}{xy}$$

(d) $\left(-\dfrac{xy^{-1}z^2}{3k^3}\right)^{-2}$

This example is an *expression to a power,* just like the examples in Section 1.1. As the expression is made up of **products** and **quotients,** the exponent properties allow us to multiply the exponents.

$\left(-\dfrac{xy^{-1}z^2}{3k^3}\right)^{-2} = \left(\dfrac{xy^{-1}z^2}{3k^3}\right)^{-2}$ Even power

$= \dfrac{x^{-2}y^2 z^{-4}}{3^{-2}k^{-6}}$ Quotient to a power and power of a power

$= \dfrac{3^2 k^6 y^2}{x^2 z^4}$ Negative exponents

$= \dfrac{9k^6 y^2}{x^2 z^4}$

(e) $\dfrac{3^{-1} + 3^{-2}}{1 - 3^{-2}}$

Be Careful!

It is **important** to note that the two rules for moving negative exponents do **not** apply here. Those rules apply **only to factors.** The 3^{-1} and the 3^{-2} in the numerator are **terms, not factors,** as are the two numbers in the denominator. The *definition* of negative exponents must be used in this problem.

$$\dfrac{3^{-1} + 3^{-2}}{1 - 3^{-2}} = \dfrac{\dfrac{1}{3^1} + \dfrac{1}{3^2}}{1 - \dfrac{1}{3^2}}$$ Definition of negative exponent

This is a complex fraction. Multiply the numerator and denominator by 9.

$$= \dfrac{\left(\dfrac{1}{3} + \dfrac{1}{9}\right) \cdot 9}{\left(1 - \dfrac{1}{9}\right) \cdot 9} = \dfrac{3 + 1}{9 - 1} = \dfrac{4}{8} = \dfrac{1}{2}$$

EXAMPLE 5. Perform the operations indicated and simplify.

(a) $(x^2 + y^2)^{-1}$ (b) $(x^{-1} + y^{-1})^2$

Solutions:

We *must not* write *either one* of these as $x^{-2} + y^{-2}$! To do so would be to make the most common mistake in elementary mathematics [writing $(A + B)^n$ as $A^n + B^n$]! Each is an expression to a power. However, we cannot multiply the exponents, as the expression in each is a *sum*, not a product or quotient.

Be Careful!

(a) $(x^2 + y^2)^{-1} = \dfrac{1}{(x^2 + y^2)^1}$ Definition of negative exponent

$= \dfrac{1}{x^2 + y^2}$

(b) $(x^{-1} + y^{-1})^2 = \left(\dfrac{1}{x} + \dfrac{1}{y}\right)^2$ Definition of negative exponent

This is like the square of a binomial.

$= \left(\dfrac{1}{x}\right)^2 + 2 \cdot \dfrac{1}{x} \cdot \dfrac{1}{y} + \left(\dfrac{1}{y}\right)^2$

$= \dfrac{1}{x^2} + \dfrac{2}{xy} + \dfrac{1}{y^2}$ □

Scientific Notation

Multiply 12345 by 12345 on a calculator and it may display an answer like

$$\boxed{1.5239903 \quad ^{08}} \quad \text{or} \quad \boxed{1.5239903 \quad 08}$$

Either one of these is short for 1.5239903×10^8, a large number written in **scientific notation**.

Divide 3 by 12345, and the calculator may display an answer like

$$\boxed{2.4301337 \quad ^{-04}} \quad \text{or} \quad \boxed{2.4301337 \quad -04}$$

This is short for 2.4301337×10^{-4}, a small number written in scientific notation. Each of the numbers is written as a number between 1 and 10 times a power of 10. As is usually the case in calculator computations, these are both approximations.

Scientific notation is a convenient way to express very large and very small numbers. Two examples that occur in physics are Planck's constant,

$$h \approx 0.0000000000000000000000000006625 \text{ erg-sec}$$

and the speed of light in a vacuum,

$$c \approx 29900000000 \text{ cm/sec}$$

Written in scientific notation, these become

$$h \approx 6.625 \times 10^{-27} \text{ erg-sec} \quad \text{and}$$

$$c \approx 2.99 \times 10^{10} \text{ cm/sec}$$

These are both approximations, as the $\approx$ symbol indicates.

Sec. 5.1   Negative Integer Exponents

> **To Write a Number in Scientific Notation**
>
> 1. Starting with the number in decimal format, move the decimal point until the number is between 1 and 10. Count the number of places moved.
> 2. Multiply the number formed in step 1 by 10 to the power equal to the number of decimal places moved. If the original number was between 0 and 1, the power of 10 is negative. If the original number was greater than 1, the power is positive.
> 3. This procedure is for a **positive** number. If the original number is **negative**, perform steps 1 and 2 on the absolute value of the original number, then attach a negative sign to the result.

EXAMPLE 6. Write each number in scientific notation.

(a) 93,000,000

(b) 0.000001554

(c) $-254,000$

Solutions:

(a) Think of 93,000,000 as 93000000.0, then move the decimal point until the number is between 1 and 10.

   | Step 1 | 9.3 0 0 0 0 0 0.0   Move decimal point.

   The decimal point was moved 7 places. Now we write

   | Step 2 | 9.3×10^7 Write in scientific notation.

   The exponent is positive because 93,000,000 is greater than 1.

(b)   0.0 0 0 0 0 1.5 5 4   Move decimal point.

   We moved the decimal point 6 places. We write

   1.554×10^{-6} Write in scientific notation.

   The exponent is negative because 0.000001554 is between 0 and 1.

(c) $-254,000$ is negative, so we work with its absolute value.

   2.5 4 0 0 0.0   Move decimal point.

   We moved the decimal point 5 places. We have now

   2.54×10^5 Write in scientific notation.

   The exponent is positive because 254,000 is greater than 1. As the original number was negative, we write

   -2.54×10^5 Adjust sign.

EXAMPLE 7. Write each number without an exponent.

(a) 8.771×10^{14} (b) 3.2×10^{-13} (c) -9.99231×10^{-3}

Solutions:

To change scientific notation to standard decimal format, reverse the steps given above.

(a) 8.7 7 1 0 0 0 0 0 0 0 0 0 0 0 0.0

$$877,100,000,000,000$$

Note that the decimal point was moved 14 places to the right (to make a number larger than 1), because the exponent 14 is positive.

(b) 0.0 0 0 0 0 0 0 0 0 0 0 0 0 3.2

$$0.00000000000032$$

This time we moved the decimal point 13 places to the left (to make a number between 0 and 1), because the exponent -13 is negative.

(c) As the given number is negative, we work with its absolute value.

0.0 0 9.9 9 2 3 1

$$0.00999231$$

We moved the decimal point 3 places to the left because *the exponent of 10, -3, is negative*. Now we append a negative sign because the *original number* was negative.

$$-0.00999231$$

CALCULATOR BOX

Scientific Notation on a Calculator

$\boxed{\text{EE}}$

As noted above, a calculator will often give answers in scientific notation. The exact value of 500^5 is given by

$$500^5 = 31{,}250{,}000{,}000{,}000 = 3.125 \times 10^{13}$$

Compute 500^5 on a calculator ($\boxed{500}$ $\boxed{x^y}$ $\boxed{5}$ $\boxed{=}$) and note how the answer is displayed. The exact value of $\dfrac{1}{8000000}$ is given by

$$\frac{1}{8000000} = 0.000000125 = 1.25 \times 10^{-7}$$

Divide 1 by 8000000 on a calculator and note how this answer is displayed.

It is sometimes convenient to *enter* numbers into a calculator in scientific notation. A scientific calculator should have a key marked $\boxed{\text{EE}}$ or $\boxed{\text{EXP}}$. (If you cannot find such a key, consult an instruction manual.) To enter 7.2×10^8,

(continued)

Sec. 5.1 Negative Integer Exponents

press $\boxed{7.2}$ $\boxed{EE}$ $\boxed{8}$. The display should show $\boxed{7.2^{08}}$ or something similar. To enter 1.25×10^{-7}, press $\boxed{1.25}$ $\boxed{EE}$ $\boxed{7}$ $\boxed{+/-}$ and the display should show $\boxed{1.25^{-07}}$ or something similar.

As an example, we can use a calculator to find the product of Planck's constant and the speed of light in a vacuum. Recall that Planck's constant in scientific notation is $h \approx 6.625 \times 10^{-27}$ erg-sec, and the speed of light in a vacuum is $c \approx 2.99 \times 10^{10}$ cm/sec. Therefore, we are to calculate

$$hc \approx (6.625 \times 10^{-27})(2.99 \times 10^{10})$$

Press the keys

$\boxed{6.625}$ $\boxed{EE}$ $\boxed{27}$ $\boxed{+/-}$ $\boxed{\times}$ $\boxed{2.99}$ $\boxed{EE}$ $\boxed{10}$ $\boxed{=}$

and read $\boxed{1.980875^{-16}}$ on the display. So

$$hc \approx 1.980875 \times 10^{-16} \quad \text{or}$$

$$hc \approx 0.0000000000000001980875 \text{ erg-cm}$$

Calculator Exercises

Perform the calculations indicated using the following constants:

Planck's constant: $\quad h \approx 1.980875 \times 10^{-27}$
Speed of light: $\quad c \approx 2.99 \times 10^{10}$
Avogadro's number: $\quad N \approx 6.02217 \times 10^{23}$

Write answers in scientific notation. (Remember to indicate approximations.)

1. c^2
2. $\sqrt{N}$
3. $\dfrac{1}{h}$
4. $\dfrac{2c^2}{h}$
5. $(5 \times 10^7)\sqrt{h}$
6. $\sqrt{\dfrac{2Nh}{\pi c}}$

Answers:

1. $c^2 \approx 8.9401 \times 10^{20}$
2. $\sqrt{N} \approx 7.7603 \times 10^{11}$
3. $\dfrac{1}{h} \approx 5.04827 \times 10^{26}$
4. $\dfrac{2c^2}{h} \approx 9.0264 \times 10^{47}$
5. $(5 \times 10^7)\sqrt{h} \approx 2.2254 \times 10^{-6}$
6. $\sqrt{\dfrac{2Nh}{\pi c}} \approx 1.5937 \times 10^{-7}$

■ PROBLEM SET 5.1

Warm-ups

In problems 1 through 12, rewrite each expression without a negative exponent, and simplify. See Example 1.

1. 2^{-3}
2. 3^{-2}
3. 6^{-1}
4. $(-2)^{-4}$
5. -2^{-4}
6. $(-2)^{-3}$
7. -2^{-3}
8. -10^{-1}
9. $\left(\dfrac{1}{3}\right)^{-2}$
10. $-2x^{-3}$
11. $3^{-2}x$
12. $-5x^{-4}z$

In problems 13 through 24, use the properties of exponents to simplify each expression. The answers should not contain negative exponents. See Example 2.

13. $2^3 \cdot 2^{-2}$
14. $3^{-3} \cdot 3^3$
15. $5^{-1} \cdot 5^4$
16. $(2^{-3})^2$
17. $(3^2)^{-2}$
18. $(2^{-2})^{-3}$
19. $(xy)^{-1}$
20. $(2x)^{-2}$
21. $(-3x)^{-3}$
22. $\dfrac{5^3}{5^6}$
23. $\dfrac{2^3}{2^{-2}}$
24. $\dfrac{-x^{-2}}{x^3}$

In problems 25 through 39, write each expression without a negative exponent. For problems 25 through 30, see Example 3.

25. $(-5^{-1})^{-2}$
26. $(-x^{-2})^{-3}$
27. $(-x^{-3})^{-2}$
28. $(-2^{-2}x)^{-3}$
29. $\left(-\dfrac{x^{-1}}{2y}\right)^{-3}$
30. $\left(-\dfrac{2x^{-2}}{y^{-3}}\right)^{-4}$

For problems 31 through 39, see Example 4.

31. $\dfrac{2^{-3}}{3}$
32. $\dfrac{3}{5^{-2}}$
33. $-\dfrac{6^{-1}}{5^{-2}}$
34. $\left(\dfrac{-3}{x^2}\right)^{-2}$
35. $-\left(-\dfrac{3x^{-2}}{yz^{-1}}\right)^{-1}$
36. $-\left(-\dfrac{(-x^{-2})y^2}{-(-z^{-2})^3}\right)^0$
37. $\dfrac{(2x^{-1}y^2)^{-2}}{(3xy^{-2})^{-1}}$
38. $\dfrac{(-2x^3yz^{-1})^{-2}}{(x^{-1}yz^{-2})^2}$
39. $\dfrac{(2x)^{-3}yz^{-2}}{2x(y^2z)^{-1}}$

In problems 40 through 45, perform the operation indicated, and simplify. See Examples 4e and 5.

40. $2^{-1} + 3^{-1}$
41. $3^{-2} - 2^{-3}$
42. $(2^{-2} + 3^{-1})^{-1}$
43. $\dfrac{2^{-1} + 2^{-2}}{2^{-1}}$
44. $\dfrac{2^{-2} + 3^{-2}}{2^{-2} - 3^{-2}}$
45. $\dfrac{2^{-3} + x^{-1}}{x^{-2}}$

In problems 46 through 49, write each number in scientific notation. See Example 6.

46. 0.000000021367
47. -0.0000012345
48. $-32,000$
49. $77,722,000,000,000$

In problems 50 through 53, write each number without an exponent. See Example 7.

50. 1.609×10^5
51. 5.43×10^{-5}
52. 1.1×10^{-9}
53. -8.0×10^{11}

Practice Exercises

In problems 54 through 86, rewrite each expression without a negative exponent, and simplify.

54. 4^{-3}
55. 2^{-2}
56. 3^{-1}
57. $(-3)^{-4}$
58. -3^{-4}
59. $(-3)^{-3}$
60. -3^{-3}
61. -10^{-2}
62. $\left(\dfrac{1}{2}\right)^{-3}$
63. $\left(\dfrac{2}{3}\right)^{-3}$
64. $\left(-\dfrac{1}{10}\right)^{-1}$
65. $-\left(\dfrac{3}{5}\right)^{-1}$
66. $3x^{-2}$
67. $2^{-3}x$
68. $4x^{-5}z$
69. $-6x^{-2}y^{-1}$
70. $2(x^{-2}yz^3)^0$
71. $3t^{-21}$
72. $\dfrac{3^{-2}}{2}$
73. $\dfrac{5}{4^{-2}}$
74. $-\dfrac{6^{-2}}{5^{-1}}$
75. $\dfrac{-2x^{-2}}{y^{-3}}$
76. $\dfrac{3x^{-1}}{y^{-2}}$
77. $4^{-1} + 4^{-2}$
78. $2^{-1} - 3^{-1}$
79. $3^{-2} + 2^{-3}$
80. $(2^{-1} - 2^{-3})^{-1}$
81. $\dfrac{3^{-1} + 3^{-2}}{3^{-1}}$

Sec. 5.1 Negative Integer Exponents

82. $\dfrac{2^{-2} - 3^{-2}}{2^{-2} + 3^{-2}}$ 83. $\dfrac{3^{-2} - x^{-1}}{x^{-2}}$ 84. $\dfrac{1}{3^{-1}} + \left(\dfrac{1}{3}\right)^{-1}$

85. $\left(\dfrac{2}{x}\right)^{-3} + x^3$ 86. $\dfrac{s}{t} + \left(\dfrac{s}{t}\right)^{-1}$

In problems 87 through 125, use the properties of exponents to simplify each expression. The answers should not contain negative exponents.

87. $3^2 \cdot 3^{-3}$
88. $2^{-2} \cdot 2^2$
89. $6^{-1} \cdot 6^4$
90. $5^3 \cdot 5^{-5}$

91. $3^{-3} \cdot 3^{-1}$
92. $2^{-3} \cdot \dfrac{1}{2}$
93. $(3^{-2})^3$
94. $(2^3)^{-3}$

95. $(3^{-3})^{-2}$
96. $-(4^{-1})^2$
97. $(-4^{-1})^2$
98. $-(4^{-1})^{-2}$

99. $(-4^{-1})^{-2}$
100. $(-x^{-4})^{-3}$
101. $(-x^{-3})^{-4}$
102. $(xy)^{-2}$

103. $(3x)^{-3}$
104. $(-2x)^{-1}$
105. $(x^{-2}y^4)^{-3}$
106. $(2^{-3}x^2)^{-3}$

107. $-(3x^{-2}y)^{-3}$
108. $(-3^{-2}x)^{-1}$
109. $\left(\dfrac{x^{-1}}{3y}\right)^{-2}$
110. $\left(-\dfrac{3x^{-3}}{y^{-2}}\right)^{-4}$

111. $\left(\dfrac{-2}{x^3}\right)^{-3}$
112. $-\left(-\dfrac{2x^{-3}}{yz^{-1}}\right)^{-1}$
113. $-\left(-\dfrac{(-x^{-3})y^3}{(-z^{-3})^2}\right)^0$
114. $\dfrac{4^3}{4^6}$

115. $\dfrac{3^2}{3^{-3}}$
116. $\dfrac{-x^{-3}}{x^2}$
117. $\dfrac{(-y)^{-4}}{(-y)^{-2}}$
118. $\dfrac{(3t)^{-3}}{(3t)^{-6}}$

119. $\dfrac{3t^{-3}}{3t^{-6}}$
120. $\dfrac{(x+y)^{-3}}{(x+y)^2}$
121. $\dfrac{7(x+2)^{-2}}{(x+2)^3}$
122. $\dfrac{-2(x-y^2)^{-3}}{3(x-y^2)^{-2}}$

123. $\dfrac{(3x^2y^2)^{-2}}{(2xy^2)^{-1}}$
124. $\dfrac{(-2x^2z^{-1})^{-3}}{(2x^{-2}z^{-1})^2}$
125. $\dfrac{(2x)^{-2}y^2z^{-3}}{2x(y^{-2}z)^{-1}}$

In problems 126 through 131, write each number in scientific notation.

126. 11,280,000
127. 0.05432
128. 0.00000000088
129. 4402
130. $-600{,}066$
131. -0.901

In problems 132 through 137, write each number without an exponent.

132. 5.678×10^{-5}
133. 1.43×10^5
134. 6.81×10^{-2}
135. 1.1×10^{-11}
136. -4.32101×10^2
137. -1.0×10^{-1}

138. The rest mass of an electron is 9.11×10^{-28} g. How many electrons are there in 1 g of electrons?

139. Light travels at the rate of 2.99×10^{10} cm/sec. A light-year is the distance light travels in a year. The nearest star, other than the sun, is 4.3 light-years away. How far away is the nearest star, in centimeters?

140. Avogadro's number is the number of molecules in a mole of a substance. It's value is approximately 6×10^{23}. How many molecules are there in a gold ring that weighs 1 g? (One gram of gold is 1/197th of a mole).

Challenge Problems

We stated that the definition of negative exponents is consistent with the five properties of exponents. In problems 141 through 145, demonstrate the truth of the properties for the given values of m and n.

141. Using $m = -4$ and $n = -1$, demonstrate that the definition

$$x^{-k} = \dfrac{1}{x^k}$$

satisfies the property $x^m \cdot x^n = x^{m+n}$.

142. Using $m = -3$ and $n = 2$, demonstrate the property

$$(x^m)^n = x^{m \cdot n}$$

143. Using $n = -5$, demonstrate the property

$$(x \cdot y)^n = x^n \cdot y^n$$

144. Using $n = -1$, demonstrate the property
$$\left(\frac{x}{y}\right)^n = \frac{x^n}{y^n}.$$

145. Using $m = 3$ and $n = -2$, demonstrate that
$$\frac{x^m}{x^n} = x^{m-n}.$$

In problems 146 through 153, perform the operations indicated. Assume that j and k are integers.

146. $x^j(x^j + x^k)$

147. $(x^j + x^k)(x^j - x^k)$

148. $(x^j + x^k)^2$

149. $(x^j - x^{-j})^2$

150. $(x^j y^k)^j$

151. $(x^j y^k)^2$

152. Factor $x^{-4} - 16$ completely.

153. Factor $x^{-2} - x^{-1} - 6$.

IN YOUR OWN WORDS . . .

154. What is the relationship between x^{-n} and x^n?

155. Explain why $(x^{-1} + y^{-1})^{-1}$ is not the same as $x + y$.

156. Make a rule that we could use to simplify a fraction to the -1 power. $\left[\text{HINT: } \left(\dfrac{p}{q}\right)^{-1} = ?\right]$

5.2 ROOTS AND RADICALS

There are two real numbers whose squares are 4. They are 2 and -2. There are two real numbers whose squares are 9. They are 3 and -3. How about 7? Are there two real numbers whose squares are 7? The answer to this question is: Yes, there are two such numbers. However, they are not integers or even rational numbers. They are irrational numbers. How do we express them? The following notation has been chosen for such numbers. We let $\sqrt{7}$ be the *positive* number whose square is 7 and $-\sqrt{7}$ the negative number whose square is 7.

$$(\sqrt{7})^2 = 7 \qquad (-\sqrt{7})^2 = 7$$

The number $\sqrt{7}$ is called the **principal square root** of 7. As $\sqrt{4}$ is the positive number whose square is 4, we have $\sqrt{4} = 2$. Is there a real number whose square is -4? The answer is no. Such a number would violate our rules for signed-number arithmetic. So expressions such as $\sqrt{-4}$ are undefined in the system of real numbers.

Square Roots of Negative Numbers

If $k > 0$, then $\sqrt{-k}$ is undefined in the system of real numbers.

There is a real number whose cube is 8. It is 2. Also, there is a real number whose cube is -8. It is -2. Is there a real number whose cube is 11? Yes, it is denoted by $\sqrt[3]{11}$. Furthermore, there is a real number whose cube is -11, $\sqrt[3]{-11}$. In fact, $\sqrt[3]{-11} = -\sqrt[3]{11}$. We call $\sqrt[3]{11}$ the **cube root** of 11. Note that $\sqrt[3]{8} = 2$ and $\sqrt[3]{-8} = -2$. Every real number has a cube root.

In general, there are roots for every natural number, n, denoted by the expression

$$\sqrt[n]{q}$$

where $(\sqrt[n]{q})^n = q$. We call $\sqrt[n]{q}$ a **radical**. $\sqrt[n]{}$ is called a **radical sign**. The real number q is called the **radicand** and the natural number n is called the **index**. If the index is omitted, the radical is a square root.

> **Definition of $\sqrt[n]{q}$**
>
> If n is an *even* natural number,
>
> 1. $\sqrt[n]{q}$ $\begin{cases} \text{is a real number when } q \geq 0 \\ \text{is } not \text{ a real number when } q < 0. \end{cases}$
> 2. $\sqrt[n]{q}$ is the *nonnegative* number such that $(\sqrt[n]{q})^n = q$.
>
> If n is an *odd* natural number,
>
> 1. $\sqrt[n]{q}$ exists for all real numbers, q.
> 2. $\sqrt[n]{q}$ is the real number such that $(\sqrt[n]{q})^n = q$.
> 3. $\sqrt[n]{-q} = -\sqrt[n]{q}$.

EXAMPLE 1. Find each root.

(a) $\sqrt{121}$ (b) $\sqrt[3]{125}$ (c) $\sqrt[4]{81}$

(d) $\sqrt[3]{-27}$ (e) $\sqrt{\dfrac{9}{16}}$ (f) $\sqrt[3]{\dfrac{8}{125}}$

Solutions:

(a) As $11^2 = 121$ and 11 is *positive*,
$$\sqrt{121} = 11$$

(b) As $5^3 = 125$,
$$\sqrt[3]{125} = 5$$

(c) As $3^4 = 81$ and 3 is *positive*,
$$\sqrt[4]{81} = 3$$

(d) As $(-3)^3 = -27$,
$$\sqrt[3]{-27} = -3$$

(e) As $\left(\dfrac{3}{4}\right)^2 = \dfrac{9}{16}$ and $\dfrac{3}{4}$ is *positive*,
$$\sqrt{\dfrac{9}{16}} = \dfrac{3}{4}$$

(f) As $\left(\dfrac{2}{5}\right)^3 = \dfrac{8}{125}$,
$$\sqrt[3]{\dfrac{8}{125}} = \dfrac{2}{5}$$

We must be *very* careful when simplifying radicals with *even* indices if there are variables in the radicand. If we *know* that x is *not negative*, we can write $\sqrt{x^2} = x$. However, if x has the value -3, then to say that $\sqrt{x^2} = x$ would be to say that $\sqrt{(-3)^2} = -3$ or $\sqrt{9} = -3$, which is incorrect! The Challenge Problems at the end of this section investigate this problem further. Unless otherwise stated, we will assume that letters appearing in the radicand represent nonnegative real numbers.

EXAMPLE 2. Find each of the following roots. Assume that all variables represent **nonnegative** real numbers.

(a) $\sqrt{25x^2}$ (b) $\sqrt{16y^4}$
(c) $\sqrt[3]{-8x^3}$ (d) $\sqrt{576x^4y^8}$

Solutions:

(a) $\sqrt{25x^2} = 5x$

[since $5x$ is nonnegative and $(5x)^2 = 25x^2$]

(b) $\sqrt{16y^4} = 4y^2$

[$4y^2$ is nonnegative and $(4y^2)^2 = 16y^4$]

(c) $\sqrt[3]{-8x^3} = -2x$

[as $(-2x)^3 = -8x^3$]

Sometimes it is difficult to recognize a perfect square. In the next example, notice how factoring helps.

(d) $\sqrt{576x^4y^8} = \sqrt{2^6 \cdot 3^2 x^4 y^8}$
$= 2^3 \cdot 3x^2 y^4$
$= 24x^2 y^4$

As radicals are real numbers, we utilize them in real number algebra. The following properties are useful.

Properties of Radicals

If $\sqrt[n]{p}$ and $\sqrt[n]{q}$ represent real numbers, then

1. $\sqrt[n]{p \cdot q} = \sqrt[n]{p} \cdot \sqrt[n]{q}$ Root of a product
2. $\sqrt[n]{\dfrac{p}{q}} = \dfrac{\sqrt[n]{p}}{\sqrt[n]{q}}$; $q \neq 0$ Root of a quotient

---

These properties are used for simplifying roots of products and quotients. Consider $\sqrt{75}$. Since 75 is not a perfect square, $\sqrt{75}$ is not an integer. However, 75 has a factor that is a perfect square. Notice how the *root of a product* property works to simplify a product.

$\sqrt{75} = \sqrt{5^2 \cdot 3}$ Factor.
$= \sqrt{5^2}\sqrt{3}$ Root of a product
$= 5\sqrt{3}$

The *root of a product* property allows us to take out perfect square **factors** of the radicand. The same idea applies to cube roots. Perfect cube factors can be taken out of the radicand. The idea works for *n*th roots in general. First we factor the radicand to determine what can be taken out.

Sec. 5.2 Roots and Radicals

EXAMPLE 3. Simplify each of the following radicals. Assume that all variables represent **nonnegative** real numbers.

(a) $\sqrt{8}$ (b) $\sqrt{18}$ (c) $2\sqrt{72x^3}$

(d) $\sqrt[3]{-54x^3}$ (e) $\sqrt[3]{\dfrac{-16}{27}}$ (f) $\sqrt[5]{96x^{12}y^{10}}$

Solutions:

(a) $\sqrt{8} = \sqrt{2^2 \cdot 2}$ Factor.
$\phantom{\sqrt{8}} = \sqrt{2^2} \cdot \sqrt{2}$ Root of a product
$\phantom{\sqrt{8}} = 2\sqrt{2}$

(b) $\sqrt{18} = \sqrt{3^2 \cdot 2}$ Factor.
$\phantom{\sqrt{18}} = \sqrt{3^2} \cdot \sqrt{2}$ Root of a product
$\phantom{\sqrt{18}} = 3\sqrt{2}$

(c) $2\sqrt{72x^3} = 2\sqrt{6^2 \cdot 2 \cdot x^2 \cdot x}$ Factoring

We rearrange the factors so that the perfect squares are first.

$\phantom{2\sqrt{72x^3}} = 2\sqrt{6^2 \cdot x^2 \cdot 2x}$
$\phantom{2\sqrt{72x^3}} = 2\sqrt{6^2} \cdot \sqrt{x^2} \cdot \sqrt{2x}$ Root of a product
$\phantom{2\sqrt{72x^3}} = 12x\sqrt{2x}$

(d) $\sqrt[3]{-54x^3} = \sqrt[3]{(-3)^3 \cdot 2x^3}$ Factoring

We place the cubes first.

$\phantom{\sqrt[3]{-54x^3}} = \sqrt[3]{(-3)^3 x^3 \cdot 2}$
$\phantom{\sqrt[3]{-54x^3}} = \sqrt[3]{(-3)^3 x^3} \cdot \sqrt[3]{2}$ Root of a product
$\phantom{\sqrt[3]{-54x^3}} = -3x\sqrt[3]{2}$

(e) $\sqrt[3]{\dfrac{-16}{27}} = \dfrac{\sqrt[3]{-16}}{\sqrt[3]{27}}$ Root of a quotient

$\phantom{\sqrt[3]{\dfrac{-16}{27}}} = \dfrac{\sqrt[3]{(-2)^3 \cdot 2}}{\sqrt[3]{27}}$

$\phantom{\sqrt[3]{\dfrac{-16}{27}}} = \dfrac{\sqrt[3]{(-2)^3} \cdot \sqrt[3]{2}}{\sqrt[3]{27}}$ Root of a product

$\phantom{\sqrt[3]{\dfrac{-16}{27}}} = \dfrac{-2\sqrt[3]{2}}{3}$

(f) $\sqrt[5]{96x^{12}y^{10}} = \sqrt[5]{2^5 \cdot 3x^{10}x^2y^{10}}$
$\phantom{\sqrt[5]{96x^{12}y^{10}}} = \sqrt[5]{2^5 x^{10} y^{10} \cdot 3x^2}$
$\phantom{\sqrt[5]{96x^{12}y^{10}}} = \sqrt[5]{2^5 x^{10} y^{10}} \cdot \sqrt[5]{3x^2}$ Root of a product
$\phantom{\sqrt[5]{96x^{12}y^{10}}} = 2x^2 y^2 \sqrt[5]{3x^2}$

Example 4 illustrates that the *root of a product* property will split up **factors** but not **terms**.

EXAMPLE 4. Simplify each of the following expressions. Assume that all variables represent *nonnegative* real numbers.

(a) $\sqrt{x^2 y^2}$ (b) $\sqrt{x^2 + y^2}$ (c) $\sqrt{(x + y)^2}$

Solutions:

(a) $\sqrt{x^2 y^2} = \sqrt{x^2} \cdot \sqrt{y^2}$ Root of a product

because x^2 and y^2 are *factors*.

$= xy$

as x and y are both nonnegative.

(b) $\sqrt{x^2 + y^2}$ does not simplify, because x^2 and y^2 are not *factors* of the radicand. They are *terms*.

(c) $\sqrt{(x + y)^2} = x + y$

as x and y are both nonnegative.

Be sure to note the difference between parts (b) and (c) in Example 4. We must be careful in part (b). If x has the value 3 and y has the value 4, we can see that $\sqrt{x^2 + y^2}$ and $\sqrt{x^2} + \sqrt{y^2}$ represent *different* numbers.

$$\sqrt{3^2 + 4^2} = \sqrt{9 + 16} = \sqrt{25} = 5$$
$$\sqrt{3^2} + \sqrt{4^2} = 3 + 4 = 7$$

We say that a radical expression is simplified if it satisfies the following conditions.

Simplified Radical Expression

1. The power of any factor in a radicand is *less* than the index of the radical.
2. No radicand contains fractions or negative numbers and there are no radicals in a denominator.

■ PROBLEM SET 5.2

Warm-ups

In problems 1 through 26, simplify each expression. Assume that all variables represent nonnegative real numbers. For problems 1 through 9, see Example 1.

1. $2\sqrt{49}$
2. $-\sqrt{64}$
3. $\sqrt{121}$
4. $\sqrt[3]{125}$
5. $3\sqrt[3]{-64}$
6. $2\sqrt{36} - \sqrt{16}$
7. $\sqrt{\dfrac{25}{49}}$
8. $\sqrt{\dfrac{256}{625}}$
9. $\sqrt[3]{\dfrac{27}{64}}$

For problems 10 through 15, see Example 2.

10. $-\sqrt{4x^2}$
11. $\sqrt[3]{-8x^6 y^3}$
12. $\sqrt{100z^6}$
13. $-\sqrt{144 j^6 t^4}$
14. $\sqrt{\dfrac{121 y^4}{x^2 z^2}}$
15. $\sqrt{\dfrac{36 x^{10}}{y^8}}$

Sec. 5.2 Roots and Radicals

For problems 16 through 24, see Example 3.

16. $\sqrt{50}$
17. $-2\sqrt{45}$
18. $\sqrt{252}$
19. $\sqrt{240}$
20. $\sqrt[3]{40}$
21. $-3\sqrt[3]{-108}$
22. $\sqrt[4]{162}$
23. $\sqrt[6]{320}$
24. $\sqrt[3]{54x^3}$

For problems 25 and 26, see Example 4.

25. (a) $\sqrt{4y^2}$ (b) $\sqrt{4+y^2}$ (c) $\sqrt{(2+y)^2}$

26. (a) $\sqrt[3]{64k^6}$ (b) $\sqrt[3]{64+k^6}$ (c) $\sqrt[3]{(2+k)^6}$

Practice Exercises

In problems 27 through 71, simplify each expression, if possible. Assume that all variables represent nonnegative real numbers.

27. $3\sqrt{36}$
28. $\sqrt{81}$
29. $-\sqrt{100}$
30. $\sqrt{169} - 2\sqrt{49}$
31. $\sqrt[3]{216}$
32. $2\sqrt[3]{-125}$
33. $-\sqrt{16x^4y^2}$
34. $\sqrt[3]{27x^9y^3}$
35. $\sqrt{72}$
36. $-3\sqrt{20}$
37. $\sqrt{288}$
38. $\sqrt{275}$
39. $\sqrt[3]{54}$
40. $-5\sqrt[3]{-192}$
41. $\sqrt[5]{-486}$
42. $\sqrt{-256}$
43. $\sqrt{\dfrac{16}{81}}$
44. $\sqrt{\dfrac{144}{169}}$
45. $\sqrt{\dfrac{27}{49}}$
46. $\sqrt{\dfrac{40}{81}}$
47. $-3\sqrt{\dfrac{8}{9}}$
48. $\sqrt[3]{\dfrac{-81}{125}}$
49. $\sqrt{9x^3y^2}$
50. $2\sqrt{5xy^3z^2}$
51. $\sqrt[3]{\dfrac{64}{27}}$
52. $\sqrt{\dfrac{8}{25}}$
53. $\sqrt{\dfrac{75}{49}}$
54. $-2\sqrt{\dfrac{7}{25}}$
55. $\sqrt[3]{\dfrac{56}{27}}$
56. $\sqrt{4x^2y^3}$
57. $3\sqrt{7x^2y^3z}$
58. $\sqrt{169z^8}$
59. $-\sqrt{450x^3y^9}$
60. $-\sqrt{100j^4k^2l^3}$
61. $\sqrt{\dfrac{196}{p^4q^2}}$
62. $\sqrt{\dfrac{135x^7}{y^{12}}}$
63. $\sqrt[3]{16z^3}$
64. $\sqrt[3]{-128x^9y^4}$
65. $\sqrt[3]{\dfrac{216a^6b^{12}c^4}{d^{27}}}$
66. $\sqrt[4]{243xy^6z^{10}}$
67. $\sqrt[5]{-96x^{15}z^{26}}$
68. $\sqrt[5]{486x^6k^{12}}$
69. $\sqrt{(13+2x)^2}$
70. $\sqrt{x^4+16}$
71. $\sqrt{36+9v^2}$

Challenge Problems

72. If x is positive (or zero), $\sqrt{x^2}$ is x. Is $\sqrt{x^2} = x$ always true? Try -2 for x and work very carefully.
73. If x is *negative*, what is the value of $\sqrt{x^2}$?
74. Write a formula for $\sqrt{x^2}$ that is valid for all real numbers. (HINT: Remember the definition of $|x|$.)

In problems 75 through 80, simplify each expression. Assume that a, b, and c are real numbers with a positive and b negative.

75. $\sqrt{a^2}$
76. $\sqrt{b^2}$
77. $\sqrt{c^2}$
78. $\sqrt{(5+a)^2}$
79. $\sqrt{(b-1)^2}$
80. $\sqrt{(a+b)^2}$

IN YOUR OWN WORDS . . .

81. Explain what is meant by the "square root of a positive number."

5.3 OPERATIONS WITH RADICALS

We continue our study of radicals with a look at addition. We know that

$$3x + 4x = 7x$$

is always true because we added like terms, whereas

$$3x + 4y$$

cannot be simplified because x and y are **not** like terms. Compare the following two expressions.

$$\boxed{1} \quad 3\sqrt{5} + 4\sqrt{5}$$
$$\boxed{2} \quad 3\sqrt{5} + 4\sqrt{7}$$

The first can be simplified,

$$3\sqrt{5} + 4\sqrt{5} = 7\sqrt{5}$$

whereas the second cannot, because $\sqrt{5}$ and $\sqrt{7}$ are not alike. Only radicals with the same index and the same radicand can be combined.

EXAMPLE 1. Simplify each expression.

(a) $\sqrt{6} - 8\sqrt{6}$ (b) $\sqrt{3} + \sqrt{27}$
(c) $\sqrt{128} - \sqrt{32}$ (d) $\sqrt[3]{16} + \sqrt[3]{54} - \sqrt[3]{2}$
(e) $3\sqrt{8x^2y} + x\sqrt{2y^3} - x\sqrt{50y}$; x and y positive

Solutions:

(a) $\sqrt{6} - 8\sqrt{6} = -7\sqrt{6}$

(b) $\sqrt{3} + \sqrt{27} = \sqrt{3} + \sqrt{9 \cdot 3}$
$\phantom{\sqrt{3} + \sqrt{27}} = \sqrt{3} + 3\sqrt{3}$
$\phantom{\sqrt{3} + \sqrt{27}} = 4\sqrt{3}$

(c) $\sqrt{128} - \sqrt{32} = \sqrt{64 \cdot 2} - \sqrt{16 \cdot 2}$
$\phantom{\sqrt{128} - \sqrt{32}} = \sqrt{64}\sqrt{2} - \sqrt{16}\sqrt{2}$
$\phantom{\sqrt{128} - \sqrt{32}} = 8\sqrt{2} - 4\sqrt{2}$
$\phantom{\sqrt{128} - \sqrt{32}} = 4\sqrt{2}$

(d) $\sqrt[3]{16} + \sqrt[3]{54} - \sqrt[3]{2} = \sqrt[3]{8 \cdot 2} + \sqrt[3]{27 \cdot 2} - \sqrt[3]{2}$
$\phantom{\sqrt[3]{16} + \sqrt[3]{54} - \sqrt[3]{2}} = \sqrt[3]{8}\sqrt[3]{2} + \sqrt[3]{27}\sqrt[3]{2} - \sqrt[3]{2}$
$\phantom{\sqrt[3]{16} + \sqrt[3]{54} - \sqrt[3]{2}} = 2\sqrt[3]{2} + 3\sqrt[3]{2} - \sqrt[3]{2}$
$\phantom{\sqrt[3]{16} + \sqrt[3]{54} - \sqrt[3]{2}} = 4\sqrt[3]{2}$

(e) $3\sqrt{8x^2y} + x\sqrt{2y^3} - x\sqrt{50y} = 3\sqrt{4 \cdot 2x^2y} + x\sqrt{2y^2y} - x\sqrt{25 \cdot 2y}$
$\phantom{3\sqrt{8x^2y} + x\sqrt{2y^3} - x\sqrt{50y}} = 3\sqrt{4}\sqrt{x^2}\sqrt{2y} + x\sqrt{y^2}\sqrt{2y} - x\sqrt{25}\sqrt{2y}$
$\phantom{3\sqrt{8x^2y} + x\sqrt{2y^3} - x\sqrt{50y}} = 3 \cdot 2x\sqrt{2y} + xy\sqrt{2y} - 5x\sqrt{2y}$
$\phantom{3\sqrt{8x^2y} + x\sqrt{2y^3} - x\sqrt{50y}} = 6x\sqrt{2y} + xy\sqrt{2y} - 5x\sqrt{2y}$
$\phantom{3\sqrt{8x^2y} + x\sqrt{2y^3} - x\sqrt{50y}} = x\sqrt{2y} + xy\sqrt{2y}$ or $(x + xy)\sqrt{2y}$ ▫

Next, we look at multiplication. The *root of a product* property in Section 5.2 shows how to multiply radicals with the same index.

Multiplication of Radicals
If $\sqrt[n]{p}$ and $\sqrt[n]{q}$ represent real numbers, then
$$\sqrt[n]{p} \cdot \sqrt[n]{q} = \sqrt[n]{pq}$$

EXAMPLE 2. Perform the operations indicated. Assume that all variables are nonnegative real numbers.

(a) $\sqrt[3]{3xy} \cdot \sqrt[3]{-9x^5y^4}$ (b) $\sqrt{x} \cdot \sqrt{x}$
(c) $\sqrt{2}(\sqrt{3} + \sqrt{5})$ (d) $(\sqrt{2} + 2\sqrt{3})(5\sqrt{2} - \sqrt{3})$
(e) $(\sqrt{2} + 3\sqrt{5})^2$ (f) $(2\sqrt{2} - \sqrt{3})(2\sqrt{2} + \sqrt{3})$

Solutions:

(a) $\sqrt[3]{3xy} \cdot \sqrt[3]{-9x^5y^4} = \sqrt[3]{-27x^6y^5}$ Multiplication of radicals
$\phantom{\sqrt[3]{3xy} \cdot \sqrt[3]{-9x^5y^4}} = -3x^2y\sqrt[3]{y^2}$

(b) $\sqrt{x} \cdot \sqrt{x} = (\sqrt{x})^2$
$\phantom{\sqrt{x} \cdot \sqrt{x}} = x$ Definition of square root

The distributive property is used in the next example.

(c) $\sqrt{2}(\sqrt{3} + \sqrt{5}) = \sqrt{2} \cdot \sqrt{3} + \sqrt{2} \cdot \sqrt{5}$
$\phantom{\sqrt{2}(\sqrt{3} + \sqrt{5})} = \sqrt{6} + \sqrt{10}$

The next example is like the product of two binomials.

(d) $(\sqrt{2} + 2\sqrt{3})(5\sqrt{2} - \sqrt{3})$
$\phantom{(\sqrt{2} + 2\sqrt{3})} = \sqrt{2} \cdot 5\sqrt{2} - \sqrt{2}\sqrt{3} + 2 \cdot 5\sqrt{2}\sqrt{3} - 2\sqrt{3}\sqrt{3}$
$\phantom{(\sqrt{2} + 2\sqrt{3})} = 5 \cdot 2 - \sqrt{6} + 10\sqrt{6} - 2 \cdot 3$
$\phantom{(\sqrt{2} + 2\sqrt{3})} = 10 + 9\sqrt{6} - 6 = 4 + 9\sqrt{6}$

(e) Notice the special product, the *square of a binomial*, $(a - b)^2 = a^2 - 2ab + b^2$.
$(\sqrt{2} - 3\sqrt{5})^2 = (\sqrt{2})^2 - 2 \cdot 3\sqrt{2}\sqrt{5} + (3\sqrt{5})^2$
$\phantom{(\sqrt{2} - 3\sqrt{5})^2} = 2 - 6\sqrt{10} + 9 \cdot 5$
$\phantom{(\sqrt{2} - 3\sqrt{5})^2} = 47 - 6\sqrt{10}$

(f) We recognize the special product, the *difference of two squares*, $(a - b)(a + b) = a^2 - b^2$.
$(2\sqrt{2} - \sqrt{3})(2\sqrt{2} + \sqrt{3}) = (2\sqrt{2})^2 - (\sqrt{3})^2$
$\phantom{(2\sqrt{2} - \sqrt{3})(2\sqrt{2} + \sqrt{3})} = 4 \cdot 2 - 3 = 8 - 3$
$\phantom{(2\sqrt{2} - \sqrt{3})(2\sqrt{2} + \sqrt{3})} = 5$ ☐

Numbers that differ only in their middle sign, such as $2\sqrt{2} - \sqrt{3}$ and $2\sqrt{2} + \sqrt{3}$, are called **conjugates** of each other.

To divide two radicals with the **same** index, we use the *root of a quotient* property of Section 5.2.

> ### Division of Radicals
> If $\sqrt[n]{p}$ and $\sqrt[n]{q}$ represent real numbers and $q \neq 0$, then
> $$\frac{\sqrt[n]{p}}{\sqrt[n]{q}} = \sqrt[n]{\frac{p}{q}}$$

EXAMPLE 3. Perform each division, and simplify.

(a) $\dfrac{\sqrt{24}}{\sqrt{2}}$ (b) $\dfrac{\sqrt[3]{54}}{\sqrt[3]{2}}$

Solutions:

(a) $\dfrac{\sqrt{24}}{\sqrt{2}} = \sqrt{\dfrac{24}{2}}$ Division of radicals

$= \sqrt{12}$

$= 2\sqrt{3}$

(b) $\dfrac{\sqrt[3]{54}}{\sqrt[3]{2}} = \sqrt[3]{\dfrac{54}{2}}$ Division of radicals

$= \sqrt[3]{27}$

$= 3$

This property does not work as well with $\dfrac{\sqrt{5}}{\sqrt{3}}$.

$$\frac{\sqrt{5}}{\sqrt{3}} = \sqrt{\frac{5}{3}}$$

However, a radical expression is not considered simplified if it contains a fraction in the radicand or a radical in the denominator. So neither of the forms above is simplified. The process of removing a radical from the denominator of a fractional expression is called **rationalizing the denominator.**

The first technique we will learn for rationalizing the denominator applies if a **square root** is a **factor** of the denominator.

EXAMPLE 4. Rationalize the denominator.

(a) $\dfrac{5}{\sqrt{6}}$ (b) $\dfrac{2\sqrt{3}}{5\sqrt{2}}$ (c) $\sqrt{\dfrac{7}{12}}$

Solutions:

(a) $\sqrt{6}$ is a factor of the denominator. Multiply numerator and denominator by this factor.

$$\frac{5}{\sqrt{6}} = \frac{5 \cdot \sqrt{6}}{\sqrt{6}\sqrt{6}} = \frac{5\sqrt{6}}{6}$$

(continued)

(b) $\sqrt{2}$ is a factor of the denominator. Multiply numerator and denominator by this factor.

$$\frac{2\sqrt{3}}{5\sqrt{2}} = \frac{2\sqrt{3}\sqrt{2}}{5\sqrt{2}\sqrt{2}} = \frac{2\sqrt{6}}{5 \cdot 2}$$

$$= \frac{2\sqrt{6}}{10} = \frac{\sqrt{6}}{5}$$

(c) First, use the division property of radicals and simplify the denominator.

$$\sqrt{\frac{7}{12}} = \frac{\sqrt{7}}{\sqrt{12}} = \frac{\sqrt{7}}{2\sqrt{3}}$$

$\sqrt{3}$ is a factor of the denominator.

$$= \frac{\sqrt{7}\sqrt{3}}{2\sqrt{3}\sqrt{3}} = \frac{\sqrt{21}}{2 \cdot 3} = \frac{\sqrt{21}}{6} \qquad \square$$

If a radical of index higher than 2 is a **factor** of the denominator, we modify the procedure as shown in the next example.

EXAMPLE 5. Rationalize the denominator.

(a) $\dfrac{1}{\sqrt[3]{4}}$ (b) $\dfrac{3\sqrt[4]{7}}{2\sqrt[4]{3}}$

Solutions:

(a) As $4 = 2^2$ we can get $\sqrt[3]{2^3}$ in the denominator by multiplying numerator and denominator by $\sqrt[3]{2}$.

$$\frac{1}{\sqrt[3]{4}} = \frac{1 \cdot \sqrt[3]{2}}{\sqrt[3]{2^2}\sqrt[3]{2}}$$

$$= \frac{\sqrt[3]{2}}{\sqrt[3]{2^3}}$$

$$= \frac{\sqrt[3]{2}}{2}$$

(b) Multiply by $\sqrt[4]{3^3}$ to make a perfect fourth power in the denominator.

$$\frac{3\sqrt[4]{7}}{2\sqrt[4]{3}} = \frac{3\sqrt[4]{7}\sqrt[4]{3^3}}{2\sqrt[4]{3}\sqrt[4]{3^3}}$$

$$= \frac{3\sqrt[4]{7 \cdot 27}}{2\sqrt[4]{3^4}}$$

$$= \frac{3\sqrt[4]{189}}{2 \cdot 3}$$

$$= \frac{\sqrt[4]{189}}{2} \qquad \square$$

The third technique for rationalizing denominators applies if the denominator is a sum or difference of two terms containing **square roots.** We take advantage of the special product, the *difference of two squares*. That is,

$$(a - b)(a + b) = a^2 - b^2$$

EXAMPLE 6. Rationalize the denominator.

(a) $\dfrac{1}{5 - \sqrt{2}}$ (b) $\dfrac{2\sqrt{3} - \sqrt{2}}{2\sqrt{3} + \sqrt{2}}$

Solutions:

(a) Multiply both numerator and denominator by the number $5 + \sqrt{2}$, which is the conjugate of the denominator.

$$\frac{1}{5 - \sqrt{2}} = \frac{1 \cdot (5 + \sqrt{2})}{(5 - \sqrt{2})(5 + \sqrt{2})}$$

Notice that the denominator is now the difference of two squares.

$$= \frac{5 + \sqrt{2}}{5^2 - (\sqrt{2})^2} = \frac{5 + \sqrt{2}}{25 - 2}$$

$$= \frac{5 + \sqrt{2}}{23}$$

(b) Again, multiply the numerator and denominator by the conjugate of the denominator.

$$\frac{2\sqrt{3} - \sqrt{2}}{2\sqrt{3} + \sqrt{2}} = \frac{(2\sqrt{3} - \sqrt{2})(2\sqrt{3} - \sqrt{2})}{(2\sqrt{3} + \sqrt{2})(2\sqrt{3} - \sqrt{2})}$$

$$= \frac{(2\sqrt{3})^2 - 2 \cdot 2\sqrt{3} \cdot \sqrt{2} + (\sqrt{2})^2}{(2\sqrt{3})^2 - (\sqrt{2})^2}$$

$$= \frac{12 - 4\sqrt{6} + 2}{4 \cdot 3 - 2}$$

$$= \frac{14 - 4\sqrt{6}}{12 - 2}$$

$$= \frac{14 - 4\sqrt{6}}{10}$$

As 2 is a factor of both the numerator and the denominator, this fraction can be simplified. However, *be careful* either to factor the numerator first or to write the fraction as two fractions.

Be Careful!

$$= \frac{14}{10} - \frac{4\sqrt{6}}{10}$$

$$= \frac{7}{5} - \frac{2\sqrt{6}}{5} \quad \square$$

Before the advent of the hand-held calculator, rationalizing denominators was an important aid in calculation. It is much easier to approximate $\frac{1}{2}\sqrt{10}$ with a table of square roots than it is to approximate $\frac{\sqrt{5}}{\sqrt{2}}$. Although today most of us have calculators, it is still important to develop these skills in manipulating radicals. In further studies, it will be necessary to rationalize the denominator *or* the numerator of a fraction.

PROBLEM SET 5.3

Warm-ups

In problems 1 through 26, perform the operations indicated, and simplify. Assume that all variables represent nonnegative real numbers. For problems 1 through 12, see Example 1.

1. $\sqrt{11} - 3\sqrt{11}$
2. $3\sqrt{12} - 2\sqrt{12}$
3. $\sqrt{121} + 2\sqrt{36}$
4. $\sqrt{576} - 2\sqrt{25} + \sqrt{625}$
5. $\sqrt[3]{54} - \sqrt[3]{-16}$
6. $\sqrt[3]{81} - \sqrt[3]{24}$
7. $\sqrt{72} - \sqrt{1250} + 11\sqrt{8}$
8. $3\sqrt{18x^3} + 2x\sqrt{2x}$
9. $5x\sqrt{27y^5} - xy\sqrt{3y^3} + 2y^2\sqrt{12x^2y}$
10. $2\sqrt[3]{81} + \sqrt[3]{375} - 5\sqrt[3]{3}$
11. $\sqrt[3]{-432} - 2\sqrt[3]{-250}$
12. $2s\sqrt[3]{16st^4} - t\sqrt[3]{54s^4t} + 3\sqrt[3]{2s^4t^4}$

For problems 13 through 22, see Example 2.

13. $\sqrt{xy} \cdot \sqrt{x^3y^5}$
14. $\sqrt[3]{2xy^4} \cdot \sqrt[3]{16x^4y^2}$
15. $\sqrt{5}(2\sqrt{3} + \sqrt{5})$
16. $3\sqrt{6}(\sqrt{2} - \sqrt{3})$
17. $(1 - \sqrt{2})(2\sqrt{2} + 3)$
18. $(3\sqrt{5} + 1)(\sqrt{2} + \sqrt{3})$
19. $(2\sqrt{x} + \sqrt{y})^2$
20. $(5\sqrt{3} - 2\sqrt{2})^2$
21. $(1 - \sqrt{2})(1 + \sqrt{2})$
22. $(2\sqrt{3} + 3)(2\sqrt{3} - 3)$

For problems 23 through 26, see Example 3.

23. $\dfrac{\sqrt{48}}{\sqrt{6}}$
24. $\dfrac{\sqrt[3]{24}}{\sqrt[3]{-3}}$
25. $\dfrac{\sqrt[3]{108}}{\sqrt[3]{4}}$
26. $\dfrac{\sqrt{45}}{\sqrt{5}}$

In problems 27 through 44, rationalize the denominators. Assume that all variables represent nonnegative real numbers. For problems 27 through 32, see Example 4.

27. $\dfrac{1}{\sqrt{5}}$
28. $\dfrac{2}{3\sqrt{3}}$
29. $\dfrac{5}{2\sqrt{5}}$
30. $\sqrt{\dfrac{5}{6}}$
31. $\dfrac{\sqrt{14}}{2\sqrt{2x}}$
32. $\sqrt{\dfrac{11x}{5y}}$

For problems 33 through 38, see Example 5.

33. $\dfrac{1}{\sqrt[3]{7}}$
34. $\dfrac{1}{\sqrt[3]{36}}$
35. $\dfrac{5}{\sqrt[3]{5}}$
36. $\dfrac{\sqrt[3]{2} + 1}{3\sqrt[3]{4}}$
37. $\dfrac{1}{\sqrt[4]{4}}$
38. $\dfrac{x}{\sqrt[5]{8}}$

For problems 39 through 44, see Example 6.

39. $\dfrac{\sqrt{2}}{1 - \sqrt{2}}$
40. $\dfrac{5}{\sqrt{5} + 2}$
41. $\dfrac{2\sqrt{3}}{\sqrt{2} - \sqrt{3}}$
42. $\dfrac{2 - \sqrt{6}}{\sqrt{3} - \sqrt{2}}$
43. $\dfrac{\sqrt{k} - 1}{\sqrt{k} + 1}$
44. $\dfrac{2\sqrt{x}}{\sqrt{x} - \sqrt{y}}$

Practice Exercises

In problems 45 through 78, perform the operations indicated, and simplify. Assume that all variables represent nonnegative real numbers.

45. $\sqrt{13} + 3\sqrt{13}$
46. $4\sqrt{18} - 3\sqrt{18}$
47. $\sqrt{72} + 2\sqrt{98}$
48. $\sqrt{27} - 3\sqrt{12} + \sqrt{48}$
49. $\sqrt[3]{108} + \sqrt[3]{32}$
50. $\sqrt[3]{-16} - \sqrt[3]{-54}$
51. $\sqrt{18} - 2\sqrt{32}$
52. $2\sqrt{24} + 3\sqrt{54} - \sqrt{6}$
53. $\sqrt{75} - \sqrt{1587} + 8\sqrt{12}$
54. $3\sqrt{27x^3} + 3x\sqrt{3x}$
55. $7x\sqrt{108y^7} + xy^2\sqrt[3]{3y^3} - 12y^2\sqrt[3]{27x^2y^3}$
56. $2\sqrt[3]{54} - \sqrt[3]{250} + 7\sqrt[3]{2}$
57. $\sqrt[3]{-192} - 2\sqrt[3]{-375}$
58. $3p^2\sqrt[3]{16pq^5} - q\sqrt[3]{54p^5q^3} + 2\sqrt[3]{128p^7q^5}$
59. $\sqrt[3]{-3a^4b} \cdot \sqrt[3]{16ab^2}$
60. $\sqrt{5xy^3} \cdot \sqrt{8xy}$
61. $\sqrt{2}(2\sqrt{2} + \sqrt{6})$
62. $\sqrt{6}(\sqrt{2} - 2\sqrt{3})$
63. $(1 + \sqrt{x})(3\sqrt{x} - 2)$
64. $(2\sqrt{6} - 1)(\sqrt{2} - \sqrt{3})$
65. $(2 + \sqrt{5})(3\sqrt{2} + 2\sqrt{3})$
66. $(\sqrt{3} - \sqrt{5})(\sqrt{3} + \sqrt{7})$
67. $(1 - \sqrt{3})^2$
68. $(\sqrt{x} + 3)^2$
69. $(2\sqrt{3} - \sqrt{2})^2$
70. $(5\sqrt{5} + 2\sqrt{2})^2$
71. $(1 - \sqrt{k})(1 + \sqrt{k})$
72. $(3\sqrt{2} + 2)(3\sqrt{2} - 2)$
73. $(\sqrt{6} - \sqrt{5})(\sqrt{6} + \sqrt{5})$
74. $(\sqrt{3} + 3\sqrt{2})(\sqrt{3} - 3\sqrt{2})$
75. $\dfrac{\sqrt{72}}{\sqrt{6}}$
76. $\dfrac{\sqrt[3]{-56}}{\sqrt[3]{-7}}$
77. $\dfrac{\sqrt[3]{32}}{\sqrt[3]{2}}$
78. $\dfrac{\sqrt{120}}{\sqrt{10}}$

In problems 79 through 98, rationalize the denominators. Assume that all variables represent nonnegative real numbers.

79. $\dfrac{1}{\sqrt{7}}$
80. $\dfrac{3}{2\sqrt{2}}$
81. $\dfrac{3}{2\sqrt{x}}$
82. $\sqrt{\dfrac{6}{7}}$
83. $\dfrac{\sqrt{22}}{2\sqrt{2x}}$
84. $\sqrt{\dfrac{13s}{7t}}$
85. $\dfrac{\sqrt{3}}{2 + \sqrt{3}}$
86. $\dfrac{2}{\sqrt{2} - 5}$
87. $\dfrac{3\sqrt{2}}{\sqrt{3} - \sqrt{2}}$
88. $\dfrac{1 - \sqrt{5}}{1 + \sqrt{5}}$
89. $\dfrac{3\sqrt{2} - 1}{\sqrt{2} + 1}$
90. $\dfrac{3 + \sqrt{6}}{\sqrt{2} - \sqrt{3}}$
91. $\dfrac{\sqrt{2k} - 3}{\sqrt{2k} + 3}$
92. $\dfrac{\sqrt{x} + \sqrt{y}}{\sqrt{x} - \sqrt{y}}$
93. $\dfrac{2\sqrt{k} + 3\sqrt{m}}{\sqrt{k} + 2\sqrt{m}}$
94. $\dfrac{1}{\sqrt[3]{5}}$
95. $\dfrac{1}{\sqrt[3]{49}}$
96. $\dfrac{\sqrt[3]{3} - 2}{2\sqrt[3]{9}}$
97. $\dfrac{1}{\sqrt[4]{9}}$
98. $\dfrac{t}{\sqrt[4]{27}}$

Challenge Problems

In problems 99 and 100, multiply each expression, and simplify. Each problem is a special product. Identify it.

99. $(\sqrt[3]{a} - \sqrt[3]{b})(\sqrt[3]{a^2} + \sqrt[3]{ab} + \sqrt[3]{b^2})$
100. $(\sqrt[3]{a} + \sqrt[3]{b})(\sqrt[3]{a^2} - \sqrt[3]{ab} + \sqrt[3]{b^2})$

In problems 101 through 103, rationalize the denominator. (HINT: Use the results of problems 99 and 100.)

101. $\dfrac{1}{\sqrt[3]{3} - \sqrt[3]{2}}$
102. $\dfrac{2}{\sqrt[3]{3} + \sqrt[3]{5}}$
103. $\dfrac{\sqrt[3]{2}}{\sqrt[3]{2} - 1}$

Sec. 5.3   Operations With Radicals

In problems 104 and 105, factor the following, using irrational numbers.

104. $x^2 - 5$ [HINT: $5 = (\sqrt{5})^2$.]

105. $x^2 - 8$

106. $3x^2 - 2$

■ IN YOUR OWN WORDS . . .

107. What does it mean to "rationalize" a denominator?

108. Explain why $(\sqrt{x} - \sqrt{y})^2$ is not the same as $(\sqrt{x-y})^2$.

■ 5.4 RATIONAL EXPONENTS

Let's see if we can define exponents as fractions in a manner consistent with our earlier definitions and the five properties of exponents. If we have such a definition, then

$$(x^{1/2})^2 = x^{(1/2)2} \quad \text{Power of a power}$$
$$= x^1 = x$$

That is, $x^{1/2}$ is a number that when squared is x. We know such a number from Section 5.2; it is $\sqrt{x}$. So $x^{1/2} = \sqrt{x}$ is a definition of the one-half power that is consistent with exponent property, *power of a power*. Notice that as x is something squared, it **cannot** be negative.

By a similar argument,

$$(x^{1/3})^3 = x^{(1/3)3} = x^1 = x$$

and we see that $x^{1/3}$ is a number that when cubed is x. $\sqrt[3]{x}$ is such a number. Notice that x **can** be negative in this case.

As this can be generalized, we state the following definition, which is consistent with the five properties of exponents.

Exponent of the form $\dfrac{1}{n}$

If n is a natural number,

$$x^{1/n} = \sqrt[n]{x}; \quad x \geq 0 \quad \text{if } n \text{ is even}$$

EXAMPLE 1. Simplify each expression.

(a) $4^{1/2}$ (b) $27^{1/3}$ (c) $-256^{1/4}$ (d) $(-32)^{1/5}$

Solutions:

(a) $4^{1/2} = \sqrt{4}$
$\qquad = 2$

(b) $27^{1/3} = \sqrt[3]{27}$
$\qquad\quad = 3$

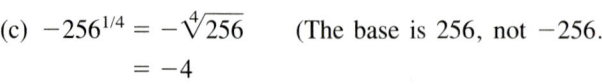

(c) $-256^{1/4} = -\sqrt[4]{256}$ (The base is 256, not -256.)
$\qquad\qquad = -4$

(d) $(-32)^{1/5} = \sqrt[5]{-32}$ (The base is -32.)
$\qquad\qquad\; = -2$

Exponent definitions and properties are summarized as follows.

Definitions

For n a natural number:

1. $x^n = \underbrace{x \cdot x \cdot x \cdots x}_{n\ x\text{'s}}$ Natural number power

2. $x^0 = 1;\ \ x \neq 0$ Zero power

3. $x^{-n} = \dfrac{1}{x^n};\ \ x \neq 0$ Negative power

4. $x^{1/n} = \sqrt[n]{x};\ \ x \geq 0$ if n even   Rational power

Properties

For x and y real numbers, and s and t reduced rational numbers:

1. $x^s \cdot x^t = x^{s+t}$ Product with same base
2. $(x^s)^t = x^{s \cdot t}$ Power of a power
3. $(xy)^s = x^s y^s$ Product to a power
4. $\left(\dfrac{x}{y}\right)^s = \dfrac{x^s}{y^s};\ \ y \neq 0$ Quotient to a power
5. $\dfrac{x^s}{x^t} = x^{s-t};\ \ x \neq 0$ Quotient with same base

With these definitions and properties, any rational number can be used as an exponent. Notice how to simplify numbers such as $4^{3/2}$.

EXAMPLE 2. Simplify each expression.

(a) $4^{3/2}$ (b) $81^{-3/4}$

Solutions:

(a) $4^{3/2} = 4^{(1/2)3}$
$\phantom{4^{3/2}} = (4^{1/2})^3$ Power of a power
$\phantom{4^{3/2}} = (\sqrt{4})^3 = 2^3$
$\phantom{4^{3/2}} = 8$

(b) $81^{-3/4} = 81^{(1/4)(-3)}$
$\phantom{81^{-3/4}} = (81^{1/4})^{-3}$ Power of a power
$\phantom{81^{-3/4}} = (\sqrt[4]{81})^{-3}$
$\phantom{81^{-3/4}} = 3^{-3} = \dfrac{1}{3^3}$
$\phantom{81^{-3/4}} = \dfrac{1}{27}$

In general, note the following result.

Sec. 5.4   Rational Exponents

> **Exponent Form ↔ Radical Form**
>
> If $\sqrt[n]{x}$ represents a real number, then
> $$x^{m/n} = (\sqrt[n]{x})^m \quad \text{and} \quad x^{m/n} = \sqrt[n]{x^m}$$

In computations, generally $(\sqrt[n]{x})^m$ is the most convenient. Consider evaluating $81^{3/4}$. The choice is between

1. $81^{3/4} = (\sqrt[4]{81})^3 = (3)^3 = 27$ or 2. $81^{3/4} = \sqrt[4]{81^3} = \sqrt[4]{531441} = 27$

Notice that in 2 we had to cube 81, then find the fourth root of 531441.

EXAMPLE 3. Simplify.

(a) $(-8)^{2/3}$ (b) $-100^{3/2}$ (c) $\left(\dfrac{256}{625}\right)^{-3/2}$ (d) $\left(\dfrac{27}{1000}\right)^{2/3}$

Solutions:

(a) $(-8)^{2/3} = (\sqrt[3]{-8})^2$ (The base is -8.)
$\qquad\qquad\;\; = (-2)^2 = 4$

(b) $-100^{3/2} = -(\sqrt{100})^3$ (The base is 100.)
$\qquad\qquad\;\;\; = -(10)^3 = -1000$

(c) $\left(\dfrac{256}{625}\right)^{-3/2} = \left(\sqrt{\dfrac{256}{625}}\right)^{-3}$

$\qquad\qquad\quad\; = \left(\dfrac{\sqrt{256}}{\sqrt{625}}\right)^{-3}$

$\qquad\qquad\quad\; = \left(\dfrac{16}{25}\right)^{-3} = \dfrac{16^{-3}}{25^{-3}} = \dfrac{25^3}{16^3}$

$\qquad\qquad\quad\; = \dfrac{4096}{15625}$

(d) $\left(\dfrac{27}{1000}\right)^{2/3} = \left(\sqrt[3]{\dfrac{27}{1000}}\right)^2$

$\qquad\qquad\quad\; = \left(\dfrac{\sqrt[3]{27}}{\sqrt[3]{1000}}\right)^2$

$\qquad\qquad\quad\; = \left(\dfrac{3}{10}\right)^2 = \dfrac{3^2}{10^2} = \dfrac{9}{100}$ ☐

EXAMPLE 4. Use the properties of exponents to simplify each expression.

(a) $3^{2/3} \cdot 3^{4/3}$ (b) $(16^{5/8})^{6/5}$ (c) $(9x^2)^{3/2}$; $x > 0$ (d) $\dfrac{32^{3/10}}{32^{1/10}}$

Solutions:

(a) $3^{2/3} \cdot 3^{4/3} = 3^{2/3+4/3}$ Product with same base
$\qquad\qquad\;\; = 3^{6/3} = 3^2$
$\qquad\qquad\;\; = 9$

(b) $(16^{5/8})^{6/5} = 16^{(5/8)\cdot(6/5)}$ Power of a power

$\qquad = 16^{3/4}$ Note: $\dfrac{5}{8} \cdot \dfrac{6}{5} = \dfrac{3}{4}$

$\qquad = (\sqrt[4]{16})^3 = 2^3$

$\qquad = 8$

(c) $(9x^2)^{3/2} = (\sqrt{9x^2})^3$

$\qquad = (3x)^3$ As $x > 0$

$\qquad = 27x^3$ Product to a power

(d) $\dfrac{32^{3/10}}{32^{1/10}} = 32^{3/10 - 1/10}$ Quotient with same base

$\qquad = 32^{2/10} = 32^{1/5}$

$\qquad = \sqrt[5]{32} = 2$ □

EXAMPLE 5. Simplify. Assume that all variables represent nonnegative real numbers. Express answers in a form containing nonnegative exponents.

(a) $(49m^3n^2)^{1/2}$ (b) $(-125x^{-3}y^6)^{-2/3}$

Solutions:

(a) $(49m^3n^2)^{1/2} = 49^{1/2}(m^3)^{1/2}(n^2)^{1/2}$ Product to a power

$\qquad = \sqrt{49}\, m^{3/2} n^{2/2}$

$\qquad = 7m^{3/2}n$

(b) $(-125x^{-3}y^6)^{-2/3} = (-125)^{-2/3}(x^{-3})^{-2/3}(y^6)^{-2/3}$

$\qquad = (\sqrt[3]{-125})^{-2} x^{-3\cdot(-2/3)} y^{6\cdot(-2/3)}$

$\qquad = (-5)^{-2} x^2 y^{-4}$

$\qquad = \dfrac{x^2}{(-5)^2 y^4} = \dfrac{x^2}{25y^4}$ □

EXAMPLE 6. Perform the operations indicated. Assume that all variables represent nonnegative real numbers.

(a) $x^{1/2}(x^{1/2} + x^{-1/2})$ (b) $5x^3(2x^{1/3} - 3x^{-1/6})$

Solutions:

(a) $x^{1/2}(x^{1/2} + x^{-1/2})$

By the distributive property,

$$x^{1/2}(x^{1/2} + x^{-1/2}) = x^{1/2} \cdot x^{1/2} + x^{1/2} \cdot x^{-1/2}$$

Next, we use the *product with same base* property.

$$= x^{1/2 + 1/2} + x^{1/2 - 1/2}$$

$$= x^1 + x^0 = x + 1$$

(b) $5x^3(2x^{1/3} - 3x^{-1/6}) = 5x^3 \cdot 2x^{1/3} - 5x^3 \cdot 3x^{-1/6}$ Distributive Property

$\qquad = 10x^{3+1/3} - 15x^{3-1/6}$ Product with same base

$\qquad = 10x^{10/3} - 15x^{17/6}$ □

Sec. 5.4   Rational Exponents

CALCULATOR BOX

Roots and Fractional Exponents with a Calculator

Calculators may be used to approximate roots and numbers to fractional powers. The following examples illustrate methods using an inexpensive scientific calculator.

EXAMPLE Use a calculator to approximate each expression to the nearest thousandth.

(a) $\sqrt{57}$ (b) $\sqrt[3]{-1192}$ (c) $38^{4/7}$ (d) $\sqrt[5]{19}$

Solutions:

(a) Every scientific calculator has a square root key. Enter 57 and press it.

$\boxed{57}$ $\boxed{\sqrt{}}$ and read $\boxed{7.5498344}$

$$\sqrt{57} \approx 7.550$$

(b) Many scientific calculators have a cube root key. If so, press

$\boxed{1192}$ $\boxed{+/-}$ $\boxed{\sqrt[3]{}}$ and read $\boxed{-10.602918}$

If there is no cube root key, use the method employed in part (d) below.

$$\sqrt[3]{-1192} \approx -10.603$$

(c) To raise 38 to the 4/7th power, we will utilize the $\boxed{y^x}$ key. (It may be necessary to press $\boxed{\text{inv}}$ or the 2nd function key before pressing the $\boxed{y^x}$ key.)

$\boxed{38}$ $\boxed{y^x}$ $\boxed{(}$ $\boxed{4}$ $\boxed{\div}$ $\boxed{7}$ $\boxed{)}$ $\boxed{=}$ and read $\boxed{7.9934357}$

$$38^{4/7} \approx 7.993$$

(d) To find $\sqrt[5]{19}$, we rewrite it in exponent form and follow the procedure described above.

$$\sqrt[5]{19} = 19^{1/5} = 19^{0.2}$$

So we press

$\boxed{19}$ $\boxed{y^x}$ $\boxed{0.2}$ $\boxed{=}$ and read $\boxed{1.8019831}$

$$\sqrt[5]{19} \approx 1.802$$

Calculator Exercises

Use a calculator to approximate each expression to the nearest thousandth.

1. $923^{3/5}$ 2. $\sqrt[6]{741}$ 3. $(0.975)^{5/9}$
4. $177^{4/7}$ 5. $\sqrt[8]{421}$ 6. $(1.175)^{7/9}$

Answers:

1. 60.134 2. 3.008 3. 0.986
4. 19.256 5. 2.128 6. 1.134

PROBLEM SET 5.4

Warm-ups

In problems 1 through 24, simplify if the expression represents a real number.
For problems 1 through 9, see Example 1.

1. $49^{1/2}$
2. $121^{1/2}$
3. $169^{1/2}$
4. $-100^{1/2}$
5. $(-100)^{1/2}$
6. $(-1000)^{1/3}$
7. $81^{1/4}$
8. $32^{1/5}$
9. $1024^{1/10}$

For problems 10 through 15, see Example 2.

10. $27^{2/3}$
11. $64^{5/6}$
12. $144^{3/2}$
13. $4^{7/2}$
14. $9^{-1/2}$
15. $27^{-2/3}$

For problems 16 through 24, see Example 3.

16. $-49^{-3/2}$
17. $(-49)^{-3/2}$
18. $2048^{-3/11}$
19. $\left(\dfrac{8}{125}\right)^{2/3}$
20. $\left(\dfrac{81}{16}\right)^{3/4}$
21. $\left(\dfrac{196}{169}\right)^{3/2}$
22. $\left(\dfrac{1}{64}\right)^{-5/6}$
23. $\left(\dfrac{100}{121}\right)^{-3/2}$
24. $\left(-\dfrac{343}{216}\right)^{-2/3}$

In problems 25 through 39, simplify, assuming that all variables represent nonnegative real numbers. Express answers in a form containing nonnegative exponents.
For problems 25 through 30, see Example 4.

25. $2^{1/2} \cdot 2^{5/2}$
26. $7^{8/3} \cdot 7^{-2/3}$
27. $(25^{3/8})^{4/3}$
28. $\left(\dfrac{4^{3/12}}{s^2}\right)^6$
29. $\dfrac{2^{7/3}}{2^{1/3}}$
30. $\dfrac{27^{11/6}}{27^{7/6}}$

For problems 31 through 39, see Example 5.

31. $(36x^2)^{1/2}$
32. $(144x^2y^4)^{1/2}$
33. $(49x^2y^{-2})^{1/2}$
34. $(8a^3b)^{1/3}$
35. $(-64x^{-6}z)^{-1/3}$
36. $-(81x^4y^{-4})^{3/4}$
37. $(8x^{-2}y)^{1/2}$
38. $-11(x^{1/2}y^{-1/2})^2$
39. $(3a^{2/3}x^{-1/3})^3$

In problems 40 through 44, perform the operation indicated. Assume that all variables represent nonnegative real numbers. Express answers in a form containing nonnegative exponents. See Example 6.

40. $x(x^{1/2} + x^{-1/2})$
41. $2x^2(3x^{1/6} - 5x^{-5/6})$
42. $x^{2/3}(2x^{2/3} + 3x^{-2/3})$
43. $(x^{1/2} + x^{-1/2})^2$
44. $(x^{1/2} + x^{-1/2})(x^{1/2} - x^{-1/2})$

Practice Exercises

In problems 45 through 74, simplify if the expression represents a real number.

45. $36^{1/2}$
46. $196^{1/2}$
47. $225^{1/2}$
48. $(-144)^{1/2}$
49. $-(100)^{1/2}$
50. $(-125)^{1/3}$
51. $81^{1/2}$
52. $81^{1/4}$
53. $-64^{1/6}$
54. $27^{1/3}$
55. $216^{1/3}$
56. $512^{1/3}$
57. $(-81)^{1/4}$
58. $(-32)^{1/5}$
59. $1024^{3/10}$
60. $8^{2/3}$
61. $32^{7/5}$
62. $100^{3/2}$
63. $9^{5/2}$
64. $25^{-1/2}$
65. $64^{-2/3}$
66. $-64^{-3/2}$
67. $(-27)^{-2/3}$
68. $4096^{-3/12}$
69. $\left(\dfrac{64}{343}\right)^{2/3}$
70. $\left(\dfrac{625}{256}\right)^{3/4}$
71. $\left(\dfrac{256}{225}\right)^{3/2}$
72. $\left(\dfrac{1}{32}\right)^{-4/5}$
73. $\left(\dfrac{169}{324}\right)^{-3/2}$
74. $\left(-\dfrac{8}{729}\right)^{-2/3}$

In problems 75 through 89, simplify, assuming that all variables represent nonnegative real numbers. Express answers in a form containing nonnegative exponents.

75. $(25y^2)^{1/2}$
76. $(100x^6y^2)^{1/2}$
77. $(16x^4y^{-6})^{1/2}$
78. $(-27a^9b^6)^{1/3}$
79. $(64x^{-3}z)^{-1/3}$
80. $-(16x^8y^{-8})^{3/4}$
81. $(18x^2y^{-2})^{1/2}$
82. $(x^{1/3}y^{-1/3})^3$
83. $-3(3a^{3/2}x^{-1/2})^2$
84. $\dfrac{x^{3/2}}{x^{1/2}y^{-1/2}}$
85. $\dfrac{5x^{-1/3}y^{1/3}}{y^{2/3}}$
86. $\dfrac{a^{3/5}b^{-1/4}}{5a^{-2/5}b}$
87. $\left(\dfrac{2x^{1/3}y}{3z^{2/3}}\right)^{-3}$
88. $\left(\dfrac{16x^4z^{-6}}{49y^{-8}}\right)^{1/2}$
89. $\left(\dfrac{25x^{-2/3}}{y^{2/3}}\right)^{-3/2}$

In problems 90 through 95, perform the operation indicated. Assume that all variables represent nonnegative real numbers. Express answers in a form containing nonnegative exponents.

90. $x(x^{-1/2} + x^{-3/2})$
91. $3x^3(2x^{1/3} - 5x^{-5/3})$
92. $x^{3/2}(3x^{3/2} + 2x^{-3/2})$
93. $2a^{-1/2}b^{3/2}(3a^{1/2}b^{-3/2} + 1)$
94. $(x^{3/2} - x^{-3/2})^2$
95. $(x^{3/2} + x^{-3/2})(x^{3/2} - x^{-3/2})$

Challenge Problems

In problems 96 through 99, perform the operation indicated. Assume that all variables represent nonnegative real numbers. Express answers in a form containing nonnegative exponents.

96. $(2x^{1/2} + x^{-1/2})(x^{3/2} - x^{1/2})$
97. $(a^{1/3} + b^{1/3})^3$
98. $(a^{1/3} + b^{1/3})(a^{2/3} - a^{1/3}b^{1/3} + b^{2/3})$
99. $(x^{1/3} - x^{-1/3})^3$
100. Factor $x^{1/3}$ from the expression $x^{2/3} + x^{1/3}$.
101. Factor $x^{2/3} - x^{1/3} - 6$.
102. Factor $z^{2/3} - 4$.
103. What is wrong with the following "proof" that $-1 = 1$?
$$-1 = (-1)^1 = (-1)^{2/2} = \sqrt{(-1)^2} = \sqrt{1} = 1$$

The property $\sqrt[n]{x^m} = x^{m/n}$ (provided that $\sqrt[n]{x}$ exists) allows us to simplify radicals when there are common factors between an exponent of the radicand and the index of the radical. Consider $\sqrt[6]{x^4}$. We can write this radical with a smaller index.

$$\sqrt[6]{x^4} = x^{4/6} = x^{2/3}$$
$$= \sqrt[3]{x^2}$$

In problems 104 through 106, write each radical with a smaller index.

104. $\sqrt[4]{x^6}$
105. $\sqrt[8]{x^4y^4}$
106. $\sqrt[6]{x^2}$

■ IN YOUR OWN WORDS . . .

107. Explain the relationship between radicals and fractional exponents.

■ 5.5 COMPLEX NUMBERS

In Section 5.2 it was noted that expressions such as $\sqrt{-4}$ have no meaning as real numbers because there is no real number that equals -4 when squared. In Chapter 6 we will solve equations like

$$x^2 + 4 = 0$$

Any solution to this equation must be a number whose square is -4. As there is no real number with this property, we will have to *enlarge* our set of numbers to include solutions to such equations. This new, larger set of numbers is called the set of **complex numbers.**

The set of complex numbers is constructed by introducing a new "number" to the set of real numbers. The new number is the number that when squared is -1. We call it "i." That is, i is a number such that

$$i^2 = -1$$

Other numbers must be included in this set so that the sum or product of two complex numbers will be a complex number. This results in the following set.

The Set of Complex Numbers

The set of complex numbers is the set of all numbers that can be written as

$$a + bi$$

where a and b are real numbers and i has the property

$$i^2 = -1$$

Some examples of complex numbers are

$$5 + 3i \qquad \frac{-6}{7} + 13i \qquad 1 - 2i$$

$$1 + \sqrt{22}\,i \qquad 6i \qquad -17$$

Notice that the set of real numbers is a subset of the set of complex numbers. (The real number -17 can be written as $-17 + 0i$; thus it belongs to the set of complex numbers.)

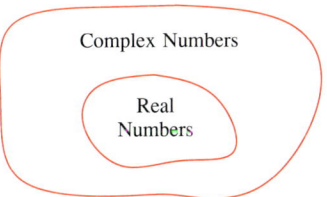

Next, let's look at $\sqrt{-5}$. If this were defined, it would be the number that when squared is -5. Of course, it cannot be a real number, but is it a complex number? To answer this question, we note that

$$(\sqrt{5}\,i)^2 = (\sqrt{5})^2 \cdot i^2 = 5 \cdot (-1) = -5$$

Thus a sensible definition is $\sqrt{-5} = \sqrt{5}\,i$. Or, in general,

---

Definition of Negative Radicand

If k is any positive real number,

$$\sqrt{-k} = \sqrt{k}\,i$$

Note that the i is **not** under the radical. To emphasize this $\sqrt{k}\,i$ may be written $i\sqrt{k}$.

Be Careful!

Sec. 5.5 Complex Numbers

EXAMPLE 1. Write each complex number in the form $a + bi$.

(a) $\sqrt{-7}$ (b) $3 + \sqrt{-4}$ (c) $1 - \sqrt{-12}$ (d) 6

Solutions:

(a) $\sqrt{-7} = \sqrt{7}i$ Definition of negative radicand

To answer the question, we put this complex number in the form $a + bi$.
$$= 0 + \sqrt{7}i$$

(b) $3 + \sqrt{-4} = 3 + \sqrt{4}i$ Definition of negative radicand
$$= 3 + 2i$$

(c) $1 - \sqrt{-12} = 1 - \sqrt{12}i$ Definition of negative radicand
$$= 1 - 2\sqrt{3}i$$

In the strictest sense, this is not quite in the form $a + bi$.
$$= 1 + (-2\sqrt{3})i$$

(d) Remember, real numbers *are* complex numbers.
$$6 = 6 + 0i$$

When dealing with symbols such as $\sqrt{-5}$ or $\sqrt{-k}$, with k positive, it is **important** to rewrite them as $\sqrt{5}i$ or $\sqrt{k}i$ **before** doing any arithmetic.

$\sqrt{-2} \cdot \sqrt{-3}$ *does not equal* $\sqrt{(-2)(-3)} = \sqrt{6}$

$\sqrt{-2}\sqrt{-3}$ equals $\sqrt{2}i \cdot \sqrt{3}i = \sqrt{6}i^2 = \sqrt{6}(-1) = -\sqrt{6}$

A complex number written in the form $a + bi$, where a and b are real numbers, is said to be in **standard form**. The real number a in the standard form is called the **real part** of the complex number, and the real number b is called the **imaginary part**.

EXAMPLE 2. Identify the real part and the imaginary part of each complex number.

(a) $5 + 17i$ (b) $3 - \sqrt{5}i$ (c) $\sqrt{-24} - 2$ (d) $\sqrt{18}$

Solutions:

(a) $5 + 17i$
   The real part is 5 and the imaginary part is 17.
(b) As $3 - \sqrt{5}i = 3 + (-\sqrt{5})i$:
   The real part is 3 and the imaginary part is $-\sqrt{5}$.
(c) First, write the number in standard form.
$$\sqrt{-24} - 2 = \sqrt{24}i - 2$$
$$= 2\sqrt{6}i - 2$$
$$= -2 + 2\sqrt{6}i$$

The real part is -2 and the imaginary part is $2\sqrt{6}$.
(d) As $\sqrt{18} = 3\sqrt{2} + 0i$:
   The real part is $3\sqrt{2}$ and the imaginary part is 0.

Let's look at some powers of i. Using the usual definitions,

$$i^0 = 1$$
$$i^1 = i$$
$$i^2 = -1$$
$$i^3 = i^2 \cdot i = (-1) \cdot i = -i$$
$$i^4 = (i^2)^2 = (-1)^2 = 1$$
$$i^5 = i^4 \cdot i = 1 \cdot i = i$$

and the pattern repeats $(1, i, -1, -i, 1, i,$ etc$)$. Can a power of i be found that fits into this pattern? Yes, for example i^{23}:

$$i^{23} = i^{20+3} = i^{20} \cdot i^3 = i^{4 \cdot 5} \cdot i^3 = (i^4)^5 \cdot i^3 = 1^5 \cdot i^3 = i^3 = -i$$

Notice that 3 is the remainder when you divide 23 by 4 *and*

$$i^{23} = i^3$$

The Principal Powers of i

$$i^0 = 1$$
$$i^1 = i$$
$$i^2 = -1$$
$$i^3 = -i$$

EXAMPLE 3. Rewrite each expression as $1, -1, i,$ or $-i$.
(a) i^{13} (b) i^{92} (c) i^{462}

Solutions:

(a) Four divides into 13 three times, with a remainder of 1, so

$$i^{13} = i^1$$
$$= i$$

(b) Division of 92 by 4 gives a remainder of 0. Thus

$$i^{92} = i^0$$
$$= 1$$

(c) If we divide 462 by 4, we get a remainder of 2.

$$i^{462} = i^2$$
$$= -1$$

Sec. 5.5   Complex Numbers

> **Sum, Difference, and Product of Complex Numbers**
>
> 1. Sum and difference
> $$(a + bi) + (c + di) = (a + c) + (b + d)i$$
> $$(a + bi) - (c + di) = (a - c) + (b - d)i$$
> 2. Product
> $$(a + bi) \cdot (c + di) = (ac - bd) + (ad + bc)i$$

Rather than memorize the definitions above, we usually do arithmetic with complex numbers the same way that we do arithmetic with polynomials, remembering to replace powers of i by 1, -1, i, or $-i$.

EXAMPLE 4. Perform the operation indicated.

(a) $(3 + 5i) + (2 - 3i)$ (b) $(3 + 5i) - (2 - 3i)$
(c) $(3 + 5i)(2 - 3i)$ (d) $\sqrt{-2}(1 - \sqrt{-2})$

Solutions:

(a) $(3 + 5i) + (2 - 3i) = (3 + 2) + (5i - 3i)$
$\qquad\qquad\qquad\qquad = 5 + 2i$
(b) $(3 + 5i) - (2 - 3i) = (3 + 5i) + (-2 + 3i)$
$\qquad\qquad\qquad\qquad = (3 - 2) + (5i + 3i)$
$\qquad\qquad\qquad\qquad = 1 + 8i$
(c) $(3 + 5i)(2 - 3i) = 6 - 9i + 10i - 15i^2$
$\qquad\qquad\qquad\quad = 6 + i - 15(-1)$
$\qquad\qquad\qquad\quad = 6 + i + 15$
$\qquad\qquad\qquad\quad = 21 + i$
(d) It is important to rewrite the square roots first.
$$\sqrt{-2}(1 - \sqrt{-2}) = \sqrt{2}i(1 - \sqrt{2}i)$$
$$= \sqrt{2}i - 2i^2$$
$$= \sqrt{2}i - 2(-1)$$
$$= 2 + \sqrt{2}i$$

Special products work with complex numbers as well.

EXAMPLE 5. Perform the operation indicated.

(a) $(2 + 3i)(2 - 3i)$ (b) $(2 + 3i)^2$ (c) $(2 - 5i)^3$

Solutions:

(a) The complex numbers $a + bi$ and $a - bi$ are called **complex conjugates.** The product of conjugates is an important special product.
$$(2 + 3i)(2 - 3i) = 2^2 - (3i)^2$$
$$= 4 - 9i^2 = 4 - 9(-1)$$
$$= 4 + 9 = 13$$

(b) $(2 + 3i)^2 = 4 + 12i + 9i^2$
$= 4 + 12i + 9(-1)$
$= 4 + 12i - 9$
$= -5 + 12i$

(c) $(2 - 5i)^3 = 2^3 - 3 \cdot 2^2 \cdot (5i) + 3 \cdot 2 \cdot (5i)^2 - (5i)^3$
$= 8 - 3 \cdot 4 \cdot 5 \cdot i + 6 \cdot 5^2 \cdot i^2 - 5^3 \cdot i^3$
$= 8 - 60i + 6 \cdot 25 \cdot (-1) - 125(-i)$
$= 8 - 60i - 150 + 125i$
$= -142 + 65i$

To do a division problem in complex numbers, write it as a fraction, then rationalize the denominator. That is, we multiply numerator and denominator by the conjugate of the denominator. Since we multiply conjugates every time, the following result saves some steps.

The Product of Complex Conjugates

$$(a + bi)(a - bi) = a^2 + b^2$$

EXAMPLE 6. Perform the operations indicated.

(a) $(2 + 3i) \div (1 + i)$ (b) $\dfrac{i}{1 - 2i}$

Solutions:

(a) First rewrite the problem as a fraction.

$$(2 + 3i) \div (1 + i) = \frac{2 + 3i}{1 + i}$$

Next, rationalize the denominator by multiplying the numerator and denominator by the **conjugate** of the **denominator.**

$$= \frac{(2 + 3i)(1 - i)}{(1 + i)(1 - i)}$$

$$= \frac{2 - 2i + 3i - 3i^2}{1^2 - i^2}$$

$$= \frac{2 - 2i + 3i + 3}{1 - (-1)}$$

$$= \frac{5 + i}{2}$$

$$= \frac{5}{2} + \frac{1}{2}i$$

(continued)

Sec. 5.5 Complex Numbers

(b) Again, rationalize the denominator.

$$\frac{i}{1 - 2i} = \frac{i(1 + 2i)}{(1 - 2i)(1 + 2i)}$$

$$= \frac{i + 2i^2}{1^2 + 2^2} \quad \text{Product of conjugates}$$

$$= \frac{i + 2(-1)}{1 + 4}$$

$$= \frac{-2 + i}{5} \quad \text{or} \quad -\frac{2}{5} + \frac{1}{5}i$$

EXAMPLE 7. Find the reciprocal of $4 - 7i$.

Solution:

$$\frac{1}{4 - 7i} = \frac{1 \cdot (4 + 7i)}{(4 - 7i)(4 + 7i)}$$

$$= \frac{4 + 7i}{4^2 + 7^2} \quad \text{Product of conjugates}$$

$$= \frac{4 + 7i}{16 + 49}$$

$$= \frac{4 + 7i}{65} \quad \text{or} \quad \frac{4}{65} + \frac{7}{65}i$$

EXAMPLE 8. Evaluate i^{-43}.

Solution:

By the definition of negative exponents,

$$i^{-43} = \frac{1}{i^{43}}$$

As 43 divided by 4 gives a remainder of 3, $i^{43} = i^3$.

$$= \frac{1}{i^3} = \frac{1}{-i} \quad \text{which we rationalize}$$

$$= \frac{1(i)}{-i(i)} = \frac{i}{-i^2} = \frac{i}{-(-1)} = \frac{i}{1} = i$$

PROBLEM SET 5.5

Warm-ups

In problems 1 through 9, write each number as a simplified complex number in standard form. See Example 1.

1. $\sqrt{-9}$
2. $\sqrt{-49}$
3. $-\sqrt{-121}$
4. $1 + \sqrt{-3}$
5. $3 - \sqrt{-4}$
6. $\sqrt{-5}$
7. $\sqrt{50}$
8. $3\sqrt{-3}$
9. $(\sqrt{-2})^2$

In problems 10 through 15, simplify the given powers of i. See Examples 3 and 8.

10. i^6 **11.** i^{11} **12.** i^{29}
13. i^{48} **14.** i^{-1} **15.** i^{-235}

In problems 16 through 39, perform the operations indicated.
For problems 16 through 25, see Example 4.

16. $(1 + i) + (2 + 3i)$ **17.** $(3 + 5i) - (2 + 3i)$
18. $(5 - 2i) + (3 - 5i)$ **19.** $(-4 - i) + (1 + 5i)$
20. $i(-7i)$ **21.** $i(14i)$
22. $i(3 + 2i)$ **23.** $3i(-2 - 5i)$
24. $(1 + i)(2 + 3i)$ **25.** $(1 - i)(5 + 2i)$

For problems 26 through 33, see Example 5.

26. $(1 + i)^2$ **27.** $(2 - 3i)^2$
28. $(\sqrt{2} - \sqrt{3}i)^2$ **29.** $(\sqrt{8} - \sqrt{-12})^2$
30. $(2 - i)(2 + i)$ **31.** $(5 + 2i)(5 - 2i)$
32. $(\sqrt{3} - \sqrt{2}i)(\sqrt{3} + \sqrt{2}i)$ **33.** $(\sqrt{7} + \sqrt{-5})(\sqrt{7} - \sqrt{-5})$

For problems 34 through 39, see Example 6.

34. $\dfrac{1}{1 + i}$ **35.** $\dfrac{1}{2 - i}$ **36.** $\dfrac{2}{1 - 3i}$

37. $\dfrac{1 + 2i}{1 + 3i}$ **38.** $\dfrac{2 - i}{3 + 2i}$ **39.** $(2 - i) \div (2 + i)$

Practice Exercises

In problems 40 through 48, write each number as a simplified complex number in standard form.

40. $\sqrt{-4}$ **41.** $\sqrt{-25}$ **42.** $-\sqrt{-169}$
43. $2 - \sqrt{-5}$ **44.** $3 + \sqrt{-9}$ **45.** $\sqrt{-7}$
46. $\sqrt{27}$ **47.** $2\sqrt{-2}$ **48.** $(\sqrt{-5})^2$

In problems 49 through 54, simplify the given powers of i.

49. i^7 **50.** i^{12} **51.** i^{30}
52. i^{49} **53.** i^{-3} **54.** i^{-237}

In problems 55 through 98, perform the operations indicated.

55. $(1 - i) + (2 - 3i)$ **56.** $(3 - 5i) - (2 - 3i)$
57. $(5 + 2i) + (3 + 5i)$ **58.** $(-4 + i) - (1 - 5i)$
59. $(3 + 3i) - (5 + 5i)$ **60.** $(6 - 7i) - (4 + 5i)$
61. $i(9i)$ **62.** $i(-11i)$
63. $i(2 - 3i)$ **64.** $2i(-3 + 4i)$
65. $(1 - i)(3 + 2i)$ **66.** $(1 + i)(2 + 4i)$
67. $(1 - 3i)(2 - 3i)$ **68.** $(3 + 2i)(2 - 5i)$
69. $(-2 - 3i)(1 - 2i)$ **70.** $(-4 - 5i)(-3 + 3i)$
71. $(1 - i)^2$ **72.** $(2 + 3i)^2$
73. $(3 + 4i)^2$ **74.** $(3 - i)^2$
75. $(\sqrt{3} + 3i)^2$ **76.** $(3 - \sqrt{3}i)^2$
77. $(\sqrt{3} + \sqrt{2}i)^2$ **78.** $(\sqrt{8} + \sqrt{-18})^2$
79. $(3 - i)(3 + i)$ **80.** $(4 + 3i)(4 - 3i)$
81. $(7 - 10i)(7 + 10i)$ **82.** $(3 + 2i)(2i - 3)$
83. $(\sqrt{3} - 2i)(\sqrt{3} + 2i)$ **84.** $(2 + \sqrt{3}i)(2 - \sqrt{3}i)$
85. $(\sqrt{2} - \sqrt{3}i)(\sqrt{2} + \sqrt{3}i)$ **86.** $(\sqrt{8} + \sqrt{-3})(\sqrt{8} - \sqrt{-3})$

Sec. 5.5 Complex Numbers

87. $\dfrac{1}{1-i}$
88. $\dfrac{1}{3+i}$
89. $\dfrac{3}{1-2i}$
90. $\dfrac{2}{3+2i}$
91. $\dfrac{i}{4+4i}$
92. $3i \div (5+i)$
93. $\dfrac{2+i}{1-3i}$
94. $\dfrac{3-i}{2+3i}$
95. $(5-i) \div (5+i)$
96. $(1-i)^3$
97. $(2+3i)^3$
98. $(1-5i)^3$

Challenge Problems

99. Show that i is a solution of $x^2 + 1 = 0$. Can you find a second solution?
100. Find two solutions of $x^2 + 4 = 0$.
101. Factor $x^2 + 9$. [HINT: $-9 = (3i)^2$.]
102. Show that $(a + bi)(a - bi) = a^2 + b^2$.

IN YOUR OWN WORDS . . .

103. Using the language of sets, describe the relationship between the set of real numbers and the set of complex numbers.
104. Why was it necessary to introduce the number i?

CHAPTER SUMMARY

GLOSSARY

The set of **complex numbers:** The set of all numbers that can be expressed as $a + bi$, where a and b are real numbers and i has the property $i^2 = -1$.

EXPONENTS

Definitions

For n a natural number:
1. $x^n = x \cdot x \cdot x \cdots x$; ($n$ x's)
2. $x^0 = 1$; $x \neq 0$
3. $x^{-n} = \dfrac{1}{x^n}$; $x \neq 0$
4. $x^{1/n} = \sqrt[n]{x}$; $x \geq 0$ if n even

Properties

For x and y real numbers and s and t reduced rational numbers:
1. $x^s \cdot x^t = x^{s+t}$
2. $(x^s)^t = x^{s \cdot t}$
3. $(xy)^s = x^s y^s$
4. $\left(\dfrac{x}{y}\right)^s = \dfrac{x^s}{y^s}$; $y \neq 0$
5. $\dfrac{x^s}{x^t} = x^{s-t}$; $x \neq 0$

DEFINITION OF nTH ROOTS OF REAL NUMBERS

If n is an *even* natural number:
1. $\sqrt[n]{x}$ is a real number *only* if $x \geq 0$.
2. $\sqrt[n]{x}$ is the *positive* number such that $(\sqrt[n]{x})^n = x$.

If n is an *odd* natural number:
1. $\sqrt[n]{x}$ is a real number for all real numbers, x.
2. $\sqrt[n]{x}$ is the real number such that $(\sqrt[n]{x})^n = x$.
3. $\sqrt[n]{-x} = -\sqrt[n]{x}$.

$$x^{m/n} = (\sqrt[n]{x})^m \quad \text{provided that } \sqrt[n]{x} \text{ is a real number}$$
$$(\sqrt[n]{x})^m = \sqrt[n]{x^m} \quad \text{provided that } \sqrt[n]{x} \text{ is a real number}$$

FRACTIONAL EXPONENTS RADICALS

If $k > 0$, then $\sqrt{-k}$ exists in the set of complex numbers and furthermore, $\sqrt{-k} = \sqrt{k}\,i$.

SQUARE ROOTS OF NEGATIVE NUMBERS

CHECKUPS

1. Use the properties of exponents to simplify.
$$\left(-\frac{xy^{-1}z^2}{3k^3}\right)^{-2}$$
Section 5.1; Example 4d

2. Write $-254{,}000$ in scientific notation. — Section 5.1; Example 6c
3. Simplify $\sqrt[3]{-54x^3}$. — Section 5.2; Example 3d
4. Simplify $3\sqrt{8x^2y} + x\sqrt{2y^3} - x\sqrt{50y}$; x and y positive. — Section 5.3; Example 1e
5. Perform the operation indicated.
$$(\sqrt{2} + 2\sqrt{3})(5\sqrt{2} - \sqrt{3})$$
Section 5.3; Example 2d

6. Rationalize the denominator.
$$\frac{2\sqrt{3} - \sqrt{2}}{2\sqrt{3} + \sqrt{2}}$$
Section 5.3; Example 6b

7. Simplify $(-125x^{-3}y^6)^{-2/3}$. — Section 5.4; Example 5b
8. Perform the operation indicated.
$$(3 + 5i)(2 - 3i)$$
Section 5.5; Example 4c

REVIEW PROBLEMS

In problems 1 through 30, simplify, if possible. Assume that all variables represent positive real numbers. The answers should not contain negative exponents.

1. 5^{-2}
2. -3^{-4}
3. $(-2)^{-6}$
4. $\left(\dfrac{11}{12}\right)^{-2}$
5. $7q^{-3}$
6. $(-2^{-2} - 2^{-1})^{-2}$
7. $6^{-4} \cdot 6^6$
8. $(3^{-2})^2$
9. $\dfrac{4x^{-2}}{y^{-1}}$
10. $(-x^2)^{-3}$
11. $\left(\dfrac{7^{-5}}{7^{-3}}\right)^{-1}$
12. $\dfrac{2x^{-1}}{3x^2y^{-2}}$
13. $\left(\dfrac{3x^2y^{-3}}{z^{-1}}\right)^{-3}$
14. $\dfrac{3^{-2} + 3^{-1}}{3^{-2} - 3^{-1}}$
15. $\dfrac{p}{q} - \left(\dfrac{p}{q}\right)^{-2}$
16. $\sqrt{75}$
17. $\sqrt[3]{-128}$
18. $\sqrt{8x^3yz^4}$
19. $\sqrt{\dfrac{24x^5}{54x}}$
20. $\sqrt[3]{\dfrac{-8x^{-2}}{xy^3}}$
21. $\sqrt{4 + x^2}$
22. $-49^{1/2}$
23. $32^{4/5}$
24. $8^{-2/3}$

Sec. 5.5 Complex Numbers

25. $\left(\dfrac{49}{121}\right)^{-3/2}$

26. $\left(\dfrac{2x^{1/2}y^{-1/3}}{z^{1/3}}\right)^6$

27. $\left(\dfrac{4x^6y^{-4}}{9z^{-2}}\right)^{-1/2}$

28. $\dfrac{x^{1/3}y^{-2/3}}{x^{-2/3}y^{1/3}}$

29. $\left(\dfrac{-8x^{3/2}y^{-1/2}}{z^{-3}}\right)^{-2/3}$

30. $\dfrac{4^{-1/2}+1}{4^{1/2}}$

In problems 31 through 40, perform the operations indicated, and simplify. Assume that all variables represent positive real numbers.

31. $\sqrt{50} - 3\sqrt{8}$
32. $\sqrt{432} - 2\sqrt{147} + 3\sqrt{3}$
33. $\sqrt[3]{16x^4} + x\sqrt[3]{2x} - \sqrt[3]{54x^4}$
34. $\sqrt{2}(\sqrt{6} - 1)$
35. $(\sqrt{3} - 1)(2 + \sqrt{2})$
36. $(\sqrt{2} - \sqrt{3})^2$
37. $(2\sqrt{5} + \sqrt{7})(2\sqrt{5} - \sqrt{7})$
38. $x^{1/2}(x^{1/2}y - x^{-1/2})$
39. $(x^{1/2} + y^{-1/2})^2$
40. $(2x^{1/2} - 3y^{-1/2})(2x^{1/2} + 3y^{-1/2})$

In problems 41 through 49, rationalize the denominator.

41. $\dfrac{2}{\sqrt{6}}$
42. $\dfrac{11}{7\sqrt{11}}$
43. $\dfrac{\sqrt{3}}{\sqrt{5}}$
44. $\dfrac{1}{\sqrt[3]{6}}$
45. $\dfrac{4}{\sqrt[5]{16}}$
46. $\dfrac{\sqrt[3]{4}+\sqrt[3]{2}}{\sqrt[3]{2}}$
47. $\dfrac{6}{\sqrt{5}-1}$
48. $\dfrac{\sqrt{2}}{\sqrt{2}+\sqrt{6}}$
49. $\dfrac{\sqrt{3}-2}{\sqrt{3}+2}$

In problems 50 through 53, write each number in scientific notation.

50. 289,000
51. 0.000007681
52. 91.02
53. -0.00000000000988

In problems 54 through 57, write each number without an exponent.

54. 1.112×10^{-6}
55. 9.006×10^{14}
56. -8.1734×10^3
57. -3.2×10^{-11}

In problems 58 through 60, write each complex number in standard form.

58. $\sqrt{-225}$
59. $\sqrt{18} + \sqrt{-18}$
60. $\sqrt{-2}(\sqrt{2} - \sqrt{-2})$

In problems 61 through 68, perform the operation indicated.

61. $(2 - 3i) + (7 - 11i)$
62. $(5 - 4i) - (2 - 6i)$
63. $4i(3 - 2i)$
64. $(1 + 2i)(5 - 3i)$
65. $(3 - 5i)^2$
66. $(6 - 7i)(6 + 7i)$
67. $\dfrac{3-7i}{5+5i}$
68. $(2 - 3i)^3$

... LET'S NOT FORGET ...

Identify the expressions that are in factored form. Factor those that are not factored.

69. $(x - 3)^2$
70. $r^2(x - y) + t^2(x - y)$
71. $(2x + 3)(4x^2 - 6x + 9)$
72. $(2x + 3)^3$
73. $8x^3 - 27$
74. $4x^2 - 12xy + 9y^2$

How many terms are in each expression? Which expressions have $1 - x$ as a factor?

75. $1 - x^2$
76. $1 - x^3$
77. $2 - x - x^2$
78. $(1 - x)^2$
79. $1 - x + x^2$

Simplify each expression, if possible, leaving only nonnegative exponents in your answer.

80. $\dfrac{8x^3 - 1}{1 - 4x^2}$

81. $\dfrac{-5x^{-1}}{y^2 z^{-3}}$

82. $-4^{-1/2}$

83. $\left(\dfrac{27a^{-6}}{8b^{-9}}\right)^{-1/3}$

84. $\dfrac{2^{-1} + 3^{-1}}{2^{-2}}$

85. $\dfrac{\sqrt{12} - \sqrt{-4}}{2}$

86. $\sqrt{(x^2 + y^2)^2}$

Find each product.

87. $\sqrt{6}(\sqrt{2} + \sqrt{6})$

88. $(x + \sqrt{2})^2$

89. $(2 - x)^3$

Reduce, if possible.

90. $\dfrac{r^2(x - a) - s^2(x - a)}{(x - a)(r + s)}$

91. $\dfrac{4 + \sqrt{8}}{4}$

The following problems can be worked by using a least common denominator. Follow the directions in each and notice how the LCD is used.

92. Solve $\dfrac{1}{x + 1} + \dfrac{1}{x - 2} = \dfrac{1}{x^2 - x - 2}$.

93. Perform the subtraction indicated: $\dfrac{4}{x + 1} - \dfrac{1}{x - 2}$.

94. Simplify $\dfrac{1 - \dfrac{1}{x + 1}}{\dfrac{1}{x - 2} + 1}$.

Label each as an equation or an expression. Solve the equations and perform the operations indicated on the expressions.

95. $\left|\dfrac{2x - 5}{3}\right| = 3$

96. $\sqrt{x + 3} = 2$

97. $\dfrac{x}{x^2 + x - 6} - \dfrac{1}{x + 3}$

98. $\dfrac{3(x + 1)}{2} - 1 = \dfrac{x + 2}{5}$

99. $(4 - 3x)^2$

100. $(3 - 2i)(1 + i)$

CHAPTER 5 TEST

In problems 1 through 9, choose the correct answer.

1. $(6^{-1/2})^{-2} = (?)$
   A. 6   B. 9   C. $\dfrac{1}{6}$ D. $-\dfrac{1}{9}$

2. $\dfrac{(b^{1/2})^6}{(b^{-1})^2} = (?)$
   A. $-b^2$ B. b^5 C. $\dfrac{1}{b^2}$ D. $\dfrac{1}{b}$

3. $\sqrt[3]{-8x^4y^2} = (?)$
   A. $-2x^3\sqrt[3]{xy^2}$ B. $-4x^2y$
   C. $-2x\sqrt[3]{xy^2}$ D. Not defined

4. $\sqrt{32} + \sqrt{18} - \sqrt{50} = (?)$
   A. 2   B. $2\sqrt{2}$ C. 0   D. 4

5. $i^{117} = (?)$
   A. 1   B. i C. -1 D. $-i$

6. $(\sqrt{5} - 2)^2 = (?)$
   A. 1   B. 21   C. $9 - 2\sqrt{5}$ D. $9 - 4\sqrt{5}$

Sec. 5.5   Complex Numbers

7. $\left(-\dfrac{729}{64}\right)^{2/3} = (?)$

    A. $\dfrac{81}{16}$ B. $-\dfrac{243}{32}$

    C. $-\dfrac{81}{16}$ D. $\dfrac{9}{4}$

8. $\sqrt{-3}(2 + \sqrt{-3}) = (?)$

    A. $3 + 2\sqrt{3}i$ B. $-3 - 6i$

    C. $-3 + 2\sqrt{3}i$ D. $3 - 2\sqrt{3}i$

9. $(5 - 2i)^2 = (?)$

    A. 29   B. 21   C. $29 - 20i$ D. $21 - 20i$

In problems 10 through 13, perform the operations indicated. Assume that all variables represent non-negative real numbers. Simplify each answer and express it in a form containing no negative exponents.

10. $\dfrac{-2x^{-1}}{y^{-2}}$

11. $\dfrac{x^{-1} + y^{-1}}{(xy)^{-1}}$

12. $\left(\dfrac{36a^4b^{-12}}{4c^{-6}}\right)^{-1/2}$

13. $\dfrac{i}{1 + i}$

In problems 14 and 15, rationalize the denominator.

14. $\dfrac{3}{2\sqrt{6}}$

15. $\dfrac{7 - \sqrt{2}}{1 + \sqrt{2}}$

CHAPTER 6

See Problem Set 6.6, Exercise 23.

Nonlinear Equations in One Variable

- **6.1** Solutions by Factoring
- **6.2** Applications
- **6.3** Completing the Square
- **6.4** The Quadratic Formula
- **6.5** Equations of Higher Degree
- **6.6** More Fractional Equations
- **6.7** Radical Equations

CONNECTIONS

During the fifteenth and sixteenth centuries mathematicians tried to solve more complicated nonlinear equations. They were hampered by their inability to accept negative numbers and complex numbers (Section 5.5).

Girolamo Cardano (1501–1576), an Italian, worked on solving the equation, $x(10 - x) = 40$. He found the two solutions, $5 + \sqrt{-15}$ and $5 - \sqrt{-15}$, but these numbers were most puzzling to him. Today, equations such as this arise as mathematical models in electrical engineering as well as in other scientific fields.

In this chapter we examine nonlinear equations in one variable. In particular, we look at the second-degree equation in detail. With the addition of an important property of our number system, we will have the tools at hand to solve all members of this important class of equations. We look at three useful methods for solving second-degree equations: factoring, completing the square, and the quadratic formula.

6.1 Solutions by Factoring

In this section second-degree equations are solved by factoring.

Definition

A **quadratic equation** in one variable is an equation of the form

$$ax^2 + bx + c = 0$$

where a, b, and c are real numbers with $a \neq 0$.

If $a > 0$, this form is called **standard form.**

If $a = 0$, the equation becomes linear. For example,

$$2x^2 - 5x + 3 = 0$$

is a quadratic equation where $a = 2$, $b = -5$, and $c = 3$.

$$3x^2 - x = 0$$

is an example where $a = 3$, $b = -1$, and $c = 0$, while

$$\frac{1}{2}x^2 - 5 = 0$$

is another example, where $a = \frac{1}{2}$, $b = 0$, and $c = -5$. The equation

$$2x^2 = -5x + 2$$

is also a quadratic equation. Rewrite it as

$$2x^2 + 5x - 2 = 0$$

to see that $a = 2$, $b = 5$, and $c = -2$.

The two tools used to solve linear equations, adding the same number to both sides and multiplying both sides by the same nonzero number, are not enough to solve higher-degree equations. However, our number system has the following important property which will provide the other tool necessary to solve higher-degree equations.

Property of Zero Products

The statement $pq = 0$ is *true* if either $p = 0$ or $q = 0$, and *false* if neither p nor q is 0.

Consider the quadratic equation
$$x^2 - 5x + 6 = 0$$
The left side of the equation will factor
$$(x - 3)(x - 2) = 0$$
This equation is certainly equivalent to the first. Notice that this is a statement that the product of two real numbers equals zero. The Property of Zero Products says that this statement will be true if *either* $(x - 3)$ *or* $(x - 2)$ is zero. That is,

$$x - 3 = 0 \quad \text{or} \quad x - 2 = 0$$
$$x = 3 \quad \text{or} \quad x = 2$$

Thus 3 and 2 are each solutions of $x^2 - 5x + 6 = 0$, so the solution set is
$$\{2, 3\}$$

EXAMPLE 1. Solve $x^2 + 3x + 2 = 0$.

Solution:

$$x^2 + 3x + 2 = 0$$
$$(x + 2)(x + 1) = 0 \qquad \text{Factor.}$$
$$x + 2 = 0 \quad \text{or} \quad x + 1 = 0 \qquad \text{Property of Zero Products}$$
$$x = -2 \quad \text{or} \quad x = -1$$
$$\{-2, -1\}$$

Solving a Quadratic Equation by Factoring

1. Write the equation in standard form.
2. Factor.
3. Apply the Property of Zero Products.
4. Check the solutions in the original equation if required.
5. Write the solution set.

Sec. 6.1 Solutions by Factoring

EXAMPLE 2. Solve $2x^2 + 5x = 3$.

Solution:

$$2x^2 + 5x - 3 = 0$$
$$(2x - 1)(x + 3) = 0 \quad \text{Factor.}$$
$$2x - 1 = 0 \quad \text{or} \quad x + 3 = 0 \quad \text{Property of Zero Products}$$
$$2x = 1 \quad \text{or} \quad x = -3$$
$$x = \frac{1}{2} \quad \text{or} \quad x = -3$$
$$\left\{\frac{1}{2}, -3\right\}$$

Notice that we *first* put the equation in standard form and *then* factored it.

EXAMPLE 3. Solve $x^2 + 2x = 0$.

Solution:

$$x^2 + 2x = 0$$
$$x(x + 2) = 0 \quad \text{Factor.}$$
$$x = 0 \quad \text{or} \quad x + 2 = 0$$
$$\{0, -2\}$$

Be Careful! A common mistake is to lose the zero solution in quadratics like the example above. Note that with only a little mental effort the solution set can be written directly after the equation has been factored.

EXAMPLE 4. Solve $(3y + 5)(2y + 1) = -1$.

Solution:

Be Careful! Be very careful to write the equation in standard form first. It is very tempting to set each factor equal to -1 and solve. *That would be wrong!*

$$(3y + 5)(2y + 1) = -1$$
$$6y^2 + 13y + 5 = -1$$
$$6y^2 + 13y + 6 = 0 \quad \text{Standard form}$$
$$(3y + 2)(2y + 3) = 0 \quad \text{Factor.}$$
$$\left\{\frac{-2}{3}, \frac{-3}{2}\right\}$$

EXAMPLE 5. Solve $x^2 + \dfrac{2}{3}x - \dfrac{8}{3} = 0$.

Solution:

Multiply both sides by 3 to clear fractions.

$$3x^2 + 2x - 8 = 0$$
$$(x + 2)(3x - 4) = 0$$
$$\left\{-2, \dfrac{4}{3}\right\}$$

EXAMPLE 6. Solve $x^2 - 4 = 0$.

Solution:

$$x^2 - 4 = 0$$
$$(x + 2)(x - 2) = 0$$
$$\{-2, 2\}$$

A common abbreviation for $\{-2, 2\}$ is $\{\pm 2\}$. Also, we often write

$$x = \pm k \quad \text{which means} \quad x = -k \quad or \quad x = k$$

EXAMPLE 7. Solve $2t^2 = 18$.

Solution:

$$2t^2 = 18$$
$$2t^2 - 18 = 0$$

Be sure to factor out the common factor first!

$$2(t^2 - 9) = 0$$
$$t^2 - 9 = 0 \quad \text{Divide by 2.}$$
$$(t - 3)(t + 3) = 0 \quad \text{Factor.}$$
$$\{-3, 3\} \quad \text{or} \quad \{\pm 3\}$$

Notice how the factor of 2 in Example 7 did not become part of the solution set.

EXAMPLE 8. Solve $-2x^2 + 9x + 35 = 0$.

Solution:

$$-2x^2 + 9x + 35 = 0$$

(continued)

Factoring is much easier if the leading coefficient is positive. So multiply both sides of the equation by -1 to write the equation in standard form.

$$2x^2 - 9x - 35 = 0$$
$$(2x + 5)(x - 7) = 0$$
$$\left\{\frac{-5}{2}, 7\right\}$$

EXAMPLE 9. Solve $z^2 - 6z + 9 = 0$.

Solution:

$$z^2 - 6z + 9 = 0$$
$$(z - 3)(z - 3) = 0$$
$$\{3\}$$

The solution set of Example 9 has only one number because both factors of the quadratic were the same. Such a single solution is called a **repeated root** or a root of **multiplicity 2**.

PROBLEM SET 6.1

Warm-ups

In problems 1 through 18, write the solution set.

1. $(x + 7)(x + 3) = 0$
2. $(x - 5)(x + 11) = 0$
3. $(x + 5)(x - 5) = 0$
4. $x(x - 3) = 0$
5. $2x(x + 5) = 0$
6. $(2x - 1)(x + 3) = 0$
7. $(x - 7)(7x + 1) = 0$
8. $(6x - 1)(2x - 3) = 0$
9. $(5x - 1)(5x + 1) = 0$
10. $-2x(3x - 1) = 0$
11. $(3x + 7)(2x - 5) = 0$
12. $(x + 4)(x + 4) = 0$
13. $x^2 = 0$
14. $(4x - 9)^2 = 0$
15. $-2(t - 3)(2t - 9) = 0$
16. $(y - \sqrt{3})(y + \sqrt{3}) = 0$
17. $(z - 2i)(z + 2i) = 0$
18. $(z + \sqrt{3}i)(z - \sqrt{3}i) = 0$

In problems 19 through 52, find the solution set by factoring. For problems 19 through 26, see Examples 1 and 2.

19. $x^2 - x - 2 = 0$
20. $x^2 - 2x - 3 = 0$
21. $x^2 + 6x = -5$
22. $x^2 - 4x + 3 = 0$
23. $x^2 - 8x + 15 = 0$
24. $3x^2 - 5x + 2 = 0$
25. $4v^2 + 5v + 1 = 0$
26. $2x^2 + 5x - 3 = 0$

For problems 27 through 30, see Example 3.

27. $x^2 - 2x = 0$
28. $x^2 - 6x = 0$
29. $x^2 = -7x$
30. $x^2 + 4x = 0$

For problems 31 through 34, see Example 4.

31. $(x + 6)(x - 1) = -10$
32. $(x - 6)(x - 2) = -3$
33. $(x + 4)^2 = 9$
34. $(x - 5)(x - 4) = 12$

For problems 35 through 40, see Example 5.

35. $\frac{2}{5}x^2 - \frac{3}{5}x - 1 = 0$ **36.** $x^2 - \frac{x}{6} - \frac{1}{3} = 0$ **37.** $\frac{3}{2}t^2 + \frac{5}{2}t = 1$

38. $\frac{4}{3}x^2 + \frac{4}{3}x - 1 = 0$ **39.** $\frac{1}{2}x^2 + \frac{7}{6}x - 1 = 0$ **40.** $t^2 + \frac{1}{6}t - \frac{1}{3} = 0$

For problems 41 through 44, see Example 6.

41. $x^2 - 4 = 0$ **42.** $x^2 = 16$
43. $x^2 - 9 = 0$ **44.** $x^2 - 25 = 0$

For problems 45 through 48, see Example 7.

45. $2x^2 - 6x = 0$ **46.** $6x^2 - 16x = 0$

47. $2x^2 - 8x = 0$ **48.** $4t^2 + 6t + 2 = 0$

For problems 49 and 50, see Example 8.

49. $-6x^2 + 5x + 1 = 0$ **50.** $-7x^2 - 4x + 3 = 0$

For problems 51 and 52, see Example 9.

51. $x^2 - 16x + 64 = 0$ **52.** $4x^2 - 4x = -1$

Practice Exercises

In problems 53 through 90, find the solution set by factoring.

53. $x^2 - 5x - 6 = 0$ **54.** $x^2 + 5x - 6 = 0$ **55.** $x^2 + 7x + 12 = 0$
56. $x^2 - 12x - 13 = 0$ **57.** $x^2 - 6x - 7 = 0$ **58.** $x^2 - 8x + 16 = 0$
59. $x^2 - 6x - 16 = 0$ **60.** $x^2 + 10x = -24$ **61.** $(x - 7)(x - 8) = 6$
62. $(x - 6)(x + 5) = 12$ **63.** $x^2 - 24x + 144 = 0$ **64.** $4y^2 - 7y - 2 = 0$
65. $5x^2 - 11x + 2 = 0$ **66.** $15w^2 + 12 = 28w$ **67.** $3x^2 + 5x + 2 = 0$
68. $5x^2 + 8x + 3 = 0$ **69.** $2x^2 - 7x + 6 = 0$ **70.** $3x^2 - 108x = 0$
71. $5x^2 - 6x - 8 = 0$ **72.** $2x^2 - 5x - 12 = 0$ **73.** $3x^2 + x = 10$
74. $7x^2 + 39x - 18 = 0$ **75.** $5x^2 - 4x - 33 = 0$ **76.** $7x^2 + 45x - 28 = 0$
77. $6x^2 + x = 12$ **78.** $9x^2 = 1$ **79.** $4x^2 + 4x - 3 = 0$
80. $6x^2 - 29x - 22 = 0$ **81.** $2s^2 - 2s - 4 = 0$ **82.** $6x^2 + 15x - 9 = 0$
83. $-x^2 + x + 12 = 0$ **84.** $-y^2 + 5y - 6 = 0$ **85.** $-2t^2 + 24t + 26 = 0$
86. $-3x^2 - 9x - 6 = 0$ **87.** $5x^2 + \frac{1}{3}x = 2$ **88.** $\frac{9}{5}t^2 - \frac{9}{5}t - 2 = 0$
89. $5x^2 + \frac{7}{4}x - \frac{3}{2} = 0$ **90.** $\frac{4}{3}w^2 - \frac{1}{3}w - \frac{1}{2} = 0$

Sec. 6.1 Solutions by Factoring

Challenge Problems

In problems 91 through 96, find the solution set by factoring.

91. $x^2 - 5 = 0$ [HINT: Find the product $(x - \sqrt{5})(x + \sqrt{5})$.]

92. $x^2 + 4 = 0$ [HINT: Find the product $(x - 2i)(x + 2i)$.]

93. $w^2 + 5 = 0$ **94.** $x^2 + 8 = 0$

95. $4x^2 + 25 = 0$ **96.** $2v^2 + 11 = 0$

In problems 97 through 102, find a quadratic equation with the solution set given.

97. $\{1, 2\}$ **98.** $\{-1, 3\}$ **99.** $\{\pm 2\}$

100. $\{\pm \sqrt{5}\}$ **101.** $\{\pm 2i\}$ **102.** $\left\{\dfrac{1}{2}, \dfrac{3}{2}\right\}$

■ IN YOUR OWN WORDS

103. Explain how factoring can be used to solve a quadratic equation.

■ 6.2 APPLICATIONS

Many word problems lead to quadratic equations. In this section, such applications are examined and the techniques developed in Section 6.1 are applied to solve them.

Often, there will be more than one solution to a quadratic equation. Each solution of the equation must be tested in the statement of the word problem to answer the original question.

Recall the procedures developed in Chapter 4 for the solution of word problems.

A Procedure to Solve Word Problems

1. Read the problem to determine what quantities are to be found.
2. Assign a variable, such as x, to represent one of the quantities to be found.
3. Express all other quantities to be found in terms of x (or the variable chosen).
4. Draw a figure or picture to illustrate the problem, if possible. Label it.
5. Read the problem and form an equation.
6. Solve the equation and find the value of all quantities to be found.
7. Check the solution(s) of the equation in the original word problem. It should make sense and answer the question.
8. Write an answer to the original question.

EXAMPLE 1. Find two consecutive positive even integers whose product is 224.

Solution:

We follow the general procedure for solving word problems, and choose a variable to represent one of the numbers.

Let x be the smaller of the two integers. Then $x + 2$ is the larger of the two. The product of the two numbers is 224.

$$x(x + 2) = 224$$

Solve the equation.

$$x^2 + 2x - 224 = 0$$

$$(x + 16)(x - 14) = 0$$

$$x = -16 \quad \text{or} \quad x = 14$$

However, the question asked for the *positive* integers. So discard -16. Checking the other solution in the problem, note that 14 and 16 are consecutive positive even integers and $14(16) = 224$. So we write the answer:

The integers are 14 and 16. ☐

EXAMPLE 2. A 3 ft by 4 ft rectangular flag is to be made with a blue rectangular center and a uniformly wide white border. If the blue part and the white part have the same area, how wide should the white border be?

Solution:

Let x be the width of the uniform border, in feet.

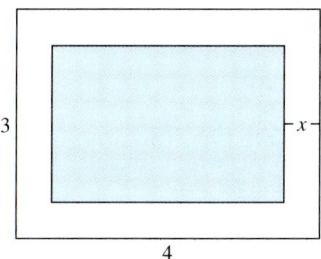

The figure shows that the area of the center (blue) rectangle is $(3 - 2x)(4 - 2x)$, and from the problem it must be the same as the area of the white border. Thus the area of the blue rectangle is one-half of the total area of the flag (3 times 4). So form the equation

$$(3 - 2x)(4 - 2x) = \frac{1}{2}(3)(4)$$

$$12 - 14x + 4x^2 = 6$$

$$4x^2 - 14x + 6 = 0$$

$$2x^2 - 7x + 3 = 0$$

$$(2x - 1)(x - 3) = 0$$

$$x = \frac{1}{2} \quad \text{or} \quad x = 3$$

A 3-ft border does not make sense since the width of the flag is 3 ft.

The border is to be $\frac{1}{2}$ ft wide. ☐

EXAMPLE 3. If the area of a triangle is 14 ft² and its base is 3 ft longer than its height, find the length of the base and the height.

Solution:

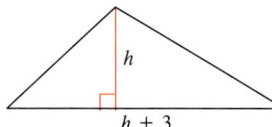

Let h be the height of the triangle in feet. Then $h + 3$ is the length of the base in feet. Remembering that the area of a triangle is one-half the base times the height (altitude), form the equation

$$14 = \frac{1}{2}(h + 3)h$$

Multiply both sides by 2, then apply the distributive law.

$$28 = (h + 3)h = h^2 + 3h$$

Then write in standard form, and solve.

$$0 = h^2 + 3h - 28$$
$$0 = (h + 7)(h - 4)$$
$$h = -7 \quad \text{or} \quad h = 4$$

Since -7 does not make sense in this problem, h must be 4.

The height is 4 ft and the base is 7 ft.

EXAMPLE 4. Find the length of the hypotenuse in a right triangle with legs of 3 m and 4 m.

Solution:

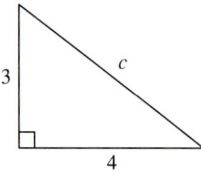

Let c be the length of the hypotenuse in meters. By the Pythagorean Theorem,

$$3^2 + 4^2 = c^2$$
$$9 + 16 = c^2$$
$$25 = c^2$$
$$0 = c^2 - 25$$
$$0 = (c - 5)(c + 5)$$
$$c = -5 \quad \text{or} \quad c = 5$$

Since -5 cannot be a length, the hypotenuse is 5 m.

EXAMPLE 5. The lengths of the sides of a right triangle are three consecutive even integers. What are they?

Solution:

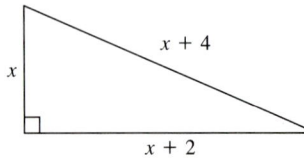

Let x be the smallest of the three consecutive even integers. Then $x + 2$ is the next one, and $x + 4$ is the largest. Since the hypotenuse is the longest side of a right triangle, its length must be given by $x + 4$. By the Pythagorean Theorem, we have the equation

$$x^2 + (x + 2)^2 = (x + 4)^2$$

Square the two binomials and simplify.

$$x^2 + x^2 + 4x + 4 = x^2 + 8x + 16$$
$$x^2 - 4x - 12 = 0$$
$$(x - 6)(x + 2) = 0$$
$$x = 6 \quad \text{or} \quad x = -2$$

Since a length cannot be negative, the smallest side must be 6. The sides of the triangle are 6, 8, and 10 units. □

PROBLEM SET 6.2

Warm-ups

For problems 1 through 6, see Example 1.

1. The two Taylor girls were born in consecutive, even years. If the product of their ages is 224, how old is each girl?
2. Find two consecutive negative odd numbers whose product is 143.
3. Two bike trails have a total length of 20 miles. If the product of the lengths of the trails is 91, find the length of each trail.
4. Find two numbers whose sum is 21 and product is 68.
5. Find three consecutive positive integers such that the sum of their squares is 77.
6. Find three consecutive negative integers such that the sum of their squares is 110.

For problems 7 through 14, see Example 2.

7. Find the dimensions of a rectangular poster of length 4 in. less than twice its width and with an area of 70 in.^{2}
8. Find the dimensions of a rectangular picture whose length is 2 in. greater than three times its width and whose area is 56 in.^{2}
9. The perimeter of a rectangular billboard is 44 m and its area is 120 m^2. What are its dimensions?
10. What are the dimensions of a rectangular placemat of perimeter 56 cm and area 187 cm^2?
11. An artist wishes to put a frame of uniform width around a painting whose dimensions are 4 ft by 6 ft. How wide should she make the frame if the total area is to be 48 ft^2?
12. A rectangular garden is to be made with a uniform path completely around the outside. If the garden is 8 ft by 12 ft and the total area of the garden and path is 221 ft^2, how wide should the path be?
13. A metal box with no top was made by cutting a 3-in. square from each corner of a square sheet of copper, then bending up the sides. What size sheet of copper was used so that the box contains 192 in.^{3}? (HINT: $V = lwh$.)
14. A rectangular field is fenced in on three sides and bordered by a river on the fourth. If the area of the field is 1250 ft^2 and the length of the fence is 100 ft, what are the dimensions of the field?

For problems 15 and 16, see Example 3.

15. The area of a triangle is 56 ft^2. If the base is 9 ft more than the altitude, find the length of the base and the altitude.
16. The area of a triangle is 25 yd^2. If the base is twice the height, find the length of the base and height.

For problems 17 through 19, see Examples 4 and 5.

17. The second grades at Main Street Elementary School are painting a big red right triangle on the gym wall. If the hypotenuse is 8m and extends from the floor to a point 4m up the wall, how far along the wall does the painting extend?
18. The hypotenuse of a right triangle is 13 ft. The longer leg is two more than twice the length of the shorter leg. Find the length of each leg.
19. The lengths of the sides of a right triangle are three consecutive integers. Find the lengths of the sides.

Sec. 6.2 Applications

Practice Exercises

20. Find two consecutive negative integers whose product is 132.
21. Jack and Bob played in their first varsity basketball game with Jack scoring two more points than Bob. If the product of their points scored is 168, how many points did each score?
22. Find two numbers whose sum is 9 and whose product is 20.
23. Find two numbers whose product is 51 and whose sum is 20.
24. Find three consecutive odd positive integers such that the sum of their squares is 35.
25. Find three consecutive even positive integers such that the sum of their squares is 200.
26. Find the dimensions of a rectangle whose length is 3 m less than twice its width and whose area is 65 m².
27. A helicopter is searching a rectangular zone of the Talledega National Forest for wreckage from a plane crash. The perimeter of the search zone is 52 miles and its area is 165 sq mi. Find the dimensions of the search zone.
28. A sidewalk of uniform width is to be built outside a square parking lot, 40 ft on each side. How wide should the sidewalk be if the total area of parking lot and sidewalk is to be 2025 ft²?
29. An open box with no top was made from a square piece of cardboard by cutting a 2-in. square from each corner and turning up the sides. If the volume of the box is 98 in.², what size was the original piece of cardboard? (HINT: $V = lwh$.)
30. The area of a triangle is 49 yd². If the base is twice the height, find the length of the base and the height.
31. The area of a triangle is 8 ft². If the base is four times the height, find the length of each.
32. The hypotenuse of a right triangle is 26 m. If the length of the longer leg is 4 more than twice the shorter leg, find the length of each leg.
33. A ball is thrown straight upward from the ground with an initial velocity of 48 ft/sec. t seconds later it will be at a height of $h = 48t - 16t^2$. How many seconds will it take the ball to reach a height of 32 ft? What is the meaning of *two* answers?
34. The profit per month of Alice's Dress Shop is given by the formula $P = x^2 + 20x - 200$, where x is the number of dresses sold. How many dresses must Alice sell this month to make a profit of $600?
35. When Rose fills her backyard reflecting pool, the volume of water in cubic feet in the pool t minutes after the water is turned on is given by the formula $V = t(3t - 14)$. If the pool holds 160 ft³ of water, how long does it take to fill it?
36. In Jefferson, Georgia, the population T years after 1985 is given by the formula $P = T^2 - 80T + 2000$. When will the population of Jefferson be 4000?

Challenge Problems

37. Two trains leave Grand Central Station at the same time. One travels south at a rate 70 mph faster than the other train, which is traveling east. After 2 hours they are 260 miles apart. Find the speed of each.
38. Winfield throws a baseball from home plate toward third base, and at the same time, Puckett throws a baseball from home plate toward first base. Winfield's ball travels 20 ft/sec faster than Puckett's. If the balls are 100 ft apart after 1 second, how fast is each ball traveling?

IN YOUR OWN WORDS

39. Make up a word problem about a right triangle that uses a quadratic equation as its mathematical model.

6.3 COMPLETING THE SQUARE

Notice the similarity in solving the following four quadratic equations.

$$\boxed{1} \quad x^2 = 4$$
$$x^2 = 2^2$$
$$x^2 - 2^2 = 0$$
$$(x + 2)(x - 2) = 0$$
$$\{\pm 2\}$$

$$\boxed{2} \quad x^2 = 5$$
$$x^2 = (\sqrt{5})^2$$
$$x^2 - (\sqrt{5})^2 = 0$$
$$(x + \sqrt{5})(x - \sqrt{5}) = 0$$
$$\{\pm \sqrt{5}\}$$

3 $x^2 = -4$ 4 $x^2 = -5$
   $x^2 = 4(-1)$ $x^2 = 5(-1)$
   $x^2 = (2i)^2$ $x^2 = (\sqrt{5}i)^2$
   $x^2 - (2i)^2 = 0$ $x^2 - (\sqrt{5}i)^2 = 0$
   $(x + 2i)(x - 2i) = 0$ $(x + \sqrt{5}i)(x - \sqrt{5}i) = 0$
   $\{\pm 2i\}$ $\{\pm \sqrt{5}i\}$

In each case note that the equation

$$x^2 = A$$

led to the solution set $\{\pm \sqrt{A}\}$. (Remember immediately to replace such expressions as $\sqrt{-17}$ with $\sqrt{17}i$.) These examples suggest the following useful result, which allows us to omit the factoring steps.

Square Root Property

$$X^2 = A$$

is equivalent to the pair of equations

$$X = \pm\sqrt{A}$$

EXAMPLE 1. Solve using the Square Root Property.

(a) $x^2 = 27$ (b) $2x^2 + 16 = 0$

Solutions:

(a) $x^2 = 27$
     $x = \pm\sqrt{27}$ Square Root Property
     $x = \pm 3\sqrt{3}$
     $\{\pm 3\sqrt{3}\}$

(b) $2x^2 + 16 = 0$
    $2x^2 = -16$
    $x^2 = -8$ Divide by 2.
    $x = \pm\sqrt{-8}$ Square Root Property
    $x = \pm\sqrt{8}i$
    $x = \pm 2\sqrt{2}i$
    $\{\pm 2\sqrt{2}i\}$

The Square Root Property says that the equation $X^2 = A$ is equivalent to the pair of equations $X = \pm\sqrt{A}$. Note that the X in this theorem need not be just a single variable, but can be an expression such as $(x + 1)$.

Sec. 6.3 Completing The Square

EXAMPLE 2. Solve $(x + 1)^2 = 9$.

Solution:

$$(x + 1)^2 = 9$$
$$x + 1 = \pm 3 \qquad \text{Square Root Property}$$
$$x = -1 \pm 3$$

This is equivalent to the *two* equations

$$x = -1 + 3 \quad \text{or} \quad x = -1 - 3$$
$$x = 2 \quad \text{or} \quad x = -4$$
$$\{2, -4\}$$

EXAMPLE 3. Solve $x^2 - 6x + 9 = 5$.

Solution:

$$x^2 - 6x + 9 = 5$$
$$(x - 3)^2 = 5 \qquad \text{Factoring}$$
$$x - 3 = \pm \sqrt{5} \qquad \text{Square Root Property}$$
$$x = 3 \pm \sqrt{5}$$
$$\{3 + \sqrt{5}, 3 - \sqrt{5}\} \quad \text{or} \quad \{3 \pm \sqrt{5}\}$$

The preceding examples suggest a general technique that will apply to *all* quadratic equations. The idea is to make one side of the equation a perfect square trinomial and then apply the Square Root Property.

EXAMPLE 4. Solve $2x^2 + 30 = 16x$.

Solution:

First write the equation in standard form.

$$2x^2 - 16x + 30 = 0$$

Divide both sides by 2, the leading coefficient.

$$x^2 - 8x + 15 = 0$$

Subtract 15 from both sides.

$$x^2 - 8x = -15$$

Before we can continue, we must learn how to *complete the square*. What must we add to $x^2 - 8x$ to make a perfect square? Notice that $x^2 - 8x + 16$ is a perfect square and 16 is $\frac{1}{2}(-8)$ squared.

Square one-half of -8 and add it to both sides.

$$x^2 - 8x + 16 = -15 + 16$$

Factor the left side and simplify the right.
$$(x - 4)^2 = 1$$
Apply the Square Root Property and simplify.
$$x - 4 = \pm\sqrt{1}$$
$$x - 4 = \pm 1$$
$$x = 4 \pm 1$$
$$\{3, 5\}$$ □

EXAMPLE 5. Calculate the term to add to each binomial to make each a perfect square.

(a) $x^2 + 8x$
(b) $x^2 - 3x$

Solutions:

(a) $x^2 + 8x$

The coefficient of x is 8. Divide 8 by 2 and square this result.
$$\left(\frac{8}{2}\right)^2 = (4)^2 = 16$$

$x^2 + 8x + 16$ is a perfect square.

(b) $x^2 - 3x$

The coefficient of x is -3.
$$\left(\frac{-3}{2}\right)^2 = \frac{9}{4}$$

$x^2 - 3x + \frac{9}{4}$ is a perfect square. □

Solving a Quadratic Equation by Completing the Square

1. Write the equation in standard form.
2. Divide both sides by the coefficient of x^2 if the coefficient is not 1.
3. Subtract the constant term from both sides.
4. Divide the coefficient of x by two, square the result, then add this number to both sides.
5. Factor the left side (it will be a perfect square) and simplify the right side.
6. Apply the Square Root Property and simplify.
7. Write the solution set.

Note the foregoing steps in the following worked examples.

EXAMPLE 6. Solve $x^2 - 5x + 3 = 0$ by completing the square.

Solution:

$$x^2 - 5x + 3 = 0 \qquad \text{Standard form}$$

$$x^2 - 5x = -3 \qquad \text{Subtract 3 from both sides.}$$

$$x^2 - 5x + \left(\frac{5}{2}\right)^2 = -3 + \left(\frac{5}{2}\right)^2 \qquad \text{Complete the square.}$$

$$\left(x - \frac{5}{2}\right)^2 = -3 + \frac{25}{4}$$

$$\left(x - \frac{5}{2}\right)^2 = \frac{-12}{4} + \frac{25}{4} = \frac{13}{4} \qquad \text{Common denominator}$$

$$x - \frac{5}{2} = \pm\sqrt{\frac{13}{4}} \qquad \text{Square Root Property}$$

$$x = \frac{5}{2} \pm \frac{\sqrt{13}}{2}$$

$$\left\{\frac{5}{2} \pm \frac{\sqrt{13}}{2}\right\} \quad \text{or} \quad \left\{\frac{5 \pm \sqrt{13}}{2}\right\} \qquad \square$$

EXAMPLE 7. Solve $2x^2 - x + 4 = 0$ by completing the square.

Solution:

$$2x^2 - x + 4 = 0$$

$$x^2 - \frac{1}{2}x + 2 = 0 \qquad \text{Divide both sides by 2.}$$

$$x^2 - \frac{1}{2}x = -2 \qquad \text{Subtract 2 from both sides.}$$

$$x^2 - \frac{1}{2}x + \frac{1}{16} = -2 + \frac{1}{16} \qquad \text{Complete the square.}$$

$$\left(x - \frac{1}{4}\right)^2 = \frac{-31}{16}$$

$$x - \frac{1}{4} = \pm\sqrt{\frac{-31}{16}} \qquad \text{Square Root Property}$$

$$x - \frac{1}{4} = \pm\sqrt{\frac{31}{16}}\,i = \pm\frac{\sqrt{31}}{4}\,i$$

$$x = \frac{1}{4} \pm \frac{\sqrt{31}}{4}\,i$$

$$\left\{\frac{1}{4} \pm \frac{\sqrt{31}}{4}\,i\right\} \qquad \square$$

Square Root Property with a Calculator

CALCULATOR BOX

Solutions to equations that can be solved with the square root property can be approximated with a calculator.

EXAMPLE Find approximate solutions to each quadratic equation accurate to the nearest thousandth.

(a) $x^2 = 17$ (b) $3t^2 - 13 = 0$ (c) $x^2 + 8 = 0$

Solutions:

(a) By the square root property, $x^2 = 17$ and $x = \pm\sqrt{17}$ are equivalent. Press $\boxed{17}$ $\boxed{\sqrt{}}$ and read $\boxed{4.1231056}$ on the display.

$$x \approx 4.123 \quad \text{or} \quad x \approx -4.123$$

The solutions are approximately ± 4.123.

(b) $3t^2 - 13\text{keys} = 0$

$\quad 3t^2 = 13$ Add 13 to both sides.

$\quad t^2 = \dfrac{13}{3}$ Divide both sides by 3.

$\quad t = \pm\sqrt{\dfrac{13}{3}}$ Square Root Property

Press the keys $\boxed{13}$ $\boxed{\div}$ $\boxed{3}$ $\boxed{=}$ $\boxed{\sqrt{}}$ and read $\boxed{2.081666}$ on the display.

$$t \approx 2.082 \quad \text{or} \quad t \approx -2.082$$

The solutions are approximately ± 2.082.

(c) $x^2 + 8 = 0$

$\quad x^2 = -8$

$\quad x = \pm\sqrt{-8} = \pm\sqrt{8}\,i$

Press the keys $\boxed{8}$ $\boxed{\sqrt{}}$ and read $\boxed{2.828427125}$ on the display.

$$x \approx 2.828i \quad \text{or} \quad x \approx -2.828i$$

The solutions are approximately $\pm 2.828i$.

Calculator Exercises.

Approximate the solutions to the nearest thousandth.

1. $x^2 = 444$
2. $x^2 - 3.4 = 0$
3. $3t^2 = -18$
4. $2z^2 = 11$
5. $4x^2 - 3.613 = 0$
6. $5x^2 - 4 = 2\pi$

Answers:

1. ± 21.071
2. ± 1.844
3. $\pm 2.449i$
4. ± 2.345
5. ± 0.950
6. ± 1.434

PROBLEM SET 6.3

Warm-ups

In problems 1 through 22, use the Square Root Property to solve each quadratic equation.

For problems 1 through 16, see Example 1.

1. $x^2 - 25 = 0$
2. $x^2 - 16 = 0$
3. $x^2 + 9 = 0$
4. $x^2 + 81 = 0$
5. $x^2 - 5 = 0$
6. $x^2 - 6 = 0$
7. $x^2 + 5 = 0$
8. $x^2 + 6 = 0$
9. $x^2 - 48 = 0$
10. $x^2 - 12 = 0$
11. $x^2 + 18 = 0$
12. $x^2 + 75 = 0$
13. $2t^2 + 16 = 0$
14. $3w^2 + 15 = 0$
15. $-6s^2 - 18 = 0$
16. $-2v^2 - 40 = 0$

For problems 17 through 22, see Example 2.

17. $(x + 3)^2 = 4$
18. $(x - 2)^2 = 1$
19. $(x - 7)^2 = 8$
20. $(y + 3)^2 = 12$
21. $(x - 5)^2 = -7$
22. $(x + 8)^2 = -18$

In problems 23 through 32, calculate the term that must be added to each binomial to make it a perfect square. See Example 5.

23. $x^2 + 4x$
24. $x^2 + 6x$
25. $x^2 - 6x$
26. $x^2 - 8x$
27. $x^2 + 7x$
28. $x^2 - 5x$
29. $x^2 + x$
30. $x^2 - x$
31. $u^2 + 11u$
32. $k^2 - 13k$

In problems 33 through 48, solve each equation by completing the square. For problems 33 through 42, see Examples 3 and 6.

33. $x^2 - 4x - 5 = 0$
34. $x^2 - 6x - 7 = 0$
35. $x^2 - 4x + 5 = 0$
36. $x^2 - 2x + 2 = 0$
37. $x^2 + x - 6 = 0$
38. $x^2 + 4x + 4 = 0$
39. $x^2 + 2x = 4$
40. $x^2 - 6x + 1 = 0$
41. $t^2 - t + 3 = 0$
42. $t^2 + 3t + 4 = 0$

For problems 43 through 48, see Examples 4 and 7.

43. $2z^2 + 2z = 7$
44. $3w^2 - 2w - 1 = 0$
45. $2x^2 - 2x + 5 = 0$
46. $4x^2 + 12x + 6 = 0$
47. $2x^2 - 5x + 1 = 0$
48. $2x^2 - 3x + 4 = 0$

Practice Exercises

In problems 49 through 68, solve each equation by completing the square.

49. $x^2 - 2x - 8 = 0$
50. $x^2 - 4x - 12 = 0$
51. $x^2 - 2x + 1 = 0$
52. $2x^2 - 4x - 6 = 0$
53. $3x^2 - 12x = -9$
54. $2x^2 - 6x + 1 = 0$
55. $x^2 - 3x - 4 = 0$
56. $x^2 - x = -2$
57. $4x^2 - 20x + 16 = 0$
58. $5z^2 - 5z + 15 = 0$
59. $2u^2 - 2u = 2$
60. $3x^2 + 2x - 5 = 0$
61. $2x^2 + 5x - 4 = 0$
62. $5y^2 - 2y + 7 = 0$
63. $3v^2 - 4v + 10 = 0$
64. $2w^2 - 3w + 1 = 0$
65. $2x^2 - 7x + 5 = 0$
66. $3x^2 - 7x + 2 = 0$

67. $4x^2 - 6x + 10 = 0$

68. $6x^2 - 8x - 2 = 0$

69. The hypotenuse of a right triangle is $3\sqrt{5}$ feet long and one leg is 6 ft long. Find the length of the other leg.

70. The hypotenuse of a right triangle is 10 m long and one leg is $5\sqrt{2}$ meters long. Find the length of the other leg.

Challenge Problems

In problems 71 through 76, use the Square Root Property to solve each equation. Assume that a and b are real numbers.

71. $x^2 = a + b$
72. $x^2 = a^2 + b^2$
73. $(x + 1)^2 = a$
74. $(x - a)^2 = -16$
75. $(x + b)^2 = 8$
76. $(x - b)^2 = -12$

In problems 77 through 81, solve each equation by completing the square. Assume that a, b, and c are real numbers.

77. $4x^2 + 4ax - a = 0$
78. $\frac{1}{2}x^2 - \frac{1}{3}x + \frac{4}{9} = 0$
79. $0.8x^2 - 0.48x + 0.01 = 0$
80. $ax^2 + 3ax = a^2; \quad a \neq 0$
81. $ax^2 + bx + c = 0; \quad a \neq 0$

■ IN YOUR OWN WORDS

82. Explain how to solve the equation $x^2 - 6x + 4 = 0$ by completing the square.

■ 6.4 THE QUADRATIC FORMULA

Completing the square is often a cumbersome method to use in solving a quadratic equation. If the quadratic can be factored, that method is easier and faster. However, not all quadratics will factor, so a general method is needed. As completing the square works for all quadratics, we can solve the general quadratic equation by completing the square and find a formula to use in solving all such equations.

We follow the procedure given in Section 6.3 for completing the square:

$ax^2 + bx + c = 0; \quad a > 0$ Standard form

$x^2 + \frac{b}{a}x + \frac{c}{a} = 0$ Divide by a.

$x^2 + \frac{b}{a}x = -\frac{c}{a}$ Subtract $\frac{c}{a}$.

$x^2 + \frac{b}{a}x + \left(\frac{b}{2a}\right)^2 = -\frac{c}{a} + \left(\frac{b}{2a}\right)^2$ Complete the square.

$\left(x + \frac{b}{2a}\right)^2 = -\frac{c}{a} + \left(\frac{b}{2a}\right)^2$ Factor left side.

$\left(x + \frac{b}{2a}\right)^2 = \frac{-4ac + b^2}{4a^2}$ Simplify right side.

$x + \frac{b}{2a} = \pm\sqrt{\frac{b^2 - 4ac}{4a^2}}$ Square Root Property

$x = -\frac{b}{2a} \pm \frac{\sqrt{b^2 - 4ac}}{2a}$ Subtract $\frac{b}{2a}$.

$x = \frac{-b \pm \sqrt{b^2 - 4ac}}{2a}$

This is an important formula and should be memorized. It is called the Quadratic Formula. Although we derived the formula for $a > 0$, it works just as well if $a < 0$.

The Quadratic Formula

The quadratic equation $ax^2 + bx + c = 0$, $a \neq 0$ is equivalent to

$$x = \frac{-b \pm \sqrt{b^2 - 4ac}}{2a}$$

EXAMPLE 1. Solve $x^2 + 6 = -5x$ by the Quadratic Formula.

Solution:

Write the equation in standard form.

$$x^2 + 5x + 6 = 0$$

Thus $a = 1$, $b = 5$, and $c = 6$. Substitute into the Quadratic Formula.

$$x = \frac{-5 \pm \sqrt{5^2 - 4(1)(6)}}{2(1)}$$

$$= \frac{-5 \pm \sqrt{25 - 24}}{2}$$

$$= \frac{-5 \pm 1}{2}$$

$$x = \frac{-5 + 1}{2} \quad \text{or} \quad x = \frac{-5 - 1}{2}$$

$$x = -2 \quad \text{or} \quad x = -3$$

$$\{-2, -3\}$$

EXAMPLE 2. Solve $2x^2 + 3x - 4 = 0$.

Solution:

Here $a = 2$, $b = 3$, and $c = -4$.

$$x = \frac{-3 \pm \sqrt{9 - 4(2)(-4)}}{2(2)}$$

$$= \frac{-3 \pm \sqrt{9 + 32}}{4}$$

$$= \frac{-3 \pm \sqrt{41}}{4} = -\frac{3}{4} \pm \frac{\sqrt{41}}{4}$$

There are several acceptable ways to write this solution set.

$$\left\{\frac{-3 \pm \sqrt{41}}{4}\right\} \quad \text{or} \quad \left\{\frac{-3 + \sqrt{41}}{4}, \frac{-3 - \sqrt{41}}{4}\right\} \quad \text{or} \quad \left\{-\frac{3}{4} \pm \frac{\sqrt{41}}{4}\right\} \quad \text{or}$$

$$\left\{-\frac{3}{4} + \frac{\sqrt{41}}{4}, -\frac{3}{4} - \frac{\sqrt{41}}{4}\right\}$$

□

EXAMPLE 3. Solve $t^2 - 3t + 3 = 0$.

Solution:

For this equation, $a = 1$, $b = -3$, and $c = 3$.
Be careful substituting -3 for b in the formula, $-b$ will become $-(-3)$.

$$t = \frac{-(-3) \pm \sqrt{(-3)^2 - 4(1)(3)}}{2(1)}$$

$$= \frac{3 \pm \sqrt{9 - 12}}{2}$$

$$= \frac{3 \pm \sqrt{-3}}{2}$$

$$= \frac{3 \pm \sqrt{3}i}{2} \qquad \text{Simplify } \sqrt{-3}.$$

$$= \frac{3}{2} \pm \frac{\sqrt{3}}{2}i \qquad \text{Standard form of complex numbers}$$

$$\left\{\frac{3}{2} \pm \frac{\sqrt{3}}{2}i\right\} \quad \text{or} \quad \left\{\frac{3}{2} + \frac{\sqrt{3}}{2}i, \frac{3}{2} - \frac{\sqrt{3}}{2}i\right\}$$

The solutions are complex conjugates.

□

The Quadratic Formula often leads to expressions such as the one below. Notice how it is simplified.

$$\frac{8 \pm \sqrt{20}}{4} = \frac{8 \pm \sqrt{4(5)}}{4}$$

$$= \frac{8 \pm 2\sqrt{5}}{4}$$

$$= \frac{8}{4} \pm \frac{2\sqrt{5}}{4}$$

$$= 2 \pm \frac{\sqrt{5}}{2}$$

This may also be written with a common denominator as $\dfrac{4 \pm \sqrt{5}}{2}$.

Sec. 6.4 The Quadratic Formula

EXAMPLE 4. Solve $3x^2 - 2x + 5 = 0$.

Solution:

With $a = 3$, $b = -2$, and $c = 5$, we have

$$x = \frac{-(-2) \pm \sqrt{4 - 4(3)(5)}}{2(3)} \qquad \text{NOTE: } -b = -(-2).$$

$$= \frac{2 \pm \sqrt{4 - 60}}{6}$$

$$= \frac{2 \pm \sqrt{-56}}{6} = \frac{2 \pm \sqrt{56}\,i}{6}$$

Immediately rewrite $\sqrt{-56}$ as $\sqrt{56}\,i$ and note that there are two complex solutions. Notice that $\sqrt{56} = \sqrt{4(14)} = 2\sqrt{14}$, so

$$x = \frac{2 \pm 2\sqrt{14}\,i}{6}$$

$$= \frac{2}{6} \pm \frac{2\sqrt{14}\,i}{6}$$

$$= \frac{1}{3} \pm \frac{\sqrt{14}\,i}{3}$$

$$= \frac{1}{3} \pm \frac{\sqrt{14}}{3}i \qquad \text{Standard form of complex numbers}$$

$$\left\{\frac{1}{3} \pm \frac{\sqrt{14}}{3}i\right\} \quad \text{or} \quad \left\{\frac{1}{3} + \frac{\sqrt{14}}{3}i, \frac{1}{3} - \frac{\sqrt{14}}{3}i\right\}$$

Notice that the solutions are complex conjugates.

EXAMPLE 5. Solve $2x^2 + 2x - 1 = 0$.

Solution:

$$x = \frac{-2 \pm \sqrt{2^2 - 4(2)(-1)}}{2(2)} \qquad \text{Substitute into formula.}$$

$$= \frac{-2 \pm \sqrt{4 + 8}}{4} = \frac{-2 \pm \sqrt{12}}{4}$$

$$= \frac{-2 \pm 2\sqrt{3}}{4} \qquad \text{Simplify } \sqrt{12}.$$

$$= \frac{-2}{4} \pm \frac{2\sqrt{3}}{4}$$

$$= -\frac{1}{2} \pm \frac{\sqrt{3}}{2} \qquad \text{Reduce.}$$

$$\left\{-\frac{1}{2} \pm \frac{\sqrt{3}}{2}\right\}$$

EXAMPLE 6. Solve $\frac{1}{6}x^2 + \frac{1}{3}x - \frac{1}{6} = 0$.

Solution:

It is a good idea to clear fractions first, then use the Quadratic Formula. Multiply by 6 to get

$$x^2 + 2x - 1 = 0$$

So $a = 1$, $b = 2$, and $c = -1$, and with the formula

$$x = \frac{-2 \pm \sqrt{4 - 4(1)(-1)}}{2(1)}$$

$$= \frac{-2 \pm \sqrt{8}}{2}$$

$$= \frac{-2 \pm 2\sqrt{2}}{2}$$

$$= -\frac{2}{2} \pm \frac{2\sqrt{2}}{2}$$

$$= -1 \pm \sqrt{2}$$

$$\{-1 \pm \sqrt{2}\}$$

Solving a Quadratic Equation Using the Quadratic Formula

1. Write the equation in standard form.
2. Identify the values of a, b, and c.
3. Substitute these values into the formula and simplify.
4. Check the solutions if required.
5. Write the solution set.

A quadratic equation must be in standard form before the Quadratic Formula can be used.

EXAMPLE 7. Solve $(2r + 1)(r - 1) = 5$ using the Quadratic Formula.

Solution:

Be careful to write the equation in standard form first.

$$(2r + 1)(r - 1) = 5$$
$$2r^2 - r - 1 = 5$$
$$2r^2 - r - 6 = 0 \quad \text{Standard form}$$

Be Careful!

(continued)

Sec. 6.4 The Quadratic Formula

Thus $a = 2$, $b = -1$, and $c = -6$.

$r = \dfrac{-(-1) \pm \sqrt{(-1)^2 - 4(2)(-6)}}{2(2)}$ Substitute into Quadratic Formula.

$= \dfrac{1 \pm \sqrt{1 + 48}}{4}$

$= \dfrac{1 \pm \sqrt{49}}{4}$

$= \dfrac{1 \pm 7}{4}$

$\left\{-\dfrac{3}{2}, 2\right\}$ ◻

Notice the important role played by the expression that appears under the radical, $b^2 - 4ac$. This expression, called the **discriminant,** gives the following information about the nature of the solutions of the quadratic equation.

The Discriminant

$b^2 - 4ac$	Nature of solutions
Positive	Two real solutions
Zero	One real solution of multiplicity 2
Negative	Two complex solutions

EXAMPLE 8. Find the discriminant and determine the nature of the solutions for each equation.

(a) $3x^2 + 2x + 4 = 0$ (b) $x^2 + 2x + 1 = 0$ (c) $3x^2 + 5x - 2 = 0$

Solutions:

(a) $b^2 - 4ac = 2^2 - 4(3)(4)$
$= 4 - 48$
$= -44$

The value of the discriminant is -44 and the equation has two complex solutions.

(b) $b^2 - 4ac = 2^2 - 4(1)(1)$
$= 4 - 4$
$= 0$

Hence the equation has one real solution of multiplicity 2.

(c) $b^2 - 4ac = 5^2 - 4(3)(-2)$
$= 25 + 24$
$= 49$

The equation has two real solutions. ◻

We have learned how to solve a quadratic equation by factoring, the Square Root Property, completing the square, and the Quadratic Formula. Which method should be used in solving a quadratic equation?

Any quadratic equation can be solved by using the Quadratic Formula or by completing the square. However, factoring is much easier if it can be used. The Square Root Property saves some steps if it can be used.

> **Choosing a Method to Use in Solving a Quadratic Equation**
>
> 1. Use the Square Root Property if the equation is of the form $X^2 = A$.
> 2. Next, try factoring.
> 3. Use the Quadratic Formula if factoring won't work.
> 4. Use completing the square when directed to do so.

Quadratic Formula with a Calculator

CALCULATOR BOX

We will be using the "memory in" key to store an intermediate result. Find the key that stores the display in memory. It may look like $\boxed{\text{STO}}$ or $\boxed{\text{Min}}$. We will also need the "recall" key. It should look like $\boxed{\text{RCL}}$ or $\boxed{\text{MR}}$.

EXAMPLE Approximate the solutions for each quadratic equation to the nearest ten-thousandth.

(a) $2x^2 + 5x + 1 = 0$ (b) $3x^2 + 2x + 4 = 0$

Solutions:

(a) Using the Quadratic Formula yields

$$x = \frac{-5 \pm \sqrt{5^2 - 4 \cdot 2 \cdot 1}}{2 \cdot 2}$$

It is convenient to calculate the solution using the $+$ of the $\pm$ first. We will store the square root in memory so that we won't have to compute it again for the other solution. We generally do simple arithmetic as we enter the problem in the calculator. Press

and read $\boxed{-0.2192235}$ on the display. This is the first solution. We now calculate the other solution.

We read $\boxed{-2.2807764}$ on the display.

$$x \approx -0.2192 \quad \text{or} \quad x \approx -2.2808$$

The solutions are approximately -2.2808 and -0.2192.

(continued)

Sec. 6.4 The Quadratic Formula

(b) $3x^2 + 2x + 4 = 0$

$$x = \frac{-2 \pm \sqrt{2^2 - 4 \cdot 3 \cdot 4}}{2 \cdot 3}$$

We follow the same procedure as before. Press $\boxed{4}\boxed{-}\boxed{48}\boxed{=}$. At this point we notice $\boxed{-44}$ on the display. This is the discriminant. It is negative, so we have two complex roots. If we press $\boxed{\sqrt{}}$ now, we will get an error signal, as the calculator expects real numbers. Since $\sqrt{-44} = \sqrt{44}i$, we continue computing the imaginary part of the complex root. Press $\boxed{+/-}\boxed{\sqrt{}}\boxed{\div}\boxed{6}\boxed{=}$ and the imaginary part of the root is on the display: $\boxed{1.1055416}$. The real part is just $\frac{-2}{6}$. To the nearest ten-thousandth,

$$x \approx -0.3333 + 1.1055i \quad \text{or} \quad x \approx -0.3333 - 1.1055i$$

The solutions are approximately $-0.3333 + 1.1055i$ and $-0.3333 - 1.1055i$. □

Calculator Exercises

Approximate the solutions to each quadratic equation to the nearest ten-thousandth.

1. $2x^2 - 5x - 2 = 0$
2. $3y^2 - y + 1 = 0$
3. $4x^2 - 8x - 1 = 0$
4. $5x^2 + x + 3 = 0$
5. $2.1x^2 + 5.2x - 0.25 = 0$
6. $t^2 - 5.742t + 3.333 = 0$

Answers:

1. $x \approx 2.8508$
   $x \approx -0.3508$
2. $x \approx 0.1667 \pm 0.5528i$
3. $x \approx 2.1180$
   $x \approx -0.1180$
4. $x \approx -0.1 \pm 7.681i$
5. $x \approx 0.0472$
   $x \approx -2.5234$
6. $x \approx 5.0868$
   $x \approx 0.6552$

■ PROBLEM SET 6.4

Warm-ups

In problems 1 through 40, solve each equation using the Quadratic Formula.
For problems 1 through 8, see Examples 1 and 2.

1. $x^2 + 2x - 8 = 0$
2. $x^2 = 4$
3. $x^2 + 2x = 0$
4. $x^2 - x - 1 = 0$
5. $x^2 + x - 1 = 0$
6. $x^2 - x - 4 = 0$
7. $x^2 + 4x + 4 = 0$
8. $y^2 - 6y + 9 = 0$

For problems 9 through 14, see Example 3.

9. $x^2 - x + 1 = 0$
10. $x^2 - 3x + 12 = 0$
11. $x^2 + 1 = 0$
12. $x^2 + 2 = 0$
13. $x^2 + 5x + 9 = 0$
14. $x^2 - x + 3 = 0$

For problems 15 through 22, see Example 4.

15. $x^2 - 2x = 1$
16. $x^2 - 4x - 2 = 0$
17. $x^2 + 2x - 1 = 0$
18. $x^2 + 4x - 2 = 0$
19. $x^2 - 6x - 2 = 0$
20. $x^2 + 6x + 10 = 0$
21. $x^2 + 2x + 2 = 0$
22. $x^2 + 4x = -10$

For problems 23 through 36, see Examples 5 and 6.

23. $4x^2 + 3 = 8x$
24. $4x^2 - x - 3 = 0$
25. $6x^2 + 5x - 6 = 0$
26. $4x^2 - 5x - 6 = 0$
27. $2s^2 - 2s - 1 = 0$
28. $2z^2 + 3z - 2 = 0$
29. $3x^2 - x + 2 = 0$
30. $5t^2 + t + 3 = 0$
31. $6t^2 - t = 5$
32. $2x^2 = 5x - 4$
33. $8x - 16 = 4x^2$
34. $2x = x^2 + 6$
35. $\frac{2}{3}x^2 + \frac{11}{3}x - 7 = 0$
36. $5v^2 - \frac{11}{2}v - 3 = 0$

For problems 37 through 40, see Example 7.

37. $r(r - 6) - (2r - 1) = 0$
38. $(x - 1)(x + 1) = x$
39. $(2y + 3)(y - 1) = 3$
40. $t(t + 3) + 2(t + 1) = 1$

In problems 41 through 44, find the discriminant for each equation and determine the nature of the solution. See Example 8.

41. $x^2 - x + 4 = 0$
42. $x^2 + 2x - 1 = 0$
43. $2x^2 + 3x = 4$
44. $x^2 - 4x + 4 = 0$

Practice Exercises

In problems 45 through 86, solve each equation using any method.

45. $x^2 + 6x + 9 = 0$
46. $x(x + 6) + 6(x + 6) = 0$
47. $2x^2 - x - 1 = 0$
48. $2x^2 + 8 = 0$
49. $x^2 - 2(x + 4) = 0$
50. $x^2 = 4(x - 3)$
51. $5x^2 = 13x + 6$
52. $\left(x - \frac{2}{3}\right)^2 = \frac{2}{9}$
53. $6x^2 + x - 2 = 0$
54. $x^2 = -x$
55. $x(x - 1) = 0$
56. $5x^2 = 2x$
57. $x^2 + 25 = 0$
58. $\left(x + \frac{1}{2}\right)^2 = \frac{13}{4}$
59. $x^2 + x - 4 = 0$
60. $x^2 - x + 4 = 0$
61. $\left(y - \frac{3}{2}\right)^2 = -\frac{11}{4}$
62. $x^2 - 10x + 5 = 0$
63. $x^2 - 2x + 2 = 0$
64. $\frac{1}{4}z^2 + z + \frac{7}{2} = 0$
65. $x^2 - 6x + 10 = 0$
66. $(x - 1)(2x + 1) = 1$
67. $(2x + 1)(x + 3) = 2$
68. $x^2 - 1.8x - 1.44 = 0$
69. $y^2 + 0.1y - 0.06 = 0$
70. $\frac{5}{2}x^2 - 7x - \frac{3}{2} = 0$
71. $x^2 + 8x + 7 = 0$
72. $\left(x - \frac{5}{2}\right)^2 = \frac{81}{4}$
73. $2x^2 - 6 = 0$
74. $2x^2 - 7x - 4 = 0$

Sec. 6.4 The Quadratic Formula

75. $3x^2 - 2x - 1 = 0$
76. $(x - 5)^2 + 5 = 0$
77. $2x^2 - x = 0$
78. $y^2 + 8 = 0$
79. $z^2 = 4$
80. $(x - 1)(x + 2) = 0$
81. $2x^2 - 2x + 1 = 0$
82. $2x^2 + 2x + 1 = 0$
83. $9x^2 + 12x + 4 = 0$
84. $9y^2 - 6y + 1 = 0$
85. $5 = 2s^2 - s$
86. $4 = 2x + 3x^2$

In problems 87 through 90, find the discriminant for each equation and determine the nature of the solutions.

87. $5x^2 - x = 0$
88. $3x^2 - x + 5 = 0$
89. $5 + x^2 = 0$
90. $4x^2 - 2x + 1 = 0$

Challenge Problems

In problems 91 through 97, solve each equation for the variables indicated using the Quadratic Formula or the Square Root Property.

91. $ax^2 + 3x - 2 = 0$, Solve for x.
92. $x^2 + 2mx - n = 0$, Solve for x. ($m > 0$ and $n > 0$)
93. $x^2 + (n - m)x - mn = 0$, Solve for x. ($m > 0$ and $n > 0$)
94. $\frac{1}{2}gx^2 - kx + L = 0$, Solve for x.
96. $d = \frac{D}{L}(L^2 - a^2)$, Solve for L.
95. $V = \frac{2}{3}pR^2h$, Solve for R.
97. $L = \frac{2}{R^2} - d^2$; $R > 0$, Solve for d.

In problems 98 through 104, find all real values of k that make each condition true.

98. $x^2 - 2x + k = 0$ has two distinct real solutions.
99. $x^2 - 2x + k = 0$ has two complex solutions.
100. $x^2 - 2x + k = 0$ has one solution of multiplicity 2.
101. $x^2 - kx - 1 = 0$ has two distinct real solutions.
102. $x^2 + kx + 4 = 0$ has one solution of multiplicity 2.
103. $x^2 + k = 0$ has two complex solutions.
104. $x^2 + k = 0$ has one solution of multiplicity 2.
105. Suppose that $b^2 - 4ac > 0$. What can be said about the solutions to $ax^2 + bx + c = 0$ if $b^2 - 4ac$ is a perfect square? If it is not a perfect square?

106. What is the discriminant for the equation
$$ax^2 + bx = 0?$$
Will the solutions be real or complex?

107. What is the discriminant for the equation
$$ax^2 + c = 0?$$
Will the solutions be real or complex?
$-4ac$; $ac < 0 \Rightarrow$ real, $ac >$

In problems 108 and 109, suppose that R_1 and R_2 are two solutions of the general quadratic equation, $ax^2 + bx + c = 0$.

108. Find the "sum of the roots formula"; that is, find a formula for $R_1 + R_2$.

109. Find the "product of the roots formula"; that is, find a formula for $R_1 R_2$.

In problems 110 through 115, find the sum of the roots and the product of the roots without solving the equation.

110. $2x^2 + 3x + 4 = 0$
111. $x^2 - 5x + 10 = 0$
112. $2x - 17 = 5x^2$
113. $3x^2 - 11 = 0$

114. $x^2 + Mx + N = 0$ **115.** $Kx^2 + (Q - P)x + Q + P = 0$

■ IN YOUR OWN WORDS

116. What is the Quadratic Formula? What does it represent?

■ 6.5 EQUATIONS OF HIGHER DEGREE

By repeated use of the Property of Zero Products, we see that the solution set of the equation $(x + 1)(x - 1)(x - 2) = 0$ is $\{-1, 1, 2\}$. Thus factoring is the key to solving some polynomial equations of higher degree. These equations represent very special situations because not all polynomial equations can be solved by factoring.

EXAMPLE 1. Solve $x^3 + 3x^2 + 2x = 0$.

Solution:

We notice that this equation is in standard form.

$$x^3 + 3x^2 + 2x = 0$$
$$x(x^2 + 3x + 2) = 0 \quad \text{Factor.}$$
$$x(x + 1)(x + 2) = 0$$

Now we use the Property of Zero Products and write the solution set.

$$\{0, -1, -2\} \qquad \square$$

Solving Higher-Degree Equations

1. Write the equation in standard form.
2. Factor, if possible.
3. Use the Property of Zero Products, the Square Root Property, or the Quadratic Formula to find the solution set.

EXAMPLE 2. Solve $x^4 - 5x^2 + 4 = 0$.

Solution:

$$x^4 - 5x^2 + 4 = 0 \qquad \text{Standard form}$$
$$(x^2 - 4)(x^2 - 1) = 0 \qquad \text{Factor.}$$
$$x^2 = 4 \quad \text{or} \quad x^2 = 1 \qquad \text{Property of Zero Products}$$
$$x = \pm 2 \quad \text{or} \quad x = \pm 1 \qquad \text{Square Root Property}$$
$$\{2, -2, 1, -1\} \quad \text{or} \quad \{\pm 2, \pm 1\} \qquad \square$$

EXAMPLE 3. Solve $x^3 = -8$.

Solution:

$$x^3 + 8 = 0 \qquad \text{Standard form}$$
$$(x + 2)(x^2 - 2x + 4) = 0 \qquad \text{Factor.}$$
$$x + 2 = 0 \quad \text{or} \quad x^2 - 2x + 4 = 0 \qquad \text{Property of Zero Products}$$
$$x = -2 \quad \text{or} \quad x = \frac{2 \pm \sqrt{4 - 4(1)(4)}}{2} \qquad \text{Quadratic Formula}$$
$$x = \frac{2 \pm \sqrt{-12}}{2}$$
$$x = \frac{2 \pm 2\sqrt{3}i}{2}$$
$$x = 1 \pm \sqrt{3}i$$
$$\{-2, 1 + \sqrt{3}i, 1 - \sqrt{3}i\}$$

PROBLEM SET 6.5

Warm-ups

In problems 1 through 34, solve each equation.

For problems 1 through 4, see Example 1.

1. $(x - 5)(x + 4)(x + 3) = 0$
2. $(x + 1)(x - 1)(5 - x) = 0$
3. $x\left(x + \dfrac{3}{2}\right)(x + 17) = 0$
4. $x\left(x - \dfrac{17}{41}\right) = 0$

For problems 5 through 24, see Example 2.

5. $x^4 - 10x^2 + 9 = 0$
6. $x^4 - 17x^2 + 16 = 0$
7. $x^4 - 2x^2 + 1 = 0$
8. $x^4 - 10x^2 + 25 = 0$
9. $x^4 - 3x^2 - 4 = 0$
10. $x^4 - x^2 - 6 = 0$
11. $x^4 + x^2 - 6 = 0$
12. $x^4 + x^2 - 2 = 0$
13. $x^4 - 13x^2 + 12 = 0$
14. $x^4 - 8x^2 - 20 = 0$
15. $x^4 + 2x^2 = 0$
16. $x^4 - 4x^2 = 0$
17. $x^4 - 4 = 0$
18. $x^4 - 1 = 0$
19. $4x^2 + 5x^2 + 6 = 0$
20. $9x^4 + 7x^2 - 2 = 0$
21. $16x^4 - 8x^2 - 15 = 0$
22. $4x^4 + 11x^2 + 6 = 0$
23. $\dfrac{1}{4}w^4 - 1 = 0$
24. $\dfrac{1}{3}x^4 - \dfrac{9}{4}x^2 - \dfrac{10}{3} = 0$

For problems 25 through 34, see Example 3.

25. $x^3 + 1 = 0$
26. $x^3 = 27$
27. $Y^3 = 8$
28. $Z^3 + 64 = 0$
29. $x^3 - 125 = 0$
30. $r^4 + 27r = 0$
31. $s^5 - s^2 = 0$
32. $3x^5 - 24x^2 = 0$

Chap. 6 Nonlinear Equations in One Variable

33. $125x^3 + 8 = 0$

34. $27t^3 = 1$

Practice Exercises

In problems 35 through 58, find the solution set.

35. $x^4 - 37x^2 + 36 = 0$
36. $x^4 - 50x^2 + 49 = 0$
37. $2x^4 - 6x^2 = 0$
38. $4x^4 - 16x^2 = 0$
39. $x^4 - x^2 - 42 = 0$
40. $x^4 - 16x^2 + 48 = 0$
41. $x^4 - x^2 - 12 = 0$
42. $x^4 + 10x^2 + 24 = 0$

43. $-x^4 + x^2 + 6 = 0$
44. $16x^4 + 32x^2 + 15 = 0$

45. $4x^4 - x^2 - 3 = 0$
46. $16x^4 - 25 = 0$

47. $9x^4 + 11x^2 - 14 = 0$
48. $-x^4 + x^2 + 2 = 0$

49. $x^4 = 16$
50. $81x^4 = 1$

51. $2x^3 + 54 = 0$
52. $x^3 - 216 = 0$

53. $2w^3 + 250 = 0$
54. $8y^3 - 8 = 0$

55. $u^6 = u^3$
56. $t^5 + 8t^2 = 0$

57. $\dfrac{1}{5}x^4 - 5 = 0$
58. $12x^4 + \dfrac{19}{3}x^2 - 2 = 0$

Challenge Problems

In problems 59 through 64, find the solution set.

59. $x^6 - 64 = 0$
60. $x^6 - 1 = 0$
61. $x^3 - 2x^2 + 4x - 8 = 0$
62. $x^3 + x^2 - x - 1 = 0$
63. $(x + 1)^4 - 5(x + 1)^2 + 4 = 0$
64. $(x - 3)^4 + 2(x - 3)^2 - 15 = 0$
65. Find an equation whose solution set is $\{4, 2, -1\}$.
66. Find an equation whose solution set is $\{\pm 3, \pm i\}$.

IN YOUR OWN WORDS

67. How can factoring be used to solve some equations?

6.6 MORE FRACTIONAL EQUATIONS

While studying linear equations, we solved equations that contained fractions. The procedure we used was to multiply both sides of the equation by the least common denominator of all the fractions in the equation. If any denominator contains a variable, there is a possibility of multiplication by zero. (Remember, we can multiply both sides of an equation by any nonzero number).

The best strategy is to clear fractions and solve the resulting equation for possible solutions. Then check to see if any of the possible solutions would make any denominator of the original equation have a value of zero. If so, that number cannot be included in the solution set.

EXAMPLE 1. Solve $\dfrac{x}{x-2} - \dfrac{7}{x+5} = \dfrac{17}{x^2+3x-10}$.

Solution:

$$\dfrac{x}{x-2} - \dfrac{7}{x+5} = \dfrac{17}{x^2+3x-10}$$

Factor all denominators.

$$\dfrac{x}{x-2} - \dfrac{7}{x+5} = \dfrac{17}{(x-2)(x+5)}$$

Multiply both sides by $(x-2)(x+5)$, the LCD.

$$x(x+5) - 7(x-2) = 17$$
$$x^2 + 5x - 7x + 14 = 17$$
$$x^2 - 2x - 3 = 0$$
$$(x-3)(x+1) = 0$$
$$x = 3 \quad \text{or} \quad x = -1$$

As neither -1 nor 3 makes a denominator in the original equation have a value of zero, they are both in the solution set.

$$\{-1, 3\}$$ ◻

EXAMPLE 2. Solve $\dfrac{1}{x-6} + \dfrac{6}{x^2-11x+30} - \dfrac{x+1}{5-x} = 0$.

Solution:

$$\dfrac{1}{x-6} + \dfrac{6}{x^2-11x+30} - \dfrac{x+1}{5-x} = 0$$

Factor all denominators.

$$\dfrac{1}{x-6} + \dfrac{6}{(x-5)(x-6)} - \dfrac{x+1}{5-x} = 0$$

Since $x-5$ and $5-x$ are opposites, replace $5-x$ with $-(x-5)$.

$$\dfrac{1}{x-6} + \dfrac{6}{(x-5)(x-6)} - \dfrac{x+1}{-(x-5)} = 0$$

$$\dfrac{1}{x-6} + \dfrac{6}{(x-5)(x-6)} + \dfrac{x+1}{x-5} = 0$$

Multiply both sides by $(x-5)(x-6)$ to clear fractions.

$$(x-5) + 6 + (x+1)(x-6) = 0$$
$$x - 5 + 6 + x^2 - 5x - 6 = 0$$
$$x^2 - 4x - 5 = 0$$
$$(x-5)(x+1) = 0$$
$$x = 5 \quad \text{or} \quad x = -1$$

In this example note that replacing x by 5 in the original equation makes two of the denominators have a value of zero, while -1 does not make any denominator have a value of zero. Thus -1 is in the solution set and 5 is not.

$$\{-1\}\qquad\square$$

> ### Solving Fractional Equations
> 1. Multiply both sides of the equation by the least common denominator to clear fractions.
> 2. Find possible solutions by solving the resulting equation.
> 3. Discard possible solutions that make any denominator in the original equation have a value of zero.
> 4. Write the solution set.

EXAMPLE 3. Solve $\dfrac{x}{x-7} = \dfrac{35}{x^2 - 9x + 14} + \dfrac{7}{x-2}$.

Solution:

$$\dfrac{x}{x-7} = \dfrac{35}{x^2 - 9x + 14} + \dfrac{7}{x-2}$$

$\dfrac{x}{x-7} = \dfrac{35}{(x-7)(x-2)} + \dfrac{7}{x-2}$ Factor.

$x(x-2) = 35 + 7(x-7)$ Multiply by $(x-7)(x-2)$.

$x^2 - 2x = 35 + 7x - 49$ Distributive Property

$x^2 - 9x + 14 = 0$ Standard form

$(x-7)(x-2) = 0$ Factor.

$x = 7$ or $x = 2$ Property of Zero Products

Neither 7 nor 2 is in the solution set, as each makes a denominator of the original equation have a value of zero. So, as they are the only possibilities, the solution set must be empty.

$$\emptyset\qquad\square$$

Recalling the definition of negative exponents, we see that equations containing them are often just equations containing fractions.

EXAMPLE 4. Solve $x^{-4} - 10x^{-2} + 9 = 0$.

Solution:

METHOD 1

$$x^{-4} - 10x^{-2} + 9 = 0$$

$$\dfrac{1}{x^4} - \dfrac{10}{x^2} + 9 = 0$$

(continued)

Multiply both sides by x^4.

$$1 - 10x^2 + 9x^4 = 0$$
$$(1 - 9x^2)(1 - x^2) = 0$$
$$1 - 9x^2 = 0 \quad \text{or} \quad 1 - x^2 = 0$$
$$x^2 = \frac{1}{9} \quad \text{or} \quad x^2 = 1$$
$$x = \pm\frac{1}{3} \quad \text{or} \quad x = \pm 1$$

As none of these values make any of the denominators in the original equation have a value of zero, all four possibilities are in the solution set.

$$\left\{\pm\frac{1}{3},\ \pm 1\right\}$$

There is another approach to this problem. The original equation will factor.

METHOD 2

$$x^{-4} - 10x^{-2} + 9 = 0$$
$$(x^{-2} - 1)(x^{-2} - 9) = 0$$

$x^{-2} = 1$ or $x^{-2} = 9$		Factor.
$\dfrac{1}{x^2} = 1$ or $\dfrac{1}{x^2} = 9$		Definition of x^{-2}
$1 = x^2$ or $\dfrac{1}{9} = x^2$		Multiply by x^2.
$x = \pm 1$ or $x = \pm\dfrac{1}{3}$		Square Root Property

$$\left\{\pm\frac{1}{3},\ \pm 1\right\}$$

EXAMPLE 5. Tony entered a 17-mile race. He ran for 15 miles and hurt his leg and was forced to walk the rest of the race. His running rate was 3 mph faster than his walking rate. If the race took 2 hours to complete, find his walking speed.

Solution:

Let x be his walking speed in mph.

We make a chart and fill in D, R, and T for the two parts, running and walking. The let statement gives us the rates.

	DISTANCE	RATE	TIME
Walking		x	
Running		$x + 3$	

The problem tells us the distances.

	DISTANCE	RATE	TIME
Walking	2	x	
Running	15	$x + 3$	

Use the formula $T = \dfrac{D}{R}$ to find the time for each part.

	DISTANCE	RATE	TIME
Walking	2	x	$\dfrac{2}{x}$
Running	15	$x + 3$	$\dfrac{15}{(x + 3)}$

We form an equation by noting that the total time is 2 hours.

$$\frac{2}{x} + \frac{15}{x + 3} = 2$$

Multiply both sides by the LCD, $x(x + 3)$.

$$2(x + 3) + 15x = 2x(x + 3)$$
$$2x + 6 + 15x = 2x^2 + 6x$$
$$0 = 2x^2 - 11x - 6$$
$$0 = (2x + 1)(x - 6)$$
$$x = \frac{-1}{2} \quad \text{or} \quad x = 6$$

We see that $\dfrac{-1}{2}$ cannot be a speed.

His walking speed was 6 mph. ☐

EXAMPLE 6. Jim and Roy can both clear their camping site in 3 hours if they work together. If each worked alone, it would take Roy 8 hours longer than Jim to clear the same site. How long would it take Jim to clear the site if he were working alone?

Solution:

Let x be the number of hours it would take Jim, working alone. Then $x + 8$ is the number of hours it would take Roy alone. So in 1 hour Jim would clear $\dfrac{1}{x}$ of the site, and in 1 hour Roy would clear $\dfrac{1}{x + 8}$ of the site, while Jim and Roy together

(continued)

Sec. 6.6   More Fractional Equations

would clear $\frac{1}{3}$ of the site. Thus in 1 hour we have

$$\frac{1}{x} + \frac{1}{x+8} = \frac{1}{3}$$

Multiply both sides by $3x(x+8)$, the LCD.

$$3(x+8) + 3(x) = x(x+8)$$
$$3x + 24 + 3x = x^2 + 8x$$
$$0 = x^2 + 2x - 24$$
$$= (x-4)(x+6)$$
$$x = 4 \quad \text{or} \quad x = -6$$

Since -6 cannot be an answer to this problem, we may write the solution.
It would take Jim 4 hours working alone. □

PROBLEM SET 6.6

Warm-ups

In problems 1 through 20, solve each equation.
For problems 1 through 7, see Example 1.

1. $\dfrac{6}{x-2} - \dfrac{3}{x} = 1$

2. $\dfrac{2}{(x-3)^2} - \dfrac{x}{x-3} = 1$

3. $\dfrac{2}{x} - 1 = \dfrac{4}{x+3}$

4. $\dfrac{x}{x+4} = \dfrac{3}{x-1}$

5. $\dfrac{1}{t-1} + \dfrac{1}{2t+1} = \dfrac{6}{5}$

6. $\dfrac{4}{x+5} - \dfrac{5}{x-2} = 3$

7. $\dfrac{Z^2 - 5}{Z^2 + Z - 20} + \dfrac{Z+1}{Z+5} = \dfrac{Z-3}{Z-4}$

For problems 8 through 11, see Example 2.

8. $\dfrac{18}{9-y^2} + \dfrac{y}{y-3} = \dfrac{1}{y+3}$

9. $\dfrac{z}{z-2} + \dfrac{2}{z+1} = \dfrac{7z+1}{z^2 - z - 2}$

10. $\dfrac{2}{w+5} - \dfrac{7}{w-4} = \dfrac{w^2 + 6w - 13}{w^2 + w - 20}$

11. $\dfrac{x}{x+1} - \dfrac{2}{1-x} = \dfrac{8x-4}{x^2 - 1}$

For problems 12 through 15, see Example 3.

12. $\dfrac{1}{3-s} + \dfrac{s}{s-1} + \dfrac{2}{s^2 - 4s + 3} = 0$

13. $\dfrac{2x^2}{x^2 - 4x + 3} + \dfrac{1}{x-1} = \dfrac{9}{x-3}$

14. $\dfrac{x+2}{x-1} = \dfrac{3}{x+1} - \dfrac{6}{1-x^2}$

15. $\dfrac{2}{x-3} + \dfrac{x^2 - 1}{x^2 - 10x + 21} = \dfrac{12}{x-7}$

For problems 16 through 20, see Example 4.

16. $x^{-2} - x^{-1} - 2 = 0$

17. $x^{-2} + 2x^{-1} - 24 = 0$

18. $y^{-2} - 3y^{-1} - 18 = 0$

19. $t^{-4} - 2t^{-2} + 1 = 0$

20. $x^{-4} - 3x^{-2} - 4 = 0$

For problems 21 through 24, see Example 5.

21. A ship sails at constant speed for 24 nautical miles. If the ship's rate was 2 knots faster, it would have saved 1 hour on the trip. What was its speed?
22. A bus travels 200 miles from Atlanta at a constant speed. It returns immediately at a constant speed, 10 miles per hour slower than it went. If the total trip took 9 hours, what speed did it travel from Atlanta?
23. Alan set out on a 23-mile bike trip to Tacoma. After 22 miles his bike broke and he walked the rest of the trip. His riding speed was 4 mph faster than his walking speed. If the total trip took 3 hours, find his walking speed.
24. John sailed for 2 miles and then rowed for 5 miles. His sailing speed was 3 mph faster than his rowing speed. If the total trip took 2 hours, what was his rowing speed?

For problems 25 and 26, see Example 6.

25. Carolyn's hot tub has two fill pipes. If both pipes are open, they can fill the hot tub in 4 hours. If only the smaller pipe is open, it takes 6 hours longer to fill the tub than if only the larger pipe were open. Find the time it would take each pipe to fill the tub if working alone.
26. Odette and Alex can clean their house in 2 hours working together. Working alone, it takes Odette 3 hours longer than it takes Alex working alone. How long does it take each to clean the house working alone?

Practice Exercises

In problems 27 through 80, find the solution set.

27. $\dfrac{2x}{5} - \dfrac{2}{x} = \dfrac{1}{5}$

28. $\dfrac{y^2 + 2y - 5}{y + 1} = 1$

29. $\dfrac{-4}{3x - 1} = 2x + 3$

30. $\dfrac{x + 2}{x - 1} = x + 2$

31. $\dfrac{-3}{x} - 1 = \dfrac{4}{1 - x}$

32. $\dfrac{3}{x - 1} = \dfrac{x}{1 - x}$

33. $\dfrac{1}{x + 1} - \dfrac{10}{x - 2} = \dfrac{3}{2}$

34. $\dfrac{25}{y + 2} - \dfrac{9}{y - 2} = 16$

35. $\dfrac{14}{x - 2} - \dfrac{35}{x - 5} = \dfrac{3}{2}$

36. $\dfrac{x + 1}{x - 2} - \dfrac{8}{x + 1} = \dfrac{9}{x^2 - x - 2}$

37. $\dfrac{2}{x^2 - 1} + \dfrac{3x + 6}{x^2 - x - 2} = \dfrac{8}{x^2 - 3x + 2}$

38. $\dfrac{2x}{(x - 3)(x + 1)} + \dfrac{3}{x + 1} = 1$

39. $\dfrac{5}{x + 2} - \dfrac{3}{x - 5} = \dfrac{x^2 - 8x - 6}{x^2 - 3x - 10}$

40. $\dfrac{3}{x} + \dfrac{x}{x + 4} = \dfrac{10}{x^2 + 4x}$

41. $\dfrac{1}{x} + \dfrac{5}{3x} = \dfrac{x + 2}{3}$

42. $\dfrac{1}{x - 5} + \dfrac{x}{x - 2} = \dfrac{3}{x^2 - 7x + 10}$

43. $\dfrac{1}{x} + \dfrac{x + 3}{7} = \dfrac{5}{7x}$

44. $\dfrac{20}{7} + \dfrac{x^2}{x + 1} = \dfrac{23}{7x + 7}$

45. $\dfrac{2}{x + 7} + \dfrac{16}{x^2 + 6x - 7} + \dfrac{x}{x - 1} = 0$

46. $\dfrac{5}{x^2 + 3x - 4} + \dfrac{1}{x + 4} = \dfrac{x}{x - 1}$

47. $\dfrac{1}{5} - \dfrac{7}{15 - 5x} = \dfrac{x^2}{x - 3}$

48. $\dfrac{4}{y - 8} + \dfrac{2}{y^2 - 11y + 24} + \dfrac{y + 1}{y - 3} = 0$

49. $\dfrac{2}{t - 3} - \dfrac{t}{2 - t} = \dfrac{2}{t^2 - 5t + 6}$

50. $\dfrac{-3}{x + 4} + \dfrac{x}{x - 3} = \dfrac{21}{x^2 + x - 12}$

51. $\dfrac{1}{x} + \dfrac{x - 3}{x + 1} = \dfrac{16}{x^2 + x}$

52. $\dfrac{5}{w - 5} + \dfrac{w}{w + 5} = \dfrac{50}{w^2 - 25}$

53. $\dfrac{1}{x^2} + \dfrac{1}{3} = \dfrac{4}{x^2}$

54. $\dfrac{3}{t - 2} + \dfrac{t + 1}{t + 3} = \dfrac{15}{t^2 + t - 6}$

55. $\dfrac{Y}{Y + 5} = \dfrac{1}{Y - 3} - \dfrac{8}{Y^2 + 2Y - 15}$

56. $\dfrac{7}{3X + 6} + 3 = \dfrac{-8}{3X - 3}$

57. $\dfrac{-2}{x + 1} - \dfrac{5}{3 - x} = 1$

58. $\dfrac{15}{4(x - 2)} - \dfrac{7}{4(x + 2)} = 1$

Sec. 6.6 More Fractional Equations

59. $\dfrac{6}{x} + \dfrac{x}{x+5} = \dfrac{30}{x^2 + 5x}$

60. $\dfrac{y}{y+3} - \dfrac{2}{y+1} = \dfrac{y-3}{y^2+4y+3}$

61. $\dfrac{t+3}{t-1} + \dfrac{1}{t+4} = \dfrac{2t^2+6t+12}{t^2+3t-4}$

62. $\dfrac{x}{x^2+3x+2} + \dfrac{1}{x^2+4x+3} = \dfrac{2}{x^2+5x+6}$

63. $\dfrac{1}{t^2-1} + \dfrac{2t}{2t^2-3t+1} = \dfrac{1}{2t^2+t-1}$

64. $\dfrac{3x}{x^2-x-2} + \dfrac{2}{x^2+4x+3} = \dfrac{14}{x^2+x-6}$

65. $\dfrac{18}{s^2+s-6} + \dfrac{s-1}{s^2+5s+6} = \dfrac{12}{s^2-4}$

66. $\dfrac{y}{y^2-y-2} + \dfrac{2}{y^2-2y-3} + \dfrac{2}{y^2-5y+6} = 0$

67. $\dfrac{1}{x^2-1} + \dfrac{x}{2x^2-3x+1} = \dfrac{1}{2x^2+x-1}$

68. $\dfrac{x}{x^2-x-2} - \dfrac{1}{2x^2+x-1} = \dfrac{1}{2x^2-5x+2}$

69. $\dfrac{2}{5y^2-15y-50} + \dfrac{y+1}{y^2-y-6} = \dfrac{-8}{5y^2-40y+75}$

70. $x^{-2} + 4x^{-1} + 4 = 0$

71. $x^{-2} - 3x^{-1} - 10 = 0$

72. $x^{-2} + 5x^{-1} - 6 = 0$

73. $2x^{-2} - 17x^{-1} + 21 = 0$

74. $3x^{-2} + x^{-1} = 4$

75. $4x^{-2} + 27 = 21x^{-1}$

76. $3 - 2x^{-1} = 5x^{-2}$

77. $x^{-2} - 10x^{-1} + 25 = 0$

78. $3x^{-4} - x^{-2} - 4 = 0$

79. $2x^{-4} - 11x^{-2} + 9 = 0$

80. $6x^{-4} - x^{-2} - 1 = 0$

81. Sam rode his bike 3 miles and then walked for 6 miles. His walking speed was 2 mph slower than his speed on his bike. If he walked 1 hour longer than he rode his bike, find his speed on his bike.

82. Susan ran 3 miles and then decreased her speed by 1 mph and ran 4 more miles. If it took 1 hour longer on the last 4 miles, find her original speed.

83. Joan rides her bicycle to Tacoma 74 miles away in 5 hours. Rain forces her to slow down by 6 mph for the last 10 miles. What was her original speed?

84. Marvin drove at a constant speed from Atlanta to Savannah, a distance of 260 miles, and returned immediately. He drove 13 mph faster on the return trip and shortened his time by 1 hour. What was his speed each way?

85. Working together, Cathy and Jim can prepare a gourmet Hunan feast in 6 hours. If Jim were to do it alone, it would take him 9 hours longer than it would take Cathy working alone. How long would it take each to prepare the feast working alone?

86. Charles can pour a concrete driveway in 16 hours less than Robert. If they work together, they can pour the driveway in 6 hours. How long does it take each to pour the driveway working alone?

Challenge Problems

In problems 87 through 103, solve each equation.

87. $4x^{-4} - 68x^{-2} + 225 = 0$

88. $4x^{-4} + 4x^{-2} - 3 = 0$

89. $9x^{-4} + 7x^{-2} - 2 = 0$

90. $4p^{-4} + 5p^{-2} - 6 = 0$

91. $16x^{-4} - 8x^{-2} - 15 = 0$

92. $4x^{-4} + 11x^{-2} + 6 = 0$

93. $6x^{-2} - x^{-1} = 0$

94. $y^{-4} + y^{-2} = 0$

95. $\dfrac{3x}{x-1} = \dfrac{x^2}{1-x}$

96. $\dfrac{x}{x+1} - \dfrac{10x}{x-2} = \dfrac{3x}{2}$

97. $\dfrac{5x}{x^2+3x-4} + \dfrac{x}{x+4} = \dfrac{x^2}{x-1}$

98. $\dfrac{2x^2}{x-3} + \dfrac{x^3}{x-2} = \dfrac{2x^2}{x^2-5x+6}$

99. $\dfrac{a-2}{x+2} + \dfrac{x}{x-1} = \dfrac{2a^2-a+2}{x^2+x-2}$

100. $a^2 x^{-4} + 3ax^{-2} - 4 = 0;\ a \ne 0$

101. $(x + 1)^{-4} - 5(x + 1)^{-2} + 4 = 0$ **102.** $(z - 2)^{-4} - 6(z - 2)^{-2} + 9 = 0$ **103.** $\dfrac{2}{3\sqrt{x}} + \dfrac{\sqrt{x}}{3} = \dfrac{x^2}{3\sqrt{x}}$

104. An airplane flies between Denver and Loveland. With a tailwind of 20 mph it arrives 4 minutes earlier than it would without the tailwind. If the cities are 40 miles apart, find the speed of the plane in still air.

105. Ron can row 30 miles down the Columbia River and back in 8 hours. If the rate of the current is 5 mph, find the rate that Ron would row in still water.

IN YOUR OWN WORDS

106. Explain how to solve a fractional equation.

6.7 RADICAL EQUATIONS

We learned in Section 4.6 that sometimes it is necessary to square both sides of an equation to clear radicals to solve using familiar techniques. Often, raising both sides to a power will lead to the types of higher order equations that we are presently studying.

Never forget that raising both sides of an equation to a power may introduce extraneous solutions. We *always* must check the solutions of the squared equation to see if they are solutions of the original equation.

EXAMPLE 1. Solve $\sqrt{4x + 1} + 5 = x$.

Solution:

$$\sqrt{4x + 1} + 5 = x$$

First, isolate the radical.

$$\sqrt{4x + 1} = x - 5$$

Then square both sides.

$$(\sqrt{4x + 1})^2 = (x - 5)^2$$
$$4x + 1 = x^2 - 10x + 25$$

Simplify, and write in standard form.

$$0 = x^2 - 14x + 24$$
$$0 = (x - 2)(x - 12)$$
$$x = 2 \quad \text{or} \quad x = 12$$

Check 2 in the original equation.

LS: $\sqrt{4(2) + 1} + 5 = \sqrt{9} + 5 = 3 + 5 = \boxed{8}$
RS: $\boxed{2}$
Does *not* check.

Check 12.

LS: $\sqrt{4(12) + 1} + 5 = \sqrt{49} + 5 = 7 + 5 = \boxed{12}$
RS: $\boxed{12}$
Checks.

Thus the solution set is {12}.

> **Solving Radical Equations**
>
> 1. Isolate one radical.
> 2. Raise both sides to the appropriate power.
> 3. Simplify, and isolate another radical if necessary. Continue steps 2 and 3 until the equation is free of radicals.
> 4. Solve the resulting equation.
> 5. Check all possible solutions in the original equation.
> 6. Write the solution set.

EXAMPLE 2. Solve $\sqrt{2x-1} - \sqrt{x-4} = 2$.

Solution:

Be Careful!

$$\sqrt{2x-1} - \sqrt{x-4} = 2$$
$$\sqrt{2x-1} = 2 + \sqrt{x-4} \qquad \text{Isolate a radical.}$$
$$(\sqrt{2x-1})^2 = (2 + \sqrt{x-4})^2 \qquad \text{Square both sides.}$$
$$2x - 1 = 4 + 4\sqrt{x-4} + (x-4)$$
$$x - 1 = 4\sqrt{x-4} \qquad \text{Isolate a radical.}$$
$$(x-1)^2 = (4\sqrt{x-4})^2 \qquad \text{Square both sides.}$$
$$x^2 - 2x + 1 = 16(x-4)$$
$$x^2 - 2x + 1 = 16x - 64 \qquad \text{Distributive Property}$$
$$x^2 - 18x + 65 = 0 \qquad \text{Standard form}$$
$$(x-5)(x-13) = 0 \qquad \text{Factor.}$$
$$x = 5 \quad \text{or} \quad x = 13 \qquad \text{Property of Zero Products}$$

Check 5 in the original equation.

LS: $\sqrt{2(5)-1} - \sqrt{5-4} = \sqrt{9} - \sqrt{1} = 3 - 1 = \boxed{2}$
RS: $\boxed{2}$

5 checks.

Check 13 in the original equation.

LS: $\sqrt{2(13)-1} - \sqrt{13-4} = \sqrt{25} - \sqrt{9} = 5 - 3 = \boxed{2}$
RS: $\boxed{2}$

Also, 13 checks.

$$\{5, 13\}$$

It may be necessary to raise both sides of an equation to a power greater than 2. In the following example it will be convenient to cube both sides.

EXAMPLE 3. Solve $x^{1/3} = 3$.

Solution:

$$x^{1/3} = 3$$
$$(x^{1/3})^3 = 3^3$$
$$x^1 = 27$$
$$x = 27 \quad \text{27 checks by inspection.}$$
$$\{27\}$$

PROBLEM SET 6.7

Warm-ups

In problems 1 through 22, find the solution set.

For problems 1 through 10, see Example 1.

1. $x = \sqrt{2x^2 - 3x + 2}$
2. $x + 1 = \sqrt{x + 3}$
3. $x - 2 = \sqrt{6 - 3x}$
4. $x = -1 + \sqrt{6x - 2}$
5. $x = \sqrt{2x + 6} - 3$
6. $x = 2 - \sqrt{3x + 4}$
7. $2x - 1 = \sqrt{2x^2 + 3x - 2}$
8. $3x - 1 = \sqrt{8x^2 - 5x + 7}$
9. $2x + 3 = \sqrt{8 + 7x - 2x^2}$
10. $x - 1 = \sqrt{2x^2 - 3x - 1}$

For problems 11 through 18, see Example 2.

11. $\sqrt{x - 3} - 1 = \sqrt{2x - 4}$
12. $\sqrt{x + 5} = 3 + \sqrt{2x - 8}$
13. $\sqrt{3x + 1} = 1 + \sqrt{2x - 1}$
14. $\sqrt{4x + 13} = 2 + \sqrt{3x}$
15. $\sqrt{2y - 5} = \sqrt{3y + 4} - 2$
16. $\sqrt{5z + 1} = \sqrt{3z - 5} + 2$
17. $\sqrt{5x + 21} = 1 - \sqrt{3x + 16}$
18. $\sqrt{2x + 4} = \sqrt{9 - 2x} - 1$

For problems 19 through 22, see Example 3.

19. $x^{1/3} = 2$
20. $x^{1/3} = -3$
21. $\sqrt[3]{2x - 1} - 3 = 0$
22. $\sqrt[3]{x^2 - 1} = 2$

Practice Exercises

In problems 23 through 48, find the solution set.

23. $\sqrt{x + 2} = x$
24. $\sqrt{9x} = x + 2$
25. $\sqrt{16x} = x + 3$
26. $\sqrt{3x - 1} = \sqrt{x + 1}$
27. $\sqrt{4 - x} = x - 4$
28. $\sqrt{x + 3} = x - 3$
29. $\sqrt{2x} = x - 4$
30. $\sqrt{18x + 10} - x = 5$
31. $\sqrt{8x + 1} - 2 = x$
32. $x = \sqrt{2x^2 - 11x + 30}$
33. $\sqrt{z + 27} - 1 = \sqrt{2z + 20}$
34. $\sqrt{1 - 8r} + r = 2$
35. $\sqrt{5w} - \sqrt{w - 4} = 4$
36. $\sqrt{2 - 7t} - 3 = \sqrt{t + 3}$
37. $\sqrt{3 - 2x} - \sqrt{x + 4} = 2$
38. $\sqrt{4x + 1} - 2 = \sqrt{x - 1}$
39. $\sqrt{t + 9} = 2 + \sqrt{2t + 1}$
40. $\sqrt{2y - 1} - 1 = \sqrt{y - 1}$
41. $\sqrt{z - 2} + 2 = \sqrt{2z}$
42. $\sqrt{2x + 7} - 1 = \sqrt{x + 3}$

Sec. 6.7 Radical Equations

43. $\sqrt{2x+1} + 1 = \sqrt{x+4}$
44. $\sqrt{5w+20} - 1 = 2\sqrt{w+3}$
45. $\sqrt[3]{x} = 1$
46. $\sqrt[3]{x} = -2$
47. $(3x+1)^{1/3} = 2$
48. $(x^2+2)^{1/3} = 3$

Challenge Problems

In problems 49 through 51, find the solution set.

49. $\sqrt{x} + \sqrt{x+5} = \sqrt{5x+5}$
50. $\sqrt{3-x} - \sqrt{2+x} = \sqrt{2x+3}$
51. $\sqrt{4x+2} = \sqrt{6x+6} - \sqrt{2-2x}$

52. Solve $x + \sqrt{x} - 2 = 0$ two ways. First, isolate the radical and square both sides, then solve it by factoring.

In problems 53 through 55, find the solution set.

53. $x^{2/3} - x^{1/3} - 2 = 0$
54. $x^{2/3} - 16 = 0$
55. $x^{2/5} - 2x^{1/5} - 3 = 0$

IN YOUR OWN WORDS

56. Explain the steps in solving an equation containing radicals.

CHAPTER SUMMARY

GLOSSARY

Quadratic equation in one variable: Equation that can be written in the form $ax^2 + bx + c = 0$. If $a > 0$, the equation is in **standard form.**

Discriminant: The value of $b^2 - 4ac$, using a, b, and c from the standard form of a quadratic equation.

If $b^2 - 4ac > 0$, the equation has two real solutions.
If $b^2 - 4ac < 0$, the equation has two complex solutions.
If $b^2 - 4ac = 0$, the equation has one real repeated solution.

SQUARE ROOT PROPERTY

$X^2 = A$

is equivalent to the pair of equations

$X = \pm\sqrt{A}$

SOLVING A QUADRATIC BY COMPLETING THE SQUARE

1. Write the equation in standard form.
2. Divide both sides by the coefficient of x^2 if the coefficient is not 1.
3. Subtract the constant term from both sides.
4. Divide the coefficient of x by 2, square the result, then add this number to both sides.
5. Factor the left side and simplify the right side.
6. Apply the Square Root Property and write the solution set.

$ax^2 + bx + c = 0; \quad a \neq 0$

is equivalent to

$x = \dfrac{-b \pm \sqrt{b^2 - 4ac}}{2a}$

QUADRATIC FORMULA

1. Is the equation of the form $X^2 = A$? If so, use the Square Root Property.
2. Try factoring.
3. Use the Quadratic Formula
4. Use completing the square when directed to do so.

SOLVING A QUADRATIC EQUATION

1. Write the equation in standard form.
2. Factor, if possible.
3. Use the Property of Zero Products, the Square Root Property, or the Quadratic Formula to find the solution set.

SOLVING HIGHER-DEGREE EQUATIONS

1. Multiply both sides by the LCD.
2. Find possible solutions by solving the resulting equation.
3. Discard possible solutions that make any denominator have a value of zero.
4. Write the solution set.

SOLVING FRACTIONAL EQUATIONS

1. Isolate one radical.
2. Raise both sides to the appropriate power.
3. Simplify and isolate another radical if necessary. Continue steps 2 and 3 until the equation is free of radicals.
4. Solve the resulting equation.
5. Check all possible solutions in the original equation.
6. Write the solution set.

SOLVING RADICAL EQUATIONS

▮ CHECKUPS

In problems 1 through 6, solve each equation.

1. $x^2 + \dfrac{2}{3}x - \dfrac{8}{3} = 0$ Section 6.1; Example 5
2. $2x^2 + 16 = 0$ Section 6.3; Example 1b
3. $t^2 - t + 3 = 0$ Section 6.4; Example 3
4. $x^3 = -8$ Section 6.5; Example 3
5. $\dfrac{1}{x-6} + \dfrac{6}{x^2 - 11x + 30} - \dfrac{x+1}{5-x} = 0$ Section 6.6; Example 2
6. $\sqrt{4x + 1} + 5 = x$ Section 6.7; Example 1
7. The lengths of the sides of a right triangle are three consecutive even integers. What are they?   Section 6.2; Example 5
8. Solve $2x^2 - x + 4 = 0$ by completing the square.   Section 6.3; Example 7
9. Tony entered a 17-mile race. He ran for 15 miles and hurt his leg and was forced to walk the rest of the race. His running rate was 3 mph faster than his walking rate. If the race took 2 hours to complete, find his walking speed.   Section 6.6; Example 5

REVIEW PROBLEMS

In problems 1 through 58, find the solution set.

1. $12x^2 + 5x - 2 = 0$
2. $2x^2 - 50 = 0$
3. $x^2 - 7x = 0$
4. $x^2 = 8$
5. $3x^2 + 4x - 3 = 0$
6. $(x - 7)(x + 3)(x - 10) = 0$
7. $\dfrac{3}{x - 5} + \dfrac{x}{3(x + 2)} = \dfrac{5}{x^2 - 3x - 10}$
8. $x^{-2} + 8x^{-1} + 15 = 0$
9. $4x^2 + 12x + 9 = 0$
10. $6x^2 + 13x + 6 = 0$
11. $4x^2 - x + 1 = 0$
12. $(x + 3)(x - 7) = 8$
13. $(3x + 2)^2 = 9$
14. $x(x - 3)(x + 2) = 0$
15. $\sqrt{6x + 1} - 1 = \sqrt{3x + 4}$
16. $\dfrac{1}{x + 3} + \dfrac{x}{x + 1} = \dfrac{-2}{x^2 + 4x + 3}$
17. $18x^2 + 9x - 5 = 0$
18. $-x^2 = 36$
19. $4x^2 - 9 = 0$
20. $x^4 - 6x^2 + 9 = 0$
21. $\dfrac{1}{x} + \dfrac{x + 1}{x^2} = \dfrac{6}{x^3}$
22. $3x^2 - 2x = 0$
23. $\dfrac{3}{x} + \dfrac{3}{x^2 - x} = \dfrac{x}{1 - x}$
24. $\sqrt{5x - 2} + 1 = \sqrt{10x - 3}$
25. $x^3 = \dfrac{1}{8}$
26. $-x^2 + 4x + 21 = 0$
27. $3x^2 + 8x + 4 = 0$
28. $4x^2 - 6x + 9 = 0$
29. $\sqrt{x + 2} - x = 0$
30. $6x^2 + 5x - 6 = 0$
31. $3x^2 - x - 2 = 0$
32. $\dfrac{3}{x + 3} - \dfrac{x}{3 - x} = \dfrac{7}{x^2 - 9}$
33. $x^4 - 50x^2 + 49 = 0$
34. $3x^2 = 27$
35. $\sqrt{2x + 1} - \sqrt{x} = 1$
36. $\dfrac{5}{x + 8} + \dfrac{x - 3}{x + 1} = \dfrac{11x - 13}{x^2 + 9x + 8}$
37. $\sqrt{2x + 5} - \sqrt{x - 1} = 2$
38. $6x^{-4} + x^{-2} = 1$
39. $\dfrac{3}{x^2 + 6x + 8} + \dfrac{x + 4}{x^2 + 3x + 2} = \dfrac{5}{x^2 + 5x + 4}$
40. $\sqrt{7x} - \sqrt{3x + 4} = 2$
41. $5x^2 + 5x + 2 = 0$
42. $\dfrac{1}{x^2 + 7x + 10} + \dfrac{7(x - 3)}{x^2 - 3x - 10} = \dfrac{29}{x^2 - 25}$
43. $\dfrac{1}{x + 1} + \dfrac{1}{3} = 2x$
44. $(x - 1)(2x + 3) = 1$
45. $(x - 2)^2 = 6$
46. $(x - 1)^2 = 10$
47. $\sqrt{2x + 7} - \sqrt{x} = 2$
48. $(x + 3)(x + 5) = 3$
49. $x^2(x^2 - 5) - 3(x^2 - 5) = 0$
50. $\dfrac{1}{x} + \dfrac{x + 1}{x^2} = \dfrac{1}{x^3}$

Chap. 6 Nonlinear Equations in One Variable

51. $(x + 3)^2 = -5$

52. $2\sqrt{x + 1} - 1 = \sqrt{3x}$

53. $16x^4 + 9x^2 - 7 = 0$

54. $\dfrac{1}{x - 1} + 6 = \dfrac{14}{x + 3}$

55. $\dfrac{9}{x - 1} + \dfrac{5}{3 + x} = 2$

56. $\dfrac{2}{x + 4} = \dfrac{x + 1}{x + 2}$

57. $(x - 7)^2 = -8$

58. $(2x - 5)^2 = -12$

59. Find two numbers whose sum is 17 and whose product is -390.

60. The perimeter of a rectangle is 84 m and its area is 425 m². Find its dimensions.

61. The hypotenuse of a right triangle is $3\sqrt{2}$ meters and one leg is 2 m. Find the length of the other leg.

62. The area of a triangle is 72 in.² If the length of the base is four times the length of the height, find the length of the height.

63. Roger drove 27 miles to Tampa on his new motorcycle. For the first 15 miles, he went 6 mph faster than he did the rest of the trip. If the trip to Tampa took 1 hour, what was his speed on the last part of the trip?

... LET'S NOT FORGET ...

Identify the expressions that are in factored form. Factor those that are not factored.

64. $-8x^3 + 125$
65. $(2x + 5)^3$
66. $\sqrt{a}(\sqrt{a} + \sqrt{b})$
67. $16 - (x + 1)^2$
68. $x^2 + 2xy + y^2$
69. $y(c - d) + x(c - d)$

How many terms are in each expression? Which expressions have $x - y$ as a factor?

70. $x^3 - y^3$
71. $x^2 - xy + y^2$
72. $a(x - y) + b(y - x)$
73. $x - y(a + b)$

Simplify each expression, if possible, leaving only nonnegative exponents in your answer.

74. $\dfrac{4 \pm \sqrt{-12}}{2}$

75. $\dfrac{-3x^{-2}}{x^{-1}y^3}$

76. $-3^2(ab^2)^{-2}$

77. $a^{-1}b^2(a^{-2} + b)$

78. -2^{-3}

79. $\sqrt[5]{a^5 + b^5}$

Find each product.

80. $(x^{1/2} + y^{1/2})^2$

81. $(3y - 1)^3$

Reduce each, if possible.

82. $\dfrac{a(r + t) - b(r - t)}{(r + t)(r - t)}$

83. $\dfrac{x(b + c) - y(b + c)}{(b + c)(b - c)}$

The following problems can be worked using a least common denominator. Follow the directions in each and notice how the LCD is used.

84. Solve $\dfrac{x}{x + 3} - \dfrac{2}{x + 4} = \dfrac{2}{x^2 + 7x + 12}$.

85. Perform the operation indicated: $\dfrac{3}{x + 4} - \dfrac{x}{x + 3}$.

86. Simplify $\dfrac{1 + \dfrac{x}{x + 3}}{1 - \dfrac{x}{x + 4}}$.

Sec. 6.7 Radical Equations

Label each as an equation or as an expression. Solve the equations and perform the operations indicated on the expressions.

87. $(1 - 2i)(1 + i)$

88. $(x + 1)(x + 2) = 12$

89. $\dfrac{x}{x^2 - 5x - 14} - \dfrac{1}{x - 7}$

90. $\sqrt{x - 1} - \sqrt{2x + 5} = -2$

91. $\left|\dfrac{1}{4}(x - 3)\right| = 3$

92. $\dfrac{2}{x + 2} - \dfrac{2}{x + 3} = 1$

93. $\dfrac{3}{5}(x + 2) - 1 = \dfrac{1}{3}(1 - x)$

CHAPTER 6 TEST

In problems 1 through 5, choose the correct answer.

1. The solution set for $x^2 - 2x - 3 = 5$ is (?)
 A. $\{4, 8\}$ **B.** $\{-2, 4\}$
 C. $\{-4, 2\}$ **D.** $\{2, 8\}$

2. $\dfrac{-3 \pm \sqrt{12}}{12} =$ (?)

 A. $-\dfrac{1}{4} \pm \dfrac{\sqrt{3}}{6}$ **B.** $-\dfrac{1}{4} \pm \dfrac{\sqrt{3}}{2}$

 C. $\dfrac{-1 \pm \sqrt{3}}{2}$ **D.** $-\dfrac{1}{4} \pm \dfrac{\sqrt{3}}{3}$

3. The solution set for $2x^2 - 2x + 3 = 0$ is (?)

 A. $\left\{\dfrac{3}{2}, 1\right\}$ **B.** $\{1 \pm \sqrt{5}i\}$

 C. $\left\{\dfrac{1}{2} \pm \dfrac{\sqrt{7}}{2}\right\}$ **D.** $\left\{\dfrac{1}{2} \pm \dfrac{\sqrt{5}}{2}i\right\}$

4. The solution set for $x = \sqrt{7 - x} + 1$ is (?)
 A. $\{-2, 3\}$ **B.** $\{-3, 2\}$
 C. $\{3\}$ **D.** $\emptyset$

5. The solution set for $4x^4 - 17x^2 + 4 = 0$ is (?)

 A. $\left\{\pm\dfrac{1}{2}i, \pm 2i\right\}$ **B.** $\left\{\dfrac{1}{4}, 4\right\}$ **C.** $\left\{\pm\dfrac{1}{2}, \pm 2\right\}$

 D. None of the above

In problems 6 through 12, find the solution set.

6. $3x^2 - 4x - 4 = 0$

7. $\dfrac{3}{x + 1} + \dfrac{2}{x + 3} = 2$

8. $(x - 3)^2 = 9$

9. $x^2 = 3x$

10. $x^3 - 8 = 0$

11. $\dfrac{1}{x^4} + \dfrac{3}{x^2} - 4 = 0$

12. $\sqrt{x + 1} + 2 = \sqrt{2x + 10}$

13. Find the solution set by completing the square.

$$3x^2 - 6x - 3 = 0$$

14. Find the discriminant for $2x^2 - x - 1 = 0$.

 The discriminant indicates that the equation has
 A. exactly one real solution
 B. two distinct real solutions
 C. two complex solutions
 D. no solutions

15. Martha, a cyclist, pedals 20 miles to Ventura at one speed and then decreases her speed by 10 mph for the next 30 miles to Santa Barbara. What was her original speed if the total trip took 4 hours?

CHAPTER 7

See Review Problems, Exercise 26.

Inequalities in One Variable

- **7.1** Linear Inequalities in One Variable
- **7.2** Compound Inequalities
- **7.3** Boundary Numbers and Absolute Value Inequalities
- **7.4** Higher-Degree and Fractional Inequalities

CONNECTIONS

Jim Bob, who has $23 in his pocket, convinces the very attractive Billie Sue to attend the Saturday night Atlanta Braves baseball game. Their tickets cost $9 each. While finding their seats, Jim Bob notices that hot dogs are $2 apiece. As the very attractive Billie Sue looks hungry, Jim Bob wonders how many hot dogs can they order safely. Clearly, two times the number of hot dogs ordered must be no greater than 23 minus 18.

Equations are statements about the equality of numbers or symbols that represent numbers. Just as equations arise when the language of mathematics is used to model real life situations, inequalities arise when something is larger or smaller than something else. For example, charges on a credit card must be less than the credit limit and the load limit on a bridge must be greater than the weight of the vehicles using it, and Jim Bob must not let Billie Sue order more than two hot dogs!

This chapter is concerned with inequalities in one variable. The properties of inequalities are developed and utilized to solve linear inequalities. The general method of boundary numbers is introduced and used for all nonlinear inequalities.

7.1 LINEAR INEQUALITIES IN ONE VARIABLE

An **inequality** is a statement that says one number is greater than another. For example,

$$5 > 3$$
$$7 > -9$$
$$-2 > -5$$

are such statements. Sometimes these statements are written the other way around, and say that one number is less than another. For example,

$$3 < 5$$
$$-9 < 7$$
$$-5 < -2$$

Often, we wish to allow equality along with inequality. For example, x **less than or equal** to 5 can be written

$$x \leq 5$$

We write x **greater than or equal** to 3 as

$$x \geq 3$$

Like equations, inequalities are interesting when they contain letters, such as x, which stand for numbers whose value is unknown. For example, the inequality,

$$x + 4 > 11$$

is true if x is 9 and false if x is 3. Note that the statement is true if x is any number greater than 7, and false if x is 7 or less.

The collection of numbers that when substituted for x make the statement true is called the **solution set** of the inequality. To **solve** an inequality is to find its solution set. Usually, the solution set for an inequality cannot be written by listing its elements. We use set-builder notation. For example, to write the solution set for the inequality above, we need to write the set of all numbers greater than 7. In set-builder notation that is

$$\{x \mid x > 7\}$$

Two inequalities with the same solution set are said to be **equivalent.**

Tools Used to Solve Inequalities

Addition Property
If the same number is added to or subtracted from both sides of an inequality, the resulting inequality is equivalent to the first inequality.

Multiplication by Positive Number
If both sides of an inequality are multiplied or divided by the same **positive** number, the resulting inequality is equivalent to the first inequality.

Multiplication by Negative Number
If both sides of an inequality are multiplied or divided by the same **negative** number, and the inequality symbol is **reversed,** the resulting inequality is equivalent to the first inequality.

Linear, or first-degree, inequalities in one variable can be solved in a manner very similar to the method used to solve first-degree equations. Our general approach is to use the tools above, forming equivalent inequalities until we have an inequality of the form $x < a$, $x \leq a$, $x > a$, or $x \geq a$. We can then write the solution set. Notice how this works in the following examples.

EXAMPLE 1. Solve $x + 5 < 7$.

Solution:

$$x + 5 < 7$$

We add -5 to both sides,

$$x + 5 - 5 < 7 - 5$$
$$x < 2$$

then write the solution set:

$$\{x \mid x < 2\}$$

Sec. 7.1 Linear Inequalities in One Variable

EXAMPLE 2. Solve $2x - 1 \leq 5$.

Solution:
$$2x - 1 \leq 5$$
We add 1 to both sides.
$$2x - 1 + 1 \leq 5 + 1$$
$$2x \leq 6$$
Divide both sides by 2 $\left(\text{multiply by } \dfrac{1}{2}\right)$.
$$\dfrac{2x}{2} \leq \dfrac{6}{2}$$
$$x \leq 3$$
$$\{x \mid x \leq 3\}$$

EXAMPLE 3. Solve $-2x + 1 < 7$.

Solution:
$$-2x + 1 < 7$$
Subtract 1 from both sides,
$$-2x < 6$$
Divide both sides by -2.

Be Careful!
$$x > -3$$
$$\{x \mid x > -3\}$$
Note that the inequality symbol is reversed.

EXAMPLE 4. Solve $3(t - 1) > 7(t + 1)$.

Solution:
$$3(t - 1) > 7(t + 1)$$
$3t - 3 > 7t + 7$ Distributive Property
$3t - 10 > 7t$ Subtract 7 from both sides.
$-10 > 4t$ Subtract $3t$ from both sides.
$-\dfrac{5}{2} > t$ Divide both sides by 4.
$$\left\{t \mid t < -\dfrac{5}{2}\right\}$$

In Example 4, notice that we wrote $t < -\dfrac{5}{2}$ rather than $-\dfrac{5}{2} > t$. Either is correct. The variable is usually written on the left.

EXAMPLE 5. Solve $x + 0.6 < 0.4(x + 2)$.

Solution:

Write the decimals as fractions.

$$x + \frac{6}{10} < \frac{4}{10}(x + 2)$$

Multiply both sides by 10 to clear fractions.

$$10x + 6 < 4(x + 2)$$
$$10x + 6 < 4x + 8 \qquad \text{Distributive Property}$$
$$10x - 4x < 8 - 6$$
$$6x < 2$$
$$x < \frac{1}{3}$$

Using decimals, $\frac{1}{3}$ is written as $0.\overline{3}$.

$$\{x \mid x < 0.\overline{3}\}$$

The inequality of Example 5 is stated with decimal numbers rather than fractions; thus it is good form to state the solution set with decimals. It is **important** to note that the number $0.\overline{3}$ is **not** the same as *approximations* of $\frac{1}{3}$ such as 0.3 or 0.333.

In the next example notice how the inequality symbol is unaffected when we multiply by a *positive* number, but is reversed when we multiply by a *negative* number.

EXAMPLE 6. Solve $\dfrac{x + 1}{3} - \dfrac{x + 2}{2} < \dfrac{1}{6}$.

Solution:

Multiply both sides by 6 to clear fractions.

$$2(x + 1) - 3(x + 2) < 1 \qquad \text{Inequality did not change.}$$
$$2x + 2 - 3x - 6 < 1 \qquad \text{Distributive Property}$$
$$-x - 4 < 1 \qquad \text{Combine like terms.}$$
$$-x < 5 \qquad \text{Add 4 to both sides.}$$

Next we multiply both sides by -1.

$$x > -5 \qquad \text{Note inequality change!}$$
$$\{x \mid x > -5\}$$

Be Careful!

Sec. 7.1 Linear Inequalities in One Variable

EXAMPLE 7. Solve $2x + 2 < 3(x + 1) - x + 3$.

Solution:

$$2x + 2 < 3(x + 1) - x + 3$$
$$2x + 2 < 3x + 3 - x + 3 \qquad \text{Distributive Property}$$
$$2x + 2 < 2x + 6 \qquad \text{Combine like terms.}$$
$$2 < 6 \qquad \text{Subtract } 2x.$$

$$\{x \mid x \text{ is a real number}\}$$

As the last statement is *always* true, the original inequality is true for *any* value of x. Thus the solution set is the set of all real numbers,

$$\{x \mid x \text{ is any real number}\} \qquad \square$$

Compare the result above with that of the following example.

EXAMPLE 8. Solve $x + 3 \leq 2(x - 3) - x + 3$.

Solution:

$$x + 3 \leq 2(x - 3) - x + 3$$
$$x + 3 \leq 2x - 6 - x + 3$$
$$x + 3 \leq x - 3$$
$$3 \leq -3$$

As the last statement is *never* true, the original inequality is false for *any* value of x. Thus the solution set is the empty set.

$$\emptyset \qquad \square$$

It is often convenient and instructive to draw a picture of the solution set of inequalities. This picture of the solution set is called the **graph** of the inequality.

To graph the set $\{x \mid x < 1\}$, we construct a copy of the real number line and locate all numbers less than 1 on it. They lie to the left of 1. We darken that portion of the line containing the solution set and indicate by an arrowhead that it continues indefinitely. We indicate with a **parenthesis** that the number 1 does *not* belong to the solution set.

Other notations use a hollow circle to indicate that a number is not included.

To graph the set $\left\{x \mid x \geq -\dfrac{1}{2}\right\}$, we draw a number line and shade the numbers to the right of $-\dfrac{1}{2}$. We use a **bracket** to show that $-\dfrac{1}{2}$ is included.

Other notations use a darkened circle to indicate that a number is included.

$$-\frac{1}{2}$$

EXAMPLE 9. Solve each inequality and graph the solution set.

(a) $\frac{3}{2}x + 2 < 3x$

(b) $2x + 9 \leq 5x + 3$

Solutions:

(a) $\frac{3}{2}x + 2 < 3x$

We multiply by 2 to clear fractions.

$$2\left(\frac{3}{2}x + 2\right) < 2(3x)$$

$3x + 4 < 6x$ Distributive Property

$4 < 3x$

$\frac{4}{3} < x$

$\frac{4}{3} < x$ can be written $x > \frac{4}{3}$. The latter form is commonly used in solution sets.

$$\left\{x \mid x > \frac{4}{3}\right\}$$

As we are asked to graph the solution set, we must shade the portion of the number line to the right of $\frac{4}{3}$. We indicate that the number $\frac{4}{3}$ does *not* belong to the solution set with a *parenthesis*.

(b) $2x + 9 \leq 5x + 3$

$9 \leq 3x + 3$ Subtract $2x$ from both sides.

$6 \leq 3x$

$2 \leq x$

$\{x \mid x \geq 2\}$

Again, draw the solution set on a number line. In this case we indicate that 2 *belongs* to the solution set with a *bracket*.

Inequalities arise in solving word problems just as equations do. The following phrases are used frequently.

WORDS	SYMBOL
is more than	$>$
is less than	$<$
is no less than	$\geq$
is no more than	$\leq$
is at least	$\geq$
is at most	$\leq$

EXAMPLE 10. Three times a number increased by 7 is at most 31. Find all such numbers.

Solution:

Let n be the number.

Writing the problem statement in the language of algebra, and inserting $\leq$ for "is at most," gives

$$3n + 7 \leq 31$$

Solving gives us

$$3n \leq 24$$
$$n \leq 8$$

The numbers must be less than or equal to 8.

■ PROBLEM SET 7.1

Warm-ups

In problems 1 through 16, solve the following inequalities.

For problems 1 through 6, see Examples 1 and 2.

1. $x - 3 \leq 4$
2. $x + 1 > -5$
3. $x - 7 \geq 3$
4. $x + 8 < 22$
5. $2x + 1 < -5$
6. $3x - 2 \geq 3$

For problems 7 and 8, see Example 3.

7. $-4x \geq 2$
8. $-\dfrac{1}{3}x \leq 1$

For problems 9 and 10, see Example 4.

9. $6(x - 2) \geq 2(x - 1)$
10. $3(x + 3) + 2(x - 1) \leq 0$

For problems 11 and 12, see Example 5.

11. $1.1x - 0.5 > 0.7x$
12. $2.5t + 1.5 \leq 3.0t - 2.5$

For problems 13 and 14, see Example 6.

13. $\dfrac{2}{3}(2x - 1) < \dfrac{1}{6}x + 1$
14. $\dfrac{x - 3}{5} - \dfrac{x - 1}{2} \leq 1$

For problems 15 and 16, see Examples 7 and 8.

15. $x + 3 \leq x + 4$ **16.** $2(x + 2) > 2(x + 1) + 2$

In problems 17 through 22, solve the inequality and graph the solution set. See Example 9.

17. $6x - 2 < 2x - 5$ **18.** $x - 1 > 2x + 3$

19. $7x - 3 \leq 5x - 1$ **20.** $3x - 1 \geq 6 - x$

21. $-4x + 2 > 2x - 4$ **22.** $\frac{1}{3}x - \frac{1}{2} < \frac{2}{3}x + 1$

The following word problems involve inequalities. Solve the inequality and write a sentence that answers each. See Example 10.

23. Twice a number plus 1 is at most 7. Find all such numbers.

24. Jim has grades of 90, 75, and 82 on his first three tests. What must he make on his fourth test in order to have an average of at least 80?

25. The perimeter of a rectangle is to be no more than 72 ft. If the length is 4 ft more than the width, find the measure of the largest possible width.

26. Twice 3 more than a number is no less than 12. Find all such numbers.

Practice Exercises

In problems 27 through 80, solve the inequality and graph the solution set.

27. $x + 7 > 10$ **28.** $x - 5 \leq 0$

29. $2 - x \geq 1$ **30.** $4x < 2$

31. $5 - x > 1$ **32.** $2x \leq 1$

33. $8x < 2x$ **34.** $-6x < 12$

35. $\frac{1}{2}x > -5$ **36.** $\frac{1}{4}x \leq 0$

37. $6x + 3 \leq 2x - 7$ **38.** $x - 7 > 5x + 4$

39. $7x - 3 > 2x$ **40.** $12x + 17 \leq 11 - 3x$

41. $\frac{1}{5}x - \frac{1}{3} \leq \frac{1}{5} + 2x$ **42.** $\frac{1}{2}y - \frac{1}{3} > \frac{1}{3}y + \frac{1}{6}$

43. $5(x + 1) < 2(1 - x)$ **44.** $3(x + 7) - (x + 2) \geq 0$

45. $\frac{1}{3}(x + 2) > \frac{1}{4}(x - 2)$ **46.** $-\frac{1}{2}(x - 6) > \frac{2}{3}(x + 6)$

Sec. 7.1 Linear Inequalities in One Variable

47. $\dfrac{x}{7} - \dfrac{1}{3} < \dfrac{x}{3}$

48. $\dfrac{x+2}{5} - \dfrac{x-3}{4} < 2$

49. $\dfrac{x-3}{6} \leq \dfrac{x}{2} - 1$

50. $\dfrac{3x+2}{-2} \leq \dfrac{5x-1}{4}$

51. $\dfrac{1-x}{3} > \dfrac{2}{3}(5-x)$

52. $\dfrac{1}{8}x - 2(x+3) < \dfrac{1}{4}$

53. $6(y+3) > 2(y-1)$

54. $3(t-7) > -5(t+3)$

55. $5(x-3) + 3 \leq -2(2x-4)$

56. $11(2x-1) - 3(3x+2) < 0$

57. $3[5-(s+2)] - 2s \geq 6s - 7$

58. $6 - 7z - 3(3z-4) > -8(2z-7)$

59. $\dfrac{3}{4}(y+1) < \dfrac{1}{8}y$

60. $\dfrac{1}{6}y - \dfrac{1}{2}(y-7) \geq -\dfrac{3}{2}$

61. $2[3t - 5 + 2(6t - 11)] < 10(2t - 5)$

62. $5(x+4) - (x+6) < 2[2x - (x+3)]$

63. $-3\{-2[(x-5) + 2(2x+1)] - 1\} \leq 5(x-7)$

64. $-2[(7t-3) - 2(3t+1)] \geq 8$

65. $\dfrac{x}{2} - \dfrac{1}{7} > \dfrac{1}{4}(2x - 9)$

66. $\dfrac{2x}{3} + \dfrac{x+3}{2} < -1 + \dfrac{7x}{6}$

67. $\dfrac{2(x-7)}{3} - \dfrac{3(x-3)}{2} \geq 1 - x$

68. $\dfrac{1}{4}x - \dfrac{3}{4}(1-x) > \dfrac{1}{2}(x+1)$

69. $0 < \dfrac{x-7}{4}$

70. $\dfrac{3t+4}{16} \geq 0$

71. $-5 < \dfrac{3(r-1)}{-2}$

72. $1 - 3(2x+5) \leq -\dfrac{1}{2}$

73. $0.4x - 20.8 \leq 8.0 - 0.8x$

74. $5.3 - 1.2x > 0.4x - 58.7$

75. $0.5 - 10.3x > 4.0x - 28.1$

76. $0.02u - 2 \geq 0.01u$

77. $2x + 5 > 2(x + 3)$

78. $5(2x - 1) < 10(x + 1) - 15$

79. $3(1 - x) \leq 3(x + 1) - 6x$

80. $2(x + 3) \geq 3(2 - x) + 5x$

The following word problems involve inequalities. Solve the indicated inequality and write a sentence that answers each.

81. Four times a number minus 2 is at most 10. Find all such numbers.

82. Gloria has grades of 82, 65, and 73 on her first three tests. What must she score on her fourth test to have an average of at least 80?

83. The perimeter of a rectangle is to be no more than 82 m. If the length is 7 m more than the width, find the measure of the largest possible width.

84. Five less than a number, multiplied by 3, is no less than 30. Find all such numbers.

Challenge Problems

In problems 85 through 92, solve for x; a, b, and c are constants.

85. $x - 5 < b$

86. $x + a \geq 6$

87. $3x - 2a > b$

88. $5x + 3b \leq a$

89. $ax < 5; \quad a > 0$

90. $ax < 5; \quad a < 0$

91. $a^2 x + x < b$

92. $(a + b)^2 x \geq c; \quad a \neq -b$

IN YOUR OWN WORDS . . .

93. What numbers are in the set $\{x \mid x < 1\}$?
94. What numbers are in the set $\{t \mid t \geq 1\}$?
95. Explain how to solve a linear inequality.

7.2 COMPOUND INEQUALITIES

A compound inequality is a statement made by combining two inequalities with the words **and** or **or.** For example, the statement

$$x < 1 \text{ or } x > 4$$

is a compound inequality. It is true if x is less than 1 *or* if x is greater than 4. Thus it is true when x is to the left of 1 on the number line *or* x is to the right of 4.

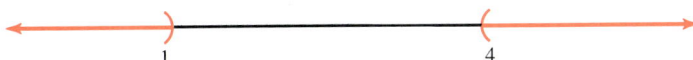

The statement

$$x < 10 \text{ and } x > 3$$

is a compound inequality that is true if x is less than 10 *and* greater than 3. Thus it is true when x is to the left of 10 *and* to the right of 3.

Notice from the graph of the inequality that if x is in the solution set, x must be *between* 3 and 10. Therefore, we usually write inequalities such as

$$x < 10 \text{ and } x > 3$$

in the **compact form**

$$3 < x < 10$$

This inequality could be written the other way around as

$$10 > x > 3$$

We prefer to write such inequalities using *less than* symbols as this puts the smaller number on the left.

The compact form will **never** be written using both $<$ and $>$ symbols. Such statements as $2 < x > 5$ are nonsense and not acceptable.

Be Careful!

EXAMPLE 1. Write a set that represents each graph.

(a)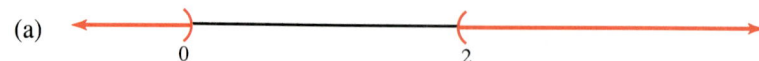

(b)

Solutions:

(a)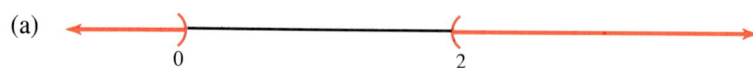

Points on the graph lie either to the left of 0 or to the right of 2. That is, $x < 0$ or $x > 2$. In set notation,

$$\{x \mid x < 0 \text{ or } x > 2\}$$

(b)

In this case, x must be *between* 1 and 3 in order to be in the given set. That is, $1 < x < 3$. Such a set is

$$\{x \mid 1 < x < 3\}$$

WARNING: Do not try to rewrite compound inequalities containing **or,** such as the solution set of Example 1a, in a more compact form such as in Example 1b. The compact form is used when x is between two numbers.

EXAMPLE 2. Write a set that represents each graph.

(a)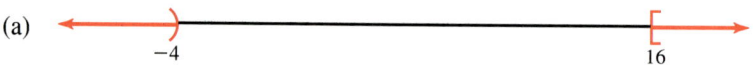

(b)

Solutions:

(a)

This is the set of all numbers less than -4 *or* greater than or equal to 16. In set notation,

$$\{x \mid x < -4 \text{ or } x \geq 16\}$$

(b)

This is the set of all numbers between -2 and 6, including -2 and 6.

$$\{x \mid -2 \leq x \leq 6\}$$

What is wrong with $2 < x < 0$? It is easy to see that we have written nonsense by noticing that $2 < x < 0$ says that $2 < 0$.

Chap. 7 Inequalities in One Variable

EXAMPLE 3. Which inequalities are nonsense?

(a) $-3 > y > 3$
(b) $2 < x < 7$
(c) $3 < t < 0$
(d) $-6 > x > -8$

Solutions:

(a) $-3 > y > 3$
This says that -3 is greater than 3, which is nonsense.
(b) $2 < x < 7$
This is all numbers that are between 2 and 7.
(c) $3 < t < 0$
This says that 3 is less than 0, which is nonsense.
(d) $-6 > x > -8$
This is all numbers between -6 and -8. This would usually be written with the smallest number on the left and the inequality signs reversed.
$$-8 < x < -6$$

EXAMPLE 4. Sketch the graph of each set.

(a) $\{x \mid x \leq -0.7 \text{ or } x > 2.3\}$
(b) $\{x \mid x \geq \frac{1}{2} \text{ and } x < \frac{3}{2}\}$

Solutions:

(a) $\{x \mid x \leq -0.7 \text{ or } x > 2.3\}$

For x to belong to this set, x must be to the left of (or equal to) -0.7, or x must be to the right of 2.3. Sketching this, we have

(b) $\{x \mid x \geq \frac{1}{2} \text{ and } x < \frac{3}{2}\}$

This set contains all numbers that are greater than (or equal to) $\frac{1}{2}$ and less than $\frac{3}{2}$, that is, all numbers between $\frac{1}{2}$ and $\frac{3}{2}$, including $\frac{1}{2}$.

To solve a compound inequality written in the compact form, work on all sides at the same time, using the tools from Section 7.1, until the variable is isolated in the middle.

Sec. 7.2 Compound Inequalities

EXAMPLE 5. Graph the solution set for $1 < 2x - 3 < 5$.

Solution:

As $1 < 2x - 3 < 5$ is shorthand for two inequalities, we can work on both at the same time. First, we add 3 to all three sides and get

$$1 + 3 < 2x - 3 + 3 < 5 + 3$$
$$4 < 2x < 8$$

Divide all three sides by 2:

$$\frac{4}{2} < \frac{2x}{2} < \frac{8}{2}$$
$$2 < x < 4$$

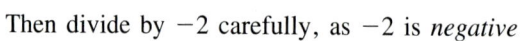

which is the graph of the set

$$\{x \mid 2 < x < 4\}$$

EXAMPLE 6. Find the solution set for $-3 \le 1 - 2x < 5$.

Solution:

$$-3 \le 1 - 2x < 5$$

Work on all three sides at one time to isolate x in the middle. First, subtract 1.

$$-4 \le -2x < 4$$

Be Careful!

Then divide by -2 carefully, as -2 is *negative*.

$$2 \ge x > -2 \quad \text{Note inequality change.}$$

We usually turn these around so that the smaller number is on the left when we write the solution set.

$$\{x \mid -2 < x \le 2\}$$

■ PROBLEM SET 7.2

Warm-ups

In problems 1 through 8, write a set to represent each graph. See Examples 1 and 2.

1.

2.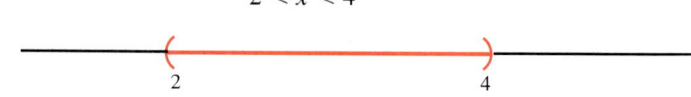

3.

4.

5.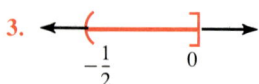

6.

7.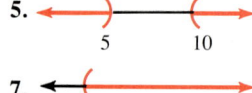

8.

In problems 9 through 14, which inequalities are nonsense? See Example 3.

9. $3 < x < 5$
10. $-2 > t \geq -5$
11. $0 \leq y \leq -3$
12. $-4 < x < -6$
13. $-1 > x > 0$
14. $-2 < x < 2$

In problems 15 through 22, sketch the graph of each set. See Example 4.

15. $\{x \mid 1 < x < 2\}$
16. $\{x \mid 2 \leq x \leq 10\}$
17. $\{x \mid -1 < x \leq 1\}$
18. $\{x \mid 0 < x < 2\}$
19. $\{x \mid x \leq -7 \text{ or } x > -2\}$
20. $\{x \mid x \leq 0 \text{ or } x > 3\}$
21. $\left\{x \mid x < \dfrac{7}{4} \text{ and } x > \dfrac{3}{4}\right\}$
22. $\{x \mid x > -2 \text{ and } x < 0\}$

In problems 23 through 28, find the solution set for each compound inequality. See Examples 5 and 6.

23. $2 \leq x + 3 \leq 7$
24. $-1 < x + 5 < 10$
25. $x < 1 \text{ and } x > -2$
26. $x \geq 3 \text{ or } x \leq -1$
27. $x \geq 4 \text{ or } x < -3$
28. $x < 2.1 \text{ and } x \geq 0$

Practice Exercises

In problems 29 through 44, sketch the graph of each set.

29. $\{x \mid 2 < x < 3\}$
30. $\{x \mid 4 \leq x < 7\}$
31. $\{x \mid -3 \leq x < 2\}$
32. $\{x \mid -9 < x < 0\}$
33. $\{x \mid x < 6 \text{ or } x > 11\}$
34. $\{x \mid x < -3 \text{ or } x > 3\}$
35. $\{x \mid x \leq -3 \text{ and } x \leq -5\}$
36. $\{x \mid x \geq 6 \text{ and } x \leq 9\}$
37. $\left\{x \mid x < \dfrac{5}{3} \text{ or } x \geq \dfrac{10}{3}\right\}$
38. $\{x \mid x > 13 \text{ and } x \leq 17\}$
39. $\{x \mid x \leq -3 \text{ or } x > 3\}$
40. $\{x \mid x < -11 \text{ or } x \geq -8\}$
41. $\{x \mid 0.5 < x \leq 1.5\}$
42. $\{x \mid 1.3 \leq x < 1.4\}$
43. $\{x \mid x \leq -0.16 \text{ and } x > -2.34\}$
44. $\{x \mid x > 51.6 \text{ and } x \leq 58.7\}$

In problems 45 through 50, which inequalities are nonsense?

45. $-3 < x < -5$
46. $-1 > t \geq 7$
47. $0 \leq y \leq 3$
48. $4 < x < -6$
49. $1 > x > 0$
50. $-2 < x < 2$

In problems 51 through 60, write a set to represent each graph.

51.

52.
```
←——————————)———————(——→
           -3        1
```

Sec. 7.2 Compound Inequalities

53.

55.

57.

59. ──[────────→
      -11

54.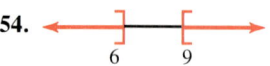
    6 9

56. ────(──[──
      -2 0

58. ─────[────→
        0

60. ─────(────→
        34

In problems 61 through 72, find the solution set for each compound inequality.

61. $3 \leq 2x + 1 \leq 5$
62. $-10 \leq 5x + 5 < 5$
63. $-5 \leq 3x - 2 < 7$
64. $0 \leq 6x + 3 < 3$
65. $4 \leq 1 - x \leq 10$
66. $-6 < 2 - 8x < 18$
67. $5 < \frac{1}{2}x - 2 \leq 8$
68. $3 < \frac{2}{3}x + \frac{1}{3} \leq \frac{15}{3}$
69. $\frac{1}{2} < 2x - 3 < 3$
70. $x \leq 2$ or $x \geq 5$
71. $x < -\frac{3}{2}$ or $x \geq -\frac{1}{2}$
72. $x < \frac{1}{2}$ and $x > -\frac{1}{2}$

Challenge Problems

In problems 73 through 78, find the solution set.

73. $x \leq a$ or $x \geq b$; $a < b$
74. $x \leq a$ and $x \leq b$; $a < b$
75. $2 \leq x - a \leq 4$
76. $1 < ax < 3$; $a > 0$
77. $1 < ax < 3$; $a < 0$
78. $a \leq ax + b \leq b$; $a < 0; a < b$

In problems 79 through 82, write a set to represent each graph.

79. ←──)────(──]──
      -1   1 2

80. ──[────]─[───
    7   8 10

81. ←─[────(──]──
    -5 0 3

82. ──(──)──[──]──
   -8 -6 -1 2

■ IN YOUR OWN WORDS . . .

83. What is a compound inequality?

84. What numbers are in the set ?

85. What numbers are in the set $\{y \mid y < -7$ or $y \geq -1\}$?

7.3 BOUNDARY NUMBERS AND ABSOLUTE VALUE INEQUALITIES

In Section 7.1 we solved linear inequalities in one variable. The tools that we used there are not sufficient to solve many other kinds of inequalities. We now examine another technique that applies to the linear inequalities as well as most other types of inequalities.

Let's solve the inequality $2x + 1 < 7$ by the method developed in Section 7.1.

$$2x + 1 < 7$$
$$2x < 6$$
$$x < 3$$
$$\{x \mid x < 3\}$$

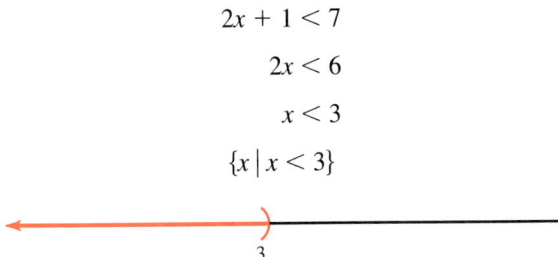

The number 3 plays an important role in this solution set. It divides the number line into two regions, labeled A and B below.

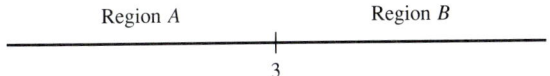

The solution set proved to be the region A. That is, the statement $2x + 1 < 7$ is true for **every** number in region A, and false for **every** number in region B.

Is there a simple way to find this number that has the property of so dividing the number line? Note that the number 3 is the solution to the equation $2x + 1 = 7$. We call 3 a **boundary number** for the inequality $2x + 1 < 7$.

Boundary Numbers

The boundary numbers of an inequality are the solutions to the equation formed by replacing the inequality symbol with an equals symbol.

Let's see how to use this technique to solve an inequality.

EXAMPLE 1. Solve $3x - 7 < 5$.

Solution:

First, we solve the associated equation in order to find the boundary number(s).

$$3x - 7 = 5$$
$$3x = 12$$
$$x = 4$$

4 is the only boundary number.

We locate 4 on a number line and note that it divides the line into two regions, A and B.

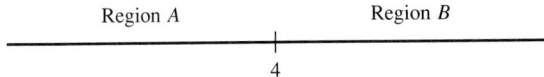

We select *any* number from region A and check to see if it is in the solution set for $3x - 7 < 5$. A particularly convenient number in region A is 0. As $3(0) - 7 < 5$ is a *true* statement, 0 must be in the solution set for the inequality. Thus *all* of region A is in the solution set.

Sec. 7.3 Boundary Numbers and Absolute Value Inequalities

Next we test *any* number from region *B*, say 5. As $3(5) - 7 < 5$ is *false*, we conclude that 5 is not in the solution set. So *none* of region *B* is in the solution set.

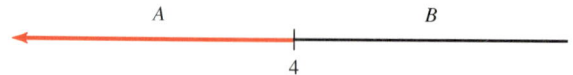

The original inequality did not include equality, so the number 4 is *not* included in the solution set.

Therefore, the solution set is $\{x \mid x < 4\}$.

> **A Procedure for Solving Inequalities**
> 1. Find the boundary numbers.
> 2. Locate the boundary numbers on a number line.
> 3. Determine which regions formed by the boundary numbers make the inequality true by testing with one number inside each region.
> 4. Shade only the regions that test true.
> 5. Check the boundary numbers themselves.
> 6. Write the solution set.

EXAMPLE 2. Solve the inequality $|x - 2| > 5$.

Solution:

Step 1 Find the boundary numbers.

We solve the associated equation.

$$|x - 2| = 5 \quad \text{(an absolute value equation)}$$
$$x - 2 = 5 \quad \text{or} \quad x - 2 = -5$$
$$x = 7 \quad \text{or} \quad x = -3$$

The boundary numbers are -3 and 7.

Step 2 Locate the boundary numbers on a number line.

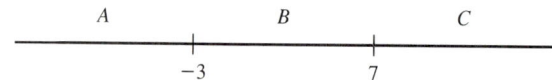

Step 3 Determine in which regions the inequality is true by testing each region.

A table is particularly convenient for this step.

REGION	NUMBER IN REGION	STATEMENT $\|x - 2\| > 5$	TRUTH OF STATEMENT	REGION IN SOLUTION SET?
A	-4	$\|-4 - 2\| > 5$	True	Yes
B	0	$\|0 - 2\| > 5$	False	No
C	8	$\|8 - 2\| > 5$	True	Yes

Step 4 Shade the regions that test true.

We shade region A and region C on the number line.

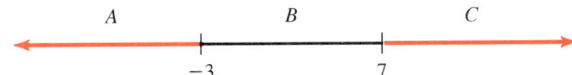

Step 5 Check the boundary numbers themselves.

The statement $|x - 2| > 5$ does not include equality, so the boundary numbers are not in the solution set.

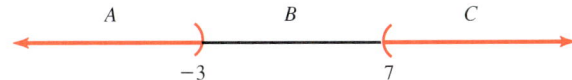

Step 6 Write the solution set.

$$\{x \mid x < -3 \text{ or } x > 7\}$$

□

EXAMPLE 3. Solve $|1 - 2x| \leq 3$.

Solution:

First we find the boundary numbers.

$$|1 - 2x| = 3$$

$1 - 2x = 3$	or	$1 - 2x = -3$
$-2x = 2$	or	$-2x = -4$
$x = -1$	or	$x = 2$

The boundary numbers are -1 and 2.

Next, we locate the boundary numbers on a number line.

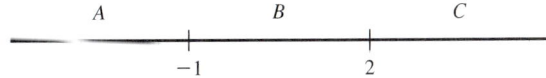

The next step is to test regions A, B, and C. A table such as the one that follows can be made or the regions can be shaded as each region is tested.

REGION	NUMBER IN REGION	STATEMENT $\|1 - 2x\| \leq 3$	TRUTH OF STATEMENT	REGION IN SOLUTION SET?
A	-2	$\|1 - 2(-2)\| \leq 3$	False	No
B	0	$\|1 - 2(0)\| \leq 3$	True	Yes
C	3	$\|1 - 2(3)\| \leq 3$	False	No

Region B is the only region in the solution set.

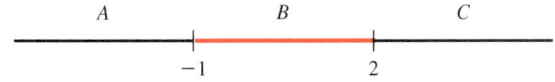

(continued)

Sec. 7.3 Boundary Numbers and Absolute Value Inequalities

The statement $|1 - 2x| \leq 3$ includes equality, so the boundary numbers *are* in the solution set.

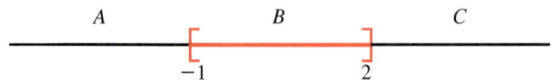

The last step is to write the solution set.

$$\{x \mid -1 \leq x \leq 2\}$$

EXAMPLE 4. Solve $|x - 6| > |2 - 3x|$.

Solution:

We find the boundary numbers by solving the equation

$$|x - 6| = |2 - 3x|$$

We recall that we must solve the two equations

$x - 6 = 2 - 3x$	or	$x - 6 = -(2 - 3x)$
$4x - 6 = 2$	or	$x - 6 = -2 + 3x$
$4x = 8$	or	$-4 = 2x$
$x = 2$	or	$-2 = x$

Thus we get the boundary numbers -2 and 2.

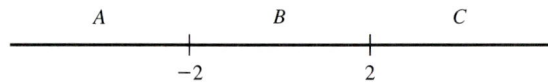

Next, we test each region. Region A contains -4. Testing with -4 gives

$$|-4 - 6| > |2 - 3(-4)|$$
$$|-10| > |2 + 12|$$
$$10 > 14$$

which is false, so region A is not shaded.

Region B contains 0, a convenient number.

$$|0 - 6| > |2 - 3(0)|$$
$$6 > 2$$

As that is a true statement, we shade region B.

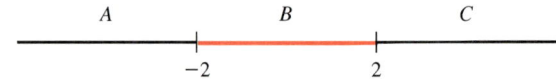

Region C contains 3. Testing with 3 gives

$$|3 - 6| > |2 - 3(3)|$$
$$|-3| > |2 - 9|$$
$$3 > 7$$

which is not true. We do not shade region C.

The statement $|x - 6| > |2 - 3x|$ does not allow equality, so the boundary numbers are not in the solution set.

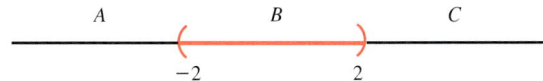

Write the solution set:
$$\{x \mid -2 < x < 2\}$$

EXAMPLE 5. Solve $|x - 5| \geq 0$.

Solution:

First we find the boundary numbers.
$$|x - 5| = 0$$
$$x - 5 = 0$$
$$x = 5$$

The only boundary number is 5.

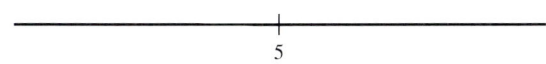

A convenient number less than 5 is 0.
$$|0 - 5| \geq 0$$
$$5 \geq 0$$

is a true statement, so the left region is in the solution set. A number to the right of 5 is 7, and
$$|7 - 5| \geq 0$$
$$2 \geq 0$$

is also a true statement, so the right region also belongs to the solution set.

Since equality is included in the original inequality, the boundary number is included in the solution set.

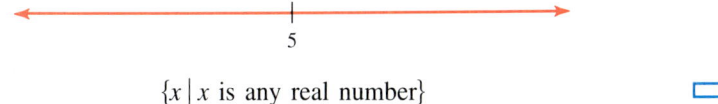

$$\{x \mid x \text{ is any real number}\}$$

EXAMPLE 6. Solve $|x| < -5$.

Solution:

To find the boundary numbers, we solve the equation
$$|x| = -5$$

As there is no solution to this equation, there are no boundary numbers. Thus there is just one region and it is the entire real line.

(continued)

Sec. 7.3 Boundary Numbers and Absolute Value Inequalities

If we test any number in the region, say 10, we find that the statement is false.
$$|10| < -5$$
$$10 < -5$$

As there are no boundary numbers to include, we get the empty set as our solution set:
$$\emptyset$$

NOTE: We could have seen that the solution set for Example 6 was the empty set because the absolute value of a number is never negative.

Absolute value inequalities such as
$$|X| < p \qquad |X| > p$$
$$|X| \leq p \qquad |X| \geq p$$

with $p > 0$, arise frequently in mathematics. Often the solution set is not needed, but rather an equivalent compound inequality without the absolute value symbol. It is useful to remember the following result.

Absolute Value Inequalities

If $p > 0$ and X is an expression, then

$$|X| < p \quad \text{is equivalent to} \quad -p < X < p$$

and

$$|X| > p \quad \text{is equivalent to} \quad X < -p \text{ or } X > p$$

The results hold if the $<$ and $>$ are replaced by $\leq$ and $\geq$.

EXAMPLE 7. Write $|x + 1| < 5$ as an equivalent compound inequality.

Solution:

We let $x + 1$ be the X in the statement
$$|X| < p \quad \text{is equivalent to} \quad -p < X < p$$
and we have
$$|x + 1| < 5 \quad \text{is equivalent to} \quad -5 < x + 1 < 5$$

Subtracting 1 from all sides yields
$$-6 < x < 4$$

EXAMPLE 8. Write $|x - 3| \geq 7$ as an equivalent compound inequality.

Solution:

We take $x - 3$ as the X and have that
$$|x - 3| \geq 7 \quad \text{is equivalent to} \quad x - 3 \leq -7 \text{ or } x - 3 \geq 7$$

Adding 3 to both sides of each part, we get the compound inequality
$$x \leq -4 \qquad \text{or} \qquad x \geq 10$$

We learned in Section 0.3 that the distance between two numbers on the number line is always given by the absolute value of their difference. This idea leads to a very important inequality.

EXAMPLE 9. Find all numbers that are less than $\frac{1}{7}$ unit from 4 on the number line.

Solution:

We let x be such a number. The distance between 4 and x is given by $|x - 4|$. As this is less than $\frac{1}{7}$, we get the inequality

$$|x - 4| < \frac{1}{7}$$

$$-\frac{1}{7} < x - 4 < \frac{1}{7}$$

$$-\frac{1}{7} + 4 < x < \frac{1}{7} + 4$$

$$\frac{27}{7} < x < \frac{29}{7}$$

So $\left\{x \mid \frac{27}{7} < x < \frac{29}{7}\right\}$ is the set of all such numbers.

▬ PROBLEM SET 7.3

Warm-ups

In problems 1 through 14, find the solution set.

For problems 1 through 6, see Examples 2 and 3.

1. $|x - 1| < 5$
2. $|x + 3| > 1$
3. $|x + 2| \geq 8$
4. $|7 - x| \leq 9$
5. $|5x - 7| - 2 < 0$
6. $|4x + 1| - 4 \geq 0$

For problems 7 through 10, see Example 4.

7. $|x + 1| > |2x + 3|$
8. $|2x| \leq |x - 2|$
9. $|2x + 3| < |5x - 3|$
10. $|3 - x| \geq |2x + 5|$

For problems 11 through 14, see Examples 5 and 6.

11. $|x - 7| \geq 0$
12. $|x + 8| > 0$
13. $|x - 5| < -7$
14. $|x + 1| > -1$

In problems 15 through 18, write each inequality as an equivalent compound inequality. See Examples 7 and 8.

15. $|x - 7| \leq 5$
16. $|2x - 5| > 6$
17. $\left|2x + \frac{1}{2}\right| \geq 7$
18. $\left|\frac{1}{2}x - 1\right| < 3$

Sec. 7.3 Boundary Numbers and Absolute Value Inequalities

The following word problems involve inequalities. Solve the indicated inequality and write a sentence that answers each. In problems 19 through 21, see Example 9.

19. Find all numbers that are within 2 units of -4 on the number line.
20. Find all numbers that are less than 1 unit from 5 on the number line.
21. Find all numbers that are less than $\frac{1}{3}$ unit from $\frac{1}{2}$ on the number line.
22. Find all numbers whose absolute value is less than or equal to 5.
23. Find all numbers whose absolute value is greater than 7.
24. Find all numbers whose absolute value is at least 10.
25. Find all numbers whose absolute value is at most 3.
26. If 6 is added to three times a number, the absolute value of the result is at least 1. Find all such numbers.

Practice Exercises

In problems 27 through 62, find the solution set.

27. $|x| \geq 8$
28. $|x| < 2$
29. $|x - 3| < 9$
30. $|x + 1| \geq 16$
31. $|x + 5| \leq 3$
32. $|x - 7| > 4$
33. $|2x + 5| > 7$
34. $|3x - 1| \leq 10$
35. $|2x - 9| - 1 \leq 0$
36. $|3x + 7| - 7 > 0$
37. $|5x - 8| > 9$
38. $|4x + 3| \leq 7$
39. $|10y + 3| - 1 \geq 0$
40. $|5 - 2z| - 3 < 0$
41. $3(|x - 1| + 3) < 15$
42. $2(3 + |2x - 1|) \geq 5$
43. $10 \leq 3(4 - |x + 3|)$
44. $0 > 5(|s| - 4)$
45. $|t - 2.5| < 6.3$
46. $\left|\frac{2}{3}x - \frac{3}{4}\right| < 6.3$
47. $|x - 2.73| > 6.72$
48. $|x - 2.4| > 7.8$
49. $\left|\frac{2x - 1}{3}\right| \leq 4$
50. $\left|\frac{3x}{5} - 2\right| \geq \frac{1}{5}$
51. $|2 - x| + 3 \geq 4$
52. $\left|3 - \frac{1}{4}x\right| + \frac{3}{8} > 1$
53. $|x + 1| \geq 0$
54. $|x - 7| > 0$
55. $|z - 3| < -5$
56. $|y + 2| > -4$
57. $|2x + 3| \leq -1$
58. $|3x + 2| \geq -4$
59. $|3x| \geq |x - 4|$
60. $|2x + 7| < |3x - 2|$
61. $|4x + 3| \leq |3x + 18|$
62. $|-3x + 2| > |x - 2|$

In problems 63 through 70, write each inequality as an equivalent compound inequality.

63. $|x + 1| \leq 3$
64. $|3x + 4| > 5$
65. $|x| > 4.3$
66. $|x| \leq 1.5$
67. $|x + 3| < 2$
68. $|2x + 3| \geq 7$
69. $\left|2x - \frac{1}{3}\right| < 2$
70. $\left|\frac{1}{3}x + 1\right| \leq 2$

The following word problems involve inequalities. Solve the indicated inequality and write a sentence that answers each.

71. Find all numbers whose absolute value is less than 4.
72. Find all numbers whose absolute value is greater than or equal to $\frac{3}{2}$.

73. Find all numbers whose absolute value is at most 8.

74. If 4 is subtracted from twice a number, the absolute value of the result is at most 5. Find all such numbers.

75. Find all numbers that are less than 2 units from 1 on the number line.

76. Find all numbers that are less than 0.05 unit from 3 on the number line.

Challenge Problems

77. Prove that $|X| < a$, $(a > 0)$ is equivalent to $-a < X < a$ using boundary numbers.

In problems 78 through 87, find the solution set. Assume that $a > 0$.

78. $|x - a| \leq 4$
79. $|x - 1| > d$; $d > 0$
80. $|x - a| < d$; $d > 0$
81. $|x - a| > d$; $d < 0$
82. $|x - a| < a$
83. $|x - a| \geq a$
84. $|x - a| \leq 0$
85. $|x - a| > 0$
86. $|x + a| < a$
87. $|x + a| \geq a$
88. Find all numbers that are within s units of a.

IN YOUR OWN WORDS . . .

89. What is a boundary number for an inequality?

90. What do we do if an inequality has no boundary numbers?

7.4 HIGHER-DEGREE AND FRACTIONAL INEQUALITIES

Just as there are fractional equations and equations of higher degree, there are fractional inequalities and inequalities of higher degree. As in Section 7.3, the method of boundary numbers reduces the problem of solving an inequality to the problem of solving an equation.

EXAMPLE 1. Solve the inequality $(x - 1)(x + 2) \leq 0$.

Solution:

First, we find the boundary numbers

$$(x - 1)(x + 2) = 0$$

The boundary numbers are -2 and 1.

We locate the boundary numbers on a number line.

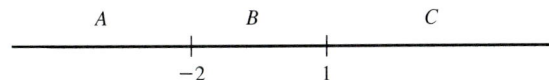

As in Section 7.3, we can make a table.

REGION	NUMBER IN REGION	STATEMENT $(x - 1)(x + 2) \leq 0$	TRUTH OF STATEMENT	REGION IN SOLUTION SET?
A	-3	$(-4)(-1) \leq 0$	False	No
B	0	$(-1)(2) \leq 0$	True	Yes
C	2	$(1)(4) \leq 0$	False	No

(continued)

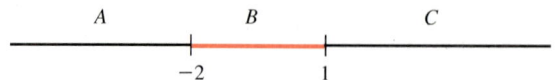

Since $(x-1)(x+2) \leq 0$ allows equality, both -2 and 1 are in the solution set.

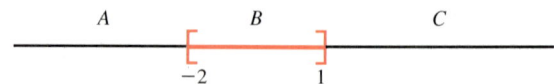

Last, we write the solution set.

$$\{x \mid -2 \leq x \leq 1\}$$

EXAMPLE 2. Solve the inequality $x(x-2)(x+3) > 0$.

Solution:

To find the boundary numbers, we solve the equation

$$x(x-2)(x+3) = 0$$

The boundary numbers are 0, 2, and -3.

Locate the boundary numbers on a number line.

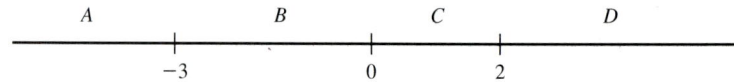

A table for step 3.

REGION	NUMBER IN REGION	STATEMENT $x(x-2)(x+3) > 0$	TRUTH OF STATEMENT	REGION IN SOLUTION SET?
A	-4	$-4(-6)(-1) > 0$	False	No
B	-1	$-1(-3)(2) > 0$	True	Yes
C	1	$1(-1)(4) > 0$	False	No
D	5	$5(3)(8) > 0$	True	Yes

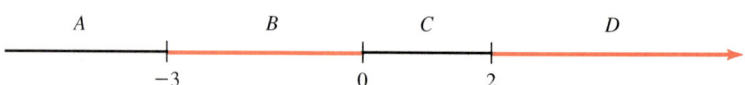

Since $x(x-2)(x+3) > 0$ does not include equality, the boundary numbers are not included in the solution set.

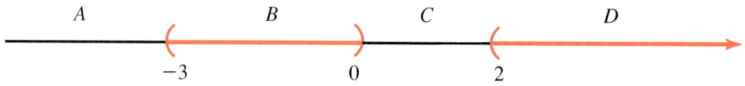

Write the solution set.

$$\{x \mid -3 < x < 0 \text{ or } x > 2\}$$

EXAMPLE 3. Solve $-x^2 + x + 2 < 0$.

Solution:

First, the boundary numbers.

$$-x^2 + x + 2 = 0$$
$$x^2 - x - 2 = 0 \quad \text{Multiply by } -1.$$
$$(x + 1)(x - 2) = 0 \quad \text{Factor.}$$

The boundary numbers are -1 and 2.

We locate the boundary numbers on a number line.

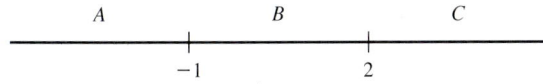

As noted in Section 7.3, it is not necessary to make a table. We check region A with any number in it, say -5.

$$-(-5)^2 + (-5) + 2 < 0$$
$$-25 - 5 + 2 < 0$$
$$-28 < 0$$

A true statement. Region A is included.

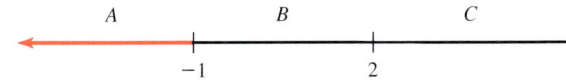

For region B, we will test 0.

$$-0^2 + 0 + 2 < 0$$
$$2 < 0$$

A false statement. Region B is not included.

For region C, we will test 3.

$$-3^2 + 3 + 2 < 0$$
$$-9 + 5 < 0$$
$$-4 < 0$$

So region C is included.

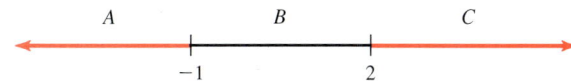

Since $-x^2 + x + 2 < 0$ does not include the equality, the boundary numbers are not included in the solution set.

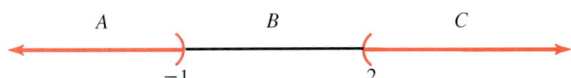

The solution set is

$$\{x \mid x < -1 \text{ or } x > 2\}$$

A fractional inequality has boundary numbers at all solutions of the associated equation just like other inequalities. In addition, a fractional inequality has a boundary number where any denominator has a value of zero. Furthermore, boundary numbers where a denominator has a value of zero are **never** in the solution set of the inequality. We call these boundary numbers **free boundary numbers.** *A free boundary number is never in a solution set.*

EXAMPLE 4. Solve $\dfrac{x+2}{x-1} > 2$.

Solution:

The number 1 is a *free* boundary number because it would make the denominator have a value of zero if x had the value 1. For other boundary numbers, we solve the equation

$$\frac{x+2}{x-1} = 2$$

$$x + 2 = 2(x - 1) \quad \text{Multiply by } x - 1.$$

$$x + 2 = 2x - 2$$

$$4 = x$$

and get a second boundary number, 4. The boundary numbers are 1 and 4.

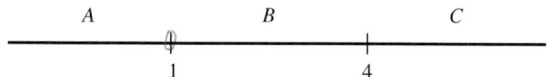

We test 0 in region A.

$$\frac{0+2}{0-1} > 2$$

$$-2 > 2 \qquad \text{False. Region } A \text{ is not included.}$$

We test 2 in region B.

$$\frac{2+2}{2-1} > 2$$

$$4 > 2 \qquad \text{True. Region } B \text{ is included.}$$

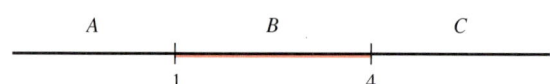

Test 5 in region C.

$$\frac{5+2}{5-1} > 2$$

$$\frac{7}{4} > 2 \qquad \text{False. Region } C \text{ is not included.}$$

As the original problem does not allow equality, the boundary number, 4, is not in the solution set. The other boundary number, 1, is a number that makes a denominator in the inequality have a value of zero. Such boundary numbers are *never* in the solution set.

$$\{x \mid 1 < x < 4\}$$

EXAMPLE 5. Solve 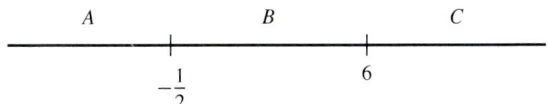 $\dfrac{3x - 5}{2x + 1} \leq 1$.

Solution:

The first step is to find the boundary numbers. We solve the associated equation,

$$\frac{3x - 5}{2x + 1} = 1$$

$$3x - 5 = 2x + 1$$

$$x = 6$$

So we have a boundary number 6, the solution of the associated equation, and a free boundary number $-\dfrac{1}{2}$, where the denominator has a value of zero.

We locate the boundary numbers on a number line.

Test region A with -1.

$$\frac{3(-1) - 5}{2(-1) + 1} \leq 1$$

$$\frac{-8}{-1} \leq 1$$

$$8 \leq 1 \qquad \text{False. Region } A \text{ is not included.}$$

Test region B with 0.

$$\frac{3(0) - 5}{2(0) + 1} \leq 1$$

$$-5 \leq 1 \qquad \text{True. Region } B \text{ is included.}$$

(continued)

Test region C with 7.

$$\frac{3(7)-5}{2(7)+1} \leq 1$$

$$\frac{16}{15} \leq 1 \quad \text{False. Region } C \text{ is not included.}$$

The boundary number 6 is in the solution set because equality is allowed in the original statement. However, the boundary number $-\frac{1}{2}$ is *not* in the solution set as it is a free boundary number (makes a denominator have a value of zero).

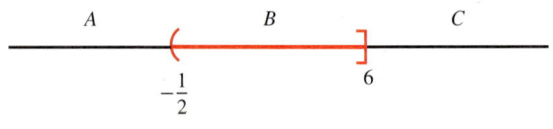

$$\left\{ x \mid -\frac{1}{2} < x \leq 6 \right\}$$

PROBLEM SET 7.4

Warm-ups

In problems 1 through 20, find the solution set for each inequality.

For problems 1 through 4, see Example 1.

1. $(x+1)(x-2) < 0$
2. $(x+4)(x-7) > 0$
3. $(x-3)(x+1) \leq 0$
4. $(2-x)(x+3) \geq 0$

For problems 5 through 8, see Example 2.

5. $x(x+2)(x-1) > 0$
6. $x(x-2)(x+5) > 0$
7. $(x+4)(x-1)(x+3) > 0$
8. $(x-1)(x-2)(x-3) \leq 0$

For problems 9 through 14, see Example 3.

9. $x^2 - 4x + 3 < 0$
10. $x^2 - x - 6 > 0$
11. $x^2 + 6x < -5$
12. $x^2 - 2x \geq 3$
13. $-x^2 - x + 2 > 0$
14. $x^2 - 4 \leq 0$

For problems 15 through 20, see Examples 4 and 5.

15. $\dfrac{x-5}{x+5} \geq 0$
16. $\dfrac{x-4}{1-x} \leq 0$
17. $\dfrac{x+5}{x-5} \geq 2$
18. $\dfrac{3-x}{x+6} \leq 2$
19. $\dfrac{3}{x+1} > \dfrac{2}{x-1}$
20. $\dfrac{1}{x-1} < \dfrac{2}{x+7}$

Practice Exercises

In problems 21 through 68, find the solution set for each inequality.

21. $(x+3)(x+4) < 0$
22. $(x-2)(x+5) > 0$
23. $(7-x)(x+1) \geq 0$
24. $(x+3)(1+x) \leq 0$

25. $(3x + 1)(x - 4) < 0$
26. $(x + 3)(5 - 2x) > 0$
27. $(x + 1)(x + 1) < 0$
28. $(x - 3)(x - 3) > 0$
29. $(5 + x)(5 + x) \leq 0$
30. $(x - 6)(x - 6) \geq 0$
31. $x^2 - 7x + 6 > 0$
32. $x^2 - 2x - 8 < 0$
33. $x^2 - 7x > 8$
34. $x^2 + 6x < -8$
35. $x^2 > 16$
36. $x^2 + 2x < -1$
37. $-t^2 - 5t \geq 0$
38. $2y^2 - y < 0$
39. $s^2 + 3 < -4s$
40. $x^2 > 0$
41. $5x + 14 \geq x^2$
42. $2x^2 + 11x - 21 < 0$
43. $10x^2 + x - 3 > 0$
44. $2z^2 + 5 > 11z$
45. $21x^2 + 19x - 12 \leq 0$
46. $-7x^2 + 5x + 2 \leq 0$
47. $12x^2 + 5x - 2 < 0$
48. $6y^2 + 5y > 6$
49. $10x^2 - 21x - 49 \geq 0$
50. $3t^2 + 17t > 6$
51. $(x - 7)(x + 3)(x - 1) \geq 0$
52. $(x - 1)(x - 2)(x - 4) < 0$
53. $x(x + 1)(x - 1) > 0$
54. $x(x + 3)(x + 2) \leq 0$
55. $x^3 - 4x \geq 0$
56. $x^3 - 2x^2 \leq 0$
57. $\dfrac{x + 1}{x - 1} > 0$
58. $\dfrac{x - 2}{x + 3} < 0$
59. $\dfrac{x - 4}{x - 5} \geq 0$
60. $\dfrac{x - 5}{x + 1} \leq 0$
61. $\dfrac{x + 2}{x} < 0$
62. $\dfrac{x}{x - 1} > 0$
63. $\dfrac{r - 5}{r + 2} \leq 2$
64. $\dfrac{x + 7}{x + 3} \geq 3$
65. $\dfrac{2x + 3}{x + 4} \geq 1$
66. $\dfrac{2}{w - 3} < \dfrac{1}{w + 2}$
67. $\dfrac{1}{x - 1} \leq \dfrac{1}{x + 1}$
68. $\dfrac{2x + 3}{2x + 1} \geq \dfrac{1}{2x + 1}$

The following word problems involve inequalities. Write a sentence that answers each.

69. The product of two consecutive integers is at most 2. Find all such integers.
70. A tablecloth in the shape of a rectangle must have an area of more than 10 ft². If the length is to be 3 ft more than the width, find all possible dimensions.
71. The reciprocal of one more than a number is negative. Find all such numbers.
72. The product of two consecutive even integers is at least 8. Find all such integers.
73. The area of a square playground must be at most 10,000 yd². How long could a side be?
74. The reciprocal of 1 less than a number is positive. Find all such numbers.
75. If one more than a number is divided by one less than the number, the quotient is nonnegative. Find all possible numbers.

Challenge Problems

In problems 76 through 83, find the solution set.

76. $x^3 + 8 > 0$
77. $x^2 + x + 1 \geq 0$
78. $x^2(x - 2) \geq 0$
79. $x(x - 1)^2 < 0$

Sec. 7.4 Higher-Degree and Fractional Inequalities

80. $(x + 2)(x - 5)^3 > 0$

81. $\dfrac{(x + 2)(x - 1)}{(x + 5)(x - 7)} \leq 0$

82. $\dfrac{x - a}{x - b} \leq 0; \quad a > 0, b < 0$

83. $\dfrac{x - a}{x - b} \leq 0; \quad a < 0, b > 0$

IN YOUR OWN WORDS . . .

84. What are free boundary numbers, and how do they differ from other boundary numbers?

CHAPTER SUMMARY

GLOSSARY

Inequality: A statement that says one number is greater or less than another number.

Solution of an inequality: A number that makes the statement true when substituted for the variable.

Solution set of an inequality: The set of all solutions to the inequality.

Solve an inequality: To find the solution set.

Equivalent inequalities: Inequalities that have the same solution set.

The **graph** of an inequality: A picture of the solution set on a number line.

Boundary numbers: Numbers that border the graph of the solution set of an inequality.

LINEAR INEQUALITIES

Linear inequalities may be solved using the following tools.

Addition Property

The same number may be added or subtracted on both sides.

Multiplication by Positive Number

Both sides may be multiplied or divided by the same positive number.

Multiplication by Negative Number

Both sides may be multiplied or divided by the same negative number, *but* the inequality symbol must be *reversed*.

THE METHOD OF BOUNDARY NUMBERS

An excellent technique for solving nonlinear inequalities is the method of boundary numbers.

1. Form an equation by replacing the inequality symbol with an equals sign. The solution(s) to this equation are the boundary number(s). Numbers that make any denominator of a fractional equation have a value of zero are also boundary numbers, called free boundary numbers.

2. Locate and label the boundary numbers on a real number line.

3. Test the original inequality with one number from the interior of each region formed by the boundary numbers. The entire region is in the solution set if the test statement is true, or none of the region is in the solution set if the test statement is false.

4. Check the boundary numbers themselves in the original inequality to see if they are included in the solution set. Free boundary numbers are *never* included in a solution set.

CHECKUPS

1. Solve $\dfrac{x+1}{3} + \dfrac{x+2}{2} < \dfrac{1}{6}$. Section 7.1; Example 6
2. Solve $-3 \leq 1 - 2x < 5$. Section 7.2; Example 6
3. Solve $|x - 2| > 5$. Section 7.3; Example 2
4. Solve $-x^2 + x + 2 < 0$. Section 7.4; Example 3
5. Solve $\dfrac{x+2}{x-1} > 2$. Section 7.4; Example 4
6. Solve $x(x - 2)(x + 3) > 0$. Section 7.4; Example 2
7. Three times a number increased by 7 is at most 31. Find all such numbers. Section 7.1; Example 10

REVIEW PROBLEMS

In problems 1 through 22, graph the solution set.

1. $2x - 3 \leq 5$
2. $1 < 2x < 4$
3. $|x + 16| > 8$
4. $(x - 1)(x - 11) \leq 0$
5. $\dfrac{x-7}{x+5} > 0$
6. $2(x - 7) + 5(3 - x) \leq 0$
7. $x \geq 5$ or $2x - 1 \leq 3$
8. $(x + 2)(x - 1)(x + 6) > 0$
9. $|2x - 3| \leq 3$
10. $\dfrac{x}{x+7} \leq 2$
11. $x^2 - 3x \leq 10$
12. $\dfrac{x-2}{x} \geq 4$
13. $\dfrac{1}{2}x - \dfrac{3}{4} < \dfrac{1}{4}x + 1$
14. $|6x - 7| \geq 0$
15. $\dfrac{x-3}{x+2} \leq 6$
16. $x^3 + 6x^2 + 8x > 0$
17. $|x - 1.5| < 4.6$
18. $2x^2 + x - 3 > 0$
19. $2x + 1 < 5$ and $x + 2 > 5$
20. $\dfrac{1}{2}x^2 - \dfrac{1}{2}x - 1 < 0$
21. $\dfrac{x-1.4}{x+2.3} \leq 0$
22. $\dfrac{1}{2}x - 2 \leq \dfrac{1}{4}(x - 3)$

Sec. 7.4 Higher-Degree and Fractional Inequalities

23. When 5 is subtracted from two times a number, the result is at most 17. Find all such numbers that fit this statement.

24. If one-third of a number is added to 15, the result is less than 25. Find all such numbers.

25. The perimeter of a square must be less than or equal to 100 ft. What could the length of a side be?

26. Amanda must score at least 30 points to qualify for the state gymnastics meet. If she has 9.2 points on the floor exercise, 8.7 points on the beam, and 7.1 points on the vault, how many points must she score on the uneven parallel bars to qualify?

27. Jo has test scores of 75, 86, 88, and 62. What must she score (out of 100 points) on her fifth test to have an average of at least 80?

28. Sharon has a pickle barrel and she sells pickles for 15 cents each. It costs her 3 cents for each pickle and she spent a total of $1.80 on wrappers. How many pickles must she sell to make a profit of more than $3.00?

... LET'S NOT FORGET ...

Identify the expressions that are in factored form. Factor those that are not factored.

29. $(a - 2)^3$
30. $8 - a^3$
31. $(x - y)^2 - 4$
32. $ab(a + c)$
33. a^2bc
34. $a^2 + 2ab + b^2$
35. $\sqrt{x}(1 + \sqrt{x})$

How many terms are in each expression? Which expressions have $x - 5$ as a factor?

36. $x^3 + 125$
37. $x^2 - 25$
38. $(x - 5)^2$
39. $a(x - 5) - b(5 - x)$

Simplify each expression, if possible, leaving only nonnegative exponents in your answer.

40. $\sqrt[3]{a^3b - a^3c}$
41. $\sqrt{a^3 + b^3}$
42. $(x^{-1} + y^{-2})^{-1}$
43. $-2x^{-1}$
44. -3^4
45. $\dfrac{(27a^3b^3)^{-1/3}}{(8a^{-2}b^3)^{-1/3}}$

Find each product.

46. $(\sqrt{x} - 2\sqrt{y})^2$
47. $(3 + 2x)^3$

Reduce, if possible.

48. $\dfrac{5 \pm 5\sqrt{5}}{10}$
49. $\dfrac{a + b + x(a + b)}{a + b}$

The following problems can be worked by using a least common denominator. Follow the directions in each and notice how the LCD is used.

50. Perform the operation indicated: $\dfrac{x}{x + 1} - \dfrac{x + 1}{x + 2}$.

51. Solve $\dfrac{x}{x + 1} = \dfrac{2}{x + 2}$.

52. Simplify $\dfrac{\dfrac{1}{x + 1} - \dfrac{2}{x + 2}}{\dfrac{-2}{x^2 + 3x + 2}}$.

Label each as an equation, inequality, or an expression. Solve the equations and inequalities, and perform the operations indicated on the expressions.

53. $2x - 5 = x - 3(4 - x)$

54. $2x - 5 < x - 3(4 - x)$

55. $x^2 - 2x - 3 = 0$

56. $|3x + 2| \leq 6$

57. $2\sqrt{x} + 3 = x$

58. $\dfrac{6}{1 - x} - \dfrac{2x}{x^2 - 1} = \dfrac{x}{x + 1}$

59. $x^2 < 4$

60. $\dfrac{1}{x - 1} \leq \dfrac{2}{x + 1}$

61. $\dfrac{x + y}{y - x} - \dfrac{(x + y)^2}{x^2 - y^2}$

CHAPTER TEST

In Problems 1 through 5, choose the correct answer.

1. The graph of the set $\{x \mid 1 < x < 3\}$ is (?)

   A.

   B.

   C.

   D.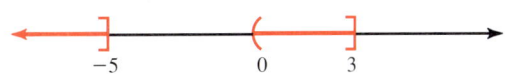

2. Which set represents the following graph?

   A. $\{x \mid x \leq -5 \text{ or } 0 < x \leq 3\}$
   B. $\{x \mid x \leq -5 \text{ or } x > 0 \text{ or } x \leq 3\}$
   C. $\{x \mid x \leq -5 \text{ or } x \leq 3\}$
   D. $\{x \mid x < -5 \text{ or } 0 < x < 3\}$

3. The solution set for $\dfrac{x - 3}{6} \leq \dfrac{x}{2} - 1$ is (?)

   A. $\left\{x \mid x \geq -\dfrac{9}{2}\right\}$
   B. $\left\{x \mid x \leq -\dfrac{9}{2}\right\}$
   C. $\left\{x \mid x \geq \dfrac{3}{2}\right\}$
   D. $\left\{x \mid x \leq \dfrac{3}{2}\right\}$

4. The solution set for $|x + 3| < 0$ is (?)
   A. $\{x \mid x < -3\}$
   B. $\{x \mid x > -3\}$
   C. $\emptyset$
   D. $\{x \mid x \text{ is a real number}\}$

5. The solution set for $(x + 2)(x + 3) = 0$ is (?)
   A. $\{-3, -2\}$
   B. $\{2, 3\}$
   C. $\{x \mid x < -2 \text{ or } x > -3\}$
   D. $\{x \mid -3 < x < -2\}$

In problems 6 through 15, find the solution set.

6. $\dfrac{x - 1}{x + 2} < 0$

7. $x(x - 2)(x + 3) > 0$

8. $\dfrac{x - 3}{x + 1} \geq 2$

9. $x^2 + 2x \leq -1$

Sec. 7.4   Higher-Degree and Fractional Inequalities

10. $1 - 2x + 2(x - 7) < 3x$

11. $\dfrac{1 - x}{x} \geq 0$

12. $2x^2 + 3x - 2 < 0$

13. $|3x + 1| > 10$

14. $(x + 3)(x + 1)(x - 2) \leq 0$

15. $\dfrac{1}{x + 2} \leq \dfrac{2}{x - 1}$

CHAPTER 8

See Problem Set 8.5, Exercise 16.

Equations in More than One Variable

- **8.1** Cartesian Coordinate System
- **8.2** Graph of $Ax + By = C$
- **8.3** Slope
- **8.4** Equations of Lines
- **8.5** Variation

CONNECTIONS

Being visual creatures, we often say "I see," meaning "I understand." With this in mind, mathematicians in the seventeenth and eighteenth centuries set about making pictures to represent abstract mathematical ideas. Their chief tool in this pursuit was our old friend the number line. Remember that there is a one-to-one correspondence between the set of real numbers and the set of points on the number line. That is, for every real number there is a single point on the number line, and for every point on the number line there is a distinct real number.

Just as each point on a number line has a coordinate, each point in a plane has a pair of coordinates. The study of such coordinates is based on the work of the French mathematician René Descartes (1596–1650).

In this chapter, we study equations of lines and their graphs. The slope of a line is defined and we learn how to find the slope of lines from an equation or from a graph.

8.1 CARTESIAN COORDINATE SYSTEM

If two copies of the number line, one horizontal and one vertical, are placed so that they intersect at the zero point of each line, a pair of **axes** is formed. The horizontal number line is called the **x-axis** and the vertical number line is called the **y-axis**. The point where the lines intersect is called the **origin**. We call this a **rectangular coordinate system,** or a **Cartesian coordinate system,** named for Descartes.

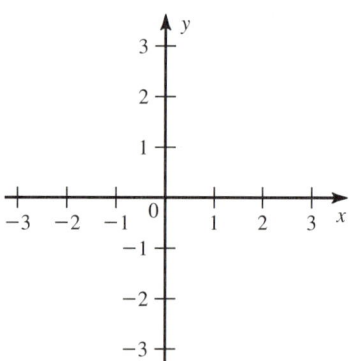

Suppose that p and q are two real numbers, written in the form

$$(p, q)$$

This is called an **ordered pair.** The order of the numbers is important.

We associate a point on a coordinate system with this ordered pair by finding the first number of the ordered pair, p, on the x-axis and the second number, q, on the y-axis. If we draw lines through these points perpendicular to the axes of the coordinate system, the lines will intersect in a single point. This unique point corresponds to the ordered pair (p, q) and is called its **graph.** To **plot** a point means to graph its ordered pair.

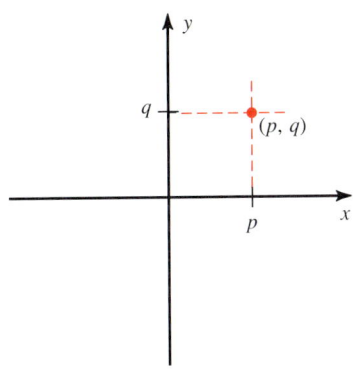

Note the graphs of the ordered pairs $(-3, 2)$, $(1, -2)$, $(-2, 1)$ and $\left(\frac{5}{2}, 1\right)$ on the coordinate system below.

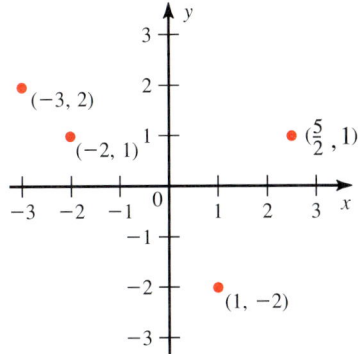

Notice that the graphs of $(-2, 1)$ and $(1, -2)$ are different points. The order in which the numbers are written in an ordered pair *is important*. The coordinates of the origin are $(0, 0)$.

The first number of an ordered pair is called the **abscissa** or **first coordinate** or **x-coordinate,** and the second number, the **ordinate** or **second coordinate** or **y-coordinate.**

The four regions made by the axes are called **quadrants.** The graph of every ordered pair is in one of the four quadrants or on one of the axes.

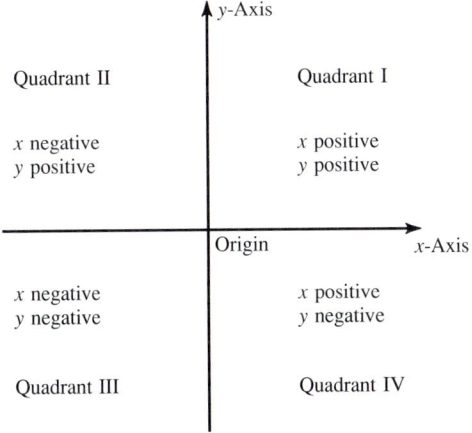

Sec. 8.1 Cartesian Coordinate System

In Section 0.3 we learned that the distance between two numbers on the number line is given by the absolute value of their difference. That is, the distance between the numbers p and q on the number line is $|p - q|$.

What about the distance between two points plotted on our coordinate system? Remembering that great discovery of the ancient Greeks, the Pythagorean Theorem, this distance can easily be found. The Pythagorean Theorem states that the square of the length of the hypotenuse of any right triangle equals the sum of the squares of the lengths of the other two sides.

Suppose that two points, P_1 and P_2, have coordinates (x_1, y_1) and (x_2, y_2). What is the distance between P_1 and P_2? Let's look at the graph of both points.

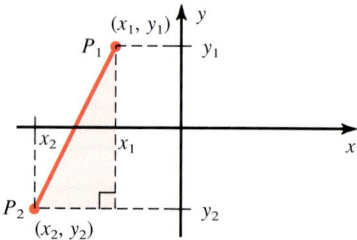

Notice that the distance from P_1 to P_2 is the length of the hypotenuse of the shaded right triangle. Also note that the length of one side is the same as the distance between x_1 and x_2 on the x-axis, which is $|x_1 - x_2|$, and the length of the other side is the same as the distance between y_1 and y_2 on the y-axis, which is $|y_1 - y_2|$.

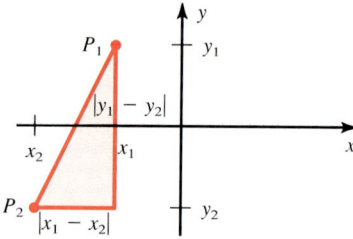

So if we let d be the distance between P_1 and P_2, by the Pythagorean Theorem,
$$d^2 = |x_1 - x_2|^2 + |y_1 - y_2|^2$$
Now, because $|q|^2 = q^2$ for any real number q, and because distance is not negative,
$$d = \sqrt{(x_1 - x_2)^2 + (y_1 - y_2)^2}$$
This is called the **distance formula.** It does not matter which point we call P_1 or P_2, because $(x_1 - x_2)^2 = (x_2 - x_1)^2$ and $(y_1 - y_2)^2 = (y_2 - y_1)^2$ (try it and see!). The distance formula says that the distance between two points is the square root of the sum of the difference in the x's squared and the difference in the y's squared.

The Distance Formula

The distance between the points (x_1, y_1) and (x_2, y_2) is given by
$$d = \sqrt{(x_1 - x_2)^2 + (y_1 - y_2)^2}$$

EXAMPLE 1. Find the distance between the points (4, 7) and (1, 3).

Solution:

Let P_1 be the point (4, 7) and P_2 be the point (1, 3). Then

$$x_1 \text{ is } 4 \quad \text{and} \quad y_1 \text{ is } 7$$
$$x_2 \text{ is } 1 \quad \text{and} \quad y_2 \text{ is } 3$$

So by the distance formula,

$$d = \sqrt{(4-1)^2 + (7-3)^2}$$
$$= \sqrt{3^2 + 4^2} = \sqrt{9 + 16} = \sqrt{25}$$
$$= 5$$

The distance is 5 units. ☐

EXAMPLE 2. Find the distance between the points (1, 6) and (4, −2).

Solution:

Let P_1 be the point (4, −2) and P_2 be the point (1, 6).

$$d = \sqrt{(4-1)^2 + (-2-6)^2}$$
$$= \sqrt{(3)^2 + (-8)^2}$$
$$= \sqrt{9 + 64} = \sqrt{73}$$

The distance is $\sqrt{73}$ units. ☐

Examples 2 and 3 show how careful we must be with signs in these computations.

EXAMPLE 3. Find the distance between the points (−2, −3) and (−10, 3).

Solution:

$$d = \sqrt{[-2-(-10)]^2 + (-3-3)^2}$$
$$= \sqrt{(-2+10)^2 + (-6)^2}$$
$$= \sqrt{64 + 36} = \sqrt{100}$$
$$= 10$$

The distance is 10 units. ☐

EXAMPLE 4. Find the distance between the points (2, 0) and (0, −3).

Solution:

$$d = \sqrt{(2-0)^2 + [0-(-3)]^2}$$
$$= \sqrt{2^2 + (-3)^2} = \sqrt{4 + 9} = \sqrt{13}$$

The distance is $\sqrt{13}$ units. ☐

Sec. 8.1 Cartesian Coordinate System

EXAMPLE 5. Find the distance between the points $\left(-\frac{1}{2}, \frac{3}{2}\right)$ and $\left(\frac{2}{3}, -\frac{1}{3}\right)$.

Solution:

$$d = \sqrt{\left(-\frac{1}{2} - \frac{2}{3}\right)^2 + \left[\frac{3}{2} - \left(-\frac{1}{3}\right)\right]^2}$$

$$= \sqrt{\left(\frac{-3-4}{6}\right)^2 + \left(\frac{9+2}{6}\right)^2} = \sqrt{\left(\frac{-7}{6}\right)^2 + \left(\frac{11}{6}\right)^2}$$

$$= \sqrt{\frac{49}{36} + \frac{121}{36}} = \sqrt{\frac{170}{36}} = \frac{\sqrt{170}}{\sqrt{36}} = \frac{\sqrt{170}}{6}$$

The distance is $\frac{\sqrt{170}}{6}$ units.

Another formula that is sometimes useful is the **Midpoint Formula,** which gives the coordinates of the point halfway between two points.

The Midpoint Formula

The midpoint between (x_1, y_1) and (x_2, y_2) is

$$\left(\frac{x_1 + x_2}{2}, \frac{y_1 + y_2}{2}\right)$$

EXAMPLE 6. Find the midpoint between the points (3, 4) and (9, 2).

Solution:

If the midpoint is (x, y), then

$$x = \frac{3+9}{2} = \frac{12}{2} = 6 \quad \text{and} \quad y = \frac{4+2}{2} = \frac{6}{2} = 3$$

The point (6, 3) is halfway between (3, 4) and (9, 2).

EXAMPLE 7. Find the coordinates of the point halfway between the points (2, −7) and (−6, 9).

Solution:

Let (x, y) be the midpoint. Then

$$x = \frac{2 + (-6)}{2} = \frac{2-6}{2} = \frac{-4}{2} = -2$$

$$y = \frac{-7+9}{2} = \frac{2}{2} = 1$$

The midpoint is (−2, 1).

EXAMPLE 8. Find the midpoint between (4, 7) and (4, −7).

Solution:

Let (x, y) be the midpoint.

$$x = \frac{4 + 4}{2} = \frac{8}{2} = 4$$

$$y = \frac{7 + (-7)}{2} = \frac{0}{2} = 0$$

The midpoint is (4, 0).

Plotting Points and Finding Distances with a Calculator

CALCULATOR BOX

Suppose that we wanted to graph the ordered pair $(-\sqrt{2}, \sqrt{3})$. We need to approximate the coordinates so that we can plot the point.

$$-\sqrt{2} \approx -1.414 \quad \text{and} \quad \sqrt{3} \approx 1.732$$

We use these approximations to help us locate the approximate position of the point.

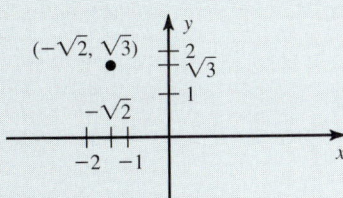

Notice that the approximations are not used as labels on the graph. We use the exact values $\sqrt{3}$ and $-\sqrt{2}$.

Distances can be approximated using a calculator.

EXAMPLE Approximate the distance between the points $(-2, 3)$ and $(5, 7)$ to the nearest hundredth.

Solution:

$$\begin{aligned} d &= \sqrt{(5 - (-2))^2 + (7 - 3)^2} \\ &= \sqrt{7^2 + 4^2} \\ &= \sqrt{49 + 16} \\ &= \sqrt{65} \\ &\approx 8.06 \end{aligned}$$

The distance is approximately 8.06 units.

(continued)

Sec. 8.1 Cartesian Coordinate System

Calculator Exercises

Plot both points, find their midpoint and plot it, and approximate the distance between the points to the nearest hundredth.

1. $(\sqrt{5}, 0), (0, 3)$
2. $(\sqrt{2}, -3), (1, 1)$
3. $(-3, \pi), \left(\frac{7}{4}, 1\right)$
4. $(-\pi, -2), (3, \sqrt{3})$

Answers:

Midpoint and approximate distance are given.

1. $\left(\frac{\sqrt{5}}{2}, \frac{3}{2}\right); 3.74$
2. $\left(\frac{\sqrt{2}+1}{2}, -1\right); 4.02$
3. $\left(-\frac{5}{8}, \frac{\pi+1}{2}\right); 5.21$
4. $\left(\frac{3-\pi}{2}, \frac{\sqrt{3}-2}{2}\right); 7.19$

PROBLEM SET 8.1

Warm-ups

In problems 1 through 9, identify the quadrant in which the points lie or on which axis they fall.

1. $(2, 5)$
2. $(-4, 7)$
3. $(11, -1)$
4. $(-5, -8)$
5. $(-3, 0)$
6. $(0, 1)$
7. $(2.4, -0.01)$
8. $\left(-\pi, -\frac{2}{3}\right)$
9. $(0, 0)$

In problems 10 through 24, plot the ordered pairs.

10. $(1, 3)$
11. $(3, 1)$
12. $(-2, 3)$
13. $(2, -3)$
14. $(3, -2)$
15. $(-3, 2)$
16. $(-2, -3)$
17. $(-3, -2)$
18. $\left(5, \frac{1}{2}\right)$
19. $\left(\frac{3}{2}, \frac{7}{2}\right)$
20. $\left(-\frac{5}{2}, \frac{11}{2}\right)$
21. $\left(\frac{1}{3}, -\frac{2}{3}\right)$

22. $\left(-\dfrac{11}{3}, -\dfrac{5}{6}\right)$ **23.** $(4.6, -2.30)$ **24.** $(-2.5, -0.1)$

In problems 25 through 37, find the distance between the points.
For problems 25 through 34, see Examples 1 through 4.

25. $(8, 6), (5, 2)$ **26.** $(1, 2), (-5, 5)$ **27.** $(-7, 3), (1, -3)$ **28.** $(-1, -2), (2, 2)$
29. $(-5, 1), (-5, 14)$ **30.** $(-3, -2), (-7, -2)$ **31.** $(-1, -2), (-3, -4)$
32. $(-4, 1), (5, 0)$ **33.** $(7, 0), (0, 7)$ **34.** $(0, 0), (3, -3)$

For problems 35 through 37, see Example 5.

35. $\left(\dfrac{1}{2}, \dfrac{5}{2}\right), \left(\dfrac{3}{2}, \dfrac{1}{2}\right)$ **36.** $\left(\dfrac{1}{2}, \dfrac{1}{3}\right), \left(\dfrac{7}{2}, -\dfrac{1}{3}\right)$ **37.** $\left(\dfrac{1}{2}, \dfrac{1}{6}\right), \left(\dfrac{1}{3}, 1\right)$

In problems 38 through 40, find the midpoint between the points. See Examples 6 through 8.

38. $(5, 1), (3, 1)$ **39.** $(5, 1), (-3, -1)$ **40.** $(-2, 10), (-6, 10)$

Practice Exercises

In problems 41 through 61, find the distance between the points.

41. $(11, 8), (3, 2)$ **42.** $(7, 5), (4, 1)$ **43.** $(2, 3), (14, 8)$
44. $(5, 1), (5, 8)$ **45.** $(10, 2), (2, 2)$ **46.** $(-3, 2), (1, 5)$
47. $(2, 3), (-4, 6)$ **48.** $(-6, 4), (2, -2)$ **49.** $(-2, -3), (1, 1)$
50. $(5, 2), (-2, -2)$ **51.** $(-5, 12), (-4, 11)$ **52.** $(8, -2), (11, -5)$
53. $(-2, 1), (-2, 10)$ **54.** $(-5, 1), (-1, 1)$ **55.** $(-4, -7), (5, -6)$
56. $(0, 4), (4, 0)$ **57.** $(0, 7), (-7, 0)$ **58.** $(-3, -6), (0, 0)$
59. $\left(\dfrac{7}{2}, \dfrac{3}{2}\right), \left(-\dfrac{5}{2}, \dfrac{1}{2}\right)$ **60.** $\left(\dfrac{6}{5}, -\dfrac{1}{5}\right), \left(\dfrac{2}{5}, \dfrac{2}{5}\right)$ **61.** $\left(\dfrac{1}{3}, \dfrac{1}{3}\right), \left(\dfrac{1}{4}, -\dfrac{1}{4}\right)$

In problems 62 through 64, find the midpoint between the points.

62. $(-1, 5), (7, -11)$ **63.** $(-5, -4), (-1, -10)$ **64.** $(1, 6), (4, -3)$

Challenge Problems

In problems 65 through 76, identify in which quadrant the points lie or on which axis they fall. For these problems only, assume that $a > 0$, $b > 0$, $c < 0$, $d < 0$.

65. (a, b) **66.** (a, c) **67.** (c, b) **68.** (d, c)
69. $(-a, c)$ **70.** $(-c, b)$ **71.** $(-b, -a)$ **72.** $(-d, -b)$
73. $(-c, -d)$ **74.** $(0, -a)$ **75.** $(-d, 0)$ **76.** $(0, d)$

77. Suppose that P_1 is the point $(2, 1)$, P_2 is $(-2, -1)$, and P_3 is $(0, 0)$. Find the distance between P_1 and P_2, between P_1 and P_3, and between P_2 and P_3. Are P_1, P_2, and P_3 collinear? (Do they all lie on the same line?) Explain the answer.

78. Suppose that Q_1 is the point $(-2, 0)$, Q_2 is $(6, 6)$, and Q_3 is $(9, 10)$. Are Q_1, Q_2, and Q_3 collinear?

79. The points $(-1, 1)$, $(-2, 5)$, and $(4, 2)$ are vertices of a triangle. Is it a right triangle? Explain the answer.

80. Are the points $(2, 4)$, $(3, -3)$, and $(5, 3)$ vertices of a right triangle?

IN YOUR OWN WORDS

81. What is a coordinate system?

82. State how to find the distance between two points without writing a formula.

8.2 GRAPH OF AX + BY = C

Let's look at some solutions for this equation in two variables,

$$2x + 3y = 12$$

A solution to the equation is a value for x and a value for y that make the equation a true statement. For example, if x has the value 0 and y has the value 4, the equation is true. We write this solution as an ordered pair with the value of x listed first and the value of y listed second.

$$(0, 4)$$

Notice that this is not the *only* ordered pair that is a solution to the equation,

$$(6, 0) \quad \text{and} \quad (3, 2)$$

are two other solutions because

$$2(6) + 3(0) = 12 \quad \text{and} \quad 2(3) + 3(2) = 12$$

How many solutions to $2x + 3y = 12$ are there? Let x have the value 12 and find y by solving the equation

$$2(12) + 3y = 12$$

We see that

$$3y = 12 - 24$$

$$y = \frac{-12}{3} \quad \text{or} \quad y = -4$$

and we have another member of the solution set, $(12, -4)$. As we can do this for *any* value of x, the solution set contains an infinite number of ordered pairs.

In Section 8.1 we learned how to graph ordered pairs. Let's graph the four ordered pairs from the solution set of $2x + 3y = 12$ found above.

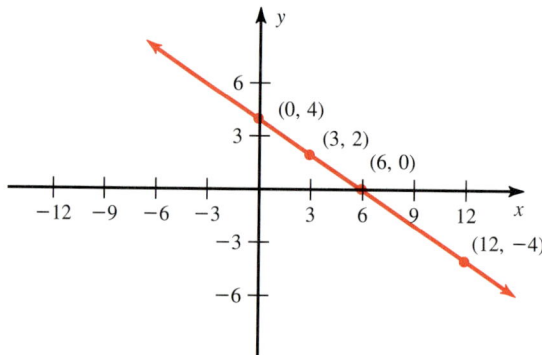

Upon close examination it appears that all these points lie on the same line. They do. As a matter of fact, the solutions of any equation of the form

$$Ax + By = C \quad A, B \text{ and } C \text{ constants;} \quad A \text{ and } B \text{ not both zero}$$

lie on the same line. Furthermore, *every* point on that line represents an ordered pair in the solution set of the equation. For this reason we call such equations **linear equations in two variables.** We call the line the **graph** of the equation. It is a picture of the solution set.

> **The Graph of an Equation of the Form**
>
> $$Ax + By = C$$
>
> is a line. A, B, C are constants with A and B not both zero. This form of the equation is called **standard form.**

As two points determine a line, we need to find two solutions of such an equation in order to draw its graph. Often the two easiest points to find are where x is 0 and where y is 0. A third point can be found as a check.

EXAMPLE 1. Draw the graph of $x - y = 2$.

Solution:

By inspection, if x is 0, then y must be -2 and if y is 0, then x must be 2, giving us the two ordered pairs

$$(0, -2) \quad \text{and} \quad (2, 0)$$

Plot these two points and then with a straightedge, draw the graph.

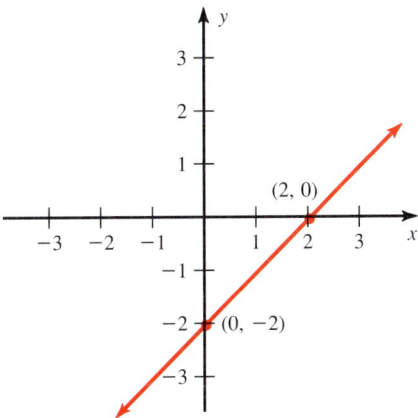

EXAMPLE 2. Draw the graph of $5x + y = 2$.

Solution:

If x is 0, then y is 2. So $(0, 2)$ is in the solution set and if y is 0, then x is $\frac{2}{5}$ and $\left(\frac{2}{5}, 0\right)$ is in the solution set.

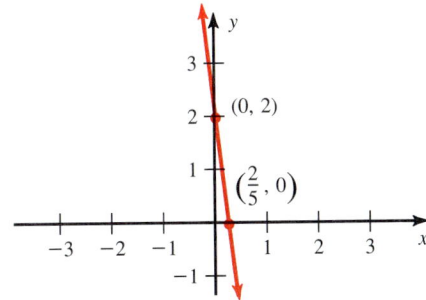

Notice that if we let x have the value 0 and find the y value that makes the equation true, we have found where the line crosses the y-axis. This y-value is called the **y-intercept**. Similarly, the **x-intercept** is the x-value where the line crosses the x-axis. Find the x-intercept by giving y the value 0 and solving for x.

To Find the Intercepts of a Line

1. To find the x-intercept, replace y with 0 and solve for x.
2. To find the y-intercept, replace x with 0 and solve for y.

Do we always get two points when we find the x-intercept and the y-intercept?

EXAMPLE 3. Draw the graph of $x + 2y = 0$.

Solution:

To find the x-intercept, replace y with 0 and solve for x. This gives a value of 0 for x. The ordered pair (0, 0) is one solution. However, when we find the y-intercept, we get (0, 0) again!

To find another point, let y be 1 and solve for x. This gives -2 for x. Thus $(-2, 1)$ is a second ordered pair in the solution set.

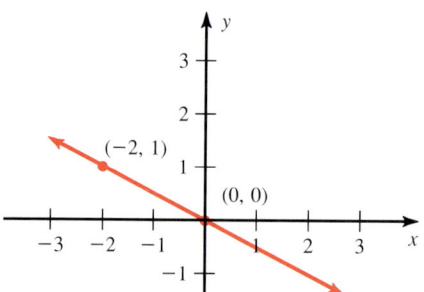

Notice that this line goes through the origin. If C is zero in the standard form, $Ax + By = C$, the graph will go through the origin.

Does every line have both an x-intercept and a y-intercept?

EXAMPLE 4. Draw the graph of $x = 2$.

Solution:

There is no y-term in this equation. Does that mean that y is 0? No, write $x = 2$ in standard form

$$x + 0y = 2$$

and we see that $A = 1$, $B = 0$, and $C = 2$.

Clearly, this equation is satisfied if x has the value 2 and y has any value at all! If we choose the pairs $(2, -1)$ and $(2, 1)$, we see that the equation $x = 2$ is true in both cases, giving the following graph.

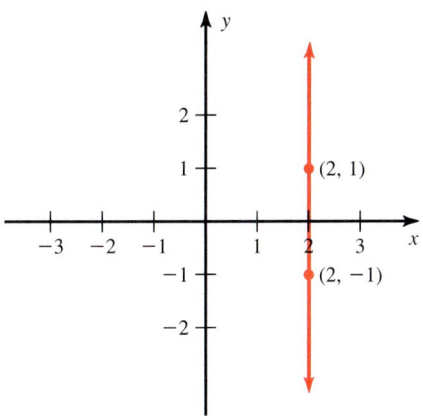

This line is called a **vertical line**.

> **The Graph of the Equation $x = p$**
>
> is a **vertical** line with x-intercept of p.

EXAMPLE 5. Draw the graph of $2y = -3$.

Solution:

In standard form, this equation is

$$0x + 2y = -3$$

Much as in the preceding example, this equation is satisfied by pairs of the form

$$\left(K, -\frac{3}{2}\right), \quad \text{where } K \text{ is } any \text{ real number!}$$

So, picking two, the graph looks as follows:

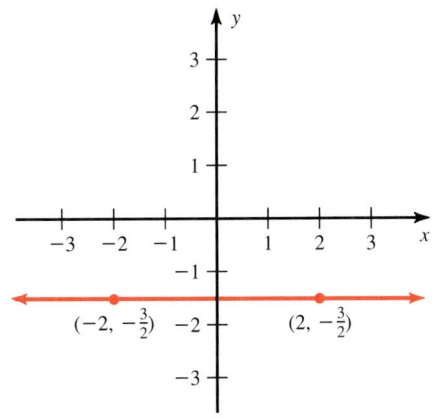

This line is called a **horizontal line**.

Sec. 8.2 Graph of $Ax + By = C$

> **The Graph of the Equation $y = q$**
>
> is a **horizontal** line with y-intercept of q.

Linear equations in two variables are used by scientists, engineers, mathematicians, and business managers to express relationships between two variables. Usually, letters that are more appropriate to the situation are used rather than x and y.

EXAMPLE 6. The distance d in miles a jogger runs at 6 mph is expressed in terms of running time t in hours by the equation $d - 6t = 0$.

(a) Graph this equation.
(b) How far has the jogger run in 2 hours?
(c) How long will it take the jogger to run 24 miles?

Solutions:

(a) Let the horizontal axis be time and the vertical axis be distance. This line goes through the origin, (0, 0). To find another point, let t be 1. This gives $d - 6(1) = 0$ or $d = 6$. So the point (1, 6) is on the graph.

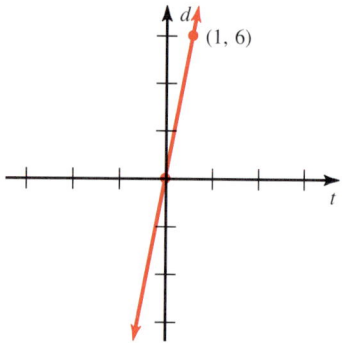

(b) If t is 2, then $d - 6(2) = 0$ or $d = 12$. The jogger has run 12 miles. On the graph this is represented by the point (2, 12).
(c) If d is 24, then $24 - 6t = 0$ or $t = 4$. It will take 4 hours to run 24 miles. This corresponds to the point (4, 24) on the graph.

PROBLEM SET 8.2

Warm-ups

In problems 1 through 24, find the intercepts, then draw the graph of the equations.
For problems 1 through 14, see Examples 1 and 2.

1. $3x + 2y = 6$
2. $3x - 2y = 6$
3. $-3x + 2y = 6$
4. $-3x - 2y = 6$

5. $x + y = 1$　　　　　**6.** $x - y = 1$　　　　　**7.** $4x + y = 4$　　　　　**8.** $5x + 3y = 10$

9. $3x - 4y = 4$　　　　**10.** $x + 3y = 1$　　　　**11.** $2x - y - 1 = 0$

12. $14x - 21y = 49$　　**13.** $10x - 15y + 25 = 0$　　**14.** $x - 3y = 6$

For problems 15 through 18, see Example 3.

15. $x + y = 0$　　　　　**16.** $x - y = 0$　　　　　**17.** $3x = -2y$　　　　　**18.** $3x = 2y$

For problems 19 through 21, see Example 4.

19. $x = 2$　　　　　　　**20.** $3x = 2$　　　　　　　**21.** $x = 0$

For problems 22 through 24, see Example 5.

22. $y = -3$　　　　　　**23.** $2y = 7$　　　　　　　**24.** $y = 0$

For problems 25 through 28, see Example 6.

25. The distance d in miles traveled by a cyclist riding at a constant speed of 10 mph is expressed in terms of time t in hours by the equation $d - 10t = 0$.
    (a) Graph this equation, using the horizontal axis as time.
    (b) How long will it take the cyclist to go 70 miles?

26. The area A in square meters of a rectangle of width 6 m is expressed in terms of its length L in meters by the equation $A - 6L = 0$.
    (a) Graph this equation, using the horizontal axis as length.
    (b) What will the length of the rectangle be if the area is 96 m²?

27. The profit P in dollars from selling x shirts is expressed by the equation $5x - P = 10$.
    (a) Graph this equation, using the horizontal axis for number of shirts sold.
    (b) How many shirts must be sold to make no profit? (This is called the *break-even point*.)
    (c) Find the profit on selling 100 shirts.
    (d) Find the profit on selling 1 shirt.

28. Madelyn can make x earrings at a cost C in dollars, expressed by the equation $2C - x = 2$.
    (a) Graph this equation, using the horizontal axis for number of earrings.
    (b) What is the cost of making 20 earrings?

Practice Exercises

In problems 29 through 55, find the intercepts, then draw the graph of each equation.

29. $4x + 3y = 12$
30. $4x - 3y = 12$
31. $-4x + 3y = 12$
32. $-4x - 3y = 12$

33. $4x + 3y = 0$
34. $4x - 3y = 0$
35. $x + 2y = 2$
36. $x = 2y + 2$

37. $x + 2y = 0$
38. $x = -2y$
39. $7x + 2y = 14$
40. $3x = 15 + 5y$

41. $x = -5$ **42.** $y = 4$ **43.** $3x + 4y = 6$ **44.** $2x - 5y = 4$

45. $4x + y = 1$ **46.** $x = 1 + 3y$ **47.** $3 + 2y = 0$ **48.** $3x - 7y = 4$

49. $2x + 5 = 3y$ **50.** $8x + 9y = 12$ **51.** $12x - 15y + 20 = 0$ **52.** $5 + 2y = 0$

53. $0 = 1 + x$ **54.** $3y = 2$ **55.** $4x + 3 = 0$

56. The distance d traveled by a cyclist riding at a constant speed of 15 mph is expressed in terms of time traveled t by the equation $d - 15t = 0$.
  (a) Graph this equation.
  (b) How long will it take the cyclist to go 75 miles?

57. The area A of a rectangle of width 5 m is expressed in terms of its length L by the equation $A - 5L = 0$.
  (a) Graph this equation.
  (b) What will the length of the rectangle be if the area is 105 m²?

58. The profit P in dollars from selling x pictures is expressed by the equation $x - P = 6$.
  (a) Graph this equation, using the horizontal axis for number of pictures sold.
  (b) How many pictures must be sold to break even?
  (c) Find the profit on selling 100 pictures.
  (d) Find the profit on selling 2 pictures.

59. Ronnie can make x belts at a cost C in dollars, expressed by the equation $C - 2x = 3$.
  (a) Graph this equation, using the horizontal axis for number of belts.
  (b) What is the cost of making 15 belts?

Challenge Problems

In problems 60 through 65, find the intercepts, then draw the graph of each equation.

60. $\dfrac{1}{3}x + \dfrac{2}{3}y = 1$

61. $\dfrac{1}{2}x - \dfrac{1}{3}y = \dfrac{1}{12}$

62. $\dfrac{1}{5}x = \dfrac{1}{3}y - \dfrac{2}{15}$

63. $\dfrac{x}{4} + \dfrac{y}{3} = \dfrac{1}{2}$

64. $\dfrac{2x}{11} - \dfrac{3y}{22} = \dfrac{4}{33}$

65. $y - 2x = \dfrac{6}{7}$

In problems 66 through 71, suppose that A and B are two real numbers with $A > 0$, $B < 0$, $C = 2A$, and $D = \dfrac{1}{2}B$. Sketch the graph of each of the following equations. Be sure to label the intercepts.

66. $Ax + Cy = C$

67. $Bx + Dy = D$

68. $Ax + By = AB$

69. $Ax - By = CD$

70. $Ax = B$

71. $C + Dy = 0$

▬ IN YOUR OWN WORDS

72. What are intercepts?

73. Describe how to graph a line whose equation is written in standard form.

8.3 SLOPE

Look at the two lines below. Notice how line A rises from left to right, while line B slopes downward. (Always look at graphs as if moving along them from *left to right*.)

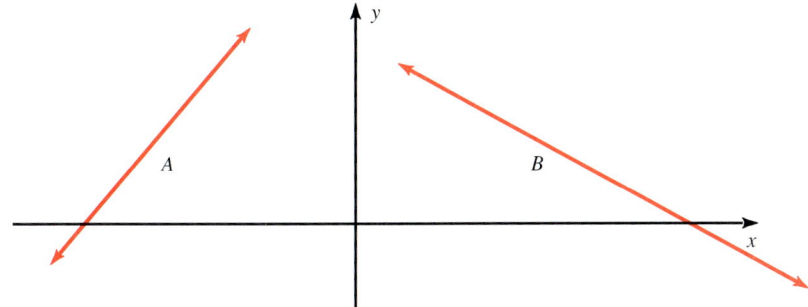

Line A is **increasing** and line B is **decreasing.** It turns out that perhaps the most important property of a line is the rate at which it is increasing or decreasing. This property is called the **slope** of the line. To measure the slope, consider any two points on the line, say P_1, whose coordinates are (x_1, y_1), and P_2, whose coordinates are (x_2, y_2).

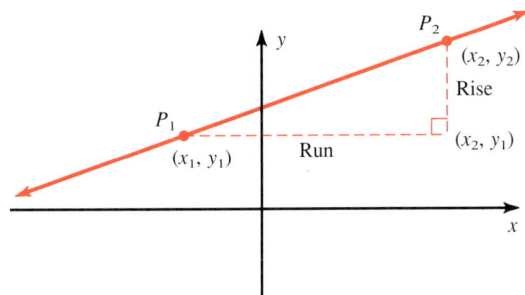

From P_1 to P_2, the vertical change or *rise* is given by the difference between the y-coordinates, while the horizontal change or *run* is given by the difference between the x-coordinates. That is,

$$\text{Rise} = y_2 - y_1$$

and

$$\text{Run} = x_2 - x_1$$

Define the slope as follows:

$$\text{Slope} = \frac{\text{Rise}}{\text{Run}}$$

Use the letter m for the slope. Therefore, the slope of a line containing the points (x_1, y_1) and (x_2, y_2) is given by the following formula.

Sec. 8.3 Slope

> ### The Slope Formula
> The slope of the line through the points (x_1, y_1) and (x_2, y_2) is given by the formula
> $$m = \frac{y_2 - y_1}{x_2 - x_1} \quad (x_1 \neq x_2)$$

EXAMPLE 1. Find the slope of the line that contains the points (1, 3) and (3, 7).

Solution:

$$m = \frac{y_2 - y_1}{x_2 - x_1}$$

Let P_1 be the point (1, 3) and P_2 be the point (3, 7); then x_1 is 1, y_1 is 3, x_2 is 3, and y_2 is 7 and the slope formula gives

$$m = \frac{7 - 3}{3 - 1} = \frac{4}{2} = 2$$

The slope of the line is 2.
The line is graphed below.

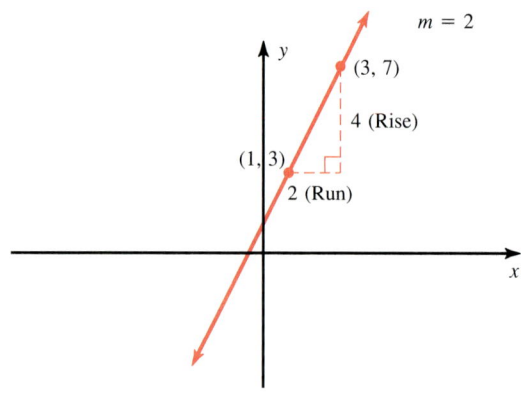

EXAMPLE 2. Find the slope of the line that contains the points (4, 3) and (8, 1).

Solution:

$$m = \frac{y_2 - y_1}{x_2 - x_1}$$

Let P_1 be (4, 3) and P_2 be (8, 1) to get

$$m = \frac{1 - 3}{8 - 4} = \frac{-2}{4}$$
$$= -\frac{1}{2}$$

The slope of the line is $-\dfrac{1}{2}$.

The line is graphed below.

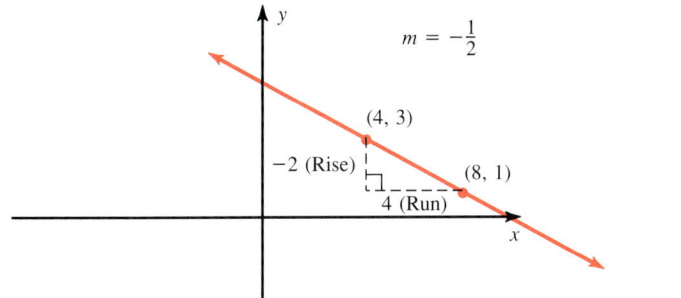

In finding the slope of the line containing the points (1, 3) and (4, 5), if P_1 is (1, 3) and P_2 is (4, 5), then using the slope formula gives

$$m = \dfrac{5 - 3}{4 - 1} = \dfrac{2}{3}$$

But if the points are turned around with P_1 as (4, 5) and (1, 3) as P_2,

$$m = \dfrac{3 - 5}{1 - 4} = \dfrac{-2}{-3} = \dfrac{2}{3}$$

Thus it does not matter which point is first, as long as we are consistent in the numerator and denominator.

In Examples 3 and 4, be careful with the signs.

EXAMPLE 3. Find the slope of the line that contains the points $(-3, 1)$ and $(3, 3)$.

Solution:

$$m = \dfrac{3 - 1}{3 - (-3)} = \dfrac{2}{6}$$
$$= \dfrac{1}{3}$$

The slope is $\dfrac{1}{3}$.

The line is graphed below.

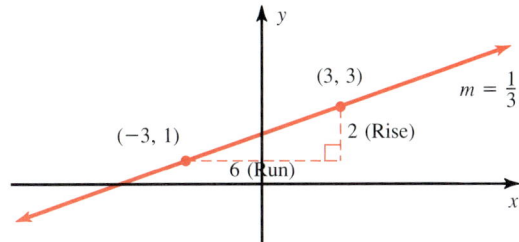

Sec. 8.3 Slope

EXAMPLE 4. Find the slope of the line that contains the points $(-1, 7)$ and $(2, -2)$.

Solution:

$$m = \frac{-2 - 7}{2 - (-1)} = \frac{-9}{3} = -3$$

The slope is -3.

The line is graphed below.

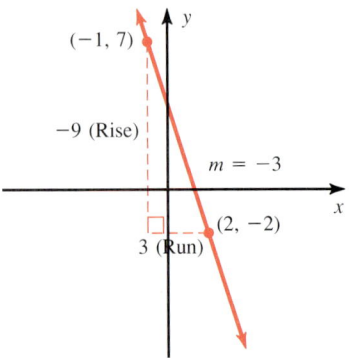

Notice in the preceding examples that *increasing* lines have *positive* slope, whereas *decreasing* lines have *negative* slope.

Let's calculate the slope of a horizontal line.

EXAMPLE 5. Find the slope of the line containing the points $(-3, 2)$ and $(4, 2)$.

Solution:

$$m = \frac{2 - 2}{4 - (-3)} = \frac{0}{7} = 0$$

The slope of the line is 0.

The graph is shown below.

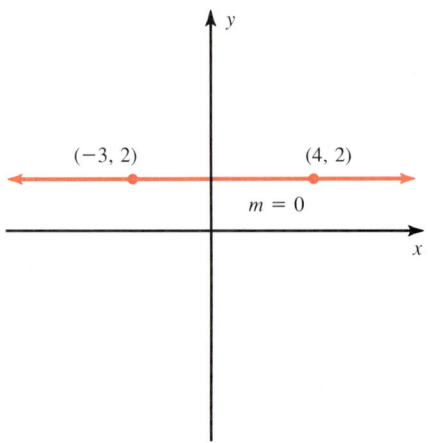

As the y-coordinates of all the points on a horizontal line are the same, the rise is zero and so the slope is zero.

EXAMPLE 6. Find the slope of the line containing the points (1, 3) and (1, −2).

Solution:

$$m = \frac{-2 - 3}{1 - 1} = \frac{-5}{0} \quad \text{which is undefined!}$$

The slope is undefined. So the line has no slope. Its graph is shown below.

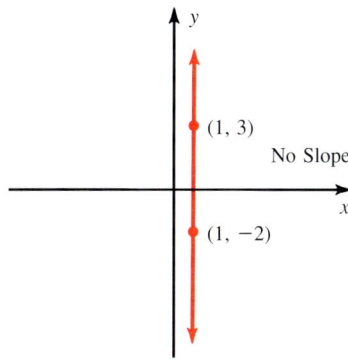

Since the x-coordinates of all the points on a vertical line are the same, the run is zero and we are unable to calculate the slope. Thus a vertical line has no slope.

Horizontal and Vertical Lines

1. Horizontal lines have equations of the form

$$y = q$$

and have a slope of zero.

2. Vertical lines have equations of the form

$$x = p$$

and have no slope.

EXAMPLE 7. Find the slope of the graph of $2x + 3y = 6$.

Solution:

The graph of $2x + 3y = 6$ is a line. Two points on that line are needed to find the slope. If the intercepts are found, we have two points. Letting y be 0 gives 3 as the x-intercept, and letting x be 0 gives 2 as the y-intercept. Using these points, (3, 0) and (0, 2), the slope is given by

$$m = \frac{2 - 0}{0 - 3} = -\frac{2}{3}$$

The slope is $-\frac{2}{3}$.

Sec. 8.3 Slope

EXAMPLE 8. Find the slope of the graph of $x - 2y = 0$.

Solution:

Again two points are needed. As this line contains the origin, we need one other point. Let y be 1. Then x is 2, giving the point (2, 1).

$$m = \frac{1 - 0}{2 - 0} = \frac{1}{2}$$

The slope is $\frac{1}{2}$.

EXAMPLE 9. Find the slope of the graph of $y = -4$.

Solution:

The equation $y = -4$ is of the form $y = q$, so its graph is a horizontal line. Therefore, the slope is zero.

EXAMPLE 10. Find the slope of the graph of $x = 16$.

Solution:

This equation is of the form $x = p$, so its graph is a vertical line and it has no slope.

Parallel and Perpendicular Lines

Suppose that l_1 is a line with slope m_1 and l_2 is a line with slope m_2, and neither line is vertical.

1. If l_1 and l_2 are **parallel** lines, their slopes are the same; that is,

$$m_1 = m_2$$

2. If l_1 and l_2 are **perpendicular** to each other, their slopes are negative reciprocals of each other; that is,

$$m_1 m_2 = -1 \quad \text{or} \quad m_1 = -\frac{1}{m_2}$$

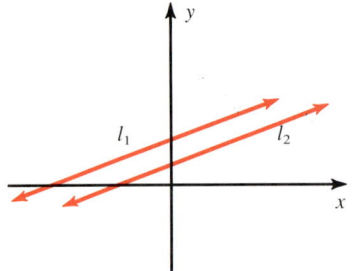

Parallel Lines; $m_1 = m_2$

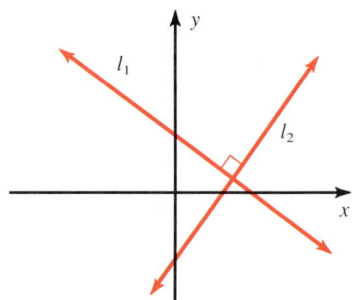

Perpendicular Lines; $m_1 = -\frac{1}{m_2}$

EXAMPLE 11. Suppose that l_1 is a line with slope $\frac{2}{3}$. What is the slope of l_2, a line parallel to l_1, and the slope of l_3, a line perpendicular to l_1?

Solution:

As l_2 is parallel to l_1, its slope is the same as l_1. So $m_2 = \frac{2}{3}$. However, l_3 is perpendicular to l_1, so its slope is the negative reciprocal of the slope of l_1.

$$m_3 = -\frac{1}{m_1}$$

$$= -\frac{1}{\frac{2}{3}} = -1 \cdot \frac{3}{2} = -\frac{3}{2}$$

EXAMPLE 12. Find the slope of *any* line parallel to the graph of $2x + y = 6$.

Solution:

First, find the slope of the graph of $2x + y = 6$. Its intercepts are $(0, 6)$ and $(3, 0)$, so the slope is

$$m = \frac{6 - 0}{0 - 3} = \frac{6}{-3} = -2$$

As parallel lines have the same slope, any line parallel to $2x + y = 6$ will have slope -2.

EXAMPLE 13. Suppose that L_1 is a line that is perpendicular to the graph of $5x - 3y = 15$. What is the slope of L_1?

Solution:

Let m_1 be the slope of L_1 and m_2 be the slope of the graph of $5x - 3y = 15$. First find m_2. The intercepts of the given line give points $(0, -5)$ and $(3, 0)$.

$$m_2 = \frac{0 - (-5)}{3 - 0} = \frac{5}{3}$$

As L_1 is perpendicular to the line whose slope is m_2,

$$m_1 = -\frac{1}{m_2}$$

$$= -\frac{1}{\frac{5}{3}} = -1 \cdot \frac{3}{5}$$

$$= -\frac{3}{5}$$

The slope of L_1 is $-\frac{3}{5}$.

Sec. 8.3 Slope

PROBLEM SET 8.3

Warm-ups

In problems 1 through 26, find the slope of the line that contains both points.
For problems 1 through 6, see Examples 1 and 2.

1. (2, 2), (3, 5)
2. (3, 2), (2, 5)
3. (2, 2), (5, 3)
4. (5, 2), (2, 3)
5. $\left(\frac{5}{2}, \frac{1}{2}\right), \left(\frac{7}{2}, \frac{5}{2}\right)$
6. $\left(\frac{1}{3}, \frac{1}{2}\right), \left(\frac{1}{4}, \frac{1}{3}\right)$

For problems 7 through 20, see Examples 3 and 4.

7. (4, −2), (2, 2)
8. (1, 1), (−1, 5)
9. (1, −1), (5, 1)
10. (2, 1), (6, −1)
11. (−1, 4), (4, 6)
12. (1, −6), (4, −1)
13. (−7, 4), (−3, 2)
14. (−3, 6), (−9, −3)
15. (−9, 1), (−1, −9)
16. (−2, −1), (−8, −9)
17. (−2, −6), (−4, −1)
18. (0, 0), (−8, −6)
19. (0, −9), (−4, 0)
20. $\left(-\frac{1}{3}, -\frac{5}{4}\right), \left(\frac{2}{3}, \frac{3}{4}\right)$

For problems 21 through 23, see Example 5.

21. (−2, 3), (−5, 3)
22. $\left(\frac{5}{7}, -\frac{2}{3}\right), \left(\frac{3}{5}, -\frac{2}{3}\right)$
23. (−2, 0), (0, 0)

For problems 24 through 26, see Example 6.

24. (8, −3), (8, 3)
25. (−3, −6), (−3, −8)
26. $\left(\frac{3}{5}, \frac{2}{5}\right), \left(\frac{3}{5}, -\frac{2}{5}\right)$

In problems 27 through 35, find the slope of the graph of each equation.
For problems 27 through 32, see Examples 7 and 8.

27. $x - y = 1$
28. $x + y = 1$
29. $2x + 3y = 6$
30. $3x - 2y = 12$
31. $4x - y = 0$
32. $2x = y - 6$

For problems 33 through 35, see Examples 9 and 10.

33. $y = 4$
34. $2x = 7$
35. $4 + 3y = 0$

For problems 36 through 41, see Examples 11 through 13.

36. Find the slope of a line parallel to the line containing the points (2, 5) and (1, −3).
37. Find the slope of a line perpendicular to the line containing the points (0, −2) and (−4, −5).
38. Find the slope of a line parallel to the line containing the points (2, −3) and (7, −3).
39. Find the slope of a line perpendicular to the line containing the points (2, −3) and (7, −3).
40. Find the slope of any line perpendicular to the graph of $4x + 3y = 24$.
41. Find the slope of a line parallel to the graph of $2x - 5y = 20$.

Practice Exercises

In problems 42 through 71, find the slope of the line that contains both points.

42. (4, 2), (5, 4)
43. (4, 4), (5, 2)
44. (2, 4), (4, 5)
45. (4, 5), (2, 4)
46. (3, 2), (6, −3)
47. (1, −2), (4, 4)
48. (0, 3), (6, −1)
49. (−2, 3), (1, 5)
50. (8, 3), (3, −1)
51. (−2, 2), (−5, 6)
52. (−3, 5), (1, −3)
53. (−6, 5), (6, −4)
54. (−6, 2), (−6, 4)
55. (1, 3), (5, 3)
56. (6, −6), (−6, 3)
57. (−5, −5), (−2, −7)
58. (−4, −3), (−8, −1)
59. (−2, −4), (−5, −4)
60. (0, 3), (−4, 0)
61. (0, −3), (0, 0)
62. (0, 0), (0, 1)

63. $\left(-\dfrac{1}{2}, \dfrac{2}{3}\right), \left(\dfrac{1}{2}, \dfrac{5}{3}\right)$
64. $\left(\dfrac{1}{3}, 6\right), \left(\dfrac{4}{3}, 4\right)$
65. $\left(-\dfrac{2}{3}, -\dfrac{5}{3}\right), \left(\dfrac{1}{6}, -\dfrac{1}{3}\right)$
66. $\left(\dfrac{2}{5}, -\dfrac{7}{2}\right), \left(\dfrac{4}{10}, -\dfrac{3}{2}\right)$
67. $\left(\dfrac{5}{2}, \dfrac{1}{6}\right), \left(2, \dfrac{2}{3}\right)$
68. $\left(\dfrac{1}{5}, -\dfrac{2}{5}\right), \left(\dfrac{1}{3}, -\dfrac{2}{5}\right)$
69. $\left(-\dfrac{3}{4}, -2\right), \left(\dfrac{1}{2}, \dfrac{1}{2}\right)$
70. $\left(2, \dfrac{2}{5}\right), \left(\dfrac{2}{7}, -2\right)$
71. $\left(\dfrac{1}{5}, \dfrac{1}{5}\right), \left(\dfrac{1}{4}, \dfrac{1}{3}\right)$

In problems 72 through 80, find the slope of the graph of each equation.

72. $2x + y = 4$
73. $2x - y = 6$
74. $x + 2y = 2$
75. $x - 2y = 8$
76. $5x - 3y = 15$
77. $3x + 5y = 0$
78. $2x + 5 = 0$
79. $11 = 5y$
80. $2x + 3y = 5$

81. Find the slope of a line parallel to the line containing the points $(1, 6)$ and $(-1, 3)$.
82. Find the slope of a line perpendicular to the line containing the points $(2, -3)$ and $(-4, 0)$.
83. Find the slope of a line parallel to the line containing the points $(5, -9)$ and $(5, -3)$.

Challenge Problems

87. Find the slope of the graph of $y = 3x + 3$.
88. Find the slope of the graph of $y = 3x + 12$.
89. Find the slope of the graph of $y = 3x + Q$ (Q any real number).

84. Find the slope of a line perpendicular to the line containing the points $(2, -1)$ and $(2, -3)$.
85. Find the slope of any line perpendicular to the graph of $2x - 3y = 24$.
86. Find the slope of a line parallel to the graph of the equation $2x + 5y = 30$.

90. Find the slope of the graph of $y = -\dfrac{2}{5}x + 10$.
91. Find the slope of the graph of $y = Px + Q$ (P and Q any real numbers).

In problems 92 through 96, find the slope of the graph of each equation by solving for y, then using the result of problem 91.

92. $2x + 3y = 6$
93. $5x - y = 7$
94. $3x - 7y = 5$
95. $4x - 2 = 5y$
96. $6 - 2y = 0$

IN YOUR OWN WORDS

97. What is slope?

8.4 EQUATIONS OF LINES

In previous sections we learned that the graph of a linear equation in two variables is a line. It is reasonable to ask: Given a line, is there a linear equation with that line as its graph? The answer is yes, and we will see how to find suitable equations in this section.

First, let's look at another form of the linear equation in two variables. Suppose that a line has slope m and y-intercept b. The equation of this line can be obtained from the slope formula. Since the y-intercept is b, the point $(0, b)$ is on the line. If (x, y) is any point on the line, the slope formula gives the following.

$$m = \dfrac{y - b}{x - 0} \quad \text{Definition of slope}$$

$$mx = y - b \quad \text{Clear fractions.}$$

$$mx + b = y$$

or

$$y = mx + b$$

We call this the **slope-intercept form.**

> **Slope-Intercept Form**
>
> An equation of a line with slope m and y-intercept of b is given by
>
> $$y = mx + b$$

EXAMPLE 1. Rewrite $5x - 6y = 18$ in slope-intercept form.

Solution:

Solve $5x - 6y = 18$ for y.

$$-6y = -5x + 18$$

$$y = \frac{-5x + 18}{-6}$$

$$y = \frac{5}{6}x - 3$$

This is the slope-intercept form, where m is $\frac{5}{6}$ and b is -3.

EXAMPLE 2. Find the slope and y-intercept of the graph of $3x + 5y = 11$.

Solution:

First, rewrite the equation in slope-intercept form by solving for y.

$$3x + 5y = 11$$

$$5y = -3x + 11$$

$$y = -\frac{3}{5}x + \frac{11}{5}$$

Now read the slope and y-intercept directly from the equation. The slope is $-\frac{3}{5}$ and the y-intercept is $\frac{11}{5}$.

It is convenient to use the slope-intercept form when drawing the graph of a linear equation. The y-intercept gives one point immediately; then use the fact that the slope is the ratio of the rise to run to get a second point.

EXAMPLE 3. Sketch the graph of the linear equation $y = \frac{2}{3}x - \frac{5}{2}$.

Solution:

As the equation is in slope-intercept form, read the y-intercept directly from the equation. As b is $-\frac{5}{2}$, this graph crosses the y axis at $-\frac{5}{2}$.

The slope is $\frac{2}{3}$. So, working from the y-intercept of $-\frac{5}{2}$, we *rise* 2 while we *run* 3 (from left to right). This gives a second point on the graph, which is enough to complete the sketch.

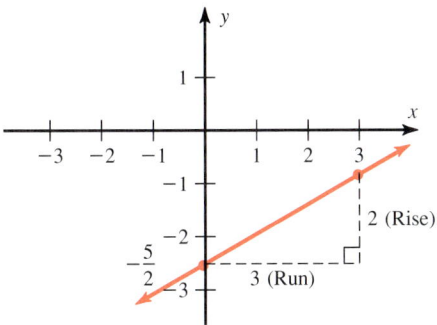

EXAMPLE 4. Graph the linear equation $y = -\frac{3}{4}x + 2$.

Solution:

This equation is in slope-intercept form. So read the slope and y-intercept directly from the equation.

$$m = \frac{-3}{4} \quad \text{and} \quad b = 2$$

Notice that we always consider the *run* to be positive. So a negative slope implies a negative *rise*. Thus, working from the y-intercept of 2, rise -3 and run 4. As a negative rise is downward, go *down* 3 and *across* 4.

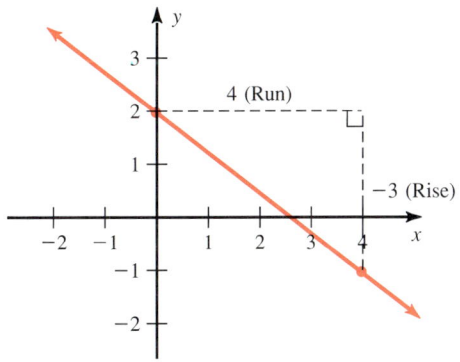

Now, suppose that we are given a line and wish to find an equation having that line as its graph. Two pieces of information are needed:

1. The slope of the line
2. The coordinates of any point on the line

Substituting this information into the slope formula and simplifying will produce the desired equation.

EXAMPLE 5. Find an equation of a line with slope 2 that contains the point $(1, -2)$, and draw its graph.

Solution:

In the slope formula we let P_1 be the point $(1, -2)$ and P_2 be any point on the line, say (x, y). As the slope is 2, the formula gives

$$2 = \frac{y - (-2)}{x - 1}$$

$$2(x - 1) = y + 2$$

$$2x - 2 = y + 2$$

$$2x - y = 4$$

The equation is $2x - y = 4$.

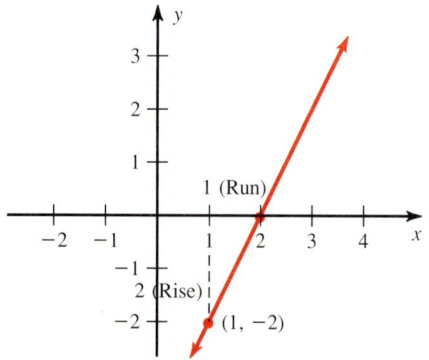

When using a given slope m, a given point (x_1, y_1), and a general point on the line (x, y), the slope formula takes the form

$$m = \frac{y - y_1}{x - x_1}$$

Multiplying both sides by $(x - x_1)$ produces the **point-slope** form,

$$m(x - x_1) = y - y_1$$

which is usually written

$$y - y_1 = m(x - x_1)$$

The Point-Slope Form

An equation of the line with slope m and containing the point (x_1, y_1) is given by

$$y - y_1 = m(x - x_1)$$

EXAMPLE 6. Find an equation for the line containing the points $(-1, 3)$ and $(2, -1)$.

Solution:

First find the slope of the line containing the two points

$$m = \frac{3 - (-1)}{-1 - 2}$$

$$= -\frac{4}{3}$$

Next, pick *either* of the two given points to be (x_1, y_1) and substitute into the point-slope form of the linear equation.

$$y - y_1 = m(x - x_1)$$

$$y - (-1) = -\frac{4}{3}(x - 2) \quad \text{Using } (2, -1) \text{ as } (x_1, y_1)$$

$$3(y + 1) = -4(x - 2)$$

$$3y + 3 = -4x + 8$$

$$4x + 3y = 5$$

This is the equation of a line in standard form. The graph of $4x + 3y = 5$ contains the points $(-1, 3)$ and $(2, -1)$. ☐

EXAMPLE 7. Find an equation for the line through the point $(4, -7)$ that is parallel to the graph of $x + 2y = 6$.

Solution:

First, find the slope of the graph of the given equation by writing it in slope-intercept form.

$$x + 2y = 6$$

$$2y = -x + 6$$

$$y = -\frac{1}{2}x + 3$$

Note that the given line has slope $-\frac{1}{2}$. The line we are looking for is parallel to the line given, so its slope is the same. As $(4, -7)$ is on our line, we have enough information to find an equation.

$$y - y_1 = m(x - x_1)$$

$$y - (-7) = -\frac{1}{2}(x - 4)$$

(continued)

Sec. 8.4 Equations of Lines

Multiply both sides by -2 to clear fractions.
$$-2(y + 7) = x - 4$$
$$-2y - 14 = x - 4$$
$$-10 = x + 2y \quad \text{or} \quad x + 2y = -10$$

This is the desired equation in standard form. ◻

EXAMPLE 8. Find an equation of the line through the origin perpendicular to the graph of $2x + 3y = 7$.

Solution:

First, find the slope. Rewriting the equation in slope-intercept form gives us
$$2x + 3y = 7$$
$$3y = -2x + 7$$
$$y = -\frac{2}{3}x + \frac{7}{3}$$

The slope of the graph of the equation is $-\frac{2}{3}$. Therefore, the slope of a line perpendicular to this line is $\frac{3}{2}$.

Thus we are looking for a line through the origin with slope $\frac{3}{2}$. As the origin has coordinates (0, 0), we have
$$y - y_1 = m(x - x_1)$$
$$y - 0 = \frac{3}{2}(x - 0)$$
$$y = \frac{3}{2}x$$
$$3x - 2y = 0$$ ◻

Forms of the Linear Equation in Two Variables

Standard Form

$Ax + By = C;$ A and B not both zero

Slope-Intercept Form

$y = mx + b$

Point-Slope Form

$y - y_1 = m(x - x_1)$

Linear equations are often stated in standard form. One of the most useful forms is the slope-intercept form. The point-slope form is convenient for writing an equation of a line given a point and its slope.

> **To Write an Equation of a Line that has Slope**
> 1. Find m, the slope of the line.
> 2. Find (x_1, y_1), the coordinates of any point on the line.
> 3. Substitute into the point-slope formula,
> $$y - y_1 = m(x - x_1)$$
> 4. Write the equation in standard or slope-intercept form, whichever is preferred.

A special case of this occurs if the slope and y-intercept are known. In this case it is easier to use the slope-intercept form.

EXAMPLE 9. Find an equation of the line with slope -1 and y-intercept of 5.

Solution:

Using the slope-intercept form, $y = mx + b$ gives
$$y = -x + 5$$

Since a vertical line has no slope, it must be treated in another way. It is usually best to treat horizontal and vertical lines as exceptions when writing their equations. They are easy to identify (slope of zero or no slope) and it is simple to write their equations.

> **Equations of Horizontal and Vertical Lines**
> 1. A line of slope 0 is *horizontal* and has the equation
> $$y = q$$
> where q is the y-coordinate of *any* point on the line.
> 2. A line with no slope is *vertical* and has the equation
> $$x = p$$
> where p is the x-coordinate of *any* point on the line.

▬ PROBLEM SET 8.4

Warm-ups

In problems 1 through 9, find the slope of the graph of each linear equation by rewriting the equation in slope-intercept form and reading the slope from the equation. See Examples 1 and 2.

1. $2x + 3y = 6$
2. $5x - y = 10$
3. $x + 7y = 14$
4. $2x + 3y = 5$
5. $5x - y = 7$
6. $x + 7y = 1$
7. $4x - 7y = 0$
8. $x = 2y - 3$
9. $y = x$

In problems 10 through 15, sketch the graph of each equation by using the information from the slope-intercept form. See Examples 3 and 4.

10. $y = x - 1$

11. $y = -x + 1$

12. $y = \dfrac{1}{2}x + \dfrac{3}{2}$

13. $y = -\dfrac{1}{3}x + \dfrac{5}{3}$

14. $y = 3x - 2$

15. $y = -4x$

In problems 16 through 21, find an equation of the line through the point having the slope given. Sketch the graph of the line first. See Example 5.

16. $(1, 2)$; $m = 3$

17. $(-1, 1)$; $m = 2$

18. $(4, 3)$; $m = -2$

19. $(-5, -2)$; $m = -1$

20. $(4, -6)$; $m = \dfrac{1}{2}$

21. $\left(\dfrac{2}{5}, 1\right)$; $m = 1$

In problems 22 through 27, find an equation of the line through the points. See Example 6.

22. $(2, 3), (4, 5)$
23. $(-1, 4), (2, 3)$
24. $(2, -3), (5, 0)$
25. $(0, -2), (5, 0)$
26. $(0, 0), (-1, -5)$
27. $(-1, -2), (-6, -9)$

In problems 28 through 36, find an equation of a line satisfying the conditions stated. See Examples 7, 8, and 9.

28. The line with y-intercept of 2 and slope $\dfrac{1}{2}$.

29. The line with slope -1 and x-intercept of -3.

30. The line with x-intercept of -2 and y-intercept of 3.

31. The line through $(1, -5)$ parallel to the graph of $y = 6x - 5$.

32. The line perpendicular to the graph of $y = -3x - 17$ containing the point $(4, -11)$.

33. The line parallel to the graph of $4x - 2y = 13$ containing the point $(-7, 0)$.
34. The line through $(1, 5)$ perpendicular to the graph of $2x + 3y = 4$.
35. The line through $(-4, -1)$ parallel to the x-axis.
36. The line through $(-4, -1)$ perpendicular to the x-axis.
37. The relationship between Fahrenheit and Celsius temperature is shown below. Write an equation of this line.

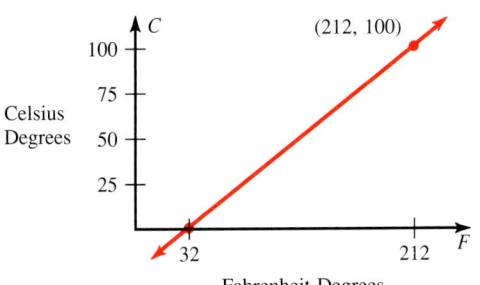

38. The marketing expenditures for Campbell Soup are graphed below. Write an equation for this line. Use the equation to find the marketing expenditures when time is 10 years.

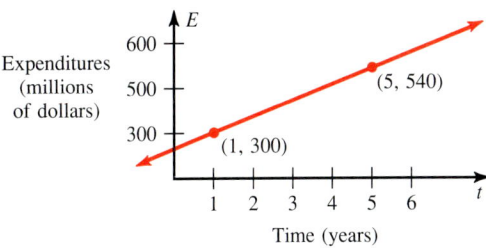

Practice Exercises

In problems 39 through 47, find the slope of the graph of each linear equation in two ways.
(a) Find two points and use the slope formula.
(b) Rewrite in slope-intercept form and read it from the equation.

39. $3x + 4y = 12$ **40.** $4x - y = 8$ **41.** $x + 5y = 10$ **42.** $3x + 4y = 5$
43. $4x - y = 7$ **44.** $x + 5y = 1$ **45.** $3x + 7y = 0$ **46.** $x = 3y - 4$ **47.** $y + 9 = 0$

In problems 48 through 53, sketch the graph of each equation by using the information from the slope-intercept form.

48. $y = x - 1$ **49.** $y = -x - 1$ **50.** $y = \frac{1}{3}x + \frac{2}{3}$

51. $y = -\frac{1}{2}x + \frac{3}{2}$ **52.** $y = -3x + 2$ **53.** $y = 4x$

In problems 54 through 59, find an equation of the line through the point having the slope given. Sketch the graph first.

54. $(2, 1); m = 2$ **55.** $(1, -1); m = -2$ **56.** $(3, 4); m = -1$
57. $(-4, -3); m = 1$ **58.** $(2, -3); m = -\frac{1}{2}$ **59.** $\left(\frac{2}{3}, \frac{3}{5}\right); m = 1$

Sec. 8.4   Equations of Lines

In problems 60 through 65, find an equation of the line through the points.

60. (1, 2), (3, 4)
61. (−2, 3), (1, 4)
62. (3, 0), (2, −3)
63. (0, 2), (−5, 0)
64. (−2, −4), (0, 0)
65. (−2, −3), (−5, −6)

In problems 66 through 80, find an equation of a line satisfying the conditions stated.

66. The line with y-intercept of 4 and slope $\frac{1}{3}$.

67. The line with slope 1 and x-intercept of $-\frac{1}{2}$.

68. The line with x and y intercepts of 1.

69. The line through $(-1, 4)$ parallel to the graph of
$$y = 4x - 3.$$

70. The line perpendicular to the graph of $y = -2x - 9$ containing the point $(4, -11)$.

71. The line parallel to the graph of $5x - 3y = 13$ containing the point $(0, -7)$.

72. The line through $(-1, 4)$ perpendicular to the graph of $x + 2y = 3$.

73. The line through $(2, -3)$ parallel to the y-axis.

74. The line through $(2, -3)$ perpendicular to the y-axis.

75. The line through the origin perpendicular to the graph of $x = -y$.

76. The line through the origin perpendicular to the line containing the points $(-5, 0)$ and $(16, 0)$.

77. The line through $(2, 0)$ parallel to the line containing the points $(-5, 3)$ and $(3, 3)$.

78. The line with x-intercept -4 and parallel to the graph of $2x + 3y = 1$.

79. The line with y-intercept 11 and perpendicular to the graph of $4x - y = 0$.

80. The line through the origin parallel to the line whose x-intercept and y-intercept are 4 and -8, respectively.

Challenge Problems

83. Find an equation of the line through (p, q) perpendicular to the graph of $px + qy = C$.

84. Find the x-intercept and y-intercept of the graph of
$$8x + 5y = 40.$$
Now, divide both sides of the original equation by 40 to

81. Total sales for Campbell Soup are shown below. Write an equation for the line. Use the equation to find sales when time is 10 years.

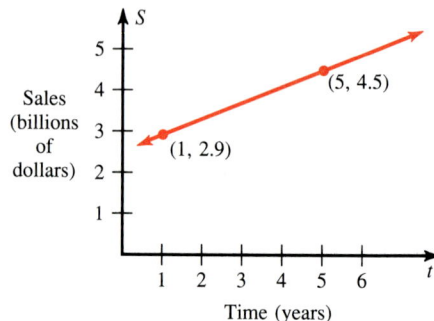

82. The depreciation on a boat is shown below. Write an equation of this line. Use the equation to find the depreciation when the boat is 3 years old.

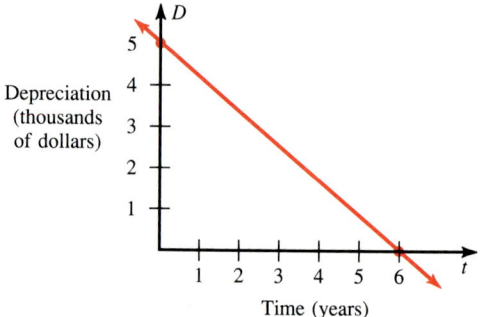

obtain the form
$$\frac{x}{a} + \frac{y}{b} = 1$$
This is called the *intercept form* of the linear equation. Why?

In problems 85 through 90, find the x-intercept and y-intercept by first rewriting the given equation in intercept form.

85. $2x + 3y = 6$
86. $5x - 4y = 20$
87. $2x + 3y = 1$
88. $5x - 7y = 11$
89. $3x = 5$
90. $2y + 9 = 0$

■ IN YOUR OWN WORDS

91. Explain how to graph a line if a point on the line and the slope of the line are given.

92. Compare the advantages of the slope-intercept form and the standard form of a line.

8.5 VARIATION

The formula $C = \pi d$ relates the diameter of a circle and its circumference. The relationship between the variables C and d is such that if d gets *larger* then C also gets *larger* and if d gets *smaller* then C also gets *smaller*. π is a constant.

We often express such a relationship by saying that "C **varies directly** as d" or "C is **directly proportional** to d."

Direct Variation

y varies directly as x means that
$$y = kx$$
where k is a nonzero constant.

The constant k is called the **constant of proportionality.**

EXAMPLE 1. y varies directly as the cube of x. Find the constant of proportionality if y is 16 when x is 2.

Solution:

From the definition,
$$y = kx^3$$
Replacing y with 16 and x with 2 gives us
$$16 = k(2)^3 = 8k$$
$$2 = k$$
The constant of proportionality is 2. Thus the relationship between x and y is given by
$$y = 2x^3$$

Boyle's Law for the expansion of gas is $V = \dfrac{C}{P}$, where V is the volume of the gas, P is the pressure, and C is a constant. This formula relates a pressure and a volume. We can see that if P gets *larger*, then V gets *smaller*, and that if P gets *smaller*, then V gets *larger*.

We say that "V **varies inversely** as P" or "V is **inversely proportional** to P."

Inverse Variation

y varies inversely as x means that
$$y = \frac{k}{x}$$
where k is a nonzero constant.

Sec. 8.5 Variation

EXAMPLE 2. y varies inversely as the square root of x. Find the constant of proportionality if y is 15 when x is 9.

Solution:

From the definition,

$$y = \frac{k}{\sqrt{x}}$$

However, y is 15 when x is 9, so

$$15 = \frac{k}{\sqrt{9}} = \frac{k}{3}$$

$$45 = k$$

Thus the constant of proportionality is 45 and the relationship between x and y is given by

$$y = \frac{45}{\sqrt{x}}$$

Joint Variation

y varies jointly as x and z means that

$$y = kxz$$

where k is a nonzero constant.

We often say that **y is jointly proportional to x and z** instead of y varies jointly as x and z.

EXAMPLE 3. Translate each statement into an equation.

(a) w varies jointly as y and the cube of x.
(b) x is directly proportional to y and inversely proportional to z.

Solutions:

(a) Using the definition,

$$w = kxy^3$$

(b) Combining direct and inverse variation, we obtain

$$x = \frac{ky}{z}$$

EXAMPLE 4. Newton's law of gravitation says that the force between two particles that are a constant distance apart varies jointly as the product of the masses of the particles. If a force of 2 units exists between two particles of masses 4 and 5 units, find the force between two particles of masses 8 and 15 units.

Solution:

If F is force and m_1 and m_2 are the masses,

$$F = km_1m_2$$

We find the constant of proportionality.

$$2 = k(4)(5)$$

$$\frac{1}{10} = k$$

The formula becomes $F = \frac{1}{10}m_1m_2$.

$$F = \frac{1}{10}(8)(15)$$

$$= 12$$

The force is 12 units. □

PROBLEM SET 8.5

Warm-ups

In problems 1 through 6, translate each statement into an equation. See Examples 1, 2, and 3.

1. The area of a circle varies directly as the square of its radius.
2. The volume of a cylinder varies jointly as the square of its radius and its height.
3. The volume of a sphere is directly proportional to the cube of its radius.
4. F is inversely proportional to G.
5. u varies directly as v and inversely as w.
6. x varies jointly as the square of y and z and inversely as w.

In problems 7 through 15, find the constant of proportionality and answer each question. See Example 4.

7. x varies directly as the square of y. If x is 12 when y is 2, find x when y is 5.
8. u varies inversely as the cube root of v. If u is 6 when v is 27, find u when v is 8.
9. s is jointly proportional to the square of t and g. If s is 36 when t is 2 and g is 3, find s when t is $\sqrt{2}$ and g is 32.2.
10. N is directly proportional to the square root of L and inversely proportional to M. If N is 2 when L is 16 and M is 8, find N when L is 25 and M is 2.
11. Hooke's Law says that the force required to stretch a spring is directly proportional to the distance stretched. If a force of 10 lb is required to stretch a spring 5 in., how much force will be required to stretch the spring 10 in.?
12. Ohm's Law says that the current, I, in a wire varies directly as the electromotive force, E, and inversely as the resistance, R. If I is 11 amperes when E is 110 volts and R is 10 ohms, find I if E is 220 volts and R is 11 ohms.
13. The distance a stone falls when dropped off a cliff is directly proportional to the square of time. If the stone falls 64.4 ft in 2 seconds, how far will the stone have fallen in 3 seconds?
14. The intensity of illumination, I, from a light is inversely proportional to the square of the distance, d, from the light. If the intensity is 120 candlepower at a distance of 10 ft from the light, what is the intensity 20 ft from the light?
15. The period of a pendulum varies directly as the square root of its length. If a pendulum of length 16 in. has a period of $\frac{1}{2}$ second, find the length of a pendulum with a period of $\frac{1}{4}$ second.

Practice Exercises

In problems 16 through 21, translate each statement into an equation.

16. The distance an object falls from the Trans Am Building (ignoring air resistance) is directly proportional to the square of the time it falls.
17. The spin on a billiards ball is directly proportional to the velocity of the stroke of the cue stick.
18. The perimeter of a square varies directly as the length of a side.
19. The volume of a box varies jointly as length, width, and height.
20. t is inversely proportional to s.
21. h is directly proportional to the square root of g and inversely proportional to d.

In problems 22 through 30, find the constant of proportionality and answer each question.

22. u varies directly as the cube of v. If u is 8 when v is 1, find u when v is 2.
23. x varies inversely as the square of y. If x is 2 when y is 3, find x when y is 5.
24. s is jointly proportional to t and g. If s is 10 when t is 3 and g is $\frac{5}{2}$, find s when t is 7 and g is 6.
25. N is directly proportional to the square of L and inversely proportional to the cube of M. If N is 9 when L is $\sqrt{3}$ and M is 2, find N when L is $\frac{\sqrt{2}}{2}$ and M is $\frac{3}{2}$.
26. Velocity varies directly as time. If velocity is 64 ft/sec at 2 seconds, find the velocity at 4 seconds.
27. If two lines are perpendicular, the slope of one is inversely proportional to the slope of the other. If two perpendicular lines have slopes of $\frac{2}{3}$ and $-\frac{3}{2}$, find the slope of a line perpendicular to a line with slope $-\frac{3}{4}$.
28. The distance traveled at a constant speed is directly proportional to time. If it takes 2 hours to go 100 miles, how many miles will be covered in 5 hours?
29. Hooke's Law (problem 11) is also true if a spring is compressed. If a force of 5 lb is required to compress a spring 2 in., how much force will be required to compress it 1 in?
30. The resistance of a wire varies directly as its length and inversely as the square of its diameter. If 50 ft of wire with a diameter of 0.01 in. has a resistance of 5 ohms, what is the resistance of 100 ft of the same kind of wire with a diameter of 0.02 in.?

IN YOUR OWN WORDS

31. Explain the difference in direct and inverse variation.

CHAPTER SUMMARY

GLOSSARY

Cartesian coordinate system: A pair of perpendicular number lines that intersect at the zero of each line. This point of intersection is called the *origin*.

Ordered pair: A pair of numbers written in the form (p, q). The first number is called the *x-coordinate* and the second number is called the *y-coordinate*.

Linear equation in two variables: An equation that can be written in the form $Ax + By = C$, where A and B are not both zero.

x-intercept of a line: The *x*-coordinate of the point where the line crosses the *x*-axis.

y-intercept of a line: The *y*-coordinate of the point where the line crosses the *y*-axis.

y varies directly as x means $y = kx$.

y varies inversely as x means $y = \frac{k}{x}$.

y varies jointly as x and z means $y = kxz$.

DISTANCE FORMULA The distance between the points (x_1, y_1) and (x_2, y_2) is given by the formula

$$d = \sqrt{(x_1 - x_2)^2 + (y_1 - y_2)^2}$$

The midpoint between (x_1, y_1) and (x_2, y_2) is

$$\left(\frac{x_1 + x_2}{2}, \frac{y_1 + y_2}{2}\right)$$

MIDPOINT FORMULA

1. *Standard form*: $Ax + By = C$; A and B not both 0.
2. *Slope-intercept form*: $y = mx + b$.
   m is the slope and b is the y-intercept.
3. *Point-slope form*: $y - y_1 = m(x - x_1)$.
   m is the slope and (x_1, y_1) is any point on the line.

FORMS OF A LINEAR EQUATION IN TWO VARIABLES

1. Using standard form, $Ax + By = C$. Find the intercepts and plot them.
2. Using slope intercept form, $y = mx + b$. Read the y-intercept from the equation and plot it. Read the slope and use it to draw the line.

GRAPHING LINES

1. The slope of the line containing the points (x_1, y_1) and (x_2, y_2) is given by the slope formula

$$m = \frac{\text{rise}}{\text{run}} = \frac{y_1 - y_2}{x_1 - x_2}; \; x_1 \neq x_2$$

2. The slope of a line can be found from the slope intercept form of the equation of the line.

SLOPE

1. Parallel lines have the same slope.
2. Perpendicular lines have slopes that are negative reciprocals.

PARALLEL AND PERPENDICULAR LINES

1. $x = p$ is a line parallel to the y-axis (*vertical*). It has no slope.
2. $y = q$ is a line parallel to the x-axis (*horizontal*). It has zero slope.

HORIZONTAL AND VERTICAL LINES

1. Find the slope of the line.
2. Find the coordinates of any point on the line.
3. Substitute into the point-slope form.

WRITING AN EQUATION OF A LINE WITH A SLOPE

CHECKUPS

1. Find the distance between the points $(1, 6)$ and $(4, -2)$.   Section 8.1; Example 2
2. Sketch the graph of each equation:
   (a) $5x + y = 2$ Section 8.2; Example 2
   (b) $x + 2y = 0$ Section 8.2; Example 3
   (c) $x = 2$ Section 8.2; Example 4
   (d) $y = \frac{2}{3}x - \frac{5}{2}$ Section 8.4; Example 3
3. Find the slope of the line that contains the points $(-1, 7)$ and $(2, -2)$.   Section 8.3; Example 4
4. Suppose that L_1 is a line perpendicular to the graph of $5x - 3y = 15$. What is the slope of L_1?   Section 8.3; Example 13
5. Find an equation of the line containing the points $(-1, 3)$ and $(2, -1)$.   Section 8.4; Example 6
6. y varies inversely as the square root of x. Find the constant of proportionality if y is 15 when x is 9.   Section 8.5; Example 2

Sec. 8.5 Variation

REVIEW PROBLEMS

In problems 1 through 6, find the distance between the points and the point midway between them.

1. (1, 1), (5, 4)
2. (6, 8), (2, 5)
3. (2, 1), (1, 3)
4. (1, −1), (−4, 2)
5. (8, −3), (9, −2)
6. (−6, −5), (−2, −8)

In problems 7 through 15, graph the equation.

7. $2x + 3y = 12$
8. $3x - 2y = 6$
9. $y = \frac{1}{2}x - 2$

10. $y = -x + 1$
11. $6x + y = 6$
12. $2x + 3y = 0$

13. $x = -3$
14. $5x = 9y$
15. $4y = 7$

In problems 16 through 24, find the slope of the line that contains both points.

16. (6, 8), (5, 3)
17. (5, 9), (3, 11)
18. (2, 6), (−2, 2)
19. (3, −2), (−2, 8)
20. (−4, −5), (−1, −7)
21. (−5, 0), (0, 3)
22. $\left(\frac{1}{2}, \frac{3}{2}\right), \left(\frac{3}{2}, -\frac{1}{2}\right)$
23. $\left(-\frac{4}{3}, \frac{1}{2}\right), \left(\frac{1}{3}, -\frac{5}{2}\right)$
24. $\left(-2, -\frac{3}{5}\right), \left(\frac{1}{5}, 3\right)$

In problems 25 through 27, find the y-intercept and the slope of the graph of the equation.

25. $2x - 3y = 6$
26. $3x = 4y - 6$
27. $2y + 3 = 0$

In problems 28 through 36, find an equation of a line satisfying the conditions given.

28. The line through (1, 2) with slope $\frac{3}{5}$.
29. The line through (4, −2) with slope −1.
30. The line containing (4, 1) and (3, −2).
31. The line containing (−9, −6) and (−2, −1).
32. The line through (−3, 6) parallel to the graph of $y = 2x - 7$.

33. The line through $(1, -7)$ parallel to the graph of
$$2x + 3y = 4.$$
34. The line perpendicular to the line $y = -3x - 9$, containing the point $(3, -5)$.
35. The line through $(3, 1)$ perpendicular to the graph of
$$3x = 4y.$$
36. The line through the origin parallel to the graph of the line containing the points $(7, -1)$ and $(2, 5)$.
37. u varies directly as v. If u is 8 when v is 4, find u when v is 12.
38. x varies inversely as the fourth power of y. If x is 5 when y is $\sqrt{2}$, find x when y is $\sqrt{3}$.
39. The gravitational force between two objects varies inversely as the square of the distance between the objects. If a force of 50 force units results from two objects that are 12 units apart, how much force results from two objects that are 20 units apart?
40. The work done by a constant force is directly proportional to the distance the object is moved. If 50 foot-pounds of work is required to move an object 10 ft, how much work is done in moving the same object 35 ft?

... LET'S NOT FORGET ...

Identify the expressions that are in factored form. Factor those that are not factored.

41. $27 + a^3$
42. $(3 - b)^3$
43. $(q + r)^2$
44. $\dfrac{a}{b} + a^2 b$
45. $(f - g)^2 - 9$
46. $a^2 bc + d^2 br$

How many terms are in each expression? Which expressions have $y - 7$ as a factor?

47. $y^3 - 7$
48. $(y - 7)^2 - x^2$
49. $2y^2 - 11y - 21$
50. $r(y - 7) - s(7 - y)$

Simplify each expression, if possible. Leave only nonnegative exponents in the answer.

51. 9^{-2}
52. -9^2
53. $(-9)^2$
54. $(-9)^{-2}$
55. $\sqrt{x^2 - 4}$
56. $\sqrt[3]{32x^3 + 64x^5}$
57. $(a^{2/3} b^{3/2})^6$
58. $\dfrac{-3a^{-4} b^2}{b^{-2}}$

Reduce, if possible.

59. $\dfrac{x^2 + y^2}{x^2 - y^2}$
60. $\dfrac{x^2 - y^2}{x^3 - y^3}$

The following problems can be worked by using a least common denominator. Follow the directions in each and notice how the LCD is used.

61. Perform the operation indicated: $\dfrac{x}{x - 1} - \dfrac{3}{1 - x}$.

62. Solve $\dfrac{3}{x - 1} + \dfrac{6}{x^2 - 1} = 1$.

63. Simplify $\dfrac{\dfrac{3}{x^2 - 1}}{\dfrac{1}{x + 1} - \dfrac{1}{x - 1}}$.

Label each problem as an expression, equation, or inequality. Solve the equations and inequalities, and perform the operations indicated on the expressions, leaving only nonnegative exponents in your answer.

64. $|5 - x| = 3$
65. $\dfrac{s - 5}{2s - 12} + \dfrac{s - 4}{12 - 2s}$
66. $(\sqrt{7} - \sqrt{5})^2$
67. $6x^2 - 5x < 6$
68. $\sqrt{x + 3} - \sqrt{x - 2} = 1$
69. $x^4 - 3x^2 - 4 = 0$

CHAPTER TEST

In problems 1 through 6, choose the correct answer.

1. The distance between the points $(2, -3)$ and $(-1, 5)$ is
   A. $\sqrt{65}$ **B.** $\sqrt{73}$
   C. $\sqrt{5}$ **D.** $\sqrt{89}$

2. The slope of the line containing the points $(5, -1)$ and $(2, 4)$ is
   A. $\dfrac{5}{3}$ **B.** $\dfrac{3}{5}$ **C.** $-\dfrac{5}{3}$ **D.** $-\dfrac{3}{5}$

3. Which of the following is the equation of a line with slope $-\dfrac{2}{3}$ that contains the point $(3, 7)$?
   A. $2x + 3y = 27$ **B.** $3x + 2y = 23$
   C. $2x - 3y = -15$ **D.** $3x - 2y = -5$

4. Which of the following is an equation of the line through the origin perpendicular to the graph of $3x - 4y = -1$?
   A. $4x - 3y = 0$ **B.** $4x + 3y = 0$
   C. $3x - 4y = 0$ **D.** $3x + 4y = 0$

5. The slope of the line whose equation is $y = -3$ is (?)
   A. $\{-3\}$ **B.** It has no slope.
   C. 0 **D.** $\emptyset$

6. The statement "x varies jointly as y and z and inversely as the square of w" may be written (?)
   A. $x = \dfrac{yz}{w^2}$ **B.** $x = \dfrac{yw^2}{z}$
   C. $x = \dfrac{kyz}{w^2}$ **D.** $x = \dfrac{kyw^2}{z}$

In problems 7 through 10, sketch the graph of each equation.

7. $x + 2y = 2$

8. $2x - y = 0$

9. $x = -\dfrac{2}{3}$

10. $y = \dfrac{3}{2}x + \dfrac{1}{2}$

11. Write an equation of the line containing the point $(-1, 0)$ and parallel to the line through the points $(1, 2)$ and $(3, 4)$.

12. Write an equation of the line perpendicular to the x-axis containing the point $(-2, -3)$.

13. Find the distance between the points $(5, -6)$ and $(-7, -1)$.

14. Find the midpoint between the points $(5, -6)$ and $(-7, -1)$.

15. Find the slope of a line whose x-intercept is 3 and whose y-intercept is 2.

16. Find the y-intercept and the slope for the line whose equation is $5x + 3y = 7$.

17. x varies directly as the square root of y and inversely as the square of z. If x is 6 when y is 9 and z is 2, find x when y is 16 and z is 4.

CHAPTER 9

See Problem Set 9.3, Exercise 19.

Systems of Equations

- **9.1** Systems of Equations in Two Variables
- **9.2** Linear Systems in More Than Two Variables
- **9.3** Applications
- **9.4** Linear Inequalities in Two Variables
- **9.5** Systems of Linear Inequalities in Two Variables
- **9.6** Determinants and Cramers Rule (Optional)
- **9.7** Matrix Methods (Optional)

CONNECTIONS

Suppose we wished to cut a 13-foot length of conduit into two pieces such that one piece is 3 feet longer than the other. We are looking for two numbers whose sum is 13 and whose difference is 3. This problem could be solved with the methods developed in Chapter 4, or it could be done by trial and error. However, it seems natural to let x be the larger of the two numbers and y be the other number. Then we immediately get two equations,

$$\begin{cases} x + y = 13 \\ x - y = 3 \end{cases}$$

We call such an arrangement a system of two equations in two variables.

In this chapter we learn two analytical methods for solving systems of equations and examine some applications. We will study inequalities in two variables and learn how to graph linear systems of inequalities. Cramer's rule and matrix methods are also discussed in two optional sections.

9.1 SYSTEMS OF EQUATIONS IN TWO VARIABLES

Let's look at the problem of finding two numbers whose sum is 13 and difference is 3. If we let x be the larger of the two numbers and y be the other number, we get two equations,

$$x + y = 13 \quad \text{and} \quad x - y = 3$$

Linked equations such as these are usually written in the form,

$$\begin{cases} x + y = 13 \\ x - y = 3 \end{cases}$$

Such an arrangement is called a **system of two equations in two variables.** Notice that x stands for the same number in both equations, as does y. So a **solution** of the system is an ordered pair that satisfies both equations. Is there such an ordered pair? Is there more than one such pair? To answer these questions, let's look at the graph of these two equations on the same coordinate system (top of next page).

The graphs of the two equations intersect in a single point. As the point of intersection lies on *both* lines, the corresponding ordered pair is in *both* solution sets. We conclude that the *system* of equations has exactly one solution.

We can look at the graph and guess that the lines intersect at (8, 5). This ordered pair is in the solution set of both equations, so it is the desired solution. This is the **graphical method** of solving systems of equations. As it requires judgment and estimation, it is not a very satisfactory technique. In this section we learn two analytical methods for solving such systems.

Let's look at our two equations again (numbered for reference).

$$\boxed{1} \begin{cases} x + y = 13 \\ \boxed{2} x - y = 3 \end{cases}$$

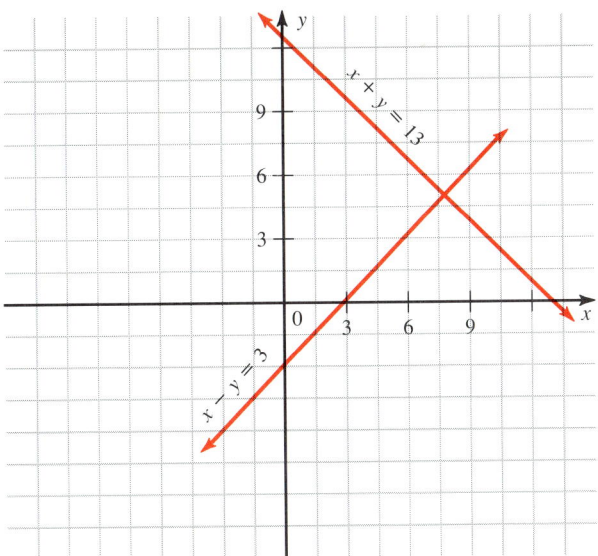

Suppose that we solve equation 1 for x, and call it equation 3.

$$\boxed{3} \quad x = 13 - y$$

Now, if x has the value $13 - y$ in equation 1, it must have the value $13 - y$ in equation 2. So we replace the x in equation 2 by $(13 - y)$.

$$(13 - y) - y = 3$$

This is an equation in one variable, which can be solved by the methods learned earlier.

$$13 - 2y = 3$$
$$-2y = -10 \quad \text{or} \quad y = 5$$

This gives the y-value. If we replace the y in equation 3 by this value, we find the corresponding x-value.

$$x = 13 - 5 = 8$$

We have found that the ordered pair $(8, 5)$ is a solution for *both* equations. Thus the solution set is

$$\{(8, 5)\}$$

This is called the **method of substitution.**

Method of Substitution

1. Solve one of the equations for one of the variables.
2. Substitute this for the same variable in the other equation.
3. Solve the resulting equation.
4. Substitute the result back into step 1 to find the other variable.
5. Check solution(s), if required.
6. Write the solution set.

Sec. 9.1 Systems of Equations in Two Variables

EXAMPLE 1. Find the solution set for each system.

(a) $\begin{cases} 2x + y = 2 \\ 6x - 4y = -1 \end{cases}$ (b) $\begin{cases} 4x + 9y = 12 \\ 2x - 3y = 1 \end{cases}$

Solutions:

(a) $\boxed{1}$ $\begin{cases} 2x + y = 2 \\ 6x - 4y = -1 \end{cases}$
    $\boxed{2}$

The most convenient variable is y, in equation $\boxed{1}$. We solve that equation for y.

$\boxed{\text{Step 1}}$ Solve one equation for one variable.

$$y = 2 - 2x$$

$\boxed{\text{Step 2}}$ Substitute for same variable in other equation.

Replace y in equation $\boxed{2}$ by $2 - 2x$.

$$6x - 4y = -1$$
$$6x - 4(2 - 2x) = -1$$

$\boxed{\text{Step 3}}$ Solve for the remaining variable.

$$6x - 8 + 8x = -1$$
$$14x = 7$$
$$x = \frac{7}{14} = \frac{1}{2}$$

$\boxed{\text{Step 4}}$ Substitute back into step 1 and find other variable.

We replace x by $\frac{1}{2}$ in the equation found in step 1, and solve.

$$y = 2 - 2x$$
$$y = 2 - 2\left(\frac{1}{2}\right)$$
$$y = 2 - 1 = 1$$

$\boxed{\text{Step 5}}$ Check.

The ordered pair $\left(\frac{1}{2}, 1\right)$ checks in both equations.

$\boxed{\text{Step 6}}$ Write the solution set.

$$\left\{\left(\frac{1}{2}, 1\right)\right\}$$

(b) $\boxed{1}$ $\begin{cases} 4x + 9y = 12 \\ 2x - 3y = 1 \end{cases}$
    $\boxed{2}$

Chap. 9 Systems of Equations

Solve one equation for one variable.

We solve equation $\boxed{2}$ for x.

$$2x = 3y + 1$$

$$x = \frac{1}{2}(3y + 1)$$

Substitute for same variable in other equation.

We replace x in equation $\boxed{1}$ by $\frac{1}{2}(3y + 1)$.

$$4\left[\frac{1}{2}(3y + 1)\right] + 9y = 12$$

Solve.

$$2(3y + 1) + 9y = 12$$
$$6y + 2 + 9y = 12$$
$$15y = 10$$
$$y = \frac{10}{15} = \frac{2}{3}$$

Substitute back into step 1 and solve.

We replace y by $\frac{2}{3}$ in the equation found in the first step to get

$$x = \frac{1}{2}\left(3 \cdot \frac{2}{3} + 1\right)$$

$$x = \frac{1}{2}(2 + 1) = \frac{3}{2}$$

Check, if required.

We have found the ordered pair $\left(\frac{3}{2}, \frac{2}{3}\right)$, which checks in both equations.

Write the solution set.

$$\left\{\left(\frac{3}{2}, \frac{2}{3}\right)\right\} \qquad \square$$

EXAMPLE 2. Find the solution set for

$$\begin{cases} -x + 2y = 2 \\ 3x - 6y = 1 \end{cases}$$

Solution:

We solve the first equation for x, as it seems to be the most convenient.

$$-x = 2 - 2y$$
$$x = -2 + 2y \qquad \text{(continued)}$$

Sec. 9.1 Systems of Equations in Two Variables

Putting this into the second equation gives

$$3(-2 + 2y) - 6y = 1$$
$$-6 + 6y - 6y = 1$$
$$-6 = 1$$

What has happened? The variable y is gone, and we are left with $-6 = 1$, a false statement!

Let's graph the two given equations to see what has happened.

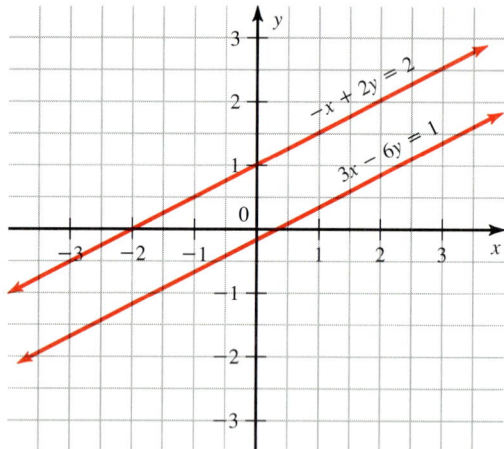

The lines look parallel. If we check the slopes, we see that the slope of each line is $\frac{1}{2}$. As each line has the same slope, they are parallel. There is *no* ordered pair common to both lines. The solution set is the empty set.

$$\emptyset$$

We call such a system **inconsistent.** If all the variables drop out when solving a system of equations, leaving a *false* statement, the system is inconsistent and the solution set is the empty set.

What if we arrive at a statement that is always *true*?

EXAMPLE 3. Solve the following system.

$$\begin{cases} x - 2y = 5 \\ 6y - 3x = -15 \end{cases}$$

Solution:

We solve the first equation for x.

$$x = 2y + 5$$

Substituting that into the second gives

$$6y - 3(2y + 5) = -15$$
$$6y - 6y - 15 = -15$$
$$-15 = -15$$

This statement is always true. Let's graph these equations.

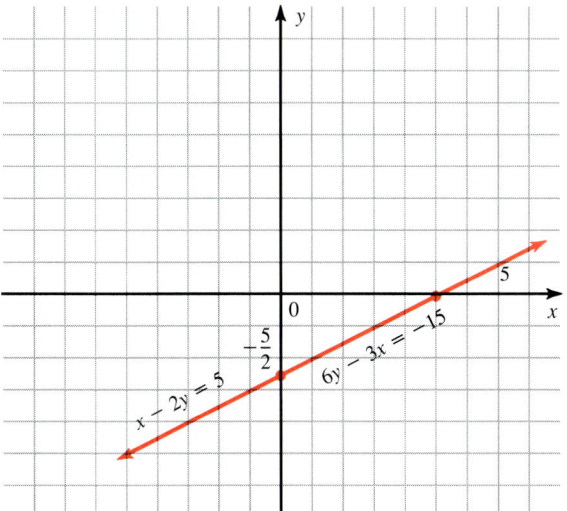

Both equations have the same line for their graph! If we look closely at the second equation, we see that it is just the first equation multiplied by -3. Thus the two equations are equivalent and have the same solution set.

Therefore, the solution set of the system is the solution set of either of the equations. In this case, the solution set is the set of ordered pairs that make the first equation true, or

$$\{(x, y) \mid x - 2y = 5\}$$

We call such systems **dependent**.

As seen above, there are three situations that can occur when solving systems of two linear equations in two variables.

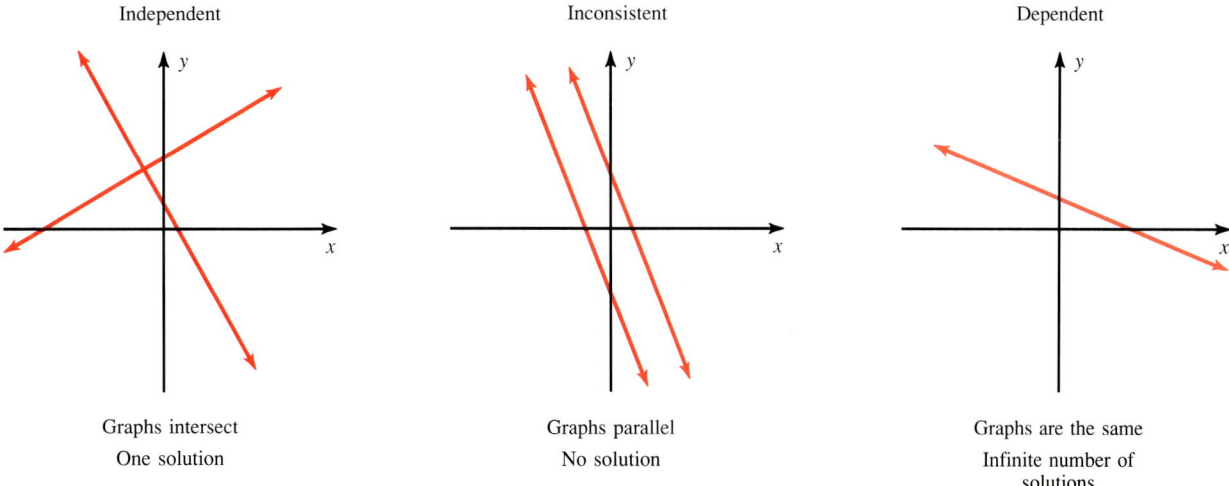

A system containing one or more *nonlinear* equations is called a nonlinear system. The method of substitution is often the best approach to use in solving nonlinear systems.

Sec. 9.1 Systems of Equations in Two Variables

EXAMPLE 4. Solve the system

$$\begin{cases} x^2 + y^2 = 5 \\ x + 2y = 3 \end{cases}$$

Solution:

Notice that the first equation is *not* linear. However, the method of substitution will still work.

We solve the second equation for *x*.

[*] $\qquad x = 3 - 2y$

Substitute this into the first equation, and solve for *y*.

$$x^2 + y^2 = 5$$
$$(3 - 2y)^2 + y^2 = 5$$
$$9 - 12y + 4y^2 + y^2 = 5$$
$$5y^2 - 12y + 4 = 0$$

This is a quadratic equation in *y*, which factors.

$$(5y - 2)(y - 2) = 0$$

$$y = 2 \quad \text{or} \quad y = \frac{2}{5}$$

We have found two *y*-values. We substitute each *y*-value into the equation marked [*]. First, if *y* is 2,

$$x = 3 - 2(2)$$
$$x = -1$$

We have found one ordered pair solution, $(-1, 2)$. To get another solution, replace *y* by $\frac{2}{5}$.

$$x = 3 - 2y$$
$$= 3 - 2\left(\frac{2}{5}\right)$$
$$= 3 - \frac{4}{5}$$
$$= \frac{11}{5}$$

Therefore, the solution set contains two ordered pairs,

$$\left\{\left(\frac{11}{5}, \frac{2}{5}\right), (-1, 2)\right\} \qquad \square$$

It is instructive to look at the graphs of the two equations in the system of Example 4.

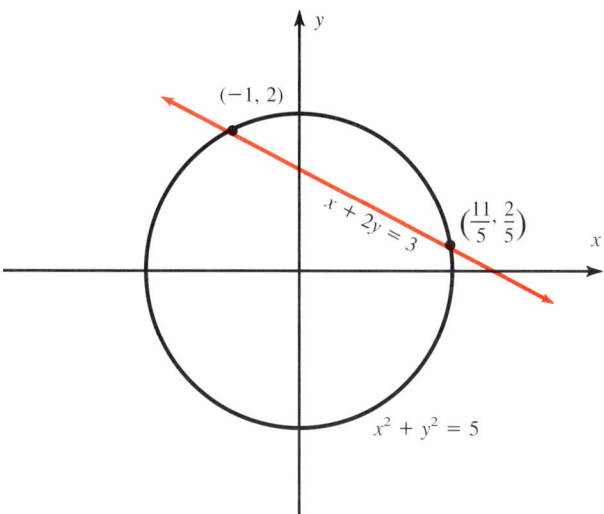

The second method for solving systems of equations works particularly well for linear systems, but may not work at all in nonlinear systems. It is called the method of **elimination.**

The addition property in Chapter 4 allows us to add two equations together. This property can be of great assistance in solving a system of equations. Some examples should clarify this.

EXAMPLE 5. Solve the system

$$\begin{cases} 3x + 2y = 2 \\ x - 2y = 14 \end{cases}$$

Solution:

Notice that if we add these equations together, the y-terms will add to zero.

$$\begin{aligned} 3x + 2y &= 2 \\ x - 2y &= 14 \\ \hline 4x \phantom{{}+2y} &= 16 \end{aligned}$$

$$x = 4$$

Now we substitute this x-value in either of the original equations and solve for y.

$$3(4) + 2y = 2$$
$$12 + 2y = 2$$
$$2y = -10$$
$$y = -5$$

We have found an ordered pair, $(4, -5)$, that satisfies both equations of the system. Therefore, the solution set is

$$\{(4, -5)\}$$

Sec. 9.1 Systems of Equations in Two Variables

EXAMPLE 6. Solve the system
$$\begin{cases} 5x - 3y = 13 \\ 2x + y = 3 \end{cases}$$

Solution:

If we add these equations together, we get the equation $7x - 2y = 16$, which does not help. But note what happens if we multiply both sides of the second equation by 3, *then* add.

$$\begin{aligned} 5x - 3y &= 13 \\ 6x + 3y &= 9 \\ \hline 11x &= 22 \\ x &= 2 \end{aligned}$$

We now find y.
$$2(2) + y = 3$$
$$4 + y = 3$$
$$y = -1.$$

The solution set is
$$\{(2, -1)\}$$

So we see that for the elimination step to work, it is sometimes necessary to multiply both sides of one of the equations by a suitable number, and both sides of the other equation by another number.

Method of Elimination

1. Write both equations in standard form, $Ax + By = C$.
2. Multiply both sides of each equation by a suitable real number so that one of the variables will be eliminated by addition of the equations. (This step may not be necessary.)
3. Add the equations and solve the resulting equation.
4. Substitute the value found in step 3 into one of the original equations, and solve this equation.
5. Check the solution, if required.
6. Write the solution set.

We say that two systems are **equivalent** if they have the same solution set.

EXAMPLE 7. Find the solution set for the following system.
$$\begin{cases} 2x + 5y = 11 \\ 3x + 7y = 15 \end{cases}$$

Solution:

Step 1 Write both equations in standard form.

Both equations are already in the form $Ax + By = C$.

Step 2 Multiply both sides of each equation by a suitable number so that a variable will be eliminated.

One way to do this step is to multiply both sides of the first equation by 3, and both sides of the second equation by -2.

$$\begin{cases} 6x + 15y = 33 \\ -6x - 14y = -30 \end{cases}$$

This system is equivalent to the first, but notice what happens if we add the two equations together.

Step 3 Add the equations.

$$\begin{aligned} 6x + 15y &= 33 \\ -6x - 14y &= -30 \\ \hline y &= 3 \end{aligned}$$

We have the y-value.

Step 4 Substitute the value found in step 3 into one of the original equations and solve.

We substitute 3 for y in the first of the original equations.

$$2x + 5y = 11$$
$$2x + 5(3) = 11$$
$$2x + 15 = 11$$
$$2x = -4$$
$$x = -2$$

Step 5 Check, if required.

As

$$2(-2) + 5(3) = -4 + 15 = 11$$

and

$$3(-2) + 7(3) = -6 + 21 = 15$$

the ordered pair $(-2, 3)$ checks in both equations.

Step 6 Write the solution set.

$$\{(-2, 3)\}$$

EXAMPLE 8. Solve

$$\begin{cases} 9x + 12y = 18 \\ 3x + 4y = 6 \end{cases}$$

Solution:

Multiply the second equation by -3.

$$\begin{cases} 9x + 12y = 18 \\ -9x - 12y = -18 \end{cases}$$

(continued)

Sec. 9.1 Systems of Equations in Two Variables

Add the two equations together.

$$9x + 12y = 18$$
$$-9x - 12y = -18$$
$$0 = 0$$

A true statement. As in the method of substitution, a true statement means that the system is **dependent** and the solution set is

$$\{(x, y) \mid 3x + 4y = 6\}$$

EXAMPLE 9. Solve the following system.

$$\begin{cases} \dfrac{1}{2}x - \dfrac{1}{3}y = \dfrac{1}{6} \\ 2x - \dfrac{4}{3}y = 1 \end{cases}$$

Solution:

First, we multiply both sides of the top equation by 6 and the bottom equation by 3 to clear fractions.

$$\begin{cases} 3x - 2y = 1 \\ 6x - 4y = 3 \end{cases}$$

Next we multiply the top equation by -2 to prepare for elimination.

$$\begin{cases} -6x + 4y = -2 \\ 6x - 4y = 3 \end{cases}$$

Now we add.

$$-6x + 4y = -2$$
$$6x - 4y = 3$$
$$0 = 1$$

A *false* statement, indicating an **inconsistent** system with an empty solution set,

$$\emptyset$$

Many word problems are easily solved using two variables.

EXAMPLE 10. Find two numbers whose sum is 15 if one of the numbers subtracted from twice the other number also is 15.

Solution:

First, we make two assignments:

Let x be one of the numbers.
Let y be the other.

From the problem statement we get that the sum of x and y is 15. Also, one of the numbers, say y, subtracted from two times the other is also 15. This gives us the two equations,

$$\begin{cases} x + y = 15 \\ 2x - y = 15 \end{cases}$$

We solve by elimination.

$$\begin{array}{r} x + y = 15 \\ 2x - y = 15 \\ \hline 3x = 30 \\ x = 10 \end{array}$$

$$10 + y = 15$$
$$y = 5$$

The numbers are 5 and 10. □

PROBLEM SET 9.1

Warm-ups

In problems 1 through 12, use the method of substitution to find the solution set of the system.
For problems 1 through 6, see Example 1.

1. $\begin{cases} x - 2y = 0 \\ 2x + y = 5 \end{cases}$
2. $\begin{cases} x + 2y = 0 \\ 2x + y = 6 \end{cases}$
3. $\begin{cases} 2x - y = 0 \\ x + 4y = 9 \end{cases}$

4. $\begin{cases} 3x + 2y = 6 \\ 4x - y = 8 \end{cases}$
5. $\begin{cases} 2x - 5y = 4 \\ 3x - 2y = -5 \end{cases}$
6. $\begin{cases} 3x - 2y = 2 \\ 2x + 3y = -3 \end{cases}$

For problems 7 through 9, see Examples 2 and 3.

7. $\begin{cases} x + y = 4 \\ 3x + 3y = 12 \end{cases}$
8. $\begin{cases} 3x - 2y = 6 \\ 15x - 10y = -2 \end{cases}$
9. $\begin{cases} \frac{1}{3}x - \frac{3}{4}y = 1 \\ 4x - 9y = 6 \end{cases}$

For problems 10 through 12, see Example 4.

10. $\begin{cases} x^2 + y^2 = 25 \\ x - 3 = 0 \end{cases}$
11. $\begin{cases} x^2 - y = -1 \\ x - y = -1 \end{cases}$
12. $\begin{cases} x^2 - y^2 = 4 \\ 2x - y = -4 \end{cases}$

In problems 13 through 27, use the method of elimination to find the solution set.
For problems 13 through 18, see Examples 5 and 6.

13. $\begin{cases} x + y = 1 \\ x - y = 3 \end{cases}$
14. $\begin{cases} 2x + 3y = 10 \\ x - 3y = -4 \end{cases}$
15. $\begin{cases} 2x - y = -5 \\ 3x + 2y = 3 \end{cases}$

16. $\begin{cases} 2x + 2y = 3 \\ -x + 6y = 2 \end{cases}$
17. $\begin{cases} 3x + 2y = 5 \\ 6x - y = 0 \end{cases}$
18. $\begin{cases} 7x - 3y = -10 \\ 5x - y = -4 \end{cases}$

For problems 19 through 24, see Example 7.

19. $\begin{cases} 2x + 3y = 5 \\ 3x - 2y = 1 \end{cases}$
20. $\begin{cases} 4x - 5y = 2 \\ 3x + 2y = 13 \end{cases}$
21. $\begin{cases} 5x - 2y = 0 \\ 7x - 3y = -1 \end{cases}$

22. $\begin{cases} \frac{1}{2}x + \frac{1}{3}y = 11 \\ \frac{1}{3}x - \frac{1}{5}y = 1 \end{cases}$
23. $\begin{cases} \frac{4}{7}x - \frac{3}{5}y = 0 \\ -\frac{2}{5}x + \frac{1}{7}y = 0 \end{cases}$
24. $\begin{cases} -\frac{3}{7}x + \frac{2}{5}y = \frac{1}{35} \\ \frac{1}{5}x + \frac{7}{3}y = -\frac{38}{15} \end{cases}$

Sec. 9.1 Systems of Equations in Two Variables

For problems 25 through 27, see Examples 8 and 9.

25. $\begin{cases} 6x - 4y = 23 \\ -9x + 6y = 12 \end{cases}$

26. $\begin{cases} \dfrac{1}{3}x - \dfrac{3}{2}y = 0 \\ \dfrac{2}{5}x - \dfrac{9}{5}y = 0 \end{cases}$

27. $\begin{cases} \dfrac{2}{3}x + 8y = \dfrac{6}{7} \\ \dfrac{3}{4}x + 9y = 3 \end{cases}$

For problems 28 through 30, see Example 10.

28. The sum of two numbers is 30. If twice one of them is three times the other, what are the numbers?

29. The difference of two numbers is 18. What are they if their sum is zero?

30. One-half the sum of two numbers equals their difference. What are the numbers if their sum is 64?

Practice Exercises

In problems 31 through 48, use the method of substitution to find the solution set of each system.

31. $\begin{cases} x + 2y = 5 \\ 2x - y = 0 \end{cases}$

32. $\begin{cases} x - 2y = 4 \\ 2x - y = 5 \end{cases}$

33. $\begin{cases} 2x + y = 5 \\ x - 4y = -2 \end{cases}$

34. $\begin{cases} 3x - 2y = 1 \\ 4x + y = 5 \end{cases}$

35. $\begin{cases} 2x + 5y = -10 \\ 3x - 2y = 4 \end{cases}$

36. $\begin{cases} 3x - 2y = -6 \\ 2x + 3y = -4 \end{cases}$

37. $\begin{cases} 4x - 5y = -8 \\ 2x + 3y = 7 \end{cases}$

38. $\begin{cases} 2x + 6y = -2 \\ 5x + 3y = -13 \end{cases}$

39. $\begin{cases} 4x + 6y = 1 \\ 2x + 3y = 2 \end{cases}$

40. $\begin{cases} x - 6y = -1 \\ 2x - 3y = 1 \end{cases}$

41. $\begin{cases} 2x + 6y = 9 \\ 4x + y = 7 \end{cases}$

42. $\begin{cases} x + 6y = 1 \\ 3x + 18y = 3 \end{cases}$

43. $\begin{cases} 2x - y = 4 \\ 6x - 3y = 8 \end{cases}$

44. $\begin{cases} 3x + 6y = 2 \\ 6x - 9y = 2 \end{cases}$

45. $\begin{cases} x - 2y = 0 \\ 3x - 4y = 0 \end{cases}$

46. $\begin{cases} x^2 - 2y^2 = -2 \\ x - y = 1 \end{cases}$

47. $\begin{cases} x + y = 3 \\ xy = 2 \end{cases}$

48. $\begin{cases} 3x - 4y = 5 \\ \dfrac{3}{2}x - 2y = 1 \end{cases}$

In problems 49 through 80, use the method of elimination to find the solution set of each system.

49. $\begin{cases} 2x - y = 7 \\ 3x + y = 8 \end{cases}$

50. $\begin{cases} x - 2y = 1 \\ 3x + 2y = 11 \end{cases}$

51. $\begin{cases} 3x + y = 5 \\ 4x - 2y = 10 \end{cases}$

52. $\begin{cases} 4x + 3y = 14 \\ -2x + 4y = 4 \end{cases}$

53. $\begin{cases} 2x - 3y = -5 \\ 3x + y = -2 \end{cases}$

54. $\begin{cases} 2x + 3y = 2 \\ x + 5y = -6 \end{cases}$

55. $\begin{cases} 2x - 3y = -3 \\ 3x + 2y = 15 \end{cases}$

56. $\begin{cases} 5x + 3y = 1 \\ 2x - 4y = 3 \end{cases}$

57. $\begin{cases} 7x + 4y = -1 \\ 9x - 6y = -5 \end{cases}$

58. $\begin{cases} 5x + 4y = 2 \\ 4x + 3y = 1 \end{cases}$

59. $\begin{cases} 5x - 2y = 0 \\ 3x - 7y = 0 \end{cases}$

60. $\begin{cases} 11x - 7y = 9 \\ 8x + 6y = 62 \end{cases}$

61. $\begin{cases} 3x + 4y = 0 \\ 4x + 3y = 0 \end{cases}$

62. $\begin{cases} 12x - 5y = -1 \\ 3x - 2y = -1 \end{cases}$

63. $\begin{cases} 6x - 5y = 6 \\ 9x + 7y = -20 \end{cases}$

64. $\begin{cases} 2x + 9y = -2 \\ 4x + 3y = 1 \end{cases}$

65. $\begin{cases} -4x + 15y = -2 \\ 12x + 5y = -4 \end{cases}$

66. $\begin{cases} 4x + 5y = 2 \\ 6x + 7y = 3 \end{cases}$

67. $\begin{cases} 3x - 4y = 5 \\ 4x - 7y = 9 \end{cases}$

68. $\begin{cases} 5x - 3y = 5 \\ -15x + 9y = 7 \end{cases}$

69. $\begin{cases} 17x - 11y = -1 \\ -34x + 22y = 2 \end{cases}$

70. $\begin{cases} \dfrac{1}{2}x + \dfrac{3}{4}y = \dfrac{7}{4} \\ \dfrac{1}{3}x - \dfrac{1}{6}y = \dfrac{1}{2} \end{cases}$

71. $\begin{cases} \dfrac{1}{5}x - \dfrac{2}{3}y = \dfrac{1}{15} \\ \dfrac{3}{4}x - \dfrac{5}{6}y = \dfrac{1}{3} \end{cases}$

72. $\begin{cases} \dfrac{2}{3}x - \dfrac{2}{5}y = 1 \\ \dfrac{5}{2}x - \dfrac{3}{2}y = 2 \end{cases}$

73. $\begin{cases} \dfrac{2}{7}x + \dfrac{2}{5}y = \dfrac{3}{7} \\ \dfrac{3}{5}x + \dfrac{7}{3}y = \dfrac{5}{2} \end{cases}$

74. $\begin{cases} \dfrac{6}{11}x - \dfrac{4}{7}y = 0 \\ -\dfrac{7}{4}x + \dfrac{11}{6}y = 0 \end{cases}$

75. $\begin{cases} \dfrac{3}{8}x - \dfrac{1}{9}y = 0 \\ \dfrac{9}{2}x - \dfrac{5}{3}y = 0 \end{cases}$

76. $\begin{cases} 2x + 3y^2 = 2 \\ x - y = 1 \end{cases}$

77. $\begin{cases} 3x^2 - 2y^2 = 4 \\ 2x - y^2 = 0 \end{cases}$

78. $\begin{cases} x + xy = -1 \\ xy = -6 \end{cases}$

79. $\begin{cases} x^2 + y^2 = 1 \\ x^2 - y = 1 \end{cases}$

80. $\begin{cases} x^2 + y^2 = 13 \\ x^2 - y^2 = 5 \end{cases}$

81. The difference between two numbers is 2. If three times the larger plus five times the smaller is 94, what are the numbers?

82. Three times one number is four times another. If the sum of the two numbers is 49, what are they?

83. The sum of three numbers is 68. One of the numbers is twice the average of the other two. The difference of the "other two" is 20. What are the three numbers?

Challenge Problems

In problems 84 through 86, the systems are not linear, but the substitutions

$$u = \dfrac{1}{x} \quad \text{and} \quad v = \dfrac{1}{y}$$

will transform them into linear equations in u and v. Find the solution sets.

84. $\begin{cases} \dfrac{2}{x} + \dfrac{3}{y} = 12 \\ \dfrac{5}{x} - \dfrac{2}{y} = 11 \end{cases}$

85. $\begin{cases} \dfrac{3}{x} + \dfrac{4}{y} = 5 \\ \dfrac{6}{x} + \dfrac{5}{y} = 7 \end{cases}$

86. $\begin{cases} \dfrac{2}{x} + \dfrac{3}{y} = 0 \\ \dfrac{3}{x} + \dfrac{2}{y} = 1 \end{cases}$

▮▮▮ IN YOUR OWN WORDS . . .

87. Describe the method of *substitution* for solving two equations in two variables.

88. Describe the method of *elimination* for solving two equations in two variables.

9.2 LINEAR SYSTEMS IN MORE THAN TWO VARIABLES

For the same reasons that there are equations in two variables, there are equations in three, four, or even more variables. Modern high-speed computers, with their ability to solve huge systems, have made the study of such systems worthwhile.

A linear equation in three variables is an equation of the form

$$Ax + By + Cz = D \quad (A, B, C \text{ not all zero})$$

A solution of such an equation is an ordered triple of real numbers, (x, y, z), that makes the statement true when substituted for the variables. An example of an equation in three variables is

$$2x + 3y - z = 5$$

The ordered triple $(2, 1, 2)$ is a solution to this equation, as are $(0, 1, -2)$ and $(1, 1, 0)$.

A linear equation in four variables is an equation of the form

$$Ax + By + Cz + Dw = E \quad (A, B, C, D \text{ not all zero})$$

Solutions to this equation are of the form (x, y, z, w).

An example of an equation in *five* variables is

$$2x_1 - x_2 - 3x_3 + 5x_4 + x_5 = 10$$

The variables are x_1, x_2, x_3, x_4, and x_5. (Subscripts are commonly used in large systems.) Some solutions to this equation are $(1, 1, -1, 1, 1)$, $(0, 0, 1, 1, 8)$, and $(2, 0, -2, 0, 0)$.

In this section we limit our discussion to *linear* systems with the same number of equations as there are variables.

EXAMPLE 1. Find the solution set for the following system. (The equations are numbered for convenience.)

|1|
|2|
|3|

$$\begin{cases} x + 2y + z = 3 \\ x - 3y - z = 2 \\ x - 2y - 2z = -2 \end{cases}$$

Solution:

Our strategy is to eliminate one variable from two of the equations, forming an equation in two variables. Next, we eliminate the *same* variable from any *other* pair of equations. This gives us another equation in the same two variables.

Add |1| to |2| to eliminate z.

|1| $\qquad x + 2y + z = 3$

|2| $\qquad \underline{x - 3y - z = 2}$

|4| $\qquad 2x - y \phantom{{}-z} = 5$

We label the result |4|. Next multiply both sides of |1| by 2, so the same variable, z, will be eliminated when we add |1| to |3|.

|1| $\qquad 2(x + 2y + z) = 2 \cdot 3$

$\qquad\qquad 2x + 4y + 2z = 6$

Now add |1| to |3| and call the result |5|.

|1| $\qquad 2x + 4y + 2z = 6$

|3| $\qquad \underline{x - 2y - 2z = -2}$

|5| $\qquad 3x + 2y \phantom{{}- 2z} = 4$

Notice that equations |4| and |5| form a system of *two* equations in the *same two* variables, x and y. We reduced a system of three equations in three variables to a system of two equations in two variables. Now to solve the new system:

[4]
[5]
$$\begin{cases} 2x - y = 5 \\ 3x + 2y = 4 \end{cases}$$

If we multiply both sides of [4] by 2 and add the two equations together, we will eliminate y.

[4]
[5]
$$4x - 2y = 10$$
$$\underline{3x + 2y = 4}$$
$$7x = 14$$
$$x = 2$$

We have the value of x. Put this value into [5] to find the value of y.

[5]
$$3x + 2y = 4$$
$$3(2) + 2y = 4$$
$$6 + 2y = 4$$
$$2y = -2$$
$$y = -1$$

Now we have the values of x and y. Put them into equation [1] to find z.

[1]
$$x + 2y + z = 3$$
$$2 + 2(-1) + z = 3$$
$$2 - 2 + z = 3$$
$$z = 3$$

We have found the only solution, $(2, -1, 3)$. We write the solution set,

$$\{(2, -1, 3)\}$$

To Solve Three Equations in Three Variables

1. Select any two of the equations and eliminate one of the variables by the methods of Section 9.1.
2. Use the remaining equation and one of the equations used in step 1 to eliminate the *same* variable.
3. Solve the system of two equations in two variables, obtained in steps 1 and 2, by the methods of Section 9.1.
4. Put the values found in step 3 into one of the original equations to find the remaining variable.
5. Check, if required.
6. Write the solution set.

Sec. 9.2 Linear systems in more than two variables

EXAMPLE 2. Solve the system

$$\boxed{1}$$
$$\boxed{2}$$
$$\boxed{3}$$
$$\begin{cases} x + 2y + z = 0 \\ 2x - 2y + z = -2 \\ 3x + 2y - 2z = 3 \end{cases}$$

Solution:

Step 1 Eliminate one variable from two of the equations.

It pays to study a system and form a strategy of attack. In this system we notice that the variable y can be eliminated easily by adding $\boxed{1}$ and $\boxed{2}$ for one equation, then adding $\boxed{2}$ and $\boxed{3}$ for a second equation.

$$\boxed{1} \qquad x + 2y + z = 0$$
$$\boxed{2} \qquad \underline{2x - 2y + z = -2}$$
$$\boxed{4} \qquad 3x + 2z = -2$$

Then,

Step 2 Eliminate the same variable from the remaining equation and one of the equations of step 1.

$$\boxed{2} \qquad 2x - 2y + z = -2$$
$$\boxed{3} \qquad \underline{3x + 2y - 2z = 3}$$
$$\boxed{5} \qquad 5x - z = 1$$

Step 3 Solve the two-variable system.

$$\boxed{4}$$
$$\boxed{5}$$
$$\begin{cases} 3x + 2z = -2 \\ 5x - z = 1 \end{cases}$$

We multiply both sides of equation $\boxed{5}$ by 2 and add.

$$\boxed{4} \qquad 3x + 2z = -2$$
$$\boxed{5} \qquad \underline{10x - 2z = 2}$$
$$\phantom{\boxed{5}} \qquad 13x = 0$$
$$\phantom{\boxed{5}} \qquad x = 0$$

Putting the value of x into $\boxed{4}$ gives

$$\boxed{4} \qquad 3x + 2z = -2$$
$$3(0) + 2z = -2$$
$$2z = -2$$
$$z = -1$$

Step 4 Substitute the values found in step 3 into one of the original equations, and solve.

We substitute the values $x = 0$ and $z = -1$ into any of the original equations to find y.

$$\boxed{1} \qquad x + 2y + z = 0$$
$$0 + 2y + (-1) = 0$$
$$2y - 1 = 0$$
$$2y = 1$$
$$y = \frac{1}{2}$$

Be Careful!

Step 5 Check, if required.

The ordered pair $\left(0, \dfrac{1}{2}, -1\right)$ satisfies all three of the original equations.

Step 6 Write the solution set.

$$\left\{\left(0, \frac{1}{2}, -1\right)\right\} \qquad \square$$

Systems in more than two variables can be dependent or inconsistent.

EXAMPLE 3. Solve

$$\begin{cases} x - y + 3z = 1 \\ -x + 2y - 2z = 1 \\ x - 3y + z = 2 \end{cases}$$

Solution:

Adding the first two equations yields

$$y + z = 2$$

Adding the second two equations yields

$$-y - z = 3$$

But if we add these two equations, we get

$$0 = 5$$

a *false* statement. The system is **inconsistent,** so the solution set is the empty set,

$$\emptyset \qquad \square$$

Sec. 9.2 Linear systems in more than two variables

EXAMPLE 4. Solve the system
$$\begin{cases} x + 2y - 2z = 1 \\ x - 2y + 3z = 1 \\ 3x - 2y + 4z = 3 \end{cases}$$

Solution:

If we add the first two equations, we get
$$2x + z = 2$$
and if we add the first and third, we get
$$4x + 2z = 4$$
We now have a system of two equations in two variables:
$$\begin{cases} 2x + z = 2 \\ 4x + 2z = 4 \end{cases}$$
If we multiply both sides of the first equation by 2, we get the system
$$\begin{cases} 4x + 2z = 4 \\ 4x + 2z = 4 \end{cases}$$

As both of these equations are exactly the same, the system is clearly dependent, so the original system is **dependent**. (We will not write the solution set for dependent systems of three or more variables.) ▭

Systems with four or more variables may be solved by an extension of the method described above. For example, a system of four linear equations in four variables may be solved by eliminating one variable from a pair of equations. Then, when the same variable is eliminated from two *other* pairs of equations, the resulting system of three equations in three variables may be solved by the three-variable method.

EXAMPLE 5. Find the solution set.

[1]
[2]
[3]
[4]
$$\begin{cases} x + y = 3 \\ 3y + 3z = 7 \\ z + w = 1 \\ 3x - 3w = 1 \end{cases}$$

Solution:

This is an example of a system of four equations in four variables. Systems this large can be tedious to solve. However, this system has quite a few "missing terms" which will aid us in our task. For example, the variable w does not occur in equation [1] or [2]. We start by eliminating w from [3] and [4]. Multiply both sides of [3] by 3, add it to [4] to get [5].

[3] $\qquad\qquad 3z + 3w = 3$

[4] $\qquad 3x \qquad - 3w = 1$

[5] $\qquad 3x + 3z \qquad = 4$

Now the system of equations ⑴, ⑵, and ⑸ is a system of three equations in the three variables x, y, and z.

⑴
⑵
⑸
$$\begin{cases} x + y = 3 \\ 3y + 3z = 7 \\ 3x + 3z = 4 \end{cases}$$

We eliminate z from this system by multiplying both sides of ⑸ by -1 and adding it to ⑵.

⑵ $\qquad 3y + 3z = 7$
⑸ $\qquad \underline{-3x - 3z = -4}$
$\qquad\qquad -3x + 3y = 3 \quad$ or
⑹ $\qquad -x + y = 1$

Finally, the system made up of equations ⑴ and ⑹ is a system of two equations in two variables. Adding them eliminates x.

⑴ $\qquad x + y = 3$
⑹ $\qquad \underline{-x + y = 1}$
$\qquad\qquad 2y = 4$
$\qquad\qquad y = 2$

We put this result in ⑴ to get x:

⑴
$$x + y = 3$$
$$x + 2 = 3$$
$$x = 1$$

and in ⑵ to get z:

⑵
$$3y + 3z = 7$$
$$3(2) + 3z = 7$$
$$3z = 1$$
$$z = \frac{1}{3}$$

The value of z in equation ⑶ yields w.

⑶
$$z + w = 1$$
$$\frac{1}{3} + w = 1$$
$$w = \frac{2}{3}$$

The solution set is
$$\left\{\left(1, 2, \frac{1}{3}, \frac{2}{3}\right)\right\}$$

Sec. 9.2 Linear systems in more than two variables

PROBLEM SET 9.2

Warm-ups

In problems 1 through 10, find the solution set for each system unless it is dependent.
If dependent, so indicate.
For problems 1 through 6, see Examples 1 and 2.

1. $\begin{cases} x + y + z = 6 \\ x + 2y - z = 2 \\ 2x - 2y + z = 1 \end{cases}$
2. $\begin{cases} 2x - y + z = 5 \\ -x - y + 2z = 4 \\ 2x + y - 3z = -5 \end{cases}$
3. $\begin{cases} 2x - y + 2z = 2 \\ -2x + 3y + 4z = 10 \\ -2x + y - z = 0 \end{cases}$

4. $\begin{cases} x + y + z = 5 \\ 2x - 3y - 2z = 12 \\ 3x + 2y - 2z = 7 \end{cases}$
5. $\begin{cases} x + 2y + 3z = 1 \\ 2x - 2y - 3z = -1 \\ 3x + y + 6z = 2 \end{cases}$
6. $\begin{cases} x + y + 3z = 0 \\ 2x - 2y - z = 0 \\ 5x - 3y - z = 0 \end{cases}$

For problems 7 through 10, see Examples 3 and 4.

7. $\begin{cases} 3x + 3y - z = 5 \\ x + y + z = 5 \\ -2x - 2y + z = -3 \end{cases}$
8. $\begin{cases} 2x + y - z = 4 \\ 4x - y + 2z = -2 \\ -2x + 2y - 3z = 6 \end{cases}$

9. $\begin{cases} 3x - 5y - 2z = 9 \\ x - y + 2z = 1 \\ 2x - 3y = 1 \end{cases}$
10. $\begin{cases} x + y + z = 1 \\ 3x - 2y + 2z = 5 \\ 3x - 7y + z = 7 \end{cases}$

Practice Exercises

In problems 11 through 32, find the solution set for each system unless it is dependent.
If dependent, so indicate.

11. $\begin{cases} x + y + z = 6 \\ x + 2y + z = 9 \\ 2x + 2y - z = 6 \end{cases}$
12. $\begin{cases} x + y + 3z = 0 \\ 2x - y + 2z = 1 \\ 3x + y + z = 6 \end{cases}$
13. $\begin{cases} 3x + y + 2z = 10 \\ -3x - 2y + 4z = -11 \\ 2y + z = 2 \end{cases}$

14. $\begin{cases} 3x - 2y + 4z = 22 \\ x + y + z = 3 \\ 2x - 2y - 3z = -1 \end{cases}$
15. $\begin{cases} 3x + 3y - z = 3 \\ 5x + y + 3z = 1 \\ 2x + 4y - 3z = 4 \end{cases}$
16. $\begin{cases} 2x + 6y + z = 5 \\ -2x - 3y + 2z = -7 \\ x + 9y - 3z = 8 \end{cases}$

17. $\begin{cases} 2x + 2y - z = 1 \\ x + 2y - 3z = 4 \\ 5x + 6y - 5z = 3 \end{cases}$
18. $\begin{cases} x - 2y + 3z = 9 \\ 3x - y - 2z = 6 \\ 4x + 4y + 2z = 2 \end{cases}$
19. $\begin{cases} 3x + y - z = 8 \\ 2x - y + 2z = 3 \\ x + 2y - 3z = 5 \end{cases}$

20. $\begin{cases} x + 2y - z = -2 \\ 2x + 2y + 2z = 3 \\ 6x - 4y - 2z = 2 \end{cases}$
21. $\begin{cases} 2x - 3y - 10z = 4 \\ 4x - 5z = 3 \\ 6y + 5z = -3 \end{cases}$

22. $\begin{cases} 3x + y = 5 \\ 2y - 4z = 7 \\ x + y - z = 4 \end{cases}$
23. $\begin{cases} 2x + 6y = 5 \\ 3y - z = 2 \\ -3x - 3z = 1 \end{cases}$
24. $\begin{cases} 2x + 3z = 2 \\ 5y - 9z = -3 \\ 6x + 7y = 3 \end{cases}$

25. $\begin{cases} 2x - 3y = 6 \\ -2y - 3z = 4 \\ 3x - 2z = 0 \end{cases}$
26. $\begin{cases} 2x + 3y + 4z = 0 \\ x + y + z = 0 \\ -x + 2y + 2z = 0 \end{cases}$
27. $\begin{cases} x + y - z = -8 \\ 4x + 5y - 6z = -2 \\ 2x + 3y - 4z = 14 \end{cases}$

28. $\begin{cases} 2x - y + 4z = -4 \\ 4x - 5y + 7z = 1 \\ -2x + 7y - 2z = 6 \end{cases}$
29. $\begin{cases} 2x + 4y - z = 3 \\ x + y = 1 \\ 2x + 3y + z = 2 \end{cases}$
30. $\begin{cases} 2x + 3y - 3z = 0 \\ x - 3y + 2z = 5 \\ 4x - 6y - 5z = 5 \end{cases}$

31. $\begin{cases} 3x - 2y - z = 4 \\ x + 4y + 2z = -1 \\ 2x - 4y - 3z = 6 \end{cases}$
32. $\begin{cases} 3x - y = 0 \\ y + 2z = -1 \\ x - z = 0 \end{cases}$

33. Suppose that the sum of a certain number and a second number is 6. Further, suppose that the sum of the certain number and a third number is 7. If the sum of all three is 12, find the three numbers.
34. Find three numbers whose sum is 8 if the sum of two of them is 7 and the difference of the same two is 3.
35. Suppose that the sum of a certain number and a second number is 1. Further, suppose that the sum of the certain number and a third number is 8. If the sum of all three is 6, find the three numbers.
36. Four numbers have a sum of 9. The sum of two of them is 8 and the difference of the same two is 2. If the difference of the other two is 3, find the four numbers.

Challenge Problems

In problems 37 through 40, find the solution set for each system.

37. $\begin{cases} x + y + z + w = 2 \\ x - y + 2z = -1 \\ 2y + 3z - 3w = -9 \\ 2x - 3y + 2w = 6 \end{cases}$

38. $\begin{cases} x - 2y - z + 2w = -2 \\ 2x - y + z - w = 0 \\ -x + 3y - w = 4 \\ 5x - 3y = -4 \end{cases}$

39. $\begin{cases} x + 2y - z = 2 \\ 2x + z + w = 9 \\ y - w = -2 \\ 3x + 4y = 11 \end{cases}$

40. $\begin{cases} x_1 + x_2 = 1 \\ x_2 - x_3 = -2 \\ 2x_3 - 2x_4 = 5 \\ 2x_4 + 2x_5 = 1 \\ x_1 - 2x_5 = 0 \end{cases}$

■ IN YOUR OWN WORDS . . .

41. How do we find out that a system of equations is dependent or inconsistent?

■ 9.3 APPLICATIONS

Many word problems become much simpler if more than one variable is used. We follow a procedure similar to that given in Section 4.3.

A Procedure to Solve Word Problems

1. Read the problem to determine what quantities are to be found.
2. Assign a variable, such as x, y, or z, to represent each quantity to be found.
3. Draw a figure or picture if possible. Label it.
4. Reread the problem and form as many equations as there are variables.
5. Solve the system of equations found in step 4.
6. Check the solution in the original word problem. It should make sense and answer the question.
7. Write an answer to the original question.

Mixture problems are particularly suited for more than one variable.

EXAMPLE 1. Perry wishes to clean his boat with a 15% soap solution. How much of a 10% soap solution and a 20% soap solution should he add to a quart of pure water to make 30 quarts of a 15% solution?

Solution:

Step 1 Read the problem and determine what is to be found.

We are to find how many quarts of a 10% soap solution and a 20% soap solution are needed.

Step 2 Assign a variable to each quantity to be found.

Let x be the number of quarts of the 10% solution.
Let y be the number of quarts of the 20% solution.

Step 3 Draw a figure.

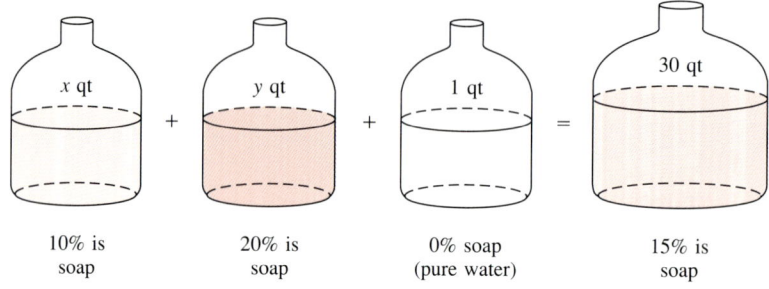

Step 4 Form as many equations as there are variables.

As x quarts + y quarts + 1 quart must be 30 quarts, our first equation is

$$x + y + 1 = 30 \quad \text{or}$$

$$\boxed{1} \quad x + y = 29$$

Now the amount of *soap* in the 10% solution added to the amount of soap in the 20% solution must equal the amount of soap in the final solution. That is,

$$0.10x + 0.20y = 0.15(30)$$

$$10x + 20y = 15(30)$$

$$10x + 20y = 450$$

$$\boxed{2} \quad x + 2y = 45 \quad \text{Divide by 10.}$$

This gives us the following system, which can be solved easily.

Step 5 Solve the system.

$$\begin{cases} x + y = 29 \\ x + 2y = 45 \end{cases}$$

$$\begin{array}{r} -x - y = -29 \\ x + 2y = 45 \\ \hline y = 16 \end{array}$$

$$x + 16 = 29$$

$$x = 13$$

Step 6 Check solution in the original word problem.

13 quarts of the 10% solution and 16 quarts of the 20% solution makes sense and answers the question.

Step 7 Write an answer to the original question.

Perry should mix 13 quarts of the 10% solution and 16 quarts of the 20% solution with 1 quart of pure water to make his solution.

EXAMPLE 2. Gloria bought 3 lb of raw peanuts and 2 lb of boiled peanuts from Big Bob's produce stand for $4.05. Jim bought 1 lb of raw and 4 lb of boiled for $4.65. What is the price per pound Big Bob charges for his raw and boiled peanuts?

Solution:

Determine what is to be found.

We are to find the price per pound of raw peanuts and the price per pound of boiled peanuts.

Assign a variable to each quantity to be found.

Let R be the price per pound of raw peanuts and B be the price per pound of boiled peanuts.

Form as many equations as there are variables.

$$\begin{cases} 3R + 2B = 4.05 \\ R + 4B = 4.65 \end{cases}$$

Solve the system.

Using elimination, we multiply the top equation by -2, then add.

$$\begin{aligned} -6R - 4B &= -8.10 \\ R + 4B &= 4.65 \\ \hline -5R &= -3.45 \\ R &= 0.69 \end{aligned}$$

$$\begin{aligned} 0.69 + 4B &= 4.65 \\ 4B &= 4.65 - 0.69 \\ 4B &= 3.96 \\ B &= 0.99 \end{aligned}$$

Check, then answer the question.

Big Bob charges 69 cents a pound for raw peanuts and 99 cents a pound for boiled. □

Often the relationships between the lengths of sides and perimeters of geometric figures are easy to express with more than one variable.

EXAMPLE 3. The perimeter of a parallelogram is 58 in. If the length of one side is 5 more than three times the length of the other side, what are the lengths of the sides?

Solution:

Let x be the length of the longest side and y be the length of the shortest side.
The perimeter of a parallelogram is given by $2p + 2q$, where p and q are the lengths of the sides. Thus we have one equation: $2x + 2y = 58$.
From the relationship between the sides we have the second equation: $x = 3y + 5$.
These two equations give us the following system:

$$\begin{cases} 2x + 2y = 58 \\ x = 3y + 5 \end{cases}$$

(continued)

As the second equation is already solved for x, we will use the method of substitution. We replace x in the first equation with its value in the second.

$$2(3y + 5) + 2y = 58$$
$$6y + 10 + 2y = 58$$
$$8y = 48$$
$$y = 6$$

Now we find x from the second equation.

$$x = 3(6) + 5$$
$$x = 18 + 5$$
$$x = 23$$

The sides of the parallelogram are 6 in. and 23 in. ☐

Motion problems can often be simplified by using more than one variable, particularly the "upstream–downstream" problems.

EXAMPLE 4. Sarah's boat takes 1 hour to travel 5 miles upstream to the Raysville Marina but only 20 minutes to return. What is the speed of the current in the river, and what is the average speed of Sarah's boat in still water?

Solution:

We let x be the average speed of the boat in still water and y be the speed of the current. Thus Sarah's speed *upstream* is

$$x - y$$

and *downstream* is

$$x + y$$

Now we make a distance–rate–time table for each part of the trip.

	DISTANCE	RATE	TIME
Upstream	5	$x - y$	1
Downstream	5	$x + y$	$\frac{1}{3}$

We need time in hours, so we must change 20 min to $\frac{1}{3}$ hour. As $D = RT$, we have the two equations

$$\begin{cases} 5 = (x - y) \cdot 1 \\ 5 = (x + y) \cdot \frac{1}{3} \end{cases}$$

Then substitute this into the second equation and solve.

$$x^2 + (7 - x)^2 = 25$$
$$x^2 + 49 - 14x + x^2 = 25$$
$$2x^2 - 14x + 24 = 0$$
$$x^2 - 7x + 12 = 0$$
$$(x - 3)(x - 4) = 0$$
$$x = 3 \quad x = 4$$

If x is 3 then y is 4, and if x is 4, y is 3.
The numbers are 3 and 4.

PROBLEM SET 9.3

Warm-ups

For problems 1 and 2, see Example 1.

1. Penny wishes to mix a tomato sauce that is 17% sugar with a sauce that is 30% sugar to obtain 26 liters of a tomato sauce that is 24% sugar. How much of each should she mix?

2. How many ounces of pure water and how many ounces of a 16% butterfat solution should Charlotte mix to obtain 32 ounces of a 10% butterfat solution?
   20 oz of 16%

For problems 3 and 4, see Example 2.

3. Martin bought 3 heads of lettuce and 2 pounds of tomatoes at the Tasty Shopette for $4.05. Beverly bought 2 heads of lettuce and 3 pounds of tomatoes for $4.35 at the same store. What is the price of a head of lettuce and a pound of tomatoes at the Tasty Shopette?

4. Latonia paid $2.39 for three Snickers and four Mr. Goodbar candy bars, while Antonio paid 3.40 for five of each of those kinds of candy bar. What was the price of Snickers and Mr. Goodbars?

For problems 5 and 6, see Example 3.

5. The perimeter of Jon's prize-winning lawn is 54 m. It is in the shape of a rectangle with the length of one side 3 m less than twice the length of the other. What are the dimensions of the lawn?

6. When not at the lake, Colombo spends his time in a rectangular dog pen that his master constructed with 28 yards of fencing. If the length of the pen is 2 yd longer than twice its width, what are its dimensions?

For problems 7 and 8, see Example 4.

7. It takes Jim's 18-year-old outboard 3 hours to travel 24 miles downstream and 5 hours to travel 10 miles upstream from his river cottage. What is the average speed of the boat in still water, and what is the speed of the current?

8. An airplane flying with the wind travels from Sioux City to Dubuque, a distance of 500 km, in 2 hours. The return trip, against the wind, takes $2\frac{1}{2}$ hours. What is the average speed of the plane in still air, and what is the wind speed?

For problems 9 and 10, see Example 5.

9. Frank and Jack working together can mow a certain field in 4 hours. One morning Frank mowed alone for 3 hours and left for the day. Jack arrived that afternoon and spent 6 hours finishing the job. How long does it take each of them, working alone, to mow the field?

10. A Texaco oil tank can be filled by two pipes in 8 hours. If the tank is filled in 10 hours when one of the pipes is shut off after 4 hours, how long will it take each of the pipes to fill the tank alone?

For problem 11, see Example 6.

11. Find two numbers whose sum is 19 if the sum of their squares is 185.

Sec. 9.3 Applications

Practice Exercises

12. The perimeter of a rectangular window is 200 cm. The length of one side is 20 cm less than 11 times the other. What are the dimensions of the window?
13. At the Farmers Market, John bought 3 pounds of grapes and 6 lemons for $3.57, and Betty bought 2 pounds of grapes and 5 lemons and paid $2.48. How much are lemons and grapes at the Farmers Market?
14. Gerry wishes to mix a 5% soap solution with an 8% soap solution to obtain 12 quarts of a 6% soap solution to wash her new Buick. How much of each should she mix?
15. Tom and Randi, working together, can mow their lawn in an hour and 12 minutes. Randi heads for the pool after working with Tom for 18 minutes and it takes Tom $1\frac{1}{2}$ hours to finish the job. How long does it take each, working alone, to mow the lawn?
16. The perimeter of a triangle is 146 ft. Twice the shortest side is 9 ft more than the longest side. The longest side is 16 ft less than the sum of the other two sides. What are the lengths of the three sides of this triangle?
17. Wayne Hilton has 1200 ft of fencing and wishes to enclose a rectangular pasture that is four times as long as it is wide. What dimensions should Wayne use for his pasture?
18. It takes a garbage scow 8 hours to travel 24 miles upstream and 8 hours to travel 88 miles downstream. What is the average speed of the scow in still water, and what is the speed of the current?

19. Carolyn walks 10 miles along the beach in the same time that Bill walks 6 miles. If Carolyn walks 1 mph less than twice Bill's rate, what is the speed at which each of them walks?
20. The sum of two positive numbers is 13, while the sum of one of them and the square of the other is 69. Find two such numbers.
21. The perimeter of an isosceles triangle is 48 inches. If the length of the shortest side is 3 inches less than either of the other two sides, what are the lengths of each of the sides of the triangle?
22. Two cars start at the same time, one from Gallop and the other from Las Cruces, 360 miles apart, and they travel toward each other. One travels 8 mph faster than the other. If they meet after 3 hours, what was the average speed of each car?
23. The perimeter of a valuable rectangular portrait is 60 in. and its area is 216 in.^{2} What are the dimensions of a frame for this painting?
24. The Smiths went soup shopping at the local Safeway. Kenneth bought 5 cans of tomato and 3 cans of mushroom for $3.09, while Margaret bought 2 cans of chicken noodle and a can of mushroom for $1.41. If Irene bought 2 cans of tomato and 3 cans of chicken noodle for $2.19, what is the price of tomato soup, mushroom soup and chicken noodle soup at the local Safeway store?

Challenge Problems

25. A $\frac{1}{4}$-in. pipe and a $\frac{1}{2}$-in. pipe, working together with the drain closed, can fill a water tank in 12 minutes. With the $\frac{1}{4}$-in. pipe closed and the drain open, it takes the $\frac{1}{2}$-in. pipe 40 minutes to fill the tank. With the $\frac{1}{2}$-in. pipe closed and the drain open it takes the $\frac{1}{4}$-in. pipe 2 hours to fill the tank. How long does it take the open drain to empty a full tank when both pipes are closed?
26. Terry, Saundra, and Linda made trips to the refreshment stand at the Cinema IV. Terry bought 2 small and 3 medium-sized drinks and one bag of popcorn for $5.25. Saundra bought a small drink and a bag of popcorn for $1.60, and Linda bought a medium-sized drink and 2 bags of popcorn for $2.90. What was the price of popcorn and drinks at the Cinema IV?
27. The sum of four numbers is 11. Twice the first plus the third is 2, while three times the first added to twice the third is 5. If three times the second plus twice the fourth is 17, what are the four numbers?
28. Two resistors R_1 and R_2 in a electrical circuit give a total resistance of R, where R is determined by the relationship

$$\frac{1}{R} = \frac{1}{R_1} + \frac{1}{R_2}$$

if they are hooked in parallel, and $R = R_1 + R_2$ if they are hooked in series. Suppose that two resistors give a total resistance of 12 ohms in parallel and 50 ohms in series. What is the number of ohms for each resistor?

IN YOUR OWN WORDS . . .

29. Write a word problem that has the following system as its mathematical model:

$$\begin{cases} x + y + z = 6 \\ x + y = 3 \\ x + z = 4 \end{cases}$$

9.4 LINEAR INEQUALITIES IN TWO VARIABLES

In Chapter 7 we studied inequalities in one variable. In this section we learn an extension of the method of boundary numbers that will allow us to graph linear inequalities in two variables. The graph of an inequality in two variables consists of all ordered pairs that make the statement of the inequality true. Inequalities in two variables, much like *equations* in two variables, required a coordinate system to display their graph.

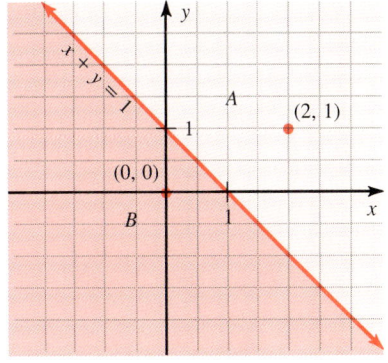

Let's look at the graph of the inequality $x + y \geq 1$. The graph is the collection of all ordered pairs (x, y) that make the statement $x + y \geq 1$ true.

Using a technique similar to the method of boundary numbers, we graph the equation $x + y = 1$.

This line is called a **boundary line.** It divides the plane into two regions, marked A and B on the sketch at right.

Region A is tested by selecting any point in the interior of the region, say $(2, 1)$, and determining whether it makes the statement of the inequality true or false. Since $2 + 1 \geq 1$ is a true statement, the point $(2, 1)$ is in the solution set and it follows that every point in region A is also in the solution set.

Similarly, we pick any point in region B, say $(0, 0)$, and see that the statement $0 + 0 \geq 1$ is false. Thus $(0, 0)$ is not in the solution set, which tells us no point in region B is in the solution set.

The boundary line itself *is* in the solution set because the original statement, $x + y \geq 1$, includes equality. The solution set is graphed below.

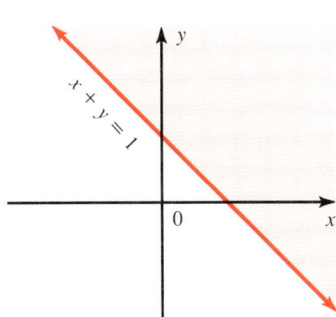

A Procedure for Graphing Linear Inequalities in Two Variables

1. Graph the boundary line:
   (a) Draw a solid line if equality is included. ($\geq$, $\leq$)
   (b) Draw a dashed line if equality is not included. ($>$, $<$)
2. Determine which region(s) formed by the line makes the inequality true by testing with one point from inside each region.
3. Shade the region(s) that makes the inequality true.

EXAMPLE 1. Graph the solution set for $x - 2y > 0$.

Solution:

Step 1 Graph the boundary line.

We draw the graph of the line $x - 2y = 0$.

Step 2 Test the regions formed by the boundary line.

A table is convenient for step 2.

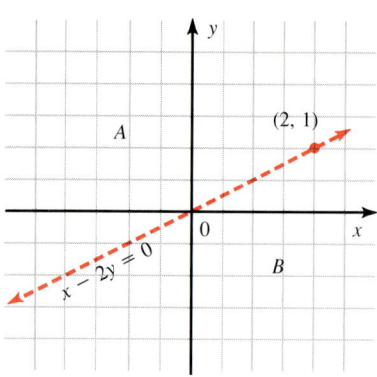

We draw a dashed line since equality is not included.

REGION	TEST POINT IN REGION	STATEMENT $x - 2y > 0$	TRUTH OF STATEMENT	REGION IN SOLUTION SET?
A	$(-1, 2)$	$-1 - 4 > 0$	False	No
B	$(1, -1)$	$1 - 2(-1) > 0$	True	Yes

Step 3 Shade the regions where the inequality is true.

Shade region B.

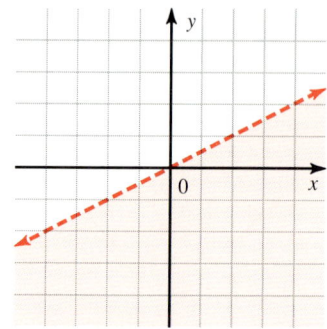

EXAMPLE 2. Graph the solution set for $x \geq 1$ (on the x,y plane).

Solution:

We graph the boundary line $x = 1$. It will be a solid line because equality is included.

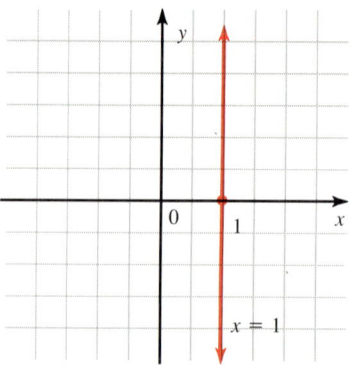

We can make a table for the next step or we can just test the regions and shade the graph.

The point (0, 0) is in the region left of the boundary line but it is not in the solution set because $0 \geq 1$ is a *false* statement. The left region *is not* shaded. The point (2, 0) is in the right region and is in the solution set as $2 \geq 1$ is a *true* statement. The right region *is* shaded.

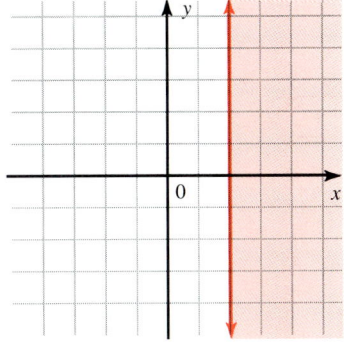

PROBLEM SET 9.4

Warm-Ups

In problems 1 through 10, graph the solution set on the x,y plane for each inequality. For problems 1 through 6, see Example 1.

1. $x - y < 2$
2. $x + 2y \leq 5$
3. $y - \frac{1}{2}x > 4$

4. $\frac{1}{3}y - x \geq 2$
5. $x - y \leq 0$
6. $y < 2x$

For problems 7 through 10, see Examples 1 and 2.

7. $x \geq 3$
8. $y > 0$
9. $y \leq -1$
10. $x < 1$

Sec. 9.4 Linear inequalities in two variables

Practice Exercises

In problems 11 through 46, graph the solution set on the x,y plane for each inequality.

11. $x + y < 3$

12. $x + 5y \leq 1$

13. $3x + y \leq 5$

14. $x + y < 0$

15. $x + y \geq 0$

16. $y \leq 3x$

17. $y > 2x - 3$

18. $x < 4 - 3y$

19. $x \leq 1$

20. $y \geq 2$

21. $y < 3x + 4$

22. $y \geq 2x - 7$

23. $x > 0$

24. $y \leq 0$

25. $2x - 3y \leq 4$

26. $3x + 2y < 6$

27. $3x - 2y \leq 5$

28. $2x + 4y > 7$

29. $3x - y < 6$

30. $2x + y \geq 0$

31. $\dfrac{1}{2}x - \dfrac{1}{6}y \leq 1$

32. $\dfrac{2}{3}x + \dfrac{3}{4}y < \dfrac{1}{6}$

33. $2.4x - 3.2y \geq 0$

34. $0.3x - 0.5y \leq 0.1$

35. $3y - 6 \geq 0$
36. $2x + y < 6$
37. $y - \frac{1}{3}x < \frac{4}{3}$
38. $\frac{1}{2}y + x > 2$

39. $\frac{1}{4}x - \frac{1}{2}y < 2$
40. $\frac{1}{2}x + \frac{1}{3}y > \frac{1}{4}$
41. $2.1x - 1.2y < 1.1$
42. $0.2x - 0.3y \leq 0.2$

43. $3x \leq 6$
44. $2x - 5 > 0$
45. $6x - y > 2$
46. $2x \leq y + 3$

Challenge Problems

In problems 47 through 56, graph the solution set on the x,y plane.

47. $y \leq mx;\ m > 0$
48. $y > mx;\ m < 0$
49. $y \leq mx + b;\ m > 0,\ b > 0$

50. $y \geq mx + b;\ m > 0,\ b < 0$
51. $y > mx + b;\ m < 0,\ b > 0$
52. $y < mx + b;\ m < 0,\ b < 0$

53. $y \leq a;\ a > 0$
54. $y \geq a;\ a < 0$

Sec. 9.4 Linear inequalities in two variables

55. $x \geq a$; $a > 0$

56. $x \leq a$; $a < 0$

■ IN YOUR OWN WORDS . . .

57. Describe the graph of the inequality $y < mx + b$.

58. Describe the graph of the inequality $x \geq k$.

■ 9.5 SYSTEMS OF LINEAR INEQUALITIES IN TWO VARIABLES

When two or more linear inequalities are considered simultaneously, a system of linear inequalities is formed. In general, the solution of such a system is a region in the plane where *all* the inequalities are true.

EXAMPLE 1. Graph the solution set of the following system of inequalities.

$$\begin{cases} x - y \leq 1 \\ x + y > 0 \end{cases}$$

Solution:

First we graph the inequality $x - y \leq 1$, as in Section 9.4.

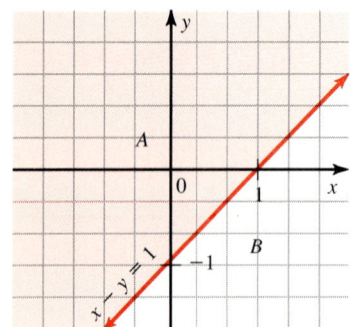

REGION	TEST POINT IN REGION	STATEMENT $x - y \leq 1$	TRUTH OF STATEMENT	REGION IN SOLUTION SET?
A	(0, 0)	$0 - 0 \leq 1$	True	Yes
B	(1, −1)	$1 - (-1) \leq 1$	False	No

So we see that the first inequality of the system is true for all points on, or above, the line $x - y = 1$. However, the solution set of the *system* is the set of points where *both* inequalities are true.

Next, we sketch the graph of $x + y > 0$.

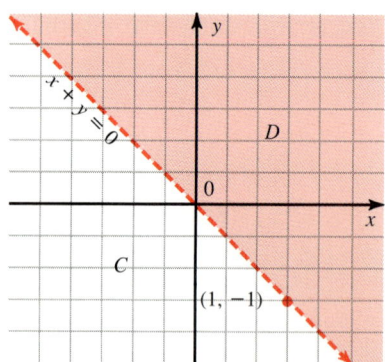

REGION	TEST POINT IN REGION	STATEMENT $x + y > 0$	TRUTH OF STATEMENT	REGION IN SOLUTION SET?
C	$(-1, 0)$	$-1 + 0 > 0$	False	No
D	$(0, 1)$	$0 + 1 > 0$	True	Yes

The second inequality of the system is true for all points above the line $x + y = 0$. Still, we are looking for points that satisfy *both* inequalities.

Notice the picture if we graph both solution sets on the *same* coordinate system.

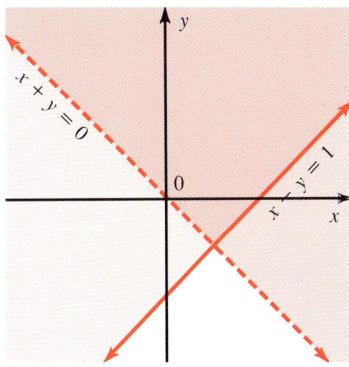

The intersection of the two solution sets is the solution set for the system. Notice that the lower segment of the solid boundary line *is not* in the solution set of the inequality, $x + y > 0$, and therefore must be changed from solid to dashed in the final sketch.

Be Careful!

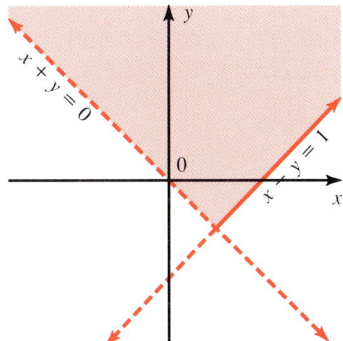

A Procedure for Graphing the Solution Set of a System of Linear Inequalities

1. Graph each inequality on the same coordinate system, lightly shading its solution set.
2. Darken the *intersection* of the lightly shaded regions.
3. Change any portion of any *solid* boundary line *not in the intersection* of the solutions from solid to dashed.

Sec. 9.5 Systems of linear inequalities in two variables

EXAMPLE 2. Graph the solution set for the following system of linear inequalities.

$$\begin{cases} x - y \geq -1 \\ x + y \geq 0 \\ 3x - y < 3 \end{cases}$$

Solution:

Sketching each of the inequalities as done in Section 9.4 gives the following graphs:

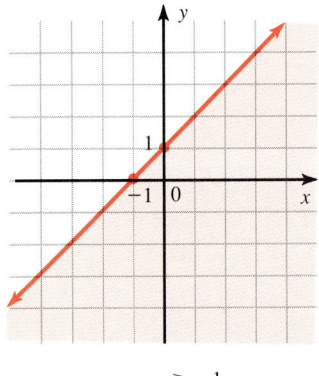

$x - y \geq -1$

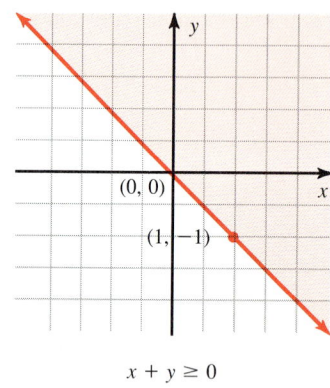

$x + y \geq 0$

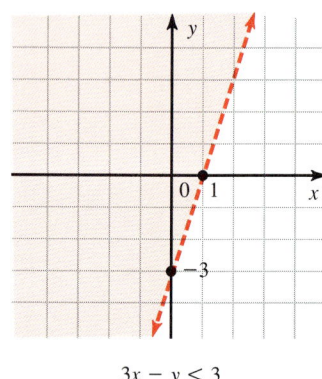

$3x - y < 3$

The solution set of the system is the intersection of all the shaded regions.

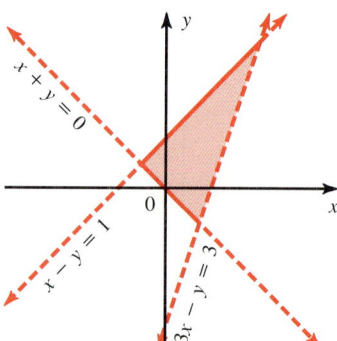

Notice the portions of the solid boundary lines that were changed from solid to dashed. □

EXAMPLE 3. The Lakeview Linen Company rents and maintains uniforms for a broad range of industrial customers. They manufacture uniforms for their service in two styles, food handling and automotive. The food-handling uniform uses 5 lb of cotton and 1 lb of polyester, while the automotive uniforms use 2 lb of cotton and 4 lb of polyester. There are 500 lb of cotton and 400 lb of polyester on hand. Write a system of inequalities that illustrate the situation and graph the system.

Solution:

Let x be the number of food handling uniforms made. Let y be the number of automotive uniforms made.

Since there is only 500 lb of cotton available and each food-handling uniform uses 5 lb and each automotive requires 2 lb, we have the inequality

$$5x + 2y \leq 500$$

Similarly, as there is 400 lb of polyester,

$$x + 4y \leq 400$$

Thus the situation is illustrated by the following system of inequalities.

$$\begin{cases} 5x + 2y \leq 500 \\ x + 4y \leq 400 \\ x \geq 0 \\ y \geq 0 \end{cases}$$

The last two inequalities are formed because x and y cannot be negative.

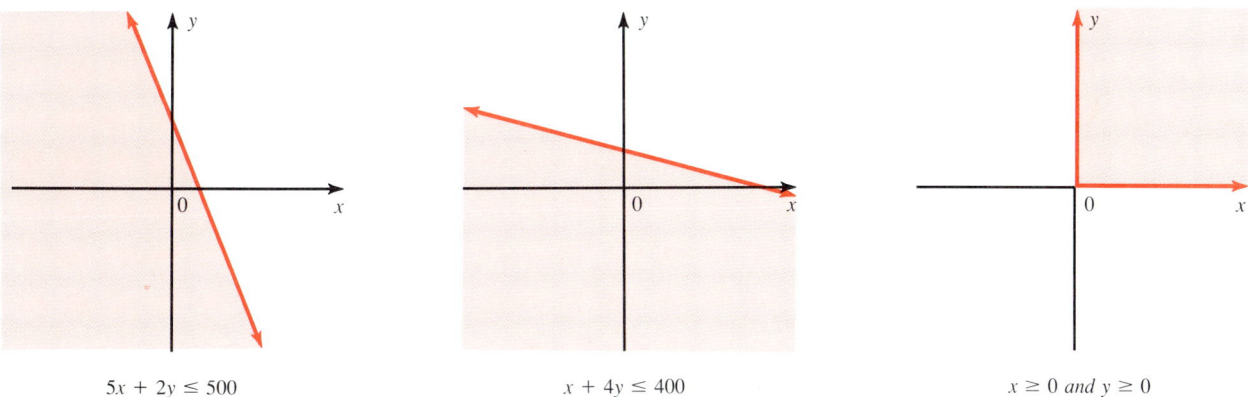

$5x + 2y \leq 500$ $x + 4y \leq 400$ $x \geq 0$ and $y \geq 0$

The solution set of the system is the intersection of all the shaded regions.

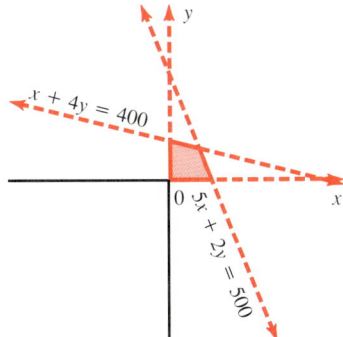

Any point (x,y) in the shaded region satisfies the system of inequalities and represents a production possibility for the Lakeside Linen Company.

Problems such as the one described above for the Lakeside Linen Company form an important branch of applied mathematics called **linear programming**.

Sec. 9.5 Systems of linear inequalities in two variables

PROBLEM SET 9.5

Warm-ups

In problems 1 through 10, graph the solution set for each system on the x,y plane.
For problems 1 through 8, see Example 1.

1. $\begin{cases} x - y \leq 4 \\ x + y \geq 2 \end{cases}$
2. $\begin{cases} x + 2y > 2 \\ x - y \leq 1 \end{cases}$
3. $\begin{cases} x + y \leq 0 \\ x - y > 0 \end{cases}$
4. $\begin{cases} x - 3y > 6 \\ 2x - y \leq 0 \end{cases}$

5. $\begin{cases} 3x - y < 1 \\ x + 2y > 5 \end{cases}$
6. $\begin{cases} 4x + y > 0 \\ x - 4y < 16 \end{cases}$
7. $\begin{cases} 2x + y \geq 3 \\ x - 2y > 4 \end{cases}$
8. $\begin{cases} \frac{1}{2}x - y < 3 \\ x + 2y > 4 \end{cases}$

For problems 9 and 10, see Example 2.

9. $\begin{cases} x - y \geq -2 \\ x + y \leq 2 \\ x - 2y \leq -2 \end{cases}$
10. $\begin{cases} 2x - y > 0 \\ 2x + y < 2 \\ y > 0 \end{cases}$

Practice Exercises

In problems 11 through 42, graph the solution set for each system on the x,y plane.

11. $\begin{cases} x - \frac{1}{3}y \leq -1 \\ 3x + y \geq -9 \end{cases}$
12. $\begin{cases} 2x - 3y \leq 4 \\ 2x - y > -2 \end{cases}$
13. $\begin{cases} x \geq 0 \\ y \geq 0 \end{cases}$

14. $\begin{cases} x > 0 \\ y < 0 \end{cases}$
15. $\begin{cases} x < 4 \\ y > 2 \end{cases}$
16. $\begin{cases} x > -1 \\ y < -1 \end{cases}$
17. $\begin{cases} 2x - y \leq 4 \\ 2x - y \geq 5 \end{cases}$

18. $\begin{cases} x + 2y < 5 \\ x + 2y < 7 \end{cases}$
19. $\begin{cases} x + 3y \geq 6 \\ 2x + 6y \leq 4 \end{cases}$
20. $\begin{cases} 4x - 3y < 15 \\ 5x + 4y > 11 \end{cases}$
21. $\begin{cases} 2x + 5y > -9 \\ 7x + 3y < 12 \end{cases}$

22. $\begin{cases} 3x + 8y \leq 7 \\ 4x - 3y \leq 23 \end{cases}$
23. $\begin{cases} x - 3y > 0 \\ x - 3y > 2 \end{cases}$
24. $\begin{cases} 4x - 3y > 4 \\ 2x + 5y > 15 \end{cases}$
25. $\begin{cases} x + 1 \leq 0 \\ y + 3 > 0 \end{cases}$

26. $\begin{cases} x - 2 > 4 \\ 2 - y < 3 \end{cases}$
27. $\begin{cases} x - y \leq 1 \\ x + 2y \leq 1 \end{cases}$
28. $\begin{cases} x + y \geq 0 \\ x - y \geq 4 \end{cases}$
29. $\begin{cases} x - 2y \leq 3 \\ 2x - 4y > 1 \end{cases}$

30. $\begin{cases} 2x - y > 4 \\ 6x - 3y > 0 \end{cases}$
31. $\begin{cases} 2x + y < 2 \\ x + \dfrac{1}{2}y < 0 \end{cases}$
32. $\begin{cases} 3x + y < 6 \\ x + \dfrac{1}{3}y > 4 \end{cases}$
33. $\begin{cases} x < 0 \\ y \leq 0 \end{cases}$

34. $\begin{cases} x \geq 1 \\ y < 2 \end{cases}$
35. $\begin{cases} x < -3 \\ y \geq 1 \end{cases}$
36. $\begin{cases} x > -4 \\ y \geq -3 \end{cases}$
37. $\begin{cases} \dfrac{1}{2}x - y > 5 \\ x + 2y < 14 \end{cases}$

Sec. 9.5 Systems of linear inequalities in two variables

38. $\begin{cases} x + \dfrac{1}{4}y \leq -2 \\ 2x - y \leq -4 \end{cases}$

39. $\begin{cases} 3x + 2y < 6 \\ 3x + 2y \geq 1 \end{cases}$

40. $\begin{cases} 4x + 4y > 4 \\ x + y > 2 \end{cases}$

41. $\begin{cases} x - y > -3 \\ x - 2y \leq 0 \\ 3x + y \leq 3 \end{cases}$

42. $\begin{cases} x + 2y > 2 \\ x + y < 3 \\ x - y \geq -1 \end{cases}$

In problems 43 and 44, write a system of inequalities that represents the situation and sketch the graph.

43. The Nut Boutique cans two styles of mixed nuts. A can of Deluxe Mixed Nuts contains 1 lb of cashews and 2 lb of peanuts, while a can of Premium Mixed Nuts contains 2 lb of cashews and 1 lb of peanuts. Only 30 lb of cashews and 40 lb of peanuts are available.

44. The High Point Furniture Company manufactures two types of desk–chair sets. The first type uses 10 board feet of pine and 5 board feet of oak, while the other requires 5 board feet of each. There is 800 board feet of pine and 500 board feet of oak available.

Challenge Problems

In problems 45 through 48, graph the solution set on the x,y plane.

45. $\begin{cases} y \leq mx; \quad m > 0 \\ y < nx; \quad n < 0 \end{cases}$

46. $\begin{cases} y \leq mx + b; \quad m > 0, b > 0 \\ y \geq nx + b; \quad n < 0 \end{cases}$

47. $\begin{cases} y < a; \quad a > 0 \\ x < b; \quad b > 0 \end{cases}$

48. $\begin{cases} y \geq a; \quad a < 0 \\ x \leq b; \quad b > 0 \end{cases}$

■ IN YOUR OWN WORDS . . .

49. Describe the steps in graphing a system of linear inequalities.

9.6 DETERMINANTS AND CRAMER'S RULE (OPTIONAL)

A square array of numbers with a bar on each side, such as

$$\begin{vmatrix} 1 & 0 & 2 \\ -3 & 1 & 5 \\ 0 & 4 & 2 \end{vmatrix}$$

is called a **determinant**. The numbers in the array are called **elements** or **entries**. The number of rows (or columns) of a determinant is called the **order** of the determinant. The example above is a third-order determinant. The following examples are second-order and fourth-order determinants.

$$\begin{vmatrix} 2 & 0 \\ 1 & 3 \end{vmatrix} \qquad \begin{vmatrix} 3 & 0 & 2 & 0 \\ 0 & 9 & 1 & 4 \\ 6 & 0 & -1 & 2 \\ 5 & 1 & 0 & 1 \end{vmatrix}$$

Second-order determinant

Fourth-order determinant

A determinant represents a number just as $\sqrt{4}$ represents a number. To evaluate any determinant, we must first learn how to evaluate a second-order determinant.

Value of a Second-Order Determinant

$$\begin{vmatrix} a & b \\ c & d \end{vmatrix} = ad - bc$$

EXAMPLE 1. Evaluate the following determinants.

(a) $\begin{vmatrix} 1 & 2 \\ 3 & 4 \end{vmatrix}$ (b) $\begin{vmatrix} 2 & 0 \\ 5 & 3 \end{vmatrix}$ (c) $\begin{vmatrix} 3 & -2 \\ 4 & 2 \end{vmatrix}$ (d) $\begin{vmatrix} -2 & 3 \\ -4 & 6 \end{vmatrix}$

Solutions:

(a) $\begin{vmatrix} 1 & 2 \\ 3 & 4 \end{vmatrix} = 1 \cdot 4 - 2 \cdot 3 = 4 - 6 = -2$

(b) $\begin{vmatrix} 2 & 0 \\ 5 & 3 \end{vmatrix} = 2 \cdot 3 - 0 \cdot 5 = 6 - 0 = 6$

(c) $\begin{vmatrix} 3 & -2 \\ 4 & 2 \end{vmatrix} = 3 \cdot 2 - (-2) \cdot 4 = 6 + 8 = 14$

(d) $\begin{vmatrix} -2 & 3 \\ -4 & 6 \end{vmatrix} = (-2) \cdot 6 - 3 \cdot (-4) = -12 + 12 = 0$ □

A determinant of order higher than 2 is evaluated by reducing it in successive steps until it is a sum of second-order determinants. To do this, two more definitions are needed.

The **minor** of an element in a determinant is the determinant that remains when we delete the row and the column of the element. For example, the minor of b in the determinant

$$\begin{vmatrix} a & b & c \\ d & e & f \\ g & h & i \end{vmatrix} \quad \text{is} \quad \begin{vmatrix} d & f \\ g & i \end{vmatrix}$$

and the minor of g in the same determinant

$$\begin{vmatrix} a & b & c \\ d & e & f \\ g & h & i \end{vmatrix} \quad \text{is} \quad \begin{vmatrix} b & c \\ e & f \end{vmatrix}$$

The **cofactor** of an element is the minor of the element with the appropriate sign attached. The appropriate sign is taken from the array of signs as shown below.

$$\begin{array}{ccc} + & - & + \\ - & + & - \\ + & - & + \end{array} \qquad \begin{array}{cccc} + & - & + & - \\ - & + & - & + \\ + & - & + & - \\ - & + & - & + \end{array} \qquad \begin{array}{ccccc} + & - & + & - & + \\ - & + & - & + & - \\ + & - & + & - & + \\ - & + & - & + & - \\ + & - & + & - & + \end{array}$$

Third Order Fourth Order Fifth Order

Notice that the signs alternate, like the squares on a checkerboard, with a "+" in the upper left-hand location.

EXAMPLE 2. Write the cofactor of b in the determinant

$$\begin{vmatrix} a & b & c \\ d & e & f \\ g & h & i \end{vmatrix}$$

Solution:

The minor of b is the determinant that remains when the row and column of b are deleted (the first row and the second column).

$$\begin{vmatrix} a & b & c \\ d & e & f \\ g & h & i \end{vmatrix}$$

$$\text{Minor of } b = \begin{vmatrix} d & f \\ g & i \end{vmatrix}$$

The position of b in the array of signs is as follows:

$$\begin{array}{ccc} + & - & + \\ - & + & - \\ + & - & + \end{array}$$

The cofactor of b is the minor with that sign.

$$\text{Cofactor of } b = -\begin{vmatrix} d & f \\ g & i \end{vmatrix}$$

> **To Evaluate a Determinant of Order Higher than 2**
> 1. Select a row or column to expand about.
> 2. For each element in the chosen row (or column), multiply the element by its cofactor.
> 3. The value of the determinant is the sum of the products found in Step 2.

EXAMPLE 3. Evaluate the following determinant by expanding about the first row.

$$\begin{vmatrix} 2 & 2 & -3 \\ 2 & -1 & 1 \\ 1 & 1 & 0 \end{vmatrix}$$

Solution:

The minors of the elements in the first row are

Step 2

$$\begin{vmatrix} -1 & 1 \\ 1 & 0 \end{vmatrix} \qquad \begin{vmatrix} 2 & 1 \\ 1 & 0 \end{vmatrix} \qquad \begin{vmatrix} 2 & -1 \\ 1 & 1 \end{vmatrix}$$

First element Second element Third element

The cofactors of the elements in the first row are

$$+\begin{vmatrix} -1 & 1 \\ 1 & 0 \end{vmatrix} \qquad -\begin{vmatrix} 2 & 1 \\ 1 & 0 \end{vmatrix} \qquad +\begin{vmatrix} 2 & -1 \\ 1 & 1 \end{vmatrix}$$

The products of the elements times their cofactors are

$$2\begin{vmatrix} -1 & 1 \\ 1 & 0 \end{vmatrix} \qquad -2\begin{vmatrix} 2 & 1 \\ 1 & 0 \end{vmatrix} \qquad -3\begin{vmatrix} 2 & -1 \\ 1 & 1 \end{vmatrix}$$

Therefore, the value of the determinant is given by

Step 3

$$\begin{vmatrix} 2 & 2 & -3 \\ 2 & -1 & 1 \\ 1 & 1 & 0 \end{vmatrix} = 2\begin{vmatrix} -1 & 1 \\ 1 & 0 \end{vmatrix} - 2\begin{vmatrix} 2 & 1 \\ 1 & 0 \end{vmatrix} - 3\begin{vmatrix} 2 & -1 \\ 1 & 1 \end{vmatrix}$$

$$= 2(0 - 1) - 2(0 - 1) - 3(2 - (-1))$$
$$= -2 + 2 - 6 - 3$$
$$= -9$$

Notice that this determinant is easier to evaluate about the *third* row.

$$\begin{vmatrix} 2 & 2 & -3 \\ 2 & -1 & 1 \\ 1 & 1 & 0 \end{vmatrix} = 1\begin{vmatrix} 2 & -3 \\ -1 & 1 \end{vmatrix} - 1\begin{vmatrix} 2 & -3 \\ 2 & 1 \end{vmatrix} + 0\begin{vmatrix} 2 & 2 \\ 2 & -1 \end{vmatrix}$$

$$= 2 - 3 - (2 + 6)$$
$$= 2 - 3 - 8 = -9$$

Sec. 9.6 Determinants and Cramer's Rule (optional)

EXAMPLE 4. Evaluate the following determinant.

$$\begin{vmatrix} 3 & 0 & 2 \\ 1 & 0 & 4 \\ 2 & 1 & 5 \end{vmatrix}$$

Solution:

First, we note the two zeros in the second column and decide to expand about that column.

$$\begin{vmatrix} 3 & 0 & 2 \\ 1 & 0 & 4 \\ 2 & 1 & 5 \end{vmatrix} = -0 + 0 - 1 \begin{vmatrix} 3 & 2 \\ 1 & 4 \end{vmatrix}$$

$$= -(12 - 2) = -10 \qquad \square$$

Evaluation of determinants larger than third order can be very tedious. It pays to look carefully for rows or columns containing zeros.

EXAMPLE 5. Evaluate

$$\begin{vmatrix} 1 & 1 & 0 & 1 \\ 0 & 1 & 1 & 0 \\ 0 & 1 & 3 & 1 \\ 2 & 0 & 1 & 1 \end{vmatrix}$$

Solution:

Notice that the first column has two zero entries. We decide to expand down that column.

$$\begin{vmatrix} 1 & 1 & 0 & 1 \\ 0 & 1 & 1 & 0 \\ 0 & 1 & 3 & 1 \\ 2 & 0 & 1 & 1 \end{vmatrix} = 1\begin{vmatrix} 1 & 1 & 0 \\ 1 & 3 & 1 \\ 0 & 1 & 1 \end{vmatrix} - 0 + 0 - 2\begin{vmatrix} 1 & 0 & 1 \\ 1 & 1 & 0 \\ 1 & 3 & 1 \end{vmatrix}$$

Now we have two third-order determinants to evaluate. We choose the first row to expand each one.

$$= 1\begin{vmatrix} 3 & 1 \\ 1 & 1 \end{vmatrix} - 1\begin{vmatrix} 1 & 1 \\ 0 & 1 \end{vmatrix} - 2\left(1\begin{vmatrix} 1 & 0 \\ 3 & 1 \end{vmatrix} - 0 + 1\begin{vmatrix} 1 & 1 \\ 1 & 3 \end{vmatrix}\right)$$

$$= 3 - 1 - (1 - 0) - 2[(1 - 0) + (3 - 1)]$$

$$= 3 - 1 - 1 - 2(1 + 2)$$

$$= 1 - 2 \cdot 3 = -5 \qquad \square$$

Determinants are an important structure found throughout mathematics. Determinants are used in a formula for the solution of systems of linear equations called **Cramer's rule.**

Cramer's rule is a general method for solving systems of linear equations that uses determinants. Cramer's rule can be used whenever there is the same number of linear equations as there are variables.

> **Cramer's Rule for Two Equations in Two Variables**
>
> The system of equations
> $$\begin{cases} Ax + By = P \\ Cx + Dy = Q \end{cases}$$
> has solution
> $$x = \frac{\begin{vmatrix} P & B \\ Q & D \end{vmatrix}}{\begin{vmatrix} A & B \\ C & D \end{vmatrix}}, \quad y = \frac{\begin{vmatrix} A & P \\ C & Q \end{vmatrix}}{\begin{vmatrix} A & B \\ C & D \end{vmatrix}} \quad \text{provided that} \quad \begin{vmatrix} A & B \\ C & D \end{vmatrix} \neq 0$$

Each equation in the system must first be written in standard form. When using Cramer's rule it is helpful to write the equations of the system carefully, lining up the same variables under one another.

Notice that the determinant in the denominator is made up of the coefficients of the variables of the original equation in their original positions. We call it the **determinant of coefficients.**

Examination of the determinant of coefficients shows that there is a *column* of numbers for each variable.

$$\text{Determinant of coefficients:} \quad \begin{vmatrix} A & B \\ C & D \end{vmatrix}$$

↓column of x-coefficients

column of y-coefficients↑

The determinant in the numerator of the formula for x is the determinant of coefficients with the x-column replaced by the column of constant terms. The formula for y is arranged in a similar manner.

If the determinant of coefficients is zero, the system is either inconsistent or dependent, and Cramer's rule will not work. Solve the system by elimination or substitution.

EXAMPLE 6. Use Cramer's rule to solve the system
$$\begin{cases} 3x + 2y = 4 \\ 4x - y = 2 \end{cases}$$

Solution:

By Cramer's rule,

$$x = \frac{\begin{vmatrix} 4 & 2 \\ 2 & -1 \end{vmatrix}}{\begin{vmatrix} 3 & 2 \\ 4 & -1 \end{vmatrix}} = \frac{4(-1) - 2 \cdot 2}{3(-1) - 2 \cdot 4} = \frac{-4 - 4}{-3 - 8} = \frac{-8}{-11}$$

$$y = \frac{\begin{vmatrix} 3 & 4 \\ 4 & 2 \end{vmatrix}}{\begin{vmatrix} 3 & 2 \\ 4 & -1 \end{vmatrix}} = \frac{3 \cdot 2 - 4 \cdot 4}{3(-1) - 2 \cdot 4} = \frac{6 - 16}{-3 - 8} = \frac{-10}{-11}$$

So the solution set is
$$\left\{\left(\frac{8}{11}, \frac{10}{11}\right)\right\}$$

Notice that the denominator is the same in the calculation of both variables. This observation saves a little work.

EXAMPLE 7. Solve
$$\begin{cases} 2x - 3y = 4 \\ 5x - 6y = 7 \end{cases}$$

Solution:

By Cramer's rule,

$$x = \frac{\begin{vmatrix} 4 & -3 \\ 7 & -6 \end{vmatrix}}{\begin{vmatrix} 2 & -3 \\ 5 & -6 \end{vmatrix}} = \frac{4(-6) - 7(-3)}{2(-6) - 5(-3)} = \frac{-24 + 21}{-12 + 15} = \frac{-3}{3} = -1$$

$$y = \frac{\begin{vmatrix} 2 & 4 \\ 5 & 7 \end{vmatrix}}{3} = \frac{2 \cdot 7 - 4 \cdot 5}{3} = \frac{14 - 20}{3} = \frac{-6}{3} = -2$$

The solution set is
$$\{(-1, -2)\} \qquad \square$$

Cramer's Rule for Three Equations in Three Variables

The system of equations
$$\begin{cases} Ax + By + Cz = P \\ Dx + Ey + Fz = Q \\ Gx + Hy + Iz = R \end{cases}$$

has solutions

$$x = \frac{\begin{vmatrix} P & B & C \\ Q & E & F \\ R & H & I \end{vmatrix}}{\begin{vmatrix} A & B & C \\ D & E & F \\ G & H & I \end{vmatrix}} \qquad y = \frac{\begin{vmatrix} A & P & C \\ D & Q & F \\ G & R & I \end{vmatrix}}{\begin{vmatrix} A & B & C \\ D & E & F \\ G & H & I \end{vmatrix}} \qquad z = \frac{\begin{vmatrix} A & B & P \\ D & E & Q \\ G & H & R \end{vmatrix}}{\begin{vmatrix} A & B & C \\ D & E & F \\ G & H & I \end{vmatrix}}$$

provided that $\begin{vmatrix} A & B & C \\ D & E & F \\ G & H & I \end{vmatrix} \neq 0.$

EXAMPLE 8. Find the solution set for the following system.
$$\begin{cases} 3x - y = 0 \\ y + 2z = -1 \\ x - z = 0 \end{cases}$$

Solution:

First, rewrite the system with the variables "lined up" and with zero coefficients where they occur.

$$\begin{cases} 3x - y + 0z = 0 \\ 0x + y + 2z = -1 \\ x + 0y - z = 0 \end{cases}$$

Next, apply Cramer's rule.

$$x = \frac{\begin{vmatrix} 0 & -1 & 0 \\ -1 & 1 & 2 \\ 0 & 0 & -1 \end{vmatrix}}{\begin{vmatrix} 3 & -1 & 0 \\ 0 & 1 & 2 \\ 1 & 0 & -1 \end{vmatrix}} = \frac{1\begin{vmatrix} -1 & 2 \\ 0 & -1 \end{vmatrix}}{3\begin{vmatrix} 1 & 2 \\ 0 & -1 \end{vmatrix} + 1\begin{vmatrix} 0 & 2 \\ 1 & -1 \end{vmatrix}} = \frac{(-1)(-1) - 2 \cdot 0}{3(-1 - 0) + 0 - 2} = \frac{1}{-5}$$

$$y = \frac{\begin{vmatrix} 3 & 0 & 0 \\ 0 & -1 & 2 \\ 1 & 0 & -1 \end{vmatrix}}{-5} = \frac{3\begin{vmatrix} -1 & 2 \\ 0 & -1 \end{vmatrix}}{-5} = \frac{3(1 - 0)}{-5} = -\frac{3}{5}$$

$$z = \frac{\begin{vmatrix} 3 & -1 & 0 \\ 0 & 1 & -1 \\ 1 & 0 & 0 \end{vmatrix}}{-5} = \frac{1\begin{vmatrix} -1 & 0 \\ 1 & -1 \end{vmatrix}}{-5} = -\frac{1}{5}$$

The solution set is

$$\left\{ \left(-\frac{1}{5}, -\frac{3}{5}, -\frac{1}{5} \right) \right\}$$

Systems with more than three variables can be solved with Cramer's rule. There must be the same number of equations as there are variables, and all the equations must be linear. The system should be written in the standard form shown above for the three-variable case, being sure to insert zeros where necessary. The determinant of coefficients can then be formed by inspection.

Examination of the determinant of coefficients shows that there is a *column* of numbers for each variable. For example, in the three-variable case

↓ column of y-coefficients

$$\begin{vmatrix} A & B & C \\ D & E & F \\ G & H & I \end{vmatrix}$$

Notice that if we were to replace the column of y-coefficients with the column of constant terms (numbers to the right of the equals sign), we would have the determinant in the numerator of the formula for y. The formulas for x and z are formed in a similar manner. By following that pattern, Cramer's rule can be applied to any linear system of n equations in n variables.

Sec. 9.6 Determinants and Cramer's rule (optional)

PROBLEM SET 9.6

Warm-ups

In problems 1 through 12, evaluate the determinants.
For problems 1 through 6, see Example 1.

1. $\begin{vmatrix} 1 & 2 \\ 2 & 5 \end{vmatrix}$
2. $\begin{vmatrix} 2 & 3 \\ 4 & 5 \end{vmatrix}$
3. $\begin{vmatrix} 3 & 2 \\ 6 & 4 \end{vmatrix}$

4. $\begin{vmatrix} -2 & 2 \\ 3 & 5 \end{vmatrix}$
5. $\begin{vmatrix} 3 & 5 \\ -2 & 4 \end{vmatrix}$
6. $\begin{vmatrix} 7 & 5 \\ -3 & -2 \end{vmatrix}$

For problems 7 through 12, see Examples 3 and 4.

7. $\begin{vmatrix} 1 & 1 & 2 \\ 2 & 3 & 1 \\ 1 & -1 & 1 \end{vmatrix}$
8. $\begin{vmatrix} 1 & -1 & 2 \\ 2 & 3 & 0 \\ 1 & 4 & 1 \end{vmatrix}$
9. $\begin{vmatrix} 4 & 2 & 1 \\ 0 & 0 & 3 \\ 3 & 5 & 3 \end{vmatrix}$

10. $\begin{vmatrix} 2 & 1 & 3 \\ 4 & 5 & 7 \\ 2 & 1 & 3 \end{vmatrix}$
11. $\begin{vmatrix} 0 & 1 & 0 \\ 0 & 0 & 1 \\ 2 & 0 & 0 \end{vmatrix}$
12. $\begin{vmatrix} 3 & 2 & -1 \\ 5 & 7 & 8 \\ 4 & -5 & 6 \end{vmatrix}$

In problems 13 through 22, use Cramer's rule to solve each system. For problems 13 through 18, work problems 13 through 18 in Problem Set 9.1. See Examples 6 and 7. For problems 19 through 22, work problems 1 through 4 in Problem Set 9.2. See Example 8.

Practice Exercises

In problems 23 through 40, evaluate the determinants.

23. $\begin{vmatrix} 2 & 1 \\ 5 & 3 \end{vmatrix}$
24. $\begin{vmatrix} 3 & 2 \\ 5 & 4 \end{vmatrix}$
25. $\begin{vmatrix} 2 & 6 \\ 3 & 9 \end{vmatrix}$
26. $\begin{vmatrix} -3 & 3 \\ 2 & 4 \end{vmatrix}$

27. $\begin{vmatrix} 4 & 5 \\ -3 & 3 \end{vmatrix}$
28. $\begin{vmatrix} -2 & -3 \\ 6 & 4 \end{vmatrix}$
29. $\begin{vmatrix} 2 & 1 & 1 \\ 1 & 3 & -1 \\ 1 & -2 & 4 \end{vmatrix}$
30. $\begin{vmatrix} 1 & 2 & 3 \\ 2 & -1 & 4 \\ 3 & 0 & 1 \end{vmatrix}$

31. $\begin{vmatrix} 1 & 2 & 0 \\ -1 & -1 & 0 \\ 3 & 4 & 2 \end{vmatrix}$
32. $\begin{vmatrix} 2 & -1 \\ -5 & -7 \end{vmatrix}$
33. $\begin{vmatrix} 2 & 0 \\ 0 & -3 \end{vmatrix}$
34. $\begin{vmatrix} 5 & 0 \\ -1 & 0 \end{vmatrix}$

35. $\begin{vmatrix} 3 & 3 & 4 \\ 2 & 2 & 1 \\ 1 & 1 & 3 \end{vmatrix}$
36. $\begin{vmatrix} 0 & 3 & 0 \\ 1 & 0 & 0 \\ 0 & 0 & 2 \end{vmatrix}$
37. $\begin{vmatrix} -2 & 3 & 5 \\ 4 & 2 & 3 \\ 7 & -3 & 2 \end{vmatrix}$

38. $\begin{vmatrix} -1 & -5 \\ 3 & -2 \end{vmatrix}$
39. $\begin{vmatrix} 4 & 0 \\ 2 & -6 \end{vmatrix}$
40. $\begin{vmatrix} 0 & 0 \\ -3 & 21 \end{vmatrix}$

For problems 41 through 60, use Cramer's rule, if it applies, to work problems 61 through 72 in Problem Set 9.1 and problems 11 through 18 in Problem Set 9.2.

Challenge Problems

In problems 61 through 63, evaluate the determinants.

61. $\begin{vmatrix} 1 & 1 & 2 & 1 \\ 2 & 1 & 0 & -3 \\ -2 & 0 & 1 & 2 \\ 4 & 0 & 5 & 1 \end{vmatrix}$
62. $\begin{vmatrix} 0 & 1 & 0 & 1 \\ 1 & 1 & 2 & 1 \\ 0 & 0 & 1 & 0 \\ 2 & 3 & 5 & 1 \end{vmatrix}$
63. $\begin{vmatrix} 1 & 1 & 0 & 0 & 0 \\ 0 & 2 & 2 & 0 & 0 \\ 0 & 0 & 3 & 3 & 0 \\ 0 & 0 & 0 & 4 & 4 \\ 5 & 0 & 0 & 0 & 5 \end{vmatrix}$

For problems 64 through 66, use Cramer's rule to work problems 37 through 39 in Problem Set 9.2.

IN YOUR OWN WORDS ...

67. What does a determinant represent?

68. How are third-order determinants evaluated.

9.7 MATRIX METHODS (OPTIONAL)

A **matrix** is a rectangular array of numbers, usually written inside brackets or parentheses. For example,

$$\begin{bmatrix} 3 & -1 & 4 \\ -5 & 0 & 7 \end{bmatrix}$$

The numbers in the array are called **elements** or **entries.** The position of elements in a matrix are identified by a row number and a column number.

$$\text{Row} \begin{array}{c} 1 \\ 2 \\ 3 \end{array} \xrightarrow{\begin{array}{cccc} \text{Column} & 1 & 2 & 3 & 4 \end{array}} \begin{bmatrix} -2 & 4 & -9 & 1 \\ 0 & 7 & 1 & -1 \\ 3 & 3 & 2 & -5 \end{bmatrix}$$

The number -9 in this matrix is in row 1, column 3. The size (or dimension) of a matrix is given by the number of rows followed by the number of columns (rows *always* before columns). The matrix above has 3 rows and 4 columns so it is a 3×4 matrix. The matrix

$$\begin{bmatrix} 3 & -1 & 4 \\ -5 & 0 & 7 \end{bmatrix}$$

is a 2×3 matrix.

There are several methods of solving systems of equations that involve matrices (the plural of "matrix"). We will examine one of them, the **method of augmented matrices.**

Suppose that we have a system of three linear equations in three variables, written in standard form.

$$Ax + By + Cz = P$$
$$Dx + Ey + Fz = Q$$
$$Gx + Hy + Iz = R$$

(Remember, some of the coefficients may be zero.) The matrix made up of the coefficients of the variables, in their proper position,

$$\begin{bmatrix} A & B & C \\ D & E & F \\ G & H & I \end{bmatrix}$$

is called the **matrix of coefficients** of the system. If we add a column on the right side of the matrix containing the constant terms of the system, we have the **augmented matrix** of the system.

$$\begin{bmatrix} A & B & C & | & P \\ D & E & F & | & Q \\ G & H & I & | & R \end{bmatrix}$$

EXAMPLE 1. Write the augmented matrix of the system
$$\begin{cases} 2x + 3y + 4z = 5 \\ 6x + 7y + 8z = 9 \\ -x + 2y - 3z = 4 \end{cases}$$

Solution:

We write the matrix by inspection.
$$\begin{bmatrix} 2 & 3 & 4 & | & 5 \\ 6 & 7 & 8 & | & 9 \\ -1 & 2 & -3 & | & 4 \end{bmatrix}$$

EXAMPLE 2. Write the augmented matrix of the system
$$\begin{cases} 2x - z = 3 \\ x + 5y = -7 \\ 6y - 5z + 4 = 0 \end{cases}$$

Solution:

First, we write the system in standard form, inserting zeros where necessary.
$$\begin{cases} 2x + 0y - z = 3 \\ x + 5y + 0z = -7 \\ 0x + 6y - 5z = -4 \end{cases}$$

Now we can write the augmented matrix.
$$\begin{bmatrix} 2 & 0 & -1 & | & 3 \\ 1 & 5 & 0 & | & -7 \\ 0 & 6 & -5 & | & -4 \end{bmatrix}$$

EXAMPLE 3. Write a system of equations that corresponds to the augmented matrix
$$\begin{bmatrix} 1 & 2 & 3 & | & 4 \\ 3 & -2 & 0 & | & -1 \\ 0 & 1 & -1 & | & 0 \end{bmatrix}$$

Solution:

We copy the coefficients from the augmented matrix and form the system
$$\begin{cases} 1x + 2y + 3z = 4 \\ 3x - 2y + 0z = -1 \\ 0x + 1y - 1z = 0 \end{cases}$$

which simplifies to
$$\begin{cases} x + 2y + 3z = 4 \\ 3x - 2y = -1 \\ y - z = 0 \end{cases}$$

We say two augmented matrices are **equivalent** if the systems they represent have the same solution set.

> ### Elementary Row Operations
> There are three operations that can be performed on an augmented matrix that will produce an equivalent matrix.
>
> 1. Exchange any two rows.
> 2. Multiply the numbers of any row by the same nonzero number.
> 3. Multiply the numbers in any row by a real number and add the result to any *other* row.

For example, the following two augmented matrices are equivalent.

$$\begin{bmatrix} 3 & 1 & -7 & | & -2 \\ 1 & 4 & 4 & | & 3 \\ 2 & -1 & 6 & | & 0 \end{bmatrix} \qquad \begin{bmatrix} 1 & 4 & 4 & | & 3 \\ 3 & 1 & -7 & | & -2 \\ 2 & -1 & 6 & | & 0 \end{bmatrix}$$

Matrix A $\qquad\qquad$ Matrix B

Matrix B was obtained from matrix A by exchanging rows 1 and 2. The following matrices are also equivalent.

$$\begin{bmatrix} 1 & 3 & | & -2 \\ -2 & -4 & | & 2 \end{bmatrix} \qquad \begin{bmatrix} 1 & 3 & | & -2 \\ 1 & 2 & | & -1 \end{bmatrix}$$

Matrix C $\qquad\qquad$ Matrix D

Matrix D was obtained from matrix C by multiplying row 2 by $-\frac{1}{2}$.

$$\begin{bmatrix} 1 & 1 & 3 & 2 & | & 2 \\ 2 & 4 & 1 & -5 & | & 3 \\ 0 & 0 & -3 & -1 & | & -1 \\ 0 & 5 & -6 & 2 & | & 2 \end{bmatrix} \qquad \begin{bmatrix} 1 & 1 & 3 & 2 & | & 2 \\ 0 & 2 & -5 & -9 & | & -1 \\ 0 & 0 & -3 & -1 & | & -1 \\ 0 & 5 & -6 & 2 & | & 2 \end{bmatrix}$$

Matrix E $\qquad\qquad\qquad$ Matrix F

Matrices E and F are also equivalent. Matrix F was obtained by multiplying the first row of matrix E by -2 and adding the result to the second row of matrix E.

Examine the following matrices. They are examples of matrices in **triangular form.**

$$\begin{bmatrix} 1 & 2 & -3 \\ 0 & 1 & 7 \end{bmatrix} \qquad \begin{bmatrix} 1 & -9 & 2 & 0 \\ 0 & 1 & 3 & -4 \\ 0 & 0 & 1 & 2 \end{bmatrix} \qquad \begin{bmatrix} 1 & -7 & -1 & 8 & 6 \\ 0 & 1 & 4 & -1 & 0 \\ 0 & 0 & 1 & 1 & -2 \\ 0 & 0 & 0 & 1 & 4 \end{bmatrix}$$

Notice the number 1 down the main diagonal. Notice that all elements *under* the main diagonal are zeros.

Sec. 9.7 Matrix Methods (optional)

The idea of the method of augmented matrices is to form the augmented matrix of the system to be solved, then write a series of equivalent matrices, ending with a matrix in triangular form. An example will illustrate this.

EXAMPLE 4. Solve the system
$$\begin{cases} 2x - y - z = 5 \\ x + 2y + 3z = -2 \\ 3x - 2y + z = 2 \end{cases}$$

Solution:

First, we write the augmented matrix of the system.

$$\begin{bmatrix} 2 & -1 & -1 & | & 5 \\ 1 & 2 & 3 & | & -2 \\ 3 & -2 & 1 & | & 2 \end{bmatrix}$$

Now, to find an equivalent matrix in the triangular form shown above, it is necessary to get a 1 in the upper left-hand corner. This could be done by multiplying row 1 by $\frac{1}{2}$. However, we can avoid fractions by exchanging rows 1 and 2.

$$\begin{bmatrix} 2 & -1 & -1 & | & 5 \\ 1 & 2 & 3 & | & -2 \\ 3 & -2 & 1 & | & 2 \end{bmatrix} \xrightarrow{\text{Exchange rows 1 and 2}} \begin{bmatrix} 1 & 2 & 3 & | & -2 \\ 2 & -1 & -1 & | & 5 \\ 3 & -2 & 1 & | & 2 \end{bmatrix}$$

The next step is to get the zeros in column 1. Notice that row 1 multiplied by -2 is

$$-2 \quad -4 \quad -6 \quad 4$$

If we add that to row 2, we get

$$\begin{bmatrix} 1 & 2 & 3 & | & -2 \\ 2 & -1 & -1 & | & 5 \\ 3 & -2 & 1 & | & 2 \end{bmatrix} \xrightarrow{\text{Multiply row 1 by } -2 \text{ and add to row 2}} \begin{bmatrix} 1 & 2 & 3 & | & -2 \\ 0 & -5 & -7 & | & 9 \\ 3 & -2 & 1 & | & 2 \end{bmatrix}$$

We get the zero in row 3 by a similar operation.

$$\begin{bmatrix} 1 & 2 & 3 & | & -2 \\ 0 & -5 & -7 & | & 9 \\ 3 & -2 & 1 & | & 2 \end{bmatrix} \xrightarrow{\text{Multiply row 1 by } -3 \text{ and add to row 3}} \begin{bmatrix} 1 & 2 & 3 & | & -2 \\ 0 & -5 & -7 & | & 9 \\ 0 & -8 & -8 & | & 8 \end{bmatrix}$$

Next, we need a 1 in row 2, column 2. Again we can avoid fractions by exchanging rows 2 and 3.

$$\begin{bmatrix} 1 & 2 & 3 & | & -2 \\ 0 & -5 & -7 & | & 9 \\ 0 & -8 & -8 & | & 8 \end{bmatrix} \xrightarrow{\text{Exchange rows 2 and 3}} \begin{bmatrix} 1 & 2 & 3 & | & -2 \\ 0 & -8 & -8 & | & 8 \\ 0 & -5 & -7 & | & 9 \end{bmatrix}$$

$$\begin{bmatrix} 1 & 2 & 3 & | & -2 \\ 0 & -8 & -8 & | & 8 \\ 0 & -5 & -7 & | & 9 \end{bmatrix} \xrightarrow{\text{Multiply row 2 by } \frac{-1}{8}} \begin{bmatrix} 1 & 2 & 3 & | & -2 \\ 0 & 1 & 1 & | & -1 \\ 0 & -5 & -7 & | & 9 \end{bmatrix}$$

We finish column 2 by multiplying row 2 by 5 and adding it to row 3.

$$\begin{bmatrix} 1 & 2 & 3 & | & -2 \\ 0 & 1 & 1 & | & -1 \\ 0 & -5 & -7 & | & 9 \end{bmatrix} \xrightarrow{\text{Multiply row 2 by 5 and add to row 3}} \begin{bmatrix} 1 & 2 & 3 & | & -2 \\ 0 & 1 & 1 & | & -1 \\ 0 & 0 & -2 & | & 4 \end{bmatrix}$$

Now, if we multiply row 3 by $-\dfrac{1}{2}$, we are finished.

$$\begin{bmatrix} 1 & 2 & 3 & | & -2 \\ 0 & 1 & 1 & | & -1 \\ 0 & 0 & -2 & | & 4 \end{bmatrix} \xrightarrow{\text{Multiply row 3 by } \dfrac{-1}{2}} \begin{bmatrix} 1 & 2 & 3 & | & -2 \\ 0 & 1 & 1 & | & -1 \\ 0 & 0 & 1 & | & -2 \end{bmatrix}$$

The reason we wanted the triangular form can be seen when we write the system for this augmented matrix.

$$\begin{cases} x + 2y + 3z = -2 \\ y + z = -1 \\ z = -2 \end{cases}$$

We automatically have the value for z, and we can find the values for y and then x in successive steps.

$$y + z = -1$$
$$y + (-2) = -1$$
$$y = 1$$

$$x + 2y + 3z = -2$$
$$x + 2(1) + 3(-2) = -2$$
$$x + 2 - 6 = -2$$
$$x = 2$$

So we have the solution set,

$$\{(2, 1, -2)\}$$

The Method of Augmented Matrices

1. Write the system in standard form, inserting zeros where necessary. Form the augmented matrix.
2. Using elementary row operations, change the matrix to triangular form. It is *important* to make the changes in the following order.
   a) Get the number 1 in row 1, column 1.
   b) Get zeros in column 1, under the 1 in row 1.
   c) Get 1 in row 2, column 2.
   d) Get zeros in column 2, under the 1 in row 2.
   e) Continue steps similar to c) and d) until the matrix is in triangular form.
3. Write the system associated with the triangular matrix.
4. Find the values of each of the variables in turn.
5. Write the solution set.

EXAMPLE 5. Solve the system

$$\begin{cases} 2x + 3y = -1 \\ 3x + 2y = 1 \end{cases}$$

Solution:

$$\begin{bmatrix} 2 & 3 & | & -1 \\ 3 & 2 & | & 1 \end{bmatrix} \quad \text{the augmented matrix}$$

$$\begin{bmatrix} 2 & 3 & | & -1 \\ 3 & 2 & | & 1 \end{bmatrix} \xrightarrow{\text{Multiply row 1 by } \frac{1}{2}} \begin{bmatrix} 1 & \frac{3}{2} & | & -\frac{1}{2} \\ 3 & 2 & | & 1 \end{bmatrix}$$

$$\begin{bmatrix} 1 & \frac{3}{2} & | & -\frac{1}{2} \\ 3 & 2 & | & 1 \end{bmatrix} \xrightarrow{\text{Multiply row 1 by } -3 \text{ and add to row 2}} \begin{bmatrix} 1 & \frac{3}{2} & | & -\frac{1}{2} \\ 0 & -\frac{5}{2} & | & \frac{5}{2} \end{bmatrix}$$

$$\begin{bmatrix} 1 & \frac{3}{2} & | & -\frac{1}{2} \\ 0 & -\frac{5}{2} & | & \frac{5}{2} \end{bmatrix} \xrightarrow{\text{Multiply row 2 by } \frac{-2}{5}} \begin{bmatrix} 1 & \frac{3}{2} & | & -\frac{1}{2} \\ 0 & 1 & | & -1 \end{bmatrix}$$

The matrix is in triangular form. The associated system is

$$x + \frac{3}{2}y = -\frac{1}{2}$$

$$y = -1$$

and x is obtained from

$$x + \frac{3}{2} \cdot (-1) = -\frac{1}{2}$$

$$x - \frac{3}{2} = -\frac{1}{2}$$

$$x = -\frac{1}{2} + \frac{3}{2} = \frac{2}{2}$$

$$x = 1$$

The solution set is

$$\{(1, -1)\} \qquad \square$$

EXAMPLE 6. Solve

$$\begin{cases} 3x - 4y + 3z = 2 \\ x - y + z = 1 \\ x - 2y + z = 0 \end{cases}$$

Solution:

Write the augmented matrix.

$$\begin{bmatrix} 3 & -4 & 3 & | & 2 \\ 1 & -1 & 1 & | & 1 \\ 1 & -2 & 1 & | & 0 \end{bmatrix} \quad \text{the augmented matrix}$$

$$\begin{bmatrix} 3 & -4 & 3 & | & 2 \\ 1 & -1 & 1 & | & 1 \\ 1 & -2 & 1 & | & 0 \end{bmatrix} \xrightarrow{\text{Exchange rows 1 and 2}} \begin{bmatrix} 1 & -1 & 1 & | & 1 \\ 3 & -4 & 3 & | & 2 \\ 1 & -2 & 1 & | & 0 \end{bmatrix}$$

$$\begin{bmatrix} 1 & -1 & 1 & | & 1 \\ 3 & -4 & 3 & | & 2 \\ 1 & -2 & 1 & | & 0 \end{bmatrix} \xrightarrow{\substack{\text{Multiply row 1 by} \\ -3 \text{ and add to row 2}}} \begin{bmatrix} 1 & -1 & 1 & | & 1 \\ 0 & -1 & 0 & | & -1 \\ 1 & -2 & 1 & | & 0 \end{bmatrix}$$

$$\begin{bmatrix} 1 & -1 & 1 & | & 1 \\ 0 & -1 & 0 & | & -1 \\ 1 & -2 & 1 & | & 0 \end{bmatrix} \xrightarrow{\substack{\text{Multiply row 1 by} \\ -1 \text{ and add to row 3}}} \begin{bmatrix} 1 & -1 & 1 & | & 1 \\ 0 & -1 & 0 & | & -1 \\ 0 & -1 & 0 & | & -1 \end{bmatrix}$$

$$\begin{bmatrix} 1 & -1 & 1 & | & 1 \\ 0 & -1 & 0 & | & -1 \\ 0 & -1 & 0 & | & -1 \end{bmatrix} \xrightarrow{\substack{\text{Multiply row} \\ 2 \text{ by } -1}} \begin{bmatrix} 1 & -1 & 1 & | & 1 \\ 0 & 1 & 0 & | & 1 \\ 0 & -1 & 0 & | & -1 \end{bmatrix}$$

$$\begin{bmatrix} 1 & -1 & 1 & | & 1 \\ 0 & 1 & 0 & | & 1 \\ 0 & -1 & 0 & | & -1 \end{bmatrix} \xrightarrow{\substack{\text{Multiply row 2 by} \\ 1 \text{ and add to row 3}}} \begin{bmatrix} 1 & -1 & 1 & | & 1 \\ 0 & 1 & 0 & | & 1 \\ 0 & 0 & 0 & | & 0 \end{bmatrix}$$

Notice the complete row of zeros, which will not allow us to finish the triangular form. This indicates that the original system was dependent.

As noted in the example above, a complete row of zeros anywhere in the matrix signals a dependent system. A row of all zeros *except* a nonzero entry in the last column signals an inconsistent system. That is, rows anywhere in the matrix, such as

$[0 \quad 0 \quad 0 \quad | \quad 5] \quad \text{or} \quad [0 \quad 0 \quad | \quad -2] \quad \text{or} \quad [0 \quad 0 \quad 0 \quad 0 \quad | \quad 19]$

indicate inconsistent systems.

PROBLEM SET 9.7

Warm-ups

In problems 1 through 6, write the augmented matrix for the system of linear equations. See Examples 1 and 2.

1. $\begin{cases} 2x + 3y = -4 \\ 5x - 6y = 1 \end{cases}$

2. $\begin{cases} 3x - y = 2 \\ x + 7y + 3 = 0 \end{cases}$

3. $\begin{cases} 2x - 3y + 4z = 1 \\ 5x - y + z = 0 \\ 7x + 6y - 2z = 5 \end{cases}$

4. $\begin{cases} x + 2y = -5 \\ 3x - 4z = 1 \\ y + 2z = 10 \end{cases}$

Sec. 9.7 Matrix Methods (optional)

5. $\begin{cases} x + 2y = z \\ -x - 11 = y \\ y - 13 = 4z \end{cases}$

6. $\begin{cases} 2x + 3y - 4z + w = 17 \\ 5x + 6z - w = 8 \\ 7x - 2y - 9w = 5 \\ y + 2z + 2w = -1 \end{cases}$

In problems 7 through 10, write the linear system associated with the matrix. See Example 3.

7. $\begin{bmatrix} 1 & 2 & | & 4 \\ 9 & -8 & | & -1 \end{bmatrix}$

8. $\begin{bmatrix} 1 & -1 & 0 & | & 7 \\ 0 & 1 & 1 & | & 1 \\ 0 & 0 & 1 & | & -4 \end{bmatrix}$

9. $\begin{bmatrix} 1 & 0 & 0 & | & 5 \\ 0 & 1 & 0 & | & -3 \\ 0 & 0 & 1 & | & -2 \end{bmatrix}$

10. $\begin{bmatrix} 1 & -2 & 3 & -4 & | & 5 \\ 4 & -1 & 1 & 0 & | & 2 \\ 2 & 1 & 0 & 0 & | & 1 \\ -1 & 1 & -1 & 1 & | & -1 \end{bmatrix}$

In problems 11 through 16, solve the systems by the method of augmented matrices. See Examples 4, 5, and 6.

11. $\begin{cases} x + 3y = 7 \\ -2x + y = 0 \end{cases}$

12. $\begin{cases} 2x + 5y = 8 \\ x + 2y = 3 \end{cases}$

13. $\begin{cases} 2x + 4y = 2 \\ 3x + 7y = 1 \end{cases}$

14. $\begin{cases} 2x + 5y = 4 \\ 3x - 6y = 33 \end{cases}$

15. $\begin{cases} x + 2y + 2z = 3 \\ 2x + 3y + 6z = 2 \\ -x + y + z = 0 \end{cases}$

16. $\begin{cases} 2x + 5y + 2z = 9 \\ x + 3y - z = 0 \\ 2x + 3y - 3z = 1 \end{cases}$

Practice Exercises

In problems 17 through 22, write the augmented matrix for the system of linear equations.

17. $\begin{cases} 3x + 2y = 5 \\ 6x - 5y = 3 \end{cases}$

18. $\begin{cases} 2x - 2y = 1 \\ -x + 1y - 2 = 0 \end{cases}$

19. $\begin{cases} 3x - 2y + 5z = 4 \\ 2x + 3y - z = 0 \\ x - y - 2z = 1 \end{cases}$

20. $\begin{cases} x - 3y = 4 \\ 2x - z = 3 \\ -y - 2z = 13 \end{cases}$

21. $\begin{cases} x + 3z = 2y \\ -x - 10 = z \\ y + 3 = 4x - z \end{cases}$

22. $\begin{cases} x - 3y + 4z + 2w = 11 \\ 4x - 5z + 4w = 7 \\ 5x - y - w = 13 \\ -x + 4y + 2z = 9 \end{cases}$

In problems 23 through 26, write the linear system associated with the matrix.

23. $\begin{bmatrix} 2 & 1 & | & 3 \\ 6 & -5 & | & 11 \end{bmatrix}$

24. $\begin{bmatrix} 1 & 0 & -1 & | & 6 \\ 0 & 1 & -1 & | & 9 \\ 0 & 0 & 1 & | & -6 \end{bmatrix}$

25. $\begin{bmatrix} 0 & 0 & 1 & | & -3 \\ 0 & 1 & 0 & | & 4 \\ 1 & 0 & 0 & | & 2 \end{bmatrix}$

26. $\begin{bmatrix} 1 & 0 & 0 & 1 & | & -3 \\ 0 & 1 & -3 & 0 & | & 2 \\ 0 & 0 & 1 & -2 & | & -5 \\ 0 & 0 & 0 & 1 & | & 2 \end{bmatrix}$

In problems 27 through 36, solve each system by the method of augmented matrices.

27. $\begin{cases} x - 2y = -1 \\ -3x + 8y = 5 \end{cases}$

28. $\begin{cases} -3x + y = 2 \\ x + 3y = -4 \end{cases}$

29. $\begin{cases} 3x - 9y = 3 \\ 2x - 5y = 4 \end{cases}$

30. $\begin{cases} 3x - 7y = 1 \\ -2x + 10y = 10 \end{cases}$

31. $\begin{cases} x + 3y - 2z = 1 \\ 2x + 5y - 2z = 6 \\ -2x - 4y + 3z = -1 \end{cases}$

32. $\begin{cases} 5x - 9y - 7z = 7 \\ 3x - 4y + z = 7 \\ x - 2y - z = 1 \end{cases}$

33. $\begin{cases} 2x - 6y + 4z = 0 \\ 3x + y - 4z = 10 \\ 4x + 3y - 9z = 13 \end{cases}$

34. $\begin{cases} 2x - 3y - 2z = -3 \\ 3x - 2y + 5z = 14 \\ 3x + 6y - 3z = 6 \end{cases}$

35. $\begin{cases} 3x + 9y - 6z = -15 \\ 2x + 3y + 2z = 2 \\ 5x + 8y + 14z = 13 \end{cases}$

36. $\begin{cases} -3x + 5y + 2z = 5 \\ 2x + 3y = 13 \\ 5x - 5y + 10z = -25 \end{cases}$

For problems 37 through 56, solve the systems in problems 61 through 72 in Problem Set 9.1, and problems 11 through 18 in Problem Set 9.2, by the method of augmented matrices.

Challenge Problems

For problems 57 through 60, use the method of augmented matrices to work problems 37 through 40 in Problem Set 9.2.

IN YOUR OWN WORDS . . .

61. How do we form the augmented matrix of a system of linear equations?

62. What are the elementary row operations?

CHAPTER SUMMARY

GLOSSARY

System of equations: Two or more equations considered simultaneously.

A **solution** of a system of equations with *n* variables: An ordered *n*-tuple of numbers that satisfies all the equations in the system.

Inconsistent system: A system of equations that has no solutions.

Dependent system: A system of equations that has an infinite number of solutions.

Boundary line: A line separating the regions of an inequality in two variables.

OPTIONAL SECTIONS

Determinant: A square array of numbers that represents a number.

Matrix: A rectangular array of numbers.

Augmented matrix: A matrix that represents a system of equations.

Sec. 9.7   Matrix Methods (optional)

SYSTEMS OF EQUATIONS

Graphical Method

Graph both equations and estimate the coordinates of their point(s) of intersection. This method provides an approximation to the solution(s).

Method of Substitution

1. Solve one of the equations for one of the variables.
2. Substitute this for the same variable in the other equation.
3. Solve the resulting equation.
4. Substitute the result back into step 1 to find the other variable.
5. Check solution(s), if required.
6. Write the solution set.

Method of Elimination

1. Write both equations in standard form, $Ax + By = C$.
2. Multiply both sides of each equation by a suitable real number so that one of the variables will be eliminated by addition of the equations. (This step may not be necessary.)
3. Add the equations and solve the resulting equation.
4. Substitute the value found in step 3 into one of the original equations, and solve this equation.
5. Check the solution, if required.
6. Write the solution set.

INEQUALITIES IN TWO VARIABLES

1. Graph the equation formed by replacing the inequality symbol with an equals symbol.
   (a) Draw a solid line if equality is included. ($\geq$, $\leq$)
   (b) Draw a dashed line if equality is not included. ($>$, $<$)
2. Determine which region(s) formed by the line make the inequality true by testing with one point from inside each region.
3. Shade the region(s) that make the inequality true.

SYSTEMS OF INEQUALITIES IN TWO VARIABLES

1. Graph each inequality on the same coordinate system, lightly shading its solution set.
2. Darken the *intersection* of the lightly shaded regions.
3. Change any portion of any solid boundary line not in the intersection of the solutions from solid to dashed.

CHECKUPS

1. Use the method of substitution to solve

$$\begin{cases} 2x + y = 2 \\ 6x - 4y = -1 \end{cases}$$

Section 9.1; Example 1a

2. Use the method of elimination to solve

$$\begin{cases} 2x + 5y = 11 \\ 3x + 7y = 15 \end{cases}$$

Section 9.1; Example 7

3. Solve the system

$$\begin{cases} x + 2y + z = 0 \\ 2x - 2y + z = -2 \\ 3x + 2y - 2z = 3 \end{cases}$$

Section 9.2; Example 2

4. Gloria bought 3 lb of raw peanuts and 2 lb of boiled peanuts from Big Bob's produce stand for $4.05. Jim bought 1 lb of raw and 4 lb of boiled for $4.65. What is the price per pound for raw and boiled peanuts at Big Bob's? Section 9.3; Example 2

5. Graph the solution set for $x - 2y > 0$. Section 9.4; Example 1

6. Graph the solution set for the system of inequalities
$$\begin{cases} x - y \le 1 \\ x + y > 0 \end{cases}$$
Section 9.5; Example 1

OPTIONAL SECTIONS

7. Use Cramer's rule to solve
$$\begin{cases} 3x + 2y = 4 \\ 4x - y = 2 \end{cases}$$
Section 9.6; Example 6

8. Solve by the method of augmented matrices.
$$\begin{cases} 2x + 3y = -1 \\ 3x + 2y = 1 \end{cases}$$
Section 9.7; Example 5

REVIEW PROBLEMS

In problems 1 through 6, use the method of substitution to find the solution set for each system.

1. $\begin{cases} 2x + y = 4 \\ 3x - 2y = -1 \end{cases}$
2. $\begin{cases} 2x - 3y = 7 \\ x - 2y = 4 \end{cases}$
3. $\begin{cases} 5x + 3y = 1 \\ 7x + 2y = 8 \end{cases}$
4. $\begin{cases} 6x - 3y = 2 \\ 9x + 4y = 3 \end{cases}$
5. $\begin{cases} x^2 - 4y^2 = 9 \\ x - y = 3 \end{cases}$
6. $\begin{cases} x - 6y = 5 \\ xy = 11 \end{cases}$

In problems 7 through 25, use the method of elimination to find the solution set for each system.

7. $\begin{cases} x + 2y = 2 \\ x - 2y = 6 \end{cases}$
8. $\begin{cases} 3x + y = 1 \\ 4x + y = -1 \end{cases}$
9. $\begin{cases} 2x + 3y = 15 \\ 2x - 7y = 5 \end{cases}$
10. $\begin{cases} 2x - 3y = -5 \\ 3x + 2y = -1 \end{cases}$
11. $\begin{cases} 5x + 6y = -3 \\ -4x + 9y = 7 \end{cases}$
12. $\begin{cases} 5x + 4y = 12 \\ -6x + 8y = -8 \end{cases}$
13. $\begin{cases} 24x - 18y = -1 \\ 6x - 10y = -3 \end{cases}$
14. $\begin{cases} -x + 3y = 4 \\ 2x - 6y = -8 \end{cases}$
15. $\begin{cases} 5x + 4y = 2 \\ 10x + 12y = 5 \end{cases}$
16. $\begin{cases} x + y - z = -3 \\ x - y + z = -1 \\ x + y + z = -1 \end{cases}$
17. $\begin{cases} x - 2y + z = 2 \\ x + y + z = 8 \\ x - y - z = 2 \end{cases}$
18. $\begin{cases} x - y - z = 0 \\ x + 2y - z = 3 \\ x + y + 2z = 5 \end{cases}$
19. $\begin{cases} 2x + 3y + 5z = 0 \\ x - 2y - z = 0 \\ 3x + 2y - 2z = 0 \end{cases}$
20. $\begin{cases} x - 2y + z = 17 \\ 2x + 3y + 2z = 13 \\ 3x + y - z = 14 \end{cases}$
21. $\begin{cases} 2x + 5y - z = -6 \\ x - y + 3z = 18 \\ 2x + 3y - 2z = -12 \end{cases}$
22. $\begin{cases} 3x - y + 4z = 9 \\ x - 2y + 3z = -2 \\ 2x - y + z = 3 \end{cases}$
23. $\begin{cases} x + 3y + z = 1 \\ 2x + y + 2z = 2 \\ 3x - y + 2z = 0 \end{cases}$
24. $\begin{cases} 2x - 3y + 4z = 4 \\ 3x - y + z = -2 \\ x + 2y - 3z = -6 \end{cases}$
25. $\begin{cases} x - y + 3z = 6 \\ 2x + y - z = -3 \\ 3x - 2y + z = -4 \end{cases}$

26. The sum of two numbers is 12. Their difference is 2. Find the numbers.

27. Jane, Brenda, and Virginia bought melons at Hasty-Mart on the same day. Jane bought 3 cantaloupes, 4 honeydews, and 1 watermelon for $11.92. Brenda bought 1 cantaloupe, 1 honeydew, and 1 watermelon for $4.77. Virginia bought 1 cantaloupe and 2 watermelons for $5.17. What was the price of each type of melon?

28. Jon and Tamra know that they can paint their house in 8 hours if they work together. This year Jon painted alone for 6 hours, then Tamra finished it in 12 hours. How long would it have taken each working alone to paint the house?

29. The sum of the squares of two numbers is 58 and their difference is 10. Find one such pair of numbers.

In problems 30 through 53, graph the solution set on the x,y plane.

30. $x - y \leq 4$

31. $x + 2y \geq 3$

32. $2x - y \geq 3$

33. $x - y < 0$

34. $2x + y < 1$

35. $y < 3x - 2$

36. $\begin{cases} x - y \leq 2 \\ x + y \geq 0 \end{cases}$

37. $\begin{cases} 2x + y < 7 \\ x - y > 5 \end{cases}$

38. $\begin{cases} 2x + y \leq 3 \\ x - 2y > 9 \end{cases}$

39. $\begin{cases} y \leq 2 \\ x \geq 0 \end{cases}$

40. $y > x$

41. $2x \leq y + 3$

42. $x + 2y < 3$

43. $y < 2x$

44. $\begin{cases} 2x + y < 7 \\ 4x + 2y < 14 \end{cases}$

45. $\begin{cases} x - y > 7 \\ x - y < 5 \end{cases}$

46. $2x + 3y \leq 6$

47. $2x - 3y \leq 4$

48. $x - 3y \leq 1$

49. $5x - 7y > 0$

50. $3x + 2y \leq 4$

51. $\dfrac{1}{2}x - y < 1$

52. $\begin{cases} 3x + 2y \leq 1 \\ x - y \leq 2 \end{cases}$

53. $\begin{cases} 2x - 5y \leq 5 \\ 4x - 10y \leq 0 \end{cases}$

Cramer's Rule

For problems 54 through 72, use Cramer's rule to solve the systems in problems 7 through 25 above.

Augmented Matrices

For problems 73 through 91, use the method of augmented matrices to solve the systems in problems 7 through 25 above.

... LET'S NOT FORGET ...

Identify the expressions that are in factored form. Factor those that are not factored.

92. $125 - 45s^2$

93. $(4 - d)^3$

94. $(x - 2)(x^2 + 2x + 4)$

95. $x^3 + 8$

How many terms are in each expression? Which expressions have $x + 2$ as a factor?

96. $x^4 - 16$

97. $1 + (x + 2)^3$

98. $8 + x^3$

99. $ax - bx + 2a - 2b$

Simplify each expression, if possible. Leave only nonnegative exponents in the answer.

100. -4^{-2}

101. -2^{-4}

102. $(-3)^{-2}$

103. $(-2)^{-3}$

104. $\dfrac{-2k^{-2}}{3b^3 c^{-1}}$

105. $\left(\dfrac{8x^{-6}y^9}{27y^3}\right)^{-1/3}$

Reduce, if possible.

106. $\dfrac{4 \pm \sqrt{8}}{2}$

107. $\dfrac{a - (x + b)^2}{a^2 + (x + b)^2}$

The following problems can be worked by using a least common denominator. Follow the directions in each and notice how the LCD is used.

108. Perform the operation indicated: $\dfrac{5}{x + 1} + \dfrac{4x - 1}{1 - x^2}$.

109. Solve $\dfrac{1}{1 - x} + \dfrac{2}{x^2 - 1} = 1$.

110. Simplify $\dfrac{\dfrac{1}{x + 1} - \dfrac{1}{x - 1}}{\dfrac{1}{x + 1} + \dfrac{1}{x - 1}}$.

Classify each problem as an expression, equation, or inequality. Solve the equations and inequalities, and perform the operations indicated on the expressions.

111. $|5 - x| < 3$

112. $(-x)^2 = -x^2 + 2$

113. $\sqrt[3]{8x^3 + 64x^6}$

114. $\dfrac{x^2 + 1}{x^2 - 2} + \dfrac{x^2 - 1}{2 - x^2}$

Sec. 9.7 Matrix Methods (optional)

CHAPTER TEST

In problems 1 through 5, choose the correct answer.

1. The solution set for the system $\begin{cases} 3x + y = 1 \\ y = x - 1 \end{cases}$ is (?)

   A. $\emptyset$ B. $\{0\}$
   C. $\{(0, -1)\}$ D. $\left\{\left(\dfrac{1}{2}, -\dfrac{1}{2}\right)\right\}$

2. The solution set for the system $\begin{cases} 3x + 2y = -1 \\ 3x - 4y = 11 \end{cases}$ is (?)

   A. $\{(-1, -2)\}$ B. $\{(1, -2)\}$
   C. $\emptyset$ D. $\left\{\left(\dfrac{13}{9}, \dfrac{5}{3}\right)\right\}$

3. The solution set for the system $\begin{cases} x + y = -1 \\ y + z = 1 \\ x + z = 4 \end{cases}$ is (?)

   A. $\emptyset$ B. $\{(0, -1, 4)\}$
   C. $\{(2, -3, 4)\}$ D. $\{(1, -2, 3)\}$

4. The system $\begin{cases} x + y = -1 \\ 2x + 4y + z = 1 \\ 2y + z = 1 \end{cases}$ is best described as (?)

   A. dependent B. inconsistent
   C. independent D. consistent

5. The solution set for the system $\begin{cases} 2x + 3y = 1 \\ 4x + 6y = 0 \end{cases}$ is (?)

   A. $\{(-1, 1)\}$ B. $\{(0, -2)\}$
   C. $\emptyset$ D. None of the above

In problems 6 through 9, sketch the graph.

6. $x + 2y \leq 2$

7. $2x - y > 0$

8. $\begin{cases} x - 2y \geq -2 \\ 2x - y \geq 0 \end{cases}$

9. $\begin{cases} x + y \geq 1 \\ x < 1 \end{cases}$

10. Solve the system
    $$\begin{cases} x + y = 3 \\ x^2 + y^2 = 17 \end{cases}$$

11. Use the method of *substitution* to solve the system
    $$\begin{cases} 3x - 5y = 6 \\ 6x + y = 1 \end{cases}$$

12. Use the method of *elimination* to solve the system
    $$\begin{cases} 7x + 2y = 3 \\ 9x + 3y = 3 \end{cases}$$

13. Find the solution set for the system
    $$\begin{cases} y = 3y - 6 \\ 9x - 3y = 18 \end{cases}$$

14. Find the solution set for the system
    $$\begin{cases} 3x + 2y = 1 \\ y - 3z = -10 \\ x + y + z = 3 \end{cases}$$

15. If 3 lb of bananas and 2 lb of apples cost $2.97, while one lb of apples and 5 lb of bananas cost $2.64, how much does a pound of bananas cost?

Optional Sections

16. Evaluate $\begin{vmatrix} 1 & 2 & 1 \\ 2 & 0 & -1 \\ 3 & 1 & 0 \end{vmatrix}$

17. Use Cramer's rule to solve
    $$\begin{cases} 2x + 3y = 1 \\ 3x + 2y = 2 \end{cases}$$

18. Use Cramer's rule to solve
    $$\begin{cases} x + 2y + z = 0 \\ -y + z = 1 \\ 2x + 3y = 2 \end{cases}$$

19. Use the method of augmented matrices to solve
    $$\begin{cases} 2x + y = 1 \\ x - 7y = -2 \end{cases}$$

20. Use the method of augmented matrices to solve
    $$\begin{cases} 2x - z = 1 \\ x + y = 2 \\ x + 3y - 4z = -2 \end{cases}$$

CHAPTER 10

See Section 10.2, The Parabola.

Conic Sections

10.1 The Circle
10.2 The Parabola
10.3 The Ellipse and the Hyperbola
10.4 The General Second-Degree Equation in Two Variables

CONNECTIONS

For reasons lost in the mists of antiquity, the early Greek mathematicians spent a great deal of time studying the figures formed when a cone is sliced by a plane. In particular, Apollonius of Perga, born about 260 B.C., earned the title "The Great Geometer" from his study of these figures. The sketches below illustrate why these curves are called conic sections.

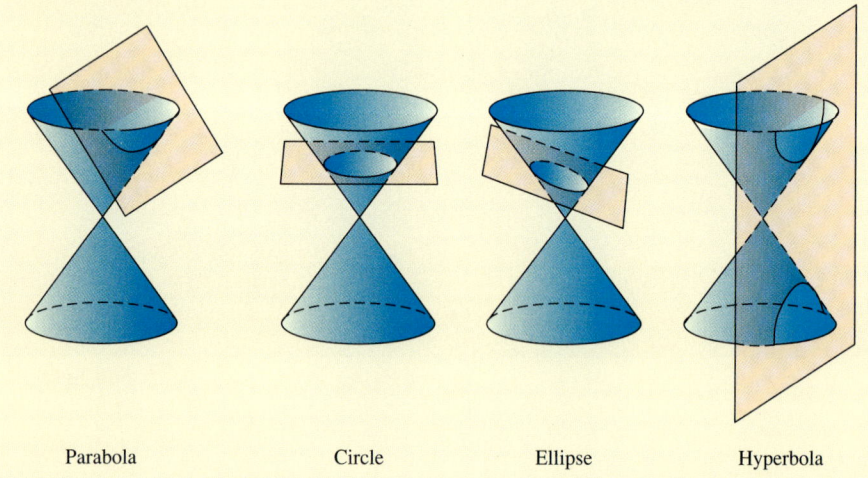

Parabola   Circle   Ellipse   Hyperbola

The conic sections have many interesting applications. For example, satellite orbits are ellipses, some comets follow hyperbolic paths as they approach the sun, and flashlight reflectors are parabolic in cross section.

This chapter begins with the study of a special figure with which we are already familiar, the circle. Next, we study the parabola and learn how to graph this conic section. In Section 10.3 we learn about the ellipse and the hyperbola and how to sketch their graphs. The last section shows the connection between the conic sections and the general second-degree equation in two variables.

10.1 THE CIRCLE

In plane geometry we learned that a circle is the set of all points the same distance from a given point. The distance is called the **radius** and the given point the **center** of the circle.

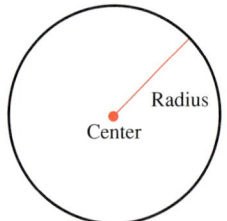

To find the basic equation for a circle with radius r, place a coordinate system so that the origin is at the center of the circle. Let (x, y) be *any* point on the circle. Then, as (x, y) must be r units from the center of the circle, the distance formula gives,

$$r = \sqrt{(x-0)^2 + (y-0)^2}$$
$$= \sqrt{x^2 + y^2}$$
$$r^2 = x^2 + y^2$$

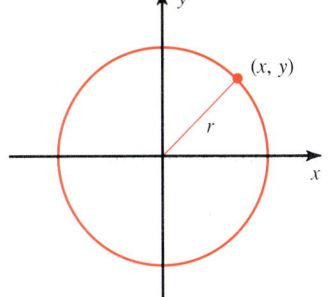

This is an equation whose graph is a circle of radius r and center at the origin. We usually write it $x^2 + y^2 = r^2$. It is the basic equation for a circle. Since r is a distance, it must be nonnegative. If r is 0, we have a circle of 0 radius which is called a point circle or a degenerate circle.

If the circle has its center at the point (h, k), the basic equation is modified as follows.

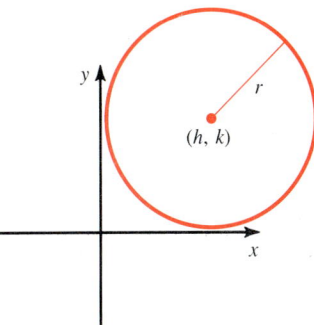

The Circle

The graph of

$$(x - h)^2 + (y - k)^2 = r^2$$

is a circle of radius r with center at the point whose coordinates are (h, k).

EXAMPLE 1. Graph each equation.

(a) $x^2 + y^2 = 16$ (b) $(x - 1)^2 + (y - 2)^2 = 16$

Solutions:

(a) This is of the form $x^2 + y^2 = r^2$, where r^2 is 16. As r is a distance and must be positive, r must be 4. So this is a circle of radius 4 with center at the origin.

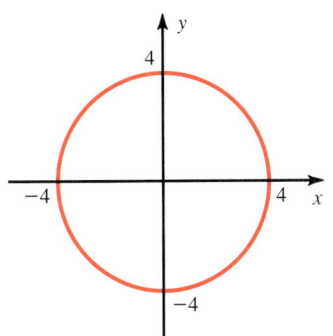

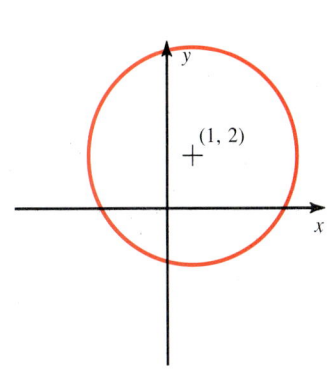

(b) This is of the form $(x - h)^2 + (y - k)^2 = r^2$, where h is 1, k is 2, and r is 4. Thus it is an equation of a circle of radius 4 centered at the point $(1, 2)$. It is the graph of part (a) *shifted* so that its center is at the point $(1, 2)$.

Sec. 10.1 The Circle

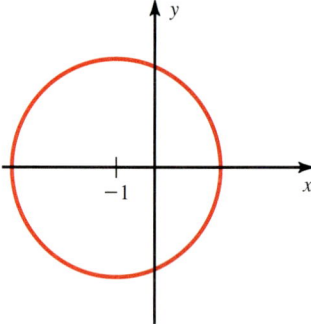

EXAMPLE 2. Graph the equation
$$(x + 1)^2 + y^2 = 7$$

Solution:

Notice that we can rewrite the equation as
$$[x - (-1)]^2 + (y - 0)^2 = (\sqrt{7})^2$$
which is of the form $(x - h)^2 + (y - k)^2 = r^2$, where h is -1, k is 0, and $r = \sqrt{7}$. Since $\sqrt{7} \approx 2.6$, we sketch the graph shown at the left. ☐

Sometimes the form of the equation is not as convenient as those in Examples 1 and 2, and we must complete the square before graphing.

EXAMPLE 3. Sketch the graph of $x^2 + y^2 - 2x = 3$.

Solution:

We note that there is *both* an x^2-term *and* an x-term in the equation. We combine them to form the $(x - h)^2$ term by completing the square.
$$x^2 + y^2 - 2x = 3$$

First we group the x-terms together.
$$(x^2 - 2x) + y^2 = 3$$

To complete the square, we divide -2 (the coefficient of x) by 2, square the result, and add this number to both sides of the equation. As $\left(\dfrac{-2}{2}\right)^2 = 1$, we add 1 to both sides.
$$(x^2 - 2x + 1) + y^2 = 3 + 1$$
$$(x^2 - 2x + 1) + y^2 = 4$$

Now $(x^2 - 2x + 1)$ is a perfect square.
$$(x - 1)^2 + y^2 = 4$$
or to put it in the form $(x - h)^2 + (y - k)^2 = r^2$,
$$(x - 1)^2 + (y - 0)^2 = 2^2$$

We can see that h is 1 and k is 0. So the center of the circle is $(1,0)$ and the radius is 2.

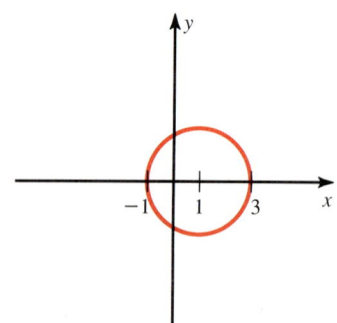

☐

Often, we must complete the square in *both* x and y.

EXAMPLE 4. Graph each equation.

(a) $x^2 + y^2 + 4x - 6y + 12 = 0$
(b) $x^2 + y^2 - 3x + 2y = 0$

Solutions:

(a) $x^2 + y^2 + 4x - 6y + 12 = 0$

We move 12 to the right side of the equation and group the x-terms together and the y-terms together.

$$(x^2 + 4x) + (y^2 - 6y) = -12$$

Now we complete the square twice, being careful to add to both sides of the equation. We must add 4 to complete the square in x and 9 to complete the square in y.

$$(x^2 + 4x + 4) + (y^2 - 6y + 9) = -12 + 4 + 9$$

$$(x + 2)^2 + (y - 3)^2 = 1$$

To determine the values of h and k, we must be careful to write the equation in the form

$$(x - h)^2 + (y - k)^2 = r^2$$

This means that $x + 2$ must be written as $x - (-2)$.

$$[x - (-2)]^2 + (y - 3)^2 = 1^2$$

So we see that the center is $(-2, 3)$ and the radius is 1.

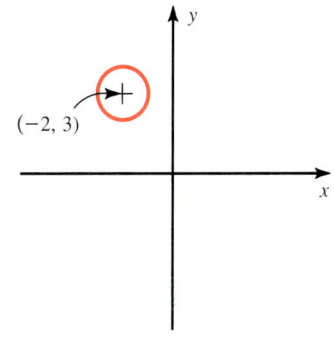

(b) $x^2 + y^2 - 3x + 2y = 0$

$(x^2 - 3x) + (y^2 + 2y) = 0$

It is necessary to add $\frac{9}{4}$ to complete the square in x and 1 to complete the square in y.

$$\left[x^2 - 3x + \frac{9}{4}\right] + [y^2 + 2y + 1] = 0 + \frac{9}{4} + 1$$

$$\left[x^2 - 3x + \frac{9}{4}\right] + [y^2 + 2y + 1] = \frac{13}{4}$$

$$\left[x - \frac{3}{2}\right]^2 + [y + 1]^2 = \frac{13}{4}$$

$$\left[x - \frac{3}{2}\right]^2 + [y - (-1)]^2 = \left[\frac{\sqrt{13}}{2}\right]^2$$

We see that the center is $\left(\frac{3}{2}, -1\right)$ and the radius is $\frac{\sqrt{13}}{2}$. To graph the circle, we must approximate $\frac{\sqrt{13}}{2}$.

$$\frac{\sqrt{13}}{2} \approx 1.8.$$

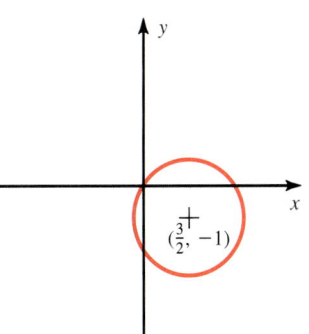

Sec. 10.1 The Circle

PROBLEM SET 10.1

Warm-ups

In problems 1 through 6, sketch the graph of the equation. See Examples 1 and 2.

1. $x^2 + y^2 = 4$
2. $x^2 + y^2 = 5$
3. $x^2 + (y - 1)^2 = 9$
4. $(x - 3)^2 + y^2 = 8$
5. $(x + 1)^2 + (y - 2)^2 = 10$
6. $(x - 1)^2 + (y + 1)^2 = 6$

In problems 7 through 16, the graph of each equation is a circle. Find the center and radius, then sketch the graph.

For problems 7 through 12, see Example 3.

7. $x^2 + y^2 - 2x = 8$
8. $x^2 + y^2 - 4y = 12$
9. $x^2 + y^2 + 8y = 4$
10. $x^2 + y^2 + 6x = -1$
11. $x^2 + y^2 - 10x + 16 = 0$
12. $x^2 + y^2 - 12y + 32 = 0$

For problems 13 through 16, see Example 4.

13. $x^2 + y^2 - 2x - 4y = 4$
14. $x^2 + y^2 - 6x - 8y = 0$
15. $x^2 + y^2 - x - 2 = 0$
16. $x^2 + y^2 - 3y - 4 = 0$

Practice Exercises

In problems 17 through 58, the graph of each equation is a circle. Find the center and radius, then sketch the graph.

17. $x^2 + y^2 = 9$

18. $x^2 + y^2 = 6$

19. $x^2 + (y - 2)^2 = 16$

20. $(x - 1)^2 + y^2 = 12$

21. $(x + 3)^2 + (y + 2)^2 = 49$

22. $\left(x + \dfrac{1}{2}\right)^2 + \left(y - \dfrac{3}{2}\right)^2 = \dfrac{9}{4}$

23. $x^2 + y^2 - 4x = 5$

24. $x^2 + y^2 - 2y = 15$

25. $x^2 + y^2 - 8y = 0$

26. $x^2 + y^2 - 6x = 16$

27. $x^2 + y^2 - 6y = 0$

28. $x^2 + y^2 - 8x = 9$

29. $x^2 + y^2 + 8x = -7$

30. $x^2 + y^2 + 4y = 0$

31. $x^2 + y^2 + 2y = 7$

32. $x^2 + y^2 + 6x = 3$

33. $x^2 + y^2 - 12x - 13 = 0$

34. $x^2 + y^2 - 10y - 11 = 0$

Sec. 10.1   The Circle

35. $x^2 + y^2 - 4x - 2y = 11$
36. $x^2 + y^2 - 8x - 6y = 0$
37. $x^2 + y^2 - 4x + 4y - 1 = 0$

38. $x^2 + y^2 + 12x - 2y + 1 = 0$
39. $x^2 + y^2 + 6x - 6y + 9 = 0$
40. $x^2 + y^2 - 10x + 2y + 10 = 0$

41. $(x + 2)^2 + (y + 3)^2 = 25$
42. $\left(x + \dfrac{3}{2}\right)^2 + \left(y - \dfrac{1}{2}\right)^2 = \dfrac{9}{4}$
43. $x^2 + y^2 + 4x = -3$

44. $x^2 + y^2 + 2y = 0$
45. $x^2 + y^2 + 12x + 4y = 0$
46. $x^2 + y^2 + 8x + 10y - 4 = 0$

47. $x^2 + y^2 - 5x - y + 4 = 0$
48. $x^2 + y^2 + x - y - \dfrac{1}{2} = 0$
49. $x^2 + y^2 + 7x + y = 0$

50. $x^2 + y^2 + 5x + 5y + \dfrac{1}{4} = 0$
51. $x^2 + y^2 + 6x + 8y + 5 = 0$
52. $x^2 + y^2 + 2x + 6y = 0$

53. $x^2 + y^2 - y - 1 = 0$

54. $x^2 + y^2 - 3x - 3 = 0$

55. $x^2 + y^2 - x - 3y - 6 = 0$

56. $x^2 + y^2 - x + y = 0$

57. $x^2 + y^2 + x + 5y = 0$

58. $x^2 + y^2 + 7x + 7y + \dfrac{3}{4} = 0$

Challenge Problems

59. Show that the equation of a circle of radius r, centered at the point whose coordinates are (h, k), is
$$(x - h)^2 + (y - k)^2 = r^2$$

60. What is the graph of $x^2 + y^2 - 2x - 4y + 5 = 0$? The point $(1, 2)$

In problems 61 through 64, the graph of each equation is a circle. Find the center and radius, then sketch the graph.

61. $2x^2 + 2y^2 - 4x - 2y + 1 = 0$

62. $2x^2 + 2y^2 + 6x - 2y - 3 = 0$

63. $4 + 2x + 4y - x^2 - y^2 = 0$

64. $5x - 3y - 2x^2 - 2y^2 + 1 = 0$

IN YOUR OWN WORDS . . .

65. Describe how completing the square is used in graphing circles.

66. Would the graph of $x^2 + y^2 = -4$ be a circle? Explain.

10.2 THE PARABOLA

In Section 8.4 we learned that the graph of the first-degree equation
$$y = mx + b$$
is a line. Here we will examine second-degree equations of the form
$$y = Ax^2 + Bx + C; \quad A \neq 0$$
and
$$x = Dy^2 + Ey + F; \quad D \neq 0$$

First, let's graph a simple example of such an equation, $y = x^2$. We begin by making a table of x and y coordinates, and then plotting the points.

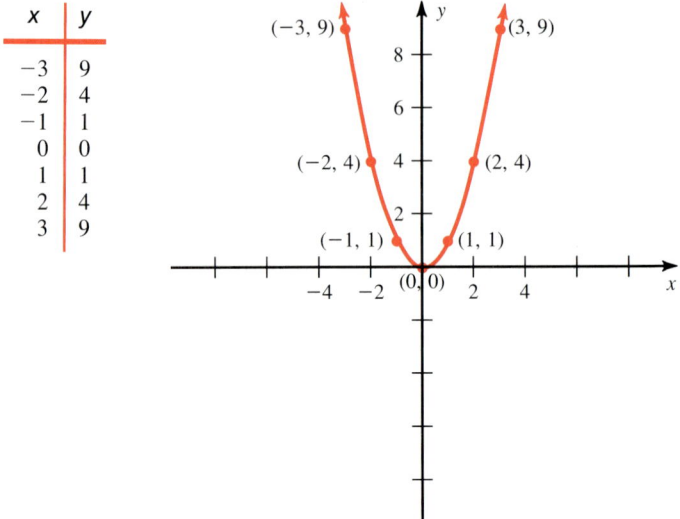

This graph is called a **parabola.** Parabolas are common in everyday life. The reflector in a flashlight is in the shape of a parabola. The cables of the George Washington Bridge hang in parabolic form and a parabola was utilized in the design of the 200-in. telescope on Mount Palomar.

Notice that the parabola is drawn as a smooth curve. There are **no sharp places.** We make an extra effort to draw parabolas as smooth curves.

The lowest point on the parabola sketched above, (0, 0), is called the **vertex.** A line through the vertex about which the parabola is symmetric is called the **axis of symmetry.** Notice that this parabola is symmetric about the y-axis.

EXAMPLE 1. Graph the equation $y = x^2 + 1$ and identify the vertex and axis of symmetry.

Solution:

Again we make a table of x, y values and sketch the graph.

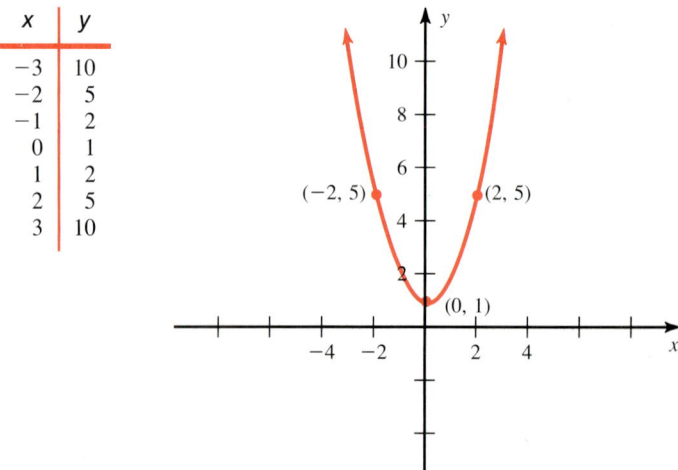

By inspection, we see that the vertex is the point (0, 1), and again the axis of symmetry is the y-axis. □

Compare this graph with the graph of $y = x^2$ that we drew earlier in the section. Notice that the size and shape of the two curves are just the same. The graph of $y = x^2 + 1$ is exactly the graph of $y = x^2$ shifted *up* 1 unit.

EXAMPLE 2. Graph $y = x^2 - 2$ and identify the vertex and axis of symmetry.

Solution:

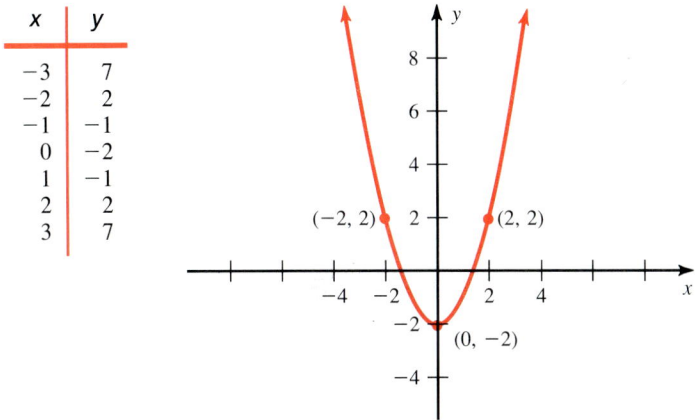

The vertex is at $(0, -2)$ and the y-axis is the axis of symmetry. □

Notice that the graph of $y = x^2 - 2$ is exactly the graph of $y = x^2$ shifted *down* 2 units. As a matter of fact, the graph of the equation $y = x^2 + k$ is a parabola with exactly the same *shape* as the graph of $y = x^2$ with its vertex at $(0, k)$.

EXAMPLE 3. Graph $y = 2x^2$, $y = x^2$, and $y = \frac{1}{4}x^2$.

Solution:

As before, we graph the three curves.

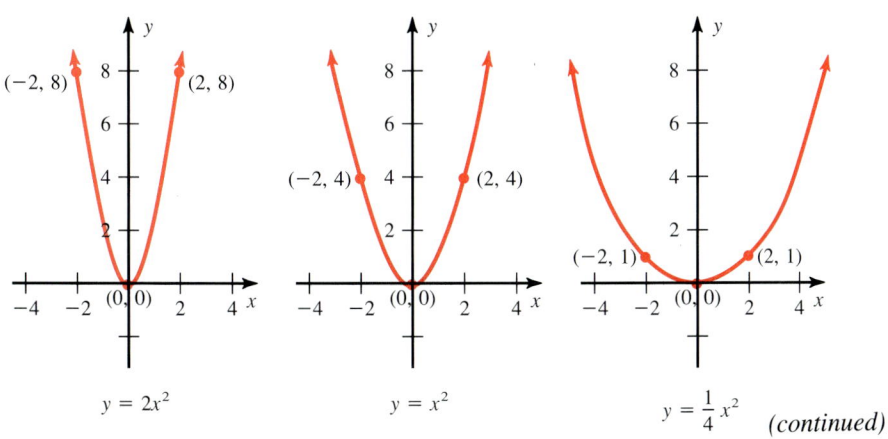

(continued)

Sec. 10.2 The Parabola

Compare these three parabolas. Notice the effect that the coefficient of x^2 has on the shape of the graph. The 2 makes a thinner parabola and the $\frac{1}{4}$ makes a wider one.

EXAMPLE 4. Graph $y = -x^2$.

Solution:

Plotting a few points gives the graph shown below. Notice that this is the graph of $y = x^2$ turned upside down.

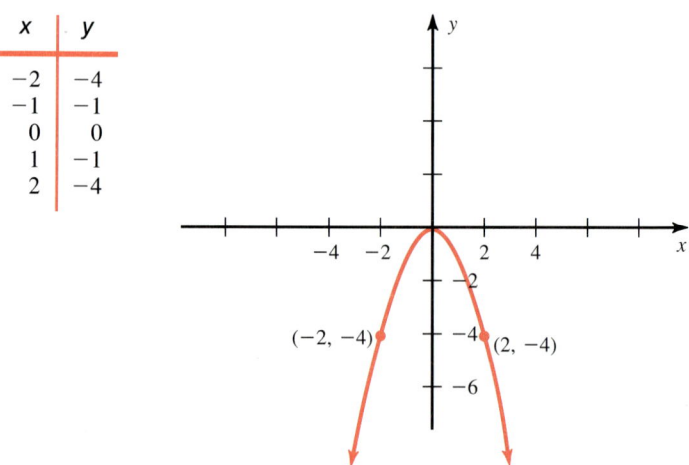

EXAMPLE 5. Graph $y = -2x^2 + 1$.

Solution:

Sketch the graph by plotting a few points.

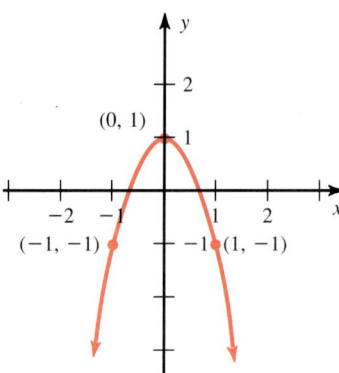

Note that this is a parabola that opens downward with vertex at (0, 1). It is thinner than the graph of $y = x^2$.

We can now make some observations about graphing parabolas. The graph of $y = px^2 + k$ is a parabola with vertex at the point (0, k). It opens downward if p is negative and it opens upward if p is positive. It is thinner than the graph of $y = x^2$ if $|p| > 1$ and wider if $|p| < 1$.

EXAMPLE 6. Graph $y = (x - 1)^2$.

Solution:

Plotting several points gives the following graph.

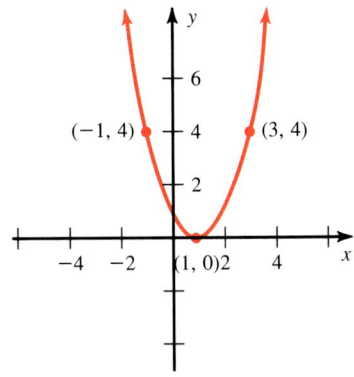

The vertex of this parabola is (1, 0). In fact, it is the graph of $y = x^2$ shifted 1 unit to the right. Notice that the axis of symmetry is the line $x = 1$. ☐

EXAMPLE 7. Graph $y = -2(x - 1)^2 + 2$.

Solution:

Plot several points to see the parabola at the right. Its vertex is (1, 2). It is thinner than the graph of $y = x^2$ and it opens downward. ☐

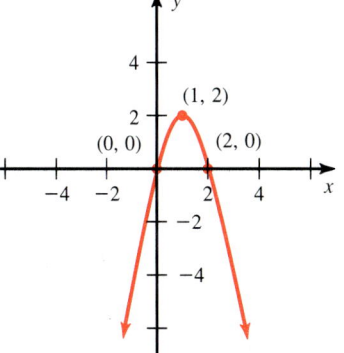

The graph of $y = p(x - h)^2 + k$ is a parabola with vertex at the point (h, k). Its axis of symmetry is the line $x = h$. The parabola opens upward if p is positive and downward if p is negative.

EXAMPLE 8. Find the vertex for each parabola.

(a) $y = (x + 1)^2 + 2$
(b) $y = x^2 - 1$
(c) $y = (x + 3)^2$

Solutions:

(a) To match the form of $y = p(x - h)^2 + k$, we see that $y = (x + 1)^2 + 2$ must be written as $y = [x - (-1)]^2 + 2$. So h has a value of -1 and k has a value of 2. Thus the vertex is $(-1, 2)$.

(b) $y = x^2 - 1$
$= (x - 0)^2 - 1$
So the vertex is $(0, -1)$.

(continued)

(c) $y = (x + 3)^2$
$ = (x + 3)^2 + 0$
$ = [x - (-3)]^2 + 0$
and the vertex is $(-3, 0)$.

As in the case of the circle, it is often necessary to complete the square in order to find the vertex and graph a parabola.

EXAMPLE 9. Complete the square and find the vertex for each parabola.

(a) $y = x^2 - 8x$ (b) $y = 3x^2 + 9x - 1$

Solutions:

(a) $y = x^2 - 8x$

We need $\left[\dfrac{-8}{2}\right]^2 = 16$ for a perfect square. We cannot simply add it on, for that would change the values of the equation. However, we can add zero without harming the equation. So we add *and subtract* 16.

$$y = x^2 - 8x + 16 - 16$$
$$y = (x - 4)^2 - 16$$

The vertex is at $(4, -16)$.

(b) $y = 3x^2 + 9x - 1$

Again we need to complete the square. First, it is necessary to factor 3 out of the x-terms.

$$y = 3(x^2 + 3x) - 1$$

Next, we complete the square *inside the parentheses*.

$$y = 3\left[x^2 + 3x + \dfrac{9}{4} - \dfrac{9}{4}\right] - 1$$

Using the distributive property, we remove the last term from inside the parentheses.

$$y = 3\left[x^2 + 3x + \dfrac{9}{4}\right] - \dfrac{27}{4} - 1$$

$$y = 3\left[x + \dfrac{3}{2}\right]^2 - \dfrac{31}{4}$$

The vertex is at $\left[-\dfrac{3}{2}, -\dfrac{31}{4}\right]$.

EXAMPLE 10. Graph $y = -2x^2 + 4x - 3$.

Solution:

To graph the equation of a parabola, we write it in the form $y = p(x - h)^2 + k$. In this form we can identify the vertex and the axis of symmetry. To write the equation in the form desired, we must complete the square.

$$y = -2x^2 + 4x - 3$$

The coefficient of x^2 must be 1 for the completing the square procedure to work. We factor -2 out of the *first two terms*.

$$y = -2(x^2 - 2x) - 3$$

Now complete the square as in Example 9b.

$$y = -2(x^2 - 2x + 1 - 1) - 3$$
$$y = -2(x^2 - 2x + 1) + 2 - 3$$
$$y = -2(x - 1)^2 - 1$$

The vertex is $(1, -1)$ and the parabola will open downward.

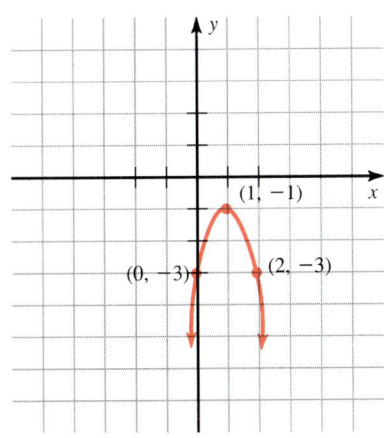

EXAMPLE 11. Graph the equation $x = y^2$.

Solution:

Make a table and plot a few points.

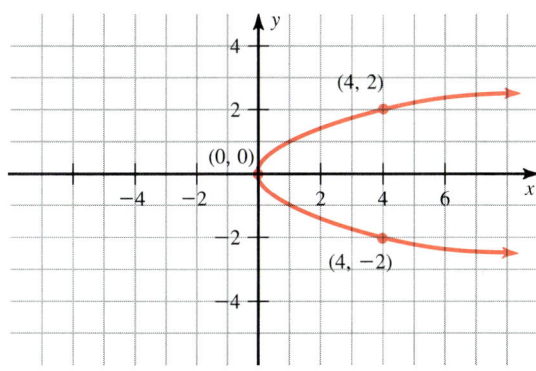

x	y
0	0
1	1
1	-1
4	2
4	-2

This is a parabola of the same general shape as the graph of $y = x^2$ except that it is lying on its side, opening to the right. In fact, if we interchange the roles of x and y (and h and k) in the equation $y = p(x - h)^2 + k$, we get similar figures, except that they open left–right instead of up–down.

Sec. 10.2 The Parabola

> ### The Parabolas
>
> $$y = p(x - h)^2 + k$$
>
> is a parabola with vertex at the point (h, k). It opens *upward* if p is positive and *downward* if p is negative. Its axis of symmetry is the line $x = h$.
>
> $$x = p(y - k)^2 + h$$
>
> is a parabola with vertex at the point (h, k). It opens *to the right* if p is positive and *to the left* if p is negative. Its axis of symmetry is the line $y = k$.

EXAMPLE 12. Graph the equation $x + 2y^2 - 8y = -5$.

Solution:

Solve the equation for x. (Note the y^2 term.)

$$x = -2y^2 + 8y - 5$$

Before completing the square it is necessary to factor -2 from the first two terms.

$$= -2(y^2 - 4y) - 5$$

Now we can complete the square inside the parentheses.

$$= -2(y^2 - 4y + 4 - 4) - 5$$
$$= -2(y^2 - 4y + 4) + 8 - 5$$
$$= -2(y - 2)^2 + 3$$

This is a parabola with vertex at $(3, 2)$, opening to the left with the line $y = 2$ as its axis of symmetry.

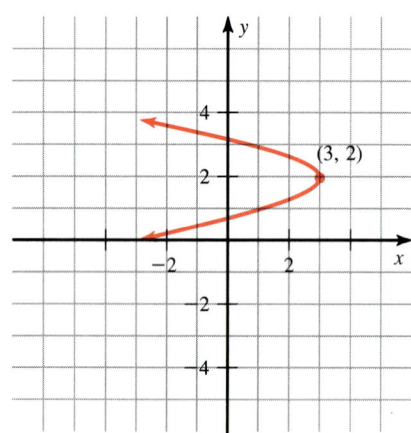

PROBLEM SET 10.2

Warm-ups

In problems 1 through 8, identify the axis of symmetry and sketch the graph of each equation.
For problems 1 through 4, see Examples 1 through 5.

1. $y = x^2 + 3$

2. $y = x^2 - 1$

3. $y = 3x^2$

4. $y = -\dfrac{1}{2}x^2 + 1$

For problems 5 through 8, see Examples 6 and 7.

5. $y = (x - 3)^2$　　　**6.** $y = -(x - 1)^2$　　　**7.** $y = (x + 2)^2 + 1$　　　**8.** $y = -(x + 3)^2 - 2$

In problems 9 through 12, find the vertex of each parabola. See Example 8.

9. $y = (x - 2)^2 + 3$　$(2, 3)$　　　　　　**10.** $y = (x + 5)^2 + 1$　$(-5, 1)$
11. $y = -(x - 4)^2 - 6$　$(4, -6)$　　　**12.** $y = 2(x + 2)^2 - 3$　$(-2, -3)$

In problems 13 through 20, the graph of each equation is a parabola. Find the vertex of each and sketch the graph.

For problems 13 through 16, see Example 10.

13. $y = -2x^2 + 4x$　　　**14.** $2x + y = x^2$　　　**15.** $y = x^2 - 4x + 2$　　　**16.** $x^2 + x = y$

For problems 17 through 20, see Examples 11 and 12.

17. $x = y^2 - 3y + 2$　　　**18.** $y^2 + 2y = x$　　　**19.** $x = -2y^2 + 4y$　　　**20.** $x = 3y^2 + 18y + 29$

Practice Exercises

In problems 21 through 56, the graph of each equation is a parabola. Find the vertex of each and sketch the graph.

21. $x = (y - 1)^2$　　　**22.** $x = -y^2 + 2$　　　**23.** $y = (x - 3)^2 + 2$　　　**24.** $y = -(x - 2)^2 - 3$

Sec. 10.2　The Parabola

25. $x = (y + 3)^2 + 3$

26. $y = -2(x + 2)^2 - 1$

27. $y = \dfrac{1}{2}\left[x + \dfrac{2}{3}\right]^2 + \dfrac{1}{3}$

28. $x = -\dfrac{2}{3}\left[y + \dfrac{1}{2}\right]^2 - \dfrac{2}{3}$

29. $y = -(x + 3)^2 + 3$

30. $x = 2(y + 2)^2 - 1$

31. $x = \dfrac{1}{3}\left[y + \dfrac{3}{2}\right]^2 - \dfrac{1}{2}$

32. $y = -\left[x + \dfrac{1}{3}\right]^2 - \dfrac{4}{3}$

33. $y^2 - 2x - 4 = 0$

34. $x^2 + 3y = 3$

35. $x = -2y^2 - 8y - 8$

36. $y = -2x^2 - 2x - 1$

37. $3x^2 - 9x - y + 6 = 0$

38. $4y^2 - 2x - 12y = -7$

39. $y^2 - x = 0$

40. $x^2 + y = 0$

41. $x^2 - 2 = y$

42. $x = -y^2 - 1$

43. $3y^2 + 3 = x$ **44.** $y = -2x^2 + 2$ **45.** $x^2 + 2y + 4 = 0$ **46.** $3y^2 - x = 3$

47. $y^2 - 8y = x - 10$ **48.** $y = -x^2 - 2x$ **49.** $x^2 - x + 2 = y$ **50.** $y^2 + 5y = x$

51. $x = -2y^2 + 8y$ **52.** $y = 3x^2 + 6x + 5$ **53.** $y = -2x^2 + 4x - 2$

54. $x = -3y^2 - 3y - 1$ **55.** $3y^2 + 9y - x + 6 = 0$ **56.** $4x^2 - 2y + 4x = 1$

Challenge Problems

In problems 57 through 64, the graph of each equation is a parabola. Find the vertex of each and sketch the graph.

57. $2y^2 - x - 3y + 1 = 0$ **58.** $3x^2 + 2x + y = 1$ **59.** $3x^2 - 4x - 2y = 0$

60. $2y^2 + 3x + 5y + 2 = 0$ **61.** $2x^2 + y + 3x + 2 = 0$ **62.** $2y^2 + y - 2x = 1$

Sec. 10.2 The Parabola

63. $4y^2 - 5y - 2x = 0$

64. $2x^2 + 5y + 7x + 3 = 0$

■ IN YOUR OWN WORDS . . .

65. Describe the procedure used to rewrite an equation of the form $y = Ax^2 + Bx + C$ into the form
$$y = p(x - h)^2 + k.$$

■ 10.3 THE ELLIPSE AND THE HYPERBOLA

In plane geometry, an **ellipse** is described as the set of all points in a plane, the sum of whose distances from two fixed points is constant.

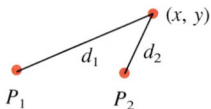

The ellipse is all points such that $d_1 + d_2 = K$, a constant. Sketching the points results in a figure like

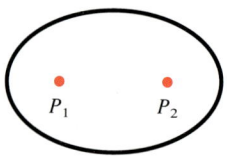

We put a coordinate system on this figure so that the points P_1 and P_2, called the *foci* (the singular is *focus*), are on the *x*-axis and the origin is halfway between the foci at the center of the ellipse.

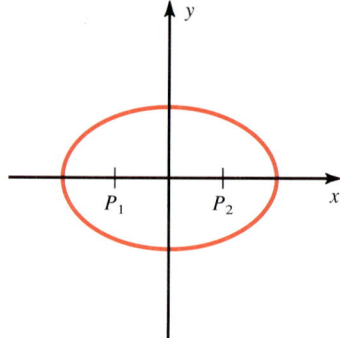

The equation whose graph is this ellipse can be written in the form

$$\frac{x^2}{a^2} + \frac{y^2}{b^2} = 1$$

where *a* and *b* are the constants shown on the following figure.

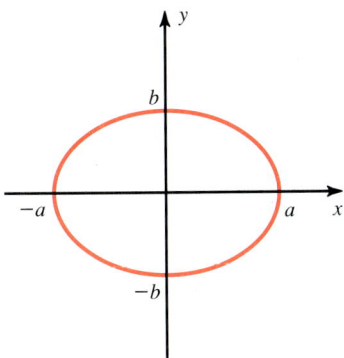

The *x*-intercepts of the graph are at $\pm a$ and the *y*-intercepts are at $\pm b$.

If *b* is larger than *a*, the foci lie on the *y*-axis and the graph looks like:

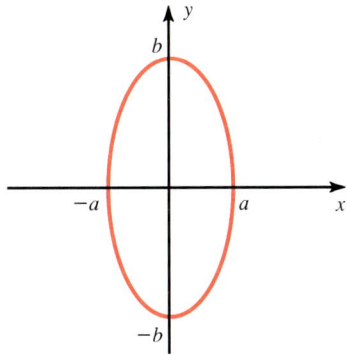

If both foci are the same point, the graph is a circle. Thus we see that a circle is a special case of an ellipse.

EXAMPLE 1. Sketch the graph of the equation

$$\frac{x^2}{9} + \frac{y^2}{4} = 1$$

Solution:

This is in the form $\frac{x^2}{a^2} + \frac{y^2}{b^2} = 1$ if *a* is 3 and *b* is 2. That is,

$$\frac{x^2}{3^2} + \frac{y^2}{2^2} = 1$$

Thus the *x*-intercepts are ± 3 and the *y*-intercepts are ± 2. The graph is a *smooth* oval through those points. ☐

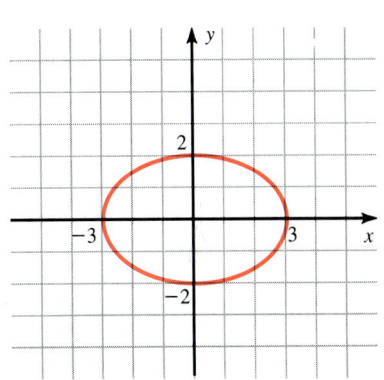

EXAMPLE 2. Sketch the graph of the equation

$$\frac{x^2}{16} + \frac{y^2}{25} = 1$$

(continued)

Sec. 10.3 The Ellipse And The Hyperbola

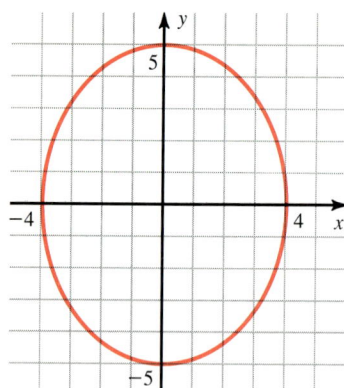

Solution:

Rewriting the equation in the form $\dfrac{x^2}{4^2} + \dfrac{y^2}{5^2} = 1$ shows that a is 4 and b is 5. The x-intercepts are ± 4 and the y-intercepts are ± 5.

The x-intercepts, a and $-a$, and the y-intercepts, b and $-b$, are sufficient to sketch the graph of an ellipse. Find these four points, then connect them with the smoothest oval possible.

Here is the basic equation for the ellipse.

The Ellipse Centered at the Origin

The graph of the equation

$$\frac{x^2}{a^2} + \frac{y^2}{b^2} = 1$$

is an ellipse with x-intercepts a and $-a$, and y-intercepts b and $-b$, centered at the origin.

The last of the conic sections, the **hyperbola,** can be described as the set of all points the difference of whose distances from two fixed points is constant. Sketching these points results in a figure like

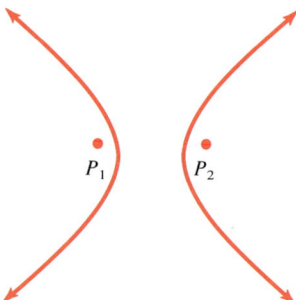

We put a coordinate system on this figure so that the foci, the points P_1 and P_2, are on the x-axis and the origin is halfway between them.

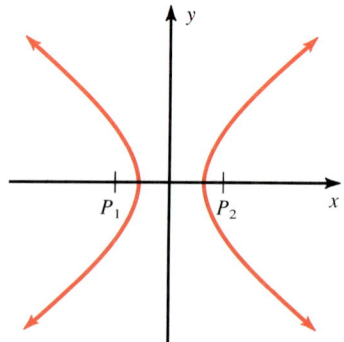

460

Chap. 10 Conic Sections

The equation whose graph is this hyperbola is

$$\frac{x^2}{a^2} - \frac{y^2}{b^2} = 1$$

where a and $-a$ are the x-intercepts as shown on the following figure.

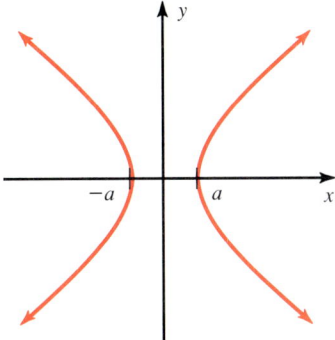

If we reverse the roles of x and y, we have an equation of the form

$$\frac{y^2}{b^2} - \frac{x^2}{a^2} = 1$$

where b and $-b$ are the *y-intercepts*, as shown on the figure below.

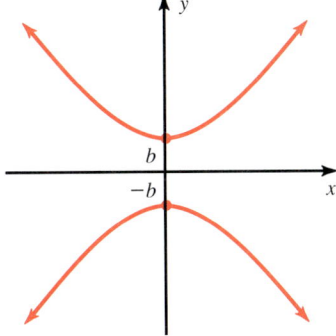

So we have hyperbolas opening left–right given by the equation

$$\frac{x^2}{a^2} - \frac{y^2}{b^2} = 1$$

and hyperbolas opening up–down given by the equation

$$\frac{y^2}{b^2} - \frac{x^2}{a^2} = 1$$

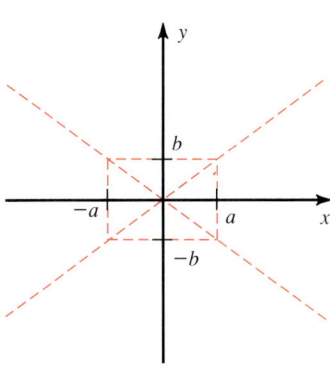

We use the values of a and b to sketch a hyperbola by drawing some construction lines. First, we draw the rectangle shown at the right using the values of a and b. Next we draw the diagonals of the rectangle, extending them as far as necessary. The two diagonals are asymptotes of the hyperbola. That is, the farther the hyperbola is from the origin, the closer it gets to the asymptote. Using these construction lines as guides, we now sketch the hyperbola, either left–right or up–down, depending on the form of the equation.

Sec. 10.3   The Ellipse And The Hyperbola

EXAMPLE 3. Sketch the graph of the equation

$$\frac{x^2}{9} - \frac{y^2}{4} = 1$$

Solution:

The equation $\frac{x^2}{9} - \frac{y^2}{4} = 1$ can be written $\frac{x^2}{3^2} - \frac{y^2}{2^2} = 1$. Therefore, a is 3 and b is 2. The construction lines are drawn as shown below. Noting that the equation is in the left–right form, we can then sketch the hyperbola.

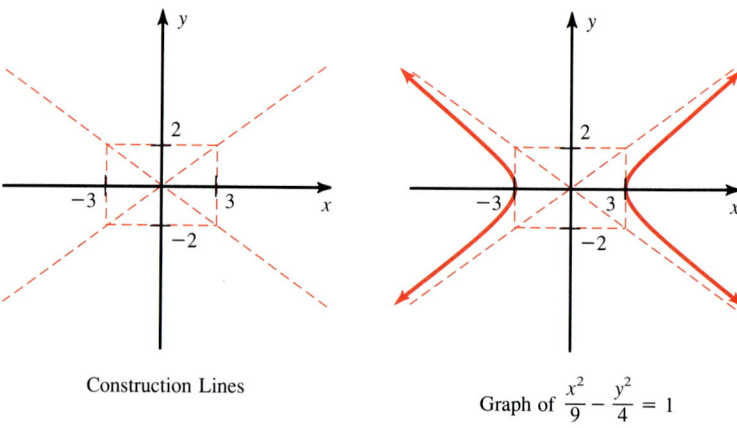

Construction Lines

Graph of $\frac{x^2}{9} - \frac{y^2}{4} = 1$

EXAMPLE 4. Sketch the graph of the equation

$$\frac{y^2}{16} - \frac{x^2}{25} = 1$$

Solution:

Here a is 5 and b is 4 and we have an up–down form.

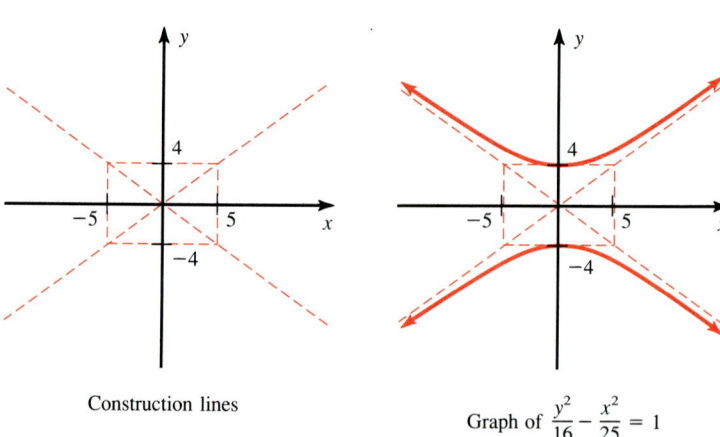

Construction lines

Graph of $\frac{y^2}{16} - \frac{x^2}{25} = 1$

We have the basic equations for the hyperbola.

> ### Hyperbolas Centered at the Origin
>
> The graph of the equation
>
> $$\frac{x^2}{a^2} - \frac{y^2}{b^2} = 1$$
>
> is a hyperbola with x-intercepts a and $-a$ centered at the origin.
> The graph of the equation
>
> $$\frac{y^2}{b^2} - \frac{x^2}{a^2} = 1$$
>
> is a hyperbola with y-intercepts b and $-b$ centered at the origin.

Like the circle, if x is replaced with $x - h$ and y with $y - k$, the basic ellipse or hyperbola will be shifted on the coordinate system.

> ### The Ellipse
>
> The graph of the equation
>
> $$\frac{(x-h)^2}{a^2} + \frac{(y-k)^2}{b^2} = 1$$
>
> is the graph of the ellipse $\frac{x^2}{a^2} + \frac{y^2}{b^2} = 1$ shifted so that its center is at the point (h, k).
>
> ### The Hyperbolas
>
> The graph of the equation
>
> $$\frac{(x-h)^2}{a^2} - \frac{(y-k)^2}{b^2} = 1$$
>
> is the graph of the hyperbola $\frac{x^2}{a^2} - \frac{y^2}{b^2} = 1$ shifted so that its center is at the point (h, k).
> The graph of the equation
>
> $$\frac{(y-k)^2}{b^2} - \frac{(x-h)^2}{a^2} = 1$$
>
> is the graph of the hyperbola $\frac{y^2}{b^2} - \frac{x^2}{a^2} = 1$ shifted so that its center is at the point (h, k).

Sec. 10.3 The Ellipse And The Hyperbola

EXAMPLE 5. Graph the equation

$$4x^2 + y^2 - 24x + 4y + 24 = 0$$

Solution:

As this equation contains both x and x^2 terms, we complete the square to form the $(x - h)^2$ term. Because it contains y and y^2 terms, we also complete the square to obtain the $(y - k)^2$ term.

$$4x^2 + y^2 - 24x + 4y + 24 = 0$$

First, group the x and y terms and factor where necessary.

$$4(x^2 - 6x) + (y^2 + 4y) + 24 = 0$$

Now we complete the square inside each set of parentheses and simplify.

$$4(x^2 - 6x + 9 - 9) + (y^2 + 4y + 4 - 4) + 24 = 0$$

$$4(x^2 - 6x + 9) - 36 + (y^2 + 4y + 4) - 4 + 24 = 0$$

$$4(x - 3)^2 + (y + 2)^2 - 16 = 0$$

Add 16 to both sides and then divide by 16.

$$4(x - 3)^2 + (y + 2)^2 = 16$$

$$\frac{4(x - 3)^2}{16} + \frac{(y + 2)^2}{16} = 1$$

$$\frac{(x - 3)^2}{4} + \frac{(y + 2)^2}{16} = 1$$

Notice that this is the ellipse, $\frac{x^2}{2^2} + \frac{y^2}{4^2} = 1$, shifted so that its center is at the point $(3, -2)$. An easy way to make such a shift is to sketch a copy of the coordinate system with its origin at the point $(3, -2)$, then graph the basic ellipse on the shifted system.

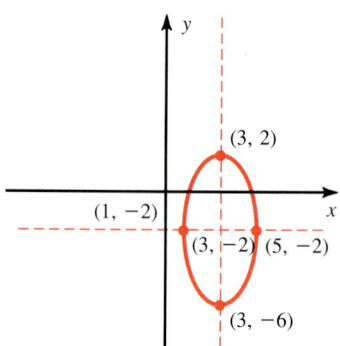

Graph of $\frac{(x - 3)^2}{2^2} + \frac{(y + 2)^2}{4^2} = 1$

EXAMPLE 6. Graph the equation

$$4x^2 - 6y^2 + 32x + 12y + 22 = 0$$

Solution:

Again we complete the square in both x and y.

$$4x^2 - 6y^2 + 32x + 12y + 22 = 0$$
$$4(x^2 + 8x) - 6(y^2 - 2y) + 22 = 0$$
$$4(x^2 + 8x + 16 - 16) - 6(y^2 - 2y + 1 - 1) + 22 = 0$$
$$4(x^2 + 8x + 16) - 64 - 6(y^2 - 2y + 1) + 6 + 22 = 0$$
$$4(x + 4)^2 - 6(y - 1)^2 = 36$$
$$\frac{4(x + 4)^2}{36} - \frac{6(y - 1)^2}{36} = 1$$
$$\frac{(x + 4)^2}{9} - \frac{(y - 1)^2}{6} = 1$$

We have a left–right hyperbola with $a = 3$ and $b = \sqrt{6}$, centered at the point $(-4, 1)$. We draw the basic hyperbola in its shifted position as we did the ellipse in Example 5.

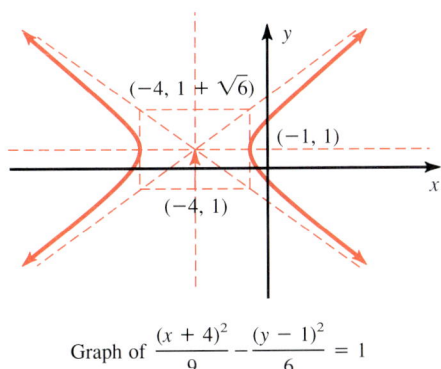

Graph of $\dfrac{(x + 4)^2}{9} - \dfrac{(y - 1)^2}{6} = 1$

EXAMPLE 7. Graph the equation

$$81x^2 - 16y^2 - 162x + 117 = 0$$

Solution:

As there is no y-term in this equation, we only need to complete the square in x.

$$81x^2 - 16y^2 - 162x + 117 = 0$$
$$81(x^2 - 2x) - 16y^2 + 117 = 0$$
$$81(x^2 - 2x + 1 - 1) - 16y^2 + 117 = 0$$
$$81(x^2 - 2x + 1) - 81 - 16y^2 + 117 = 0$$
$$81(x - 1)^2 - 16y^2 = -36$$
$$\frac{81(x - 1)^2}{-36} - \frac{16y^2}{-36} = 1$$
$$-\frac{9(x - 1)^2}{4} + \frac{4y^2}{9} = 1$$

(continued)

Sec. 10.3 The Ellipse And The Hyperbola

We rearrange the terms, then divide numerator and denominator of the *x*-term by 9 and the *y*-term by 4 to put it in the form

$$\frac{(y-k)^2}{b^2} - \frac{(x-h)^2}{a^2} = 1.$$

$$\frac{y^2}{\frac{9}{4}} - \frac{(x-1)^2}{\frac{4}{9}} = 1$$

This is an up-down hyperbola centered at (1, 0), where *a* is $\frac{2}{3}$ and *b* is $\frac{3}{2}$. We are now ready to sketch the graph. Again, we move the *coordinate axes* so that they are centered at (1, 0) and draw our construction lines there.

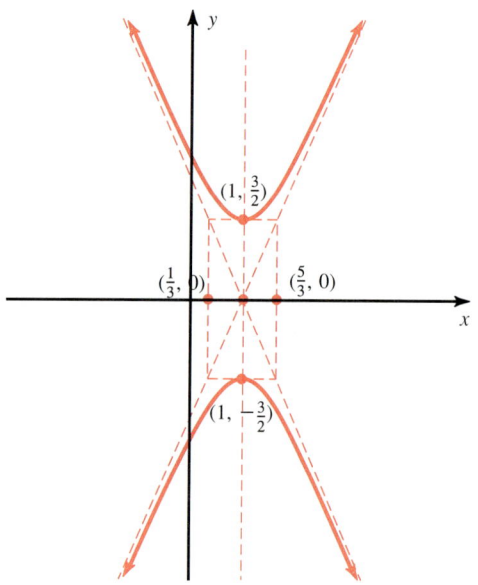

PROBLEM SET 10.3

Warm-ups

In problems 1 through 16, using the techniques developed in this section, sketch the graph of each equation.

For problems 1 through 4, see Examples 1 and 2.

1. $\dfrac{x^2}{36} + \dfrac{y^2}{16} = 1$
2. $\dfrac{x^2}{25} + \dfrac{y^2}{64} = 1$
3. $x^2 + \dfrac{y^2}{9} = 1$
4. $\dfrac{x^2}{49} + \dfrac{y^2}{36} = 1$

For problems 5 through 8, see Examples 3 and 4.

5. $\dfrac{x^2}{16} - \dfrac{y^2}{4} = 1$

6. $\dfrac{y^2}{9} - \dfrac{x^2}{16} = 1$

7. $\dfrac{y^2}{4} - x^2 = 1$

8. $x^2 - y^2 = 1$

For problems 9 through 12, see Example 5.

9. $\dfrac{(x-1)^2}{36} + \dfrac{(y-2)^2}{49} = 1$

10. $\dfrac{16(x-1)^2}{25} + \dfrac{9(y-1)^2}{4} = 1$

11. $9x^2 + 4y^2 - 18x - 8y - 23 = 0$

12. $x^2 + 9y^2 - 4x + 36y + 31 = 0$

For problems 13 through 16, see Examples 6 and 7.

13. $\dfrac{(y-1)^2}{9} - \dfrac{(x+1)^2}{4} = 1$

14. $\dfrac{(x+3)^2}{9} - \dfrac{(y+2)^2}{9} = 1$

15. $4x^2 - y^2 - 16x + 2y + 11 = 0$

16. $x^2 - y^2 - 2x + 2y - 4 = 0$

Practice Exercises

In problems 17 through 48, using the techniques developed in this section, sketch the graph of each equation.

17. $\dfrac{x^2}{36} - \dfrac{y^2}{16} = 1$

18. $\dfrac{x^2}{4} + \dfrac{y^2}{16} = 1$

19. $\dfrac{y^2}{9} + \dfrac{x^2}{16} = 1$

Sec. 10.3 The Ellipse And The Hyperbola

20. $\dfrac{y^2}{25} - \dfrac{x^2}{64} = 1$

21. $\dfrac{x^2}{4} + y^2 = 1$

22. $\dfrac{(x-1)^2}{36} - \dfrac{(y-2)^2}{49} = 1$

23. $\dfrac{(x+1)^2}{9} + \dfrac{(y-1)^2}{4} = 1$

24. $\dfrac{(y+3)^2}{9} - \dfrac{(x+2)^2}{9} = 1$

25. $\dfrac{16(x-1)^2}{25} - \dfrac{9(y-1)^2}{4} = 1$

26. $y^2 - x^2 = 1$

27. $2x^2 + y^2 = 8$

28. $x^2 - 2y^2 = 12$

29. $x^2 - 4y^2 + 16y = 0$

30. $25x^2 + 9y^2 + 50x - 200 = 0$

31. $4x^2 + 9y^2 + 32x - 36y + 64 = 0$

32. $x^2 - y^2 + 6x + 4y - 4 = 0$

33. $16x^2 + y^2 + 64x - 8y + 64 = 0$

34. $4x^2 + 5y^2 - 8x + 20y + 4 = 0$

35. $9x^2 - 36y^2 + 72y = 0$

36. $y^2 - 2x^2 = 8$

37. $x^2 - y^2 - 2x + 2y + 4 = 0$

38. $4x^2 + 9y^2 + 8x + 18y - 23 = 0$ **39.** $x^2 - 4y^2 + 16y - 32 = 0$ **40.** $9x^2 + y^2 + 36x - 4y + 31 = 0$

41. $25x^2 - 9y^2 + 50x - 200 = 0$ **42.** $4x^2 + 9y^2 + 32x + 36y + 64 = 0$ **43.** $x^2 - y^2 + 6x + 4y + 14 = 0$

44. $x^2 + 25y^2 + 10x + 50y + 25 = 0$ **45.** $4x^2 - 5y^2 - 8x - 20y - 36 = 0$ **46.** $5x^2 + 3y^2 - 20x + 6y + 8 = 0$

47. $9x^2 + 36y^2 + 72y + 32 = 0$ **48.** $100x^2 - 144y^2 + 600x + 1125 = 0$

Challenge Problems

In problems 49 through 52, sketch the graph of each equation.

49. $4x^2 + 9y^2 - 4x - 6y - 34 = 0$ **50.** $64x^2 - 36y^2 + 64x + 108y - 641 = 0$

51. $9x^2 - 4y^2 - 3x - 2y + 144 = 0$ **52.** $4x^2 + 36y^2 + 12x + 36y + 9 = 0$

▬ IN YOUR OWN WORDS . . .

53. Explain how to graph $\dfrac{(x-h)^2}{a^2} + \dfrac{(y-k)^2}{b^2} = 1$.   **54.** Why is a circle a special case of an ellipse?

Sec. 10.3   The Ellipse And The Hyperbola

10.4 THE GENERAL SECOND-DEGREE EQUATION IN TWO VARIABLES

Our interest in the conic sections stems from their relationship to the general second-degree equation in two variables,

$$Ax^2 + Bxy + Cy^2 + Dx + Ey + F = 0 \quad \text{where } A, B, \text{ and } C \text{ are not } \textit{all } 0$$

Except for some special cases, the graphs of these equations turn out to be conic sections. The term Bxy, in the general equation, will *rotate* the conic if $B \neq 0$. We will only consider the cases where $B = 0$. Certain characteristics of the equation made some graphs parabolas, some ellipses, and some hyperbolas. It is possible to detect which conic section will be graphed by looking at the original equation.

The Graph of $Ax^2 + Cy^2 + Dx + Ey + F = 0$

1. If A or C is zero, the graph is a parabola.
   (a) A *horizontal* axis of symmetry if $A = 0$.
   (b) A *vertical* axis of symmetry if $C = 0$.
2. If $A = C$, the graph is a circle.
3. If A and C are both positive or both negative, the graph is an ellipse.
4. If A and C are of opposite sign, the graph is a hyperbola.
5. There are certain degenerate forms that *may* result from these equations, such as lines, points, and no graph at all.

Some examples of the degenerate forms mentioned in 5, above, are:

Equation	*Graph*
$x^2 + y^2 = 0$	The point $(0, 0)$
$x^2 - 1 = 0$	Two parallel lines
$x^2 - y^2 = 0$	Two intersecting lines
$x^2 + y^2 + 1 = 0$	No graph

EXAMPLE 1. Identify each of the following as the equation of a parabola, circle, ellipse, or hyperbola. (None are degenerate.)

(a) $3x^2 - 4y^2 - 7x + 8y + 14 = 0$ (b) $5x^2 + 4y^2 + x - 10 = 0$
(c) $5x^2 + 4y + x - 10 = 0$ (d) $-3x^2 - 3y^2 + 6x + 5y - 11 = 0$

Solutions:

(a) Neither the coefficient of x^2 nor the coefficient of y^2 is zero, and they are of opposite sign. Therefore, the graph of this equation is a hyperbola.
(b) The coefficient of x^2 and the coefficient of y^2 are both positive; thus the graph is an ellipse.
(c) There is no y^2 term in this equation, so its coefficient must be zero. The graph is a parabola opening left or right.
(d) The coefficients of x^2 and y^2 are equal. The graph is a circle. Of course, it is also an ellipse.

> **Graphing Conic Sections**
>
> 1. If the center (vertex, in the case of a parabola) is not at the origin, lightly sketch a coordinate system at the point where the center is shifted. Work from this "new" coordinate system.
> 2. **Parabola.** Plot the vertex and a point or two on either side of the vertex. Draw the smoothest "parabola shape" possible through those points.
> 3. **Ellipse.** Mark the x- and y-intercepts and connect them with the smoothest oval possible.
> 4. **Hyperbola.** Draw the construction lines lightly. Check to see if it is an up–down or left–right hyperbola, then draw both parts as smooth curves.
>
>   **Please Note!** The conic sections are all very smooth curves without sharp turns or pointy places. Draw them as evenly as possible.

■ PROBLEM SET 10.4

Warm-ups

In problems 1 through 8, identify each equation as that of a parabola, circle, ellipse, or hyperbola. (Assume that none are degenerate.) See Example 1.

1. $x^2 - y^2 - 3x + y - 4 = 0$
2. $22x^2 + 7y^2 + 41x - 234 = 0$
3. $8x^2 + 5y^2 - 19x - 7y - 25 = 0$
4. $9y^2 + 32x - 36y + 64 = 0$
5. $3x^2 + 3y^2 + 13y = 0$
6. $2x^2 - 3y^2 + 7x + 14y - 4 = 0$
7. $x^2 - 4x + 36y + 31 = 0$
8. $64x + 8y - 16x^2 - 16y^2 + 64 = 0$

Practice Exercises

In problems 9 through 20, identify each equation as that of a parabola, circle, ellipse, or hyperbola. (Assume that none are degenerate.)

9. $13 - 2y^2 + 3x - 7y = 0$
10. $5x^2 - 11x + 7y^2 + 9y - 44 = 0$
11. $3x^2 + 2y^2 - 2x - 2y - 14 = 0$
12. $11x^2 + 11y^2 - 13x - 5 = 0$
13. $11x^2 + 11y - 13x - 5 = 0$
14. $11x^2 - 11y^2 - 13x - 5 = 0$
15. $5x^2 + 6x - 6y^2 - 5y = 0$
16. $5x^2 + 5y - 7x + y^2 - 17 = 0$
17. $2x - 3y + 4x^2 + 5y^2 - 6 = 0$
18. $x^2 + 3x - 4y = 16 + 3y^2$
19. $7x - 8y - 4x^2 - 4y^2 = 13$
20. $17x = 21y^2$

Challenge Problems

In problems 21 and 22, sketch the graph of each, if it exists.

21. $4x^2 + 3y^2 - 8x + 6y + 7 = 0$
22. $x^2 - 4y^2 = 0$

■ IN YOUR OWN WORDS . . .

23. Describe the kind of equations that graph as conic sections.

CHAPTER SUMMARY

SECOND-DEGREE EQUATION IN TWO VARIABLES

The graph of the general second-degree equation in two variables

$$Ax^2 + Bxy + Cy^2 + Dx + Ey + F = 0; \quad A, B, C \text{ not all zero}$$

is a conic section or a degenerate form of a conic section.

THE CONIC SECTIONS

The basic forms of the conic sections are:

1. Parabola

   $y = ax^2$ $x = by^2$

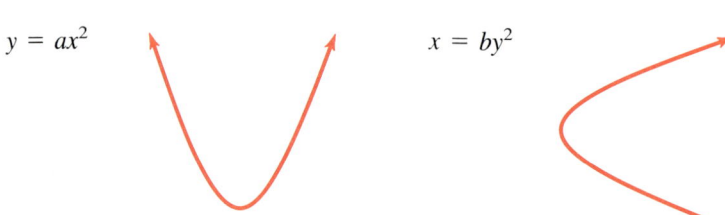

2. Ellipse

   $$\frac{x^2}{a^2} + \frac{y^2}{b^2} = 1$$

3. Hyperbola

   $$\frac{x^2}{a^2} - \frac{y^2}{b^2} = 1 \qquad \frac{y^2}{b^2} - \frac{x^2}{a^2} = 1$$

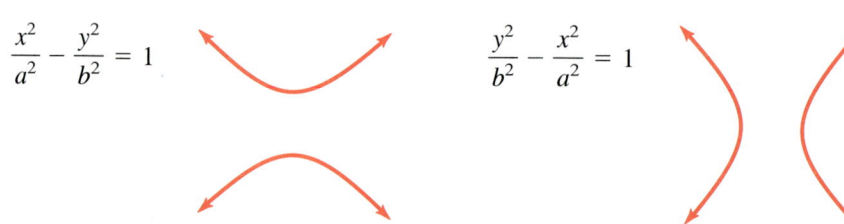

4. Circle (a special ellipse)

   $$x^2 + y^2 = r^2$$

TRANSLATION OF AXES

The substitution of $(x - h)$ for x and $(y - k)$ for y in one of the basic forms will shift the graph so that its center (or vertex in the case of a parabola) is at the point (h, k).

CHECKUPS

In Checkups 1 through 4, sketch the graph of each equation.

1. $(x - 1)^2 + (y - 2)^2 = 16$ Section 10.1; Example 1b
2. $y = -2(x - 1)^2 + 2$ Section 10.2; Example 7
3. $4x^2 + y^2 - 24x + 4y + 24 = 0$ Section 10.3; Example 5

4. $4x^2 - 6y^2 + 32x + 12y + 22 = 0$ Section 10.3; Example 6
5. Identify each of the following as the equation of a parabola, circle, ellipse, or hyperbola.   Section 10.4; Example 1
   (a) $3x^2 - 4y^2 - 7x + 8y + 14 = 0$
   (b) $5x^2 + 4y^2 + x - 10 = 0$
   (c) $5x^2 + 4y + x - 10 = 0$
   (d) $-3x^2 - 3y^2 + 6x + 5y - 11 = 0$

REVIEW PROBLEMS

In problems 1 through 10, identify the conic section.

1. $x^2 - y^2 - 2x - 6y = 0$
2. $x^2 + y^2 - 4x + 8y = 0$
3. $x^2 + 5y^2 - 10x - 13 = 0$
4. $x^2 - y - 2x = 0$
5. $2x^2 + 3y^2 - 4x + 6y = 21$
6. $x - 2y^2 - 8y - 2 = 0$
7. $3x^2 - 5y^2 - 18x - 20y - 100 = 0$
8. $5x + 10y - 5x^2 - 5y^2 + 33 = 0$
9. $2x^2 + x - y + 8 = 0$
10. $3x^2 + 4y^2 + 2x - 3y = 0$

In problems 11 through 20, sketch the graph of each equation.

11. $\dfrac{x^2}{4} + \dfrac{y^2}{25} = 1$
12. $(x - 3)^2 + (y + 2)^2 = 16$
13. $x + 2 = 2(y - 1)^2$

14. $\dfrac{(y + 3)^2}{9} - \dfrac{(x + 1)^2}{36} = 1$
15. $x^2 - y^2 - 4x - 4y - 16 = 0$
16. $4x^2 + 5y^2 + 24x - 10y + 21 = 0$

17. $16x^2 + 9y^2 - 32x + 18y - 119 = 0$
18. $16x^2 - 4y^2 + 96x - 16y + 96 = 0$

19. $x^2 + y^2 - 4x + 6y + 12 = 0$
20. $y = 2x^2 + 8x + 7$

... LET'S NOT FORGET ...

Identify the expressions that are in factored form. Factor those that are not factored.

21. $(2x - 1)(4x^2 + 2x + 1)$
22. $(3x - 2)^3$
23. $8 + z^3$
24. $(11x - 17)^2$
25. $63x^2 - 42x + 7$
26. $5 - 20x^2$

How many terms are in each expression? Which expressions have $x - 3$ as a factor?

27. $(x - 3)^2 - (x - 3)$ 2;
28. $x - 3 + y$
29. $x^2 - 6x + 9$ 3;
30. $x^3 - 3$

Simplify each expression, if possible, leaving only nonnegative exponents in your answer.

31. -4^{-4}
32. $(-4)^{-1/2}$
33. $\dfrac{4x^2 - 9}{27 - 8x^3}$
34. $\dfrac{2^{-2} + x^{-2}}{2x^{-2}}$
35. $\dfrac{-3x^{-2}}{y^{-3}z^2}$
36. $\sqrt[5]{x^5 + y^5}$

Find each product.

37. $[\sqrt{x} + 2]^2$
38. $(6 - 5x)^3$
39. $a^2b(abc)$

Reduce, if possible.

40. $\dfrac{2(x - 7) - b(7 - x)}{(x + 7)(x - 7)}$
41. $\dfrac{x(a + b) + y(a - b)}{(a + b)(a - b)}$

The following problems can be worked by using a least common denominator. Follow the directions in each and notice how the LCD is used.

42. Perform the operation indicated: $\dfrac{x}{5 - x} + \dfrac{2}{x - 5}$.

43. Solve $\dfrac{x^2}{x^2 - 25} + \dfrac{1}{5 - x} = \dfrac{x}{5 + x}$.

44. Simplify $\dfrac{\dfrac{x + 1}{5 - x}}{\dfrac{x - 2}{x - 5}}$.

Label each as an equation or an expression. Solve the equations in one variable, graph the equations in two variables, and perform the operations indicated on the expressions.

45. $(x - 7)^2$
46. $y = (x - 7)^2$
47. $\left|\dfrac{7 - 5x}{11}\right| = 3$
48. $\dfrac{1}{2x^2 - 3x - 5} + \dfrac{1}{10 + x - 2x^2}$
49. $\dfrac{3 - 2i}{1 + i}$
50. $\sqrt{x + 1} = x - 1$

CHAPTER TEST

In problems 1 through 9, circle the correct response.

1. The radius of the circle given by the equation $x^2 + y^2 - 2x + 4y = 4$ is (?)
   A. 2   B. 3
   C. 9   D. $\sqrt{14}$

2. The vertex of the parabola given by the equation $x = y^2 - 4y + 3$ is (?)
   A. $(-1, 2)$ B. $(1, -2)$
   C. $(3, -4)$ D. The origin

3. The center of the graph of the equation
   $5x^2 + 6y^2 + 50x - 24y + 119 = 0$ is (?)
   A. $(5, -2)$ **B.** $(-5, 2)$
   C. $(10, -4)$ **D.** $(-5, 4)$
4. The graph of the equation $x^2 + y + 2x - 3 = 0$ is (?)
   A. an ellipse, but not a circle
   B. a circle
   C. a parabola opening upward
   D. a parabola opening downward
5. The graph of the equation $2x^2 + 2y^2 + 2x - 3 = 0$ is (?)
   A. an ellipse, but not a circle
   B. a circle
   C. a parabola opening upward
   D. a parabola opening downward
6. The graph of the equation $4x^2 + 8x - 4y^2 + 8y - 4 = 0$ is (?)
   A. an ellipse, but not a circle   **B.** a circle
   C. a parabola   **D.** a hyperbola
7. The graph of the equation $2x^2 + 2y + y^2 = 0$ is (?)
   A. an ellipse, but not a circle   **B.** a circle
   C. a parabola   **D.** a hyperbola
8. The graph of the equation
   $$2x^2 + 3y^2 - 8x + 18y + 29 = 0$$
   most nearly resembles (?)

   A.
   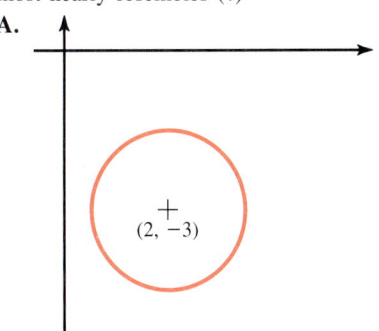

   B.
   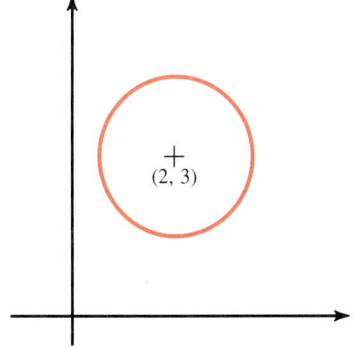

   8. C.
   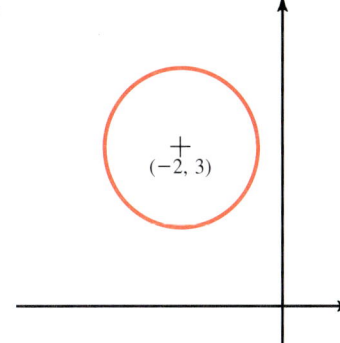

   D.
   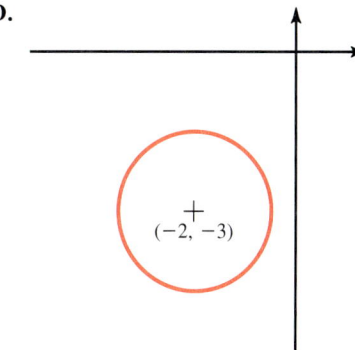

9. The graph of the equation $y^2 - x - 4y + 5 = 0$ most nearly resembles (?)

   A.
   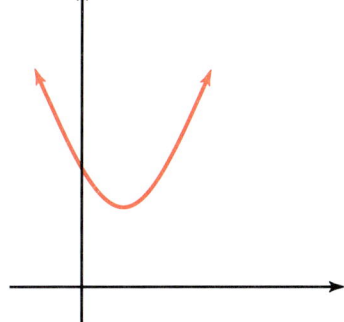

   B.
   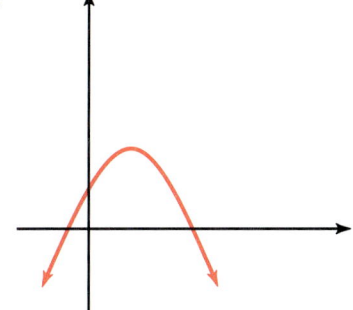

   (*see p. 476 for parts C and D*)

C.

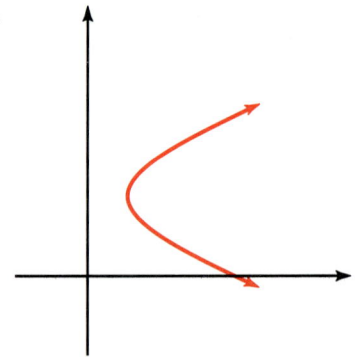

D.

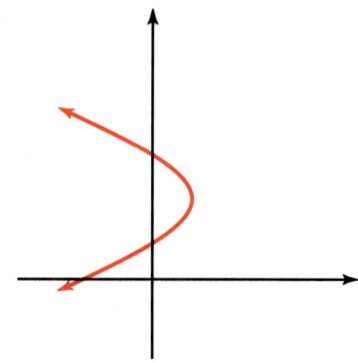

10. Find the center and radius of the circle given by the equation $x^2 + y^2 - 4y = 1$.

11. Where is the vertex of the parabola given by the equation $y = -2x^2 - 4x - 3$, and in what direction does it open?

In problems 12 through 15, sketch the graph of the equation.

12. $x^2 - y^2 + 4x + 2y + 4 = 0$

13. $4x^2 + 9y^2 - 16x - 18y - 11 = 0$

14. $x^2 + y^2 - 6x - 6y + 9 = 0$

15. $x = y^2 - 6y + 9$

CHAPTER 11

See Problem Set 11.5, Exercise 48.

Functions

11.1 Definition of Function and Functional Notation
11.2 Graphs of Functions
11.3 Algebra of Functions
11.4 Inverse Functions
11.5 Exponential and Logarithmic Functions
11.6 Properties of Logarithms

CONNECTIONS

In our everyday lives, objects are frequently paired together. A person is given a name. A football player is assigned a jersey with a number on it. A student is assigned a grade. A rectangle has a perimeter. The telephone company sends each customer a bill.

There are several ways of showing these pairings. Tables or charts that list all the pairs are very common. The formula

$$F = \frac{9}{5}C + 32$$

pairs a Celsius reading with a Fahrenheit reading. Ordered pairs are often used.

The idea of pairing is essential to the concept of a function. The study of functions is one of the most important aspects of mathematics. In this chapter we focus on an introduction to functions.

11.1 DEFINITION OF FUNCTION AND FUNCTIONAL NOTATION

In this section the concept of a function is defined and functional notation is introduced.

Function

A **function** is a rule that assigns to each member in one set (the **domain**) exactly one member from another set (the **range**).

Both a rule and a domain must be specified when a function is defined.

EXAMPLE 1. Consider the assignments made by the following sketches. Is each a function?

(a)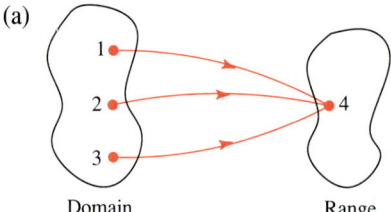

(b)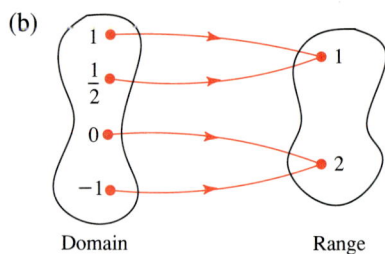

(c)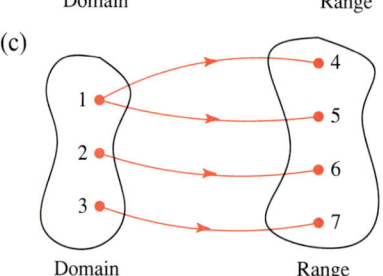

Solutions:

(a) This is a function. Its rule assigns 4 to every number in the domain. To be a function, a rule must assign to each number in the domain *exactly one number* in the range. This rule does that. It does not matter that more than one number in the domain is paired with the same number in the range.

(b) This is also a function. Its domain is $\left\{-1, 0, \frac{1}{2}, 1\right\}$ and its range is $\{1, 2\}$. It does not matter that both 1 and $\frac{1}{2}$ are paired with 1 or that both -1 and 0 are paired with 2. Each number in the domain is assigned to *exactly one* number in the range.

(c) This is *not* a function. The number 1 is assigned to two different numbers, 4 and 5. This violates the definition of a function which states that *each* number in the domain must be assigned to *exactly one* number in the range. □

There are several ways of giving the rule for a function. Using pictures as in Example 1 is one way. Sometimes the assignments are listed as ordered pairs or in tables with the first number being from the domain.

EXAMPLE 2. The domain of a function is $\{1, 2, 3, 4\}$ and the rule is: Assign to each number its square.

(a) Draw a picture to show the assignment of numbers in the domain to numbers in the range.
(b) Write the assignments as ordered pairs.
(c) Find the range.

Solutions:

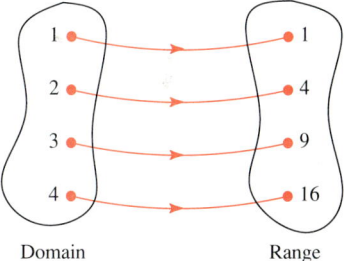
Domain   Range

(a) Draw a picture to indicate the assignment of numbers in the domain to numbers in the range.
(b) The assignments can be listed with the ordered pairs: (1, 1), (2, 4), (3, 9), (4, 16).
(c) So the range is $\{1, 4, 9, 16\}$.
Notice that *each* number in the domain is assigned to *exactly one* number in the range. □

In mathematical applications of functions the rule is often given using **functional notation.** We use letters such as f, g, and h to name functions.

If f is a function and x is a number in its domain, $f(x)$ is a symbol used to indicate the corresponding number in the range. We read $f(x)$ as "f of x" or "f evaluated at x." This could be represented by writing the ordered pair $(x, f(x))$ or by the following sketch.

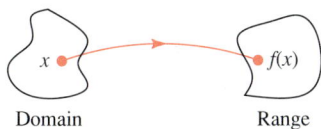
Domain   Range

NOTE: In using functional notation, remember that $f(x)$ may look like f times x, but it does *not* mean that.

Be Careful!

Sec. 11.1 Definition of Function and Functional Notation

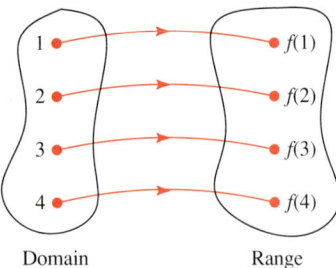

Domain   Range

The rule in Example 2 is: Assign to each number its square. Written in functional notation, this would be

$$f(x) = x^2$$

If x is a number in the domain, it is paired with $f(x)$ in the range. The illustration to the left shows the pairings.

Remember that $f(1)$ is a name for the number in the range paired with 1 from the domain. To calculate the value of $f(1)$, we use the rule $f(x) = x^2$ and replace x with 1. Thus

$$f(1) = 1^2 = 1$$
$$f(2) = 2^2 = 4$$
$$f(3) = 3^2 = 9$$
$$f(4) = 4^2 = 16$$

EXAMPLE 3. If $h(x) = x^2 - 2x$, and the domain is $\mathbb{R}$, find

(a) $h(0)$ (b) $h(-5)$
(c) $h(\sqrt{3})$ (d) $h(\pi)$

Solutions:

(a) $h(0) = 0^2 - 2(0)$ Replace x with 0 in $h(x)$.
$= 0 - 0$
$= 0$

(b) $h(-5) = (-5)^2 - 2(-5)$ Replace x with -5 in $h(x)$.
$= 25 + 10$
$= 35$

(c) $h(\sqrt{3}) = (\sqrt{3})^2 - 2\sqrt{3}$ Replace x with $\sqrt{3}$ in $h(x)$.
$= 3 - 2\sqrt{3}$

(d) $h(\pi) = \pi^2 - 2\pi$ Replace x with π in $h(x)$.

EXAMPLE 4. If $f(x) = \sqrt{x + 1}$ with domain $x \geq 0$ and $g(x) = 3x - 4$ with domain $\mathbb{R}$, find

(a) $f(3) + g(2)$ (b) $f(0) \cdot g(0)$
(c) $\sqrt{g(2)}$ (d) $[g(1)]^2$

Solutions:

(a) $f(3) + g(2) = \sqrt{3 + 1} + 3(2) - 4$
$= 2 + 6 - 4$
$= 4$

(b) $f(0) \cdot g(0) = \sqrt{0 + 1} \cdot (3 \cdot 0 - 4)$
$= 1(-4)$
$= -4$

(c) $\sqrt{g(2)} = \sqrt{3(2) - 4}$
$= \sqrt{2}$

(d) $[g(1)]^2 = (-1)^2$
$= 1$

EXAMPLE 5. If $g(x) = x^2 + 3$ with domain $\mathbb{R}$, find

(a) $g(a)$ (b) $g(b)$ (c) $g(a + b)$ (d) $g(a) + g(b)$

Solutions:

(a) $g(a) = a^2 + 3$ Replace x with a in $g(x)$.

(b) $g(b) = b^2 + 3$ Replace x with b in $g(x)$.

(c) $g(a + b) = (a + b)^2 + 3$ Replace x with $a + b$ in $g(x)$.

(d) $g(a) + g(b) = (a^2 + 3) + (b^2 + 3)$
$$= a^2 + b^2 + 6$$

It is common practice to state the rule for a function and omit the domain. When this happens, we use the natural domain of the rule. The **natural domain** is the largest subset of real numbers for which the rule has meaning.

EXAMPLE 6. Give the natural domain of the function $k(x) = 2x - 3$.

Solution:

$k(x)$ is a real number for any value of x. So the natural domain of k is $\{x \mid x$ is a real number$\}$.

EXAMPLE 7. Find the natural domain of the function

$$g(x) = \frac{5}{x + 3}$$

Solution:

We can see that $g(x)$ is a real number for any value of x except when x is -3. $g(-3)$ is $\frac{5}{-3 + 3}$ or $\frac{5}{0}$, which is undefined. So $g(-3)$ is not a real number. Thus the natural domain of g is $\{x \mid x \neq -3\}$.

EXAMPLE 8. Find the natural domain of the function $g(x) = \sqrt{x}$.

Solution:

$g(x)$ is a real number for all nonnegative values of x, that is, for $x \geq 0$. [If x is -5, then $g(-5)$ is $\sqrt{-5}$ which is not a real number.] So the natural domain of g is $\{x \mid x \geq 0\}$.

EXAMPLE 9. Give the natural domain of the function

$$h(x) = \sqrt{x^2 - x - 2}$$

Solution:

$h(x)$ will be a real number if $x^2 - x - 2 \geq 0$. We solve the inequality by the technique of boundary numbers developed in Chapter 7.

(continued)

The equation $x^2 - x - 2 = 0$ becomes

$$(x - 2)(x + 1) = 0$$

and gives boundary numbers of -1 and 2.

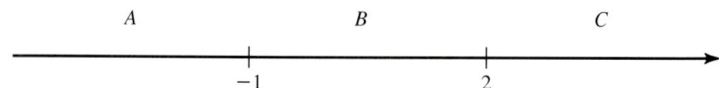

Testing each of the regions A, B, and C, the solution set for $x^2 - x - 2 \geq 0$ is $\{x \mid x \leq -1 \text{ or } x \geq 2\}$.

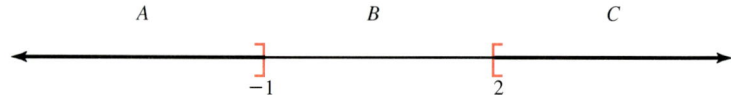

So the natural domain of h is $\{x \mid x \leq -1 \text{ or } x \geq 2\}$. ☐

A function can be evaluated only at numbers in its domain. If $f(x) = \sqrt{x}$, then $f(-2)$ is undefined since $\sqrt{-2}$ is not a real number.

PROBLEM SET 11.1

Warm-ups

In problems 1 through 4, identify which rules below establish functions. Write the assignments in the pictures as ordered pairs. See Examples 1 and 2.

1.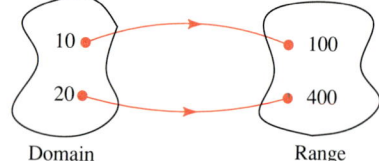

2.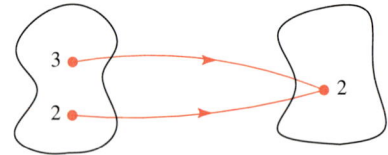

3.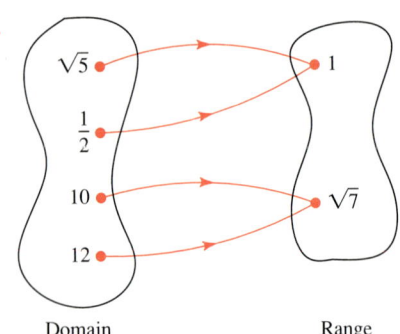

4.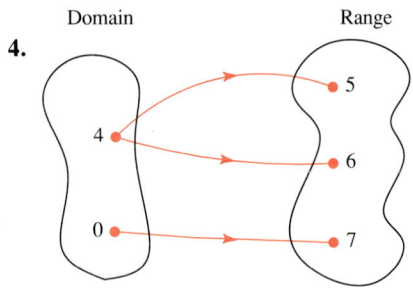

For problems 5 through 11, see Example 3.
If $f(x) = 2x - 5$, find

5. $f(2)$ **6.** $f(-3)$ **7.** $f(0)$ **8.** $f\left(\dfrac{1}{2}\right)$

If $g(x) = \sqrt{x + 1}$, find

9. $g(3)$ **10.** $g(0)$ **11.** $g(-1)$

For problems 12 through 16, see Example 4.

If $f(x) = x^2$ and $g(x) = 2x - 3$, find

12. $f(3) + g(3)$ **13.** $f(3) - g(3)$ **14.** $f(3)g(3)$

15. $\dfrac{f(3)}{g(3)}$ **16.** $(g(3))^2$

For problems 17 through 21, see Example 5.

If $f(x) = 2x^2 + x - 1$, find

17. $f(t)$ **18.** $f(a + b)$ **19.** $f(a) + f(b)$
20. $f(2t)$ **21.** $f(\sqrt{a})$

In problems 22 through 33, find the natural domain of each function.
For problems 22 through 25, see Example 6.

22. $g(x) = 3x - 1$ **23.** $h(x) = x^2 - 4$ **24.** $d(x) = \sqrt[3]{x}$ **25.** $f(x) = |x|$

For problems 26 through 29, see Example 7.

26. $r(x) = \dfrac{7}{x + 2}$ **27.** $f(x) = \dfrac{3}{x} + 2$

28. $h(x) = \dfrac{x + 7}{x - 1}$ **29.** $g(x) = \dfrac{6}{(x + 2)(x - 7)}$

For problems 30 through 33, see Examples 8 and 9.

30. $f(x) = \sqrt{x + 2}$ **31.** $g(x) = \sqrt{x - 1}$
32. $f(x) = \sqrt{x^2 - x - 6}$ **33.** $p(x) = \sqrt{x^2 + x - 2}$

Practice Exercises

In problems 34 through 40, find the range for each function.

34. Domain is $\{-1, 0, 1\}$; $f(x) = x^3$.
35. Domain is $\{1, 4, 7, 11, 14\}$; $f(x) = x - 7$.
36. Domain is $\{x \mid x \text{ is an odd natural number}\}$; $f(x) = x + 1$.
37. Domain is $\{x \mid x \text{ is an even natural number}\}$; $f(x) = 2x - 1$.
38. Domain is $\{x \mid x \text{ is an integer}\}$; $f(x) = 2x$.
39. Domain is $\{x \mid x \text{ is a real number}\}$; $f(x) = 11$.
40. Domain is $\{x \mid x \geq 0\}$; $f(x) = \sqrt{x}$.

In problems 41 through 60, find the natural domain for each function.

41. $g(x) = x - 1$ **42.** $f(x) = \dfrac{1}{3}x + 1$

43. $h(x) = \dfrac{x}{3} - \dfrac{1}{4}$ **44.** $g(x) = \dfrac{1}{x + 2}$

45. $r(x) = \dfrac{3}{x - 1}$ **46.** $f(x) = \dfrac{1}{x} + 2$

47. $h(x) = \dfrac{x}{x - 3}$ **48.** $g(x) = \dfrac{3}{(x - 2)(x + 3)}$

49. $h(x) = \dfrac{x}{x + 3}$ **50.** $t(x) = \dfrac{x}{(x - 2)(x + 1)}$

51. $f(x) = \sqrt{x - 2}$ **52.** $g(x) = \sqrt{x + 1}$
53. $h(x) = x^2 - 3$ **54.** $f(x) = \sqrt{x^2 - 2x - 3}$
55. $d(x) = \sqrt[3]{x} + 1$ **56.** $p(x) = \sqrt{x^2 + 2x - 15}$
57. $g(x) = \sqrt{x^2 + 1}$ **58.** $f(x) = |x + 1|$
59. $f(x) = |x - 5|$ **60.** $w(x) = |2 - x|$

Sec. 11.1 Definition of Function and Functional Notation

If $f(x) = 3x + 1$, find

61. $f(2)$ **62.** $f(-1)$ **63.** $f(0)$ **64.** $f\left(\dfrac{1}{3}\right)$

65. $f(\sqrt{3})$ **66.** $f(a)$ **67.** $f(a - b)$ **68.** $f(a) - f(b)$

69. $f(2t)$ **70.** $f(\sqrt{b})$ **71.** $\sqrt{f(a)}$ **72.** $f(-a)$

If $g(x) = \sqrt{x - 1}$, find

73. $g(6)$ **74.** $g(0)$ **75.** $g(2)$ **76.** $g(-2)$

77. $2g(5) - 1$ **78.** $g(b - 1)$ **79.** $g(2k + 3)$

80. $g(a^2)$ **81.** $[g(a)]^2$ **82.** $g(\pi)$

If $f(x) = 1 - 2x^2$, find

83. $f(t + h)$ **84.** $f(t) + f(h)$ **85.** $f(t + h) - f(t)$

86. $f(2t)$ **87.** $2f(t)$ **88.** $f(kt)$ **89.** $kf(t)$

90. $\dfrac{f(3t)}{3}$ **91.** $f(3t) - f(2t)$ **92.** $f(t) + 2f(k)$

If $g(x) = \dfrac{1}{x - 1}$, find

93. $g(0)$ **94.** $g(-1)$ **95.** $g(1)$ **96.** $2g(5)$

97. $\dfrac{3}{g(7)}$ **98.** $g\left(\dfrac{1}{2}\right)$ **99.** $\dfrac{g(1)}{g(2)}$ **100.** $g\left(\dfrac{-3}{2}\right)$

If $f(x) = 2x^2$ and $g(x) = x + 5$, find

101. $f(3) + g(3)$ **102.** $f(3) - g(3)$ **103.** $f(3)g(3)$

104. $\dfrac{f(3)}{g(3)}$ **105.** $f(g(3))$ **106.** $g(f(3))$

107. The Yellow Cab Company's fare (in dollars) for riding x miles is given by

$$F(x) = 0.25x + 1.5$$

(a) What is the fare for a 10-mile ride?
(b) What would $F(3)$ represent?
(c) How far would $3 take you?

108. The cost (in dollars) of manufacturing n radios is given by

$$C(n) = n^2 + n + 100$$

(a) What is the cost of manufacturing 1 radio?
(b) What would $C(100)$ represent?
(c) How many radios could be built for $210?

109. The number of ants on a picnic table after t minutes is given by

$$N(t) = t^3 - 2t^2 + t$$

(a) How many ants are on the picnic table after 1 minute?
(b) How many ants are on the table after 2 minutes?
(c) What does $N(10)$ represent?

Challenge Problems

110. Find the natural domain for $f(x) = \dfrac{1}{\sqrt{x^2 - x - 2}}$.

111. Find the natural domain of $f(x) = \dfrac{\sqrt{x + 1}}{\sqrt{x - 2}}$.

■ IN YOUR OWN WORDS

112. What is a function?

113. What is the domain of a function?

114. Explain the relationship between x and $f(x)$.

11.2 GRAPHS OF FUNCTIONS

To **graph** a function f means to graph the equation $y = f(x)$ for every x in the domain of the function. We begin by graphing some special kinds of functions: linear, quadratic, and absolute value.

Let's consider linear functions first.

Linear Functions

A function of the form

$$f(x) = mx + b$$

where m and b are real numbers, is called a **linear function.** The graph of a linear function is a line.

EXAMPLE 1. Graph the function $g(x) = 3x - 1$.

Solution:

We graph the equation $y = 3x - 1$. Its graph is the line below.

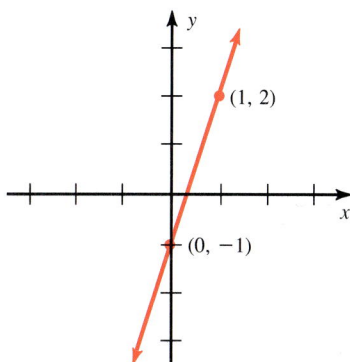

EXAMPLE 2. Graph $f(x) = x$.

Solution:

Graph $y = x$. The graph is the line below.

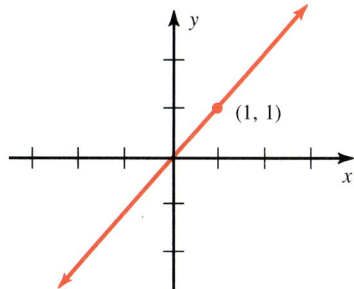

This linear function is called the **identity function.**

Sec. 11.2 Graphs and Functions

EXAMPLE 3. Graph the function $g(x) = 2$.

Solution:

Graph the equation $y = 2$. It is the horizontal line below.

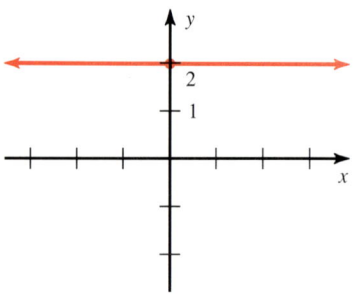

Such a function is called a **constant function.**

Now let's look at quadratic functions.

Quadratic Functions

A function of the form
$$f(x) = ax^2 + bx + c$$
where a, b, and c are real numbers and $a \neq 0$, is called a **quadratic function.** The graph of a quadratic function is a **parabola.**

EXAMPLE 4. Graph each function.

(a) $g(x) = 2 - x^2$ (b) $f(x) = x^2 - 2x + 3$

Solutions:

(a) The graph of $y = 2 - x^2$ is a parabola with vertex at $(0, 2)$ opening down. The graph is shown below.

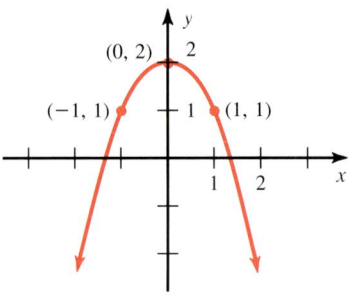

(b) The graph of $y = x^2 - 2x + 3$ will be a parabola. The equation must be in the form $y = p(x - h)^2 + k$ in order to find the vertex. To do this, complete the square.

$$y = x^2 - 2x + 3$$

First, group the x-terms together.
$$y = (x^2 - 2x) + 3$$
Then complete the square by taking $\frac{1}{2}$ the coefficient of x and squaring it. Add and subtract this number.
$$y = (x^2 - 2x + 1) + 3 - 1$$
$$y = (x - 1)^2 + 2$$

The vertex is (1, 2) and the parabola will open upward. The graph is shown below.

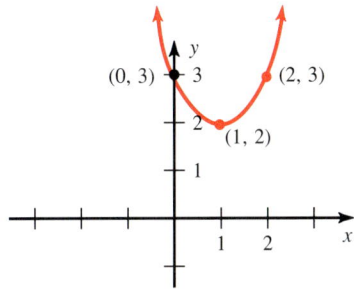

The graph of an absolute value function behaves much like the graph of a quadratic function.

EXAMPLE 5. Graph $f(x) = |x|$.

Solution:

Plotting several points shows us that the graph is in the shape of a "V."

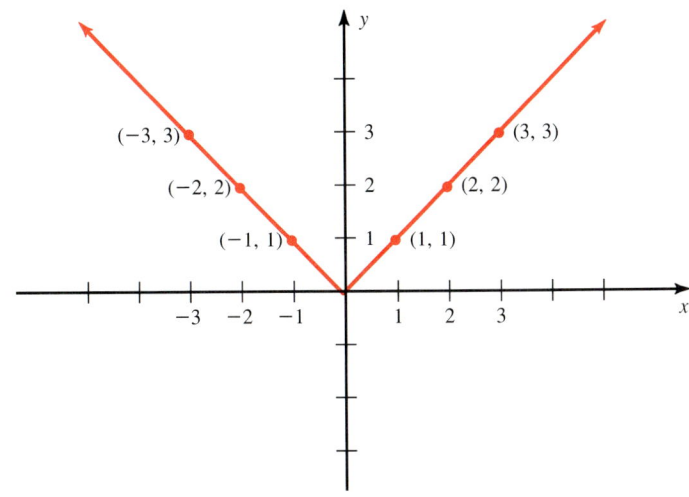

Absolute value graphs can be shifted up and down or right and left just as we did with the parabola.

> ## Absolute Value Functions
> A function in the form
> $$f(x) = p|x - h| + k$$
> is called an **absolute value** function. The graph is a "V" with vertex at the point (h, k). The "V" opens upward if p is positive and downward if p is negative.

EXAMPLE 6. Give the vertex of the graph of each function.

(a) $f(x) = |x| + 1$ (b) $g(x) = |x + 2|$ (c) $h(x) = |x - 2| - 1$

Solutions:

(a) $f(x) = |x| + 1$
$\quad\quad = |x + 0| + 1$
So the vertex is $(0, 1)$.

(b) $g(x) = |x + 2|$
$\quad\quad = |x + 2| + 0$
So the vertex is $(-2, 0)$.

(c) $h(x) = |x - 2| - 1$
The vertex is $(2, -1)$.

EXAMPLE 7. Graph $g(x) = |x + 2| - 1$.

Solution:

The vertex will be $(-2, -1)$ and the graph will open upward.

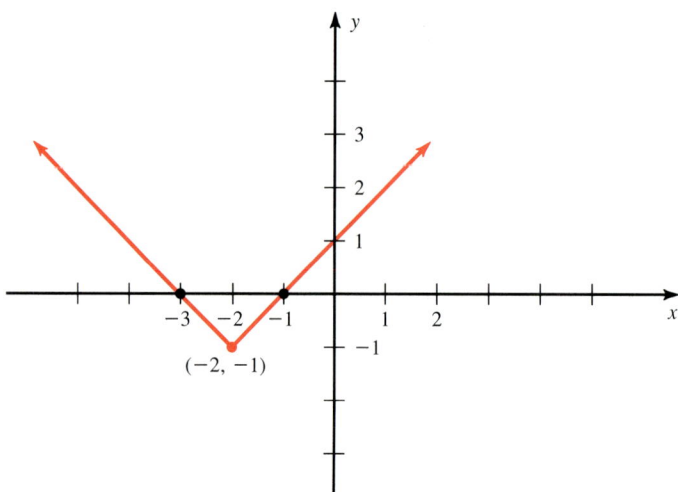

A graph of a function is very helpful to determine its range. The range consists of all possible y-values. Consider the linear function f graphed below. If a value on the y-axis is chosen, is there a point on the line having it as a y-coordinate?

Consider 3 on the y-axis.

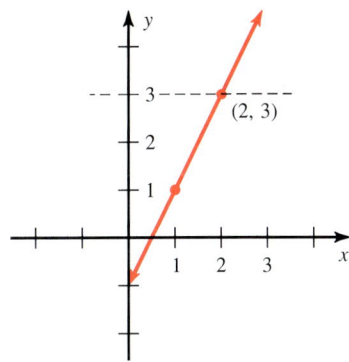

The dashed line shows that there is a point on the line with a y-coordinate of 3. Notice that (2, 3) is a point on the line. So 3 is in the range of f. Since any value on the y-axis can be chosen, the range of f is $\{y \mid y \text{ is a real number}\}$.

EXAMPLE 8. The graph of a function g is shown below. Give its domain and range.

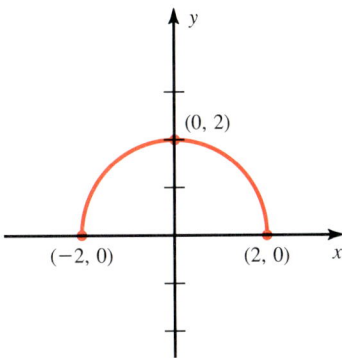

Solution:

The points on the graph of g have x-coordinates which are between -2 and 2 and also including -2 and 2. So the domain of f is $\{x \mid -2 \leq x \leq 2\}$. The y-coordinates of points on the graph of g are between 0 and 2, including 0 and 2. So the range of g is $\{y \mid 0 \leq y \leq 2\}$. □

EXAMPLE 9. The graph of h is at the right. Give its domain and range.

Solution:

The points on the graph of h use every real number as an x-coordinate. So the domain of h is $\{x \mid x \text{ is a real number}\}$. The only y-coordinate for points on the graph of h is 2. So the range of h is $\{2\}$. □

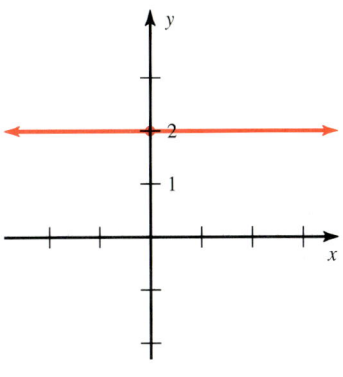

How can we look at a graph and determine whether or not it is the graph of a function? When will one number in the domain be paired with more than one number?

Sec. 11.2 Graphs and Functions

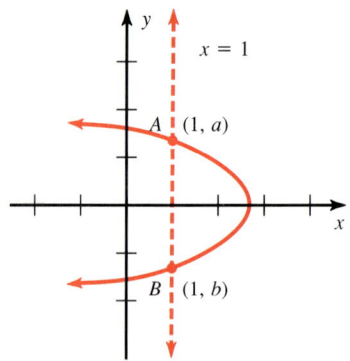

Look at the graph at the left. Draw the line $x = 1$.

The line crosses the graph at points A and B. The x-coordinate of A and B is 1. However, A and B have different y-coordinates, say a and b. This means that the number 1 in the domain is paired with the two numbers, a and b. So the graph could *not* be the graph of a function.

> ### Vertical Line Test
> If any vertical line crosses a graph in more than one point, the graph is **not** the graph of a function.

Is each graph the graph of a function?

(a)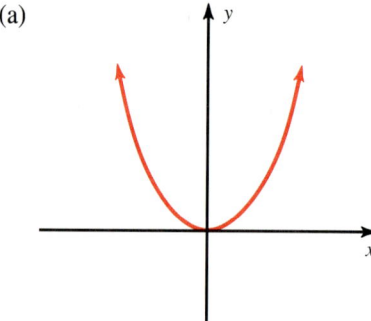

(b)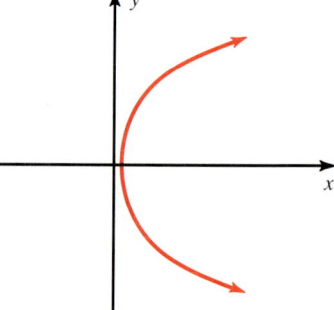

(c)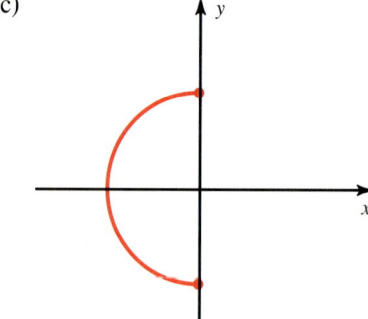

(d)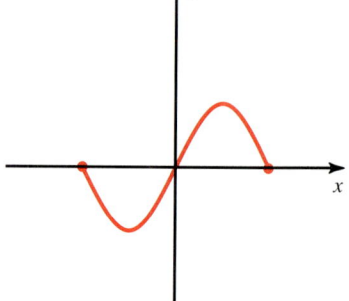

(e)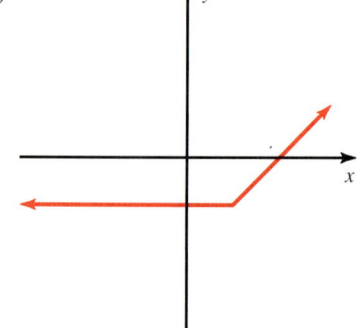

490   Chap. 11   Functions

Solution:

The graphs (a), (d), and (e) could be graphs of functions because they pass the vertical line test, but (b) and (c) are *not* graphs of functions because the vertical lines shown below cross each graph in more than one point.

(b) 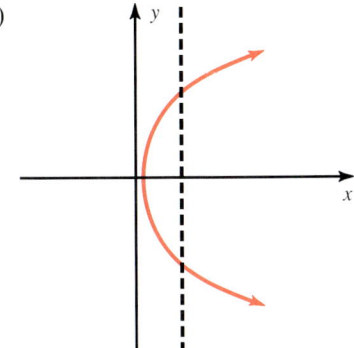   (c)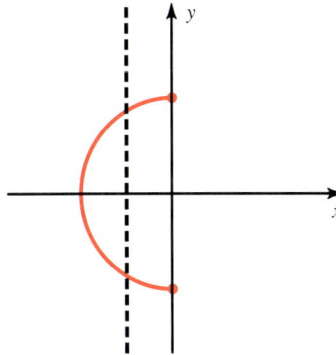

PROBLEM SET 11.2

Warm-ups

In problems 1 through 18, sketch the graph of each function. Label the vertex of parabolas and absolute value functions.
For problems 1 through 4, see Examples 1, 2, and 3.

1. $f(x) = 2x + 1$
2. $g(x) = 5x$
3. $h(x) = \frac{1}{2}x - 1$
4. $f(x) = -1$

For problems 5 through 12, see Example 4.

5. $f(x) = 3x^2$
6. $g(x) = 2x^2 + 3$
7. $h(x) = \frac{1}{2}x^2 - 2$
8. $f(x) = 1 - x^2$

9. $g(x) = (x + 4)^2$
10. $f(x) = -(x - 2)^2 + 1$
11. $g(x) = x^2 - 2x + 3$
12. $g(x) = 2x^2 + 4x - 1$

For problems 13 through 18, see Examples 5, 6, and 7.

13. $g(x) = |x - 4|$
14. $h(x) = |x + 3| + 2$
15. $f(x) = 2|x - 1| - 2$

16. $g(x) = -2|x + 2| - 1$
17. $h(x) = |x| + 2$
18. $h(x) = |x - 1| + 3$

In problems 19 through 24, give the domain and range for each function graphed. See Examples 8 and 9.

19.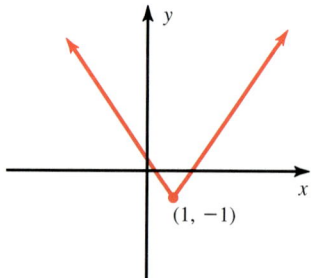

20.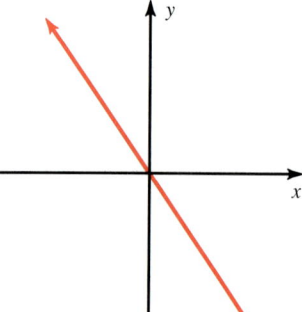

21.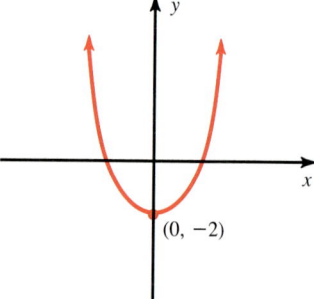

22.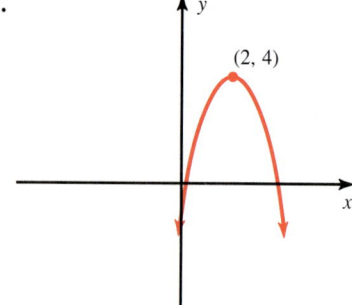

23.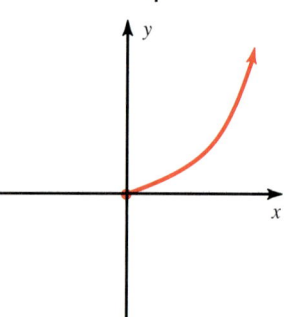

24.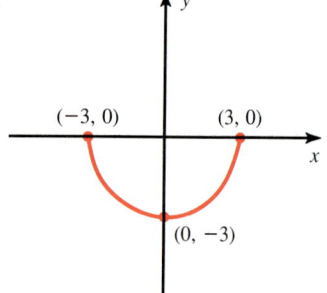

In problems 25 through 30, determine which graphs are graphs of functions. See Example 10.

25.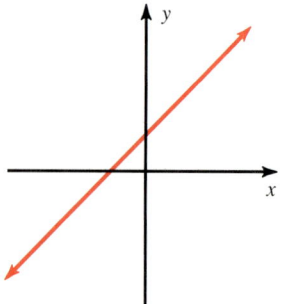

26.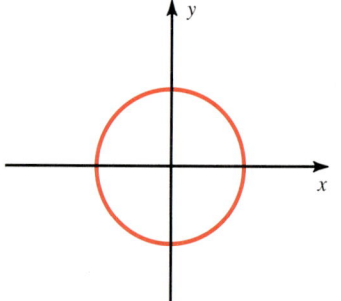

27.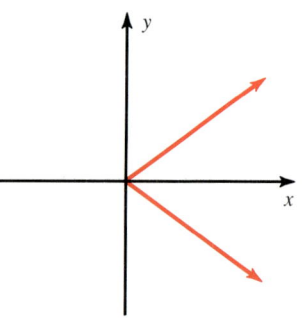

28. 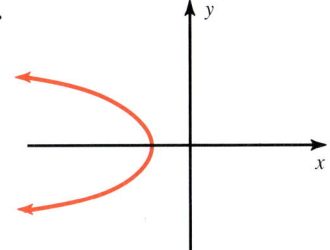   **29.** 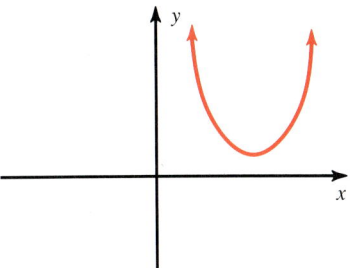   **30.**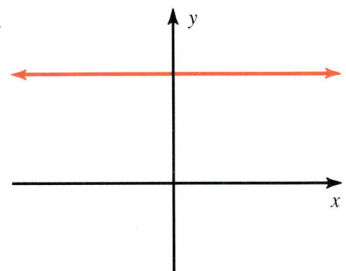

Practice Exercises

In problems 31 through 60, sketch the graph of each function and give its domain and range. Label the vertex of parabolas and absolute value functions.

31. $f(x) = 3x + 2$

32. $g(x) = 3x$

33. $h(x) = \dfrac{1}{4}x + 2$

34. $f(x) = 1$

35. $f(x) = 4x^2$

36. $g(x) = x^2 + 3$

37. $h(x) = \dfrac{1}{3}x^2 - 1$

38. $f(x) = 2 - x^2$

39. $f(x) = x^2 - 4$

40. $g(x) = \dfrac{1}{2}x^2 + 3$

41. $g(x) = (x - 3)^2$

42. $h(x) = (x + 1)^2 + 5$

43. $h(x) = \dfrac{1}{2}(x - 2)^2 + 1$

44. $f(x) = 3(x - 2)^2 + 4$

45. $f(x) = -(x - 1)^2 + 4$

46. $f(x) = -2(x + 3)^2 + 1$

Sec. 11.2 Graphs and Functions

47. $g(x) = x^2 - 2x + 4$ **48.** $h(x) = x^2 + 6x + 1$ **49.** $f(x) = x^2 - 6x$ **50.** $g(x) = x^2 - 8x$

51. $g(x) = 3x^2 + 6x + 5$ **52.** $h(x) = 2x^2 + 12x + 8$ **53.** $f(x) = -x^2 - 2x - 3$ **54.** $f(x) = -2x^2 + 8x - 7$

55. $g(x) = |x + 7|$ **56.** $h(x) = |x| + 3$ **57.** $f(x) = |x - 1| + 2$ **58.** $g(x) = -|x + 2| + 6$

59. $h(x) = 2|x| + 1$ **60.** $h(x) = -2|x - 1| + 1$

61. Graph $f(x) = x^2 + 1$. What is the smallest y-coordinate? This number is called the *minimum* value of f.

62. Graph $f(x) = 2 - x^2$. What is the *maximum* value of f?

63. The profit at Mary's Dress Shop is given by
$$P(x) = -(x - 250)^2 + 20{,}000$$
where x represents the number of dresses sold. Find how many dresses she must sell to receive a maximum profit.

64. The height (in feet) of a ball thrown upward from a building 100 ft high with an initial velocity of 96 ft/sec is given by
$$s(t) = -16t^2 + 96t + 100$$
where t is the time in seconds. Find how high up the ball will go.

Challenge Problems

In problems 65 through 74, graph each function. Give the domain and range for each.

65. $f(x) = \sqrt{x}$

66. $f(x) = \sqrt{x - 1}$

67. $f(x) = \sqrt{4 - x}$

68. $f(x) = -\sqrt{x}$

69. $f(x) = -\sqrt{x - 1}$

70. $f(x) = \sqrt{1 - x^2}$

71. $f(x) = x^3$

72. $f(x) = x^3 + 1$

73. $f(x) = x^3 - 1$

74. $f(x) = x^4$

■ IN YOUR OWN WORDS

75. Explain how to graph a function.

11.3 ALGEBRA OF FUNCTIONS

If f and g are functions, we can make new functions by adding, subtracting, multiplying, and dividing the rules for f and g.

Operations with Functions

1. $(f + g)(x) = f(x) + g(x)$ Sum
2. $(f - g)(x) = f(x) - g(x)$ Difference
3. $(fg)(x) = f(x)g(x)$ Product
4. $\left(\dfrac{f}{g}\right)(x) = \dfrac{f(x)}{g(x)};\ g(x) \neq 0$ Quotient

EXAMPLE 1. If $f(x) = x^2$ and $g(x) = x + 2$, find

(a) $(f + g)(x)$ (b) $(f - g)(x)$

(c) $(fg)(x)$ (d) $\left(\dfrac{f}{g}\right)(x)$

Solutions:

(a) $(f + g)(x) = f(x) + g(x)$
$= x^2 + x + 2$

(b) $(f - g)(x) = f(x) - g(x)$
$= x^2 - (x + 2)$
$= x^2 - x - 2$

(c) $(fg)(x) = f(x)g(x)$
$= x^2(x + 2)$
$= x^3 + 2x^2$

(d) $\left(\dfrac{f}{g}\right)(x) = \dfrac{f(x)}{g(x)}$
$= \dfrac{x^2}{x + 2}; \quad x \neq -2$

EXAMPLE 2. If $g(x) = \sqrt{x}$ and $h(x) = x^2$, find

(a) $(g + h)(2)$ (b) $\left(\dfrac{h}{g}\right)(4)$

Solutions:

(a) $(g + h)(2) = g(2) + h(2)$
$= \sqrt{2} + 4$

(b) $\left(\dfrac{h}{g}\right)(4) = \dfrac{h(4)}{g(4)}$
$= \dfrac{16}{\sqrt{4}}$
$= \dfrac{16}{2}$
$= 8$

Composition of Functions

If f and g are functions, then

$$(f \circ g)(x) = f(g(x))$$

for all x in the domain of g such that $g(x)$ is in the domain of f. $f \circ g$ is called the **composition** of f with g or f **composed with** g.

EXAMPLE 3. If $f(x) = 2x - 5$ and $g(x) = x + 3$, find $(f \circ g)(x)$.

Solution:

$$(f \circ g)(x) = f(g(x)) \quad \text{Definition}$$
$$= f(x + 3)$$
$$= 2(x + 3) - 5$$
$$= 2x + 6 - 5$$
$$= 2x + 1$$

Another way to evaluate $(f \circ g)(x)$ in this example is

$$(f \circ g)(x) = f(g(x))$$
$$= 2g(x) - 5$$
$$= 2(x + 3) - 5$$
$$= 2x + 6 - 5$$
$$= 2x + 1 \quad \square$$

EXAMPLE 4. If $f(x) = x^2$ and $g(x) = x - 2$, find $(f \circ g)(x)$ and $(g \circ f)(x)$.

Solution:

$$(f \circ g)(x) = f(g(x))$$
$$= f(x - 2)$$
$$= (x - 2)^2$$

$$(g \circ f)(x) = g(f(x))$$
$$= g(x^2)$$
$$= x^2 - 2 \quad \square$$

EXAMPLE 5. If $f(x) = x^2$ and $g(x) = 2x - 1$, find $f(g(3))$.

Solution:

$$f(g(3)) = f(2 \cdot 3 - 1)$$
$$= f(5)$$
$$= 25 \quad \square$$

PROBLEM SET 11.3

Warm-ups

In problems 1 through 4, find $(f + g)(x)$, $(f - g)(x)$, $(fg)(x)$, and $\left(\dfrac{f}{g}\right)(x)$. See Example 1.

1. $f(x) = 2x; \quad g(x) = x - 5$

2. $f(x) = x - 7; \quad g(x) = \dfrac{2}{3}x - 3$

3. $f(x) = 3 - x^2$; $g(x) = x + 1$

4. $f(x) = x^2 - x - 3$; $g(x) = x^2 - x$

In problems 5 through 8, $f(x) = x - 1$ and $g(x) = 1 - x^2$. Find the value of each. See Example 2.

5. $(g - f)(0)$ 6. $(fg)(-1)$ 7. $\left(\dfrac{f}{g}\right)(2)$ 8. $(f + g)(-2)$

In problems 9 through 12, find $(f \circ g)(x)$ and $(g \circ f)(x)$. See Examples 3 and 4.

9. $f(x) = 5x + 8$; $g(x) = 7x - 4$
10. $f(x) = 6x + 9$; $g(x) = \dfrac{1}{4}x$
11. $f(x) = x^2 + 2$; $g(x) = x - 3$
12. $f(x) = x^3$; $g(x) = x - 1$

In problems 13 through 16, $f(x) = x - 1$ and $g(x) = 1 - x^2$. Find the value of each. See Example 5.

13. $(f \circ g)(-1)$ 14. $(g \circ f)(0)$ 15. $(f \circ f)(1)$ 16. $(f \circ g)(1)$

Practice Exercises

In problems 17 through 28, find $(f + g)(x)$, $(f - g)(x)$, $(fg)(x)$, $\left(\dfrac{f}{g}\right)(x)$, and $(f \circ g)(x)$.

17. $f(x) = 3x$; $g(x) = x + 8$

18. $f(x) = x + 7$; $g(x) = x + 4$

19. $f(x) = 4x + 2$; $g(x) = 3x - 9$

20. $f(x) = 6x - 1$; $g(x) = \dfrac{1}{2}x$

21. $f(x) = x - 3$; $g(x) = \dfrac{1}{3}x + 3$

22. $f(x) = x^2 + 1$; $g(x) = x - 5$

23. $f(x) = 2x^2 + x + 3$; $g(x) = 2x - 5$

24. $f(x) = 3x^2 - x - 2$; $g(x) = 3 - x$

25. $f(x) = 1 - x^2$; $g(x) = x + 3$

26. $f(x) = x^2 + 1$; $g(x) = 3x^2 - 1$

27. $f(x) = x^2 - x - 2$; $g(x) = x^2 - 2x$

28. $f(x) = x^3$; $g(x) = x + 1$

In problems 29 through 36, $f(x) = 2x - 5$ and $g(x) = 4 - x^2$. Find the value of each.

29. $(g - f)(0)$
30. $(fg)(-1)$
31. $\left(\dfrac{f}{g}\right)(1)$
32. $(f + g)(-2)$
33. $(f \circ g)(-1)$
34. $(g \circ f)(0)$
35. $(f \circ f)(1)$
36. $(f \circ g)(1)$

Challenge Problems

In problems 37 and 38, find $(f \circ g)(x)$ and $(g \circ f)(x)$.

37. $f(x) = x + 3$; $g(x) = x - 3$
38. $f(x) = x^3$; $g(x) = \sqrt[3]{x}$
39. Is 1 in the domain of $\dfrac{f}{g}$ if $f(x) = x - 1$ and $g(x) = 1 - x^2$?

IN YOUR OWN WORDS

40. Explain the composition of two functions.

11.4 INVERSE FUNCTIONS

Let's consider the functions shown below.

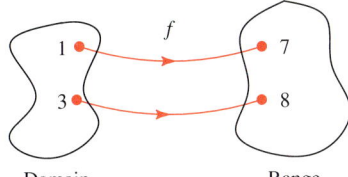
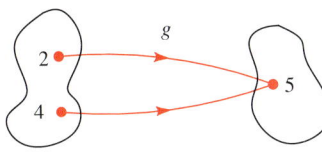

Notice that each number in the domain of f is assigned to a different number in the range of f. However, this is not true for each number in the domain of g.

One-to-One Function

A function is said to be **one-to-one** if every number in the *range* is paired with exactly one number in the domain.

EXAMPLE 1. The illustration below describes a function. Is it one-to-one?

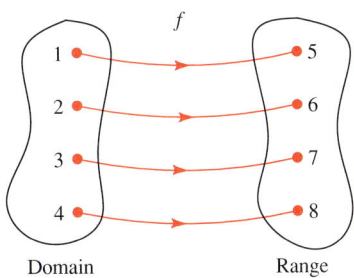

Solution:

Notice that each number in the range is paired with *one* number in the domain. So f is one-to-one. ☐

Sec. 11.4 Inverse Functions

499

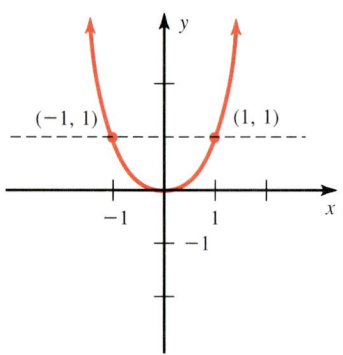

EXAMPLE 2. A function f is graphed at the left. Is f one-to-one?

Solution:

Notice that 1 is a number in the range. It is paired with -1 and 1. So f is *not* one-to-one. □

> ### Horizontal Line Test
> If any horizontal line crosses the graph of a function in more than one point, the function is *not* one-to-one.

In many applications of functions, it is desirable to interchange the domain and range of a function f. In other words, make a function that would "reverse" what f does. Such a function is named f^{-1} (read "f inverse").

NOTE: f^{-1} does *not* mean $\dfrac{1}{f}$.

Consider f and f^{-1} as shown below.

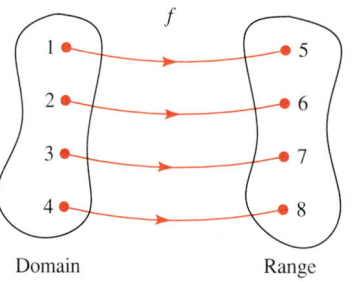 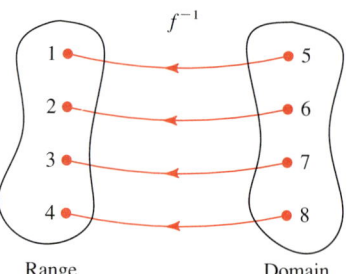

Notice that the domain of f^{-1} is the range of f and the range of f^{-1} is the domain of f. In other words, the domain and range are interchanged. We can see that f defines ordered pairs of (1, 5), (2, 6), (3, 7), and (4, 8), while f^{-1} defines (5, 1), (6, 2), (7, 3), and (8, 4). Notice that in each case, the x- and y- coordinates are interchanged.

Functional notation is very useful in giving names to the numbers in the range.

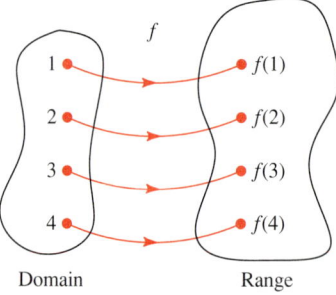 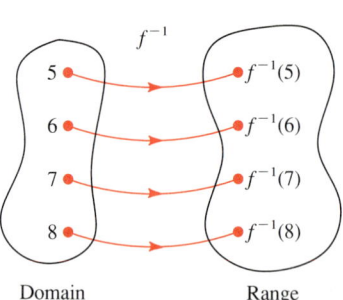

The *value* of $f(1)$ is 5 and the *value* of $f^{-1}(5)$ is 1.

To see how f^{-1} undoes what f does, consider the following. The function f assigns 1 to $f(1)$, whose value is 5.

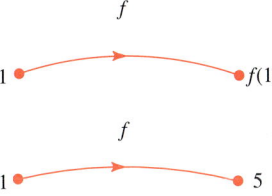

The function f^{-1} assigns 5 to $f^{-1}(5)$, whose value is 1.

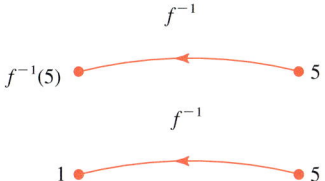

Combining these into one picture shows that f and f^{-1} undo each other.

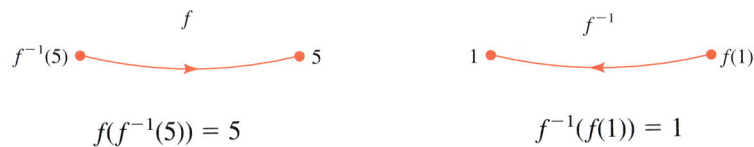

Definition of f^{-1}

If f is a one-to-one function, f^{-1} is the function such that
$$f(f^{-1}(x)) = x \text{ for } x \text{ in the domain of } f^{-1}$$
and
$$f^{-1}(f(x)) = x \text{ for } x \text{ in the domain of } f.$$
f^{-1} is called the **inverse** of f.

Notice that f must be one-to-one in order for f^{-1} to exist. There is an algorithm that finds the inverse of simple one-to-one functions.

To Find A Rule for f^{-1}

1. Replace $f(x)$ with y.
2. Swap x and y.
3. Solve this for y.
4. Replace y with $f^{-1}(x)$.

Sec. 11.4 Inverse Functions

EXAMPLE 3. If $f(x) = 2x - 3$, find $f^{-1}(x)$.

Solution:

$$f(x) = 2x - 3$$
$$y = 2x - 3 \quad \text{Replace } f(x) \text{ with } y.$$
$$x = 2y - 3 \quad \text{Swap } x \text{ and } y.$$
$$x + 3 = 2y \quad \text{Solve for } y.$$
$$\frac{x + 3}{2} = y$$
$$f^{-1}(x) = \frac{x + 3}{2} \quad \text{Replace } y \text{ with } f^{-1}(x).$$

EXAMPLE 4. If $f(x) = 5x + 4$, find a rule for f^{-1}.

Solution:

$$y = 5x + 4 \quad \text{Replace } f(x) \text{ with } y.$$
$$x = 5y + 4 \quad \text{Swap } x \text{ and } y.$$
$$\frac{x - 4}{5} = y \quad \text{Solve for } y.$$
$$f^{-1}(x) = \frac{x - 4}{5} \quad \text{Replace } y \text{ with } f^{-1}(x).$$

PROBLEM SET 11.4

Warm-ups

In problems 1 through 8, determine whether or not each function graphed is one-to-one. See Examples 1 and 2.

1.
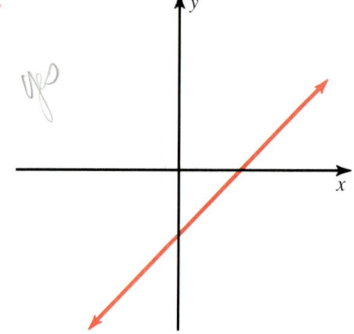

2.
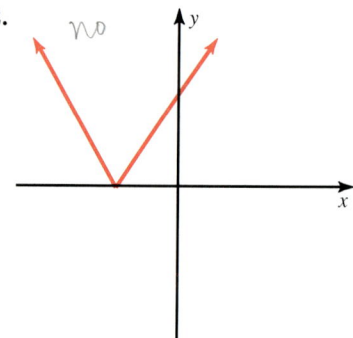

3.
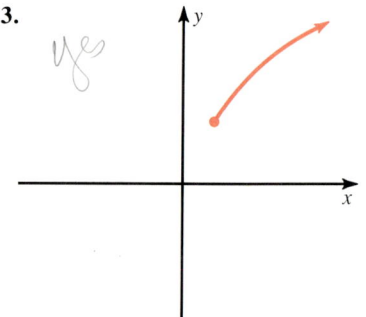

4.
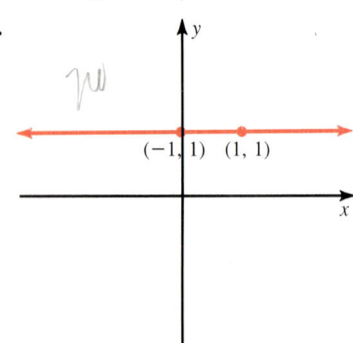

5.
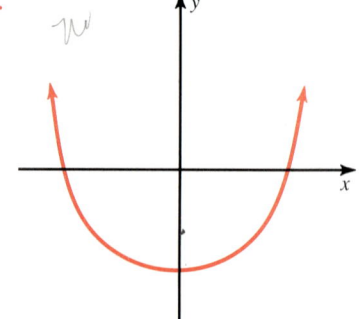

6.
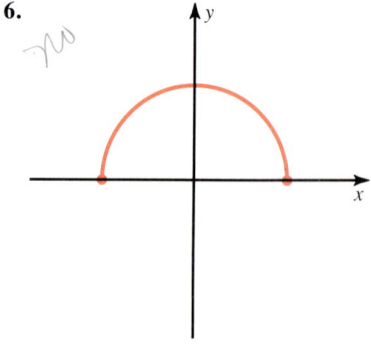

7.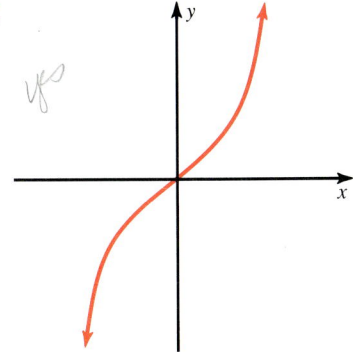
8.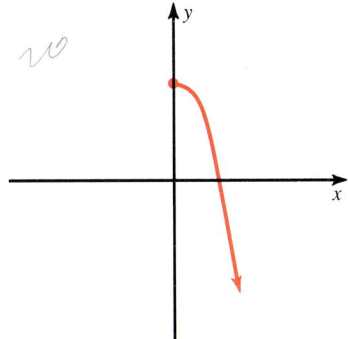

In problems 9 through 14, find $f^{-1}(x)$. See Examples 3 and 4.

9. $f(x) = 3x - 1$
10. $f(x) = x - 4$
11. $f(x) = \frac{1}{2}x + 3$
12. $f(x) = \frac{2}{3}x + \frac{1}{2}$
13. $f(x) = x^3$
14. $f(x) = x^3 + 1$

Practice Exercises

In problems 15 through 22, sketch the graph of each function and determine whether or not each function is one-to-one.

15. $f(x) = 3x + 1$
16. $h(x) = 5$
17. $f(x) = x^2 + 1$
18. $f(x) = -x^2$
19. $g(x) = (x + 1)^2 - 1$
20. $g(x) = |x| + 2$
21. $h(x) = |x + 1|$
22. $f(x) = -\frac{1}{2}x^2$

In problems 23 through 28, find $f^{-1}(x)$.

23. $f(x) = 5x - 1$
24. $f(x) = x + 2$
25. $f(x) = \frac{1}{3}x - 1$
26. $f(x) = \frac{2}{5}x - \frac{1}{2}$
27. $f(x) = x^3 + 2$
28. $f(x) = x^3 - 1$

In problems 29 through 32, f and f^{-1} are given. Verify that $f(f^{-1}(x)) = f^{-1}(f(x)) = x$ for each.

29. $f(x) = 2x + 1;\ f^{-1}(x) = \frac{x - 1}{2}$
30. $f(x) = x^3 + 1;\ f^{-1}(x) = \sqrt[3]{x - 1}$
31. $f(x) = x + 1;\ f^{-1}(x) = x - 1$
32. $f(x) = x^3;\ f^{-1}(x) = \sqrt[3]{x}$

Sec. 11.4   Inverse Functions

Challenge Problems

33. Sketch the graph of $f(x) = x + 2$ and $f^{-1}(x) = x - 2$ on the same set of axes. Draw the graph of $y = x$ as a dashed line. Is there a relationship between the graphs of f, f^{-1}, and $y = x$? (HINT: Imagine folding the paper along the graph of $y = x$.)

34. The graph of g is

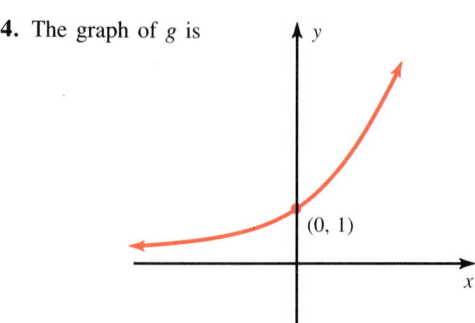

Draw the graph of g^{-1}.

IN YOUR OWN WORDS

35. What is a one-to-one function?

36. If f is a one-to-one function, explain what f^{-1} is.

11.5 EXPONENTIAL AND LOGARITHMIC FUNCTIONS

Let's look at the graph of $f(x) = 2^x$. This function is different from others we have seen because it has a variable as an exponent. To help in graphing the function f, we make a table of some coordinates for $y = 2^x$.

$\left(\text{Remember: } 2^{-3} = \dfrac{1}{2^3}\right)$

x	y
−3	$\frac{1}{8}$
−2	$\frac{1}{4}$
−1	$\frac{1}{2}$
0	1
1	2
2	4
3	8

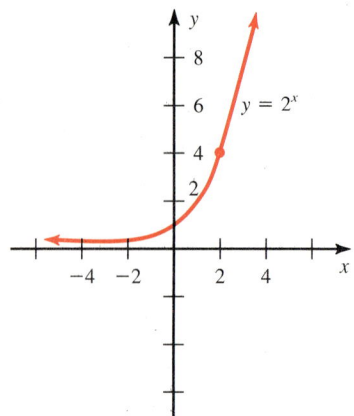

Plot the points and connect them with a smooth curve. The graph of $f(x) = 2^x$ is as shown (left). As x continues in the *negative* direction, the graph gets closer and closer to the x-axis, but it never touches it. Such a line is called an **asymptote**. Notice that the domain is $\{x \mid x \text{ is a real number}\}$ and that the range is $\{y \mid y > 0\}$.

Exponential Function

A function of the form

$$f(x) = b^x$$

where $b > 0$ and $b \neq 1$, is called an **exponential function**.

EXAMPLE 1. Graph $f(x) = \left(\dfrac{1}{2}\right)^x$.

Solution:

Make a table. As the domain is $\{x \mid x \text{ is a real number}\}$, plot the points in the table and connect them to get the graph below.

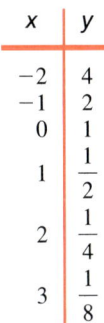
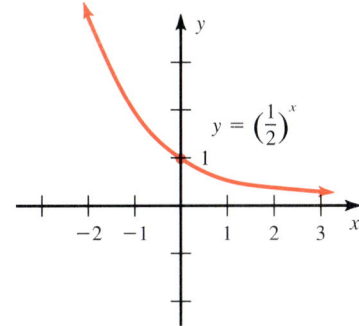

The range is $\{y \mid y > 0\}$ and the graph is the graph of $f(x) = 2^x$ turned around. ☐

EXAMPLE 2. Graph $f(x) = -2^x$.

Solution:

We remember that -2^x is not the same as $(-2)^x$. That is, the base is 2 not -2. However, the domain is still $\{x \mid x \text{ is a real number}\}$. Making a table and plotting points gives the graph below.

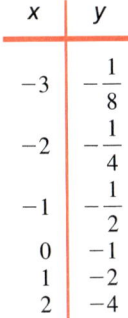
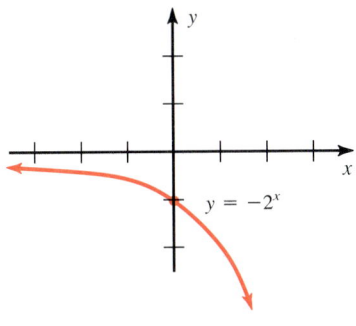

The range is $\{y \mid y < 0\}$. This is the graph of $f(x) = 2^x$ turned upside down. ☐

The number e whose value is approximately 2.71828 is a number that occurs naturally just as the number π does. It is used in physics, engineering, business, and other fields. Because it arises so often, the exponential function

$$f(x) = e^x$$

is often called *the* exponential function. Its graph is shown at the right.

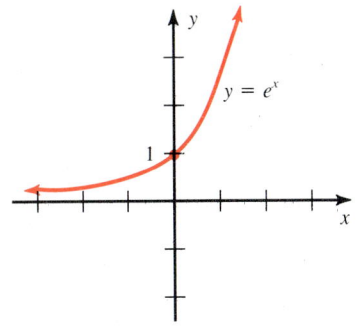

Sec. 11.5 Exponential and Logarithmic Functions

Notice that an exponential function is a one-to-one function. Because of this, we have the following result.

A Property of Exponential Functions

If $b^x = b^y$ where $b > 0$ and $b \neq 1$, then $x = y$.

EXAMPLE 3. Solve $4^x = 32$.

Solution:

$$4^x = 32$$
$$(2^2)^x = 2^5$$
$$2^{2x} = 2^5$$
$$2x = 5 \qquad \text{Exponential Property}$$
$$x = \frac{5}{2}$$

$$\left\{\frac{5}{2}\right\}$$

□

Since an exponential function is a one-to-one function, an exponential function has an inverse. To find a rule for the inverse of $f(x) = b^x$, we follow the procedure developed in Section 11.4.

$$y = b^x \qquad \text{Replace } f(x) \text{ with } y.$$
$$x = b^y \qquad \text{Swap } x \text{ and } y.$$

We do not have the tools to solve this for y, but since f has an inverse, we know such a function exists. The name for this function is $\log_b x$. So write

$$y = \log_b x \text{ (read ''log base } b \text{ of } x\text{'')}$$
$$f^{-1}(x) = \log_b x \qquad \text{Replace } y \text{ with } f^{-1}(x).$$

"Log" is short for "logarithm." The term "$\log_b$" is the name of a function, just as f, g, h, and f^{-1} are names of functions.

A logarithmic function is the inverse of an exponential function. If $f(x) = b^x$, then $f^{-1}(x) = \log_b x$.

Logarithmic Function

A function of the form

$$f(x) = \log_b x; \qquad x > 0$$

where $b > 0$ and $b \neq 1$, is called a **logarithmic function.**

We have seen that the statements $x = b^y$ and $y = \log_b x$ are equivalent. There are many times when the form $y = \log_b x$ (logarithmic form) must be changed to the equivalent form $x = b^y$ (exponential form).

Logarithmic and Exponential Forms

$y = \log_b x$ is equivalent to $x = b^y$
(logarithmic)   (exponential)

When we write $y = \log_b x$, we mean that y is the exponent of b that gives x. Thus logarithms can be thought of as exponents.

To graph a logarithmic function, calculate some coordinates using the equivalent exponential form.

EXAMPLE 4. Graph $f(x) = \log_2 x$.

Solution:

$y = \log_2 x$ is equivalent to $x = 2^y$. Make a table by choosing a value for y *first* and calculating the corresponding value of x.

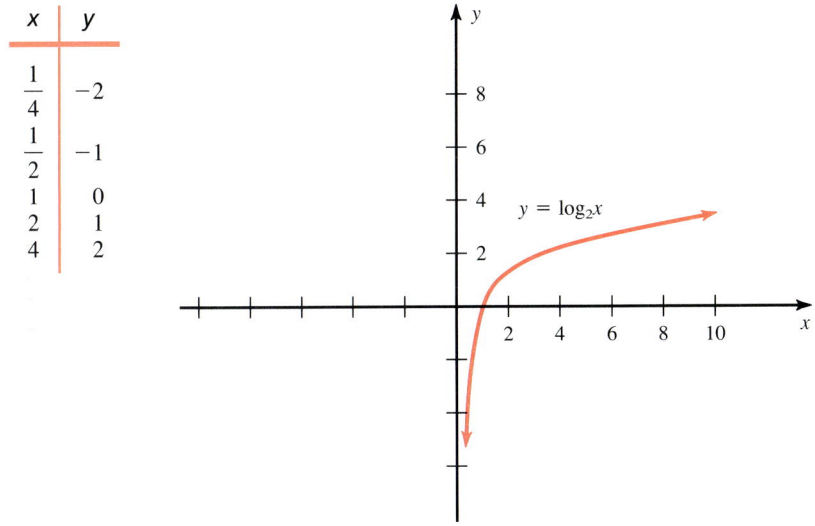

x	y
$\frac{1}{4}$	-2
$\frac{1}{2}$	-1
1	0
2	1
4	2

Notice that the domain is $\{x \mid x > 0\}$.

This means that we can find the logarithm of *positive* real numbers only. The range is $\{y \mid y \text{ is a real number}\}$, which says that the logarithm of a number can be *negative, zero,* or *positive*.

Since logarithmic and exponential functions are inverses, there is an interesting relationship between the graphs of $y = 2^x$ and $y = \log_2 x$. The graphs of both functions are shown below.

Sec. 11.5 Exponential and Logarithmic Functions

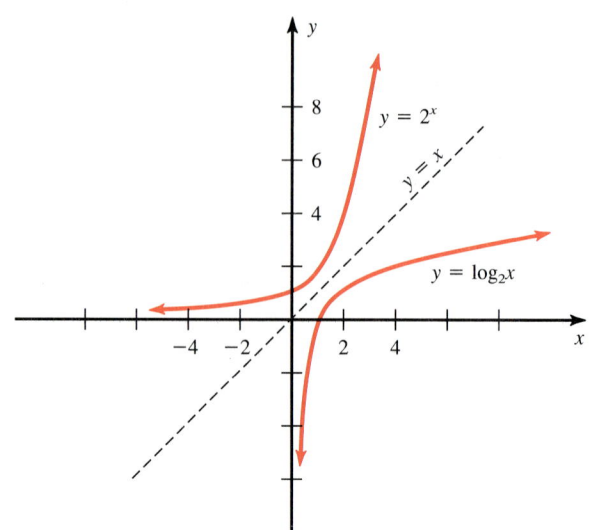

Notice that each is a reflection of the other about the line $y = x$. If the page were folded along the graph of $y = x$, the two curves would fall on top of one another.

The inverse of the exponential function $f(x) = e^x$ is an important function. Instead of writing $f^{-1}(x) = \log_e x$, we write $f^{-1}(x) = \ln x$. Logarithms with base e are called **natural logarithms.** The graphs of these functions are shown below.

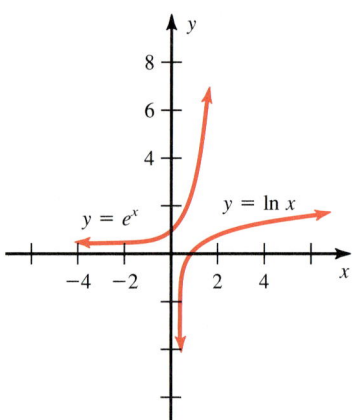

Many problems require changing from exponential form to logarithmic form, and vice versa.

EXAMPLE 5. Change $2 = \log_3 9$ to exponential form.

Solution:

$y = \log_b x$ is equivalent to $x = b^y$. So $2 = \log_3 9$ is equivalent to $9 = 3^2$. ☐

EXAMPLE 6. Change $25 = 5^2$ to logarithmic form.

Solution:

$x = b^y$ is equivalent to $y = \log_b x$. So $25 = 5^2$ is equivalent to $2 = \log_5 25$. ☐

EXAMPLE 7. Solve $\log_2 x = 4$.

Solution:

This is in logarithmic form and if changed to exponential form, it becomes

$$x = 2^4$$
$$x = 16$$
$$\{16\}$$

EXAMPLE 8. Find each logarithm, if possible.

(a) $\log_5 25$ (b) $\log_2(-8)$

Solutions:

(a) Let $y = \log_5 25$.

$$25 = 5^y \qquad \text{Change to exponential form.}$$
$$5^2 = 5^y$$
$$y = 2 \qquad \text{Property of Exponential Functions}$$

So $\log_5 25 = 2$.

(b) Let $y = \log_2(-8)$

$$-8 = 2^y \qquad \text{Change to exponential form.}$$

Since there is no power of 2 that is -8, there is no value of y that will make this statement true. So $\log_2(-8)$ does not exist. It is not defined.

Remember that a logarithm is defined for positive numbers only.

Be Careful!

The next examples show how the exponential function is used in applications.

EXAMPLE 9. The atmospheric pressure, $P(\text{lb/in.}^2)$, at an elevation of x feet above sea level is given by the formula

$$P = 14.7e^{-0.00004x}$$

Approximate (to the nearest tenth) the atmospheric pressure at an elevation of 1000 ft.

Solution:

$$P = 14.7e^{-0.00004x}$$
$$= 14.7e^{-0.00004(1000)}$$
$$= 14.7e^{-0.04}$$

Approximate the value of $e^{-0.04}$ with a calculator. Enter $-.04$ and press the keys that find e^x. (Sometimes $\boxed{\text{inv}}$ followed by $\boxed{\text{ln}}$.)

$$P \approx 14.1$$

The atmospheric pressure is approximately 14.1 lb/in.^{2}.

If a quantity grows exponentially, the growth rate is very rapid.

Sec. 11.5 Exponential and Logarithmic Functions

EXAMPLE 10. A strain of *Bacillus anthracis* bacteria initially contains 10,000 bacteria. The formula

$$A = (10{,}000)3^{0.3t}$$

gives the number of bacteria present after t hours. Approximate to the nearest integer how many bacteria will be present after $\frac{1}{2}$ hour.

Solution:

$$A = (10{,}000)3^{0.3t}$$
$$= (10{,}000)3^{0.3(0.5)}$$
$$= (10{,}000)3^{0.15}$$

With a calculator, approximate the value of $3^{0.15}$ by entering

$\boxed{3}\ \boxed{y^x}\ \boxed{.15}\ \boxed{=}\ $ and read $\boxed{1.1791476}$

$$A \approx 11{,}791$$

After $\frac{1}{2}$ hour, there will be approximately 11,791 bacteria.

CALCULATOR BOX

Logarithms with a Calculator

$\boxed{\log}\quad \boxed{\ln}$

Calculators make finding the logarithm of a number very easy to do. Log tables or slide rules were used to find logarithms before calculators were invented.

Most calculators have a $\boxed{\log}$ key. This key will find a logarithm with base 10. A $\log_{10}$ is called a **common logarithm** and very often the 10 is omitted. This is because our number system is a base 10 system.

Many calculators have an $\boxed{\ln}$ key. This key will find a natural logarithm, that is, with base e.

EXAMPLE Use a calculator to approximate the following correct to four decimal places.

(a) log 3.21   (b) ln 345.69

Solutions:

(a) log means $\log_{10}$. So enter

$\boxed{3.21}\quad \boxed{\log}\ $ and read $\boxed{0.506505032}$

log 3.21 ≈ .5065

(b) ln means the natural log. Enter

$\boxed{345.69}\quad \boxed{\ln}\ $ and read $\boxed{5.84554242}$

ln 345.69 ≈ 5.8455

Sometimes the logarithm of a number x is known and we want to find the number x itself.

EXAMPLE Use a calculator to approximate x correct to two decimal places.

(a) $\log x = -0.0987$ (b) $\ln x = 24$

Solutions:

(a) $\log x = -0.0987$

$$x = 10^{-0.0987} \quad \text{Change to exponential form.}$$

Enter $\boxed{0.0987}$ $\boxed{+/-}$ $\boxed{\text{inv}}$ $\boxed{\log}$ and read $\boxed{0.796709507}$.

$$x \approx 0.80$$

(b) $\ln x = 24$

$$x = e^{24} \quad \text{Change to exponential form.}$$

Enter $\boxed{24}$ $\boxed{\text{inv}}$ $\boxed{\ln}$ and read $\boxed{2.648912212^{10}}$.

$$x \approx 2.65 \times 10^{10}$$

Calculator Exercises.

In Problems 1 through 6, approximate each to four decimal places.

1. $\log 98.7$ 2. $\log 1.009$ 3. $\log 0.0016$
4. $\log 234.89$ 5. $\ln 0.98$ 6. $\ln 87,953$

In problems 7 through 10, find x correct to three decimal places.

7. $\log x = 3.2897$ 8. $\log x = -1.3648$
9. $\ln x = 10.23445$ 10. $\ln x = -5.6730$

Answers:

1. 1.9943   2. 0.0039   3. -2.7959 4. 2.3709
5. -0.0202 6. 11.3846   7. 1948.4982   8. 0.04317
9. 27,846.1501   10. 0.0034

▬ PROBLEM SET 11.5

Warm-ups

In problems 1 through 4, sketch the graph of each function. Give the domain and range for each. See Examples 1 and 2.

1. $f(x) = 4^x$ 2. $f(x) = \left(\dfrac{1}{5}\right)^x$ 3. $f(x) = -\left(\dfrac{1}{3}\right)^x$ 4. $f(x) = -5^x$

In problems 5 through 8, solve each equation. See Example 3.

5. $3^x = 81$ **6.** $2^{2x} = 32$ **7.** $3^{x-2} = 1$ **8.** $3^{x-7} = 9$

In problems 9 and 10, sketch the graph of each function. See Example 4.

9. $f(x) = \log_6 x$ **10.** $f(x) = \log_8 x$

In problems 11 through 18, write each equation in exponential form. See Example 5.

11. $\log_2 8 = 3$ **12.** $\log_{1/2} 8 = -3$ **13.** $\log_8 64 = 2$ **14.** $\log_3 3 = 1$

15. $\log_5 1 = 0$ **16.** $\log_{16} 2 = \dfrac{1}{4}$ **17.** $\log_9 27 = \dfrac{3}{2}$ **18.** $\log_2 \dfrac{1}{4} = -2$

In problems 19 through 26, write each equation in logarithmic form. See Example 6.

19. $2^3 = 8$ **20.** $4^{-2} = \dfrac{1}{16}$ **21.** $\left(\dfrac{1}{3}\right)^2 = \dfrac{1}{9}$ **22.** $3^0 = 1$

23. $3^{-1} = \dfrac{1}{3}$ **24.** $10^3 = 1000$ **25.** $4^3 = 64$ **26.** $4^1 = 4$

In problems 27 through 38, solve each equation. See Example 7.

27. $\log_4 x = 2$ **28.** $\log_3 81 = x$ **29.** $\log_5 x = 3$ **30.** $\log_5 x = 0$

31. $\log_2 x = 1$ **32.** $\log_3 3^x = 4$ **33.** $\log_{1/3} x = -1$ **34.** $\log_3 x = -1$

35. $\log_7 7 = x$ **36.** $\log_{16} 4 = x$ **37.** $\log_{10}(x + 2) = 2$ **38.** $\log_3(2x - 1) = 1$

In problems 39 through 46, find the value of each logarithm. See Example 8.

39. $\log_5 125$ **40.** $\log_5 625$ **41.** $\log_3(-9)$ **42.** $\log_{1/3}(-81)$

43. $\log_3 3^2$ **44.** $\log_{10} 10^3$ **45.** $\log_{100} 10{,}000$ **46.** $\log_2 1$

For problems 47 through 49, see Examples 9 and 10.

47. Use the formula given in Example 9 to approximate the atmospheric pressure to the nearest tenth at 1 mile (5280 ft).

48. Rabbits multiply exponentially according to the formula
$$N = N_0 e^{0.3t}$$
where N_0 is the number present initially and t is the time in months. How many rabbits will there be in 3 months if 20 are present initially? (To the nearest rabbit.)

49. If P dollars is invested at an interest rate of $100i\%$ compounded annually, the amount, A, after t years is given by the formula
$$A = P(1 + i)^t$$
Find A to the nearest cent if $3000 is invested at 12% ($i = 0.12$) for 5 years.

Practice Exercises

In problems 50 through 83, sketch the graph of each function. Give the domain and range for each.

50. $f(x) = 5^x$ **51.** $f(x) = \left(\dfrac{1}{3}\right)^x$

52. $f(x) = -3^x$ **53.** $f(x) = 10^x$ **54.** $f(x) = -\left(\dfrac{1}{2}\right)^x$ **55.** $f(x) = \log_2 x$

56. $f(x) = -\log_2 x$ **57.** $f(x) = \log_{10} x$ **58.** $f(x) = \log_5 x$ **59.** $f(x) = -\log_4 x$

Solve each equation.

60. $\log_5 x = 2$ **61.** $\log_3 9 = x$ **62.** $\log_3 27 = x$ **63.** $\log_x 1 = 0$

64. $\log_2 x = 4$ **65.** $\log_4 x = \dfrac{3}{2}$ **66.** $\log_{1/2} x = -1$ **67.** $\log_2 x = 5$

68. $\log_5 5 = x$ **69.** $\log_{27} 3 = x$ **70.** $3^x = 9$ **71.** $2^{2x} = 16$

72. $3^{x-1} = 1$ **73.** $3^{x-7} = 27$ **74.** $\log_{10}(x + 1) =$ **75.** $\log_3(2x - 1) = 2$

76. $\log_3 81 = x$ **77.** $\log_5 25 = x$ **78.** $\log_{1/2} 4 = x$ **79.** $\log_{1/3} 9 = x$

80. $\log_5 5^2 = x$ **81.** $\log_{10} 10 = x$ **82.** $\log_{100} 10 = x$ **83.** $\log_{25} 5 = x$

84. Use the formula in Example 9 to approximate to the nearest tenth the atmospheric pressure at $\dfrac{1}{2}$ mile (2640 ft).

85. Use the formula in problem 48 and find how many rabbits there will be in 3 months if 50 are present initially. (To the nearest rabbit.)

86. Use the formula in problem 49 to find A to the nearest cent if $10,000 is invested at 8% ($i = 0.08$) for 10 years.

87. The population of a city is growing exponentially. The city had a population of 50,000 in 1970. If the population is given by the formula
$$P = P_0 e^{0.05t}$$
where t is the number of years since 1970 and P_0 is the population in 1970, what will the population be in the year 2000? (To the nearest person.)

88. Use the formula in problem 87 to find the population in the year 1990. (To the nearest person.)

89. Use the formula in problem 48 to find how many months it would take 20 rabbits to multiply to 100 rabbits. (To the nearest month.)

90. Use the formula in problem 48 to find out how many months it would take the rabbit population to double if 50 were present initially. (To the nearest month.)

Challenge Problems

91. Graphs of exponential and logarithmic functions can be shifted just as graphs of quadratic and absolute value functions. Graph each of the following and make a rule for shifting these functions.
   (a) **1.** $f(x) = 2^x$ **2.** $f(x) = 2^{x+1}$ **3.** $f(x) = 2^{x-1}$ **4.** $f(x) = 2^x + 1$ **5.** $f(x) = 2^x - 1$

Sec. 11.5 Exponential and Logarithmic Functions

91. (b) 1. $f(x) = \log_2 x$ 2. $f(x) = \log_2(x + 1)$ 3. $f(x) = \log_2(x - 1)$ 4. $f(x) = \log_2 x + 1$ 5. $f(x) = \log_2 x - 1$

92. Graph each function. Find the domain before graphing.
   (a) 1. $f(x) = 2^{-x}$ 2. $f(x) = 2^{x^2}$ 3. $f(x) = 2^{|x|}$

   (b) 1. $f(x) = \log_2(-x)$ 2. $f(x) = \log_2|x|$ 3. $f(x) = \log_2 x^2$ 4. $f(x) = 2\log_2 x$

93. Why do we restrict the base in an exponential function to be positive and not equal to 1? [Look at $(-2)^x$ and 1^x.]

94. Use a calculator to approximate log 101,000,000,000.

■ IN YOUR OWN WORDS

95. Explain the relationship between a logarithmic function and an exponential function.

■ 11.6 PROPERTIES OF LOGARITHMS

Logarithms are very useful tools that have properties that help simplify calculations. Logarithms were even more useful before calculators were invented.

Let $\log_b M = x$ and $\log_b N = y$, where b, M, and N are positive and $b \neq 1$. Changing to exponential form gives us

$$M = b^x \quad \text{and} \quad N = b^y$$

$$MN = b^x b^y = b^{x+y} \quad \text{is equivalent to}$$

$$\log_b(MN) = x + y$$

$$\log_b(MN) = \log_b M + \log_b N$$

An important property of logarithms has been proved. This property says: "The log of a product is the sum of the logs." It allows us to add instead of multiply.

Now consider a quotient.

$$\frac{M}{N} = \frac{b^x}{b^y} = b^{x-y} \quad \text{is equivalent to}$$

$$\log_b \frac{M}{N} = x - y$$

$$\log_b \frac{M}{N} = \log_b M - \log_b N$$

This proves that the log of a quotient is the difference in the logs. This property allows us to subtract instead of divide.

Properties of Logarithms

If $M > 0$, $N > 0$, $b > 0$, $b \neq 1$, and r is real:

1. $\log_b(MN) = \log_b M + \log_b N$ Log of a product
2. $\log_b(M/N) = \log_b M - \log_b N$ Log of a quotient
3. $\log_b M^r = r \log_b M$ Log of a power
4. $\log_b b = 1$ Log of the base
5. $\log_b 1 = 0$ Log of 1

EXAMPLE 1. If $\log_3 5 \approx 1.5$ and $\log_3 2 \approx 0.6$, approximate each of the following.

(a) $\log_3 10$ (b) $\log_3 5/2$ (c) $\log_3 25$
(d) $\log_3 \sqrt{2}$ (e) $\log_3 50$

Solutions:

(a) $\log_3 10 = \log_3(2)(5)$
$= \log_3 2 + \log_3 5$ Log of a product
$\approx 0.6 + 1.5 \approx 2.1$

(b) $\log_3 5/2 = \log_3 5 - \log_3 2$ Log of a quotient
$\approx 1.5 - 0.6 \approx 0.9$

(c) $\log_3 25 = \log_3 5^2$
$= 2 \log_3 5$ Log of a power
$\approx 2(1.5) \approx 3.0$

(d) $\log_3 \sqrt{2} = \log_3 2^{1/2}$
$= \frac{1}{2} \log_3 2$ Log of a power
$\approx \frac{1}{2}(0.6) \approx 0.3$

(continued)

Sec. 11.6 Properties of Logarithms

(e) $\log_3 50 = \log_3(2)(5^2)$
$= \log_3 2 + \log_3 5^2$ Log of a product
$= \log_3 2 + 2\log_3 5$ Log of a power
$\approx 0.6 + 2(1.5)$
$\approx 0.6 + 3.0 \approx 3.6$

To solve an equation containing logarithms, use the properties of logarithms and try to write the equation in the form

$$\log_b M = N$$

and then change to exponential form,

$$M = b^N$$

EXAMPLE 2. Solve $\log_2 x - \log_2 3 = 5$.

Solution:

$\log_2 x - \log_2 3 = 5$

$\log_2 \dfrac{x}{3} = 5$ Log of a quotient

$\dfrac{x}{3} = 2^5$ Change to exponential form.

$\dfrac{x}{3} = 32$

$x = 96$

Check 96 in the original equation.

$\{96\}$

Since logarithmic functions are one-to-one functions, we have the following useful result.

A Property of Logarithmic Functions

If $\log_b x = \log_b y$, then $x = y$ ($x > 0$, $y > 0$, $b > 0$, $b \neq 1$).

EXAMPLE 3. Solve $\log_2(3x - 1) = \log_2 8$.

Solution:

$\log_2(3x - 1) = \log_2 8$

$3x - 1 = 8$ Property of Logarithms

$3x = 9$

$x = 3$

Check 3 in the original equation.

$\{3\}$

Logarithms are defined for positive numbers only. So we must be careful to check possible solutions in the original equation.

EXAMPLE 4. Solve $\log_3 x + \log_3(x - 2) = 1$.

Solution:

$$\log_3 x + \log_3(x - 2) = 1$$
$$\log_3 x(x - 2) = 1 \qquad \text{Log of a Product Property}$$
$$x(x - 2) = 3^1 \qquad \text{Change to exponential form.}$$
$$x^2 - 2x - 3 = 0$$
$$(x - 3)(x + 1) = 0$$
$$x = 3 \quad \text{or} \quad x = -1$$

We must be careful to check before writing the solution set.
Check 3.

LS: $\log_3 3 + \log_3(3 - 2) = \log_3 3 + \log_3 1 = \boxed{1}$
RS: $\boxed{1}$

Thus 3 checks and is in the solution set.
Check -1.

LS: $\log_3(-1) + \log_3(-1 - 2)$ is undefined because we cannot find the log of a negative number. Thus -1 is not in the solution set.

$\qquad \{3\} \qquad$ Write the solution set.

Finding Logs with Other Bases with a Calculator

CALCULATOR BOX

Computers have made bases of 2, 8, and 16 very important. Most calculators do not have log keys for these bases. However, we can use the formula

$$\log_b x = \frac{\log_c x}{\log_c b}$$

to find a logarithm with any base.

EXAMPLE Approximate $\log_2 5$ correct to four decimal places.

Solution:

We can use the formula above and the $\boxed{\log}$ key.

$$\log_2 5 = \frac{\log 5}{\log 2} \qquad \text{(using the formula above)}$$

(continued)

Sec. 11.6 Properties of Logarithms

> Enter [5] [log] [÷] [2] [log] [=] and read 2.321928095 on the display.
>
> $$\log_2 5 \approx 2.3219$$
>
> We could also use the formula and the [ln] key.
>
> $$\log_2 5 = \frac{\ln 5}{\ln 2}$$
>
> Enter [5] [ln] [÷] [2] [ln] [=] and read 2.321928095.
>
> $$\log_2 5 \approx 2.3219$$
>
> **Calculator Exercises.**
>
> *Approximate the value of each logarithm correct to four decimal places.*
>
> 1. $\log_2 10$ 2. $\log_2 e$ 3. $\log_2 (38)^2$ 4. $\log_2 0.876$
> 5. $\log_8 \sqrt{57}$ 6. $\log_8 (68)^{-2}$ 7. $\log_{16} 127.975$ 8. $\log_{16}(7)$
>
> **Answers:**
>
> 1. 3.3219   2. 1.4427   3. 10.4959   4. −0.1910
> 5. 0.9721   6. −4.0583   7. 1.7499   8. 0.7018

PROBLEM SET 11.6

Warm-ups

In problems 1 through 8, approximate each logarithm if $\log_b 2 \approx 1.32$ and $\log_b 3 \approx 1.62$. See Example 1.

1. $\log_b 6$
2. $\log_b \frac{3}{2}$
3. $\log_b 2^2$
4. $\log_b \sqrt{3}$
5. $\log_b \frac{1}{2}$
6. $\log_b 18$
7. $\log_b \frac{4}{3}$
8. $\log_b \sqrt{6}$

In problems 9 through 22, solve each equation. For problems 9 through 15, see Example 2.

9. $\log_3 x + \log_3 2 = 2$
10. $\log_2 3x - \log_2(x-1) = 2$
11. $\log x + \log 2 = 0$
12. $\log_2 x - \log_2 3 = 3$
13. $\log_{16} x + \log_{16} 4 = 1$
14. $\log_3 x + \log_3 2 - \log_3 5 = 0$
15. $\log_7(x+1) - \log_7 49 = 1$

For problems 16 through 18, see Example 3.

16. $\ln(x-1) = \ln e$
17. $\log_8(2x+2) = \log_8 8$
18. $\ln x - \ln e = 0$

For problems 19 through 22, see Example 4.

19. $\log x + \log(x-3) = 1$
20. $\log x(x-3) = 1$
21. $\log_2 x^2 = 2$
22. $2 \log_2 x = 2$

Practice Exercises

In problems 23 through 30, approximate each logarithm if $\log_b 3 \approx 1.62$ and $\log_b 5 \approx 2.52$.

23. $\log_b 15$
24. $\log_b \frac{3}{5}$
25. $\log_b 5^2$
26. $\log_b \sqrt{5}$
27. $\log_b \frac{1}{3}$
28. $\log_b 45$
29. $\log_b \frac{9}{5}$
30. $\log_b \sqrt{15}$

In problems 31 through 37, classify each statement as true or false. $b > 0$; $b \neq 1$

31. $\log_b 14 = \log_b 7 + \log_b 2$
32. $\log_b 14 = (\log_b 7)(\log_b 2)$
33. $\log_b 13 = \dfrac{\log_b 26}{\log_b 2}$
34. $\log_b 13 = \log_b 26 - \log_b 2$
35. $\log_b(17^2) = (\log_b 17)^2$
36. $\log_b(17^2) = 2\log_b 17$
37. $\sqrt{\log_b 5} = \log_b \sqrt{5}$

In problems 38 through 53, solve each equation.

38. $\log_5 x + \log_5 2 = 2$
39. $\log_2 5x = \log_2(x + 1)$
40. $\log x + \log 2 = 3$
41. $\log_2 x - \log_2 3 = 4$
42. $\log_6(2x + 2) = \log_6 6$
43. $\log_{16} 2x + \log_{16} 2 = 2$
44. $\log_5 x - \log_5 3 + \log_5 2 = 0$
45. $\log_2(x - 1) - \log_2 16 = 1$
46. $\ln x^2 - \ln e^2 = 0$
47. $\ln(2x + 1) = \ln e$
48. $\log_2 x + \log_2(x + 1) = 1$
49. $\log_3 |x| = 1$
50. $\log_6 x + \log_6(x - 1) = 1$
51. $\log_2 x(x - 3) = 2$
52. $\log x + \log x = 0$
53. $\log_5 x + \log_5(x - 4) = 1$

Challenge Problems

54. Prove that $\log_b \dfrac{1}{x} = -\log_b x$.

■ IN YOUR OWN WORDS

55. What properties does a logarithmic function have that other functions do not have?

CHAPTER SUMMARY

GLOSSARY

Function: A rule that assigns to each member of the **domain** exactly one member in the **range.**

The **vertical line test:** Determines whether or not a graph is a function.

Functional notation: Used in giving the rule for a function.

A **one-to-one** function: Each number in the range is paired with exactly one number in the domain.

The **horizontal line test:** Determines whether or not a graph is one-to-one. The linear, exponential, and logarithmic functions that we have studied are one-to-one.

The **inverse** of a function: A function that "undoes" the function. A function must be one-to-one to have an inverse. A logarithmic function is the inverse of an exponential function.

FUNCTIONS

1. Linear $f(x) = mx + b$
2. Quadratic $f(x) = ax^2 + bx + c$; $a \neq 0$
3. Absolute value $f(x) = p|x - h| + k$; $a \neq 0$
4. Exponential $f(x) = b^x$; $b > 0$; $b \neq 1$
5. Logarithmic $f(x) = \log_b x$; $b > 0$; $b \neq 1$; $x > 0$

PROPERTIES OF LOGARITHMS

$\log_b MN = \log_b M + \log_b N$ Log of a product

$\log_b \dfrac{M}{N} = \log_b M - \log_b N$ Log of a quotient

$\log_b M^r = r \log_b M$ Log of a power

$\log_b b = 1$ Log of the base

$\log_b 1 = 0$ Log of 1

LOGARITHMIC AND EXPONENTIAL FORMS

$x = b^y$ and is equivalent to $y = \log_b x$.

CHECKUPS

1. If $h(x) = x^2 - 2x$, find $h(-5)$.   Section 11.1; Example 3b
2. Find the natural domain of $g(x) = \dfrac{5}{x+3}$.   Section 11.1; Example 7
3. Graph each function.
   (a) $f(x) = 3x - 1$ Section 11.2; Example 1
   (b) $g(x) = 2 - x^2$ Section 11.2; Example 4a
   (c) $g(x) = |x + 2| - 1$ Section 11.2; Example 7
   (d) $f(x) = \left(\dfrac{1}{2}\right)^x$ Section 11.5; Example 1
   (e) $f(x) = \log_2 x$ Section 11.5; Example 4
4. If $g(x) = \sqrt{x}$ and $h(x) = x^2$, find $(g + h)(2)$.   Section 11.3; Example 2a
5. If $f(x) = 2x - 5$ and $g(x) = x + 3$, find $(f \circ g)(x)$.   Section 11.3; Example 3
6. If $f(x) = 2x - 3$, find $f^{-1}(x)$.   Section 11.4; Example 3
7. Solve each equation.
   (a) $4^x = 32$ Section 11.5; Example 3
   (b) $\log_2 x = 4$ Section 11.5; Example 7
   (c) $\log_2 x - \log_2 3 = 5$ Section 11.6; Example 2
   (d) $\log_3 x + \log_3(x - 2) = 1$ Section 11.6; Example 4

REVIEW PROBLEMS

1. Find the range of the function whose domain is $\{x \mid x \text{ is a positive integer}\}$; rule is $f(x) = -x$.

In problems 2 through 7, find the natural domain of each function.

2. $f(x) = 2x - 8$
3. $f(x) = |x - 7|$
4. $f(x) = \sqrt{x + 2}$
5. $f(x) = \sqrt{x^2 - x}$
6. $f(x) = \dfrac{3}{x - 7}$
7. $f(x) = \dfrac{x + 3}{(x + 1)(x - 2)}$

If $f(x) = \dfrac{3}{x + 5}$, find:

8. $f(1)$
9. $f(-5)$
10. $f(0)$
11. $f(f(1))$
12. $f(a)$
13. $f(-3)$
14. $f(a) + f(b)$
15. $f(a^2)$

In problems 16 through 31, sketch the graph of each function. Give the domain and range for each and determine whether or not each is one-to-one.

16. $f(x) = \dfrac{1}{2}x - \dfrac{5}{2}$

17. $f(x) = \dfrac{1}{4}x^2 + 3$

18. $f(x) = 4^x$

19. $f(x) = 5x - 2$

20. $f(x) = \log_7 x$

21. $f(x) = (x - 2)^2 + 3$

22. $f(x) = |x - 3| + 1$

23. $f(x) = -8$

24. $f(x) = -4^x$

25. $f(x) = |x + 7|$

26. $f(x) = 2(x + 1)^2 - 5$

27. $f(x) = -\dfrac{1}{2}x^2$

28. $f(x) = \log_5 x$

29. $f(x) = x^2 - 4x + 3$

30. $f(x) = 1 - |x|$

31. $f(x) = -2x^2 + 12x + 4$

If $f(x) = 3x - 5$ and $g(x) = x^2$, find:

32. $(f + g)(x)$

33. $\left(\dfrac{f}{g}\right)(x)$

34. $(f - g)(x)$

35. $(f \circ g)(x)$

36. $(fg)(x)$

37. $(g \circ f)(x)$

38. If $f(x) = 6x - 5$, find $f^{-1}(x)$.

In problems 39 through 50, solve each equation.

39. $2^x = 128$
40. $\log_x 4 = 2$
41. $\log_{10} 10^x = 1$
42. $\log_{10} x = \log_{10} 10^3$
43. $\ln(x + e) - \ln e = 1$
44. $\ln e + \ln e^3 - \ln e^2 = x$
45. $3^{x+1} = 9$
46. $\ln x = 1$
47. $(25)^x = 625$
48. $\log_x b = 1$
49. $\log x + \log(x + 3) = 1$
50. $\log_2 x^2 = 0$

... LET'S NOT FORGET ...

Identify the expressions that are in factored form. Factor those that are not factored.

51. $(a + 2)^2$
52. $125 - 8a^3$
53. $4 - (a + b)^2$
54. $ab^2(a^2 + b^2)$
55. $a^3 b + c^2 b$
56. $(a + 2b)^3$

How many terms are in each expression? Which expressions have $a - b$ as a factor?

57. $a^3 - b^3$
58. $a - b + 7$
59. $w(a - b) + y(b - a)$
60. $(a - b)^2 - 4$

Simplify each expression, if possible, leaving only nonnegative exponents in your answer.

61. $\sqrt[5]{64x^5 - 128y^5}$
62. $(x^{1/2} y^{1/2})^4$
63. $\sqrt[3]{(x - 3)^3}$
64. $(xy^{-4})^{-2}$
65. $(2^{-1} + 4^{-1})^{-1}$
66. $\dfrac{4 \pm \sqrt{-12}}{2}$
67. $\dfrac{-5x^{-3}}{y}$
68. $\left(\dfrac{16a^{-8}}{9b^4}\right)^{-1/2}$

Find each product.

69. $(3x - 1)^3$
70. $(2x - 3y)^2$

Reduce each, if possible.

71. $\dfrac{2(x - 1) + a(x + 1)}{4(x - 1) + a(x + 1)}$
72. $\dfrac{2(x - 1) + a(x - 1)}{4(x - 1) + a(x - 1)}$

The following problems can be worked by using a least common denominator. Follow the directions in each and notice how the least common denominator is used.

73. Perform the operation indicated: $\dfrac{3x}{x^2 + 6x + 9} + \dfrac{1}{x + 3}$.

74. Solve $\dfrac{x}{x + 3} - \dfrac{3}{x^2 + 6x + 9} = 1$.

75. Simplify $\dfrac{\dfrac{1}{x + 3} - 1}{\dfrac{x + 2}{x^2 + 6x + 9}}$.

Label each problem as an expression, equation, or inequality. Solve the equations and inequalities. Perform the operations indicated on the expressions, leaving only nonnegative exponents in your answer.

76. $4^x = 16$
77. $\dfrac{1}{x - 2} = \dfrac{4}{2 - x}$
78. $(-2)^2$
79. -2^2

80. $(\sqrt{3} + \sqrt{5})^2$
82. $\sqrt{x - 5} = 4$
84. $(x + 2)(x - 7) > 0$
86. $\dfrac{a}{a - b} + \dfrac{b}{b - a}$
88. $\dfrac{1}{x - 3} < 1$
90. $\sqrt{x + 1} - \sqrt{x + 6} = 1$

81. $2x + 3(x - 7) \leq 5$
83. $|x + 11| = 0$
85. $\sqrt{48} - \sqrt{12}$
87. $a^2(a^{-3}b)^3$
89. $ab^{-3}(ab^4 - b^5)$
91. $|x + 7| \geq 11$

CHAPTER TEST

In problems 1 through 5, choose the correct answer.

1. If $\log_b x = y$, which of the following statements is true?
   A. $b > 0$ B. $b \neq 1$
   C. $x > 0$ D. All are true
2. If $g(x) = x^2$ and the domain of g is $\{-2, -1, 0, 1, 2\}$, the range of g is (?)
   A. $\{1, 4\}$ B. $\{-4, -1, 0\}$
   C. $\{0, 1, 4\}$ D. $\{-2, -1, 0, 1, 2\}$
3. If $f(x) = 1 - 2x + x^2$, then $f(-3) = $ (?)
   A. -2 B. 16
   C. -14 D. 13
4. The vertex of the graph of $f(x) = x^2 + 2x + 2$ is (?)
   A. $(-1, 1)$ B. $(1, 1)$
   C. $(1, -1)$ D. $(-1, -1)$
5. Which of the following illustrations is the graph of a function?
   A.
   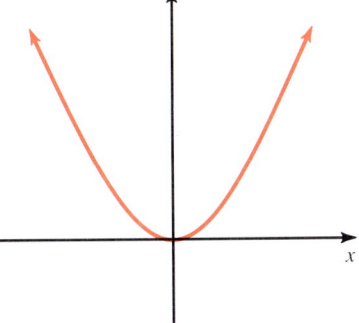
   B.
   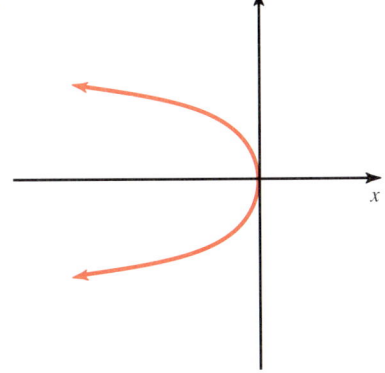
   C.
   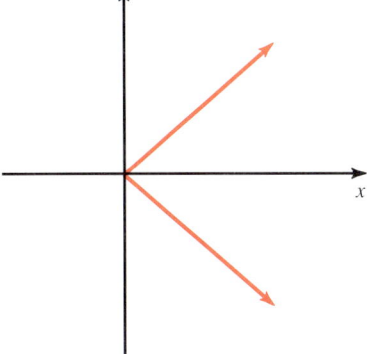
   D.
   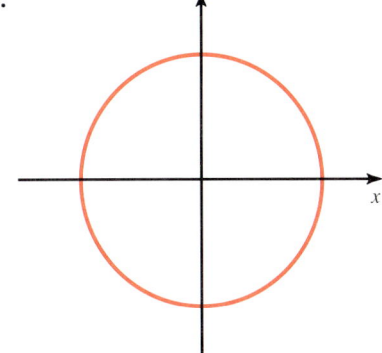

6. Solve the following equations.
   a. $\log_3 x = 3$ b. $5^x = 125$
   c. $\log_2 2 + \log_2 4 = x$ d. $\log 10 = x$
   e. $\log_5 1 = x$
7. Approximate x to two decimal places in each of the following.
   a. $\log 4.73 = x$
   b. $\log x = -0.8174$
   c. $\ln 10.1 = x$
8. If $f(x) = x - 3$ and $g(x) = x^2$, find:
   a. $(f + g)(x)$
   b. $(g \circ f)(x)$

9. If $f(x) = 2x + 3$, find $f^{-1}(x)$.

10. Find the natural domain of each function.
    a. $f(x) = \sqrt{(x+1)(x-2)}$
    b. $g(x) = \dfrac{1}{x-7}$

11. Sketch the graph of each function.
    a. $f(x) = x^2 - 1$
    b. $f(x) = 3^x$
    c. $f(x) = |x - 3|$
    d. $f(x) = -2(x-1)^2 + 3$

CHAPTER 12

See Section 12.5, Example 2.

Other Topics

- **12.1** Sequences
- **12.2** Summation Notation
- **12.3** Factorials and Binomial Coefficients
- **12.4** The Binomial Theorem
- **12.5** Permutations and Combinations

CONNECTIONS

We complete our study of intermediate algebra with an introduction to topics that will become important in future mathematics courses.

First, we look at sequences or ordered collections of numbers such as

$$1, 2, 4, 8, 16, \ldots$$

which are used in the study of calculus.

Next, we look at sums of related numbers and introduce a very convenient notation for such sums. We use this new notation to state the binomial theorem.

Finally, we define permutations and combinations and learn how to compute them. These ideas form a cornerstone in the study of probability.

12.1 SEQUENCES

Ordered lists of numbers such as

$$1, 3, 5, 7, 9, \ldots$$
$$-2, 4, -8, 16, \ldots$$
$$1, \frac{1}{2}, \frac{1}{3}, \frac{1}{4}, \frac{1}{5}, \ldots$$

are called **sequences.** The three dots, called an ellipsis, mean continue on, in the pattern that has been established. The expressions separated by commas are called the **terms** of the sequence. Often, we can determine the **general term** that expresses every term of the sequence. For example, the three sequences given above have general terms given by

$$2n - 1$$
$$(-2)^n$$
$$\frac{1}{n}$$

where n has the values $1, 2, 3, 4, \ldots$. This leads to the formal definition of sequence.

Definition of Sequence

A sequence is a function whose domain is the set of natural numbers.

Since the domain of a sequence is the set of natural numbers, we use the general term rather than the usual functional notation. Instead of writing

$$f(x) = 2x - 1; \quad x = 1, 2, 3, \ldots$$

for the first sequence, we use n as an element of the domain and a_n instead of $f(n)$, with the understanding that n is a natural number. So we write the sequence as

$$a_n = 2n - 1$$

EXAMPLE 1. Write the first five terms of the sequence $a_n = 4n + 5$.

Solution:

Evaluate the terms of a sequence just like we evaluate a function. We evaluate the first five terms.

$$a_1 = 4 \cdot 1 + 5 = 9$$
$$a_2 = 4 \cdot 2 + 5 = 13$$
$$a_3 = 4 \cdot 3 + 5 = 17$$
$$a_4 = 4 \cdot 4 + 5 = 21$$
$$a_5 = 4 \cdot 5 + 5 = 25$$

and see that the sequence is

$$9, 13, 17, 21, 25, \ldots$$

EXAMPLE 2. Write the first five terms of the sequence $b_n = n(n + 1)$.

Solution:

Evaluate the first five terms.

$$b_1 = 1(1 + 1) = 1 \cdot 2 = 2$$
$$b_2 = 2(2 + 1) = 2 \cdot 3 = 6$$
$$b_3 = 3(3 + 1) = 3 \cdot 4 = 12$$
$$b_4 = 4(4 + 1) = 4 \cdot 5 = 20$$
$$b_5 = 5(5 + 1) = 5 \cdot 6 = 30$$

So this sequence is

$$2, 6, 12, 20, 30, \ldots$$

Often we are given some of the terms of a sequence and are asked to provide a suitable general term.

EXAMPLE 3. Find a suitable general term for each of the following sequences.

(a) $2, 6, 18, 54, 162, \ldots$ (b) $2, 5, 8, 11, 14, \ldots$
(c) $\sqrt{2}, 2, \sqrt{6}, 2\sqrt{2}, \sqrt{10}, 2\sqrt{3}, \ldots$ (d) $1, -\dfrac{1}{2}, \dfrac{1}{4}, -\dfrac{1}{8}, \dfrac{1}{16}, \ldots$

Solutions:

(a) Notice that each term, after the first, is three *times* the previous term. Thus

$$a_2 = a_1 \cdot 3 = 2 \cdot 3$$

(continued)

By the same reasoning

$$a_3 = a_2 \cdot 3 = 2 \cdot 3 \cdot 3 = 2 \cdot 3^2$$
$$a_4 = 2 \cdot 3^2 \cdot 3 = 2 \cdot 3^3$$
$$a_5 = 2 \cdot 3^3 \cdot 3 = 2 \cdot 3^4$$

Now we see the pattern, $a_n = 2 \cdot 3^{n-1}$.

(b) In this sequence, each term, after the first, is three *added to* the previous term. So

$$b_2 = b_1 + 3 = 2 + 3$$
$$b_3 = b_2 + 3 = 2 + 3 + 3 \quad = 2 + 2 \cdot 3$$
$$b_4 = b_3 + 3 = 2 + 2 \cdot 3 + 3 = 2 + 3 \cdot 3$$
$$b_5 = b_4 + 3 = 2 + 3 \cdot 3 + 3 = 2 + 4 \cdot 3$$

Look closely to see the pattern:

$$b_n = 2 + (n - 1) \cdot 3$$

(c) Rewriting this sequence as unsimplified square roots,

$$\sqrt{2}, \sqrt{4}, \sqrt{6}, \sqrt{8}, \sqrt{10}, \sqrt{12}, \ldots$$

and each term is the square root of $2n$,

$$c_n = \sqrt{2n}$$

(d) Since the terms alternate in sign, think of a negative number raised to the *n*th power. Since powers of 2 are in the denominators, notice that

$$d_1 = 1 = \text{(any nonzero number)}^0$$
$$d_2 = -\frac{1}{2} = \left(-\frac{1}{2}\right)^1$$
$$d_3 = \frac{1}{4} = \left(-\frac{1}{2}\right)^2$$
$$d_4 = -\frac{1}{8} = \left(-\frac{1}{2}\right)^3$$

and the pattern is

$$d_n = \left(-\frac{1}{2}\right)^{n-1}$$

A sequence such as

$$1, 4, 7, 10, 13, \ldots$$

where every term is greater than the preceding term is called an **increasing** sequence. A sequence such as

$$1, \frac{2}{3}, \frac{4}{9}, \frac{8}{27}, \frac{16}{81}, \ldots$$

where every term is less than the preceding term is called a **decreasing** sequence.

EXAMPLE 4. Label each of the following sequences as *increasing, decreasing,* or *neither increasing nor decreasing*.

(a) $a_n = 3n - 2$
(b) $b_n = 4 - n$
(c) $c_n = 5 + (-2)^n$

Solutions:

(a) A few terms of this sequence,

$$1, 4, 7, 10, \ldots$$

show that each term is 3 more than the preceding term. The sequence is *increasing*.

(b) As the value of n increases by one in each term, the value of the term decreases by one.

$$4, 3, 2, 1, 0, -1, \ldots$$

The sequence is *decreasing*.

(c) Notice

$$c_1 = 5 + (-2)^1 = 5 + (-2) = 3$$
$$c_2 = 5 + (-2)^2 = 5 + 4 = 9$$
$$c_3 = 5 + (-2)^3 = 5 - 8 = -3$$
$$c_4 = 5 + (-2)^4 = 5 + 16 = 21$$
$$c_5 = 5 + (-2)^5 = 5 - 32 = -27$$

and the sequence is *neither increasing nor decreasing*.

$$3, 9, -3, 21, -27, \ldots$$

Any sequence of the form

$$a_n = a_1 + (n - 1)d$$

is called an **arithmetic sequence** or **arithmetic progression.** The number d is called its **common difference.** Some examples of arithmetic sequences are

$1, 5, 9, 13, 17, \ldots$ Common difference is 4.
$-3, 6, 15, 24, \ldots$ Common difference is 9.
$2, -1, -4, -7, \ldots$ Common difference is -3.

We could write these sequences as

$$a_n = 1 + (n - 1) \cdot 4$$
$$b_n = -3 + (n - 1) \cdot 9$$
$$c_n = 2 + (n - 1)(-3)$$

Notice that d is constant for each arithmetic sequence.

EXAMPLE 5. For each of the following *arithmetic sequences,* find the common difference and write each in the form $a_n = a_1 + (n-1)d$.

(a) 1, 3, 5, 7, 9, . . .   (b) -2, 6, 14, 22, . . .

(c) 3, $\frac{7}{2}$, 4, $\frac{9}{2}$, 5, $\frac{11}{2}$, . . .   (d) $-4, -7, -10, -13,$ . . .

Solutions:

If the sequence is an arithmetic sequence, find d by subtracting any term from the following term.

(a) $d = a_2 - a_1 = 3 - 1 = 2$. So, as a_1 is 1,
$$a_n = 1 + (n-1) \cdot 2$$

(b) $d = 6 - (-2) = 8$.
$$a_n = -2 + (n-1) \cdot 8$$

(c) $d = \frac{7}{2} - 3 = \frac{7-6}{2} = \frac{1}{2}$.
$$a_n = 3 + (n-1) \cdot \frac{1}{2}$$

(d) $d = -7 - (-4) = -7 + 4 = -3$.
$$a_n = -4 + (n-1)(-3)$$

Any sequence of the form
$$a_n = a_1 r^{n-1}$$

is called a **geometric sequence** or **geometric progression.** The number r is called its **common ratio.** Some examples of geometric sequences are

$\qquad$ 1, 2, 4, 8, 16, . . . $\qquad$ Common ratio is 2.

$\qquad$ 2, 10, 50, 250, . . . $\qquad$ Common ratio is 5.

$\qquad$ 4, -2, 1, $-\frac{1}{2}$, $\frac{1}{4}$, . . . $\qquad$ Common ratio is $-\frac{1}{2}$.

We could write these sequences as
$$a_n = 1 \cdot 2^{n-1} \quad \text{(or simply } a_n = 2^{n-1}\text{)}$$
$$b_n = 2 \cdot 5^{n-1}$$
$$c_n = 4\left(-\frac{1}{2}\right)^{n-1}$$

The common ratio r is a constant for each geometric sequence.

EXAMPLE 6. For each of the following *geometric sequences,* find the common ratio and write each in the form $a_n = a_1 r^{n-1}$.

(a) 2, 4, 8, 16, . . .   (b) 4, 12, 36, 108, . . .

(c) 3, -2, $\frac{4}{3}$, $-\frac{8}{9}$, $\frac{16}{27}$, . . .

Solutions:

If a sequence is a geometric sequence, find the common ratio by dividing any term into the following term.

(a) $r = \dfrac{a_2}{a_1} = \dfrac{4}{2} = 2$. Therefore, the common ratio is 2.

$$a_n = 2 \cdot 2^{n-1} \quad \text{(or simply } a_n = 2^n\text{)}$$

(b) $r = \dfrac{12}{4} = 3$.

$$a_n = 4 \cdot 3^{n-1}$$

(c) $r = \dfrac{-2}{3} = -\dfrac{2}{3}$.

$$a_n = 3\left(-\dfrac{2}{3}\right)^{n-1}$$

PROBLEM SET 12.1

Warm-ups

In problems 1 through 9, write the first five terms of each sequence. See Examples 1 and 2.

1. $a_n = n$
2. $a_n = 2n + 1$
3. $a_n = 2(n + 1)$
4. $b_n = 14 - 3n$
5. $c_n = n^2$
6. $h_n = 2(n - 1)^2$
7. $a_n = 3$
8. $z_n = (-1)^{n-1} n^3$
9. $k_n = (-1)^n (1 - n)$

In problems 10 through 15, find a suitable general term for each sequence. See Example 3.

10. 2, 4, 6, 8, ...
11. 1, 4, 9, 16, ...
12. 2, 5, 8, 11, ...
13. −1, 2, −3, 4, ...
14. 1, 9, 25, 49, ...
15. 2, −4, 8, −16, ...

In problems 16 through 21, label each sequence as increasing, decreasing, or neither increasing nor decreasing. See Example 4.

16. $a_n = 3 + 2n$
17. $a_n = 3 - 2n$
18. $b_n = n^{-2}$
19. $a_n = (-n)^{-1}$
20. $v_n = (-n)^{-n}$
21. $k_n = 1 - n^{-1}$

Problems 22 through 27 are arithmetic sequences. In each, find the common difference and write in the form $a_n = a_1 + (n - 1)d$. See Example 5.

22. 3, 5, 7, ...
23. 2, 3, 4, ...
24. 1, 5, 9, ...
25. 5, 15, 25, ...
26. 0, −2, −4, ...
27. $\dfrac{1}{2}, 0, -\dfrac{1}{2}, \ldots$

Problems 28 through 33 are geometric sequences. In each, find the common ratio and write in the form $a_n = a_1 r^{n-1}$. See Example 6.

28. 3, 6, 12, ...
29. 1, 5, 25, ...
30. 3, 9, 27, ...
31. 2, −4, 8, ...
32. 4, 2, 1, ...
33. $1, -\dfrac{1}{4}, \dfrac{1}{16}, \ldots$

Practice Exercises

In problems 34 through 42, write the first five terms of each sequence.

34. $a_n = n + 1$
35. $a_n = 3n - 1$
36. $a_n = 2(n - 1)$

37. $A_n = 8 - 2n$ **38.** $b_n = n^2 - 1$ **39.** $h_n = n^2 - n + 1$
40. $c_n = -4$ **41.** $j_n = (-1)^{n-1}n^2$ **42.** $a_n = (-1)^n(n - n^2)$

In problems 43 through 48, find a suitable general term for the sequence.

43. 4, 6, 8, 10, . . . **44.** 0, 3, 8, 15, . . . **45.** -1, 3, 7, 11, . . .
46. 1, -2, 3, -4, . . . **47.** 0, 1, 8, 27, . . . **48.** -1, 2, -4, 8, . . .

In problems 49 through 54, label each sequence as increasing, decreasing, or neither increasing nor decreasing.

49. $a_n = -4 + n$ **50.** $a_n = -4 - n$ **51.** $b_n = n^{-2} - 2$
52. $c_n = 2 - n^{-2}$ **53.** $d_n = (-2)^{-2n}$ **54.** $k_n = (-1 - n)^{-1}$

Problems 55 through 60 are arithmetic sequences. In each, find the common difference and write in the form $a_n = a_1 + (n - 1)d$.

55. 1, 4, 7, . . . **56.** -1, 4, 9, . . .
57. 10, 8, 6, . . . **58.** 0, 11, 22, . . .
59. $-\frac{1}{2}, 0, \frac{1}{2}, \ldots$ **60.** $\frac{2}{3}, \frac{1}{3}, 0, \ldots$

Problems 61 through 66 are geometric sequences. In each, find the common ratio and write in the form $a_n = a_1 r^{n-1}$.

61. 1, 3, 9, . . . **62.** 2, 4, 8, . . . **63.** 2, -6, 18, . . .
64. 5, 25, 125, . . . **65.** 27, 9, 3, . . . **66.** 128, -64, 32, . . .

IN YOUR OWN WORDS

67. What is a sequence?

12.2 SUMMATION NOTATION

Ellen Adams's new job pays $1000 for the first month and includes a $10 raise each month. How much will Ellen earn the first year?

Ellen will earn the sum of 12 monthly payments as follows:

$$S = (1000) + (1000 + 10) + (1000 + 2 \cdot 10) + \cdots + (1000 + 11 \cdot 10)$$

Notice that the first two payments can be written as $(1000 + 0 \cdot 10)$ and $(1000 + 1 \cdot 10)$. Therefore, Ellen will get 12 monthly payments of the form $(1000 + j \cdot 10)$, where j has values 0, 1, 2, . . . , 11. Such sums can be written with a convenient shorthand called *summation notation*.

$$S = \sum_{j=0}^{11} (1000 + j \cdot 10)$$

(Ellen will earn $12,660.)

The symbol Σ is the Greek capital letter sigma, which is used to mean sum. The assignment, $j = 0$, under the sigma, is the first value of the variable j to be used in the sum. The integer 11, at the top of the sigma, is the final value of the variable j to be used in the sum. There is a term in the sum for every integer value of j from 0 to 11. The variable j is called the **index**.

EXAMPLE 1. Write out the terms of the following sum.

$$\sum_{j=1}^{6} (2j - 1)$$

Solution:

$$\sum_{j=1}^{6} (2j-1) = (2 \cdot 1 - 1) + (2 \cdot 2 - 1) + (2 \cdot 3 - 1)$$
$$+ (2 \cdot 4 - 1) + (2 \cdot 5 - 1) + (2 \cdot 6 - 1)$$
$$= 1 + 3 + 5 + 7 + 9 + 11 \qquad \square$$

EXAMPLE 2. Write out the terms of the following sum.

$$\sum_{j=2}^{5} \frac{j-1}{j+1}$$

Solution:

$$\sum_{j=2}^{5} \frac{j-1}{j+1} = \frac{2-1}{2+1} + \frac{3-1}{3+1} + \frac{4-1}{4+1} + \frac{5-1}{5+1}$$
$$= \frac{1}{3} + \frac{2}{4} + \frac{3}{5} + \frac{4}{6} + \frac{5}{7} \qquad \square$$

EXAMPLE 3. Write out the terms of the following sum.

$$\sum_{j=1}^{5} 2^{j-1} \cdot j$$

Solution:

$$\sum_{j=1}^{5} 2^{j-1} \cdot j = 2^0 \cdot 1 + 2^1 \cdot 2 + 2^2 \cdot 3 + 2^3 \cdot 4 + 2^4 \cdot 5$$
$$= 1 + 2 \cdot 2 + 4 \cdot 3 + 8 \cdot 4 + 16 \cdot 5$$
$$= 1 + 4 + 12 + 32 + 80 \qquad \square$$

EXAMPLE 4. Write out the terms of the following sum.

$$\sum_{j=0}^{3} (-1)^j X^j$$

Solution:

$$\sum_{j=0}^{3} (-1)^j X^j = (-1)^0 X^0 + (-1)^1 X^1 + (-1)^2 X^2 + (-1)^3 X^3$$
$$= 1 \cdot 1 + (-1)X + 1 \cdot X^2 + (-1)X^3$$
$$= 1 - X + X^2 - X^3$$

In this example, notice how the factor $(-1)^j$ caused the signs to alternate. $(-1)^j$ and $(-1)^{j-1}$ are often used for this purpose. $\square$

EXAMPLE 5. Write the following sum in summation notation.
$$2 + 4 + 6 + 8 + 10 + 12 + 14 + 16$$

Solution:

Notice that the terms can be written
$$2 \cdot 1 + 2 \cdot 2 + 2 \cdot 3 + 2 \cdot 4 + \cdots + 2 \cdot 8$$

So we can write
$$\sum_{j=1}^{8} 2j$$

EXAMPLE 6. Write the following sum in summation notation.
$$x + 2x^2 + 3x^3 + 4x^4 + 5x^5$$

Solution:

This sum is of the form jx^j as j goes from 1 to 5.
$$\sum_{j=1}^{5} jx^j$$

PROBLEM SET 12.2

Warm-ups

In problems 1 through 9, write out the terms of each sum. For problems 1 through 6, see Examples 1 through 3.

1. $\sum_{j=1}^{5} 2j$
2. $\sum_{j=1}^{6} 3(j-1)$
3. $\sum_{j=0}^{7} j^2$
4. $\sum_{j=1}^{7} 2^j$
5. $\sum_{j=0}^{4} (3-j)^2$
6. $\sum_{j=0}^{5} (j-1)^2$

For problems 7 through 9, see Example 4.

7. $\sum_{j=1}^{6} (-1)^{j-1} j$
8. $\sum_{j=1}^{5} (-1)^j 2^j$
9. $\sum_{j=1}^{5} \frac{j-1}{j+1}(-1)^{j+1}$

In problems 10 through 15, write each sum in summation notation. For problems 10 through 12, see Example 5.

10. $1 + 2 + 3 + 4 + 5 + 6 + 7 + 8$
11. $1 - 3 + 5 - 7 + 9 - 11$
12. $1 - \frac{2}{3} + \frac{3}{5} - \frac{4}{7} + \frac{5}{9} - \frac{6}{11} + \frac{7}{13} - \frac{8}{15}$

For problems 13 through 15, see Example 6.

13. $1 + \frac{x}{2} + \frac{x^2}{3} + \frac{x^3}{4} + \frac{x^4}{5} + \frac{x^5}{6} + \frac{x^6}{7}$
14. $A_1 + A_2 + A_3 + \cdots + A_{100}$
15. $1 - \sqrt{2}x^2 + \sqrt{3}x^4 - 2x^6 + \sqrt{5}x^8 - \sqrt{6}x^{10} + \sqrt{7}x^{12}$

Practice Exercises

In problems 16 through 24, write out the terms of each sum.

16. $\sum_{j=1}^{6} 3j$
17. $\sum_{j=0}^{4} 2(j-5)$
18. $\sum_{j=0}^{7} (j^2 - 1)$
19. $\sum_{j=1}^{7} 2^{j-1}$
20. $\sum_{j=0}^{4} (3 - 2j)^2$
21. $\sum_{j=0}^{5} (j-1)^3$
22. $\sum_{j=1}^{6} (-1)^{j-1} Q_j$
23. $\sum_{j=1}^{5} (-1)^j 2^{j-1}$
24. $\sum_{j=1}^{5} \frac{2j-1}{2j+1}$

In problems 25 through 30, write each sum in summation notation.

25. $3 + 4 + 5 + 6 + 7 + 8 + 9 + 10 + 11 + 12 + 13 + 14$
26. $2 - 4 + 6 - 8 + 10 - 12 + 14 - 16 + 18$
27. $1 + \frac{x^2}{3} + \frac{x^4}{5} + \frac{x^6}{7} + \frac{x^8}{9}$
28. $A_0 - A_1 + A_2 - A_3 + \cdots + A_{52}$
29. $-\pi + \frac{1-\pi}{10} + \frac{2-\pi}{100} + \frac{3-\pi}{1000} + \frac{4-\pi}{10{,}000} + \frac{5-\pi}{100{,}000} + \frac{6-\pi}{1{,}000{,}000}$
30. $1 + 2 + 4 + 8 + \cdots + 2048$

Challenge Problems

31. Write out the terms of $\sum_{i=1}^{5} i$, $\sum_{j=1}^{5} j$, and $\sum_{k=1}^{5} k$.
32. Why is the index in summation notation sometimes called a "dummy"?
33. Compare $\sum_{j=1}^{4} 5A_j$ with $5\sum_{j=1}^{4} A_j$. Write a property of summation notation.
34. Compare $\sum_{j=1}^{4} (A_j + B_j)$ with $\sum_{j=1}^{4} A_j + \sum_{j=1}^{4} B_j$. Write another property of summation notation.

▬▬ IN YOUR OWN WORDS

35. Why is summation notation useful?
36. Explain how the index works in summation notation.

▬▬ 12.3 FACTORIALS AND BINOMIAL COEFFICIENTS

Just as x^4 is shorthand notation for the product $x \cdot x \cdot x \cdot x$, shorthand notation exists for products such as $1 \cdot 2 \cdot 3 \cdot 4$ and $1 \cdot 2 \cdot 3 \cdot 4 \cdot 5 \cdot 6 \cdot 7$. Such products are called **factorials**. Write them as follows:

$$4! = 1 \cdot 2 \cdot 3 \cdot 4$$

$$7! = 1 \cdot 2 \cdot 3 \cdot 4 \cdot 5 \cdot 6 \cdot 7$$

Read 4! as "four factorial," $K!$ as "K factorial," and $(N-1)!$ as "N minus one factorial."

> **Factorial Definitions**
>
> For N a nonnegative integer,
> $$N! = 1 \cdot 2 \cdot 3 \cdots (N-1) \cdot N$$
> $$0! = 1$$

Factorials grow quickly, as illustrated below.

$$\begin{aligned}
0! &= 1 & &= 1 \\
1! &= 1 & &= 1 \\
2! &= 1 \cdot 2 & &= 2 \\
3! &= 1 \cdot 2 \cdot 3 & &= 6 \\
4! &= 1 \cdot 2 \cdot 3 \cdot 4 & &= 24 \\
5! &= 1 \cdot 2 \cdot 3 \cdot 4 \cdot 5 & &= 120 \\
&\vdots \\
9! &= 1 \cdot 2 \cdot 3 \cdot 4 \cdot 5 \cdot 6 \cdot 7 \cdot 8 \cdot 9 & &= 362{,}880 \\
10! &= 1 \cdot 2 \cdot 3 \cdot 4 \cdot 5 \cdot 6 \cdot 7 \cdot 8 \cdot 9 \cdot 10 &&= 3{,}628{,}800
\end{aligned}$$

Notice from the table above that 5! is $4! \cdot 5$ and 10! is $9! \cdot 10$.

> **Property of Factorials**
>
> If N is a natural number,
> $$N! = (N-1)! \cdot N$$

EXAMPLE 1. Calculate 11!.

Solution:

By the Property of Factorials,
$$11! = 10! \cdot 11$$
From the table, 10! is 3,628,800. So
$$11! = (3{,}628{,}800) \cdot 11$$
$$= 39{,}916{,}800$$

Quotients of factorials are interesting and quite common. They often simplify with vast cancellations.

EXAMPLE 2. Evaluate each expression.

(a) $\dfrac{16!}{17!}$ (b) $\dfrac{20!}{18!}$

(c) $\dfrac{10!}{5! \cdot 5!}$ (d) $\dfrac{8!}{8! \cdot 0!}$

Solutions:

(a) By the Property of Factorials, 17! is 16! · 17, so

$$\frac{16!}{17!} = \frac{16!}{16! \cdot 17} = \frac{1}{17}$$

Note that 16 factors were divided out from the numerator and denominator!

(b) Write 20! as 18! · 19 · 20.

$$\frac{20!}{18!} = \frac{18! \cdot 19 \cdot 20}{18!} = 19 \cdot 20 = 380$$

(c) By definition,

$$\frac{10!}{5! \cdot 5!} = \frac{1 \cdot 2 \cdot 3 \cdot 4 \cdot 5 \cdot 6 \cdot 7 \cdot 8 \cdot 9 \cdot 10}{(1 \cdot 2 \cdot 3 \cdot 4 \cdot 5) \cdot (1 \cdot 2 \cdot 3 \cdot 4 \cdot 5)}$$

Divide out the first five factors, leaving

$$= \frac{6 \cdot 7 \cdot 8 \cdot 9 \cdot 10}{1 \cdot 2 \cdot 3 \cdot 4 \cdot 5}$$

$$= \frac{7 \cdot 2 \cdot 9 \cdot 2}{1} \qquad \text{Reduce.}$$

$$= 252$$

(d) The 8!s divide out.

$$\frac{8!}{8! \cdot 0!} = \frac{1}{0!}$$

Remember, 0! is 1, not 0.

$$= \frac{1}{1} = 1 \qquad \square$$

Factorial expressions of the form $\frac{7!}{3! \cdot 4!}$ or $\frac{10!}{8! \cdot 2!}$ are quite common in mathematics. In the next section such expressions are called **binomial coefficients** and are written

$$\binom{7}{3} = \frac{7!}{3! \cdot 4!}$$

$$\binom{10}{8} = \frac{10!}{8! \cdot 2!}$$

The Binomial Coefficients

For n and r nonnegative integers, $n \geq r$,

$$\binom{n}{r} = \frac{n!}{r! \cdot (n-r)!}$$

Sec. 12.3 Factorials and Binomial Coefficients

EXAMPLE 3. Calculate each binomial coefficient.

(a) $\binom{7}{5}$ (b) $\binom{9}{3}$

(c) $\binom{6}{6}$ (d) $\binom{11}{0}$

Solutions:

(a) $\binom{7}{5} = \dfrac{7!}{5! \cdot (7-5)!} = \dfrac{7!}{5! \cdot 2!}$

$= \dfrac{1 \cdot 2 \cdot 3 \cdot 4 \cdot 5 \cdot 6 \cdot 7}{(1 \cdot 2 \cdot 3 \cdot 4 \cdot 5)(1 \cdot 2)} = \dfrac{6 \cdot 7}{1 \cdot 2} = 21$

(b) $\binom{9}{3} = \dfrac{9!}{3! \cdot (9-3)!} = \dfrac{9!}{3! \cdot 6!}$

$= \dfrac{1 \cdot 2 \cdot 3 \cdot 4 \cdot 5 \cdot 6 \cdot 7 \cdot 8 \cdot 9}{(1 \cdot 2 \cdot 3)(1 \cdot 2 \cdot 3 \cdot 4 \cdot 5 \cdot 6)} = \dfrac{7 \cdot 8 \cdot 9}{1 \cdot 2 \cdot 3} = 7 \cdot 4 \cdot 3 = 84$

(c) $\binom{8}{8} = \dfrac{8!}{8! \cdot (8-8)!} = \dfrac{8!}{8! \cdot 0!} = \dfrac{1}{0!} = \dfrac{1}{1} = 1$

(d) $\binom{11}{0} = \dfrac{11!}{0! \cdot (11-0)!} = \dfrac{11!}{0! \cdot 11!} = 1$

CALCULATOR BOX

Factorials with a Calculator

Factorials can be calculated with a calculator easily.

EXAMPLE Find 8!.

Solution:

Press the keys [8] [n!] and read [40320] on the display. The [n!] key may require a second function or an inverse, and on some calculators it may be [x!].

$$8! = 40{,}320$$

Calculator Exercises.

Evaluate each using a calculator.

1. 12!   2. 0!   3. 6!   4. $\dfrac{20!}{18!}$ 5. $\dfrac{6!}{3!2!}$

Answers:

1. 479,001,600   2. 1   3. 720   4. 380   5. 60

PROBLEM SET 12.3

Warm-ups

In problems 1 through 24, evaluate each expression.
For problems 1 through 4, see Example 1.

1. $6!$
2. $8!$
3. $12!$
4. $10!$

For problems 5 through 12, see Example 2.

5. $\dfrac{14!}{13!}$
6. $\dfrac{29!}{31!}$
7. $\dfrac{8!}{4! \cdot 4!}$
8. $\dfrac{8!}{3! \cdot 5!}$
9. $\dfrac{8!}{2! \cdot 6!}$
10. $\dfrac{8!}{1! \cdot 7!}$
11. $\dfrac{20!}{1! \cdot 19!}$
12. $\dfrac{20!}{2! \cdot 18!}$

For problems 13 through 24, see Example 3.

13. $\binom{8}{5}$
14. $\binom{5}{2}$
15. $\binom{5}{3}$
16. $\binom{6}{0}$
17. $\binom{6}{1}$
18. $\binom{6}{2}$
19. $\binom{6}{3}$
20. $\binom{6}{4}$
21. $\binom{6}{5}$
22. $\binom{6}{6}$
23. $\binom{500}{0}$
24. $\binom{69}{69}$

Practice Exercises

In problems 25 through 48, evaluate each expression.

25. $7!$
26. $13!$
27. $11!$
28. $5!$
29. $\dfrac{17!}{18!}$
30. $\dfrac{29!}{27!}$
31. $\dfrac{9!}{4! \cdot 5!}$
32. $\dfrac{9!}{3! \cdot 6!}$
33. $\dfrac{9!}{2! \cdot 7!}$
34. $\dfrac{9!}{1! \cdot 8!}$
35. $\dfrac{30!}{1! \cdot 29!}$
36. $\dfrac{30!}{2! \cdot 28!}$
37. $\binom{9}{6}$
38. $\binom{4}{3}$
39. $\binom{4}{1}$
40. $\binom{7}{0}$
41. $\binom{7}{1}$
42. $\binom{7}{2}$
43. $\binom{7}{3}$
44. $\binom{7}{4}$
45. $\binom{7}{5}$
46. $\binom{7}{6}$
47. $\binom{7}{7}$
48. $\binom{39}{38}$

Challenge Problems

49. $\binom{7}{2}$ has the same value as $\binom{7}{5}$ and $\binom{6}{1}$ has the same value as $\binom{6}{5}$. Use the definition of binomial coefficient to show that $\binom{n}{r} = \binom{n}{n-r}$.

IN YOUR OWN WORDS

50. Explain $n!$ without using a formula.

12.4 THE BINOMIAL THEOREM

In Chapter 1 we learned that $(a + b)^2$ is $a^2 + 2ab + b^2$. Let's look at a few powers of the binomial $(a + b)$.

$(a + b)^0 = 1$
$(a + b)^1 = a + b$
$(a + b)^2 = a^2 + 2ab + b^2$
$(a + b)^3 = a^3 + 3a^2b + 3ab^2 + b^3$
$(a + b)^4 = a^4 + 4a^3b + 6a^2b^2 + 4ab^3 + b^4$
$(a + b)^5 = a^5 + 5a^4b + 10a^3b^2 + 10a^2b^3 + 5ab^4 + b^5$

The Binomial Theorem provides a general formula for the expansion of $(a + b)^n$, for n a natural number. Some of the formula is quite easy to see. The powers of a, for example, start at n and reduce by one for each term until the power becomes zero (remember, $a^0 = 1$). Similarly, b starts with an exponent of zero and increases by one until b^n is reached. Except for coefficients, $(a + b)^7$ can be written

$$\underline{}a^7 + \underline{}a^6b + \underline{}a^5b^2 + \underline{}a^4b^3 + \underline{}a^3b^4 + \underline{}a^2b^5 + \underline{}ab^6 + \underline{}b^7$$

In Section 1.5 we found that the coefficients are the seventh row of Pascal's triangle.

```
0th row ⟶                              1
1st row ⟶                           1     1
                                 1     2     1
                              1     3     3     1
                           1     4     6     4     1
                        1     5    10    10     5     1
                     1     6    15    20    15     6     1
7th row ⟶         1     7    21    35    35    21     7     1
```

However, for large values of n, this is an awkward way to find these coefficients. Let's look at some binomial coefficients from Section 12.3.

$$\binom{7}{0} = \frac{7!}{0! \cdot 7!} = \frac{1}{0!} = 1$$

$$\binom{7}{1} = \frac{7!}{1! \cdot 6!} = \frac{6! \cdot 7}{6!} = 7$$

$$\binom{7}{2} = \frac{7!}{2! \cdot 5!} = \frac{1 \cdot 2 \cdot 3 \cdot 4 \cdot 5 \cdot 6 \cdot 7}{(1 \cdot 2)(1 \cdot 2 \cdot 3 \cdot 4 \cdot 5)} = \frac{6 \cdot 7}{1 \cdot 2} = 21$$

Continuing, we have

$$\binom{7}{3} = 35 \quad \binom{7}{4} = 35 \quad \binom{7}{5} = 21 \quad \binom{7}{6} = 7 \quad \binom{7}{7} = 1$$

Notice these are exactly the elements on the seventh row of Pascal's triangle. That is, the expansion of $(a + b)^7$ could be written as

$$\binom{7}{0}a^7 + \binom{7}{1}a^6b + \binom{7}{2}a^5b^2 + \binom{7}{3}a^4b^3 + \binom{7}{4}a^3b^4 + \binom{7}{5}a^2b^5 + \binom{7}{6}ab^6 + \binom{7}{7}b^7$$

Remembering that $x^0 = 1$ and $x^1 = x$, and using summation notation, we have

$$(a + b)^7 = \sum_{j=0}^{7} \binom{7}{j} a^{7-j} b^j$$

The Binomial Theorem

For n a natural number,

$$(a + b)^n = \sum_{j=0}^{n} \binom{n}{j} a^{n-j} b^j$$

EXAMPLE 1. Write the first four terms in the expansion of $(x + y)^{11}$.

Solution:

By the Binomial Theorem,

$$(x + y)^{11} = \sum_{j=0}^{11} \binom{11}{j} x^{11-j} y^j$$

So the first four terms are

$$\binom{11}{0} x^{11} y^0 + \binom{11}{1} x^{10} y^1 + \binom{11}{2} x^9 y^2 + \binom{11}{3} x^8 y^3$$

Now

$$\binom{11}{0} = \frac{11!}{0! \cdot 11!} = 1$$

$$\binom{11}{1} = \frac{11!}{1! \cdot 10!} = \frac{10! \cdot 11}{1! \cdot 10!} = \frac{11}{1!} = 11$$

$$\binom{11}{2} = \frac{11!}{2! \cdot 9!} = \frac{1 \cdot 2 \cdot 3 \cdot 4 \cdot 5 \cdot 6 \cdot 7 \cdot 8 \cdot 9 \cdot 10 \cdot 11}{(1 \cdot 2)(1 \cdot 2 \cdot 3 \cdot 4 \cdot 5 \cdot 6 \cdot 7 \cdot 8 \cdot 9)}$$

$$= \frac{10 \cdot 11}{1 \cdot 2} = 55$$

$$\binom{11}{3} = \frac{11!}{3! \cdot 8!} = \frac{1 \cdot 2 \cdot 3 \cdot 4 \cdot 5 \cdot 6 \cdot 7 \cdot 8 \cdot 9 \cdot 10 \cdot 11}{(1 \cdot 2 \cdot 3)(1 \cdot 2 \cdot 3 \cdot 4 \cdot 5 \cdot 6 \cdot 7 \cdot 8)}$$

$$= \frac{9 \cdot 10 \cdot 11}{1 \cdot 2 \cdot 3} = 165$$

So the first four terms of $(x + y)^{11}$ are

$$1 \cdot x^{11} y^0 + 11 x^{10} y^1 + 55 x^9 y^2 + 165 x^8 y^3 \quad \text{or}$$

$$x^{11} + 11 x^{10} y + 55 x^9 y^2 + 165 x^8 y^3$$

EXAMPLE 2. Find the fourteenth term of $(2 + z)^{16}$.

Solution:

From the Binomial Theorem,

$$(2 + z)^{16} = \sum_{j=0}^{16} \binom{16}{j} 2^{16-j} z^j$$

(continued)

The fourteenth term is when j has the value 13.

$$\binom{16}{13} 2^{16-13} z^{13}$$

But

$$\binom{16}{13} = \frac{16!}{13! \cdot 3!} = \frac{14 \cdot 15 \cdot 16}{1 \cdot 2 \cdot 3} = 7 \cdot 5 \cdot 16 = 560$$

and

$$2^{16-13} = 2^3 = 8$$

so the fourteenth term is $560 \cdot 8z^{13}$ or $4480z^{13}$.

PROBLEM SET 12.4

Warm-ups

In problems 1 through 6, find the first three terms of the binomial expansion. See Example 1.

1. $(a + b)^{12}$
2. $(x + y)^{20}$
3. $(s + 2)^{15}$
4. $(2 + t)^9$
5. $(2a + 3b)^6$
6. $(x^2 + 1)^{25}$

In problems 7 through 10, see Example 2.

7. Find the fifth term of $(a + b)^{13}$.
8. Find the ninth term of $(x + 1)^{12}$.
9. Find the fourth term of $\left(3x + \frac{1}{2}\right)^{11}$.
10. Find the middle term of $\left(\frac{x}{2} + \frac{1}{3}\right)^{10}$.

Practice Exercises

In problems 11 through 22, use the Binomial Theorem to expand each binomial.

11. $(a + b)^6$
12. $(x + y)^8$
13. $(x + 1)^9$
14. $(A + 2)^7$
15. $(2x + y)^6$
16. $(3x + 2y)^5$
17. $(a + b)^9$
18. $(x + 1)^8$
19. $(x + 1)^9$
20. $(A + 2)^6$
21. $(2x + y)^7$
22. $(2x + 3y)^6$

In problems 23 through 34, find the last three terms of the binomial expansion.

23. $(a + b)^{13}$
24. $(x + y)^{21}$
25. $(s + 3)^5$
26. $(3 + t)^{17}$
27. $(3x + 2y)^7$
28. $(x^2 + y^2)^{30}$
29. $(a + b)^{12}$
30. $(x + y)^{24}$
31. $(s + 3)^6$
32. $(2 + t)^{19}$
33. $(2x + 3y)^7$
34. $(2x^2 + y^2)^{23}$

In problems 35 through 40, find the first two terms of the binomial expansion.

35. $(a + b)^{50}$
36. $(x + 1)^{100}$
37. $(s + 3)^{500}$
38. $(a + b)^{150}$
39. $(1 + x)^{100}$
40. $(s + 5)^{400}$

41. Find the fourth term of $\left(\dfrac{x}{2} + \dfrac{1}{3}\right)^{12}$

42. Find the middle term of $\left(\dfrac{x}{3} + 1\right)^{12}$

Challenge Problems

43. Notice that $1 = 1^n = \left(\dfrac{1}{2} + \dfrac{1}{2}\right)^n$. Use the Binomial Theorem to expand $\left(\dfrac{1}{2} + \dfrac{1}{2}\right)^6$, then show that it equals 1.

44. As $1.02 = 1 + 0.02$, powers of 1.02 can be found with the Binomial Theorem. Use this technique to find the exact value of $(1.02)^5$.

45. The expansion of $(a - b)^n$ is the same as the expansion of $(a + b)^n$, except that the signs alternate, starting with positive. Write a form of the binomial theorem for $(a - b)^n$.

In problems 46 through 51, use the form of the Binomial Theorem developed in problem 45 to expand each binomial.

46. $(a - b)^7$
47. $(x - y)^6$
48. $(x - 1)^9$
49. $(q - 2)^8$
50. $(2x - 3y)^5$
51. $(3x^2 - 2y^2)^4$

52. Use the results of problem 45 with the technique of problem 44 to find the exact value of $(0.98)^5$.

IN YOUR OWN WORDS

53. State the Binomial Theorem without using a formula.

33. $20{,}412x^2y^5 + 10{,}206xy^6 + 2187y^7$

12.5 PERMUTATIONS AND COMBINATIONS

In this section we examine two methods of counting the number of ways that various events can occur.

An arrangement of a collection of objects is called a **permutation.** Suppose that three books were to be arranged on a bookshelf. How many different ways could this be done? If the books are labeled A, B and C, the possibilities are:

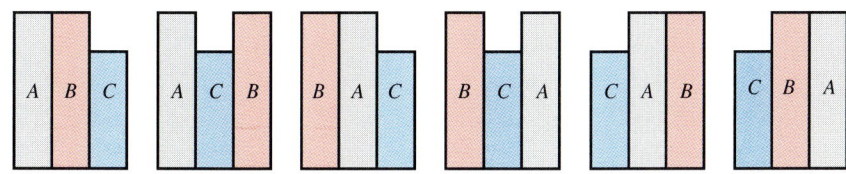

As there are no other possibilities, we conclude that there are six different ways to arrange the three books. Now, suppose that there is room for only two books on our bookshelf. Again, we list the various ways to place the books.

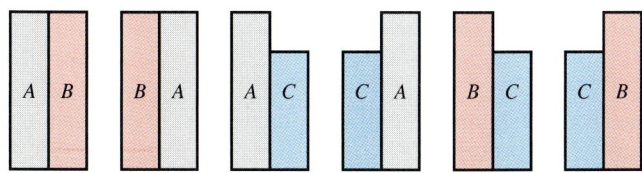

There are six ways to do this task. We say that there are six permutations of *three things taken two at a time*. A common notation used for this is $P(3,2)$. As there are also six permutations of three things taken three at a time (the first arrangement), we see that

$$P(3, 3) = 6 \quad \text{and} \quad P(3, 2) = 6$$

If there were only one space on our shelf, we could put either book *A*, book *B*, or book *C* in that space. Thus

$$P(3, 1) = 3$$

Now suppose that there are 12 books and only room for 8. If we start listing the possibilities we will soon tire of the task (or run out of paper). Let's consider the 8 spaces.

Any one of the 12 books can be placed in the first space, so there are 12 ways to fill it. Suppose that we put book *G* there.

Now there are only 11 books left to fill the next space. Thus there are $12 \cdot 11$ ways to fill the first two spaces. Suppose that book *D* is chosen for the second space.

Now there are only 10 books left to fill the third space. So there are $12 \cdot 11 \cdot 10$ ways to fill the first three spaces. If we continue placing books until we have one space left,

| G | D | H | F | A | J | B | |

we find that we have 5 books left to fill the last space, so we can fill it 5 different ways. Therefore, there are

$$12 \cdot 11 \cdot 10 \cdot 9 \cdot 8 \cdot 7 \cdot 6 \cdot 5 = 19{,}958{,}400$$

different ways to place 12 books in 8 spaces. That is,

$$P(12, 8) = 19{,}958{,}400$$

We extend these ideas to a formula for the number of permutations of *n* things taken *r* at a time.

$$P(n, r) = n \cdot (n - 1) \cdot (n - 2) \cdots (n - (r - 1)); \quad 0 \leq r \leq n$$

Two other commonly used notations for the number of permutations of *n* things taken *r* at a time are $_nP_r$ and P_r^n.

The factorial notation, introduced in Section 12.3, is very useful for expressing this formula. However, it is more convenient to rearrange the factors using the Commutative Property for Multiplication.

Factorial Definitions

For n a nonnegative integer:

1. $n! = n \cdot (n-1) \cdots 4 \cdot 3 \cdot 2 \cdot 1$
2. $0! = 1$

The formula for $P(n, r)$ can now be written using factorials. To see how, look at $P(12, 8)$ again.

$$P(12, 8) = 12 \cdot 11 \cdot 10 \cdot 9 \cdot 8 \cdot 7 \cdot 6 \cdot 5$$

If we multiply and divide by $4 \cdot 3 \cdot 2 \cdot 1$, we get

$$P(12, 8) = \frac{12 \cdot 11 \cdot 10 \cdot 9 \cdot 8 \cdot 7 \cdot 6 \cdot 5 \cdot 4 \cdot 3 \cdot 2 \cdot 1}{4 \cdot 3 \cdot 2 \cdot 1}$$

or

$$P(12, 8) = \frac{12!}{4!}$$

$$P(12, 8) = \frac{12!}{(12-8)!}$$

We now have a general formula for permutations:

Number of Permutations of n Things taken r at a Time

For integers n and r, where $0 \leq r \leq n$,

$$P(n, r) = \frac{n!}{(n-r)!}$$

EXAMPLE 1. Find

(a) $P(6, 2)$ (b) $P(5, 3)$
(c) $P(8, 0)$ (d) $P(7, 7)$

Solutions:

(a) $P(6, 2) = \dfrac{6!}{(6-2)!} = \dfrac{6!}{4!}$

$= \dfrac{6 \cdot 5 \cdot 4 \cdot 3 \cdot 2 \cdot 1}{4 \cdot 3 \cdot 2 \cdot 1} = 6 \cdot 5 = 30$

(b) $P(5, 3) = \dfrac{5!}{(5-3)!} = \dfrac{5!}{2!}$

$= \dfrac{5 \cdot 4 \cdot 3 \cdot 2 \cdot 1}{2 \cdot 1} = 3 \cdot 4 \cdot 5 = 60$

(continued)

(c) $P(8, 0) = \dfrac{8!}{(8-0)!} = \dfrac{8!}{8!}$
$= 1$

(d) $P(7, 7) = \dfrac{7!}{(7-7)!} = \dfrac{7!}{0!}$ (remember, $0! = 1$, not 0)

$= \dfrac{7 \cdot 6 \cdot 5 \cdot 4 \cdot 3 \cdot 2 \cdot 1}{1}$

$= 5040$ ☐

EXAMPLE 2. If 9 horses are entered in The Motor City Handicap, how many different win, place, and show results are possible? (Win is first place, place is second place, and show is third place.)

Solution:

As each possible selection of first-, second-, and third-place horses is a permutation of 9 horses taken 3 at a time, we must compute $P(9, 3)$.

$$P(9, 3) = \dfrac{9!}{(9-3)!} = \dfrac{9!}{6!}$$

$$= 7 \cdot 8 \cdot 9 = 504$$

There are 504 different win, place, and show possibilities. ☐

Suppose that we have 5 books and only have room for 3 of them on the bookshelf, and it does not matter in what order they are placed. This is not a permutation problem, as the *order* of an arrangement is part of a permutation. In this problem the collection of books *A C D* is the same as the collection *D A C*. We call such collections **combinations,** and use the notation $C(5, 3)$ for the number of combinations of 5 things taken 3 at a time. We list the different combinations of the 5 books, *A*, *B*, *C*, *D*, and *E*, taken 3 at a time.

| A B C | A B D | A B E | A C D | A C E |
| A D E | B C D | B C E | B D E | C D E |

Any other combination is an arrangement of one of the 10 combinations listed. Therefore, $C(5, 3) = 10$.

The formula for combinations also involves factorials.

Number of Combinations of *n* Things taken *r* at a Time

For integers *n* and *r*, where $0 \le r \le n$,

$$C(n, r) = \dfrac{n!}{(n-r)! \cdot r!}$$

Chap. 12 Other Topics

Notice that this is *exactly* the formula for the binomial coefficients that was developed in Section 12.3.

$$C(n, r) = \binom{n}{r}$$

Other commonly used notations for the number of combinations of n things taken r at a time are $_nC_r$ and C_r^n.

EXAMPLE 3. Find

(a) $C(6, 2)$ (b) $C(5, 3)$
(c) $C(8, 0)$ (d) $C(7, 7)$

Solutions:

(a) $C(6, 2) = \dfrac{6!}{(6-2)! \cdot 2!} = \dfrac{6!}{4! \cdot 2!}$

$= \dfrac{1 \cdot 2 \cdot 3 \cdot 4 \cdot 5 \cdot 6}{1 \cdot 2 \cdot 3 \cdot 4 \cdot 1 \cdot 2} = 5 \cdot 3 = 15$

(b) $C(5, 3) = \dfrac{5!}{(5-3)! \cdot 3!} = \dfrac{5!}{2! \cdot 3!}$

$= \dfrac{1 \cdot 2 \cdot 3 \cdot 4 \cdot 5}{1 \cdot 2 \cdot 1 \cdot 2 \cdot 3} = 2 \cdot 5 = 10$

(c) $C(8, 0) = \dfrac{8!}{(8-0)! \cdot 0!} = \dfrac{8!}{8! \cdot 0!}$

$= \dfrac{1}{0!} = \dfrac{1}{1} = 1$

(d) $C(7, 7) = \dfrac{7!}{(7-7)! \cdot 7!} = \dfrac{7!}{0! \cdot 7!}$

$= 1$ □

EXAMPLE 4. The Clarkston Womens club has 16 members. A steering committee made up of 4 members is to be formed. How many different ways can this be done?

Solution:

As there is no particular order to this committee, the number of different committees is the number of combinations of 16 things taken 4 at a time.

$$C(16, 4) = \dfrac{16!}{(16-4)! \cdot 4!} = \dfrac{16!}{12! \cdot 4!}$$

$$= \dfrac{1 \cdot 2 \cdot 3 \cdots 12 \cdot 13 \cdot 14 \cdot 15 \cdot 16}{1 \cdot 2 \cdot 3 \cdots 12 \cdot 1 \cdot 2 \cdot 3 \cdot 4}$$

$$= 13 \cdot 7 \cdot 5 \cdot 4 = 1820$$

There are 1820 different ways to form the steering committee.   □

EXAMPLE 5. How many different 5-card poker hands can be dealt from an ordinary deck of 52 cards?

Solution:

Since the order in which the cards are dealt does not matter, we need to find the number of combinations of 52 things taken 5 at a time.

$$C(52, 5) = \frac{52!}{(52-5)! \cdot 5!} = \frac{52!}{47! \cdot 5!}$$

$$= \frac{1 \cdot 2 \cdot 3 \cdots 47 \cdot 48 \cdot 49 \cdot 50 \cdot 51 \cdot 52}{1 \cdot 2 \cdot 3 \cdots 47 \cdot 1 \cdot 2 \cdot 3 \cdot 4 \cdot 5}$$

$$= 2 \cdot 49 \cdot 10 \cdot 51 \cdot 52 = 2,598,960$$

There are 2,598,960 different 5-card poker hands.

CALCULATOR BOX

Permutations and Combinations with a Calculator

[nPr]   [nCr]

Permutations and combinations can be done on a scientific calculator. The method used depends on the capability of the calculator.

EXAMPLE Calculate $C(27, 4)$ *without* using special keys.

Solution:

First, we do as much pencil-and-paper simplification as we can.

$$C(27, 4) = \frac{27!}{(27-4)! \cdot 4!} = \frac{27!}{23! \cdot 4!}$$

$$= \frac{1 \cdot 2 \cdot 3 \cdots 23 \cdot 24 \cdot 25 \cdot 26 \cdot 27}{1 \cdot 2 \cdot 3 \cdots 23 \cdot 1 \cdot 2 \cdot 3 \cdot 4}$$

$$= 25 \cdot 26 \cdot 27$$

Now, the calculator: [25] [×] [26] [×] [27] [=] and read [17550] on the display.

$$C(27, 4) = 17,550$$

EXAMPLE Find $C(18, 8)$ using a *factorial* key.

Solution:

Write

$$C(18, 8) = \frac{18!}{(18-8)! \cdot 8!} = \frac{18!}{10! \cdot 8!}$$

Press the keys

$$\boxed{18}\ \boxed{n!}\ \boxed{\div}\ \boxed{10}\ \boxed{n!}\ \boxed{\div}\ \boxed{8}\ \boxed{n!}\ \boxed{=}$$

and read `43758` on the display.

$$C(18,8) = 43{,}758$$

EXAMPLE Use $_nP_r$ and $_nC_r$ keys to calculate each expression.

(a) $P(13, 6)$ (b) $C(52, 13)$

Solutions:

(a) Press $\boxed{13}\ \boxed{nPr}\ \boxed{6}\ \boxed{=}$ and read `1235520` on the display.

$$P(13, 6) = 1{,}235{,}520$$

(b) Press $\boxed{52}\ \boxed{nCr}\ \boxed{13}\ \boxed{=}$ and read `6.3501356`^{11} on the display.

This is a very large number in scientific notation. We give our answer in scientific notation, being careful to indicate that it is an *approximation*.

$$C(52, 13) \approx 6.3501 \times 10^{11}$$

The last example, the number of combinations of 52 things taken 13 at a time, is the total number of possible distinct bridge hands.

Calculator Exercises.

Evaluate each expression using a calculator.

1. $P(10, 8)$
2. $C(14, 7)$
3. $_{12}P_7$
4. $_{44}C_5$
5. $\binom{16}{12}$
6. $C(65, 60)$

Answers:

1. 1,814,400
2. 3432
3. 3,991,680
4. 1,086,008
5. 1820
6. 8,259,888

PROBLEM SET 12.5

Warm-ups

In problems 1 through 14, evaluate each expression.
For problems 1 through 7, see Example 1.

1. $P(4, 2)$
2. $P(5, 0)$
3. $P(5, 1)$
4. $P(5, 2)$
5. $P(5, 3)$
6. $P(5, 4)$
7. $P(5, 5)$

For problems 8 through 14, see Example 3.

8. $C(4, 2)$
9. $C(9, 0)$
10. $C(9, 1)$
11. $C(9, 2)$
12. $C(9, 3)$
13. $C(9, 9)$
14. $C(9, 8)$

Sec. 12.5 Permutations and Combinations

For problems 15 through 18, see Examples 2 and 4.

15. How many different three-letter "words" can be made with the letters of the word TIMERS?
16. How many different ways can first through fourth place be determined in the Kentucky Derby if 11 horses run?
17. Seven players are to be chosen from a roster of 13. In how many ways can this be done?
18. The local Ace Hardware carries 14 different shades of interior latex paint. How many different combinations of four different shades can be selected?

Practice Exercises

In problems 19 through 42, evaluate each expression.

19. $P(8, 4)$
20. $P(7, 5)$
21. $P(6, 0)$
22. $P(6, 1)$
23. $P(6, 2)$
24. $P(6, 3)$
25. $P(6, 4)$
26. $P(6, 5)$
27. $P(6, 6)$
28. $C(8, 4)$
29. $C(7, 5)$
30. $C(9, 5)$
31. $C(8, 0)$
32. $C(8, 1)$
33. $C(8, 2)$
34. $C(8, 3)$
35. $C(8, 8)$
36. $C(8, 7)$
37. $C(8, 6)$
38. $C(8, 5)$
39. $P(14, 4)$
40. $P(14, 10)$
41. $C(14, 4)$
42. $C(14, 10)$

43. How many different four-letter "words" can be made with the letters of the word OUTFIELD?
44. The officers of the Spin-around Dance Club are president, vice–president, and secretary–treasurer. No person is allowed to hold more than one office. If there are 28 members, how many ways can a slate of officers be selected?
45. Nine players are to be chosen from a roster of 16. In how many ways can this be done?
46. In how many ways can six algebra books be taken from a pile of 13 algebra books?

Challenge Problems

In problems 47 through 55, suppose that N is an integer greater than 1. Simplify each expression.

47. $P(N, N)$
48. $P(N, 0)$
49. $P(N, N - 2)$
50. $C(N, N)$
51. $C(N, 0)$
52. $C(N, N - 2)$
53. $C(N + 2, N)$
54. $P(N, 1)$
55. $C(N + 1, N - 1)$

IN YOUR OWN WORDS . . .

56. Describe permutations without using a formula.
57. Describe combinations without using a formula.

CHAPTER SUMMARY

GLOSSARY

Sequence: A function whose domain is the set of natural numbers.

Terms of a sequence: The numbers in the range of the function that defines the sequence.

Arithmetic sequence: A sequence with a general term in the form $a_1 + (n - 1)d$.

Geometric sequence: A sequence with a general term in the form $a_1 r^{n-1}$.

Increasing sequence: A sequence in which every term of the sequence is larger than the preceding term.

Decreasing sequence: A sequence in which every term of the sequence is smaller than the preceding term.

Permutation: An arrangement of objects in a particular order.

Combination: An arrangement of objects where order is not important.

$$A_1 + A_2 + A_3 + \cdots + A_n = \sum_{j=1}^{n} A_j$$ **SUMMATION NOTATION**

$$1 \cdot 2 \cdot 3 \cdots (n-1) \cdot n = n!$$ **FACTORIAL NOTATION**

$$\binom{n}{r} = \frac{n!}{r! \cdot (n-r)!}; \quad n, r \text{ nonnegative integers, with } n \geq r$$ **BINOMIAL COEFFICIENTS**

$$(a+b)^n = \sum_{j=0}^{n} \binom{n}{j} a^{n-j} b^j$$ **BINOMIAL THEOREM**

CHECKUPS

1. Write the first five terms of the sequence
   $$a_n = 4n + 5$$
   Section 12.1; Example 1

2. Find a suitable general term for the sequence,
   $$1, -\frac{1}{2}, \frac{1}{4}, -\frac{1}{8}, \frac{1}{16}, \ldots$$
   Section 12.1; Example 3d

3. Label the sequence as increasing, decreasing, or neither.
   $$a_n = 4 - n$$
   Section 12.1; Example 4b

4. Write out the terms of each sum.

   (a) $\sum_{j=1}^{6} (2j - 1)$ Section 12.2; Example 1

   (b) $\sum_{j=0}^{3} (-1)^j x^j$ Section 12.2; Example 4

5. Write the following sum in summation notation:
   $$2 + 4 + 6 + 8 + 10 + 12 + 14 + 16$$
   Section 12.2; Example 5

6. Evaluate $\dfrac{10!}{5!5!}$.   Section 12.3; Example 2c

7. Calculate $\binom{7}{5}$.   Section 12.3; Example 3a

8. Write the first four terms of $(x + y)^{11}$.   Section 12.4; Example 1
9. Find $P(6, 2)$.   Section 12.5; Example 1a
10. Find $C(6, 2)$.   Section 12.5; Example 3a

REVIEW PROBLEMS

In problems 1 through 6, write the first five terms of each sequence.

1. $a_n = 3n + 1$
2. $a_n = 2^n$
3. $a_n = 3 \cdot 5^{n-1}$
4. $b_n = 11 + 5n$
5. $c_n = (-1)^n \cdot 3n$
6. $d_n = (-1)^{n-1} A_n x^n$

Sec. 12.5 Permutations and Combinations

In problems 7 through 9, label each sequence as increasing, decreasing, or neither increasing nor decreasing.

7. $a_n = 10 + 3n$ **8.** $b_n = 10 - 3n$ **9.** $c_n = 3 - n^{-1}$

Problems 10 through 12 are arithmetic progressions. In each, find the common difference and write the next term.

10. 7, 9, 11, . . . **11.** −2, 0, 2, . . . **12.** 5, $\frac{13}{2}$, 8, . . .

Problems 13 through 15 are geometric progressions. In each, find the common ratio and write the next term.

13. 1, 4, 16, . . . **14.** 16, −8, 4, . . . **15.** 40, 60, 90, . . .

In problems 16 through 21, write out the terms of each sum.

16. $\sum_{j=0}^{5}(2j+1)$ **17.** $\sum_{j=1}^{6}(1-j)^2$ **18.** $\sum_{j=1}^{7}(2^j + j^2)$

19. $\sum_{j=0}^{5}(2j)^2$ **20.** $\sum_{j=1}^{6}(j-6)^2$ **21.** $\sum_{j=1}^{4}(-1)^{j+1}j^j$

In problems 22 through 24, write each sum in summation notation.

22. $1 + 3 + 5 + 7 + 9 + 11 + 13 + 15$

23. $1 + 2k + 3k^2 + 4k^3 + 5k^4 + 6k^5 + 7k^6$

24. $\frac{1}{2} - \frac{2}{3} + \frac{3}{4} - \frac{4}{5} + \frac{5}{6} - \frac{6}{7}$

In problems 25 through 33, evaluate each expression.

25. $9!$ **26.** $3! \cdot 5!$ **27.** $0! \cdot 1! \cdot 2! \cdot 3!$

28. $\frac{14!}{13!}$ **29.** $\frac{41!}{39!}$ **30.** $\frac{10!}{5!}$

31. $\binom{9}{3}$ **32.** $\binom{17}{15}$ **33.** $\binom{10}{5}$

In problems 34 through 39, use the Binomial Theorem to expand each binomial.

34. $(a+b)^5$
35. $(x+1)^7$
36. $(y+2)^8$
37. $(2+3y)^6$
38. $(4x+3y)^4$
39. $(1+s)^{10}$

40. Find the third term of $(a+b)^{11}$.
41. Find the first three terms of $(2+5t)^8$.
42. Find the fifteenth term of $(2+k)^{17}$.
43. Find the middle term of $(a+b)^{14}$.
44. Simplify $P(6, 5)$.
45. Simplify $C(10, 4)$.

46. How many different ways can six balls be drawn from an urn containing 11 numbered balls if their order *is not* important?

47. How many different ways can six balls be drawn from an urn containing 11 numbered balls if their order *is* important?

... LET'S NOT FORGET ...

Identify the expressions that are in factored form. Factor those that are not factored.

48. $(2x - 1)^5$
49. $8x^3 - 27$
50. $96x^2 - 4x - 15$
51. $(2x - 3)^3$
52. $63 + 7x^3$
53. $x^3 - 3x^2 + 3x - 1$

How many terms are in each expression? Which expressions have $7 - y$ as a factor?

54. $49 - y^2$
55. $t(7 - y) + y - 7$
56. $7 - 14y + y^2$ 3;

Simplify each expression, if possible, leaving only nonnegative exponents in your answer.

57. -2^{-4}
58. $\dfrac{-2x^{-3}}{y^2 z^{-2}}$
59. $\dfrac{16x^2 - 25}{125 - 64x^3}$
60. $|3x - 2|$
61. $\dfrac{2^{-3} + 8}{x^{-3} + 2^{-2}}$
62. $\sqrt[3]{(a + b)^6}$

Find each product.

63. $(3x + 2y)^3$
64. $(2\sqrt{2} + \sqrt{3})^2$

Reduce, if possible.

65. $\dfrac{-14 + \sqrt{-28}}{14}$
66. $\dfrac{z - w - r(w - z)}{w - z}$

The following problems can be worked by using a least common denominator. Follow the directions in each problem and notice how the LCD is used.

67. Perform the operation indicated: $\dfrac{1}{x} + \dfrac{1}{x^2}$.

68. Solve $\dfrac{4}{x} - \dfrac{1}{x^2} = 0$.

69. Simplify $\dfrac{\dfrac{1}{x} - 1}{\dfrac{1}{x^2} - 1}$.

Label each as an equation or an expression. Solve the equations in one variable, graph the equations in two variables, and perform the operations indicated on the expressions.

70. $\sqrt{4x - 3} = 2x - 3$
71. $\dfrac{1 - \dfrac{x}{x + 1}}{1 + \dfrac{1}{x + 1}}$

72. $(2x + y)^2$
73. $\left|\dfrac{2 - 3x}{4}\right| = -3$

74. $\dfrac{1}{2x^2 + 3x - 2} - \dfrac{1}{4x^2 - 4x + 1}$
75. $\dfrac{i}{2 - 3i}$

CHAPTER TEST

In problems 1 through 9, choose the correct response.

1. The sequence $a_n = (-1)^n n^2$ can be written
  A. $1, -4, 9, \ldots$
  B. $-1, -4, -9, \ldots$
  C. $1, 4, 9, \ldots$
  D. $-1, 4, -9, \ldots$

2. The sequence $a_n = 3 \cdot 2^{n-1}$ is
  A. an increasing, arithmetic progression
  B. an increasing, geometric progression
  C. a decreasing, arithmetic progression
  D. a decreasing, geometric progression

3. $\sum_{j=0}^{4} j \cdot (j+1)$ is another way of writing the sum
   A. $0 \cdot 1 + 1 \cdot 2 + 2 \cdot 3 + 3 \cdot 4$
   B. $0 \cdot 1 + 2 \cdot 3 + 4 \cdot 5$
   C. $0 + 1 + 2 + 3 + 4 + 5$
   D. $0 \cdot 1 + 1 \cdot 2 + 2 \cdot 3 + 3 \cdot 4 + 4 \cdot 5$

4. $\sum_{j=1}^{6} (-1)^{j+1} j = (?)$
   A. 21
   B. -3
   C. 3
   D. -21

5. $6! = (?)$
   A. 21 B. 30
   C. 36 D. 720

6. $\dfrac{12!}{10!} = (?)$
   A. $\dfrac{6!}{5!}$ B. 132
   C. $2!$ D. $\dfrac{6}{5}$

7. $\binom{9}{3} = (?)$
   A. 84
   B. $6!$
   C. $7 \cdot 8 \cdot 9$
   D. $\dfrac{1}{2}$

8. $(x+2)^5 = (?)$
   A. $x^5 + 32$
   B. $x^5 + 5x^4 + 10x^3 + 10x^2 + 5x + 2$
   C. $x^5 + 10x^4 + 40x^3 + 80x^2 + 80x + 32$
   D. $5x + 10$

9. The middle term of the expansion of $(a+b)^{12}$ is
   A. $a^6 b^6$ B. $7 \cdot 8 \cdot 9 \cdot 10 \cdot 11 \cdot 12 a^6 b^6$
   C. $6! a^6 b^6$ D. $924 a^6 b^6$

10. Find a suitable general term for the sequence
    $$4, 7, 10, 13, \ldots$$

11. Write the first five terms of the sequence
    $$a_n = (-1)^{n+1} 2^n.$$

12. Evaluate $\sum_{j=1}^{5} (2j - 1)$.

13. How many different five-letter "words" can be made from the letters of the word COMPUTER?

14. Use the Binomial Theorem to expand $(3+t)^4$.

15. Find the last three terms of the expansion of $(a+b)^{18}$.

Answers to Selected Problems

CHAPTER 0

PROBLEM SET 0.1

Warm-ups

1. T **3.** T **5.** F **7.** F **9.** F **11.** −2, 0, 2, 4, 6, 8 **13.** {−2, 0, 2}, {−2, 0}, {−2, 2}, {0, 2}, {−2}, {0}, {2}, ∅ **15.** red, white, blue **17.** a, b, c, d, e, f, g, h, i, j **19.** $\{x \mid x$ is an even natural number less than 13$\}$ **21.** $\{x \mid x$ is a day of the week$\}$ **23.** $0.75\overline{0}$ **25.** $0.625\overline{0}$ **27.** Rational **29.** Irrational **31.** Rational

Practice Exercises

33. F **35.** T **37.** T **39.** T **41.** T **43.** $\frac{1}{2}, 2, \frac{9}{4}, 5$ **45.** $\left\{\frac{9}{4}, 5\right\}, \left\{\frac{9}{4}\right\}, \{5\}, \emptyset$ **47.** Friday **49.** q, r, s, t, u, v, w, x, y, z **51.** $\{x \mid x$ is an integer between −3 and 3$\}$ **53.** $\{x \mid x$ is a suit in a deck of cards$\}$ **55.** Rational **57.** Irrational **59.** Rational **61.** $0.6\overline{0}$ **63.** $0.875\overline{0}$

Challenge Problems

65. $\frac{3}{10}$ **67.** $\frac{655}{1000}$

PROBLEM SET 0.2

Warm-ups

1. $-\sqrt{3}$,

$\xleftarrow{\qquad\quad|\quad\ \ |\quad\ \ |\qquad\quad}$
$-\sqrt{3}\ \ 0\ \ \sqrt{3}$

3. $\frac{1}{3}$,

$\xleftarrow{\qquad\quad|\quad\ \ |\quad\ \ |\qquad\quad}$
$-\frac{1}{3}\ \ 0\ \ \frac{1}{3}$

5. 0,

$\xleftarrow{\qquad\quad\quad|\qquad\quad\quad}$
0

7. $-q$,

$\xleftarrow{\qquad\quad|\quad\ \ |\quad\ \ |\qquad\quad}$
$q\ \ 0\ \ -q$

9. $\frac{1}{7}$ **11.** $\frac{3}{2}$ **13.** $\frac{5}{8}$ **15.** Reflexive **17.** Associative **19.** Additive Identity **21.** Multiplicative Identity **23.** Multiplicative inverse **25.** Distributive **27.** Substitution **29. (a)** Commutative property **(b)** Distributive property **(c)** Addition **(d)** Commutative property

Practice Exercises

31. Commutative **33.** Symmetric **35.** Commutative **37.** Commutative **39.** Additive inverse **41.** Multiplication by zero **43.** Transitive **45.** Associative

47. 2, [number line with points at $-2, 0, 2$] **49.** $-\frac{2}{3}$, [number line with points at $-\frac{2}{3}, 0, \frac{2}{3}$] **51.** p, [number line with points at $-p, 0, p$]

53. 1 **55.** 4 **57.** $\frac{7}{3}$ **59. (a)** Associative property **(b)** Addition

Challenge Problems

61. [number line with points at $0, p-q$]

PROBLEM SET 0.3

Warm-ups

1. 46 **3.** 4 **5.** 0 **7.** 38 **9.** -13 **11.** -60 **13.** a **15.** Won't simplify **17.** a **19.** $|c|$ **21.** $5|y|$ **23.** $-53|k|$ **25.** $\frac{6}{|z|}$

27. [number line with points at 2, 10] **29.** [number line with points at $\sqrt{7}, \pi$]

31. x and y can be anywhere on a number line, $|x-y|$ units.

Practice Exercises

33. 16 **35.** 3 **37.** 13 **39.** -47 **41.** -3 **43.** -108 **45.** $21|x|$ **47.** $-81|g|$ **49.** $\frac{|u|}{9}$ **51.** Won't simplify **53.** r **55.** $-s$ **57.** r **59.** $-r$ **61.** $-r$ **63.** $|-(-3)| = -3$, False **65.** $|-(-3)| = -(-3)$, True **67.** $|7| = 7$, True

69. [number line with points at $-1, 7$] **71.** [number line with points at $-9, -5$] **73.** [number line with points at $-\sqrt{7}, -\sqrt{5}$]

75. 13 **77.** -2 **79.** $\sqrt{2} + \pi$

Challenge Problems

81. $\pi - 3$ **83.** $\pi + 3$

PROBLEM SET 0.4

Warm-ups

1. 9 **3.** -4 **5.** Undefined **7.** 2 **9.** $\frac{3}{2}$ **11.** 5 **13.** 1 **15.** $-\frac{1}{7}$ **17.** -10 **19.** -17 **21.** 81 **23.** 12

Practice Exercises

25. -15 **27.** 42 **29.** -4 **31.** -8 **33.** -11 **35.** -7 **37.** 42 **39.** $-\frac{4}{27}$ **41.** -24 **43.** $\frac{25}{16}$ **45.** $-\frac{10}{3}$ **47.** 15 **49.** 1 **51.** -6 **53.** -1 **55.** 13

Challenge Problems

57. $\frac{1}{2}$

PROBLEM SET 0.5

Warm-ups

1. $\angle a = 45°$, $\angle b = 135°$, $\angle c = 135°$, $\angle d = 45°$, $\angle e = 135°$, $\angle f = 45°$, $\angle g = 45°$, $\angle h = 135°$, $\angle i = 90°$, $\angle j = 90°$, $\angle k = 90°$, $\angle l = 90°$, $\angle m = 135°$, $\angle n = 45°$, $\angle o = 45°$, $\angle p = 135°$, $\angle q = 45°$, $\angle r = 135°$, $\angle s = 135°$, $\angle t = 45°$
3. 120° **5.** 30°
7. (a) Two angles of one triangle are equal to two angles of the other triangle.   **(b)** Two angles of one triangle are equal to two angles of the other triangle.

Practice Exercises

9. F   **11.** F   **13.** T   **15.** F   **17.** F   **19.** T   **21.** 50°   **23.** 140°   **25.** 68°
27. $10^2 = 100$; $6^2 + 8^2 = 36 + 64 = 100$ **29.** Corresponding sides are proportional.   **31.** T   **33.** T   **35.** F
37. T   **39.** T

Challenge Problems

41. $(n - 2)180°$

PROBLEM SET 0.6

Warm-ups

1. 12 in.   **3.** 34 cm   **5.** 74 yd   **7.** 360 ft   **9.** 28 sq m   **11.** 60 sq cm   **13.** 125 cubic units
15. 5,250 cu ft   **17.** 1.452π cu in.   **19.** $64,000,000\pi$ sq mi $\approx 200,960,000$ sq mi   **21.** $325\frac{1}{2}$ sq in.

Practice Exercises

23. 7.5 in.   **25.** 62 ft   **27.** 75.06 yd   **29.** 8000π mi   **31.** $\frac{9}{8}\pi$ sq in.   **33.** $\frac{343}{27}$ cu in.   **35.** $26\frac{13}{16}$ cu in.
37. 240π cu yd   **39.** $\frac{256,000,000,000}{3}\pi$ cu mi   **41.** $\frac{3}{2}$ cu ft   **43.** $131,250\pi$ sq in.   **45.** 2 cu ft

Challenge Problems

47. 174 sq ft

REVIEW PROBLEMS CHAPTER 0

1. 35   **3.** $\frac{-21}{13}$ **5.** 1   **7.** -41 **9.** 7   **11.** 7   **13.** 56   **15.** Commutative   **17.** Distributive
19. Multiplicative identity   **21.** Irrational   **23.** Rational   **25.** Rational   **27.** c **29.** 1, 3, 5   **31.** -20
33. -25 **35.** -21 **37.** $-b$ **39.** a **41.** The perimeter is 40 inches. The area is 104 square inches.   **43.** The circumference is 5π meters. The area is 6.25π square meters.   **45.** The volume is 12π cubic units. The surface area is 24π square units.   **47.** The volume is $\frac{500\pi}{3}$ cubic units. The surface area is 100π square units.   **49.** Corresponding sides are proportional.

CHAPTER 0 TEST

1. B   **2.** D   **3.** B   **4.** C   **5.** C   **6.** D   **7.** C   **8.** C   **9.** B   **10.** 62   **11.** $\frac{1}{27}$ **12.** 49
13. -21 **14.** 0   **15.** 1   **16.** 18   **17.** $-7 + 6$ **18.** $x(6) + xy$ **19.** 0   **20.** $(x + t)(2)$
21. $\angle a = 130°$, $\angle b = 50°$, $\angle c = 100°$, $\angle d = 30°$, $\angle e = 150°$, $\angle f = 50°$, $\angle g = 130°$ **22.** The volume is 192π cubic feet.
23. The amount of fencing needed is 100 meters.   **24.** The surface area of the ball is 49π square centimeters.
25. The area is 308 square yards.

CHAPTER 1

PROBLEM SET 1.1

Warm-ups

1. 16 3. -16 5. -64 7. -12 9. x^8 11. $-s^{11}$ 13. 2^5 15. $-x^5$ 17. 2^6 19. $-x^{10}$ 21. $\dfrac{9}{4}$
23. x^8 25. x^6 27. $8x^3y^6$ 29. $(x+y)^{20}$ 31. $\dfrac{8x^3}{y^6}$ 33. $\dfrac{x^4}{y^8}$ 35. $(x-1)^5$ 37. $\dfrac{-x^{20}y^5z^{15}}{2^5s^{10}}$ 39. $\dfrac{5x}{z^7}$
41. $-\dfrac{y}{xz^5}$ 43. $-\dfrac{1}{16}$

Practice Exercises

45. 8 47. -27 49. -9 51. $\dfrac{1}{4}$ 53. 54 55. 3^{10} 57. $-\dfrac{1}{4}$ 59. -72 61. -36 63. -108
65. -2^9 67. 16 69. 216 71. 36 73. $-\dfrac{27}{8}$ 75. $\dfrac{36}{25}$ 77. 2 79. $\dfrac{1}{2^8}$ 81. $-\dfrac{1}{27}$ 83. 7
85. -1 87. x^7 89. $-x^5$ 91. $-x^5$ 93. x^3 95. $-x^6$ 97. $x^{35}y^7$ 99. -4^3x^3 101. $(x+7)^{16}$
103. $-\dfrac{2^5x^3}{y^{10}}$ 105. $(x+4)^5$ 107. $\dfrac{3^4x^3}{2^9y^8}$ 109. $\dfrac{4}{7x^4z}$ 111. $-\dfrac{1}{x^4yz^7}$ 113. $\dfrac{4z^{10}}{9}$ 115. $9x^4y^6$
117. $\dfrac{121x^{10}z^2}{49y^6}$ 119. $9x^2$ ft^2 121. $64t^3$ in.^{3}

Challenge Problems

123. 6^{n+2} 125. 8^{2n} 127. 2^{1+n} 129. x^{3n}

PROBLEM SET 1.2

Warm-ups

1. $x+6$; 1 3. x^2-3x+4; 2 5. $-3x^3+4x^2+\dfrac{1}{2}x-7$; 3 7. 1 9. -16 11. 5 13. -7
15. $8x^3-8x^2$ 17. $2z^2+2z$ 19. Won't combine 21. 6

Practice Exercises

23. $x-4$; 1; binomial 25. x^2-x+1; 2; trinomial 27. $-2x^5+9x^4+\dfrac{2}{5}x-9$; 5 29. x^5+y^5; 5; binomial
31. $x^7y^2-x^4y^3+y^7$; 9; trinomial 33. -2 35. 11 37. -8 39. 0 41. 1 43. -5 45. $x-4$
47. $4y-4z$ 49. $3x^3+x^2+2$ 51. 0 53. 48 ft; 0 ft 55. 55; 91

Challenge Problems

57. 2 59. -2

PROBLEM SET 1.3

Warm-ups

1. $7x-14$ 3. $5x^2-3xy-3y^2$ 5. $-5x^6+3x^4-3$ 7. $-15x^4y+12x^3y^2-7x^2y^2+12$ 9. $-2x-12$
11. $-4x^2-8xy+11y^2$ 13. $3x-3$ 15. $4x^3y-3x^2y^2+11xy+12$

Practice Exercises

17. $24x+4$ 19. $6y^2+4y-17$ 21. $-2x^2-11xy+4y^2$ 23. $15x^3+2x^2+4x-8$
25. $8x^5-3x^4+13x^3-x^2+7x+12$ 27. $6x^3y-9x^2y^2+15xy-13$ 29. $24x^5+14x^4-4x^3-10x$
31. $x^2-15x+10$ 33. z^2-2z-7 35. $-2x^2-3x+17$ 37. $2v^4-2v^3-4v^2+6v-12$ 39. $-7z^2-1$
41. s^2+4s-8 43. x^2+x+1 45. $2x-11y$ 47. $7x^2+5x-5$ 49. $6W-6$; 36 ft 51. $119x+27$ mi
53. $42-x$ 55. $4x+26$ dollars 57. $150-x^2+2x$ degrees

Answers to Selected Problems

Challenge Problems

59. $7x^{2n} + x^n + 4$ **61.** $3x^m - 4x^m y^n - 4y^n$

PROBLEM SET 1.4

Warm-ups

1. $x^2 + 3x + 2$ **3.** $\frac{1}{4}x^2 + \frac{3}{2}x + 2$ **5.** $x^2 - 13x + 42$ **7.** $x^2 + 2x - 35$ **9.** $-12a^3b^2$ **11.** rs^4
13. $-3x^2 + 3x^3$ **15.** $15x^2 - 61x + 56$ **17.** $x^3 - 2x^2y^2 - x^2y + 3xy^3 - y^4$ **19.** $x^2 + 10x + 25$
21. $9x^2 - 24xy + 16y^2$ **23.** $x^2 - 4$ **25.** $25x^2 - 49$ **27.** $x^3 - 8$ **29.** $x^3 + 125$ **31.** $x^3 + 6x^2 + 12x + 8$
33. $27a^3 - 27a^2b + 9ab^2 - b^3$ **35.** $3x^2 - 12x - 3$ **37.** $-8x^2 - 2x - 1$

Practice Exercises

39. $-8a^4b^2$ **41.** $2rs^5$ **43.** $-2^3x^5y^4$ **45.** $-3^7a^3b^8c^3$ **47.** $-r^{20}t^5$ **49.** $9r^{17}s^3t^{15}$ **51.** $a^{26}b^{16}c^{16}$
53. $3a^4bc + 5a^4b^4c^7 - 4a^3b^2c$ **55.** $-3x^6yz^2 + \frac{9}{5}x^5y^6z^8 + \frac{3}{5}x^5yz$ **57.** $x^{11} - 3x^{10} + \frac{1}{3}x^7 - x^4$ **59.** $x^2 + 10x + 9$
61. $x^2 + 4x - 21$ **63.** $35 + 12x + x^2$ **65.** $x^2 - 10x + 24$ **67.** $x^2 + \frac{1}{3}x - \frac{2}{9}$ **69.** $6 - 5x + x^2$
71. $28 + 3x^2 - x^4$ **73.** $8x^2 + 30x + 25$ **75.** $33x^2 + 26x + 5$ **77.** $20 - 21x + 4x^2$ **79.** $5x^4 - 41x^2 + 42$
81. $30x^2 - 71x + 35$ **83.** $(u - z)^2 - 2(u - z) - 15$ **85.** $2x^3 - 7x^2y + 12xy^2 - 9y^3$ **87.** $6x^3 - 19x^2 - 154x - 49$
89. $x^2 - 8x + 16$ **91.** $121x^2 + 264x + 144$ **93.** $25x^2 - 40x + 16$ **95.** $x^2 - 25y^2$ **97.** $49x^2 + 84x + 36$
99. $a^2x^2 - 4abxy + 4b^2y^2$ **101.** $343a^3 + 27b^3$ **103.** $t^3 + 64r^3$ **105.** $x^3 - 216$ **107.** $x^3 - 15x^2 + 75x - 125$
109. $x^3 + 6x^2y + 12xy^2 + 8y^3$ **111.** $64s^3 - 144s^2t + 108st^2 - 27t^3$ **113.** $(x - y)^2 - 49$ **115.** $(s + t)^2 + 6(s + t) + 9$
117. $-2x^2 + 18x - 18$ **119.** $-6x^2 - 3x - 2$ **121.** $N^2 + 6N + 8$ **123.** $N^2 + 3N + 2$
125. $\frac{4}{3}\pi x^3 - 20\pi x^2 + 100\pi x - \frac{500}{3}\pi$ m^3 **127.** $\frac{3}{2}W^3$ in^3

Challenge Problems

129. $2x^{2+2n} + 6x^{2+n} + 2x^2$ **131.** $x^{2n} - 4$ **133.** $x^{3n} + 3x^{2n} + 3x^n + 1$

PROBLEM SET 1.5

Warm-ups

1. $x^3 + 3x^2 + 3x + 1$ **3.** $x^4 + 8x^3 + 24x^2 + 32x + 16$
5. $64x^6 + 576x^5y + 2160x^4y^2 + 4320x^3y^3 + 4860x^2y^4 + 2916xy^5 + 729y^6$
7. $u^6 - 12u^5w + 60u^4w^2 - 160u^3w^3 + 240u^2w^4 - 192uw^5 + 64w^6$
9. $x^{10} - 20x^9 + 180x^8 - 960x^7 + 3360x^6 - 8064x^5 + 13{,}440x^4 - 15{,}360x^3 + 11{,}520x^2 - 5120x + 1024$

Practice Exercises

11. $x^3 - 3x^2 + 3x - 1$ **13.** $y^4 - 8y^3 + 24y^2 - 32y + 16$ **15.** $b^5 - 15b^4 + 90b^3 - 270b^2 + 405b - 243$
17. $16x^4 - 32x^3 + 24x^2 - 8x + 1$ **19.** $a^5 + 5a^4 + 10a^3 + 10a^2 + 5a + 1$ **21.** $16x^4 + 96x^3 + 216x^2 + 216x + 81$
23. $27x^3 - 54x^2 + 36x - 8$ **25.** $243x^5 + 810x^4 + 1080x^3 + 720x^2 + 240x + 32$
27. $x^6 - 6x^5y + 15x^4y^2 - 20x^3y^3 + 15x^2y^4 - 6xy^5 + y^6$
29. $x^8 - 16x^7 + 112x^6 - 448x^5 + 1120x^4 - 1792x^3 + 1792x^2 - 1024x + 256$ **31.** $-1{,}082{,}565x^8y^3$ **33.** $700{,}000z^4$

PROBLEM SET 1.6

Warm-ups

1. a^4b **3.** $(x - y)^2z$ **5.** $x + \frac{2y}{x}$ **7.** $z - 6$ **9.** $x - 3$ **11.** $x - 1 + \frac{-1}{x + 3}$
13. $x^3 - x^2 - 4x - 7 + \frac{3x + 25}{x^2 - 2x + 3}$ **15.** $x^2 - xy + y^2$ **17.** $x^2 - x + 2 + \frac{-3 + x}{x^2 - 1}$

Practice Exercises

19. $\dfrac{3a}{2b}$ 21. $\dfrac{x^2}{2y} + \dfrac{x}{y} - \dfrac{3}{2y}$ 23. $\dfrac{x}{y} - 1 + \dfrac{y}{x}$ 25. $3x - 1$ 27. $y + 3$ 29. $x^2 - x - 1 + \dfrac{4}{2x + 1}$

31. $2x^2 - x + 1 + \dfrac{-7}{3x + 2}$ 33. $2x + 1$ 35. $3x - y$ 37. $x^2 - x + 1 + \dfrac{-4x + 5}{3x^2 + x - 1}$ 39. $3x^2 - x + 1 + \dfrac{4}{7x + 4}$

41. $x^3 - 3x^2 + 9x - 28 + \dfrac{86}{3 + x}$ 43. $x^3 + 2 + \dfrac{-1}{x^3 - 1}$ 45. $2x^3 - 4x + 1 + \dfrac{4x}{2x^2 + 1}$ 47. $x^2 + 3x - 4$ 49. 27

Challenge Problems

51. $x^n - 3 + \dfrac{7}{x^n + 1}$ 53. 12

PROBLEM SET 1.7

Warm-ups

1. $2x + 1$ 3. $x^2 + 2x + 1$ 5. Remainder is 0

Practice Exercises

7. $x^3 + x^2 - x - 1$ 9. $4x^2 - x + 3 + \dfrac{-1}{x + 2}$ 11. $2x - 3$ 13. $x^3 + 2x^2 + 2x - 1 + \dfrac{1}{x - 2}$

15. $2x^2 + 3x + 3 + \dfrac{2}{x - 2}$ 17. $y^3 - y^2 + y - 1 + \dfrac{2}{y + 1}$ 19. $x^3 + x^2 + 2x + 2$ 21. $x^3 + 2x^2 + 3x + 4 + \dfrac{5}{x - 1}$

23. Remainder is 0

REVIEW PROBLEMS CHAPTER 1

1. -35 3. 25 5. -108 7. $x^4 y^{12}$ 9. $(a + b)^5$ 11. $\dfrac{4x^2}{y}$ 13. $-\dfrac{1}{10}$ 15. $-\dfrac{1}{6}$ 17. $\dfrac{125}{16}$

19. $a^3 b^4 c^2 - a^2 bc^5$ 21. $6p^3 - 4p^2 - 6p - 2$ 23. $\dfrac{y^2 z}{x}$ 25. $r^6 s^3 t^5$ 27. $-x^4 - 3x^3 y + 4x^2 y^2 - 12xy + y^5$

29. $x^2 - 2x + 4 + \dfrac{-8}{x + 7}$ 31. $t^3 - 15t^2 + 75t - 125$ 33. $14r^2 - 11r + 2$ 35. $c^2 - 9d^2$ 37. $-2y$

39. $x^2 y^3 - xy^2$ 41. $4s^2 - 6s + 9$ 43. $10x + 5$ 45. $x^2 + 3x + \dfrac{-6x + 6}{x^2 - x + 2}$ 47. $5r^2 - rt - t^2$

49. $-x^2 + 3x + 12$ 51. x^2 sq ft

Let's Not Forget

52. 2 53. 3 54. 3 55. $4x^2$ 56. $2x^2 + 4x$ 57. $x^2 - 4$ 58. $x^2 - 4x + 4$ 59. $x^3 + 3x^2 + 3x + 1$
60. -9 61. -27 62. 9 63. b^2 64. $-b^3$

CHAPTER 1 TEST

1. A   2. D   3. A   4. B   5. B   6. $6 - 4x$ 7. $10x^2 - x - 21$ 8. $2x^3 y^2 - 3x^2 y - 4xy$

9. $16x^2 - 56x + 49$ 10. $8x^3 + 27$ 11. $-x^4 - x^3 - x^2 - 3$ 12. $x^3 + 9x^2 + 27x + 27$ 13. $6x^2 - 8x + 9 + \dfrac{-11}{x + 1}$

14. $4x^3 - x^2 + 2x + 11$ 15. $x^2 - 2x - 5$

CHAPTER 2

PROBLEM SET 2.1

Warm-ups

1. 2^6 **3.** $3^2 \cdot 5$ **5.** $2 \cdot 3^4$ **7.** 3^5 **9.** $2^3 \cdot 3^3$ **11.** 24   **13.** 18   **15.** x^3y^2 **17.** st^2 **19.** $16x$
21. $24x^2y^3$ **23.** $16a^2b^2c(3b + 4a)$ **25.** $4p^2q^3(5q^4 - 7p^3q^2 + 9p^6)$ **27.** $15t^2(t - 1)$ **29.** $24m^2n^3(5 - 3m^2n + 2mn^2)$
31. $-3x(2x - 1)$ **33.** $-3v(2u - 1)$ **35.** $(p - q)(m + n)$ **37.** $(x + y)(x + y + 3)$
39. $(1 - r)^2[5(1 - r)^2 + 3(1 - r) - 1]$ **41.** $(s - t)(r + u)$ **43.** $(3u - v)[w(3u - v)^2 - 3u(3u - v) - v]$
45. $(x + y)(a + 7)$ **47.** $(a - b)(a + 2)$ **49.** $(a + b)(a + 1)$ **51.** $(r - s)(5 - s)$ **53.** $(u + w)(v - 1)$
55. $a(a - c)(a + b)$ **57.** $z(x - 1)(z + 1)$ **59.** $5a(x - y)(3b - 5c)$

Practice Exercises

61. $6a^2b^2c(7a^2 + 9b^3)$ **63.** $24r^3s^3(rs - 2)$ **65.** $9ab(3a^3 - 2b^4 + 1)$ **67.** $-8x^3(2 + 3x^3)$ **69.** $12t^3(t - 1)$
71. $a^3(x - 2y + w)$ **73.** $12m^3n^3(3 - 4m^2 + 5mn^3)$ **75.** $(d - 2)(c + b)$ **77.** $(s - t)(a - u)$
79. $(x + y)^2(x + y - 5)$ **81.** $(1 - r)^2[3(1 - r) - 3(1 - r)^2 - 4]$ **83.** $3t(a - x)$
85. $(5 + b)^2[c^2(5 + b)^2 - d^2(5 + b)^3 - 2]$ **87.** $5(a - b)(3(a - b)^3 + 9(a - b) + 7)$ **89.** $(x + 6)(y + 1)$
91. $(a - b)(r - s)$ **93.** $(a - b)(1 - b)$ **95.** $(s + 3)(4 - b)$ **97.** $(t + 1)(t^4 + s^4)$ **99.** $(q^2 - t)(q^2 + 1)$
101. $x(a + b^2)(x - 1)$ **103.** $8c(r + 2s)(2c - 1)$ **105.** $w^2(w - t)(1 - w)$

Challenge Problems

107. $x^{n+2}(1 - x)$ **109.** $x^{2mn}(1 + x^{2mn})$

PROBLEM SET 2.2

Warm-ups

1. $(x - 3)(x + 3)$ **3.** $(x^2 - 5)(x^2 + 5)$ **5.** $(3x^4 - 2y^2)(3x^4 + 2y^2)$ **7.** $(x + 1)(x^2 - x + 1)$
9. $(4 - x)(16 + 4x + x^2)$ **11.** $(1 - 2x)(1 + 2x + 4x^2)$ **13.** $(r - s)(r^2 + rs + s^2)$ **15.** $(x^2 + y)(x^4 - x^2y + y^2)$
17. $(3a - 2b)(9a^2 + 6ab + 4b^2)$ **19.** $(x + y - 2)(x + y + 2)$ **21.** $(9 - r - t)(9 + r + t)$
23. $(4 - a + b)[16 + 4(a - b) + (a - b)^2]$ **25.** $(a + b - x - y)(a + b + x + y)$
27. $[s + t - u - v][(s + t)^2 + (s + t)(u + v) + (u + v)^2]$ **29.** $(a - b)(a^2 + ab + b^2)(a^6 + a^3b^3 + b^6)$
31. $(x + 1)(5 - r)(5 + r)$ **33.** $5(x + y - 2)(x + y + 2)$ **35.** $4(x - 4)(x + 4)$

Practice Exercises

37. $(x - 9)(x + 9)$ **39.** $(x^2 - 6)(x^2 + 6)$ **41.** $(2x^3 - 5y^2)(2x^3 + 5y^2)$ **43.** $(x + 3)(x^2 - 3x + 9)$
45. $(3 - x)(9 + 3x + x^2)$ **47.** $(1 - 3x)(1 + 3x + 9x^2)$ **49.** $(r - 2s)(r^2 + 2rs + 4s^2)$ **51.** $(x^2 + 2y)(x^4 - 2x^2y + 4y^2)$
53. $(5a - 4b)(25a^2 + 20ab + 16b^2)$ **55.** $(x - y - 4)(x - y + 4)$ **57.** $[2 - r + t][4 + 2(r - t) + (r - t)^2]$
59. $[5 - a + b][25 + 5(a - b) + (a - b)^2]$ **61.** $(a - b - x + y)(a - b + x - y)$ **63.** $(a + b)(2 + y)(4 - 2y + y^2)$
65. $3(x - 5)(x + 5)(x^2 + 25)$ **67.** $(1 - x)(1 + x + x^2)(1 + x^3 + x^6)$ **69.** $5(t - 5s)(t^2 + 5st + 25s^2)$
71. $2[2 + s - t][4 - 2(s - t) + (s - t)^2]$ **73.** Prime   **75.** 16   **77.** $\pi h(x - y)(x + y)$ **79.** $\pi(y - x)(y + x)$

Challenge Problems

81. $(x^p - 1)(x^p + 1)$ **83.** $(x^p + 1)(x^{2p} - x^p + 1)$
85. $(x - 1)(x + 1)(x^2 - x + 1)(x^2 + x + 1)$; $(x - 1)(x + 1)(x^4 + x^2 + 1)$

PROBLEM SET 2.3

Warm-ups

1. $(x + 2)(x + 1)$ **3.** $(x - 7)^2$ **5.** $(r - 5s)(r + 3s)$ **7.** Prime   **9.** $(z - 5)(z + 2)$ **11.** $(y - 12)(y + 5)$
13. $(2z + 1)(z + 1)$ **15.** $(5y + 3z)(4y + 3z)$ **17.** $(4x - 5)(2x - 3)$ **19.** $(2 - 7x)(1 - 3x)$ **21.** $(6z - 1)(2z + 1)$
23. $4(a - 2)(a + 1)$ **25.** $-6(b - 2c)(b + c)$ **27.** $z(z + 1)^2$ **29.** $(x^2 + 5)(x^2 + 1)$ **31.** $(x^3 + 7)(x^3 - 2)$
33. $(3y^2 - 2)(2y^2 + 1)$ **35.** $(x - 1)^2(x + 1)^2$ **37.** $(x^2 + 2)(x - 1)(x + 1)$

Practice Exercises

39. $(x + 8)(x + 1)$ **41.** $(x + 2)^2$ **43.** $(x - 4)(x - 2)$ **45.** Prime **47.** $(r - 7s)(r + 2s)$ **49.** $(x - 4)(x + 3)$
51. $(z - 9)(z + 6)$ **53.** $(y - 8)(y + 7)$ **55.** $(2z + 3)(z + 1)$ **57.** $(6y + z)(y + 4z)$ **59.** $(3x - 4)(2x - 3)$
61. $(7 - 2x)(5 - 3x)$ **63.** $(3z + 5)(z - 2)$ **65.** $(3t + 1)(t - 1)$ **67.** $(7s + 5)(2s - 3)$ **69.** $(7r - 5)(4r + 3)$
71. $5(a - 3)(a + 2)$ **73.** $-3(2b - c)(b + c)$ **75.** $z^2(z - 4)(z + 2)$ **77.** $x^2(x - 12)(x + 5)$ **79.** $x^2(x + 4)(x - 3)$
81. $(x^3 - 5)(x^3 - 3)$ **83.** $4(y^3 - 2)(y + 1)(y^2 - y + 1)$ **85.** $(x + 2)(x - 1)(x^2 - 2x + 4)(x^2 + x + 1)$
87. $(x - 2)(x + 2)(x^2 + 2)$ **89.** 6 or -6 **91.** 2 or -2

Challenge Problems

93. $(x^n + 4)(x^n - 2)$ **95.** $(x^n + y^m)^2$

PROBLEM SET 2.4

Warm-ups

1. $16(x - 1)(x + 1)$ **3.** $x(x - 1)$ **5.** Factored **7.** $45a^2b^2c(5a^3 - 2)$ **9.** $(2x - 1)(2x + 1)$ **11.** $(4 - 5x)(4 + 5x)$
13. $[z + t + 1][(z + t)^2 - z - t + 1]$ **15.** $(2w - 3)(4w^2 + 6w + 9)$ **17.** $(x^5 - 4)(x^5 + 4)$ **19.** Prime
21. $(2a - y)(b + x)$ **23.** $(4 - y)(4 + y)(1 + y^2)$ **25.** $(2x - 7)(x + 1)$ **27.** $(x^3 + 2)(x - 1)(x^2 + x + 1)$
29. $x^3(x + 1)(x^2 + 1)$ **31.** Factored **33.** Prime **35.** Prime **37.** $(x - z + 3)(x + z + 3)$

Practice Exercises

39. $(x + 7)(x + 2)$ **41.** $45r^2t^4(3rt + 5)$ **43.** $(1 - 3ab)(1 + 3ab + 9a^2b^2)$ **45.** $(x + y)(x - y + 1)$ **47.** Factored
49. $2(4 - x)(4 + x)$ **51.** $(x - t)(r - 1)(r^2 + r + 1)$ **53.** $(z - xy + 4)(z + xy + 4)$ **55.** $(ab + 1)(a + b)$
57. $(x^2 + 2)(x^2 + 1)$ **59.** $(2x^3 + y)(4x^6 - 2x^3y + y^2)$ **61.** $(x + 6)^2$ **63.** Prime
65. $[t + z + 8][(t + 8)^2 - z(t + 8) + z^2]$ **67.** $(x^2 + 5)(x - 2)(x + 2)$ **69.** $(9 - x)^2$ **71.** $(a^3 + 2)(a^6 - 2a^3 + 4)$
73. $a^5b^5c^3(a^2c - abc^3 + b)$ **75.** $(4 - x - y)(4 + x + y)$

Challenge Problems

77. $(x^n - 1)^2(x^n + 1)^2$ **79.** $x^{2n}(x^n - 1)(x^n + 1)$ **81.** (a) $(x^2 - x + 2)(x^2 + x + 2)$ (b) $(x^2 - 2x + 3)(x^2 + 2x + 3)$
(c) $(x^2 - x + 4)(x^2 + x + 4)$ (d) $(x^2 - 2x - 2)(x^2 + 2x - 2)$ (e) $(x^2 - x - 5)(x^2 + x - 5)$

REVIEW PROBLEMS CHAPTER 2

1. $2(x - 9)(x + 9)$ **3.** $(2z + a)(t - 2z)$ **5.** $(x + 4z - 2)(x + 4z + 2)$ **7.** $2(3x + 1)(2x - 7)$
9. $6abc(2ab - 16ab^2 + 9)$ **11.** $(x - 4)(x + 4)(x^2 + 1)$ **13.** $(x + 3)(x^2 - 3x + 9)(x + 1)(x^2 - x + 1)$
15. $a(b + 2y)(2a - 3t)$ **17.** $2(a + 2)(x + 1)$ **19.** $(x + 9)^2$ **21.** $(w - x - 2)(w + x + 2)$
23. $(x^2 - 3)(x^2 + 3)(x^4 + 9)$ **25.** Prime **27.** $(3a - 1)(9a^2 + 3a + 1)$ **29.** $(x + z + 2)(x - z + 2)$

Let's Not Forget

31. Factored **32.** $(2x + 5)(4x^2 - 10x + 25)$ **33.** $(s + t)^2(s + t - a)$ **34.** Factored **35.** Factored
36. $(b - 1)(1 - a)$ **37.** Two terms; $y + 7$ is a factor. **38.** One term; $y + 7$ is a factor.
39. Three terms; $y + 7$ is a factor. **40.** Three terms; $y + 7$ is a factor. **41.** Two terms; $y + 7$ is a factor.
42. $-27x^3$ **43.** $9x^2 - 12x + 4$ **44.** $x^3 - 9x^2 + 27x - 27$ **45.** -25 **46.** -125 **47.** 25 **48.** x^2 **49.** $-x^3$

CHAPTER 2 TEST

1. B **2.** C **3.** B **4.** D **5.** D **6.** D **7.** $(2a - 3)(4a^2 + 6a + 9)$ **8.** $2(2x - 1)(x + 5)$
9. $(a - b)(x + 2)$ **10.** $(p - q)(x + y)$ **11.** $(a - 2)(a + 2)(a^2 + 2)$ **12.** $(x + 7)^2$ **13.** $(x + y - 3)(x + y + 3)$
14. $(2y - 5)(2y + 5)$ **15.** $9a^2b^2(3b - 4a + 5a^2b^2)$

CHAPTER 3

PROBLEM SET 3.1

Warm-ups

1. 3 **3.** 5 **5.** None **7.** None **9.** 2 **11.** -3 and 1 **13.** 0 and 2 **15.** ± 2 **17.** $-\dfrac{3}{x}; \dfrac{3}{-x}$

19. $\dfrac{-7}{x}; \dfrac{7}{-x}$ **21.** $-\dfrac{2x}{x+5}; \dfrac{-2x}{x+5}$ **23.** $\dfrac{-(x-1)}{x^2}; -\dfrac{x-1}{x^2}$ **25.** $\dfrac{-(x-7)}{x+4}; \dfrac{x-7}{-(x+4)}$

Practice Exercises

27. -6 **29.** 0 **31.** 0 **33.** 7 **35.** -9 **37.** 2 and 5 **39.** -1 and 2 **41.** 3 **43.** $-\dfrac{x^2}{2}; \dfrac{-x^2}{2}$
45. $-\dfrac{x+1}{x}; \dfrac{x+1}{-x}$ **47.** $\dfrac{-x}{x-3}; \dfrac{x}{-(x-3)}$ **49.** $\dfrac{x+5}{-(x+1)}; -\dfrac{x+5}{x+1}$ **51.** $\dfrac{160x}{x+12}$ hrs **53.** $\dfrac{2N}{8-N}$ hrs

Challenge Problems

55. -3 and -1 **57.** -2 and 2

PROBLEM SET 3.2

Warm-ups

1. x^5 **3.** $\dfrac{4t}{3}$ **5.** $3a$ **7.** $\dfrac{x}{2(x+y)}$ **9.** $(x+y)^3$ **11.** $\dfrac{2(x-1)}{3x}$ **13.** $x+4$ **15.** $\dfrac{x+2}{x+3}$ **17.** $\dfrac{s-1}{s^2-s+1}$
19. $\dfrac{2x-1}{x-2}$ **21.** $\dfrac{c+d}{c-d}$ **23.** -1 **25.** $\dfrac{1-2x}{x+2}$ **27.** $\dfrac{2-x}{x+3}$ **29.** $\dfrac{50}{10x}$ **31.** $\dfrac{r^2(m+n)}{6m+6n}$ **33.** $\dfrac{12}{4a}$
35. $\dfrac{-3x(x-1)}{x^2-1}$ **37.** $\dfrac{-x(x+1)}{x-x^2}$ **39.** $\dfrac{-x(x+3)}{x^2-9}$

Practice Exercises

41. x^5 **43.** $3t^2$ **45.** $\dfrac{x}{4}$ **47.** $\dfrac{-3}{x^2 y^2}$ **49.** $\dfrac{5x}{7y^2}$ **51.** $\dfrac{-n}{(n-m)^3}$ **53.** $\dfrac{x+2}{(x-1)(x+5)}$ **55.** $-\dfrac{a+3}{a+1}$
57. $-a$ **59.** $\dfrac{1}{5-3x}$ **61.** $\dfrac{3(4y^3+1)}{5(3y^2-5)}$ **63.** $\dfrac{ab}{a+1}$ **65.** $\dfrac{y^2+5}{y+3y^3}$ **67.** 4 **69.** $x-5$ **71.** $x+4$
73. $\dfrac{p^2+2p+4}{p-3}$ **75.** $\dfrac{b+1}{b+2}$ **77.** $\dfrac{x+3}{x+1}$ **79.** $-\dfrac{x+5}{x+4}$ **81.** $\dfrac{x^2+x+1}{2(x+1)}$ **83.** $\dfrac{x^2+1}{x^2+3}$ **85.** $-\dfrac{2x+3}{x+1}$
87. x^2+5 **89.** $\dfrac{-(m+n)}{m^2+mn+n^2}$ **91.** $\dfrac{a+3}{c+6}$ **93.** $\dfrac{q-p}{q+p}$ **95.** $\dfrac{x+y}{x-y}$ **97.** $\dfrac{60x}{12x^2}$ **99.** $\dfrac{2r^2(x+y)}{6x+6y}$
101. $\dfrac{12a}{4a^2}$ **103.** $\dfrac{-100axy^2}{30x^2y^3}$ **105.** $\dfrac{4a^2}{4a+4b}$ **107.** $\dfrac{2p}{2p-8}$ **109.** $\dfrac{-3x(x+1)}{x^2-1}$ **111.** $\dfrac{2z^2+6z}{z^2+6z+9}$
113. $\dfrac{x^2-5x+6}{x^2-3x}$ **115.** $\dfrac{-x^2-2x}{x^2-4}$ **117.** $\dfrac{2x(x^2-xy+y^2)}{x^3+y^3}$ **119.** $\dfrac{2x(x^2+xy+y^2)}{x^3-y^3}$ **121.** $\dfrac{w^3-27}{w^2-9}$ **123.** $\dfrac{2x^3z}{2z}$

Challenge Problems

125. $x+2$

PROBLEM SET 3.3

Warm-ups

1. $\dfrac{yz}{x^3}$ **3.** $\dfrac{-2}{m^2}$ **5.** $\dfrac{-xy^2}{z^2}$ **7.** $-\dfrac{13}{3}$ **9.** $\dfrac{-4(a^2+2)}{a^2-3}$ **11.** $\dfrac{1}{2}$ **13.** $\dfrac{2(a-2b)}{a+2b}$ **15.** $\dfrac{3x-1}{2x-1}$ **17.** $-\dfrac{1}{t}$
19. $-\dfrac{x+4}{x+1}$ **21.** $\dfrac{5y}{2x}$ **23.** $\dfrac{-2a}{x^3}$ **25.** $\dfrac{6}{rs^2 t}$ **27.** $\dfrac{(5-x)(2-x)}{2x}$ **29.** $\dfrac{(x+3)(x-1)}{x^2(x+1)}$ **31.** $\dfrac{x(x-7)}{2(x-1)(x+1)}$
33. $\dfrac{(p-q)^2}{(r-s)(r+s)}$ **35.** $-y$ **37.** $-\dfrac{(x+3)(x+1)}{(x-2)(x+2)^2}$

Practice Exercises

39. $\dfrac{y^2 z^3}{x}$ **41.** $\dfrac{-3}{m^3}$ **43.** $-\dfrac{y^2}{xz^5}$ **45.** $-\dfrac{c^5 b^9}{x^2}$ **47.** $abcxy$ **49.** $-\dfrac{1}{t^2}$ **51.** $\dfrac{q}{p^2}$ **53.** $\dfrac{rt(rs^2-1)}{s^2(1+t^2 u^2)}$

Chapter 3

55. $-(s + t)$ **57.** $-x^3(x^2 + 5x + 25)$ **59.** $\dfrac{r + s}{r - s}$ **61.** $\dfrac{(m - n)(m + 2n)}{2(m + n)(m - 2n)}$ **63.** $-\dfrac{s + 1}{s - 2}$ **65.** $\dfrac{c^2y^2}{bz^4}$
67. $\dfrac{m^2n^3r^2s}{pt}$ **69.** $\dfrac{-a^2}{5bxy}$ **71.** $-\dfrac{b^2rs^{11}}{a^2}$ **73.** $-\dfrac{p}{q}$ **75.** $\dfrac{2x}{ay}$ **77.** -1 **79.** $\dfrac{4x + 1}{(x + 4)(16x^2 + 4x + 1)}$
81. $\dfrac{(x + 2)(x - 2)}{(x + 1)(x + 3)}$ **83.** $-\dfrac{(x + 2)^2}{(x + 1)^2}$ **85.** $\dfrac{(x + 3)(x - 1)}{x^2(x - 3)}$ **87.** $\dfrac{(b - 2)(b + 3)}{(b + 2)(b^2 - 3)}$ **89.** $\dfrac{(a + b)(a + 2b)}{(a + d)(a - b)}$
91. $\dfrac{x - 2}{x + 1}$ **93.** $\dfrac{(2p + t)(p + t)}{(p - 2t)(2p - t)}$ **95.** $\dfrac{c^3}{c + d}$

Challenge Problems

97. $\dfrac{(x^n - 3)(x^p + 1)}{(x^n + 2)(x^p + 3)}$

PROBLEM SET 3.4

Warm-ups

1. $\dfrac{5 - x}{x^2}$ **3.** $\dfrac{2}{y}$ **5.** $\dfrac{2x + y}{x + y}$ **7.** $\dfrac{m + 1}{m - 2}$ **9.** $\dfrac{5y}{x - 5}$ **11.** $\dfrac{p^2 + q^2}{p^2q^2}$ **13.** $\dfrac{3r + 1}{(r + 2)(r - 3)}$ **15.** $\dfrac{41}{15(a + 2)}$
17. $\dfrac{x^2 + 4x - 4}{(x - 2)^2(x + 2)}$ **19.** $\dfrac{1}{(x + 3)(x - 1)}$ **21.** $\dfrac{7y^2 + 20y + 28}{(y + 2)(2 - y)(y^2 + 2y + 4)}$ **23.** $\dfrac{x^2 + 2}{x}$ **25.** $\dfrac{a^2 + ab + 3}{a + b}$
27. $\dfrac{2}{a(a + 1)}$ **29.** $\dfrac{2(p + 1)}{p + 3}$ **31.** $\dfrac{3 - 7p}{p(p - 1)}$ **33.** $\dfrac{5x + 3}{(x - 2)(x + 2)}$

Practice Exercises

35. $\dfrac{-2}{x + 1}$ **37.** $\dfrac{2t^3 - r}{r^2t^4}$ **39.** $\dfrac{yz + xz - xy}{xyz}$ **41.** $\dfrac{3r - 1}{2(r + 2)}$ **43.** $\dfrac{2s - 4}{(s + 1)(s - 1)}$ **45.** $-\dfrac{2p + 70}{(p + 5)(p - 7)}$
47. $\dfrac{v^2 + 6v + 3}{(v - 3)(v + 7)}$ **49.** $\dfrac{6}{x - y}$ **51.** $\dfrac{x + 2}{x^2}$ **53.** $\dfrac{-1}{1 + r}$ **55.** $\dfrac{a - 2b}{a - b}$ **57.** $\dfrac{2(x + 2)}{x + 4}$ **59.** $\dfrac{y}{(y - 5)(y + 1)}$
61. $\dfrac{t^2 - 11t}{(t - 3)(t + 1)(t - 5)}$ **63.** $\dfrac{3b - 9}{(b - 1)(b + 2)}$ **65.** $\dfrac{6x - 9}{(x + 2)(x - 1)}$ **67.** $\dfrac{6t^2 - 6t + 3}{(t + 2)(t - 1)(t + 5)}$ **69.** $\dfrac{p^2 + p + 5}{p + 1}$
71. $\dfrac{r^2 + 2rt}{r + t}$ **73.** $\dfrac{-x + 8}{x(x + 2)}$ **75.** $\dfrac{-(x^2 + 8x + 14)}{(x + 4)(x + 3)(x + 2)}$ **77.** $\dfrac{1}{x + 3}$ **79.** $\dfrac{n^3 + mn^2 - m^3}{m^2(m + n)}$ **81.** $\dfrac{3r - 21}{(r + 3)(r - 3)}$
83. $\dfrac{2x + 5}{x(x + 5)}$ **85.** $\dfrac{3x - 2}{(x + 1)(2x - 3)}$

Challenge Problems

87. $\dfrac{4x - 1}{x - 1}$ **89.** $\dfrac{(9 - 6x - x^2)(x^2 + 6x - 9)}{(x - 3)^2(x + 3)^2}$

PROBLEM SET 3.5

Warm-ups

1. $\dfrac{a}{2}$ **3.** $\dfrac{-s^2}{6r}$ **5.** y **7.** $-\dfrac{1}{2}$ **9.** $\dfrac{-2}{15}$ **11.** $\dfrac{r + 2}{r}$ **13.** $\dfrac{n + m}{n - m}$ **15.** $t - s$ **17.** $\dfrac{3n - m}{3n + m}$
19. $w + 3$ **21.** $\dfrac{-2ab}{a^2 + b^2}$

Practice Exercises

23. $\dfrac{z^3t^2}{x}$ **25.** $\dfrac{4a^3n}{-3b^4m^2}$ **27.** $\dfrac{s^2}{4}$ **29.** $-\dfrac{3}{7}$ **31.** $\dfrac{a}{4(c + d)}$ **33.** $\dfrac{2(u + v)}{u - v}$ **35.** $\dfrac{1}{3}$ **37.** $\dfrac{1}{x + 2}$
39. $\dfrac{3 - 2y}{1 + 4y}$ **41.** $-\dfrac{a - 1}{a - 3}$ **43.** $\dfrac{2}{a^2 - ab + b^2}$ **45.** 1 **47.** $\dfrac{1}{t - 4}$ **49.** $\dfrac{1}{x + 3}$ **51.** $\dfrac{1}{x - 1}$ **53.** 0

55. $\dfrac{1}{x-1}$ **57.** $\dfrac{3yz - 2xz - 4xy}{yz - xz - xy}$

Challenge Problems

59. $\dfrac{5}{3}$ **61.** $\dfrac{2x^2 + x - 1}{x - 1}$

REVIEW PROBLEMS CHAPTER 3

1. 0   **3.** None   **5.** -3 and -1 **7.** $\dfrac{8p^2 - 16p}{6p^2(p-2)}$ **9.** $\dfrac{x^2 - x - 12}{x^2 - 8x + 16}$ **11.** $\dfrac{1}{m^2 n^2}$ **13.** $\dfrac{x-5}{x^2 - x + 1}$
15. $t(s-1)$ **17.** $\dfrac{-3(x+1)(x+2)}{(x+3)(x-2)}$ **19.** $\dfrac{a(a+b)}{(a-b)(a+3)}$ **21.** $-\dfrac{rt^7 v^2}{u^3}$ **23.** $\dfrac{u-v}{(u+2w)^2}$ **25.** $\dfrac{y+4}{3y+1}$
27. $\dfrac{4x+1}{(x+y)(4x+3)}$ **29.** $\dfrac{3t^3(2t-3)}{(t^2-2)(t^4+4)(t-3)}$ **31.** $-\dfrac{4t^3 s^5}{9}$ **33.** $\dfrac{-24}{a^5 b^2 (a-b)}$
35. $\dfrac{-4(m+4n)}{(2m^2+1)(16n^2+4mn+m^2)}$ **37.** xy **39.** $\dfrac{y-2}{2y+1}$ **41.** t **43.** $\dfrac{1}{2(x^2+1)}$ **45.** s

Let's Not Forget

46. Factored   **47.** $(3x-1)(9x^2+3x+1)$ **48.** $(a-b)(x+y)$ **49.** Factored   **50.** Factored
51. Two terms; not a factor.   **52.** One term; $x-1$ is a factor.   **53.** Three terms; not a factor.
54. Three terms; $x-1$ is a factor.   **55.** Two terms; $x-1$ is a factor.   **56.** Two terms; $x-1$ is a factor.
57. $x^2 + 4xy + 4y^2$ **58.** $c^3 - 6c^2 + 12c - 8$ **59.** -4 **60.** 4   **61.** -8 **62.** a^4 **63.** $-a^5$
64. Won't reduce   **65.** $\dfrac{a-b}{y-z}$ **66.** $\dfrac{5x-6}{(x-2)(x+2)}$ **67.** $\dfrac{(x+2)^2}{(x-2)^2}$

CHAPTER 3 TEST

1. C   **2.** B   **3.** A   **4.** C   **5.** B   **6.** $\dfrac{a^2 + b^2}{ab(a-b)}$ **7.** $\dfrac{a-4}{(a+4)(a+2)}$ **8.** $\dfrac{x(x+3)}{(x-3)(x-1)}$ **9.** $\dfrac{-1}{y+3}$
10. $\dfrac{3x-2}{x-1}$ **11.** $\dfrac{(a+2b)(a+b)}{r}$ **12.** $\dfrac{3t+s}{3t-s}$ **13.** $\dfrac{2-x}{4x^2(2+x)}$ **14.** $\dfrac{3w^4}{15w^3}$ **15.** (A) -1 and 2 (B) None

CHAPTER 4

PROBLEM SET 4.1

Warm-ups

1. $\{6\}$ **3.** $\{5\}$ **5.** $\{3\}$ **7.** $\{3\}$ **9.** $\{5\}$ **11.** $\{x \mid x \text{ is a real number}\}$ **13.** $\{6\}$ **15.** $\left\{-\dfrac{3}{2}\right\}$
17. $\left\{-\dfrac{4}{3}\right\}$ **19.** $\emptyset$ **21.** $\{3\}$ **23.** $\{6\}$ **25.** $\left\{\dfrac{21}{2}\right\}$ **27.** $\left\{-\dfrac{10}{3}\right\}$ **29.** $\{3\}$ **31.** $\{-1\}$ **33.** $\emptyset$ **35.** $\{1.1\}$
37. $\{-17\}$ **39.** $\{-3.8\overline{3}\}$ **41.** The strawberry patch is 17 meters by 22 meters.   **43.** The numbers are 2, 10, and 12.

Practice Exercises

45. $\{4\}$ **47.** $\left\{\dfrac{19}{2}\right\}$ **49.** $\{1\}$ **51.** $\left\{\dfrac{4}{3}\right\}$ **53.** $\emptyset$ **55.** $\{-3\}$ **57.** $\left\{-\dfrac{1}{2}\right\}$ **59.** $\{8\}$ **61.** $\{-2\}$
63. $\{-1\}$ **65.** $\{2\}$ **67.** $\{x \mid x \text{ is a real number}\}$ **69.** $\left\{-\dfrac{5}{3}\right\}$ **71.** $\emptyset$ **73.** $\left\{\dfrac{3}{2}\right\}$ **75.** $\left\{\dfrac{1}{10}\right\}$ **77.** $\{14\}$
79. $\{18\}$ **81.** $\{14\}$ **83.** $\{22\}$ **85.** $\{-21\}$ **87.** $\{2\}$ **89.** $\{-3\}$ **91.** $\{-0.87\}$ **93.** $\{3.\overline{3}\}$ **95.** $\{-0.4\}$
97. The dimensions are 22 feet by 30 feet.   **99.** The numbers are 6, 11, and 12.
101. The acute angles have measures of 30° and 60°.

Challenge Problems

103. $\{a + b\}$ **105.** $\left\{\dfrac{3a + 8}{5}\right\}$

PROBLEM SET 4.2

Warm-ups

1. $r = \dfrac{C}{2\pi}$ **3.** $C = \dfrac{5F - 160}{9}$ **5.** $i = \dfrac{Am - Pm}{P}$ **7.** $w = \dfrac{S - 2lh}{2h + 2l}$ **9.** $x = \dfrac{a + 7}{2}$ **11.** $x = \dfrac{3a + 1}{2}$
13. $x = \dfrac{-3b}{a}$ **15.** $x = \dfrac{5b + 6}{5}$ **17.** $x = 10.0 - 10b$ **19.** $x = \dfrac{b}{b - a}$ **21.** $x = \dfrac{ab + b^2}{a - b}$ **23.** $x = \dfrac{12a + 12b}{9b - 4}$
25. $x = \dfrac{42}{2a + 3b}$ **27.** $x = \dfrac{-1}{2b}$ **29.** $x = \dfrac{2a^2 - 3a}{3b - 3a}$

Challenge Problems

31. $l = \dfrac{A}{w}$ **33.** $w = \dfrac{V}{lh}$ **35.** $t = \dfrac{I}{Pr}$ **37.** $R = \dfrac{R_1 R_2}{R_1 + R_2}$ **39.** $t = \dfrac{A - P}{Pr}$ **41.** $l = \dfrac{P - 2w}{2}$
43. $r = \dfrac{a}{1 - s^2}$ **45.** $R = \dfrac{E}{I}$ **47.** $a = S - Sr$ **49.** The length is 7.0$\overline{75}$ ft.   **51.** His rate is $\dfrac{2}{11}$ km/min.
53. 100° C is 212° F.   **55.** $x = 6 + b$ **57.** $x = \dfrac{7 - a}{4b}$ **59.** $x = \dfrac{b - ab}{a}$ **61.** $x = 3 - a$ **63.** $x = \dfrac{2b - a}{b}$
65. $x = \dfrac{-14}{3a}$ **67.** $x = \dfrac{2}{a - b}$ **69.** $x = \dfrac{3a + 2b}{a + 1}$ **71.** $x = \dfrac{a^2 - 4}{b - a}$ **73.** $x = \dfrac{2ab}{a - b}$ **75.** $x = \dfrac{15a + 15b}{6 - 10b}$
77. $x = \dfrac{10a - 6b^2}{a - 5b}$ **79.** $x = \dfrac{3a}{a - 1}$

Challenge Problems

81. **(a)** $x = a - 1$, if $a \neq -1$ **(b)** $x = -b - 2$, if $b \neq 2$ **(c)** $x = b - 3$, if $b \neq -2$ **(d)** $x = a + 1$, if $a \neq 1$

PROBLEM SET 4.3

Warm-ups

1. One should add 45 gal of 60% solution.   **3.** The chemist should add 5/3 gal of pure alcohol.   **5.** It will take 4/3 hr to fill the pool.   **7.** It should take 12 min.   **9.** They will meet in 4 hr. Frank walked 8 mi and Chuck walked 12 mi.
11. They will be 38 miles apart in 3 hours.   **13.** $4,000 invested at 5% and $6,000 invested at 6%.   **15.** She needs a yield of 15.5%.   **17.** The legs are 5 m and 10 m.   **19.** Yes

Practice Exercises

21. Virginia should add 5 oz of 50% alloy.   **23.** Virginia should add 10 ml of water.   **25.** The glass manufacturer should add 28 kg of pure lead.   **27.** It will take $\dfrac{30}{11}$ hours.   **29.** It will take $7\dfrac{1}{2}$ minutes.   **31.** It will take $\dfrac{6}{7}$ minutes.
33. It will take 10 hours.   **35.** It will take them 35 minutes.   **37.** It is 150 miles.   **39.** The slow freight has traveled 70 miles.   **41.** John started 1 min after Jim.   **43.** $8,000 is invested at 5.5% and $4,000 is invested at 6.5%.
45. They invested $70,000 at 18% and $30,000 at 12%.   **47.** He should invest $75,000 in AAAA bonds and $25,000 in AA bonds.   **49.** He invested $180,000 in the 20% deal and $60,000 in the 5% deal.   **51.** The side is 11 inches.
53. Glen gave away 10 tickets.   **55.** The number is 18.   **57.** The number is 30.

Challenge Problems

59. The speed of the current is $\dfrac{4}{3}$ km/hr.

PROBLEM SET 4.4

Warm-ups

1. {1}   **3.** {2}   **5.** {3}   **7.** $\left\{\dfrac{1}{3}\right\}$ **9.** {2}   **11.** $\left\{\dfrac{1}{2}\right\}$ **13.** {−2.5}   **15.** {5}   **17.** {10}   **19.** {−3}
21. $\left\{-\dfrac{1}{2}\right\}$ **23.** ∅   **25.** {−2}   **27.** {−10}   **29.** His average speed is 16 mph.   **31.** The numbers are 3 and 6.
33. There are 40 girls.   **35.** There are 30.48 cm in a foot.

Practice Exercises

37. {−12}   **39.** {−40}   **41.** $\left\{\dfrac{4}{3}\right\}$ **43.** {−2}   **45.** {1}   **47.** {6}   **49.** {11.5$\overline{90}$}   **51.** {−1.$\overline{518}$}   **53.** ∅
55. {−4}   **57.** ∅   **59.** $\left\{-\dfrac{1}{2}\right\}$ **61.** {3}   **63.** ∅   **65.** ∅   **67.** Joyce's rate is 8 mph.
69. His speed going to work is 30 mph.   **71.** The number is 2.   **73.** There are 120 republicans.
75. There are 19 liters in 5 gallons.

Challenge Problems

77. ∅   **79.** $\left\{\dfrac{2a + b}{2c}\right\}$

PROBLEM SET 4.5

Warm-ups

1. {−3, 3}   **3.** $\left\{-\dfrac{5}{2}, \dfrac{5}{2}\right\}$ **5.** $\left\{-\dfrac{3}{2}, \dfrac{5}{2}\right\}$ **7.** {−4, 4}   **9.** $\left\{\dfrac{1}{6}, \dfrac{1}{2}\right\}$ **11.** ∅   **13.** {0}   **15.** $\left\{\dfrac{5}{2}\right\}$
17. {−1, 4}   **19.** {1, 4}   **21.** ∅   **23.** {−1}   **25.** {1}   **27.** $\left\{-\dfrac{2}{3}, 4\right\}$ **29.** $\left\{-\dfrac{1}{4}, \dfrac{1}{2}\right\}$

Practice Exercises

31. {−6, 6}   **33.** {0}   **35.** {2, 6}   **37.** ∅   **39.** {−20, −6}   **41.** $\left\{-\dfrac{13}{4}, \dfrac{3}{4}\right\}$ **43.** $\left\{\dfrac{3}{10}\right\}$ **45.** $\left\{-4, -\dfrac{2}{5}\right\}$
47. {−0.8, 4}   **49.** {−2.2, 6.8}   **51.** {−1, 3}   **53.** $\left\{-\dfrac{3}{2}\right\}$ **55.** $\left\{-3, -\dfrac{1}{2}\right\}$ **57.** $\left\{0, -\dfrac{2}{3}\right\}$ **59.** $\left\{-4, \dfrac{1}{3}\right\}$
61. $\left\{-16, -\dfrac{2}{3}\right\}$ **63.** $\left\{\dfrac{1}{5}, 1\right\}$ **65.** $\left\{-2, -\dfrac{2}{3}\right\}$ **67.** The numbers are ± 3.   **69.** The numbers are $\dfrac{1}{2}$ and 1.

Challenge Problems

71. {a + 5, 5 − a}   **73.** {a + b, a − b}   **75.** $\left\{a, \dfrac{a}{3}\right\}$ **77.** $\left\{\dfrac{2}{3}, 4\right\}$ **79.** ∅

PROBLEM SET 4.6

Warm-ups

1. {3}   **3.** {11}   **5.** {−7}   **7.** {0}   **9.** ∅   **11.** {1}   **13.** ∅   **15.** {3}   **17.** ∅   **19.** {17}   **21.** {3}
23. {1}   **25.** {0}   **27.** {0}   **29.** {−2}   **31.** ∅

Practice Exercises

33. ∅   **35.** {4}   **37.** {−24}   **39.** ∅   **41.** {2}   **43.** {5}   **45.** $\left\{-\dfrac{1}{2}\right\}$ **47.** ∅   **49.** ∅   **51.** {−10}
53. {1}   **55.** {1}   **57.** {3}   **59.** {5}   **61.** {−1}   **63.** The height is $\dfrac{1}{2}$ foot.

REVIEW PROBLEMS CHAPTER 4

1. $\left\{\dfrac{33}{5}\right\}$ 3. $\left\{-\dfrac{1}{2}, \dfrac{7}{2}\right\}$ 5. $\{2\}$ 7. $\emptyset$ 9. $\left\{-\dfrac{2}{3}\right\}$ 11. $\{-13, 7\}$ 13. $\{-4\}$ 15. $\{0\}$ 17. $\left\{\dfrac{4}{3}\right\}$
19. $\left\{\dfrac{2}{3}\right\}$ 21. $\left\{\dfrac{13}{19}\right\}$ 23. $\{-1\}$ 25. $\left\{\dfrac{1}{5}\right\}$ 27. $\emptyset$ 29. $\{-2\}$ 31. $\emptyset$ 33. $\{-1\}$ 35. $\left\{-\dfrac{11}{4}\right\}$
37. $\{-7\}$ 39. $\{-15\}$ 41. $\{2\}$ 43. $\{4\}$ 45. $\emptyset$ 47. $\{-4\}$ 49. $\emptyset$ 51. $\emptyset$ 53. $\left\{-\dfrac{4}{3}\right\}$ 55. $\{3, 31\}$
57. $\left\{\dfrac{6}{7}\right\}$ 59. $\{8\}$ 61. $\{-3, 5\}$ 63. $\emptyset$ 65. $\{6\}$ 67. $\emptyset$ 69. $\{2\}$ 71. $\emptyset$ 73. $\{1\}$ 75. $\{-10, 6\}$
77. $\left\{-\dfrac{3}{4}\right\}$ 79. $\left\{\dfrac{7}{6}\right\}$ 81. They are 13, 15, and 17. 83. She should add 9 gal of water.
85. The dimensions are 15 m by 28 m.

Let's Not Forget

87. Factored 88. $(x - 3)(x^2 + 3x + 9)$ 89. $(x + y)(1 - 4x - 4y)$ 90. Factored 91. Factored
92. Two; $x + 2$ is a factor. 93. One term; $x + 2$ is a factor. 94. Two; $x + 2$ is a factor.
95. One term; $x + 2$ is a factor. 96. Two; $x + 2$ is a factor. 97. Three terms; $x + 2$ is a factor. 98. -64
99. 64 100. $x^2 - 4xy + 4y^2$ 101. $a^3 + 6a^2b + 12ab^2 + 8b^3$ 102. x^4y^2 103. $x^5y - x^4y$
104. Reduces to $\dfrac{3}{x + 3}$ 105. Won't reduce; simplifies to $\dfrac{x - 9}{x^2 - 9}$ 106. $\left\{-\dfrac{3}{2}\right\}$ 107. $\dfrac{3 - 2x}{6x}$ 108. $\dfrac{45}{2xt}$
109. Expression; $\dfrac{2 - x^2}{(x + 1)(x + 2)}$ 110. Equation; $\left\{\dfrac{9}{2}\right\}$ 111. Equation; $\{-11\}$ 112. Equation; $\left\{\dfrac{17}{2}\right\}$
113. Equation; $\{-9, 3\}$ 114. Equation; $\{14\}$ 115. Expression; $\dfrac{12}{x - y}$

CHAPTER 4 TEST

1. C 2. A 3. B 4. D 5. C 6. $\{-5, 2\}$ 7. $\emptyset$ 8. $\left\{-\dfrac{7}{2}\right\}$ 9. $\left\{-\dfrac{5}{2}\right\}$ 10. $\emptyset$
11. $x = \dfrac{a + 2y}{1 + 2a}$ 12. $C = \dfrac{4L}{4L^2F^2 + R}$ 13. The Honda traveled at a rate of 30 mph.
14. The largest frame will be 31 in. × 36 in. 15. She should add 8 quarts of 20% solution.

CHAPTER 5

PROBLEM SET 5.1

Warm-ups

1. $\dfrac{1}{8}$ 3. $\dfrac{1}{6}$ 5. $-\dfrac{1}{16}$ 7. $-\dfrac{1}{8}$ 9. 9 11. $\dfrac{x}{9}$ 13. 2 15. 125 17. $\dfrac{1}{81}$ 19. $\dfrac{1}{xy}$ 21. $-\dfrac{1}{27x^3}$
23. 32 25. 25 27. x^6 29. $-8x^3y^3$ 31. $\dfrac{1}{24}$ 33. $-\dfrac{25}{6}$ 35. $\dfrac{x^2y}{3z}$ 37. $\dfrac{3x^3}{4y^6}$ 39. $\dfrac{y^3}{16x^4z}$
41. $-\dfrac{1}{72}$ 43. $\dfrac{3}{2}$ 45. $\dfrac{x^2 + 8x}{8}$ 47. -1.2345×10^{-6} 49. 7.7722×10^{13} 51. 0.0000543
53. $-800{,}000{,}000{,}000$

Practice Exercises

55. $\dfrac{1}{4}$ 57. $\dfrac{1}{81}$ 59. $-\dfrac{1}{27}$ 61. $-\dfrac{1}{100}$ 63. $\dfrac{27}{8}$ 65. $-\dfrac{5}{3}$ 67. $\dfrac{x}{8}$ 69. $-\dfrac{6}{x^2y}$ 71. $\dfrac{3}{t^{21}}$ 73. 80
75. $\dfrac{-2y^3}{x^2}$ 77. $\dfrac{5}{16}$ 79. $\dfrac{17}{72}$ 81. $\dfrac{4}{3}$ 83. $\dfrac{x^2 - 9x}{9}$ 85. $\dfrac{9x^3}{8}$ 87. $\dfrac{1}{3}$ 89. 6^3 91. $\dfrac{1}{81}$ 93. $\dfrac{1}{3^6}$

95. 3^6 **97.** $\dfrac{1}{16}$ **99.** 16   **101.** x^{12} **103.** $\dfrac{1}{27x^3}$ **105.** $\dfrac{x^6}{y^{12}}$ **107.** $-\dfrac{x^6}{27y^3}$ **109.** $9x^2y^2$ **111.** $-\dfrac{x^9}{8}$
113. -1 **115.** 3^5 **117.** $\dfrac{1}{y^2}$ **119.** t^3 **121.** $\dfrac{7}{(x+2)^5}$ **123.** $\dfrac{2}{9x^3y^2}$ **125.** $\dfrac{1}{8x^3z^2}$ **127.** 5.432×10^{-2}
129. 4.402×10^3 **131.** -9.01×10^{-1} **133.** 143,000   **135.** 0.000000000011   **137.** -0.1
139. $\approx 4.05 \times 10^{18}$ cm

Challenge Problems

147. $x^{2j} - x^{2k}$ **149.** $x^{2j} - 2 + \dfrac{1}{x^{2j}}$ **151.** $x^{2j}y^{2k}$ **153.** $(x^{-1} - 3)(x^{-1} + 2)$

PROBLEM SET 5.2

Warm-ups

1. 14   **3.** 11   **5.** -12 **7.** $\dfrac{5}{7}$ **9.** $\dfrac{3}{4}$ **11.** $-2x^2y$ **13.** $-12j^3l^2$ **15.** $\dfrac{6x^5}{y^4}$ **17.** $-6\sqrt{5}$
19. $4\sqrt{15}$ **21.** $9\sqrt[3]{4}$ **23.** $2\sqrt[6]{5}$ **25. a)** $2y$ **b)** Won't simplify   **c)** $2 + y$

Practice Exercises

27. 18   **29.** -10 **31.** 6   **33.** $-4x^2y$ **35.** $6\sqrt{2}$ **37.** $12\sqrt{2}$ **39.** $3\sqrt[3]{2}$ **41.** $-3\sqrt[5]{2}$ **43.** $\dfrac{4}{9}$
45. $\dfrac{3\sqrt{3}}{7}$ **47.** $-2\sqrt{2}$ **49.** $3xy\sqrt{x}$ **51.** $\dfrac{4}{3}$ **53.** $\dfrac{5\sqrt{3}}{7}$ **55.** $\dfrac{2\sqrt[3]{7}}{3}$ **57.** $3xy\sqrt{7yz}$ **59.** $-15xy^4\sqrt{2xy}$
61. $\dfrac{14}{p^2q}$ **63.** $2z\sqrt[3]{2}$ **65.** $\dfrac{6a^2b^4c\sqrt[3]{c}}{d^9}$ **67.** $-2x^3z^5\sqrt[5]{3z}$ **69.** $13 + 2x$ **71.** $3\sqrt{4 + v^2}$

Challenge Problems

73. $-x$ **75.** a **77.** $|c|$ **79.** $1 - b$

PROBLEM SET 5.3

Warm-ups

1. $-2\sqrt{11}$ **3.** 23   **5.** $5\sqrt[3]{2}$ **7.** $3\sqrt{2}$ **9.** $18xy^2\sqrt[3]{3y}$ **11.** $4\sqrt[3]{2}$ **13.** x^2y^3 **15.** $2\sqrt{15} + 5$
17. $-1 - \sqrt{2}$ **19.** $4x + 4\sqrt{xy} + y$ **21.** -1 **23.** $2\sqrt{2}$ **25.** 3   **27.** $\dfrac{\sqrt{5}}{5}$ **29.** $\dfrac{\sqrt{5}}{2}$ **31.** $\dfrac{\sqrt{7x}}{2x}$
33. $\dfrac{\sqrt[3]{49}}{7}$ **35.** $\sqrt[3]{25}$ **37.** $\dfrac{\sqrt[4]{4}}{2}$ **39.** $-\sqrt{2} - 2$ **41.** $-2\sqrt{6} - 6$ **43.** $\dfrac{k - 2\sqrt{k} + 1}{k - 1}$

Practice Exercises

45. $4\sqrt{13}$ **47.** $20\sqrt{2}$ **49.** $5\sqrt[3]{4}$ **51.** $-5\sqrt{2}$ **53.** $-2\sqrt{3}$ **55.** $7xy^3\sqrt[3]{3y}$ **57.** $6\sqrt[3]{3}$ **59.** $-2ab\sqrt[3]{6a^2}$
61. $4 + 2\sqrt{3}$ **63.** $\sqrt{x} - 2 + 3x$ **65.** $6\sqrt{2} + 4\sqrt{3} + 3\sqrt{10} + 2\sqrt{15}$ **67.** $4 - 2\sqrt{3}$ **69.** $14 - 4\sqrt{6}$
71. $1 - k$ **73.** 1   **75.** $2\sqrt{3}$ **77.** $2\sqrt[3]{2}$ **79.** $\dfrac{\sqrt{7}}{7}$ **81.** $\dfrac{3\sqrt{x}}{2x}$ **83.** $\dfrac{\sqrt{11x}}{2x}$ **85.** $2\sqrt{3} - 3$
87. $3\sqrt{6} + 6$ **89.** $7 - 4\sqrt{2}$ **91.** $\dfrac{2k - 6\sqrt{2k} + 9}{2k - 9}$ **93.** $\dfrac{2k - \sqrt{km} - 6m}{k - 4m}$ **95.** $\dfrac{\sqrt[3]{7}}{7}$ **97.** $\dfrac{\sqrt[4]{9}}{3}$

Challenge Problems

99. $a - b$; the difference of two cubes   **101.** $\sqrt[3]{9} + \sqrt[3]{6} + \sqrt[3]{4}$ **103.** $2 + \sqrt[3]{4} + \sqrt[3]{2}$ **105.** $(x + 2\sqrt{2})(x - 2\sqrt{2})$

PROBLEM SET 5.4

Warm-ups

1. 7   **3.** 13   **5.** Not a real number   **7.** 3   **9.** 2   **11.** 32   **13.** 128   **15.** $\dfrac{1}{9}$ **17.** Not a real number

19. $\dfrac{4}{25}$ **21.** $\dfrac{2744}{2197}$ **23.** $\dfrac{1331}{1000}$ **25.** 8 **27.** 5 **29.** 4 **31.** $6x$ **33.** $\dfrac{7x}{y}$ **35.** $-\dfrac{x^2}{4z^{1/3}}$ **37.** $\dfrac{2\sqrt{2}y^{1/2}}{x}$
39. $\dfrac{27a^2}{x}$ **41.** $6x^{13/6} - 10x^{7/6}$ **43.** $x + 2 + \dfrac{1}{x}$

Practice Exercises

45. 6 **47.** 15 **49.** -10 **51.** 9 **53.** -2 **55.** 6 **57.** Not a real number **59.** 8 **61.** 128 **63.** 243
65. $\dfrac{1}{16}$ **67.** $\dfrac{1}{9}$ **69.** $\dfrac{16}{49}$ **71.** $\dfrac{4096}{3375}$ **73.** $\dfrac{5832}{2197}$ **75.** $5y$ **77.** $\dfrac{4x^2}{y^3}$ **79.** $\dfrac{x}{4z^{1/3}}$ **81.** $\dfrac{3\sqrt{2}x}{y}$
83. $-\dfrac{27a^3}{x}$ **85.** $\dfrac{5}{x^{1/3}y^{1/3}}$ **87.** $\dfrac{27z^2}{8xy^3}$ **89.** $\dfrac{xy}{125}$ **91.** $6x^{10/3} - 15x^{4/3}$ **93.** $6 + \dfrac{2b^{3/2}}{a^{1/2}}$ **95.** $x^3 - \dfrac{1}{x^3}$

Challenge Problems

97. $a + 3a^{2/3}b^{1/3} + 3a^{1/3}b^{2/3} + b$ **99.** $x - 3x^{1/3} + \dfrac{3}{x^{1/3}} - \dfrac{1}{x}$ **101.** $(x^{1/3} - 3)(x^{1/3} + 2)$ **103.** $a^{m/n}$ does *not* equal $\sqrt[n]{a^m}$ unless $\sqrt[n]{a}$ is a real number, and $\sqrt{-1}$ is not a real number. **105.** $\sqrt{xy}$

PROBLEM SET 5.5

Warm-ups

1. $0 + 3i$ **3.** $0 + (-11)i$ **5.** $3 + (-2)i$ **7.** $5\sqrt{2} + 0i$ **9.** $-2 + 0i$ **11.** $-i$ **13.** 1 **15.** i
17. $1 + 2i$ **19.** $-3 + 4i$ **21.** -14 **23.** $15 - 6i$ **25.** $7 - 3i$ **27.** $-5 - 12i$ **29.** $-4 - 8\sqrt{6}i$ **31.** 29
33. 12 **35.** $\dfrac{2}{5} + \dfrac{1}{5}i$ **37.** $\dfrac{7}{10} - \dfrac{1}{10}i$ **39.** $\dfrac{3}{5} - \dfrac{4}{5}i$

Practice Exercises

41. $0 + 5i$ **43.** $2 + (-\sqrt{5})i$ **45.** $0 + \sqrt{7}i$ **47.** $0 + 2\sqrt{2}i$ **49.** $-i$ **51.** -1 **53.** i **55.** $3 - 4i$
57. $8 + 7i$ **59.** $-2 - 2i$ **61.** -9 **63.** $3 + 2i$ **65.** $5 - i$ **67.** $-7 - 9i$ **69.** $-8 + i$ **71.** $-2i$
73. $-7 + 24i$ **75.** $-6 + 6\sqrt{3}i$ **77.** $1 + 2\sqrt{6}i$ **79.** 10 **81.** 149 **83.** 7 **85.** 5 **87.** $\dfrac{1}{2} + \dfrac{1}{2}i$
89. $\dfrac{3}{5} + \dfrac{6}{5}i$ **91.** $\dfrac{1}{8} + \dfrac{1}{8}i$ **93.** $-\dfrac{1}{10} + \dfrac{7}{10}i$ **95.** $\dfrac{12}{13} - \dfrac{5}{13}i$ **97.** $-46 + 9i$

Challenge Problems

99. LS: $i^2 + 1 = -1 + 1 = 0$ RS: 0 If x is $-i$, then LS: $(-i)^2 + 1 = i^2 + 1$ **101.** $(x + 3i)(x - 3i)$

REVIEW PROBLEMS CHAPTER 5

1. $\dfrac{1}{25}$ **3.** $\dfrac{1}{64}$ **5.** $\dfrac{7}{q^3}$ **7.** 36 **9.** $\dfrac{4y}{x^2}$ **11.** 49 **13.** $\dfrac{y^9}{27x^6z^3}$ **15.** $\dfrac{p^3 - q^3}{p^2q}$ **17.** $-4\sqrt[3]{2}$ **19.** $\dfrac{2x^2}{3}$
21. Will not simplify. **23.** 16 **25.** $\dfrac{1331}{343}$ **27.** $\dfrac{3y^2}{2x^3z}$ **29.** $\dfrac{y^{1/3}}{4xz^2}$ **31.** $-\sqrt{2}$ **33.** 0
35. $2\sqrt{3} + \sqrt{6} - \sqrt{2} - 2$ **37.** 13 **39.** $x + \dfrac{2x^{1/2}}{y^{1/2}} + \dfrac{1}{y}$ **41.** $\dfrac{\sqrt{6}}{3}$ **43.** $\dfrac{\sqrt{15}}{5}$ **45.** $2\sqrt[5]{2}$ **47.** $\dfrac{3\sqrt{5} + 3}{2}$
49. $4\sqrt{3} - 7$ **51.** 7.681×10^{-6} **53.** -9.88×10^{-12} **55.** $900{,}600{,}000{,}000{,}000$ **57.** -0.000000000032
59. $3\sqrt{2} + 3\sqrt{2}i$ **61.** $9 - 14i$ **63.** $8 + 12i$ **65.** $-16 - 30i$ **67.** $-\dfrac{2}{5} - i$

Let's Not Forget

69. Factored **70.** $(x - y)(r^2 + t^2)$ **71.** Factored **72.** Factored **73.** $(2x - 3)(4x^2 + 6x + 9)$ **74.** $(2x + 3y)^2$
75. 2, yes **76.** 2, yes **77.** 3, yes **78.** 1, yes **79.** 3, no **80.** $-\dfrac{4x^2 + 2x + 1}{1 + 2x}$ **81.** $-\dfrac{5z^3}{xy^2}$ **82.** $-\dfrac{1}{2}$
83. $\dfrac{2a^2}{3b^3}$ **84.** $\dfrac{10}{3}$ **85.** $\sqrt{3} - i$ **86.** $x^2 + y^2$ **87.** $2\sqrt{3} + 6$ **88.** $x^2 + 2\sqrt{2}x + 2$ **89.** $8 - 12x + 6x^2 - x^3$

90. $r - s$ **91.** $1 + \dfrac{\sqrt{2}}{2}$ **92.** $\{1\}$ **93.** $\dfrac{3x - 9}{(x + 1)(x - 2)}$ **94.** $\dfrac{x^2 - 2x}{x^2 - 1}$ **95.** Equation, $\{-2, 7\}$
96. Equation, $\{1\}$ **97.** Expression, $\dfrac{2}{x^2 + x - 6}$ **98.** Equation, $\left\{-\dfrac{1}{13}\right\}$ **99.** Expression, $16 - 24x + 9x^2$
100. Expression, $5 + i$

CHAPTER 5 TEST

1. A **2.** B **3.** C **4.** B **5.** B **6.** D **7.** A **8.** C **9.** D **10.** $-\dfrac{2y^2}{x}$ **11.** $x + y$ **12.** $\dfrac{b^6}{3a^2c^3}$
13. $\dfrac{1}{2} + \dfrac{1}{2}i$ **14.** $\dfrac{\sqrt{6}}{4}$ **15.** $8\sqrt{2} - 9$

CHAPTER 6

PROBLEM SET 6.1

Warm-ups
1. $\{-7, -3\}$ **3.** $\{\pm 5\}$ **5.** $\{-5, 0\}$ **7.** $\left\{-\dfrac{1}{7}, 7\right\}$ **9.** $\left\{\pm \dfrac{1}{5}\right\}$ **11.** $\left\{-\dfrac{7}{3}, \dfrac{5}{2}\right\}$ **13.** $\{0\}$ **15.** $\left\{3, \dfrac{9}{2}\right\}$
17. $\{\pm 2i\}$ **19.** $\{-1, 2\}$ **21.** $\{-5, -1\}$ **23.** $\{3, 5\}$ **25.** $\left\{-1, -\dfrac{1}{4}\right\}$ **27.** $\{0, 2\}$ **29.** $\{-7, 0\}$
31. $\{-4, -1\}$ **33.** $\{-7, -1\}$ **35.** $\left\{-1, \dfrac{5}{2}\right\}$ **37.** $\left\{-2, \dfrac{1}{3}\right\}$ **39.** $\left\{-3, \dfrac{2}{3}\right\}$ **41.** $\{\pm 2\}$ **43.** $\{\pm 3\}$
45. $\{0, 3\}$ **47.** $\{0, 4\}$ **49.** $\left\{-\dfrac{1}{6}, 1\right\}$ **51.** $\{8\}$

Practice Exercises
53. $\{-1, 6\}$ **55.** $\{-4, -3\}$ **57.** $\{-1, 7\}$ **59.** $\{-2, 8\}$ **61.** $\{5, 10\}$ **63.** $\{12\}$ **65.** $\left\{\dfrac{1}{5}, 2\right\}$
67. $\left\{-1, -\dfrac{2}{3}\right\}$ **69.** $\left\{\dfrac{3}{2}, 2\right\}$ **71.** $\left\{-\dfrac{4}{5}, 2\right\}$ **73.** $\left\{-2, \dfrac{5}{3}\right\}$ **75.** $\left\{-\dfrac{11}{5}, 3\right\}$ **77.** $\left\{-\dfrac{3}{2}, \dfrac{4}{3}\right\}$
79. $\left\{-\dfrac{3}{2}, \dfrac{1}{2}\right\}$ **81.** $\{-1, 2\}$ **83.** $\{-3, 4\}$ **85.** $\{-1, 13\}$ **87.** $\left\{-\dfrac{2}{3}, \dfrac{3}{5}\right\}$ **89.** $\left\{-\dfrac{3}{4}, \dfrac{2}{5}\right\}$

Challenge Problems
91. $\{\pm\sqrt{5}\}$ **93.** $\{\pm\sqrt{5}i\}$ **95.** $\left\{\pm\dfrac{5}{2}i\right\}$ **97.** $x^2 - 3x + 2 = 0$ **99.** $x^2 - 4 = 0$ **101.** $x^2 + 4 = 0$

PROBLEM SET 6.2

Warm-ups
1. The girls are 14 and 16 years old. **3.** The trails are 7 and 13 miles. **5.** The integers are 4, 5, and 6.
7. The poster is 7 in. by 10 in. **9.** The rectangle is 10 m by 12 m. **11.** The frame should be 1 ft wide.
13. The sheet of copper should be 14 in. by 14 in. **15.** The base is 16 ft and the altitude is 7 ft.
17. The painting extends $4\sqrt{3}$ m along the wall. **19.** The sides have lengths of 3, 4, and 5 units.

Practice Exercises
21. Jack scored 14 points and Bob scored 12 points. **23.** The numbers are 3 and 17.
25. The integers are 6, 8, and 10. **27.** The search zone is 11 miles by 15 miles.
29. The original piece of cardboard was 11 in. by 11 in. **31.** The height is 2 ft and the base is 8 ft. **33.** The ball will be at a height of 32 ft after 1 sec (going up) and again after 2 sec (going down). **35.** It takes 10 minutes to fill the pool.

Challenge Problems

37. The southbound train is traveling at the rate of 120 mph and the other train is traveling at the rate of 50 mph.

PROBLEM SET 6.3

Warm-ups

1. $\{\pm 5\}$ **3.** $\{\pm 3i\}$ **5.** $\{\pm\sqrt{5}\}$ **7.** $\{\pm\sqrt{5}i\}$ **9.** $\{\pm 4\sqrt{3}\}$ **11.** $\{\pm 3\sqrt{2}i\}$ **13.** $\{\pm 2\sqrt{2}i\}$ **15.** $\{\pm\sqrt{3}i\}$
17. $\{-5, -1\}$ **19.** $\{7 \pm 2\sqrt{2}\}$ **21.** $\{5 \pm \sqrt{7}i\}$ **23.** 4 **25.** 9 **27.** $\dfrac{49}{4}$ **29.** $\dfrac{1}{4}$ **31.** $\dfrac{121}{4}$
33. $\{-1, 5\}$ **35.** $\{2 \pm i\}$ **37.** $\{-3, 2\}$ **39.** $\{-1 \pm \sqrt{5}\}$ **41.** $\left\{\dfrac{1}{2} \pm \dfrac{\sqrt{11}}{2}i\right\}$ **43.** $\left\{-\dfrac{1}{2} \pm \dfrac{\sqrt{15}}{2}\right\}$
45. $\left\{\dfrac{1}{2} \pm \dfrac{3}{2}i\right\}$ **47.** $\left\{\dfrac{5}{4} \pm \dfrac{\sqrt{17}}{4}\right\}$

Practice Exercises

49. $\{-2, 4\}$ **51.** $\{1\}$ **53.** $\{1, 3\}$ **55.** $\{-1, 4\}$ **57.** $\{1, 4\}$ **59.** $\left\{\dfrac{1}{2} \pm \dfrac{\sqrt{5}}{2}\right\}$ **61.** $\left\{-\dfrac{5}{4} \pm \dfrac{\sqrt{57}}{4}\right\}$
63. $\left\{\dfrac{2}{3} \pm \dfrac{\sqrt{26}}{3}i\right\}$ **65.** $\left\{1, \dfrac{5}{2}\right\}$ **67.** $\left\{\dfrac{3}{4} \pm \dfrac{\sqrt{31}}{4}i\right\}$ **69.** The leg is 3 feet.

Challenge Problems

71. $\{\pm\sqrt{a+b}\}$ **73.** $\{-1 \pm \sqrt{a}\}$ **75.** $\{-b \pm 2\sqrt{2}\}$ **77.** $\left\{-\dfrac{a}{2} \pm \dfrac{\sqrt{a^2+a}}{2}\right\}$ **79.** $\{0.3 \pm \sqrt{0.0775}\}$
81. $\left\{\dfrac{-b \pm \sqrt{b^2-4ac}}{2a}; a \neq 0\right\}$

PROBLEM SET 6.4

Warm-ups

1. $\{-4, 2\}$ **3.** $\{-2, 0\}$ **5.** $\left\{-\dfrac{1}{2} \pm \dfrac{\sqrt{5}}{2}\right\}$ **7.** $\{-2\}$ **9.** $\left\{\dfrac{1}{2} \pm \dfrac{\sqrt{3}}{2}i\right\}$ **11.** $\{\pm i\}$ **13.** $\left\{-\dfrac{5}{2} \pm \dfrac{\sqrt{11}}{2}i\right\}$
15. $\{1 \pm \sqrt{2}\}$ **17.** $\{-1 \pm \sqrt{2}\}$ **19.** $\{3 \pm \sqrt{11}\}$ **21.** $\{-1 \pm i\}$ **23.** $\left\{\dfrac{3}{2}, \dfrac{1}{2}\right\}$ **25.** $\left\{-\dfrac{3}{2}, \dfrac{2}{3}\right\}$
27. $\left\{\dfrac{1}{2} \pm \dfrac{\sqrt{3}}{2}\right\}$ **29.** $\left\{\dfrac{1}{6} \pm \dfrac{\sqrt{23}}{6}i\right\}$ **31.** $\left\{-\dfrac{5}{6}, 1\right\}$ **33.** $\{1 \pm \sqrt{3}i\}$ **35.** $\left\{-7, \dfrac{3}{2}\right\}$ **37.** $\{4 \pm \sqrt{15}\}$
39. $\left\{-2, \dfrac{3}{2}\right\}$ **41.** -15; two complex solutions **43.** 41; two real solutions

Practice Exercises

45. $\{-3\}$ **47.** $\left\{-\dfrac{1}{2}, 1\right\}$ **49.** $\{-2, 4\}$ **51.** $\left\{-\dfrac{2}{5}, 3\right\}$ **53.** $\left\{-\dfrac{2}{3}, \dfrac{1}{2}\right\}$ **55.** $\{0, 1\}$ **57.** $\{\pm 5i\}$
59. $\left\{-\dfrac{1}{2} \pm \dfrac{\sqrt{17}}{2}\right\}$ **61.** $\left\{\dfrac{3}{2} \pm \dfrac{\sqrt{11}}{2}i\right\}$ **63.** $\{1 \pm i\}$ **65.** $\{3 \pm i\}$ **67.** $\left\{-\dfrac{7}{4} \pm \dfrac{\sqrt{41}}{4}\right\}$ **69.** $\{-0.3, 0.2\}$
71. $\{-7, -1\}$ **73.** $\{\pm\sqrt{3}\}$ **75.** $\left\{-\dfrac{1}{3}, 1\right\}$ **77.** $\left\{0, \dfrac{1}{2}\right\}$ **79.** $\{\pm 2\}$ **81.** $\left\{\dfrac{1}{2} \pm \dfrac{1}{2}i\right\}$ **83.** $\left\{-\dfrac{2}{3}\right\}$
85. $\left\{\dfrac{1}{4} \pm \dfrac{\sqrt{41}}{4}\right\}$ **87.** 1; two real solutions **89.** -20; two complex solutions

Challenge Problems

91. $x = \dfrac{-3 \pm \sqrt{9+8a}}{2a}$ **93.** $x = -n$ or $x = m$ **95.** $R = \pm\sqrt{\dfrac{3V}{2ph}}$ **97.** $d = \pm\dfrac{\sqrt{2-LR^2}}{R}$ **99.** $k > 1$

101. All real numbers   **103.** $k > 0$ **105.** If $b^2 - 4ac$ is a perfect square, the equation will have rational solutions. If $b^2 - 4ac$ is not a perfect square, the equation will have irrational solutions.   **107.** $-4ac$; real if ac is negative and complex if ac is positive.   **109.** $\dfrac{c}{a}$ **111.** Sum is 5. Product is 10.   **113.** Sum is 0. Product is $-\dfrac{11}{3}$.
115. Sum is $\dfrac{P-Q}{K}$. Product is $\dfrac{Q+P}{K}$.

PROBLEM SET 6.5

Warm-ups
1. $\{-4, -3, 5\}$ **3.** $\{-17, -\dfrac{3}{2}, 0\}$ **5.** $\{\pm 1, \pm 3\}$ **7.** $\{\pm 1\}$ **9.** $\{\pm 2, \pm i\}$ **11.** $\{\pm \sqrt{2}, \pm \sqrt{3}i\}$
13. $\{\pm 1, \pm 2\sqrt{3}\}$ **15.** $\{0, \pm \sqrt{2}i\}$ **17.** $\{\pm \sqrt{2}, \pm \sqrt{2}i\}$ **19.** $\left\{\pm \dfrac{\sqrt{3}}{2}, \pm \sqrt{2}i\right\}$ **21.** $\left\{\pm \dfrac{\sqrt{5}}{2}, \pm \dfrac{\sqrt{3}}{2}i\right\}$
23. $\{\pm \sqrt{2}, \pm \sqrt{2}i\}$ **25.** $\left\{-1, \dfrac{1}{2} \pm \dfrac{\sqrt{3}}{2}i\right\}$ **27.** $\{2, -1 \pm \sqrt{3}i\}$ **29.** $\left\{5, -\dfrac{5}{2} \pm \dfrac{5\sqrt{3}}{2}i\right\}$
31. $\left\{0, 1, -\dfrac{1}{2} \pm \dfrac{\sqrt{3}}{2}i\right\}$ **33.** $\left\{-\dfrac{2}{5}, \dfrac{1}{5} \pm \dfrac{\sqrt{3}}{5}i\right\}$

Practice Exercises
35. $\{\pm 1, \pm 6\}$ **37.** $\{0, \pm \sqrt{3}\}$ **39.** $\{\pm \sqrt{7}, \pm \sqrt{6}i\}$ **41.** $\{\pm 2, \pm \sqrt{3}i\}$ **43.** $\{\pm \sqrt{3}, \pm \sqrt{2}i\}$
45. $\left\{\pm 1, \pm \dfrac{\sqrt{3}}{2}i\right\}$ **47.** $\left\{\pm \dfrac{\sqrt{7}}{3}, \pm \sqrt{2}i\right\}$ **49.** $\{\pm 2, \pm 2i\}$ **51.** $\left\{-3, \dfrac{3}{2} \pm \dfrac{3\sqrt{3}}{2}i\right\}$ **53.** $\left\{-5, \dfrac{5}{2} \pm \dfrac{5\sqrt{3}}{2}i\right\}$
55. $\left\{0, 1, -\dfrac{1}{2} \pm \dfrac{\sqrt{3}}{2}i\right\}$ **57.** $\{\pm \sqrt{5}, \pm \sqrt{5}i\}$

Challenge Problems
59. $\{\pm 2, 1 \pm \sqrt{3}i, -1 \pm \sqrt{3}i\}$ **61.** $\{2, \pm 2i\}$ **63.** $\{-3, -2, 0, 1\}$ **65.** $(x-4)(x-2)(x+1) = 0$

PROBLEM SET 6.6

Warm-ups
1. $\{-1, 6\}$ **3.** $\{-6, 1\}$ **5.** $\left\{-\dfrac{1}{4}, 2\right\}$ **7.** $\{2, 3\}$ **9.** $\{5\}$ **11.** $\{6\}$ **13.** $\emptyset$ **15.** $\emptyset$ **17.** $\left\{-\dfrac{1}{6}, \dfrac{1}{4}\right\}$
19. $\{\pm 1\}$ **21.** The ship's speed is 6 knots.   **23.** His walking speed was 4 mph.
25. The larger pipe will take 6 hrs and the smaller pipe 12 hrs.

Practice Exercises
27. $\left\{-2, \dfrac{5}{2}\right\}$ **29.** $\left\{-1, -\dfrac{1}{6}\right\}$ **31.** $\{-1, 3\}$ **33.** $\{-3, -2\}$ **35.** $\{-5, -2\}$ **37.** $\{-2, 3\}$ **39.** $\emptyset$
41. $\{-4, 2\}$ **43.** $\{-2, -1\}$ **45.** $\{-2\}$ **47.** $\left\{-\dfrac{4}{5}, 1\right\}$ **49.** $\{-2\}$ **51.** $\{-3, 5\}$ **53.** $\{\pm 3\}$ **55.** $\{1\}$
57. $\{-2, 7\}$ **59.** $\{-6, 0\}$ **61.** $\emptyset$ **63.** $\left\{-\dfrac{3}{2}, 0\right\}$ **65.** $\{-1\}$ **67.** $\{-2, 0\}$ **69.** $\{-1\}$ **71.** $\left\{-\dfrac{1}{2}, \dfrac{1}{5}\right\}$
73. $\left\{\dfrac{1}{7}, \dfrac{2}{3}\right\}$ **75.** $\left\{\dfrac{4}{9}, \dfrac{1}{3}\right\}$ **77.** $\left\{\dfrac{1}{5}\right\}$ **79.** $\left\{\pm 1, \pm \dfrac{\sqrt{2}}{3}\right\}$ **81.** His speed on his bike was 6 mph.
83. Her original speed was 16 mph.   **85.** It will take Kathy 9 hrs and Jim 18 hrs.

Challenge Problems
87. $\left\{\pm \dfrac{\sqrt{2}}{3}, \pm \dfrac{\sqrt{2}}{5}\right\}$ **89.** $\left\{\pm \dfrac{3\sqrt{2}}{2}, \pm i\right\}$ **91.** $\left\{\pm \dfrac{2\sqrt{5}}{5}, \pm \dfrac{2\sqrt{3}}{3}i\right\}$ **93.** $\{6\}$ **95.** $\{-3, 0\}$ **97.** $\{0\}$

99. $\{-2a, a\}$ **101.** $\left\{-2, -\dfrac{3}{2}, -\dfrac{1}{2}, 0\right\}$ **103.** $\{-1, 2\}$ **105.** His rate in still water is 10 mph.

PROBLEM SET 6.7

Warm-ups
1. $\{1, 2\}$ **3.** $\{2\}$ **5.** $\{-3, -1\}$ **7.** $\left\{\dfrac{1}{2}, 3\right\}$ **9.** $\left\{-\dfrac{1}{2}, -\dfrac{1}{3}\right\}$ **11.** $\emptyset$ **13.** $\{1, 5\}$ **15.** $\{7, 15\}$ **17.** $\emptyset$
19. $\{8\}$ **21.** $\{14\}$

Practice Exercises
23. $\{2\}$ **25.** $\{1, 9\}$ **27.** $\{4\}$ **29.** $\{8\}$ **31.** $\{1, 3\}$ **33.** $\{-2\}$ **35.** $\{5\}$ **37.** $\{-3\}$ **39.** $\{0\}$
41. $\{2, 18\}$ **43.** $\{0\}$ **45.** $\{1\}$ **47.** $\left\{\dfrac{7}{3}\right\}$

Challenge Problems
49. $\{0, 4\}$ **51.** $\left\{\pm\dfrac{1}{2}\right\}$ **53.** $\{-1, 8\}$ **55.** $\{-1, 243\}$

REVIEW PROBLEMS CHAPTER 6

1. $\left\{-\dfrac{2}{3}, \dfrac{1}{4}\right\}$ **3.** $\{0, 7\}$ **5.** $\left\{\dfrac{-2 \pm \sqrt{13}}{3}\right\}$ **7.** $\{-3, -1\}$ **9.** $\left\{-\dfrac{3}{2}\right\}$ **11.** $\left\{\dfrac{1}{8} \pm \dfrac{\sqrt{15}}{8}i\right\}$ **13.** $\left\{-\dfrac{5}{3}, \dfrac{1}{3}\right\}$
15. $\{4\}$ **17.** $\left\{-\dfrac{5}{6}, \dfrac{1}{3}\right\}$ **19.** $\left\{\pm\dfrac{3}{2}\right\}$ **21.** $\left\{-2, \dfrac{3}{2}\right\}$ **23.** $\{-3\}$ **25.** $\left\{\dfrac{1}{2}, -\dfrac{1}{4} \pm \dfrac{\sqrt{3}}{4}i\right\}$ **27.** $\left\{-2, -\dfrac{2}{3}\right\}$
29. $\{2\}$ **31.** $\left\{-\dfrac{2}{3}, 1\right\}$ **33.** $\{\pm 7, \pm 1\}$ **35.** $\{0, 4\}$ **37.** $\{2, 10\}$ **39.** $\{-3\}$ **41.** $\left\{-\dfrac{1}{2} \pm \dfrac{\sqrt{15}}{10}i\right\}$
43. $\left\{\dfrac{1}{2}, -\dfrac{4}{3}\right\}$ **45.** $\{2 \pm \sqrt{6}\}$ **47.** $\{1, 9\}$ **49.** $\{\pm\sqrt{3}, \pm\sqrt{5}\}$ **51.** $\{-3 \pm \sqrt{5}i\}$ **53.** $\left\{\pm\dfrac{\sqrt{7}}{4}, \pm i\right\}$
55. $\{-2, 7\}$ **57.** $\{7 \pm 2\sqrt{2}i\}$ **59.** The numbers are -13 and 30.   **61.** The leg is $\sqrt{14}$ m.
63. His speed was 24 mph.

Let's Not Forget
64. $(5 - 2x)(25 + 10x + 4x^2)$ **65.** Factored   **66.** Factored   **67.** $(4 - x - 1)(4 + x + 1)$ **68.** $(x + y)^2$
69. $(c - d)(y + x)$ **70.** Two; $x - y$ is a factor.   **71.** Three; $x - y$ is not a factor.   **72.** Two; $x - y$ is a factor.
73. Two; $x - y$ is not a factor.   **74.** $2 \pm \sqrt{3}i$ **75.** $\dfrac{-3}{xy^3}$ **76.** $\dfrac{-9}{a^2b^4}$ **77.** $\dfrac{b^2}{a^3} + \dfrac{b^3}{a}$ **78.** $-\dfrac{1}{8}$
79. Cannot be simplified.   **80.** $x + 2x^{1/2}y^{1/2} + y$ **81.** $27y^3 - 27y^2 + 9y - 1$ **82.** Will not reduce   **83.** $\dfrac{x - y}{b - c}$
84. $\{2\}$ **85.** $\dfrac{-x^2 - x + 9}{(x + 4)(x + 3)}$ **86.** $\dfrac{2x^2 + 11x + 12}{4x + 12}$ **87.** Expression; $3 - i$ **88.** Equation; $\{-5, 2\}$
89. Expression; $\dfrac{-2}{(x - 7)(x + 2)}$ **90.** Equation; $\{2, 10\}$ **91.** Equation; $\{-9, 15\}$ **92.** Equation: $\{-4, -1\}$
93. Equation; $\left\{\dfrac{1}{7}\right\}$

CHAPTER 6 TEST

1. B   **2.** A   **3.** D   **4.** C   **5.** C   **6.** $\left\{-\dfrac{2}{3}, 2\right\}$ **7.** $\left\{-\dfrac{5}{2}, 1\right\}$ **8.** $\{0, 6\}$ **9.** $\{0, 3\}$
10. $\{2, -1 \pm \sqrt{3}i\}$ **11.** $\left\{\pm 1, \pm\dfrac{1}{2}i\right\}$ **12.** $\{3\}$ **13.** $\{1 \pm \sqrt{2}\}$ **14.** $9; b$
15. Her original speed was 20 mph.

CHAPTER 7

PROBLEM SET 7.1

Warm-ups

1. $\{x \mid x \leq 7\}$ **3.** $\{x \mid x \geq 10\}$ **5.** $\{x \mid x < -3\}$ **7.** $\left\{x \mid x \leq -\dfrac{1}{2}\right\}$ **9.** $\left\{x \mid x \geq \dfrac{5}{2}\right\}$

11. $\{x \mid x > 1.25\}$ **13.** $\left\{x \mid x < \dfrac{10}{7}\right\}$ **15.** {All reals}

17. $\left\{x \mid x < -\dfrac{3}{4}\right\}$ **19.** $\{x \mid x \leq 1\}$ **21.** $\{x \mid x < 1\}$

23. All numbers three or less. **25.** The largest width is 16 ft.

Practice Exercises

27. $\{x \mid x > 3\}$ **29.** $\{x \mid x \leq 1\}$ **31.** $\{x \mid x < 4\}$ **33.** $\{x \mid x < 0\}$

35. $\{x \mid x > -10\}$ **37.** $\left\{x \mid x \leq -\dfrac{5}{2}\right\}$ **39.** $\left\{x \mid x > \dfrac{3}{5}\right\}$ **41.** $\left\{x \mid x \geq -\dfrac{8}{27}\right\}$

43. $\left\{x \mid x < -\dfrac{3}{7}\right\}$ **45.** $\{x \mid x > -14\}$ **47.** $\left\{x \mid x > -\dfrac{7}{4}\right\}$ **49.** $\left\{x \mid x \geq \dfrac{3}{2}\right\}$

51. $\{x \mid x > 9\}$ **53.** $\{y \mid y > -5\}$ **55.** $\left\{x \mid x \leq \dfrac{20}{9}\right\}$ **57.** $\left\{s \mid s \leq \dfrac{16}{11}\right\}$

59. $\left\{y \mid y < -\dfrac{6}{5}\right\}$ **61.** $\left\{t \mid t < \dfrac{2}{5}\right\}$ **63.** $\left\{x \mid x \leq -\dfrac{4}{5}\right\}$ **65.** {All reals}

67. $\{x \mid x \geq 7\}$ **69.** $\{x \mid x > 7\}$ **71.** $\left\{r \mid r < \dfrac{13}{3}\right\}$ **73.** $\{x \mid x \leq 24\}$

75. $\{x \mid x < 2\}$ **77.** $\emptyset$ **79.** {All Reals}

81. All numbers, 3 or less. **83.** The largest width is 17 m.

Challenge Problems

85. $x < b + 5$ **87.** $x > \dfrac{b + 2a}{3}$ **89.** $x < \dfrac{5}{a}$ **91.** $x < \dfrac{b}{a^2 + 1}$

PROBLEM SET 7.2

Warm-ups

1. $\{x \mid 2 < x < 3\}$ 3. $\{x \mid -\frac{1}{2} < x \leq 0\}$ 5. $\{x \mid x < 5 \text{ or } x > 10\}$ 7. $\{x \mid x > 5\}$
9. Makes sense   11. Nonsense   13. Nonsense
15. ——(———)—— 1 2   17. ——(———]—— −1 1   19. ←——]———(——→ −7 −2   21. ——(———)—— 3/4 7/4
23. $\{x \mid -1 \leq x \leq 4\}$ 25. $\{x \mid -2 < x < 1\}$ 27. $\{x \mid x < -3 \text{ or } \geq 4\}$

Practice Exercises

29. ——(———)—— 2 3   31. ——[———)—— −3 2   33. ←——)———(——→ 6 11   35. ←——]——+—— −5 −3
37. ←——)———[—— 5/3 10/3   39. ←——]———(——→ −3 3   41. ——(———]—— 0.5 1.5   43. ——(———]—— −2.34 −0.16
45. Nonsense   47. OK   49. Makes sense   51. $\{x \mid 4 \leq x \leq 6\}$ 53. $\{x \mid -11 < x < -8\}$ 55. $\{x \mid x \leq -3 \text{ or } x > 3\}$
57. $\{x \mid x < 0\}$ 59. $\{x \mid x \geq -11\}$ 61. $\{x \mid 1 \leq x \leq 2\}$ 63. $\{x \mid -1 \leq x < 3\}$ 65. $\{x \mid -9 \leq x \leq -3\}$
67. $\{x \mid 14 < x \leq 20\}$ 69. $\{x \mid \frac{7}{4} < x < 3\}$ 71. $\{x \mid x < -\frac{3}{2} \text{ or } x \geq -\frac{1}{2}\}$

Challenge Problems

73. $\{x \mid x \leq a \text{ or } x \geq b\}$ 75. $\{x \mid 2 + a < x < 4 + a\}$ 77. $\{x \mid \frac{3}{a} < x < \frac{1}{a}\}$ 79. $\{x \mid x < -1 \text{ or } 1 < x < 2\}$
81. $\{x \mid x \leq -5 \text{ or } 0 < x \leq 3\}$

PROBLEM SET 7.3

Warm-ups

1. $\{x \mid -4 < x < 6\}$ 3. $\{x \mid x \leq -10 \text{ or } x \geq 6\}$ 5. $\{x \mid 1 < x < \frac{9}{5}\}$ 7. $\{x \mid -2 < x < -\frac{4}{3}\}$
9. $\{x \mid x < 0 \text{ or } x > 2\}$ 11. {All real numbers}   13. ∅   15. $2 \leq x \leq 12$ 17. $x \leq -\frac{15}{4} \text{ or } x \geq \frac{13}{4}$
19. All numbers between −6 and −2, inclusive.   21. All numbers between $\frac{1}{6}$ and $\frac{5}{6}$.
23. All numbers less than −7 or greater than 7.   25. All numbers between −3 and 3, inclusive.

Practice Exercises

27. $\{x \mid x \leq -8 \text{ or } x \geq 8\}$ 29. $\{x \mid -6 < x < 12\}$ 31. $\{x \mid -8 \leq x \leq -2\}$ 33. $\{x \mid x < -6 \text{ or } x > 1\}$
35. $\{x \mid 4 \leq x \leq 5\}$ 37. $\{x \mid x < -\frac{1}{5} \text{ or } x > \frac{17}{5}\}$ 39. $\{y \mid y \leq -\frac{2}{5} \text{ or } y \geq -\frac{1}{5}\}$ 41. $\{x \mid -1 < x < 3\}$
43. $\{x \mid -\frac{11}{3} \leq x \leq -\frac{7}{3}\}$ 45. $\{t \mid -3.8 < t < 8.8\}$ 47. $\{x \mid x < -3.99 \text{ or } x > 9.45\}$ 49. $\{x \mid -\frac{11}{2} \leq x \leq \frac{13}{2}\}$
51. $\{x \mid x \leq 1 \text{ or } x \geq 3\}$ 53. $\{x \mid x \text{ is a real number}\}$ 55. ∅   57. ∅   59. $\{x \mid x \leq -2 \text{ or } x \geq 1\}$
61. $\{x \mid -3 \leq x \leq 15\}$ 63. $-4 \leq x \leq 2$ 65. $x < -4.3 \text{ or } x > 4.3$ 67. $-5 < x < -1$ 69. $-\frac{5}{6} < x < \frac{7}{6}$
71. All numbers between −4 and 4.   73. All numbers between −8 and 8, inclusive.
75. All numbers between −1 and 3.

Challenge Problems

79. $\{x \mid x < 1 - d \text{ or } x > 1 + d\}$ 81. $\{x \mid x \text{ is a real number}\}$ 83. $\{x \mid x \leq 0 \text{ or } x \geq 2a\}$ 85. $\{x \mid x \neq a\}$
87. $\{x \mid x \leq -2a \text{ or } x \geq 0\}$

PROBLEM SET 7.4

Warm-ups

1. $\{x \mid -1 < x < 2\}$ 3. $\{x \mid -1 \leq x \leq 3\}$ 5. $\{x \mid -2 < x < 0 \text{ or } x > 1\}$ 7. $\{x \mid -4 < x < -3 \text{ or } x > 1\}$
9. $\{x \mid 1 < x < 3\}$ 11. $\{x \mid -5 < x < -1\}$ 13. $\{x \mid -2 < x < 1\}$ 15. $\{x \mid x < -5 \text{ or } x \geq 5\}$
17. $\{x \mid 5 < x \leq 15\}$ 19. $\{x \mid -1 < x < 1 \text{ or } x > 5\}$

Practice Exercises

21. $\{x \mid -4 < x < -3\}$ 23. $\{x \mid -1 \leq x \leq 7\}$ 25. $\left\{x \mid -\dfrac{1}{3} < x < 4\right\}$ 27. $\emptyset$ 29. $\{-5\}$
31. $\{x \mid x < 1 \text{ or } x > 6\}$ 33. $\{x \mid x < -1 \text{ or } x > 8\}$ 35. $\{x \mid x < -4 \text{ or } x > 4\}$ 37. $\{t \mid -5 \leq t \leq 0\}$
39. $\{s \mid -3 < s < -1\}$ 41. $\{x \mid -2 \leq x \leq 7\}$ 43. $\left\{x \mid x < -\dfrac{3}{5} \text{ or } x > \dfrac{1}{2}\right\}$ 45. $\left\{x \mid -\dfrac{4}{3} \leq x \leq \dfrac{3}{7}\right\}$
47. $\left\{x \mid -\dfrac{2}{3} < x < \dfrac{1}{4}\right\}$ 49. $\left\{x \mid x \leq -\dfrac{7}{5} \text{ or } x \geq \dfrac{7}{2}\right\}$ 51. $\{x \mid -3 \leq x \leq 1 \text{ or } x \geq 7\}$
53. $\{x \mid -1 < x < 0 \text{ or } x > 1\}$ 55. $\{x \mid -2 \leq x \leq 0 \text{ or } x \geq 2\}$ 57. $\{x \mid x < -1 \text{ or } x > 1\}$ 59. $\{x \mid x \leq 4 \text{ or } x > 5\}$
61. $\{x \mid -2 < x < 0\}$ 63. $\{r \mid r \leq -9 \text{ or } r > -2\}$ 65. $\{x \mid x < -4 \text{ or } x \geq 1\}$ 67. $\{x \mid -1 < x < 1\}$
69. They are -2 and -1, -1 and 0, 0 and 1, and 1 and 2. 71. All numbers less than -1. 73. 100 yds, or less
75. All numbers less than or equal to -1 or all numbers greater than 1.

Challenge Problems

77. {all real numbers} 79. $\{x \mid x < 0\}$ 81. $\{x \mid -5 < x \leq -2 \text{ or } 1 \leq x < 7\}$ 83. $\{x \mid a \leq x < b\}$

REVIEW PROBLEMS CHAPTER 7

1. 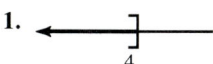 3. 5. 7.
   4   −24 −8   −5 7   2 5

9. 11. 13. 15.
   0 3   −2 5   7   −3 −2

17. 19. 21.
   −3.1 6.1   −2.3 1.4

23. Any number less than or equal to 11. 25. The side must be 25 ft or less. 27. She must make 89 or higher.

Let's Not Forget

29. Factored 30. $(2 - a)(4 + 2a + a^2)$ 31. $(x - y + 2)(x - y - 2)$ 32. Factored 33. Factored
34. $(a + b)^2$ 35. Factored 36. 2, no 37. 2, yes 38. 1, yes 39. 2, yes 40. $a\sqrt[3]{b - c}$
41. Will not simplify. 42. $\dfrac{xy^2}{x + y^2}$ 43. $\dfrac{-2}{x}$ 44. -81 45. $\dfrac{2}{3a^{5/3}}$ 46. $x - 4\sqrt{xy} + 4y$
47. $27 + 54x + 36x^2 + 8x^3$ 48. $\dfrac{1}{2} \pm \dfrac{\sqrt{5}}{2}$ 49. $x + 1$ 50. $\dfrac{-1}{(x + 1)(x + 2)}$ 51. $\{\pm\sqrt{2}\}$ 52. $\dfrac{x}{2}$
53. Equation; $\left\{\dfrac{7}{2}\right\}$ 54. Inequality; $\left\{x \mid x > \dfrac{7}{2}\right\}$ 55. Equation; $\{-1, 3\}$ 56. Inequality; $\left\{x \mid -\dfrac{8}{3} \leq x \leq \dfrac{4}{3}\right\}$
57. Equation; $\{9\}$ 58. Equation; $\{-6\}$ 59. Inequality; $\{x \mid -2 < x < 2\}$ 60. Inequality; $\{x \mid -1 < x < 1 \text{ or } x \geq 3\}$
61. Expression; $\dfrac{2x + 2y}{y - x}$

CHAPTER 7 TEST

1. B 2. A 3. C 4. C 5. A 6. $\{x \mid -2 < x < 1\}$ 7. $\{x \mid -3 < x < 0 \text{ or } x > 2\}$ 8. $\{x \mid -5 \leq x < -1\}$
9. $\{-1\}$ 10. $\left\{x \mid x > -\dfrac{13}{3}\right\}$ 11. $\{x \mid 0 < x \leq 1\}$ 12. $\left\{x \mid -2 < x < \dfrac{1}{2}\right\}$ 13. $\left\{x \mid x < -\dfrac{11}{3} \text{ or } x > 3\right\}$
14. $\{x \mid x \leq -3 \text{ or } -1 \leq x \leq 2\}$ 15. $\{x \mid -5 \leq x < -2 \text{ or } x > 1\}$

CHAPTER 8

PROBLEM SET 8.1

Warm-ups

1. I **I 3.** IV **5.** x-axis **7.** IV **9.** Both x and y axis

11. 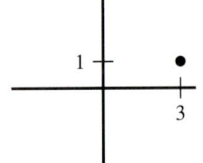 **13.** 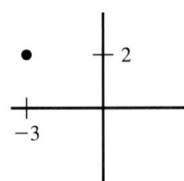 **15.**   **17.**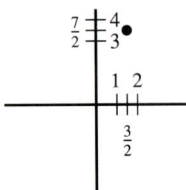

11. 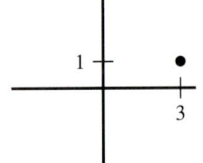 **13.** 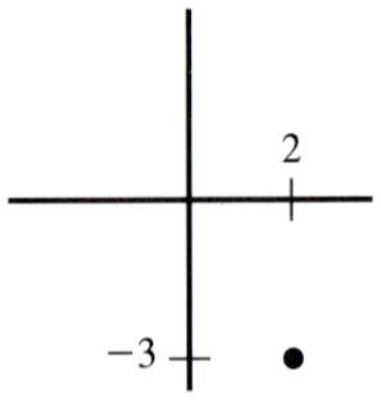 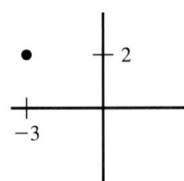 **15.**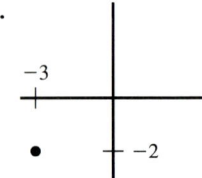

17.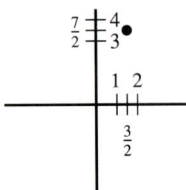

19. 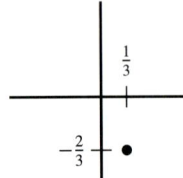 **21.** 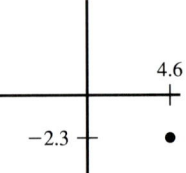 **23.**

25. 5 **27.** 10 **29.** 13 **31.** $2\sqrt{2}$ **33.** $7\sqrt{2}$ **35.** $\sqrt{5}$ **37.** $\dfrac{\sqrt{26}}{6}$ **39.** $(1, 0)$

Practice Exercises

41. 10 **43.** 13 **45.** 8 **47.** $3\sqrt{5}$ **49.** 5 **51.** $\sqrt{2}$ **53.** 9 **55.** $\sqrt{82}$ **57.** $7\sqrt{2}$ **59.** $\sqrt{37}$ **61.** $\dfrac{5\sqrt{2}}{12}$ **63.** $(-3, -7)$

Challenge Problems

65. I **67.** II **69.** III **71.** III **73.** I **75.** x-axis **77.** They are collinear because the distance between P_1 and P_2 is $2\sqrt{5}$ units, the distance between P_1 and P_3 is $\sqrt{5}$ units, and the distance between P_2 and P_3 is $\sqrt{5}$ units. **79.** It is not a right triangle because the Pythagorean Theorem does not hold.

PROBLEM SET 8.2

(The x-intercept is listed first, then the y-intercept.)

Warm-ups

1. 2, 3 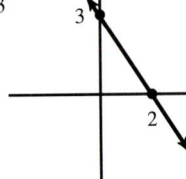 **3.** -2, 3 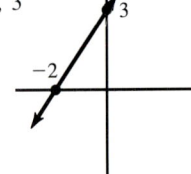 **5.** 1, 1 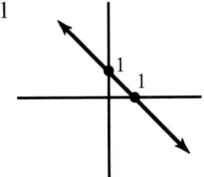 **7.** 1, 4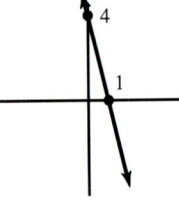

9. $\dfrac{4}{3}$, -1 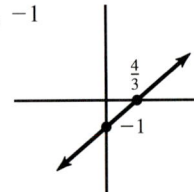 **11.** $\dfrac{1}{2}$, -1 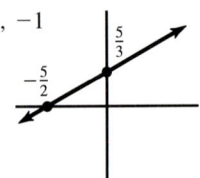 **13.** $-\dfrac{5}{2}$, $\dfrac{5}{3}$ 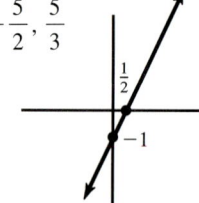 **15.** 0, 0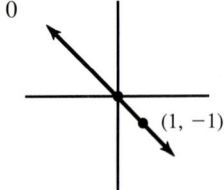

Answers to Selected Problems

17. 0, 0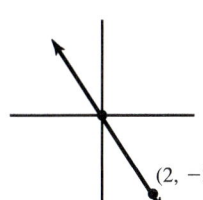

19. 2, No y-int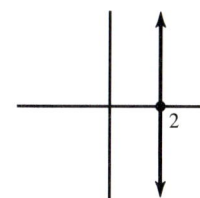

21. 0, The y-axis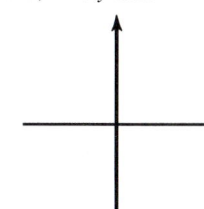

23. No x-int, $\frac{7}{2}$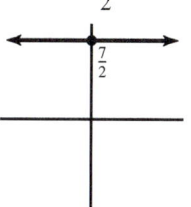

25. (a) 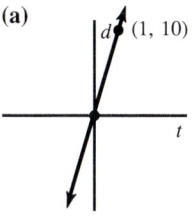 **(b)** It will take 7 hours.

27. (a) 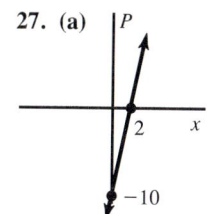 **(b)** Two shirts must be sold to break even.
(c) The profit is $490.
(d) The profit is a loss of $5.

Practice Exercises

29. 3, 4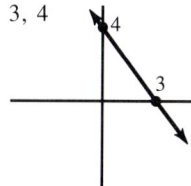

31. −3, 4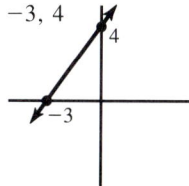

33. 0, 0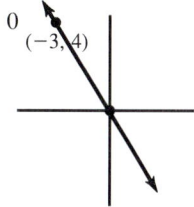

35. 2, 1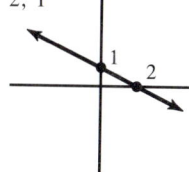

37. 0, 0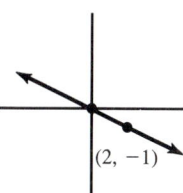

39. 2, 7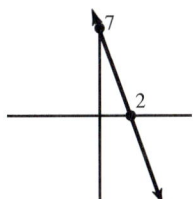

41. −5, No y-int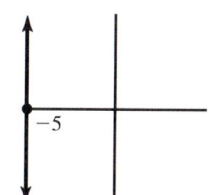

43. 2, $\frac{3}{2}$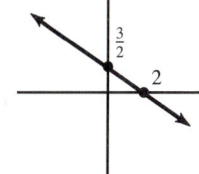

45. $\frac{1}{4}$, 1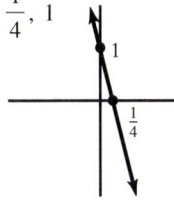

47. No x-int, $-\frac{3}{2}$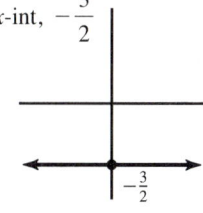

49. $-\frac{5}{2}$, $\frac{5}{3}$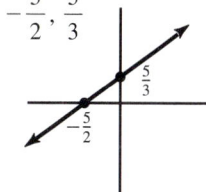

51. $-\frac{5}{3}$, $\frac{4}{3}$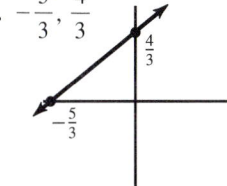

53. −1, No y-int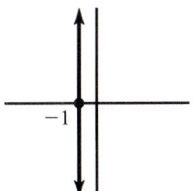

55. $-\frac{3}{4}$, No y-int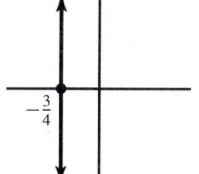

Chapter 8

57. (a) 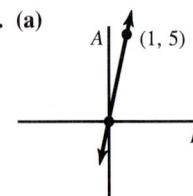 (b) The length will be 21 meters.

59. (a) 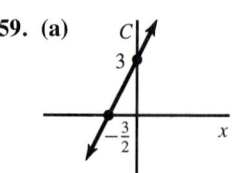 (b) The cost is $33.

Challenge Problems

61. $\frac{1}{6}, -\frac{1}{4}$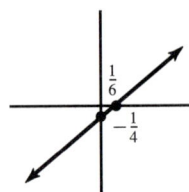

63. $2, \frac{3}{2}$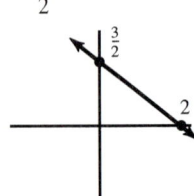

65. $-\frac{3}{7}, \frac{6}{7}$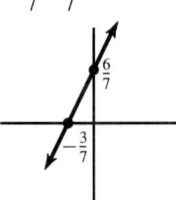

67. $\frac{1}{2}, 1$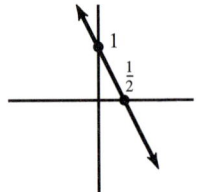

69. $B, -A$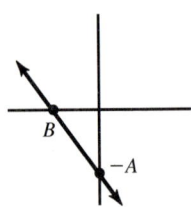

71. No x-int, $-\frac{C}{D}$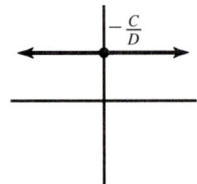

■ PROBLEM SET 8.3

Warm-ups

1. 3   **3.** $\frac{1}{3}$ **5.** 2   **7.** -2 **9.** $\frac{1}{2}$ **11.** $\frac{2}{5}$ **13.** $-\frac{1}{2}$ **15.** $-\frac{5}{4}$ **17.** $-\frac{5}{2}$ **19.** $-\frac{9}{4}$ **21.** 0

23. 0   **25.** No slope   **27.** 1   **29.** $-\frac{2}{3}$ **31.** 4   **33.** 0   **35.** 0   **37.** $-\frac{4}{3}$ **39.** No slope   **41.** $\frac{2}{5}$

Practice Exercises

43. -2 **45.** $\frac{1}{2}$ **47.** 2   **49.** $\frac{2}{3}$ **51.** $-\frac{4}{3}$ **53.** $-\frac{3}{4}$ **55.** 0   **57.** $-\frac{2}{3}$ **59.** 0   **61.** No slope

63. 1   **65.** $\frac{8}{5}$ **67.** -1 **69.** 2   **71.** $\frac{8}{3}$ **73.** 2   **75.** $\frac{1}{2}$ **77.** $-\frac{3}{5}$ **79.** 0   **81.** $\frac{3}{2}$ **83.** No slope

85. $-\frac{3}{2}$

Challenge Problems

87. 3   **89.** 3   **91.** P   **93.** 5   **95.** $\frac{4}{5}$

■ PROBLEM SET 8.4

Warm-ups

1. $-\frac{2}{3}$ **3.** $-\frac{1}{7}$ **5.** 5   **7.** $\frac{4}{7}$ **9.** 1

580   Answers to Selected Problems

11. $m = -1, b = 1$ **13.** $m = -\frac{1}{3}, b = \frac{5}{3}$ **15.** $m = -4, b = 0$ **17.** $2x - y = -3$

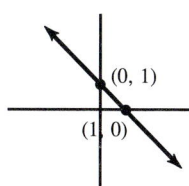 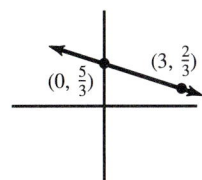 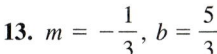 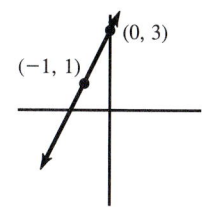

19. $x + y = -7$ **21.** $5x - 5y = -3$

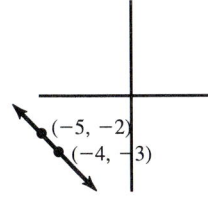 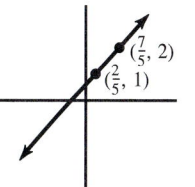

23. $x + 3y = 11$ **25.** $2x - 5y = 10$ **27.** $7x - 5y = 3$ **29.** $x + y = -3$ **31.** $6x - y = 11$ **33.** $2x - y = -14$
35. $y = -1$ **37.** $C = \frac{5}{9}(F - 32)$

Practice Exercises

39. $-\frac{3}{4}$ **41.** $-\frac{1}{5}$ **43.** 4 **45.** $-\frac{3}{7}$ **47.** 0

49. $m = -1, b = -1$ **51.** $m = -\frac{1}{2}, b = \frac{3}{2}$ **53.** $m = 4, b = 0$

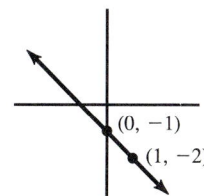 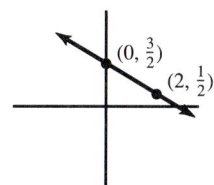 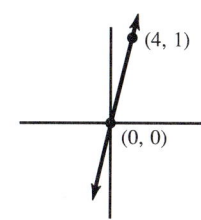

55. $2x + y = 1$ **57.** $x - y = -1$ **59.** $15x - 15y = 1$ **61.** $x - 3y = -11$ **63.** $2x - 5y = -10$
65. $x - y = 1$ **67.** $2x - 2y = -1$ **69.** $4x - y = -8$ **71.** $5x - 3y = 21$ **73.** $x = 2$ **75.** $x - y = 0$
77. $y = 0$ **79.** $x + 4y = 44$ **81.** $4t - 10s = -25$; Sales will be $6.5 billion.

Challenge Problems

83. $qx - py = 0$ **85.** $3; 2$ **87.** $\frac{1}{2}; \frac{1}{3}$ **89.** $\frac{5}{3}$; no y-intercept

■ PROBLEM SET 8.5

Warm-ups

1. $A = kr^2$ **3.** $V = kr^3$ **5.** $u = \frac{kv}{w}$ **7.** $k = 3$; x is 75 when y is 5.
9. $k = 3$; s is 193.2 when t is $\sqrt{2}$ and g is 32.2.   **11.** $k = 2$; A force of 20 lbs is required.
13. $k = 16.1$; The stone will have fallen 144.9 feet.   **15.** $k = \frac{1}{8}$; A 4-inch pendulum has a period of $\frac{1}{4}$ sec.

Practice Exercises

17. $s = kv$ **19.** $V = klwh$ **21.** $h = \frac{k\sqrt{g}}{d}$ **23.** $k = 18$; x is $\frac{18}{25}$.

25. $k = 24$; N is $\dfrac{32}{9}$ when l is $\dfrac{\sqrt{2}}{2}$ and M is $\dfrac{3}{2}$. **27.** $k = -1$; Its slope is $\dfrac{4}{3}$.

29. $k = \dfrac{5}{2}$; A force of $\dfrac{5}{2}$ lbs is required.

REVIEW PROBLEMS CHAPTER 8

1. $5, \left(3, \dfrac{5}{2}\right)$ **3.** $\sqrt{5}, \left(\dfrac{3}{2}, 2\right)$ **5.** $\sqrt{2}, \left(\dfrac{17}{2}, -\dfrac{5}{2}\right)$

In problems 7–15 the x-intercept is listed first.

7. 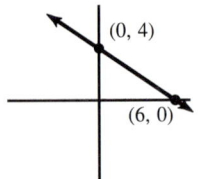   **9.** 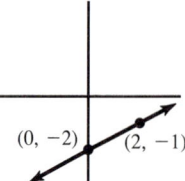   **11.**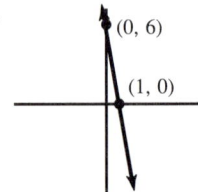

13. 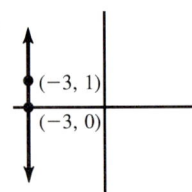   **15.**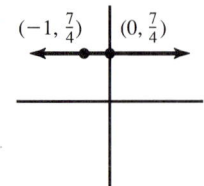

17. -1 **19.** -2 **21.** $\dfrac{3}{5}$ **23.** $-\dfrac{9}{5}$ **25.** Slope $\dfrac{2}{3}$; y-intercept -2 **27.** Slope 0; y-intercept $-\dfrac{3}{2}$

29. $x + y = 2$ **31.** $5x - 7y = -3$ **33.** $2x + 3y = -19$ **35.** $4x + 3y = 15$ **37.** u is 24. **39.** The force is 18 units.

Let's Not Forget

41. $(3 + a)(9 - 3a + a^2)$ **42.** Factored **43.** Factored **44.** $a\left(\dfrac{1}{b} + ab\right)$ **45.** $(f - g - 3)(f - g + 3)$

46. $b(a^2c + d^2r)$ **47.** Two; not a factor. **48.** Two; not a factor. **49.** Three; $y - 7$ is a factor.

50. Two; $y - 7$ is a factor. **51.** $\dfrac{1}{81}$ **52.** -81 **53.** 81 **54.** $\dfrac{1}{81}$ **55.** Will not simplify. **56.** $2x\sqrt[3]{4 + 8x^2}$

57. $a^4 b^9$ **58.** $\dfrac{-3b^4}{a^4}$ **59.** Will not reduce. **60.** $\dfrac{x + y}{x^2 + xy + y^2}$ **61.** $\dfrac{x + 3}{x - 1}$ **62.** $\{-2, 5\}$ **63.** $-\dfrac{3}{2}$

64. Equation; $\{2, 8\}$ **65.** Expression; $\dfrac{-1}{2s - 12}$ **66.** Expression; $12 - 2\sqrt{35}$ **67.** Inequality; $\left\{x \,\middle|\, -\dfrac{2}{3} < x < \dfrac{3}{2}\right\}$

68. Equation; $\{6\}$ **69.** Equation; $\{\pm 2, \pm i\}$

CHAPTER 8 TEST

1. B **2.** C **3.** A **4.** B **5.** C **6.** C

7. 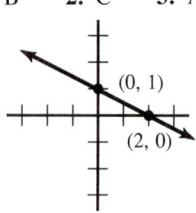 **8.** 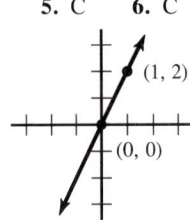 **9.** 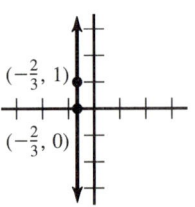 **10.**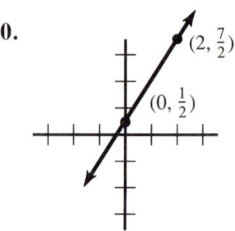

11. $y = x + 1$ **12.** $x = -2$ **13.** 13 **14.** $\left(-1, -\dfrac{7}{2}\right)$ **15.** $-\dfrac{2}{3}$ **16.** Slope $-\dfrac{5}{3}$; y-intercept $\dfrac{7}{3}$ **17.** x is 2.

CHAPTER 9

PROBLEM SET 9.1

Warm-ups

1. $\{(2, 1)\}$ **3.** $\{(1, 2)\}$ **5.** $\{(-3, -2)\}$ **7.** $\{(x, y) \mid x + y = 4\}$ **9.** $\emptyset$ **11.** $\{(0, 1), (1, 2)\}$ **13.** $\{(2, -1)\}$
15. $\{(-1, 3)\}$ **17.** $\left\{\left(\frac{1}{3}, 2\right)\right\}$ **19.** $\{(1, 1)\}$ **21.** $\{(2, 5)\}$ **23.** $\{(0, 0)\}$ **25.** $\emptyset$ **27.** $\emptyset$ **29.** 9 and -9

Practice Exercises

31. $\{(1, 2)\}$ **33.** $\{(2, 1)\}$ **35.** $\{(0, -2)\}$ **37.** $\left\{\left(\frac{1}{2}, 2\right)\right\}$ **39.** $\emptyset$ **41.** $\left\{\left(\frac{3}{2}, 1\right)\right\}$ **43.** $\emptyset$ **45.** $\{(0, 0)\}$
47. $\{(1, 2), (2, 1)\}$ **49.** $\{(3, -1)\}$ **51.** $\{(2, -1)\}$ **53.** $\{(-1, 1)\}$ **55.** $\{(3, 3)\}$ **57.** $\left\{\left(-\frac{1}{3}, \frac{1}{3}\right)\right\}$
59. $\{(0, 0)\}$ **61.** $\{(0, 0)\}$ **63.** $\left\{\left(-\frac{2}{3}, -2\right)\right\}$ **65.** $\left\{\left(-\frac{1}{4}, -\frac{1}{5}\right)\right\}$ **67.** $\left\{\left(-\frac{1}{5}, -\frac{7}{5}\right)\right\}$
69. $\{(x, y) \mid 17x - 11y = -1\}$ **71.** $\left\{\left(\frac{1}{2}, \frac{1}{20}\right)\right\}$ **73.** $\left\{\left(0, \frac{15}{14}\right)\right\}$ **75.** $\{(0, 0)\}$ **77.** $\{(2, 2), (2, -2)\}$
79. $\{(0, -1), (1, 0), (-1, 0)\}$ **81.** 11 and 13
83. 7, 27, and 34

Challenge Problems

85. $\{(3, 1)\}$

PROBLEM SET 9.2

Warm-ups

1. $\{(1, 2, 3)\}$ **3.** $\{(-1, 0, 2)\}$ **5.** $\left\{\left(0, 0, \frac{1}{3}\right)\right\}$ **7.** $\emptyset$ **9.** $\emptyset$

Practice Exercises

11. $\{(1, 3, 2)\}$ **13.** $\{(3, 1, 0)\}$ **15.** Dependent **17.** $\emptyset$ **19.** Dependent **21.** $\left\{\left(\frac{1}{2}, -\frac{1}{3}, -\frac{1}{5}\right)\right\}$ **23.** $\emptyset$
25. $\{(0, -2, 0)\}$ **27.** Dependent **29.** $\left\{\left(\frac{2}{3}, \frac{1}{3}, -\frac{1}{3}\right)\right\}$ **31.** $\left\{\left(1, \frac{1}{2}, -2\right)\right\}$ **33.** 1, 5 and 6 **35.** -2, 3 and 5

Challenge Problems

37. $\{(1, 0, -1, 2)\}$ **39.** $\{(1, 2, 3, 4)\}$

PROBLEM SET 9.3

Warm-ups

1. She should mix 12 liters of 17% solution and 14 liters of 30% solution. **3.** A pound of tomatoes costs 99 cents and a head of lettuce costs 69 cents. **5.** The dimensions are 10 m by 17 m. **7.** The boat's rate is 5 mph and the current 3 mph. **9.** It takes Frank 6 hr and Jack 12 hr. **11.** They are 8 and 11.

Practice Exercises

13. Lemons are 10 cents each and grapes are 99 cents per pound. **15.** It takes Tom 2 hr and Randy 3 hr.
17. 120 by 480 ft **19.** Carolyn's rate is 5 mph and Bill's is 3 mph. **21.** The sides are 14″, 17″, and 17″.
23. 12 by 18 in.

Challenge Problems

25. It takes 40 min. **27.** The numbers are -1, 1, 4, and 7.

PROBLEM SET 9.4

Warm-ups

1.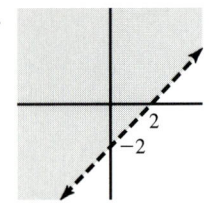

3.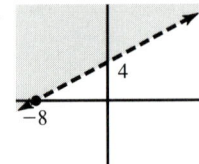

5.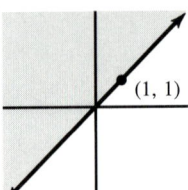

7.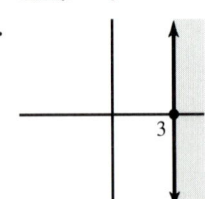

9.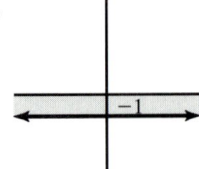

Practice Exercises

11.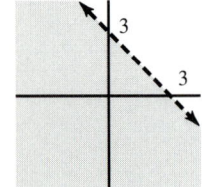

13.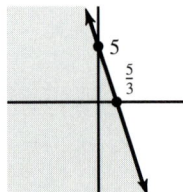

15.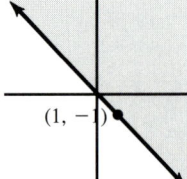

17.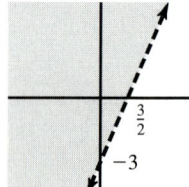

19.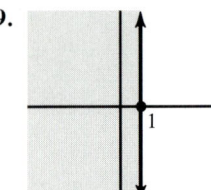

21.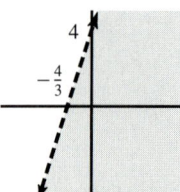

23.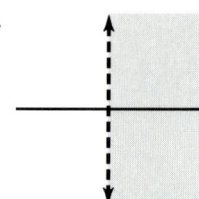

25.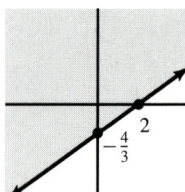

27.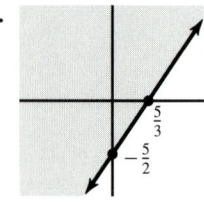

29.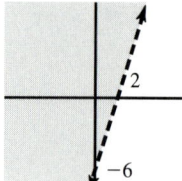

31.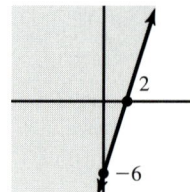

33.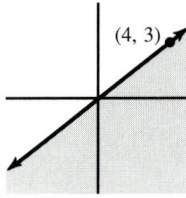

35.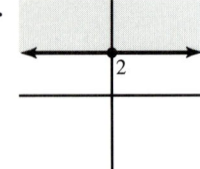

37.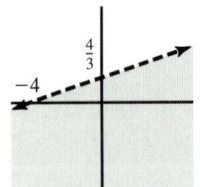

39.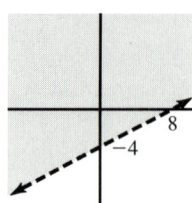

41.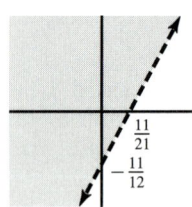

43.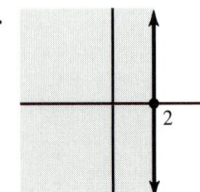

45.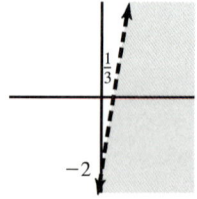

Challenge Problems

47.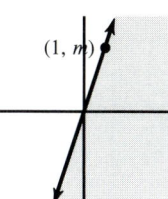

49.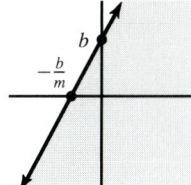

51.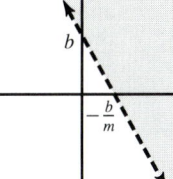

53.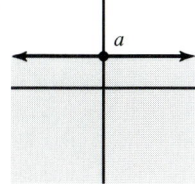

55.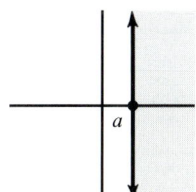

■ PROBLEM SET 9.5

Warm-ups

1.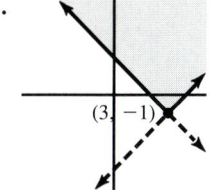

3.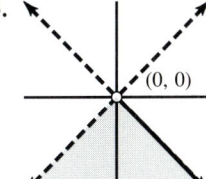

5.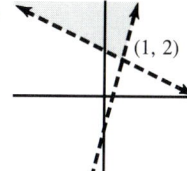

7.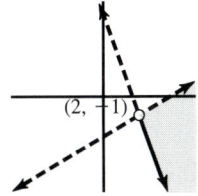

9.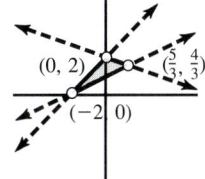

Practice Exercises

11.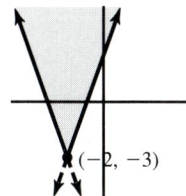

13.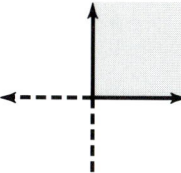

15.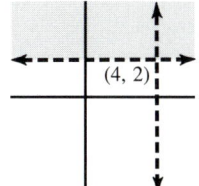

17.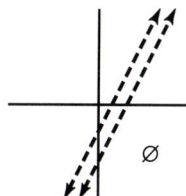

19.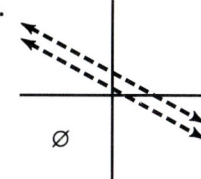

21.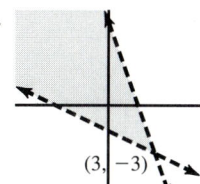

23.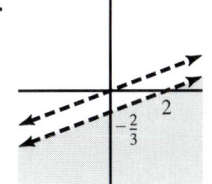

25.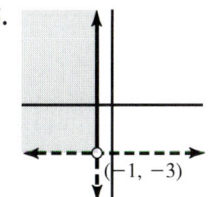

27. 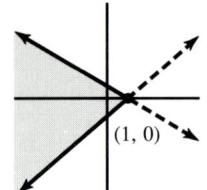 **29.** 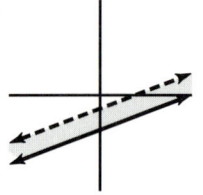 **31.** 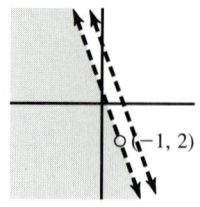 **33.**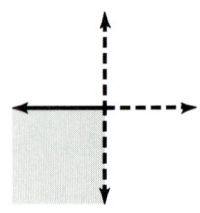

35. 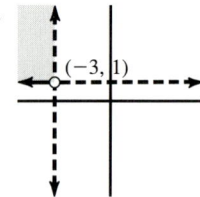 **37.** 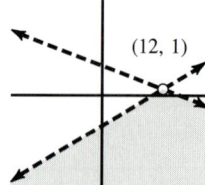 **39.** 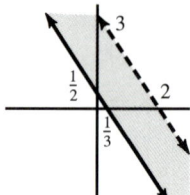 **41.**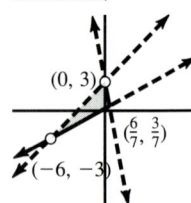

43. $D + 2P \leq 30$
$2D + P \leq 40$
$P \geq 0$
$D \geq 0$

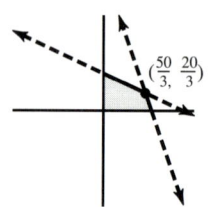

Challenge Problems

45. 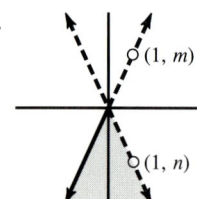   **47.**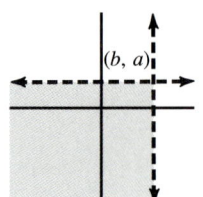

PROBLEM SET 9.6

Warm-ups

1. 1   **3.** 0   **5.** 22   **7.** -7 **9.** -42 **11.** 2

Practice Exercises

23. 1   **25.** 0   **27.** 27   **29.** 10   **31.** 2   **33.** -6 **35.** 0   **37.** -117 **39.** -24

Challenge Problems

61. -27 **63.** 240

PROBLEM SET 9.7

Warm-ups

1. $\begin{pmatrix} 2 & 3 & -4 \\ 5 & -6 & 1 \end{pmatrix}$ **3.** $\begin{pmatrix} 2 & -3 & 4 & 1 \\ 5 & -1 & 1 & 0 \\ 7 & 6 & -2 & 5 \end{pmatrix}$ **5.** $\begin{pmatrix} 1 & 2 & -1 & 0 \\ -1 & -1 & 0 & 11 \\ 0 & 1 & -4 & 13 \end{pmatrix}$ **7.** $\begin{cases} x + 2y = 4 \\ 9x - 8y = -1 \end{cases}$ **9.** $\begin{cases} x = 5 \\ y = -3 \\ z = -2 \end{cases}$

11. $\{(1, 2)\}$ **13.** $\{(5, -2)\}$ **15.** $\{(1, 2, -1)\}$

Practice Exercises

17. $\begin{pmatrix} 3 & 2 & 5 \\ 6 & -5 & 3 \end{pmatrix}$ **19.** $\begin{pmatrix} 3 & -2 & 5 & 4 \\ 2 & 3 & -1 & 0 \\ 1 & -1 & -2 & 1 \end{pmatrix}$ **21.** $\begin{pmatrix} 1 & -2 & 3 & 0 \\ -1 & 0 & -1 & 10 \\ -4 & 1 & 1 & -3 \end{pmatrix}$ **23.** $\begin{cases} 2x + y = 3 \\ 6x - 5y = 11 \end{cases}$ **25.** $\begin{cases} z = -3 \\ y = 4 \\ x = 2 \end{cases}$

27. $\{(1, 1)\}$ **29.** $\{(7, 2)\}$ **31.** $\{(1, 2, 3)\}$ **33.** $\{(4, 2, 1)\}$ **35.** $\{(3, -2, 1)\}$

CHAPTER 9 REVIEW PROBLEMS

1. {(1, 2)}   **3.** {(2, −3)}   **5.** {(3, 0), (5, 2)}   **7.** {(4, −1)}   **9.** {(6, 1)}   **11.** $\left\{\left(-1, \frac{1}{3}\right)\right\}$ **13.** $\left\{\left(\frac{1}{3}, \frac{1}{2}\right)\right\}$
15. $\left\{\left(\frac{1}{5}, \frac{1}{4}\right)\right\}$ **17.** {(5, 2, 1)}   **19.** {(0, 0, 0)}   **21.** {(0, 0, 6)}   **23.** {(−2, 0, 3)}   **25.** {(−1, 2, 3)}
27. Cantaloupes are $1.19, honeydews are $1.59, and watermelons are $1.99.   **29.** 3 and −7, or 7 and −3

31. **33.** 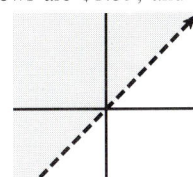 **35.** 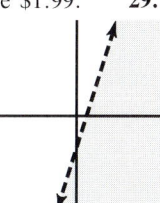 **37.**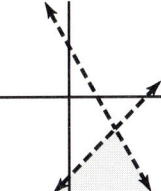

39. 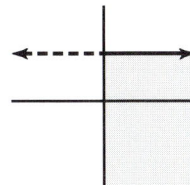 **41.** 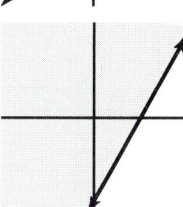 **43.** 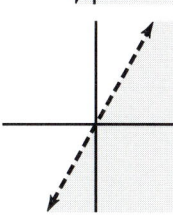 **45.**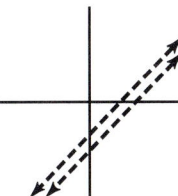

47. 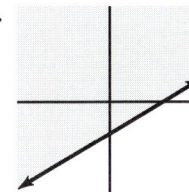 **49.** 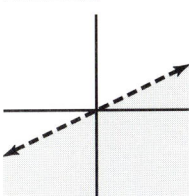 **51.** 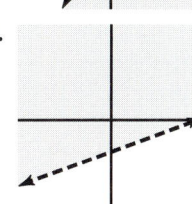 **53.**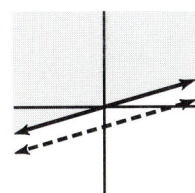

Let's Not Forget

92. $5(5 + 3s)(5 − 3s)$ **93.** Factored   **94.** Factored   **95.** $(x + 2)(x^2 − 2x + 4)$ **96.** 2; $x + 2$ is a factor.
97. 2; not a factor   **98.** 2; $x + 2$ is a factor.   **99.** 4; $x + 2$ is a factor.   **100.** $-\frac{1}{16}$ **101.** $-\frac{1}{16}$ **102.** $\frac{1}{9}$
103. $-\frac{1}{8}$ **104.** $\frac{-2c}{3b^3k^2}$ **105.** $\frac{3x^2}{2y^2}$ **106.** $2 \pm \sqrt{2}$ **107.** Won't reduce   **108.** $\frac{4-x}{1-x^2}$ **109.** {−2}
110. $-\frac{1}{x}$ **111.** Inequality; $\{x \mid 2 < x < 8\}$ **112.** Equation; $\{\pm 1\}$ **113.** Expression; $2x\sqrt[3]{1+8x^3}$
114. Expression; $\frac{2}{x^2-2}$

CHAPTER 9 TEST

1. D   **2.** B   **3.** D   **4.** B   **5.** C

6. 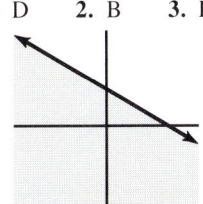 **7.** 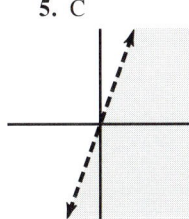 **8.** 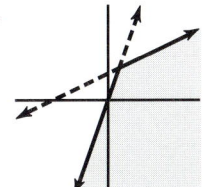 **9.**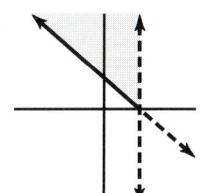

10. {(−1, 4), (4, −1)}   **11.** $\left\{\left(\frac{1}{3}, -1\right)\right\}$ **12.** {(1, −2)}   **13.** $\{(x, y) \mid y = 3x − 6\}$
14. {(1, −1, 3)}   **15.** Bananas cost 33¢ a pound and apples 99¢ a pound.   **16.** −3
17. $\left\{\left(\frac{4}{5}, -\frac{1}{5}\right)\right\}$ **18.** $\left\{\left(3, -\frac{4}{3}, -\frac{1}{3}\right)\right\}$ **19.** $\left\{\left(\frac{1}{3}, \frac{1}{3}\right)\right\}$ **20.** $\left\{\left(\frac{6}{5}, \frac{4}{5}, \frac{7}{5}\right)\right\}$

CHAPTER 10

PROBLEM SET 10.1

Warm-ups

1.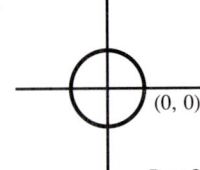
3.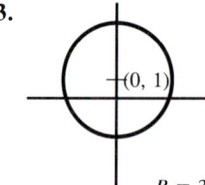
5.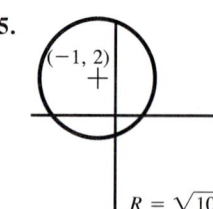
7.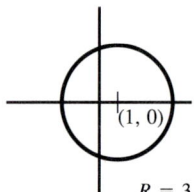

9.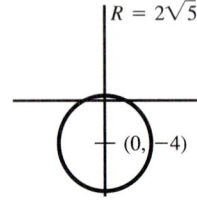
11.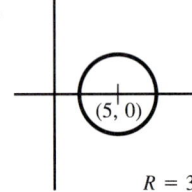
13.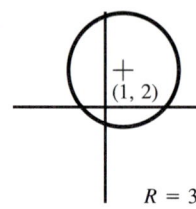
15.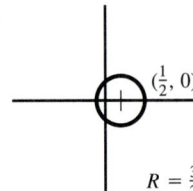

Practice Problems

17.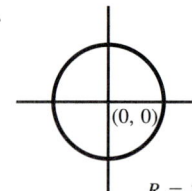
19.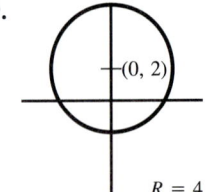
21.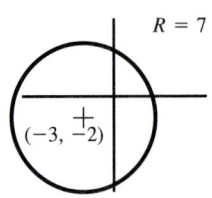
23.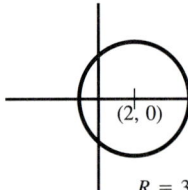

25.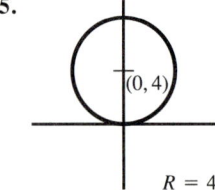
27.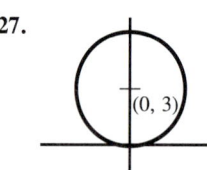
29.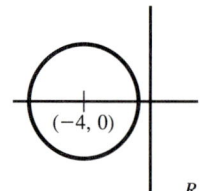
31.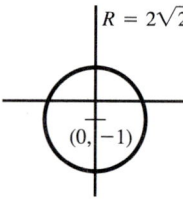

33.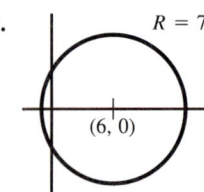
35.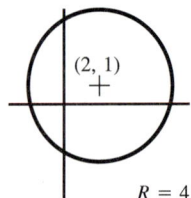
37.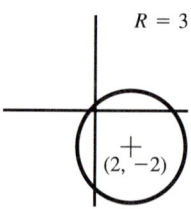
39.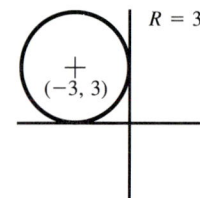

41.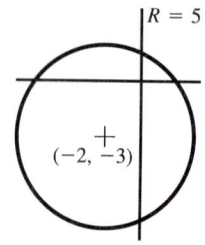
43.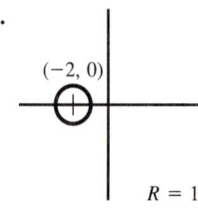
45. $R = 2\sqrt{10}$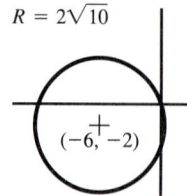
47.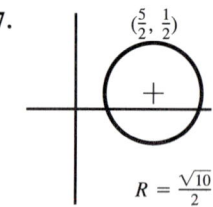

49. $R = \frac{5\sqrt{2}}{2}$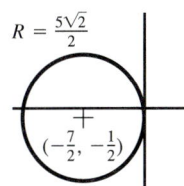

51. $R = 2\sqrt{5}$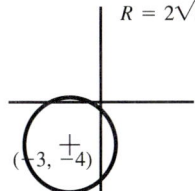

53. $R = \frac{\sqrt{5}}{2}$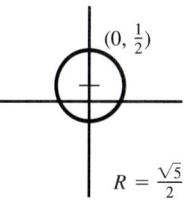

55. $R = \frac{\sqrt{34}}{2}$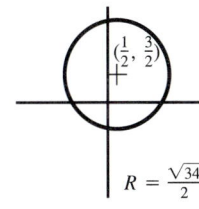

57. $R = \frac{\sqrt{26}}{2}$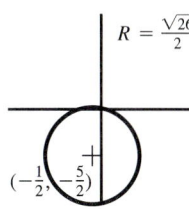

Challenge Problems

61. $R = \frac{\sqrt{3}}{2}$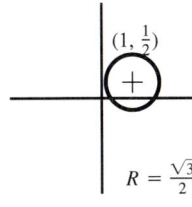

63. $R = 3$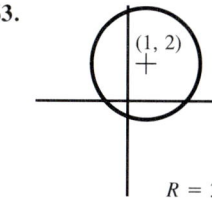

▰ PROBLEM SET 10.2

Warm-ups

In problems 1 through 8, the axis of symmetry is given.

1. y-axis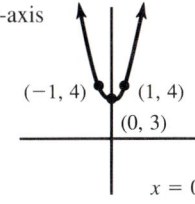

3. y-axis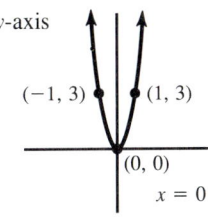

5. $x = 3$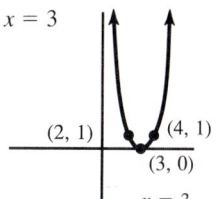

7. $x = -2$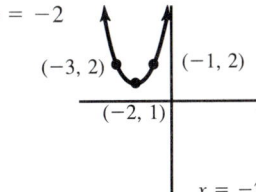

9. (2, 3)   **11.** (4, −6)

13.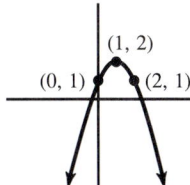

15.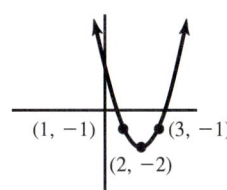

17.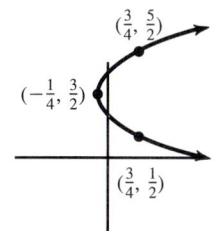

19.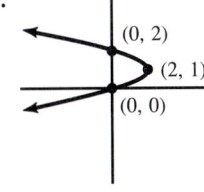

Practice Problems

21.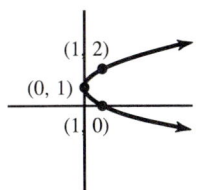

23.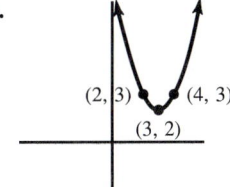

25.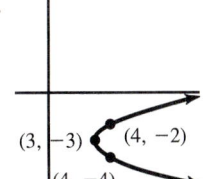

27.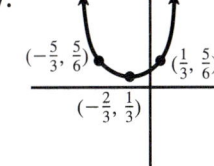

29.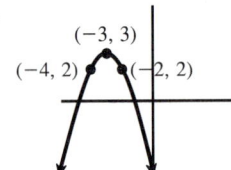
31.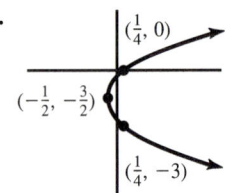
33.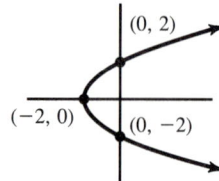
35.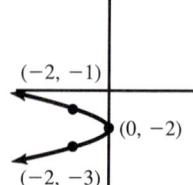

37.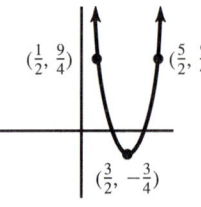
39.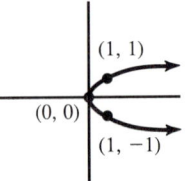
41.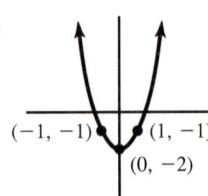
43.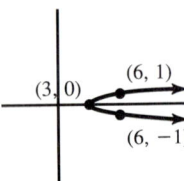

45.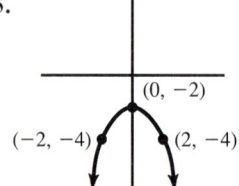
47.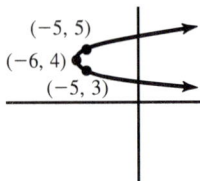
49.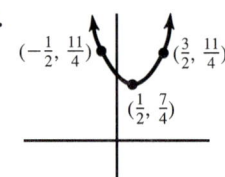
51.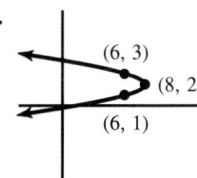

53.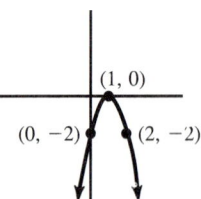
55.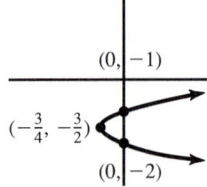

Challenge Problems

57.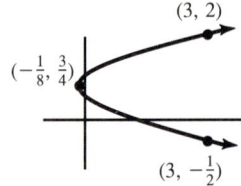
59.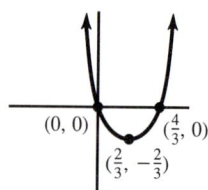
61.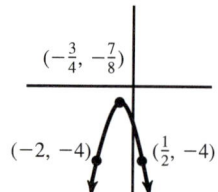
63.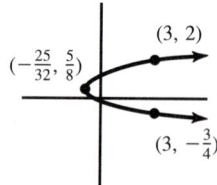

▮ PROBLEM SET 10.3

Warm-ups

1.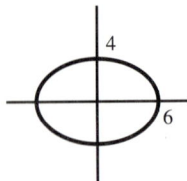
3.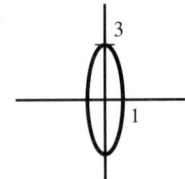
5.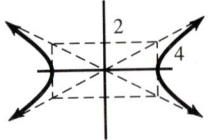
7.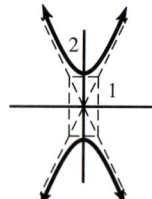

9.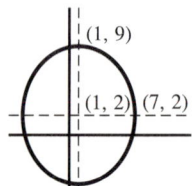
11.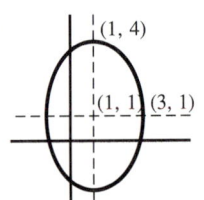
13.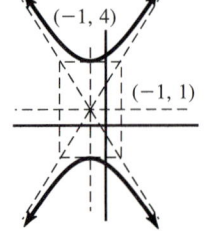
15.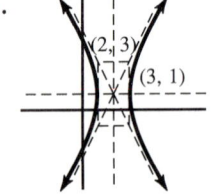

Practice Exercises

17.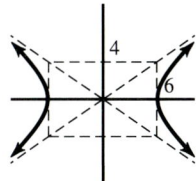
19.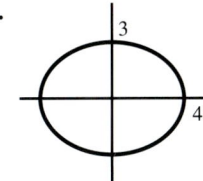
21.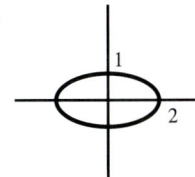
23.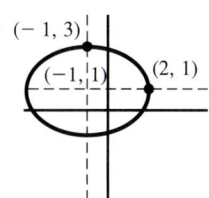

25.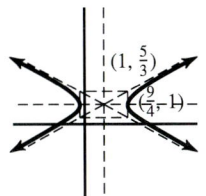
27.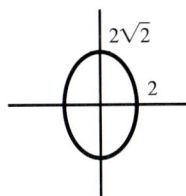
29.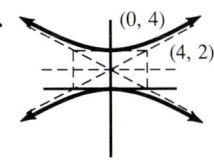
31.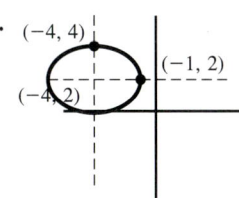

33.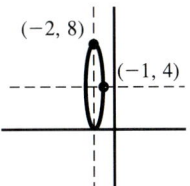
35.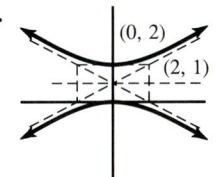
37.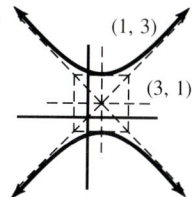
39.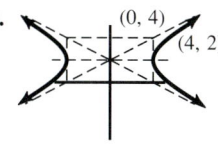

41.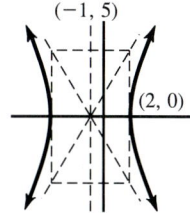
43.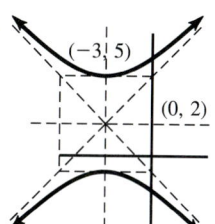
45.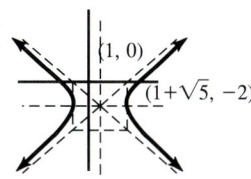
47.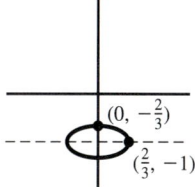

Challenge Problems

49.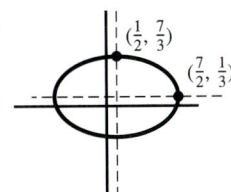
51.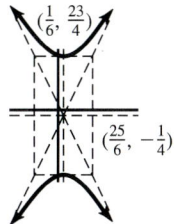

PROBLEM SET 10.4

Warm-ups

1. Hyperbola **3.** Ellipse **5.** Circle **7.** Parabola

Practice Exercises

9. Parabola **11.** Ellipse **13.** Parabola **15.** Hyperbola **17.** Ellipse **19.** Circle

Challenge Problems

21. degenerate ellipse

REVIEW PROBLEMS CHAPTER 10

1. Hyperbola **3.** Ellipse **5.** Ellipse **7.** Hyperbola **9.** Parabola

11. 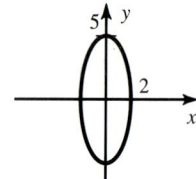   13. 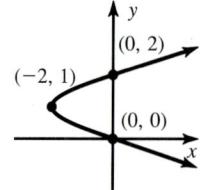   15.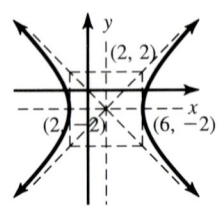

17. 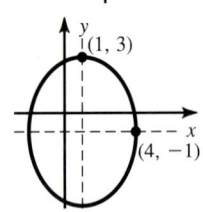   19.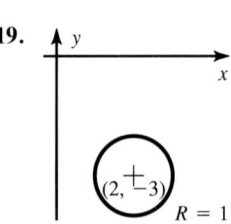

Let's Not Forget

21. Factored **22.** Factored **23.** $(2 + z)(4 - 2z + z^2)$ **24.** Factored **25.** $7(3x - 1)^2$ **26.** $5(1 - 2x)(1 + 2x)$
27. 2; $x - 3$ is a factor. **28.** 3; not a factor **29.** 3; $x - 3$ is a factor. **30.** 2; not a factor **31.** $-\dfrac{1}{256}$
32. Not a real number **33.** $-\dfrac{2x + 3}{4x^2 + 6x + 9}$ **34.** $\dfrac{x^2 + 4}{8}$ **35.** $\dfrac{-3y^3}{x^2 z^2}$ **36.** Can't be simplified.
37. $x + 4\sqrt{x} + 4$ **38.** $216 - 540x + 450x^2 - 125x^3$ **39.** $a^3 b^2 c$ **40.** $\dfrac{2 + b}{x + 7}$ **41.** Won't reduce. **42.** $\dfrac{2 - x}{x - 5}$
43. $\left\{\dfrac{5}{4}\right\}$ **44.** $-\dfrac{x + 1}{x - 2}$ **45.** Expression; $x^2 - 14x + 49$ **46.** Equation; parabola opening up with vertex at $(7, 0)$
47. Equation; $\left\{-\dfrac{26}{5}, 8\right\}$ **48.** Expression; $\dfrac{1}{(2x - 5)(x + 1)(x + 2)}$ **49.** Expression; $\dfrac{1}{2} - \dfrac{5}{2}i$ **50.** Equation; $\{3\}$

CHAPTER 10 TEST

1. B **2.** A **3.** B **4.** D **5.** B **6.** D **7.** A **8.** A **9.** C **10.** Center is $(0, 2)$; radius is $\sqrt{5}$.
11. Vertex is $(-1, -1)$; opens down.
12. 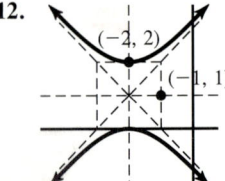   **13.** 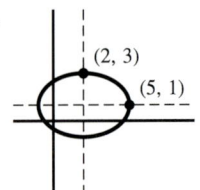   **14.** 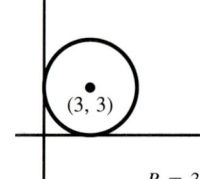   **15.**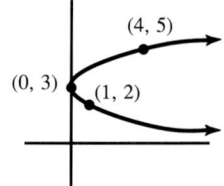

CHAPTER 11

PROBLEM SET 11.1

Warm-ups

1. Yes; $(10, 100)$, $(20, 400)$ **3.** Yes; $(\sqrt{5}, 1)\left(\dfrac{1}{2}, 1\right)(10, \sqrt{7})$, $(12, \sqrt{7})$ **5.** -1 **7.** -5 **9.** 2 **11.** 0
13. 6 **15.** 3 **17.** $2t^2 + t - 1$ **19.** $2a^2 + a + 2b^2 + b - 2$ **21.** $2a + \sqrt{a} - 1$ **23.** R **25.** R
27. $\{x \mid x \neq 0\}$ **29.** $\{x \mid x \neq -2 \text{ and } x \neq 7\}$ **31.** $\{x \mid x \geq 1\}$ **33.** $\{x \mid x \leq -2 \text{ or } x \geq 1\}$

Practice Exercises

35. $\{-6, -3, 0, 4, 7\}$ **37.** $\{y \mid y \text{ is an odd natural number}\}$ **39.** $\{11\}$ **41.** R **43.** R **45.** $\{x \mid x \neq 1\}$

47. $\{x \mid x \neq 3\}$ 49. $\{x \mid x \neq -3\}$ 51. $\{x \mid x \geq 2\}$ 53. R   55. R   57. R   59. R   61. 7   63. 1
65. $3\sqrt{3} + 1$ 67. $3(a - b) + 1$ 69. $6t + 1$ 71. $\sqrt{3a + 1}$ 73. $\sqrt{5}$ 75. 1   77. 3   79. $\sqrt{2k + 2}$
81. $a - 1$ 83. $1 - 2(t^2 + 2th + h^2)$ 85. $-4th - 2h^2$ 87. $2 - 4t^2$ 89. $k - 2kt^2$ 91. $-10t^2$ 93. -1
95. Undefined   97. 18   99. Undefined   101. 26   103. 144   105. 128   107. (a) $4.00
(b) $F(3)$ is $2.25 which is the fare for a 3 mile ride.   (c) 6 miles   109. (a) 0   (b) 2   (c) $N(10)$ is 810 which is the number of ants after 10 minutes.

Challenge Problems

111. $\{x \mid x > 2\}$

PROBLEM SET 11.2

Warm-ups

1.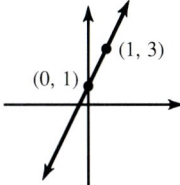
D: {all reals}
R: {all reals}

3.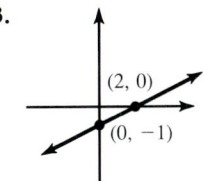
D: {all reals}
R: {all reals}

5.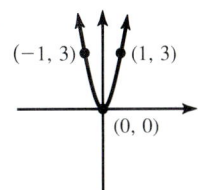
D: {all reals}
R: $\{y \mid y \geq 0\}$

7.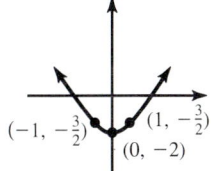
D: {all reals}
R: $\{y \mid y \geq -2\}$

9.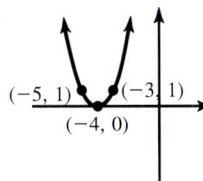
D: {all reals}
R: $\{y \mid y \geq 0\}$

11.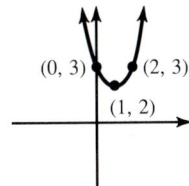
D: {all reals}
R: $\{y \mid y \geq 2\}$

13.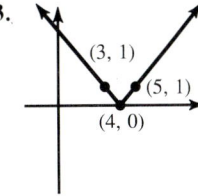
D: {all reals}
R: $\{y \mid y \geq 0\}$

15.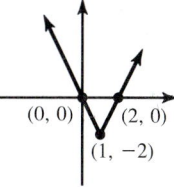
D: {all reals}
R: $\{y \mid y \geq -2\}$

17.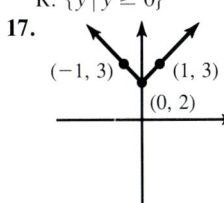
D: {all reals}
R: $\{y \mid y \geq 2\}$

19. D: {all reals}; R: $\{y \mid y \geq -1\}$ 21. D: {all reals}; R: $\{y \mid y \geq -2\}$ 23. D: $\{x \mid x \geq 0\}$; R: $\{y \mid y \geq 0\}$
25. Yes   27. No   29. Yes

Practice Exercises

31.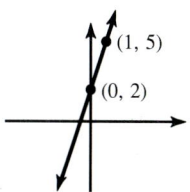
D: {all reals}
R: {all reals}

33.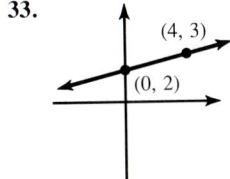
D: {all reals}
R: $\{y \mid y$ is real$\}$

35.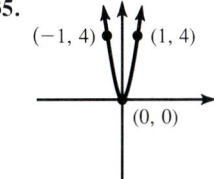
D: {all reals}
R: $\{y \mid y \geq 0\}$

37.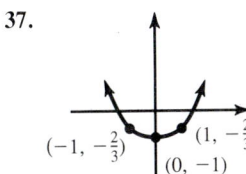
D: {all reals}
R: $\{y \mid y \geq -1\}$

39.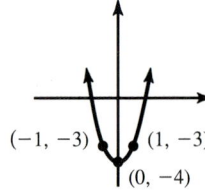
D: {all reals}
R: {y | y ≥ −4}

41.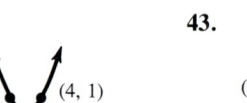
D: {all reals}
R: {y | y ≥ 0}

43.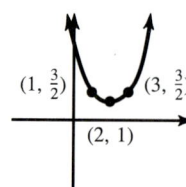
D: {all reals}
R: {y | y ≥ 1}

45.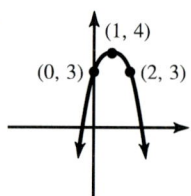
D: {all reals}
R: {y | y ≤ 4}

47.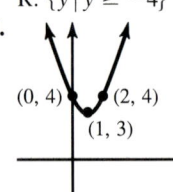
D: {all reals}
R: {y | y ≥ 3}

49.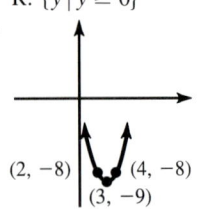
D: {all reals}
R: {y | y ≥ −9}

51.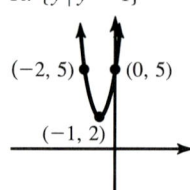
D: {all reals}
R: {y | y ≥ 2}

53.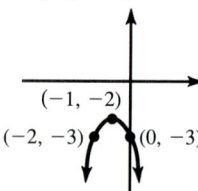
D: {all reals}
R: {y | y ≤ −2}

55.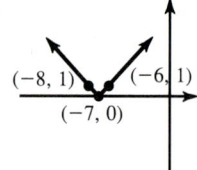
D: {all reals}
R: {y | y ≥ 0}

57.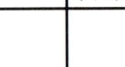
D: {all reals}
R: {y | y ≥ 2}

59.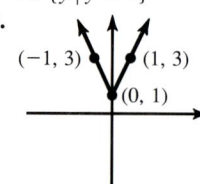
D: {all reals}
R: {y | y ≥ 1}

61.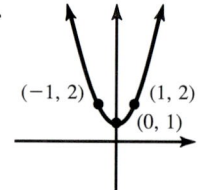
Minimum y value is 1.

63. She must sell 20,000 dresses to maximize her profit.

65.
D: {x | x ≥ 0}
R: {y | y ≥ 0}

67.
D: {x | x ≤ 4}
R: {y | y ≥ 0}

69.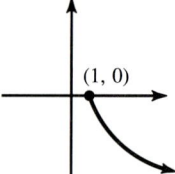
D: {x | x ≥ 1}
R: {y | y ≤ 0}

71.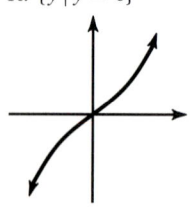
D: {all reals}
R: {all reals}

73.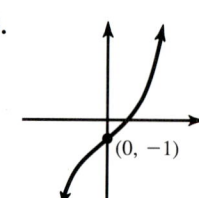
D: {all reals}
R: {all reals}

PROBLEM SET 11.3

Warm-ups

In problems 1 through 4, answers given in the order: $(f + g)(x)$, $(f - g)(x)$, $(fg)(x)$, $\left(\dfrac{f}{g}\right)(x)$.

1. $3x - 5$, $x + 5$, $2x^2 - 10x$, $\dfrac{2x}{x - 5}$ **3.** $-x^2 + x + 4$, $-x^2 - x + 2$, $-x^3 - x^2 + 3x + 3$, $\dfrac{3 - x^2}{x + 1}$ **5.** 2   **7.** $-\dfrac{1}{3}$
9. $(f \circ g)(x) = 35x - 12$; $(g \circ f)(x) = 35x + 52$ **11.** $(f \circ g)(x) = x^2 - 6x + 11$; $(g \circ f)(x) = x^2 - 1$ **13.** −1   **15.** −1

Practice Exercises

Answers given in the order: $(f + g)(x)$, $(f - g)(x)$, $(fg)(x)$, $\left(\dfrac{f}{g}\right)(x)$, $(f \circ g)(x)$.

17. $4x + 8$, $2x - 8$, $3x^2 + 24$, $\dfrac{3x}{x + 8}$, $3x + 24$ **19.** $7x - 7$, $x + 11$, $12x^2 - 30x - 18$, $\dfrac{4x + 2}{3x - 9}$, $12x - 34$ **21.** $\dfrac{4}{3}x$, $\dfrac{2}{3}x - 6$, $\dfrac{1}{3}x^2 + 2x - 9$, $\dfrac{3(x - 3)}{x + 9}$, $\dfrac{1}{3}x$ **23.** $2x^2 + 3x - 2$, $2x^2 - x + 8$, $4x^3 - 8x^2 + x - 15$, $\dfrac{2x^2 + x + 3}{2x - 5}$, $8x^2 - 38x + 48$ **25.** $-x^2 + x + 4$, $-x^2 - x - 2$, $-x^3 - 3x^2 + x + 3$, $\dfrac{1 - x^2}{x + 3}$, $-x^2 - 6x - 8$ **27.** $2x^2 - 3x - 2$, $x - 2$, $x^4 - 3x^3 + 4x$, $\dfrac{x + 1}{x} + x \neq 2$, $x^4 - 4x^3 + 3x^2 + 2x - 2$ **29.** 9 **31.** -1 **33.** 1 **35.** -11

Challenge Problems

37. $(f \circ g)(x) = x$, $(g \circ f)(x) = x$ **39.** No

PROBLEM SET 11.4

Warm-ups

1. One-to-one **2.** Not one-to-one **3.** One-to-one **4.** Not one-to-one **5.** Not one-to-one **6.** Not one-to-one **7.** One-to-one **8.** One-to-one **9.** $\dfrac{1}{3}(x + 1)$ **10.** $x + 4$ **11.** $2(x - 3)$ **12.** $\dfrac{3}{2}\left(x - \dfrac{1}{2}\right)$ **13.** $\sqrt[3]{x}$ **14.** $\sqrt[3]{x - 1}$

Practice Exercises

15.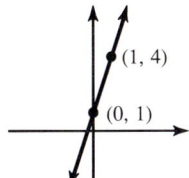
One-to-one

17.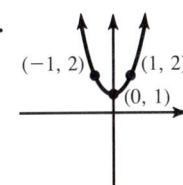
Not one-to-one

19.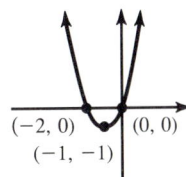
Not one-to-one

21.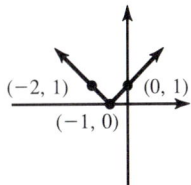
Not one-to-one

23. $\dfrac{1}{5}(x + 1)$ **25.** $3(x + 1)$ **27.** $\sqrt[3]{x - 2}$

Challenge Problems

33.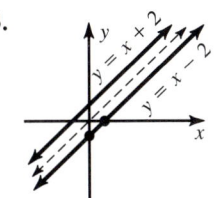

The graph of f^{-1} is the graph of f reflected about the line $y = x$.

PROBLEM SET 11.5

Warm-ups

1.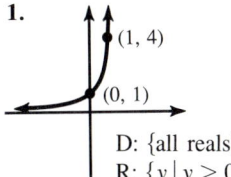
D: {all reals}
R: $\{y \mid y > 0\}$

3.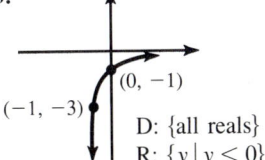
D: {all reals}
R: $\{y \mid y < 0\}$

5. {4}   **7.** {2}

9.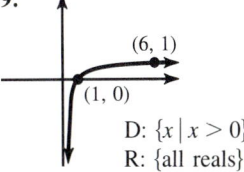
D: $\{x \mid x > 0\}$
R: {all reals}

Chapter 11

11. $2^3 = 8$ **13.** $8^2 = 64$ **15.** $5^0 = 1$ **17.** $9^{3/2} = 27$ **19.** $\log_2 8 = 3$ **21.** $\log_{1/3} \frac{1}{9} = 2$ **23.** $\log_3 \frac{1}{3} = -1$ **25.** $\log_4 64 = 3$ **27.** {16} **29.** {125} **31.** {2} **33.** {3} **35.** {1} **37.** {98} **39.** 3 **41.** Undefined **43.** 2 **45.** 2 **47.** The pressure is approximately 11.9 lbs/sq in. **49.** $A \approx \$5287.03$

Practice Exercises

51.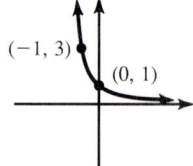
D: {all reals}
R: $\{y \mid y > 0\}$

53.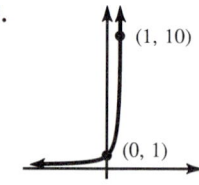
D: {all reals}
R: $\{y \mid y > 0\}$

55.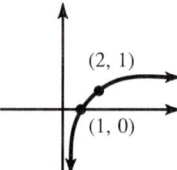
D: $\{x \mid x > 0\}$
R: {all reals}

57.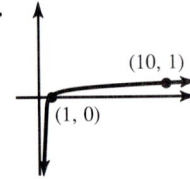
D: $\{x \mid x > 0\}$
R: {all reals}

59.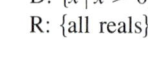
D: $\{x \mid x > 0\}$
R: {all reals}

61. {2} **63.** $\{x \mid x > 0 \text{ and } x \neq 1\}$ **65.** {8} **67.** {32} **69.** $\left\{\frac{1}{3}\right\}$ **71.** {2} **73.** {10} **75.** {5} **77.** {2}

79. {−2} **81.** {1} **83.** $\left\{\frac{1}{2}\right\}$ **85.** There are approximately 123 rabbits.

87. The population is approximately 224,084. **89.** It will take approximately 5 months.

Challenge Problems

91. a) 1. 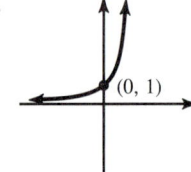 **2.** 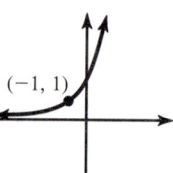 **3.** 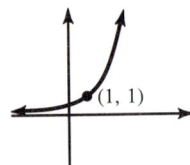 **4.**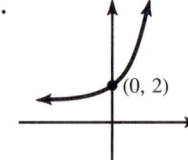

5. 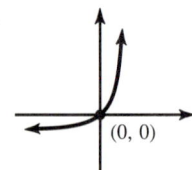 **b) 1.** 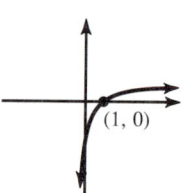 **2.** 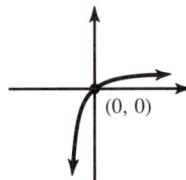 **3.**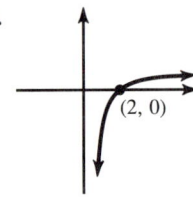

4. 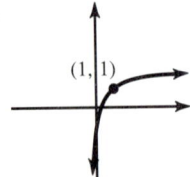 **5.**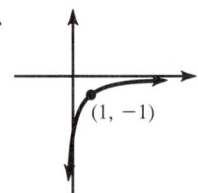

93. $(-2)^{1/2}$ is not a real number.

PROBLEM SET 11.6

Warm-ups
1. 2.94   **3.** 2.64   **5.** -1.32 **7.** 1.02   **9.** $\left\{\dfrac{9}{2}\right\}$ **11.** $\left\{\dfrac{1}{2}\right\}$ **13.** $\{4\}$ **15.** $\{342\}$ **17.** $\{3\}$ **19.** $\{5\}$
21. $\{\pm 2\}$

Practice Exercises
23. 4.14   **25.** 5.04   **27.** -1.62 **29.** 0.72   **31.** T   **33.** F   **35.** F   **37.** F   **39.** $\left\{\dfrac{1}{4}\right\}$ **41.** $\{48\}$
43. $\{64\}$ **45.** $\{33\}$ **47.** $\left\{\dfrac{e-1}{2}\right\}$ **49.** $\{\pm 3\}$ **51.** $\{-1, 4\}$ **53.** $\{5\}$

Challenge Problems
54. $\log_b \dfrac{1}{x} = \log_b 1 - \log_b x = -\log_b x$

REVIEW PROBLEMS CHAPTER 11

1. $\{y \mid y \text{ is a negative integer}\}$ **3.** $\{\text{all reals}\}$ **5.** $\{x \mid x \leq 0 \text{ or } x \geq 1\}$ **7.** $\{x \mid x \neq -1 \text{ and } x \neq 2\}$
9. Undefined   **11.** $\dfrac{6}{11}$ **13.** $\dfrac{3}{2}$ **15.** $\dfrac{3}{a^2+5}$

17.

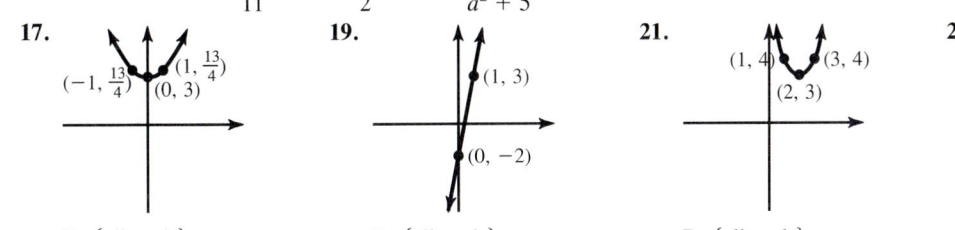

25.

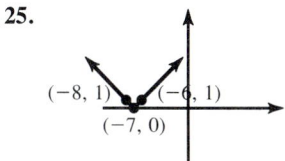

D: {all reals}
R: $\{y \mid y \geq 0\}$
not one-to-one

27.

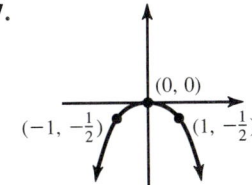

D: {all reals}
R: $\{y \mid y \leq 0\}$
not one-to-one

29.

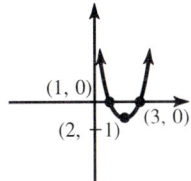

D: {all reals}
R: $\{y \mid y \geq -1\}$
not one-to-one

31.

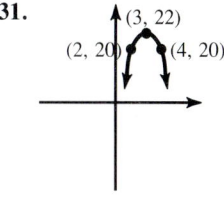

D: {all reals}
R: $\{y \mid y \leq 22\}$
not one-to-one

33. $\dfrac{3x-5}{x^2}$ **35.** $3x^2 - 5$ **37.** $(3x-5)^2$ **39.** $\{7\}$ **41.** $\{1\}$ **43.** $\{e^2 - e\}$ **45.** $\{1\}$ **47.** $\{2\}$ **49.** $\{2\}$

Let's Not Forget
51. Factored   **52.** $(5 - 2a)(25 + 10a + 4a^2)$ **53.** $(2 + a + b)(2 - a - b)$ **54.** Factored   **55.** $b(a^3 + c^2)$
56. Factored   **57.** 2; Yes   **58.** 3; No   **59.** 2; Yes   **60.** 2; No   **61.** $2\sqrt[5]{2x^5 - 4y^5}$ **62.** x^2y^2 **63.** $x - 3$
64. $\dfrac{y^8}{x^2}$ **65.** $\dfrac{4}{3}$ **66.** $2 \pm \sqrt{3}i$ **67.** $\dfrac{-5}{x^3y}$ **68.** $\dfrac{3a^4b^2}{4}$ **69.** $27x^3 - 27x^2 + 9x - 1$ **70.** $4x^2 - 12xy + 9y^2$
71. Won't reduce   **72.** $\dfrac{2+a}{4+a}$ **73.** $\dfrac{4x+3}{(x+3)^2}$ **74.** $\{-4\}$ **75.** $-(x+3)$ **76.** Equation; $\{2\}$ **77.** Equation; $\emptyset$
78. Expression; 4   **79.** Expression; -4 **80.** Expression; $8 + 2\sqrt{15}$ **81.** Inequality; $\left\{x \mid x \leq \dfrac{26}{5}\right\}$

82. Equation; {21} **83.** Equation; {−11} **84.** Inequality; $\{x \mid x < -2^- \text{ or } x > 7\}$
85. Expression; $2\sqrt{3}$ **86.** Expression; 1 **87.** Expression; $\dfrac{b^3}{a^7}$ **88.** Inequality; $\{x \mid x < 3 \text{ or } x > 4\}$
89. Expression; $a^2b - ab^2$ **90.** Equation; $\emptyset$ **91.** Inequality; $\{x \mid x \leq -18 \text{ or } x \geq 4\}$

CHAPTER 11 TEST

1. D **2.** C **3.** B **4.** A **5.** A **6. (a)** {27} **(b)** {3} **(c)** {3} **(d)** {1} **(e)** {0}
7. (a) $x \approx 0.67$ **(b)** $x \approx 0.15$ **(c)** $x \approx 2.31$ **8. (a)** $x^2 + x - 3$ **(b)** $(x - 3)^2$ **9.** $f^{-1}(x) = \dfrac{x - 3}{2}$
10. (a) $\{x \mid x \leq -1 \text{ or } x \geq 2\}$ **(b)** $\{x \mid x \neq 7\}$
11. (a) **(b)** **(c)** **(d)**

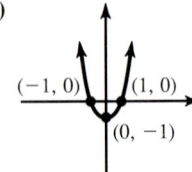
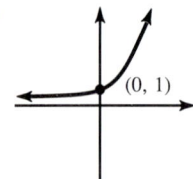
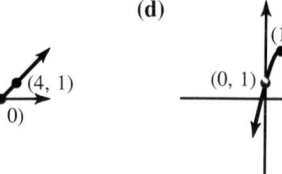

PROBLEM SET 12.1

Warm-up

1. 1, 2, 3, 4, 5 **3.** 4, 6, 8, 10, 12 **5.** 1, 4, 9, 16, 25 **7.** 3, 3, 3, 3, 3 **9.** 0, −1, 2, −3, 4 **11.** n^2
13. $(-1)^n n$ **15.** $(-1)^{n+1} 2^n$ **17.** Decreasing **19.** Increasing **21.** Increasing **23.** $d = 1; a_n = 2 + (n - 1)1$
25. $d = 10; a_n = 5 + (n - 1)10$ **27.** $d = -\dfrac{1}{2}; a_n = \dfrac{1}{2} + (n - 1)\left(-\dfrac{1}{2}\right)$ **29.** $r = 5; a_n = 1 \cdot 5^{n-1}$
31. $r = -2; a_n = 2 \cdot (-2)^{n-1}$ **33.** $r = -\dfrac{1}{4}; a_n = 1 \cdot \left(-\dfrac{1}{4}\right)^{n-1}$

Practice Exercises

35. 2, 5, 8, 11, 14 **37.** 6, 4, 2, 0, −2 **39.** 1, 3, 7, 13, 21 **41.** 1, −4, 9, −16, 25 **43.** $2n + 2$
45. $4n - 5$ **47.** $(n - 1)^3$ **49.** Increasing **51.** Decreasing **53.** Decreasing **55.** $d = 3; a_n = 1 + (n - 1)3$
57. $d = -2; a_n = 10 + (n - 1)(-2)$ **59.** $d = \dfrac{1}{2}; a_n = -\dfrac{1}{2} + (n - 1)\dfrac{1}{2}$ **61.** $r = 3; a_n = 1 \cdot 3^{n-1}$
63. $r = -3; a_n = 2 \cdot (-3)^{n-1}$ **65.** $r = \dfrac{1}{3}; a_n = 27 \cdot \left(\dfrac{1}{3}\right)^{n-1}$

PROBLEM SET 12.2

Warm-ups

1. $2 + 4 + 6 + 8 + 10$ **3.** $0 + 1 + 4 + 9 + 16 + 25 + 36 + 49$ **5.** $9 + 4 + 1 + 0 + 1$ **7.** $1 - 2 + 3 - 4 + 5 - 6$
9. $0 - \dfrac{1}{3} + \dfrac{1}{2} - \dfrac{3}{5} + \dfrac{2}{3}$ **11.** $\sum_{j=1}^{6}(-1)^{j+1}(2j - 1)$ **13.** $\sum_{j=1}^{7} \dfrac{x^{j-1}}{j}$ **15.** $\sum_{k=1}^{7}(-1)^{k+1}\sqrt{k}x^{2k-2}$

Practice Exercises

17. $-10 - 8 - 6 - 4 - 2$ **19.** $1 + 2 + 4 + 8 + 16 + 32 + 64$ **21.** $-1 + 0 + 1 + 8 + 27 + 64$
23. $-1 + 2 - 4 + 8 - 16$ **25.** $\sum_{k=3}^{14} k$ **27.** $\sum_{k=0}^{4} \dfrac{x^{2k}}{2k + 1}$ **29.** $\sum_{k=0}^{6} \dfrac{k - \pi}{10^k}$

Challenge Problems

31. $1 + 2 + 3 + 4 + 5$ All are the same. The index does not affect the sum.
33. They are the same. $\sum_{j=1}^{n} kA_j = k\sum_{j=1}^{n} A_j$.

PROBLEM SET 12.3

Warm-ups

1. 720　**3.** 479,001,600　**5.** 14　**7.** 70　**9.** 28　**11.** 20　**13.** 56　**15.** 10　**17.** 6　**19.** 20　**21.** 6
23. 1

Practice Exercises

25. 5040　**27.** 39,916,800　**29.** $\dfrac{1}{18}$　**31.** 126　**33.** 36　**35.** 30　**37.** 84　**39.** 4　**41.** 7　**43.** 35
45. 21　**47.** 1

Challenge Problems

49. $\binom{n}{r} = \dfrac{n!}{(n-r)!r!}$

$\binom{n}{n-r} = \dfrac{n!}{(n-(n-r))!(n-r)!} = \dfrac{n!}{r!(n-r)!}$

PROBLEM SET 12.4

Warm-ups

1. $a^{12} + 12a^{11}b + 66a^{10}b^2$　**3.** $s^{15} + 30s^{14} + 420s^{13}$　**5.** $64a^6 + 576a^5b + 2160a^4b^2$　**7.** $715a^9b^4$　**9.** $\dfrac{1082565}{8}x^8$

Practice Exercises

11. $a^6 + 6a^5b + 15a^4b^2 + 20a^3b^3 + 15a^2b^4 + 6ab^5 + b^6$
13. $x^9 + 9x^8 + 36x^7 + 84x^6 + 126x^5 + 126x^4 + 84x^3 + 36x^2 + 9x + 1$
15. $64x^6 + 192x^5y + 240x^4y^2 + 160x^3y^3 + 60x^2y^4 + 12xy^5 + y^6$
17. $a^9 + 9a^8b + 36a^7b^2 + 84a^6b^3 + 126a^5b^4 + 126a^4b^5 + 84a^3b^6 + 36a^2b^7 + 9ab^8 + b^9$
19. $x^9 + 9x^8 + 36x^7 + 84x^6 + 126x^5 + 126x^4 + 84x^3 + 36x^2 + 9x + 1$
21. $128x^7 + 448x^6y + 672x^5y^2 + 560x^4y^3 + 280x^3y^4 + 84x^2y^5 + 14xy^6 + y^7$
23. $78a^2b^{11} + 13ab^{12} + b^{13}$　**25.** $270s^2 + 405s + 243$　**27.** $6048x^2y^5 + 1344xy^6 + 128y^7$　**29.** $66a^2b^{10} + 12ab^{11} + b^{12}$
31. $1215s^2 + 1458s + 729$　**33.** $20{,}412x^2y^5 + 10{,}206xy^6 + 2187y^7$　**35.** $a^{50} + 50a^{49}b$　**37.** $s^{500} + 1500s^{499}$
39. $1 + 100x$　**41.** $\dfrac{55}{3456}x^9$

Challenge Problems

43. $\left(\dfrac{1}{2}\right)^6 + 6\left(\dfrac{1}{2}\right)^5\left(\dfrac{1}{2}\right) + 15\left(\dfrac{1}{2}\right)^4\left(\dfrac{1}{2}\right)^2 + 20\left(\dfrac{1}{2}\right)^3\left(\dfrac{1}{2}\right)^3 + 15\left(\dfrac{1}{2}\right)^2\left(\dfrac{1}{2}\right)^4 + 6\left(\dfrac{1}{2}\right)\left(\dfrac{1}{2}\right)^5 + \left(\dfrac{1}{2}\right)^6$

$= \dfrac{1}{64} + \dfrac{6}{64} + \dfrac{15}{64} + \dfrac{20}{64} + \dfrac{15}{64} + \dfrac{6}{64} + \dfrac{1}{64} = \dfrac{64}{64} = 1$

45. $\sum_{j=0}^{n}(-1)^j\binom{n}{j}a^{n-j}b^j$　**47.** $x^6 - 6x^5y + 15x^4y^2 - 20x^3y^3 + 15x^2y^4 - 6xy^5 + y^6$
49. $q^8 - 16q^7 + 112q^6 - 448q^5 + 1120q^4 - 1792q^3 + 1792q^2 - 1024q + 256$
51. $81x^8 - 216x^6y^2 + 216x^4y^4 - 96x^2y^6 + 16y^8$

PROBLEM SET 12.5

Warm-ups

1. 12　**3.** 5　**5.** 60　**7.** 120　**9.** 1　**11.** 36　**13.** 1　**15.** 120　**17.** 1716

Practice Exercises

19. 1680　**21.** 1　**23.** 30　**25.** 360　**27.** 720　**29.** 21　**31.** 1　**33.** 28　**35.** 1　**37.** 28
39. 24,024　**41.** 1001　**43.** 1680　**45.** 11,440

Challenge Problems

47. $N!$ **49.** $\dfrac{N!}{2}$ **51.** 1   **53.** $\dfrac{(N+2)(N+1)}{2}$

REVIEW PROBLEMS CHAPTER 12

1. 4, 7, 10, 13, 16   **3.** 3, 15, 75, 375, 1875   **5.** $-3, 6, -9, 12, -15$ **7.** Increasing   **9.** Increasing
11. $d = 2$; 4   **13.** $r = 4$; 64   **15.** $r = \dfrac{3}{2}$; 135   **17.** $0 + 1 + 4 + 9 + 16 + 25$ **19.** $0 + 4 + 16 + 36 + 64 + 100$
21. $1 - 4 + 27 - 256$ **23.** $\displaystyle\sum_{j=0}^{6}(j+1)k^j$ **25.** 362,880   **27.** 12   **29.** 1640   **31.** 84   **33.** 252
35. $x^7 + 7x^6 + 21x^5 + 35x^4 + 35x^3 + 21x^2 + 7x + 1$
37. $64 + 576y + 2160y^2 + 4320y^3 + 4860y^4 + 2916y^5 + 729y^6$
39. $1 + 10s + 45s^2 + 120s^3 + 210s^4 + 252s^5 + 210s^6 + 120s^7 + 45s^8 + 10s^9 + s^{10}$
41. $256 + 7040t + 44800t^2$ **43.** $3432a^7b^7$ **45.** 210   **47.** 332,640

Let's Not Forget

48. Factored   **49.** $(2x - 3)(4x^2 + 6x + 9)$ **50.** $(12x - 5)(8x + 3)$ **51.** Factored   **52.** $7(9 + x^3)$ **53.** $(x - 1)^3$
54. 2; factor   **55.** 3; factor   **56.** 3; not a factor   **57.** $-\dfrac{1}{16}$ **58.** $\dfrac{-2z^2}{x^3y^2}$ **59.** $-\dfrac{4x + 5}{16x^2 + 20x + 25}$
60. Won't simplify.   **61.** $\dfrac{65x^3}{2x^3 + 8}$ **62.** $(a + b)^2$ **63.** $27x^3 + 54x^2y + 36xy^2 + 8y^3$ **64.** $11 + 4\sqrt{6}$
65. $-1 + \dfrac{\sqrt{7}}{7}i$ **66.** $-1 - r$ **67.** $\dfrac{x + 1}{x^2}$ **68.** $\left\{\dfrac{1}{4}\right\}$ **69.** $\dfrac{x}{x + 1}$ **70.** Equation; $\{3\}$ **71.** Expression; $\dfrac{1}{x + 2}$
72. Expression; $4x^2 + 4xy + y^2$ **73.** Equation; $\emptyset$ **74.** Expression; $\dfrac{x - 3}{(2x - 1)^2(x + 2)}$ **75.** Expression; $-\dfrac{3}{13} + \dfrac{2}{13}i$

CHAPTER 12 TEST

1. D   **2.** B   **3.** D   **4.** B   **5.** D   **6.** B   **7.** A   **8.** C   **9.** D   **10.** $a_n = 4 + (n - 1)3$
11. $2, -2^2, 2^3, -2^4, 2^5$ **12.** 25   **13.** 6720   **14.** $81 + 108t + 54t^2 + 12t^3 + t^4$ **15.** $153a^2b^{16} + 18ab^{17} + b^{18}$

Index

Abscissa, 333
Absolute value, 16, 17, 187
   equation, 187, 190
   function, 488
   inequality, 312
   properties, 18
Addition
   associative property, 10
   commutative property, 10
   complex numbers, 240
   equality property, 155
   functions, 495
   inequality property, 297
   polynomials, 66
   radical expressions, 223
   rational expressions, 136
   real numbers, 21
Additive inverse, 11
Algebra of functions, 495
Algebraic expression, 9
Angle
   acute, 28
   alternate exterior, 29
   alternate interior, 29
   complementary, 28
   exterior, 31
   interior, 31
   obtuse, 28
   right, 28
   supplementary, 28
   vertical, 29

Area
   circle, 37
   parallelogram, 37
   rectangle, 37
   square, 37
   trapezoid, 37
   triangle, 37
Arithmetic progression, 529
Array of signs, 418
Associative property, 10
Asymptote, 504
   hyperbola, 461
Augmented matrix, 425
Axes, 332
Axis of symmetry, 448

Base, 23, 50, 52
Binomial, 62
   coefficient, 537
   cube of, 74
   expansion, 80
   special products, 71, 75
   square of, 71
   theorem, 539, 541
Boundary
   line, 405
   number, 311, 322

Calculator boxes
   calculators, 5

Calculator boxes (*cont.*)
  evaluating polynomials, 64
  exponents, 58
  factorials, 538
  finding logs with other bases, 517
  logarithms, 510
  operations, 26
  permutations & combinations, 548
  plotting points and finding distance, 337
  quadratic formula, 273
  roots and fractional exponents, 234
  scientific notation, 213
  solving equations, 162
  square root equations, 196
  square root property, 265
Cartesian coordinate system, 332
Checking
  equations, 155, 193
  word problems, 171
Circle, 34
  area, 37
  center, 34
  chord, 34
  circumference, 37
  diameter, 34
  equation, 441
Coefficient, 61
Cofactor, 418
Combinations, 546
Combining like terms, 63
Common difference, 529
Common factor, 99
Common ratio, 530
Commutative property, 10
Completing the square, 260
Complex fraction, 143
Complex numbers, 236
  addition, 240
  conjugate, 240
  division, 241
  imaginary part, 238
  multiplication, 240
  powers of i, 239
  real part, 238
  standard form, 238
  subtraction, 240
Composition of functions, 496
Cone
  surface area, 39
  volume, 39
Conic sections, 440
  circle, 440
  ellipse, 458
  hyperbola, 460
  parabola, 447
  summary, 470
Consecutive numbers, 162
Constant
  of proportionality, 367
  term, 62

Contradiction, 159
Coordinate
  axes, 332
  on number line, 7
  system, 332
Counting numbers, 4
Cramer's rule, 420–422
Cube
  of a binomial, 74
  root, 217
  surface area, 39
  volume, 39
Cylinder
  surface area, 39
  volume, 39

Decreasing line, 349
Degenerate form, 470
Degree, 61
Dependent system, 381, 386
Descartes, Rene, 332
Determinant, 417
  of coefficients, 421
  cofactor, 418
  Cramer's rule, 420, 422
  element, 417
  minor, 418
  order, 417
  sign array, 418
Difference
  in two cubes, 73, 104
  in two squares, 72, 104
Direct variation, 367
Discriminant, 272
Distance between two points
  on the number line, 18
  in the plane, 334
Distance formula, 334
Distributive property, 11
Dividend, 85
Divisibility tests, 97
Division
  complex numbers, 241
  functions, 495
  polynomials, 82
  radicals, 225
  rational expressions, 132
  real numbers, 21
  synthetic, 88
  with zero, 22
Divisor, 85
Domain, 478
  natural, 481

e, 505
Elementary row operations, 427
Elimination method, 383
Ellipse, 458
  equation, 460, 463

Ellipsis, 2
Empty set, 2
Equality
    addition property, 155
    multiplication property, 155
    properties of, 9
Equation, 154
    absolute value, 187, 190
    addition property, 155
    circle, 441
    contradiction, 159
    ellipse, 463
    equivalent, 154
    fractional, 180, 181, 279
    higher degree, 277
    hyperbola, 463
    identity, 159
    linear, one variable, 154, 157
    linear, two variables, 340
    literal, 165, 167
    multiplication property, 155
    nonlinear, 155, 249
    quadratic, 250
    radical, 192, 194, 287
    second degree, 250, 470
    solution, 154
    solve, 154
    systems, 376
Equations of a line, 362
    point-slope form, 360
    slope-intercept form, 358
    standard form, 341
Equiangular, 30
Equilateral, 30
Equivalent, 154, 297, 384
Evaluate polynomial, 62
Exponent, 23, 50
    base, 23, 50, 52
    natural number, 23, 50
    negative integer, 204
    properties, 54, 206, 231
    rational, 230
    zero, 57
Exponential function, 504
Expression, 9
Exterior angle, 29, 31

Factor, 9, 96
Factored completely, 96, 99
Factorial, 536, 545
Factoring, 96
    common factors, 99
    difference in cubes, 104
    difference in squares, 104
    GCF, 97
    perfect trinomial squares, 111
    sum of cubes, 104
    sum of squares, 104
    trinomials, 108
FOIL, 70

Formula, 165
Fractional equations, 180, 279
Fractions
    addition, 22, 136
    complex, 143
    division, 22, 132
    fundamental principle, 22
    least common denominator, 137
    multiplication, 22, 130
    rational expression, 120
    signs in, 23
Function, 478
    absolute value, 488
    algebra of, 495
    composition, 496
    constant, 486
    domain, 478
    exponential, 504
    graph, 485
    horizontal line test, 500
    inverse, 501
    linear, 485
    logarithmic, 506
    notation, 479
    one-to-one, 499
    operations with, 495
    quadratic, 486
    range, 478
    vertical line test, 490

General term, 526
Geometric progression, 530
Geometry
    acute angle, 28
    alternate exterior, 29
    alternate interior, 29
    angle, 28
    area, 37
    chord, 34
    circle, 34
    complementary angle, 28
    cone, 39
    cube, 39
    cylinder, 39
    diagonals, 30
    equiangular, 30
    equilateral, 30
    exterior angle, 31
    hypotenuse, 32
    interior angle, 30
    legs, 32
    obtuse angle, 28
    parallel lines, 29
    parallelogram, 34
    perimeter, 37
    perpendicular lines, 29
    polygon, 30
    pythagorean theorem, 32
    quadrilateral, 33
    radius, 34

Geometry (cont.)
  rectangle, 34
  rhombus, 34
  right angle, 28
  similar triangles, 32
  sphere, 39
  square, 34
  supplementary angle, 28
  surface area, 39
  tangent, 34
  transversal, 29
  trapezoid, 33
  triangle, 31
  vertical angles, 29
  volume, 39
Graph, 332
  absolute value function, 488
  circle, 441
  conic section, 471
  ellipse, 460, 463
  exponential function, 504
  function, 485
  horizontal line, 344
  hyperbola, 463
  linear equation, 340
  linear function, 485
  linear inequality, one variable, 300, 405
  logarithmic function, 507
  parabola, 454
  point, 332
  system of equations, 376
  systems of inequalities, 410
  vertical line, 343
Graphing method, 376
Greater than, 8, 296
Greatest common factor, 97
Grouping, 101
Grouping symbols, 24

Horizontal line, 344, 353, 363
Horizontal line test, 500
Hyperbola, 460
  equation, 463

i, 237
  powers, 239
Identity
  addition, 11
  equation, 159
  multiplication, 11
Inconsistent system, 380, 386
Increasing line, 349
Index
  radical, 217
  summation, 532
Inequality, 296
  absolute value, 312, 316
  boundary number, 311

  compact form, 305
  compound, 305
  equivalent, 297
  fractional, 322
  free boundary number, 322
  higher degree, 319
  linear, one variable, 297
  linear, two variables, 405
  properties, 297
  quadratic, 319
  solution, 297
  solution set, 297
  solve, 297
  systems, 410
  two variables, 405
Integer, 3
Intercept, 342
Interior angle, 29, 30
Intersection, 3
Inverse
  additive, 11
  function, 501
  multiplicative, 11
  variation, 367
Irrational numbers, 3

Joint variation, 368

Leading coefficient, 62
Least common denominator, 137
Less than, 8, 296
Like terms, 63
Linear
  function, 485
  inequality, 297
Linear equation
  one variable, 154, 157
  point-slope form, 360
  slope-intercept form, 358
  standard form, 341
  two variables, 340
Linear programming, 413
Linear system
  Cramer's rule, 421
  elimination method, 384
  graphing, 376
  matrix method, 429
  more than two variables, 391
  substitution method, 377
  two variables, 376
Literal equation, 165
Logarithm
  base, 506
  function, 506
  graph, 507
  natural, 508
  properties, 515
Long division, 82
Lowest terms, 124

Matrix, 425
    augmented, 425
    coefficients, 425
    dimension, 425
    element, 425
    triangular form, 427
Method of augmented matrices, 429
Method of boundary numbers, 312
Mid-point formula, 336
Minor, 418
Monomial, 61
Multiplication
    associative property, 10
    commutative, 10
    complex numbers, 240
    functions, 495
    polynomials, 69
    property of equality, 155
    property of inequality, 297
    radicals, 224
    rational expressions, 130
    real numbers, 21
    by zero, 13
Multiplicative inverse, 11

Negative exponent, 204
Negative radicand, 217, 237
Notation
    combination, 547
    permutation, 544
    scientific, 211
    set-builder, 2
    summation, 532
Null set, 2
Number
    complex, 236
    counting, 4
    integer, 3
    irrational, 3
    natural, 3
    rational, 3
    real, 3
    whole, 3

Obtuse
    angle, 28
    triangle, 31
One-to-one, 499
Opposite, 11, 14, 66
Order of operations, 24, 76
Ordered pair, 332
Ordinate, 333
Origin, 7, 332

Parabola, 447
Parallel lines, 29, 30, 354
Parallelogram
    area, 37
    perimeter, 37
Parentheses, 24
Pascal's triangle, 79, 540
Perfect trinomial square, 111
Perimeter
    circle, 37
    parallelogram, 37
    rectangle, 37
    square, 37
    trapezoid, 37
    triangle, 37
Permutation, 543, 545
Perpendicular lines, 29, 354
Point-slope form, 360
Polygon, 30
Polynomial, 62
    addition, 66
    binomial, 62
    coefficient, 61
    constant term, 62
    degree, 61
    division, 82
    evaluate, 62
    factor, 99
    factored completely, 99, 113
    leading coefficient, 62
    like terms, 63
    monomial, 61
    multiplication, 69
    opposite, 66
    prime, 99
    properties, 63
    standard form, 62
    subtraction, 67
    synthetic division, 88
    term, 62
    trinomial, 62
Power rules, 53, 54, 56
Powers, odd or even, 52, 208
Prime, 96, 99
Principal square root, 217
Progression
    arithmetic, 529
    geometric, 530
Properties of real numbers, 7
    associative, 10
    commutative, 10
    distributive, 10
    equality, 9
    identities, 11
    inverses, 11
    transitive, 9
    trichotomy, 8
Property of zero products, 251
Proportion, 185
Pythagorean theorem, 32, 176, 258

Quadrant, 333
Quadratic
    discriminant, 272

Quadratic (*cont.*)
  equation, 250
  formula, 268
  function, 486
  inequality, 319
  standard form, 250
Quadrilateral, 33
  parallelogram, 34
  rectangle, 34
  rhombus, 34
  square, 34
  trapezoid, 33
Quotient, 85

Radical, 217
  equation, 192, 287
  negative radicand, 217, 237
  operations, 223
  properties, 219
  sign, 217
  simplified, 221
Radical expression
  addition, 223
  division, 225
  multiplication, 224
Radicand, 217
Radius, 34
Range, 478
Ratio, 185
Rational exponent, 230
Rational expression, 120
  addition, 136
  complex fraction, 143
  division, 132
  Fundamental Principal, 123
  LCD, 137
  multiplication, 130
  reducing, 124
  sign property, 121
  subtraction, 136
  undefined, 121
Rational number, 3
Rationalizing the denominator, 225
Real number, 3
Reciprocal, 11
Rectangle, 34
  area, 37
  perimeter, 37
Rectangular coordinates, 332
Rectangular solid
  surface area, 39
  volume, 39
Reflexive property, 9
Remainder, 85
Repeated root, 254
Rhombus, 34
Right angle, 28
Right triangle
  hypotenuse, 32
  legs, 32

Rise, 349
Root, 154, 217, 254
Run, 349

Scientific notation, 211
Sequence, 526
  arithmetic, 529
  decreasing, 528
  general term, 526
  geometric, 530
  increasing, 528
  term, 526
Set, 2
  element, 2
  empty, 2
  finite, 2
  infinite, 2
  intersection, 3
  member, 2
  null, 2
  solution, 154, 297, 376
  subset, 2
  union, 3
Set-builder notation, 2
Signs in fractions, 23
Similar triangles, 32
Slope, 349
  formula, 350
  horizontal line, 353
  parallel lines, 354
  perpendicular lines, 354
  rise, 349
  run, 349
  vertical line, 353
Slope-intercept form, 358
Solution
  equation, 154
  inequality, 297
Solution set, 154, 297, 376
Solve, 154, 297, 376
Special products, 75
  cube of a binomial, 74
  difference of cubes, 73
  difference of squares, 72
  square of a binomial, 71
  sum of two cubes, 73
Sphere
  surface area, 39
  volume, 39
Square
  area, 37
  perimeter, 37
Square of a binomial, 71
Square root, 217
Square root property, 261
Squaring property, 192
Standard form
  complex number, 238
  linear equation in two variables, 341
  polynomial, 62

quadratic equation, 250
Subset, 2
Substitution method, 377
Subtraction
   complex numbers, 240
   functions, 495
   polynomials, 67
   rational expressions, 136
   real numbers, 21
Summation
   index, 532
   notation, 532
Surface area
   cube, 39
   rectangular solid, 39
   right circular cone, 39
   right circular cylinder, 39
   sphere, 39
Symmetric property, 9
Synthetic division, 88
Systems, 376, 389
   dependent, 381, 386
   elimination, 383
   equations, 376
   inconsistent, 380, 386
   inequalities, 405
   solution, 376
   substitution, 377

Tangent, 34
Term, 9, 62
Transitive property, 9
Transversal, 29, 30
Trapezoid
   area, 37
   perimeter, 37
Triangle
   acute, 31
   area, 37
   equilateral, 31
   isosceles, 31
   obtuse, 31
   perimeter, 37
   right, 31

Triangular matrix, 427
Trichotomy property, 8
Trinomial, 62, 108

Union, 3
Unlike terms, 63

Variable, 61, 154
Variation, 367
   constant of proportionality, 367
   direct, 367
   inverse, 367
   joint, 368
Vertex, 448
Vertical line, 343, 353, 363
Vertical line test, 490
Volume
   cube, 39
   rectangular solid, 39
   right circular cone, 39
   right circular cylinder, 39
   sphere, 39

Whole numbers, 3

x-axis, 332
x-coordinate, 333
x-intercept, 342

y-axis, 332
y-coordinate, 333
y-intercept, 342

Zero
   division, 22
   exponent, 57
Zero product property, 251

A SCIENTIFIC CALCULATOR